Engineering and Food for the 21st Century

EDITED BY

Jorge Welti-Chanes, Ph.D.
Universidad de las Américas-Puebla, Mexico
Cholula, Puebla

Gustavo V. Barbosa-Cánovas, Ph.D.
Washington State University
Pullman, Washington

José Miguel Aguilera, Ph.D.
Pontificia Universidad Católica de Chile
Santiago, Chile

CRC Press
Taylor & Francis Group
Boca Raton London New York

CRC Press is an imprint of the
Taylor & Francis Group, an **informa** business

CRC Press
Taylor & Francis Group
6000 Broken Sound Parkway NW, Suite 300
Boca Raton, FL 33487-2742

First issued in paperback 2019

ISBN-13: 978-1-56676-963-1 (hbk)
ISBN-13: 978-0-367-39625-1 (pbk)

Visit the Taylor & Francis Web site at
http://www.taylorandfrancis.com

and the CRC Press Web site at
http://www.crcpress.com

FOOD PRESERVATION TECHNOLOGY SERIES

Series Editor
Gustavo V. Barbosa-Cánovas

Innovations in Food Processing
Editors: Gustavo V. Barbosa-Cánovas and Grahame W. Gould

Trends in Food Engineering
Editors: Jorge E. Lozano, Cristina Añón, Efrén Parada-Arias,
and Gustavo V. Barbosa-Cánovas

**Pulsed Electric Fields in Food Processing:
Fundamental Aspects and Applications**
Editors: Gustavo V. Barbosa-Cánovas and Q. Howard Zhang

Engineering and Food for the 21st Century
Editors: Jorge Welti-Chanes, Gustavo V. Barbosa-Cánovas,
and José Miguel Aguilera

Unit Operations in Food Engineering
Editors: Albert Ibarz and Gustavo V. Barbosa-Cánovas

Part I

Vision

1   Food Engineering for the 21st Century

D. B. Lund

CONTENTS

Abstract

As a discipline, the application of engineering to food has resulted in the ability of societies to avoid widespread malnutrition, provide a healthy workforce, and have access to a varied, nutritious, low-cost food supply. With projections of the world population leveling out at slightly less than 10 billion by 2050, food engineering is more important than ever to ensure food availability. Some of the new technologies applicable to food are reviewed in this chapter.

1.1   INTRODUCTION

Although estimates vary somewhat as to both total equilibrium population of the Earth and the time when that will be achieved (year of zero population growth), none of the projections are for fewer than 10 billion people by the first quarter of the twenty-second century.[1] Although it has been demonstrated that science and its application can enhance our ability to produce food, the capacity to preserve food is equally important. It is interesting to note that population expansion has greatly increased with man's ability to cultivate food crops and domesticate animals and, ultimately, to preserve the nutritional value of these materials as foodstuffs (Figure 1.1).

With the introduction of industrial food preservation in England and France at the beginning of the nineteenth century (canning and appertization) and with subsequent knowledge of the causes of food spoilage and deterioration (microbiology and chemistry), it was possible to develop large cities with industrial complexes.

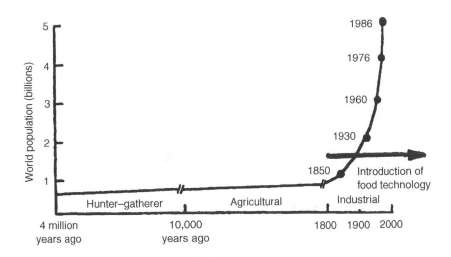

Figure 1.1 Foraging to farming to food technology. (Reprinted from Reference 2 with permission from The Nutrition Society.)

Although society has benefited greatly from the application of industrial technology to the preservation of food, we have not eliminated problems arising from lack of food (starvation) or nutritional deficiencies. FAO/WHO[3] estimated that 12.9 million children under the age of five died in 1990; 13 million preschool aged children were affected by zerophthalmia, with 500,000 becoming partially or totally blind; 28% of the world's population suffered from iron deficiency anemia; 1 billion people lived in iodine-deficient areas; 217 million people suffered from goiter; 1 billion people were at risk of malnourishment; 192 million children suffered from protein-energy malnutrition; 2 billion people had micronutrient deficiencies and/or diet-related noncommunicable diseases in developing countries. Poor diets can result in high infant and childhood mortality, low birth weight, substandard growth, reduced life expectancy, and increased healthcare expenditures.[4]

Given the outlook for food security, it is clear that the food system must be approached in its totality to ensure rational solutions. Tweeten[1] estimated that, given current scenarios of increase in yield from plant food materials, and assuming that fish and livestock production will continue to increase at the same rate (approximately 1.5% increase per year), food demand will require between 50% and 100% growth in production by the year of zero population growth (late twenty-first century to the first quarter of the twenty-second century).

It has been suggested that transgenic plants (and, perhaps, ultimately, transgenic animals) will allow us to keep pace with the growth in food demand. Interestingly, acreage of transgenic plants on a global basis was 4.3 million in 1996, 27.5 million in 1997, and 69.5 million acres in 1998 (excluding China).[5] Most of this acreage has been in soybeans and corn. Although yield is an important trait, transgenic plants have been used primarily for herbicide tolerance and insect resistance (or a combination of the two) and, to a much lesser extent, quality traits.[5] However, consumer resistance to acceptance of transgenic plants has been vocal and must be overcome if transgenic plants are to be widely adopted.

It is well understood that the food chain consists of a series of intricately linked activities, from farm input and production through processing, wholesaling/retailing, and, ultimately, consumption in eating establishments and homes. The food chain results in the availability of nutrients for sustenance of human life and human health, utilizing natural resources. This set of activities occurs within a framework of economic, biological, social, and political contexts. An important and integral part of this system is food processing, where technologies are applied to assembling, processing, and distributing food.

1.2 FOOD PROCESSING

The art of food processing has been known from the earliest written records that include reference to fermentation and drying processes. The abilities to store food from the time of harvest and plenty to other seasons of the year, and to make it transportable for long journeys and voyages, were essential for the continued development of society and the improvement of mankind.

In addition to extending shelf life, there are other important reasons for processing food. Food processing may render a foodstuff more digestible, inactivate naturally occurring toxins, or result in the formation of novel food products (for example, ice cream). Development and advances in food process technology are supported significantly by the private sector. However, there is clearly a rationale for public investment in the advancement of knowledge and application of technologies for the preservation and processing of food. Clearly, a healthy, well-fed population results in less strife within society and a more productive workforce and a more enlightened populace.

The "older" methods of food preservation and processing include drying, canning, chemical preservation, and refrigeration/freezing. The use of physical preservation methods (heating, freezing, dehydration, and packaging) and chemical methods (pH and preservatives) continue to be used extensively, and technological advances to improve the efficiency and effectiveness of these processes are being made at a rapid rate. The basis for these traditional methods is to reduce microbiological growth and metabolism to prevent undesirable chemical changes in food.

Minimizing microbial spoilage not only relies on processes for inactivating microorganisms but also is dependent on good manufacturing practices, sanitation, and hygiene. Furthermore, the application of several methods to stabilize food, either in parallel or sequentially, has been advanced through the concept of hurdle technology.[6] Once the concept of hurdle technology was articulated, it was immediately recognized as having already been broadly utilized in the preservation of food. "Hurdles" may be thought of as barriers to microbial growth and deterioration of food or chemical destabilization of food. Examples of hurdles include heating, chilling, water activity, pH, oxidation-reduction potential, preservatives, and irradiation. Each hurdle has a specific inhibitory effect on a mode of deterioration, and their combination will result in preservation of the food material. Nearly all preserved foods are examples of the application of hurdle technology. A guiding principle in the application of hurdle technology is to understand the main cause of spoilage and to know the effects of each hurdle on the root cause of spoilage (Table 1.1).

TABLE 1.1
An Overview of Different Types of Hurdle-Preserved Food Products

	Cottage cheese	Ham	MAP packaged salad
Main cause of spoilage			
Microbiological	X	X	X
Biochemical		X	X
Physical	X		
Sensory			
Type of hurdle			
High temperature		X	
Low temperature	X	X	X
High acidity (low pH)	X		
Low water activity (a_w)		X	
Low redox potential (Eh)			X
Preservative(s)	X	X	
Competitive flora		X	
Modified gas atmosphere			X
Packaging film			
Ultrahigh pressure			
Extrusion			
Product origin			
Traditional	X	X	
Recently developed			X
Country in which developed/marketed	UK, USA	USA	FRANCE, UK, USA
Shelf life (weeks)	2	>2	1

Reprinted from Reference 6 with permission from Elsevier Science.

1.3 NEW TECHNOLOGIES

Most of the newer technologies being applied to food are nonthermal in nature (i.e., they do not involve the significant elevation or reduction of product temperature). These methods include pulsed electric fields, pulsed light, ionizing radiation, high hydrostatic pressure, and active-smart packaging.

Pulsed electric fields have recently received considerable attention, although the use of pulsed electric fields to inactivate enzymes and microorganisms was convincingly demonstrated in the 1960s. The mode of action is primarily through lysis of the microbial cell (Figure 1.2).

Pulses are generally on the order of microseconds with rapid cycling (5–10 Hz). Figure 1.3 demonstrates the inactivation of *Escherichia coli* in liquid eggs subjected

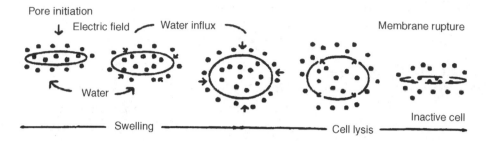

Figure 1.2 Mechanism of cell inactivation by pulsed electric fields. (Reprinted from Reference 7 with permission from Elsevier Science.)

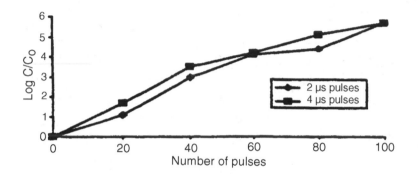

Figure 1.3 Inactivation of *Escherichia coli* suspended in liquid egg at 26 kV/cm and 37°C using 2 µs and 4 µs pulses [$Log(C/C_0)$ = plate count after electric pulse treatment (colony-forming units/ml)/plate count before electric pulse treatment (colony-forming units/ml)]. Reprinted from Reference 7 with permission from Elsevier Science.

to electric field pulses. This technology has been widely demonstrated in a variety of products against a number of microorganisms (Table 1.2).

Pulsed light technology is an intense exposure of a product to simulated sunlight. The technique requires the direct exposure of a microbial cell to a light pulse. Consequently, its greatest effect is on surfaces of packaging material or on other smooth, regular surfaces. The technique has inhibitory effects on a variety of microorganisms and has been shown to extend the shelf life of baked goods, seafood, meats, and water.[8] The technology is currently being commercialized, particularly for sterilizing food packaging materials (Figures 1.4, 1.5, and 1.6).

Whenever new technology is discussed, it is likely that irradiation of food will be highlighted. This analysis is no exception to that rule. Despite a long history of active promotion as an effective method for increasing the shelf life of foods and ensuring food safety, irradiation has never met expectations for its application in food. The most important aspect is that it does not have wide acceptance with consumers. Irradiation has received increased attention recently with outbreaks of illness associated with *E. coli* O157:H7 and species of *Salmonella*. Approval of the

TABLE 1.2
Applications of High-Intensity Pulsed Electric Fields in Food Processing

Product	Treatment regime	Inoculate	Maximum inactivation: log reduction (D)	Chamber characteristics
Orange juice	33.6–35.7 kV/cm 42–65 C; 35 pulses 1–100 μs	Natural microflora	Shelf life extended from 3 d to 1 week (5 D)	Static chamber; parallel stainless steel electrodes; 2 cm gap; volume of 157 cm^3
Milk	36.7 kV/cm; 63°C; 40 pulses; 100 μs	Salmonella dublin	3 D, resulting in 0 cfu/ml Salmonella dublin after treatment	Static chamber; parallel stainless steel electrodes; 2 cm gap; volume of 157 cm^3
Liquid egg	25.8 kV/cm; 37°C; 100 pulses; 4 μs	Escherichia coli	6 D	Continuous chamber; coaxial stainless steel electrodes; 0.6 cm gap; volume of 11.87 cm^3

Reprinted from Reference 7 with permission from Elsevier Science.

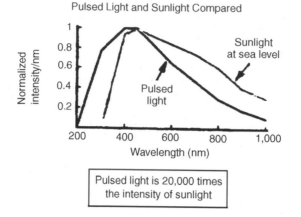

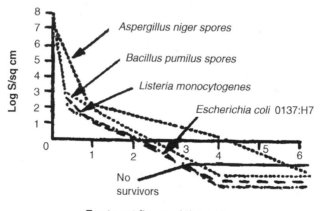

Figure 1.4 Pulsed light resembles sunlight—both have similar spectral peaks and distribution. Pulsed light contains some UV wavelengths filtered from sunlight by the atmosphere, is pulsed, and is much more intense. (Reprinted from Reference 8 with permission of the Institute of Food Technologists.)

Figure 1.5 Pulsed-light kill kinetics for a variety of microorganisms. (Reprinted from Reference 8 with permission of the Institute of Food Technologists.)

application of irradiation to meat and other food products will accelerate acceptance of this technology. Food irradiation has been applied for many years to the inactivation of microorganisms in spices.

Although the idea is nearly a century old, it is only now that hydrostatic pressure is being seriously applied to the preservation of food. The technology is relatively straightforward, and there are many applications in food processing (Table 1.3). Although the commercial adoption of this process technology has been slow, the Japanese have aggressively pursued its application to food. High-pressure processed

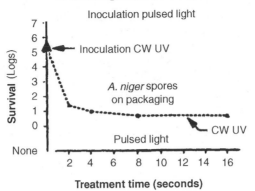

Figure 1.6 Comparison of kill kinetics for pulsed light and continuous-wave UV. (Reprinted from Reference 8 with permission of the Institute of Food Technologists.)

foods have been available in Japan since 1990. Pressures for processing are on the order of 100 to 1000 MPa (1 to 10 kbar).

One of the most promising technologies for extending the shelf life of food products is active/smart packaging. Ensuring that the food product is protected from external environments during storage and distribution is essential to maintaining shelf life. Active packaging is defined as the development of packages that actively and constantly change the internal atmospheric composition during storage and distribution. These techniques rely on oxygen absorbers, moisture absorbers, moisture regulators, CO_2 absorbers, ethylene absorbers, CO_2 emitters, and ethanol emitters. Furthermore, chemicals may be added to or incorporated in packaging materials that serve as antimicrobial agents, antioxidants, or that provide enzymatic activity (Table 1.4). Active packaging is now being used to extend the shelf life of minimally processed foods (chef-like foods) as well as for long-term shelf-stable foods (e.g., cereals).

Smart packaging systems are systems in which information is provided to the consumer; this information can be used to indicate product abuse or product quality (Table 1.5). A good example of smart packaging is external time temperature indicators. Some indicators are based on "go/no-go" abuse of the product (integrators), whereas others may indicate whether the product has experienced abuse temperatures. Although increased commercialization of active/smart packages will need to take into account a number of problems, none of these is expected to decrease the intense interest in ultimate application of active/smart packaging (Table 1.6).

1.4 FUTURE

The adoption of new technologies for application to food preservation and processing is, indeed, slow. Some would argue that it should be a slow process because of the

TABLE 1.3
Applications of High-Pressure Technology to Food Processing

Effect	Solid food					Liquid food	
	Fish	Meat	Eggs	Rice, starches	Soybean protein	Milk	Natural juice
Prolongation of storage time						☆	☆
Prevention of microbial contamination	✔	✔	✔	✔	✔	✔	☆
Development of new foodstuffs	☆	☆	☆	☆	☆		
Manufacture of partially cooked food	☆	☆	☆	☆			

Note: ☆ = Applicability is large; ✔ = Application exists.

Reprinted from Reference 9, courtesy of Marcel Dekker, Inc.

TABLE 1.4
Examples of Active Compounds Added to Packaging Materials

Function	Reagent	Application
Antimicrobial effects	Antibiotics (nisin)	Fish
	Chitosan	Fresh fruits
	Horseradish-derivative	
	Imazalil	
	Ceramic compounds:	
	Aluminum silicate	
	Silver	
	Copper	
	Synthetic zeolite	
	Manganese	
	Nickel	
	Zinc oxide	
	Magnesium oxide	
Antioxidative effects	BHT	Cereal
	BHA	
Enzymatic effects	Cholesterol reductase	Milk
	Glucose oxidase	All kinds of foods

Reprinted from Reference 10, with permission from Taylor & Francis, Inc., http://www.routledge-ny.com.

TABLE 1.5
Examples of External or Internal Indicators that Can Be Used in Smart Packaging

Technique	Principal reagents	Application
Time-temperature indicators (external)	Mechanical Chemical Enzymatic	Foods stored under chilled and frozen conditions
O_2 indicators (internal)	Redox dyes pH dyes	Foods stored in packages with reduced O_2 concentration (tamper evidence)
Microbial growth indicators (internal)	pH dyes All dyes reacting with certain metabolites	Aseptic products

Reprinted from Reference 10, with permission from Taylor & Francis, Inc., http://www.routledge-ny.com.

potential impact on humankind. Given that adoption of new technologies proceeds at a snail's pace, it is interesting to note the observations of the Executive Advisory Panel of *Food Engineering* magazine. They identified the top ten manufacturing trends (Table 1.7). Automation, information, and integration were the number one trends, followed by new products, processes, and technology. Consequently, leadership of the food industry is seeking new technologies to enhance the safety, nutritional quality, and sensory quality of food products. The same panel of 201 respondents identified the manufacturing trend priorities for 1998–2003. In addition to manufacturing trends, high priority was also placed on total quality management, outsourcing, and formation of partnerships and alliances.

Finally, the same group was asked to look at the commercial potential of new and unique food process technologies.

It is notable (Table 1.8) that irradiation had higher potential in their opinion than in the past, along with microwave-assisted processing. Other technologies, including high pressure, pulsed light, and pulsed electric fields, have less potential, in their view. Advances in automation, new products, and unit operations require a fundamental knowledge of food systems. Chemical, physical, and transport properties must be understood to transform raw materials into complex foodstuffs for the consumer. Consequently, an understanding of structure-function relationships in foods is essential.

In addition to transport properties, it is imperative that reactions that govern rates of process operations and rates of chemically, physically and biologically induced changes in foods be understood. Quantifying these reactions continues to be a significant research goal. Collectively, these data represent food quality quantification in which reactions, which dictate microbial, sensory, nutritional, physical, and chemical quality of food ingredients or raw or manufactured foods, are described as kinetic parameters that depend on environmental conditions. Having such data will ensure progress in developing new technologies for assembling, processing, and distributing food.

TABLE 1.6
Problems and Solutions Encountered with Introducing New Products Using Active and/or Smart Packaging Techniques

Problem/Fear	Solution
Consumer attitude	• Consumer research; education and information
Doubts about performance	• Storage tests before launching • Consumer education and information
Increased packaging costs	• Use in selected, high-quality products • Marketing tool for increased quality and quality assurance
False sense of security, ignorance of date markings	• Consumer education and information
Mishandling and abuse	• Active compound incorporated into label or packaging film • Consumer education and information
False complaints and returns of packs	• Color automatically readable at the point with indicators of purchase
Difficulty in checking every color indicator at the point of purchase	• Bar code labels: intended for quality assurance for retailers only

Reprinted from Reference 10, with permission from Taylor & Francis, Inc., http://www.routledge-ny.com.

TABLE 1.7
Top Ten Food Manufacturing Trends

1	Automation/information/integration
2	New products/processes/technologies
3	Process flexibility/efficiency/productivity
4	HACCP/food safety/compliance
5	Outsourcing/co-packing
6	Teams/training issues
7	New/improved packaging
8	Alliance/partnerships
9	Acquisition/consolidation
10	Improved maintenance

Reprinted from Reference 11, Copyright 1998, *Food Engineering,* Cahners Business Information, a Division of Reed Elsevier, Inc. All rights reserved.

REFERENCES

1. Tweeten, L. 1998. "Anticipating a Tighter Global Food Supply—Demand Balance in the Twenty-First Century," *Choices,* 3rd Quarter, 8–12.
2. Henry, C. J. K. 1997. "New Food Processing Technologies: From Foraging to Farming to Food Technology," *Proc. Nutr. Soc.,* 56: 855–863.

TABLE 1.8
Commercial Potential of New/Unique Process Technologies (Percent of Respondents Familiar with Each Technology)

	High potential	Moderate potential	Low potential
Ohmic heating	10	36	54
Electron beam radiation	27	38	35
Gamma irradiation	33	35	32
High pressure	19	38	43
Radio frequency cooking	17	50	33
Microwave pasteurization/sterilization	37	36	27
Pulsed light	7	44	49
Pulsed electrical field	4	36	60
CO_2 drying	18	40	42
Microwave drying	20	43	37
Low-acid aseptic particulars	23	50	27
Magnetic resonance imaging	22	35	43
Predictive process control	44	44	12

3. FAO/WHO. 1992. International Conference on Nutrition. Nutrition and Development—A Global Assessment, Food and Agriculture Organization/World Health Organization, Rome, Italy.

4. Wotkei, C.E. 1998. Impacts of diet on health in North America, in *Creating Healthful Food Systems: Linking Agriculture to Human Needs*. G. F. Combs, Jr. and R. M. Welch, eds. Ithaca, NY: Cornell University.

5. James, C. 1998. Global review of commercialized transgenic crops: 1998, *International Service for the Acquisition of Agri-Biotech Applications*. Brief No. 8. Ithaca, NY.

6. Leistner, L. and L.G.M. Gorris. 1995. "Food preservation by hurdle technology, *Trends Food Sci. Technol.*, 6: 41–46.

7. Vega-Mercado, H. et al. 1997. Non-thermal food preservation: Pulsed electric fields, *Trends Food Sci. Technol.*, 8: 151–157.

8. Dunn, J. et al. 1995. Pulsed light treatment of food and packaging, *Food Technol.*, 49(9): 95.

9. Barbosa-Cánovas, G.V. et al. 1997. *Non-thermal Preservation of Foods*. New York, NY: Marcel Dekker, Inc.

10. Ahvenianen, R. and E. Hurme. 1997. Active and smart packaging for meeting consumer demands for quality and safety, *Food Addit. Contam.*, 14: 753–763.

11. Morris, C.E. 1998. 1998 survey of food manufacturing trends: A clear direction, *Food Eng.*, 70(3): 77–86.

2 Trends in Food Engineering

G. Trystram
J. J. Bimbenet

CONTENTS

Abstract

Technological innovation is presented as one of the answers to the constraints in the food industry. Examples of innovations or research, mainly in the fields of preservation (thermal and nonthermal techniques), manufacturing operations (operations on individual pieces), automatic control, etc., are presented. New tools and concepts used in product and process development are described, such as product, material, and reaction engineering. Emphasis is placed on heterogeneous and composite products. Finally, the authors present their ideas about challenges to the food industry in coming years.

The food industry is at the same time the instigator and the subject of change in society. When the food industry is subjected to several kinds of constraints, it has to modify its structures, and these evolutions have an impact on its technology.

2.1 CONSTRAINTS, CONTEXT, AND REASONS FOR INNOVATION

2.1.1 How Innovative is the Food Industry?

Table 2.1 presents our views of constraints to which the food industry is being subjected. Due to the context described here, the food industry has to manage innovation to modify and adapt its technologies. The objectives of this adaptation have varied with time, as shown in Figure 2.1. Presently, the accent is placed on safety and on an increase in quality homogeneity. But, when we consider the

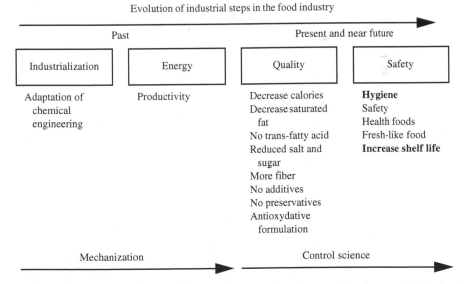

Figure 2.1 Some important steps for the evolution of objectives and requirements in the food industries.

TABLE 2.1

Constraints and Context of Evolution in the Food Industries

Consumer power	New challenges are coming; the food industry has to establish and demonstrate its ability to control its production in terms of quality, of course, but also, nowadays, more and more in terms of safety, nutritive value, and natural image of the product. People also require more convenience and information about the food they buy.
Distribution power	The competition in pricing and quality is increasing, mainly due to the increasing weight of distribution chains. The increasing pressure of retailers pushes industry to modify its way of distribution. Internet market may become an important trend.
Shareholders' power	On the international level, shares are bought and sold at the speed of electrons, often on the basis of short-term profits. Whole companies may similarly change owners in a very short time.
Respect for water, air, and environment	Like other human activities in developed countries, the food industry is asked by society to be "environmentally correct" concerning air, solid wastes, packages, landscape, and water. The pressure on water supplies is likely to become a major problem in many countries in the coming years.
Evolution in the structure of the food industry	To fulfill all of the previously described constraints at the same time, the food industry has a tendency to split itself into two entities: • *First transformation industry* (e.g., production of sugar, starch, oils, malt, etc.)—close to agriculture and international raw materials markets, making basic products for consumers and more and more ingredients for the food industry. • *Second transformation industry*—close to distribution and consumers, which mixes, assembles, and shapes ingredients to make complex products of various origins (as cereals + meals, dairy + fruits, etc.) This evolution tends to break the traditional organization of commodities from field to consumers (cereals, meat, dairy, etc.). The fast-growing activity in food ingredients is evidence of this trend.

introduction of new technologies in the food industries during the past few decades, the number of real innovations turns out to be rather low:

- There have been few new unit operations, except extrusion cooking, membrane separations, irradiation, high-pressure treatments, and, in a sense, manufacturing operations.
- Some new processes have been required to make new products (i.e., prepared salads, new composite desserts, osmo-dehydrated products, etc.) and new operations (membranes, extrusion cooking).
- Many new techniques are used in unit operations: aseptic techniques, super critical extraction and osmotic dehydration (both being new forms of solvent extraction), ohmic heating, RF heating, water-jet cutting, associative packaging, image analysis, etc.

This shows that most of the existing technologies have been in use for a long time; there is more progressive evolution than striking innovation.

2.1.2 WHAT ARE THE REASONS FOR INNOVATION?

A set, probably not exhaustive, of reasons for innovation consists of the following:

- Although heat is the most common method for transformation, sanitation, and preservation, it is well known today that the consequences of heating are not necessarily good for the product. Therefore, nonthermal processing is an important objective. On the other hand, the ability to perform accurate separations of biomolecules becomes more and more important. The consequences of such progress are the lengthening of preservation time and an increase in the consumer's perception of the food as being "natural."
- Another driving force for innovation is probably the attainment of new properties (texture or aroma, for example), which may require new technologies. The design of new products is a matter of competitiveness for industry. In such new products, safety considerations become very important (Figure 2.1).
- The competition between companies and the relative ease in the design of "me-too" products imply firms' increasing focus on technologies involved in the process.
- Innovation is evidently the direct result of research and development within the firm. But it is also the consequence of research made elsewhere. Transfer from one industrial domain to another is a frequent path of innovation. Screw extrusion was used in the plastics industry before being transferred to the food industry to be utilized in extrusion cooking. Today, one of the most promising directions of research and innovation is certainly derived from rapid progress made in the field of biology.

Another important point is the acceptance of the new technology by the user. It is obvious that this mechanism of acceptance is not easy to implement. Figure 2.2 represents different points that have to be considered during the evaluation of a new technology for industrial purposes.[1]

2.2 TECHNOLOGICAL TRENDS

2.2.1 EMERGING PROCESSING AND PRESERVATION METHODS

An exhaustive presentation of emerging methods for the processing of food products is quite difficult. In the case of preservation, some points are summarized in Table 2.2. In some cases, technologies have already been transferred to industry. The main idea for such research is to process food without heat. In fact, in many of these technologies, some heating occurs during processing. Except in some specific applications that are highlighted in Table 2.2, it becomes obvious that a combination of technologies is preferred.

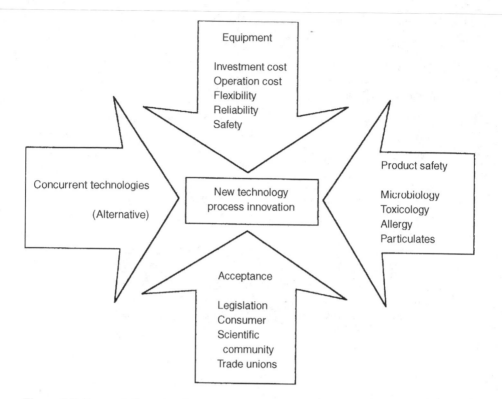

Figure 2.2 Factors influencing the success of new technologies at the industrial level.

Also, in the heat processing of food, major improvements can be provided using a combination of classical heating methods (convection in air or liquid) and new technologies as presented in brief in Table 2.3. The way to combine these technologies is not well established at present. Nevertheless, numerous applications are available.

An interesting point to discuss is the aseptic processing principle. It must be indicated here that it is not really a new unit operation or principle of processing, but it is a new set of technologies that permit work to take place in a safe and hygienic climate. The next question for such processes that researchers and engineers have to face concerns process optimization. But the introduction of the hygiene point of view probably will be very important to the future of food industries (see Figure 2.1).

2.2.2 BIOTECHNOLOGY

The use of biotechnology and of biologic steps during processing has increasing importance. A comprehensive discussion would be necessary to describe this aspect. Some of this information is reported in Reference 2. Table 2.4 summarizes a few

TABLE 2.2
New Nonthermal Principles of Preservation Treatments

	Principle	Applications (functions)	Advantages	Products
High pressure	Increase of pressure (until 8000 bars)—the stress applied modifies the behavior of microorganisms, proteins, etc.	Inactivation of microorganisms; thawing; freezing; diffusion of solutions (impregnation); and protein denaturation	Low-temperature process	Packaged product (in batch)—may be continuous for liquids
Pulsed electric field	A high-electric field is applied. If the value is higher t an a critical value, pores appear in the cell membranes. The deterioration of membranes is i reversible.	Sanitation, extraction	Nonthermal processing; different effects on parameters such as pH, temperature, etc.; Inactivation at lower temperature, alternative for pasteurization	Effect on spores is unknown—only pumpable products
Pulsed light	Very strong and fast light pulsations; effect of high peak power and broad spectrum of flash; real principle unkncwn	Disinfection, sterilization	Surface inactivation—all types of microorganisms are inactivated (spores and viruses included)	Pumpable foods; packages; in package possible; water processing
Pulsed magnetic field	High-density pulsed magnetic field provokes a dra natic decrease of microorganisms.	Sterilization, but the principle is not exactly known	Number of studies is still too low	

TABLE 2.3
New Principles of Heat Treatments

	Principle	Applications	Advantages	Products
Microwave/high frequency	*In situ* heat generation	Heating and thawing	Direct *in situ* heating—often combined with air	Numerous applications
Infrared	Infrared radiation heats the surfaces— small penetration depth.	Heating, surface treatment, and pasteurization	Function of optical properties of food	Liquid or pieces of products; application available for in-package products
Direct ohmic heating	Electric current is directly injected in the pipe; Joule effect, also in pieces of products, if present.	Heating, pasteurization	Direct *in situ* heating	Pumpable products, even if particles are present
Indirect ohmic heating	Electric current sent in exchanger walls.	Heating, pasteurization	Excellent control of exchanger walls temperature	Pumpable products
Superheated steam	Superheated steam heats the product like a hot gas.	Baking, drying and roasting	High-intensity drying; high heating rate	Thin product pieces
Immersion	Pieces of product are immersed in a concentrated solution, heated or not heated.	Dehydration, soaking, salting, and pickling	Increase of transfer coefficient; control of impregnation or drying is possible through operating conditions	Fruits, meat, fish products, and vegetables
Immersion frying	Pieces of product are immersed in oil at a high temperature (180°C).	Drying, baking, frying and extraction of oil	Increase of transfer coefficient; fast drying	Meat, cereal products, fruits, vegetables, and fat products
Under vacuum	Vacuum permits the reduction of temperature.	Every kind of heat treatment	Lower temperature	Every product
"*Sous-vide*" cooking	Heating of tightly packaged products.	Cooking and pasteurization	Quality, safety and less losses	Precooked prepared meals for cold storage
Heat processing improvement	The most used method. The combination of techniques is made to improve heat transfer. The main ways to improve heat transfer are to increase convection (better heat transfer coefficient), introduce controlled radiation and combine steam injection with an exchanger.			

TABLE 2.4

Examples of Perspectives of Biotechnology Application for Innovation in Food Industries

Processing on the field	Genetic and agronomical engineering for agricultural products permits the design of new raw materials. The incidence on processing is not established.
The cell factory	The use of microorganisms is a way to "do things that can more rapidly lead to new products." Bioreaction engineering (including enzymatic engineering) is an important area of progress.
New analytical tools	Probably, this is the major area of interest, with the design of new specific probes for measuring pathogens, unwanted xenobiotic, etc. Specific kits and at-line sensors become available.

directions of progress in biotechnology that, in our opinion, offer important contributions to the evolution of food industries. When more generally speaking of biology, nutrition must also be quoted as becoming a major incentive for the creation of new products.

Other technological evolutions have been described above. We now present as separate topics two specific trends: food manufacturing operations and automatic control.

2.3   FOOD MANUFACTURING OPERATIONS

A visit to almost any food plant will show two types of operations:

1. Treatment of product in bulk, mainly liquids or solid particles, corresponding to the classical unit operations of chemical and food engineering (centrifugation, heating/cooling in exchangers, distillation, milling, etc.)
2. Treatments on "objects," i.e., products like pizzas, cakes, pieces of meat or fish, and packaged products (cans, bottles, etc.); examples of such operations include the deposition of fruits on a pie, cutting, molding, assembling of several parts, packaging, etc.

In fact, most products sold to consumers (excluding most ingredients sent to secondary transformation) in industrial countries for many years have no longer been commercialized in bulk. Products are packaged, and often shaped, either traditionally (e.g., bread, sausages, cheeses, and biscuits) or in new shapes (e.g., fish fingers and frozen hamburgers). Furthermore, people consume more and more composite objects such as two-layer dessert creams, multilayer cakes, ice creams in cones, pizzas, industrial sandwiches, and prepared dishes. All of this means that growing parts of food plants are devoted to forming, assembling, conveying, and otherwise processing such objects.

2.3.1   WHAT ARE FOOD MANUFACTURING OPERATIONS?

We have proposed[3] to define a new category of unit operations of food engineering that we could call *food manufacturing operations*. It could be defined as operations considering objects individually or starting from a bulk product to make such individual objects. These objects are generally "large," but their size is not the relevant criterion. If fruits are peeled by a knife, they receive an individual treatment, and the position of each of them is determined; this may be considered a manufacturing operation. When potatoes are peeled by abrasion, the position of each of them is not controlled; there is random treatment on a bulk product. Similarly, the color sorting of coffee granules by high-velocity optical machines considers each grain individually, and this operation can be considered food manufacturing.

2.3.2 Can We Define Food Manufacturing Unit Operations?

We can tentatively make a list of such unit operations (Table 2.5) as follows:

- Many heat and mass transfer operations (can sterilization and drying) are based on principles that do not differ from the principles of operations performed on bulk products, and they are classically studied.
- The same situation exists for reactions in food objects; their rates are determined by heat and/or mass transfer and/or by reactions kinetics—all classical concepts.
- Transportation of objects is not specific to the food industry if these objects are packages or packaged products; however, a specificity exists if it concerns bare food objects, because problems of stickiness, hygiene, and deformation may be encountered if the products are semisolids.

This means that the most interesting and original of these unit operations consists of shaping, separation, and assembly, including packaging.

2.3.3 Characteristic Features of Food Manufacturing Operations

These manufacturing operations have other features that justify special interest. Many of them treat products on open conveyors or in open vessels or equipment, and they include human handling or the close proximity of humans. This means that hygienic questions are often critical for such operations and justify the use of microbiological control of the atmosphere. Clean rooms or aerobic protection of equipment must be employed.

In many instances, heterogeneity is part of product quality, especially when a composite object consists, for example, of the combination of soft and crispy layers. The problem is then to control the transfer of water (and/or other molecules) between these layers.

In matters of quality, each piece must fulfill certain requirements (weight, composition, contamination, etc.), whereas bulk products are sold by total weight and average characteristics with certain variation allowable among samples.

Many of these operations are results of the industrialization of manual operations developed in kitchens. Mechanization may be difficult due to the complex nature of

TABLE 2.5
Examples of Food Manufacturing Unit Operations

Principle	Unit operation	Example of products
Forming	Molding	Bread, biscuits, sugar
	Extrusion	Biscuits, ham, sugar
	Lamination	Pasta
Separation	Disassembling	Meat carcasses
	Cutting	Meat, fish
	Peeling	Fruits
	Sorting	Fruits, vegetables, coffee grains
Assembling	Filling	• of liquids
		• of pasty products
		• of powders and particles
	Dosing, deposition, powdering, coating	On pizzas, cakes, prepared meals, all composite products
	Arrangement	Candies, cookies in boxes
	Closing, sealing	Bottles, cans
	Labeling	On bottles, cans, etc.
	Wrapping, cartoning	Bottles, cans, etc.
Transportation	Conveying	Everything
Heat and mass transfer	Cooking	Bread, biscuits, meat
	Canning	Preserves
	Roasting	Meat, bread
	Cooling, freezing	Meat
	Salting	Ham, cheese
	Drying	Ham, cheese, sausages
Reactions	Biological and enzymatic	Cheese, yogurt, dry sausage

many products (thick liquids, pastes, or semisolids—fragile, deformable, often sticky) to the composite structure of many of them and to their complicated shapes or dispositions. For example, think of how we could mechanize the deposition of four anchovies on a pizza. In such operations, the mechanical design of the machine must be related to a knowledge of the mechanical behavior of the product.

Even then, repetitive mechanization, which requires constant human supervision, is real automation and is difficult to realize, because it supposes some real-time measurements. The weight of pieces is fairly easy to measure (as is color), but the determination of shape may require image analysis techniques. Furthermore, automating the control of plants that include such operations is very different from the case of bulk products; it supposes the control of waiting lines, of flow rates measured in objects per minute, etc., which are all techniques that have been more highly developed in mechanical industries than in food processing.

In many cases, the same plant has to produce a succession of several batches of products in the same day, which means that flexibility is necessary, and this may require robotization (in the sense of programmable mechanization). This may limit our interest in investments in mechanization, robotization, and automation.

To all of these specifics may be added the fact that these operations so far have not been the objects of education and academic research commensurate with their importance in industrial investments and with the concerns of food plants operators. One important exception is packaging, which has received some attention in recent decades. This deficiency of research, however, has been alleviated by the transfer of technology from mechanical industries (products, robotization). These considerations raise our awareness of the need to give more importance to the topic of mechanics in the food industry.

2.4 AUTOMATIC CONTROL

It is well recognized today that control science is one of the important avenues for progress in the food industries.[4] A review of applications and the potential of control science in the food industries has been presented.[5] The main points and ideas are as follows. In parallel with heat and mass fluxes, which are classic for food engineers, the complexity of flow sheets implies that fluxes of information are essential and must be taken into account. As it may appear from Figure 2.1, the new objectives of production imply a necessary evolution from mechanization (important for productivity criteria) to control (important for quality and safety criteria). Without control, many processes cannot work.

The consequence is that numerous studies have proposed the introduction of new sensors. Figure 2.3 proposes a set of available methods for sensor design. It is important to remark that most of the progress today is being made using classical (simple-to-use) sensors in combination with computer-based applications.

The direction of algorithm design for control purposes is still under active development. Nevertheless, an important gap remains between the level of laboratory

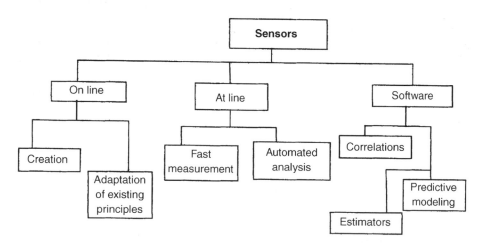

Figure 2.3 Principle of sensor development for food process control applications.

development and real industrial applications. Most of the implemented algorithms at the industrial plant level are simple (PID-like) controllers. But the efficiency of many tools actually developed becomes very attractive, especially for nonlinear and multivariable situations (which is often the case in product quality-based control).

The second important and actual direction is the search to establish relationships between product qualities and process operating conditions using data analysis tools and statistical methods. One subject of such approaches is the on-line application of these methods. Statistical process control (SPC) is one approach. Good results are obtained, although with poor predictive abilities, which means that the importance of the supervisory level probably increases.[6]

The last but not least important significant trend is the control of product during and after production. The goal is to ensure the control of the desired properties of the product. It must be established by performing numerous studies so as to develop mechanisms to measure consumer-related properties: "electronic noses" for odor and aroma, image processing for color and shape, etc. This is an important direction and, although we still are not able to design simulation tools that are able to predict the influence of processing on every property of the product, progress has been made in the on-line control of such properties.

Probably one of the most important directions for future applications is the combination of process design and automatic control tools. New processes could be designed, and the process operations will become safer and more reproducible.

2.5 ADVANCES IN TOOLS AND CONCEPTS IN FOOD ENGINEERING

2.5.1 TOOLS AND CONCEPTS IN PRODUCT DESIGN

As described, the objectives for food production are constantly evolving.[7] Nowadays, more natural, fresher, and easier-to-use products are expected by consumers.

A recent consideration is the necessity of a product engineering approach. At the recent European Conference on Chemical Engineering (ECCE 2, Montpellier, France, October 1999), it was illustrated that, in every industrial domain, there is an important need for improved tools and methods in the field of product engineering. The key issue is the ability to predict the consequences of the process on the overall quality of the final product.

Classical research in such topics is abundant. Different points of view could be highlighted for modeling approaches (Figure 2.4). The first is the black box study of formulation and process. Numerous works have been published, and the limitations of such methods are now clearly established. Nevertheless, there is still work in progress, even if the prediction ability of the resulting tools is poor.

More convincing works try to adopt the reaction engineering point of view. The classical reaction engineering toolbox[8] permits the combination of chemical or biochemical reactions and the evolution of important factors such as temperature, moisture content, etc. Very efficient tools exist that permit the inclusion of product quality into design or control simulation tools (see Reference 9, for example). Based

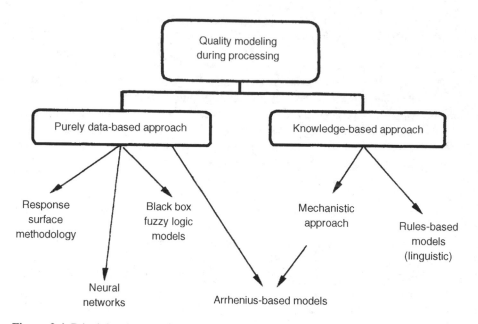

Figure 2.4 Principle of modeling approaches for product engineering purposes.

on kinetics experiments and dynamic modeling, the optimization of product properties becomes possible. Numerous modeling approaches have been described in the literature at this stage. (See Tréléa et al.[10] for the artificial neural network approach, Perrot et al.[11] for the use of fuzzy logic, etc.).

Nevertheless, there are some limitations. The first is due to the fact that not all of the desirable properties of a food product are necessarily related to a set of reactions. In many cases, there are transformations due to physical mechanisms. Even if some equivalent reaction engineering approaches are available,[12,13] for biscuit spring for example,[14] the design of simulation tools is still difficult.

The other reason for limitations is that a food product is generally complex and heterogeneous, whereas the available reaction engineering tools are valid for homogeneous properties. No tools are available to design heterogeneous products. As highlighted by Feillet,[15] the manufacture of a food product with uniform characteristics is more related to complex systems engineering. It has probably been ignored for many years that food products are complex systems with strong interactions, and that an understanding of the properties of each element (or component) is not sufficient for controlling the overall properties.[16] From this point of view, it is important to manage research dealing with formulation and in the field of process engineering. Today, most of the work is performed in one of these areas only. Integrated approaches are necessary. Another important point is related to the real objective of product formulation. Consumer demands must be taken into account, including the dimensions of the desired product.[16] A product design mechanism probably will be necessary in the future.

Increased interest in the reaction engineering approach is nevertheless very important. Our knowledge of and our ability to predict the kinetics of biochemical deterioration of food during storage are progressively improving.[17] Predictive microbiology predicts microbiological risks as a function of food processing and of storage conditions. A second emerging approach is material science and engineering. Material science may be defined as the understanding of material properties evolution due to stresses or constraints (from processing, for example). Material engineering is the application of this new field of knowledge to texturization and preservation problems. It is well known today that significant progress is made using, for example, the concept of glass transition.[18,19] The association of state diagrams, the relationship between glass transition and transport properties on one hand and biochemical reactions on the other hand, is recognized as an efficient tool. But it is necessary to reiterate that such criteria are based on the assumption of a homogeneous material. A food product is often heterogeneous. At the present time, no methods for studying these problems are readily available.

However, it must be considered that the choice of processes in food production is subject to regulation and limitations of social acceptance. A typical example is the difficulty encountered in obtaining authorization for the use of irradiation and achieving consumer acceptance. More recently, proponents of ohmic heating and high-pressure techniques had to demonstrate the safety of these technologies.

A key direction in the future will be the creation of new tools, and certainly of new ways of thinking, including consideration of the food product in its complexity, so as to be able to design a product and an appropriate processing line from the laboratory level to the plant level.

2.5.2 TOOLS AND CONCEPTS IN PROCESS DESIGN

It first must be emphasized that, due to the increasingly sophisticated properties that consumers desire in products, it is increasingly necessary to integrate process design into product design,[7] as mentioned above.

One of the objectives of process design is the *scale-up* procedure. In terms of emerging technologies, appropriate tools for that purpose are not obvious, and constraints are important considerations. The main method in process engineering for scaling up is simulation. But, before one can use simulation tools, it is necessary to choose the appropriate types of equipment. Equipment design software programs have become typical for many operations, but expert systems for the selection of equipment are more difficult to develop, because they must integrate unformulated knowledge. Some have appeared in recent years.[20]

A limiting factor for simulation is related to databases of thermophysical (and sometimes thermodynamic) and transport properties. These databases may also be necessary at this step for calculating energy requirements (and therefore of utilities) for heating and cooling products, for determining heat or mass transfer coefficients between phases (which determine equipment design), and for determining heat and mass penetration times to and from the center of products. Because most of these data are not available (but some correlations have been validated), different approaches could be adopted for modeling. The direct transfer of the principle of the model library (for example, available in the Aspen Tech library) is not necessarily the best way. To us, a library of modeling methods is probably better. Figure 2.5 illustrates some approaches that we have validated for numerous applications.

The integration of unit operations in the whole plant is now typically done by flow sheeting in the chemical industry. In the food industry, its use is less frequent, possibly because of the greater complexity caused by the coexistence of continuous and discontinuous operations and because products are treated in bulk and in pieces. However, considering that food industry needs to reduce production costs and shorten product launch times, engineering and reengineering tools are becoming more and more necessary, but they are not always available or well known in the food industry. In some instances, these considerations will lead to alternative operations or processes.

The most fascinating tool that has been applied in the food industry in the last years is computational fluid dynamics (CFD). The result is spectacular in aerodynamics; for example, it allows the simulation (and therefore the improvement) of the circulation of airflow in meat carcass chilling tunnels, in dryers,[21] in cheese ripening rooms, in stacked products, etc. An important need for such a tool also arises from stricter hygiene requirements; for example, the design of air curtains around machines becomes possible. CFD may also help to optimize the shape of heat exchangers for liquids. More recently, it has proved to be efficient in the simulation of velocities and stresses in the flow of complex (non-Newtonian, thixotropic) fluids. However, so far, it remains unable to predict transfer coefficients between phases.

Flows of utilities and circulation of all products (not only food products, ingredients, and additives, but also packages, CIP chemicals, and effluents, etc.) within the plant, as well as entrance to or exit from the plant, are part of the general design of the flow sheets.

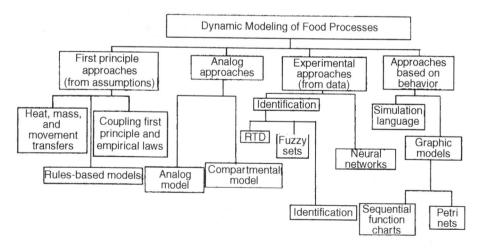

Figure 2.5 Some methods for modeling and simulating food unit operations and processes.

At this point, it must also be mentioned that, due to the increasing complexity of food plants, it is more important to integrate, as much as possible, the design of automatic control to the flow sheet design itself, including all utility flows, CIP, and product circulation. Simulation tools are very useful here for optimization of such designs.

The weight of hygiene constraints in the plant design has often become considerable. For example, the general layout of the plant may be based on the principle of avoiding proximity between contaminated and clean products, or on the optimal installation of clean rooms of different classes, on the circulation of people, or on a plant's ability to be visited by outside groups without allowing contact with critical products.

Finally, but not less important, environmental questions have to be taken into account as early as possible in plant design. Theoretically, the best concept is that of a "clean process," by which we mean that the process is designed to minimize flows of effluents (mainly by internal recycles and treatments), thus reducing (mainly) the amount of wastewater to be treated as such.

Figure 2.2 presents different points that have to be considered in the evaluation of a new technology for industrial purposes.

2.5.3 TOOLS AND CONCEPTS IN PLANT OPERATION

The food industry, especially secondary transformation, may be considered as an assembly line, comparable in that aspect to car manufacturing. Think of a dessert made of a layer of fresh cheese over a layer of jam. The items needed to assemble it are the plastic cup, the lid (plastic and metal), possibly the label(s), the jam, the fresh cheese, the plastic wrapping, and the carton. In the assembly process, these six to eight pieces are placed together. Comparable assembly procedures are used for many of the products we buy every day. Logistics has become a key function in many food plants, as well explained by, for example, Pearson.[22]

As noted above, the necessary conveying equipment inside the plant, along with appropriate control systems, must have been laid out as part of the overall plant design. But daily operation presupposes a permanent supervision of machines and conveyors, not only by computers but also by people. The temporary lack of one of the components is enough to stop an assembly machine or even a whole section of the line. Observation of food plants shows that incidents of mechanical origin are frequent. At an upper level, flux management and the concept of "just-in-time" operation makes this management more critical. The connection between sales, sales forecasts, raw materials and other supplies, production, and storage (i.e., logistics) has become a strategic factor in food firms (especially with regard to perishable products) and a part of modern food engineering in a broad sense, as much as it is a part of industrial engineering. We may also consider the logistics of distribution as related to food engineering, since transportation problems and cold chain control may require various mechanics, heat and mass transfer, chemical reactions, etc. And the logistics of distribution may be approached by the concept of residence time distribution (RTD).

Some of the tools come from the computer integrated manufacturing (CIM) concept and application. These considerations mean that organization of information fluxes are as important and necessary as heat and mass fluxes.

Concerning quality management, the food industry, like others in recent years, has rapidly developed and used more formal procedures of quality assurance: recommendations related to hygiene, HACCP, ISO certification, etc. The concept of traceability, imposed by consumers, has led to new systems to identify and track product batches.

2.6 WHAT CHALLENGES EXIST FOR FOOD ENGINEERING?

Most of the advances in this field will come from multidisciplinary research and the transfer of knowledge and technology from other sectors. In addition to the fields of present evolution described above, examples of challenges for progress and research in the upcoming years are as described below.

2.6.1 INTEGRATING RAPID PROGRESS IN BIOLOGY INTO FOOD ENGINEERING

- Can we apply the laws of heat and mass transfer with predictive microbiology? For example, is it possible to associate the prediction of temperature and water content evolution at the surface of a product to predict the development of microorganisms?
- What reliable and hygienic sensors can we integrate in chains of automatic control?
- Will rapid detection of contaminants be integrated in automatic control systems?
- Can the recent advances in biofilms bring about progress in equipment cleaning and disinfection?
- Mass transfer barriers associated with nutrition science can be used to design foods with controlled release.
- Progress in bioreactors creates new possibilities, especially in making more biological additives or performing biological transformations to food products.
- Progress in genetic engineering will have to be assimilated in the processes if they are to be accepted by the general public.

2.6.2 INTEGRATING PROGRESS IN PHYSICO-CHEMISTRY INTO FOOD ENGINEERING

- Magnetic resonance imaging (MRI), in the form of nuclear magnetic resonance (NMR) imaging, is being used to determine water content profiles in products during and after processing. Will this method be applicable to small objects (<1 mm)?

- New, sophisticated microscopy techniques increasingly will be used to observe the structure of foods and determine the effects of treatments on very small scales.
- Can we integrate the concept of glass transition in the process (i.e., to determine the path of a point of a product on a temperature-water content graph, which will allow the location of that product to be compared to the glass transition curve)?[23]
- How can the advances in interfaces be used to improve designs of emulsifiers or to help solve fouling problems?
- Can we move toward "intelligent CIP systems" using physico–chemical sensors and such contamination analyzers?

2.6.3 INTEGRATE MORE MECHANIZATION IN FOOD ENGINEERING

- Much remains to be done in the use of mechanical characterization of complex fluids and semisolids for equipment design, for example in operations incorporating product deposition or forming.
- New materials in packaging, films, membranes, sensors, etc., offer new opportunities for food engineering.
- Techniques of production management used in the mechanical industry (and other industries) must be assimilated to improve the "manufacturing" sectors of the food industry.

2.6.4 USE A MORE FUNCTIONAL APPROACH IN THE DESIGN OF NEW PRODUCTS AND PROCESSES

Analyze the transformation of a raw material to a (traditional or new) finished product as a list of functions (e.g., mix, separate, thicken, stabilize, shape, package, etc.) and, correspondingly, list the possible unit operations with the functions they perform (for example, a cooking operation on a cereal product coagulates proteins, gives porosity, dries the product, modifies the shape, provides a color, etc.). Such analyses may lead to a more creative and rational process selection. Furthermore, if we know all the properties and phenomena occurring in a product (the "virtual food"), we can simulate and compare various combinations of treatments.

2.6.5 USE MORE MATHEMATICS TO INTEGRATE MORE COMPLEXITY

New (and future) mathematical tools, if we master them and can "feed" them with data, will enable us to take more complexity into account, thereby describing and predicting reality more accurately. Examples such as fuzzy logic, fractals, multivariable and nonlinear data treatments (e.g. neuronal networks) show us the way.

2.6.6 BETTER INTEGRATION OF THE HUMAN FACTOR IN PLANT DESIGN AND OPERATION

In addition to the obvious necessity to take into account the social, communication, training, and other problems in the companies, can we make better use of man's specific abilities in food lines and control systems? For example, can advisory or sensorial perceptions be used in a semiautomatic decision chain, associating man with machine rather than replacing him?

It seems that many experiments in expert systems, in which man's opinion could be used in the design or the operation of food plants, have been disappointing: can these difficulties be overcome?

2.6.7 CAN WE INTEGRATE ALL SCALES?

Will we be able to use all of these advances to integrate all dimensional scales and all aspects in the resolution of problems? An example could be the following. An environmental question (the "megascale") has to be related to problems with cleaning-disinfection-rinsing operations (the "macroscale") and in turn has to be connected with the hygienic design of the equipment (the "mesoscale") and to the formation of biofilms and surface properties (the "micro- or nanoscale").

Finally, let us not forget that all of this future progress must be put in the perspective of challenges for the food industry:

- Feed low-income people in developing and of industrialized countries.
- Reconcile the food industry and society.

REFERENCES

1. Volanschi, A. 1999. "Viewpoint of a Potential User of the Pulsed Electric Fields Preservation Technique," Colloque "La conservation de demain," Pessac, France, October 13–14.
2. Bimbenet J. J., J. Boudrant, and G. Trystram. 1996. "Le Génie des Procédés Alimentaires et Biotechnologiques," *Ind. Alim. Agr.*, 946–951.
3. Bimbenet J. J. and G. Trystram. 1993. "Evolution du Génie Industriel Alimentaire," *Ind. alim. agr.,* (11–12): 811–817.
4. Labuza, T. 1997. "A 2020 Vision of Food Preservation," *Proceedings Rencontres,* Agoral, Nancy.
5. Trystram, G. and F. Courtois. 1994. "Food Processing Control, Reality and Problems," *Food Research International*, 27: 173–185.
6. Trystram, G., M. Danzart, R. Treillon, and B. O'Connor. 1996. "Supervision of Food Processes," in *Automatic Control and Optimisation of Food Processes*, J. J. Bimbenet, E. Dumoulin, and G. Trystram (eds.), Elsevier Sc. Publishers, pp. 441–448.
7. Bruin, S. 1997. "Frontiers in Food Engineering," *Entropie.*, 2038, 13–21.
8. Villermaux, J. 1982. *Génie de la Réaction Chimique — Conception et Fonctionnement des Réacteurs. Tech et Doc,* Lavoisier, Paris, 394 pp.

9. Courtois, F. A. Lebert, J. J. Bimbenet, and J. C. Lasserand. 1991. "Modelling of Drying in Order to Improve Quality of Maize," *Drying Technology*, 9: 927–945.
10. Tréléa, I. C., F. Courtois, and G. Trystram. 1997. "Dynamic Models for Drying and Wetmilling Quality Degradation of Corn Using Neural Networks," *Drying Technology*, 15(3): 1095–1102.
11. Perrot, N., C. Bonazzi, and G. Trystram. 1998. "Application of Fuzzy Rules Based Models to Prediction of Quality Degradation of Rice and Maize During Hot Air Drying," *Drying Technology*, 16(8): 1533–1565.
12. Broyart B., G. Trystram, and A. Duquenoy. 1998. "Predicting Color Kinetics during Cracker Baking," *Journal of Food Engineering*, 35(3): 351–368.
13. Broyart, B. 1998. *Modélisation des Phénomènes de Transferts et des Modifications de Qualité Induites lors de la Cuisson d'un Biscuit Sec en Continu.* Thèse de l'ENSIA, 21 Novembre 1998, Massy, France.
14. Abud, M., F. Courtois, C. Bonazzi, and J. J. Bimbenet. 1998. "Kinetics of Mechanical Degradation of Paddy Rice during Drying: Influence of Process Operations Conditions and Modelling for Control Design," in *Drying '98*, Proceedings of the 11th International Drying Symposium (IDS '98), Haldiki, Greece, A.S. Mujumdar, vol. B, 1303–1310.
15. Feillet, P. 1998. "Les Priorités de la Recherche sur les Industries Alimentaires," *Ind. Alim. Agr.*, 12: 14–20.
16. Raoult-Wack, A. L. and N. Bricas. 1998. "Controllable Development of the Food Sector in Tropical Areas: Main Challenges, Fields of Research and Research Procedures," *Outlook on Agriculture*, 27(4): 225–235.
17. Taoukis P. S., T. P. Labuza, and I. S. Saguy. 1997. "Kinetics of Food Deterioration and Shelf-Life Prediction," in *Food Engineering Practice*, K. J. Valentas, E. Rotstein, and P. Singh (eds). Boca Raton, FL: CRC Press.
18. Roos, Y.H.1992. "Phase Transitions and Transformations in Food Systems," *Handbook of Food Engineering*, D. R. Heldman and D. B. Lund (eds). New York: Marcel Dekker.
19. Willis, B., M. Okos, and O. Campanella. 1999. "Effects of Glass Transition on Stress Development during Drying of a Shrinking Food System," in *Proc. 6th Conf. Food Engineering (COFE '99)*, AIChE Annual Meeting, Oct. 31–Nov. 5, Dallas, TX, USA.
20. Kemp, I. C. 1998. "Progress in Dryer Selection Techniques," in *Drying '98*, Proceedings of the 11th International Drying Symposium (IDS '98), Haldiki, Greece, A.S. Mujumdar, vol. A, 668–675.
21. Frydman, A., J. Vasseur, J. Moureh, M. Sionneau, and P. Tharrault. 1998. "Comparison of Superheated Steam and Air Operated Spray Dryers Using Computational Fluid Dynamics," *Drying Technol.*, 16(7): 1305–1338.
22. Pearson, C. A. 1998. "Control of Assembly and Packaging Processes in the Food Industry," in *AcoFop IV Conference Proceedings*, SIK edition.
23. Vuataz, G. 1999. "Prévention des Transitions de Phases dans les Systèmes Déshydratés Pendant le Traitement et le Stockage," in *Les produits alimentaires et l'eau*, Recueil des conférences Agoral 99, Tec & Doc, Paris, pp. 75–86.

3   Challenges for the Process Specialist in the 21st Century

K. R. Swartzel

CONTENTS

Abstract

An overview of the challenges facing the food processing and packaging specialists of the twenty-first century will be given. Commercially emerging thermal and nonthermal processing methods will be discussed relative to public health and product quality considerations. Visions for tomorrow's processing and packaging environments are presented with special attention given to increased capabilities leading to increased system accountability and continuous validation.

3.1   INTRODUCTION

Much literature exists describing scheduled process filing protocol for traditional canning operations. As industry demand has developed for continuous-flow thermal processing systems, new challenges and protocols have been developed. The driving force has always been for methods that proved to be accurate yet conservative relative to public health considerations. These challenges have become overwhelming in the U.S. industry related to processing of shelf-stable, low-acid, continuous-flow multiphase foods. In 1995, in response to these challenges, the National Center for Food

Safety and Technology, Summit-Argo, IL; and the Center for Advanced Processing and Packaging Studies (formerly the Center for Aseptic Processing and Packaging Studies); North Carolina State University, Raleigh, NC; the University of California-Davis, Davis, CA; and The Ohio State University, Columbus, OH, organized a series of workshops. The objective of the workshops was to establish a dialogue and a mechanism for exchange on issues surrounding the application and implementation of aseptic processing of multiphase foods and to come to a consensus relative to process filing protocol. The invited expert participants recognized that each process would be unique, and that it is the processors' responsibility to demonstrate the ability of a process to commercially sterilize every portion of the product-contact surface of the system and all food products produced. The protocol adopted was presented at an Institute of Food Technologists symposium.[1–4]

The case study of these workshops[5] was later used as a guide in the U.S. Food and Drug Administration (FDA) filing that led to the first (May, 1997) no-rejection letter for a low-acid continuously flowing multiphase aseptic process.[6] Future successful filings will undoubtedly be dependent on the processor's use of scientifically sound principles and documentation of the lethality delivered.

The aim of this chapter is to discuss commercially emerging thermal and non-thermal processes in light of the workshop's developed approach to establishing acceptable filing protocol. A further aim is to provide a vision of tomorrow's processing and packaging environment relative to accountability and validation requirements.

3.2 BACKGROUND

Process evaluations and regulatory process filings have always focused on determining the system's minimum F_o value. This represents the number of equivalent minutes at $T = 250°F$ delivered to a container or unit of product calculated using a z-value of 18°F (or at $T = 121.1°C$ with a z-value of 10°C). The development of the F_o concept is based on using a z-value, which is more conservative than the z of the most heat resistant pathogen that can grow and multiply at the storage conditions of the end product. It is further assumed that the F_o is determined at the point in the product receiving the least thermal treatment (known as the cold spot). This concept has served the canning industry well over the past 70 years. The standard method based on the above concept for determining the value of a process is by the General Method.[7–11] New methods of thermal evaluation have emerged, and they provide broader understanding of the overall thermal treatment to product constituents.[12]

With the advent of commercial use of continuous flow sterilization processes, including those with particulates, new techniques were developed to identify the cold spot and ensure a safe product.[4,5,13,14] Additionally, several investigators have reported that the resistance of organisms was influenced by the method of heat treatment, even when such treatments provide equivalent total heat.[15–18] This implies that physical factors may be involved.

3.3   EMERGING THERMAL SYSTEMS

Today, a variety of methods exist for the thermal processing of biomaterials. Beyond the traditional canning and conventional continuous-flow processing (heated with steam or hot water under pressure), there exist ohmic (electric resistance heating), microwave, infrared radiation, and radio frequency processes. In these systems, conventional evaluation techniques are applicable, as long as appropriate time and temperatures can be measured. Claims of additional destruction due to the method of delivery over the thermal energy delivered would have to be documented. Destruction kinetics of the most resistant pathogen must be determined to support these claims.

3.4   EMERGING NONTHERMAL SYSTEMS

Nonthermal processing techniques have long interested the industry and university researchers. These techniques include radiation, high-pressure, filtration, chemical, and, more recently, electric pulse and oscillating magnetic waves. Traditional evaluation methods do not apply. Now, instead of integrating a time-temperature curve in pursuit of an F_o value, one must examine radiation dose levels, pressure levels, degree and depth of filtration bed, chemical concentration, electric power levels, and pulse durations, all as a factor of time. The most resistant pathogen for the treatment must be identified, and the appropriate kinetics for the full extent of the process operating range must be known. Then, and only then, can appropriate evaluation models be developed and used with reliability. In all cases, correlations to a traditional inoculate study must be carried out.

3.5   PACKAGE SYSTEM VALIDATION

Aseptic processing and aseptic packaging have added additional process evaluation challenges for the specialists. Aseptic packaging evaluations have been limited to biovalidation techniques. No general method for determining an F_o value-like parameter exists. For many aseptic packaging units, a variety of process steps exist, such as ambient temperature chemical sterilization followed by hot chemical sterilization followed by dry heat sterilization. Lacking this information, packaging manufacturers are left with mathematical models and final biovalidation as the only system checks. Biovalidation provides a needed final assurance, but it neither provides detailed system-to-system and stage-to-stage sterilization comparisons nor definitive information useful to aid design models.

3.6   FUTURE CAPABILITIES

Biomaterial processing will have many additional challenges for the process specialist. As we look to the future, what will be the processing breakthroughs of tomorrow? What will man be able to do tomorrow relative to product processing that he is not able to do today? In the future, systems will be available to track every particle in the multiphase streams. Each particle will have a thermal profile. System cold spots will be known and tagged for each run. Treatment (thermal or other) distributions will be controllable, and system constituent changes will be designable. Systems will be available to not only monitor each component contribution to the mass of the final package but to also control it. Data acquisition, measurement, and simulation capabilities of the on-line monitoring systems will allow for multiple-scenario simulations in better than real time. Decision-making capabilities of these systems will enable on-the-fly product stream diversion according to the highest benefit criteria before the examined stream reaches the filling station. Components that need to be cooked well will be so, and those that require mild thermal treatment resulting in rigid structure will be so. Initial raw component monitoring, along with the treatment log, will allow for precise shelf-life projections under controlled distribution conditions. The high, continuous throughput of these lines and the need to maintain it will eliminate process downtimes for cleaning and maintenance. Processing systems will incorporate backup elements, which the processor will flexibly incorporate, remove, clean, and sterilize without interrupting the process continuity.

The package of tomorrow will be available to the user on an as-needed basis. Barrier properties will be even more custom made than they are today. Protective agents within the product contact film will guard against distribution abuse. As abuse is accelerated, so will be the release of the protective agent.

Package seals will be made at speeds not thought possible by current standards, providing filling and sealing rates far beyond those of today's cans. Contributing to this success, there are rapid contact surface sterilization techniques. These techniques not only sterilize but also monitor the presence of any surviving spores. No product will be allowed downstream that is not sterile or that has an improper seal.

The package of tomorrow will be designed with a trigger mechanism. Once activated, the package will either begin to biodegrade or to better restructure itself such that it becomes a better and easier raw ingredient for new package material. Beyond these achievements, where will technology take us? Cell generation techniques offer the greatest potential ever in ridding the world of its endless dependency on the land and water for its nourishment.

Today, we are growing cells (plants and animals) in our labs. Most animal cell generation experiments are medical in scope—i.e., skin tissue for burn victims. The knowledge and the technology we are developing may, however, be a source for the greatest change in the ways we produce our food supply and in the way we apply aseptic technology. Imagine plant or animal tissue being grown in a controlled environment in a completely controlled atmosphere (sterile), without toxins, and with a highly controlled nutrient delivery system, little or no waste (no more lagoons), and accelerated growth, anytime, anywhere. No more planting seeds, and no more animal rights issues; the product may be so pure that it may not even need processing (at least not in the formal sense). Plant and animal tissue grown in a sterile reactor (potato tissue or turkey breast tissue, continuously growing, continuously being harvested, processed, and packaged all at the same location) envision a bright new era for food processing. However, these systems must still be designed and validated with conservative public safety protocol, a protocol well founded in the use of scientifically sound principles and a high level of accountability.

Where will the technologies take us? Bioengineering and biotechnology will open new worlds to all of us. In food processing, the new products and the new markets may be unimaginable today. Along with them will be new processing and packaging systems. The process specialist of tomorrow will need new tests and new methods. With increased capabilities will come increased demands for validation. Wherever these new technologies take us, the process specialist will be there to provide system confidence, accountability, and validation.

3.7 ACKNOWLEDGMENTS

Paper No. FSR-99-39 of the *Journal Series of the Department of Food Science,* NCSU, Raleigh, NC 27695-7624. Support from the North Carolina Agricultural Research Service (NCARS) and the Center for Advanced Processing and Packaging Studies (CAPPS) is gratefully acknowledged. The use of trade names in this publication does not imply endorsement by NCARS or CAPPS of the products named or criticism of similar ones not mentioned.

REFERENCES

1. Larkin, J. W. 1997. "Workshop Targets Continuous Multiphase Aseptic Processing of Foods," *Food Technol.,* 51(10): 43–44.
2. Sastry, S. K. 1997. "Measuring Residence Time and Modeling the System," *Food Technol.,* 51(10): 44–48.
3. Marcy, J. E. 1997. "Biological Validation," *Food Technol.,* 51(10): 48–52.
4. Digeronimo, M., W. Garthright, and J. W. Larkin. 1997. "Statistical Design and Analysis," *Food Technol.,* 51(10): 52–56.
5. Lechowich, R.V. and K. R. Swartzel. 1997. *Case Study for Condensed Cream of Potato Soup from the Aseptic Processing of Multiphase Foods Workshop.* The National Center for Food Safety and Technology, Illinois Institute of Technology, Summit-Argo, IL, and the Center for Aseptic Processing and Packaging Studies, Raleigh, NC, and Davis, CA.
6. Palaniappan, S. and C. E. Sizer. 1997. "Aseptic Process Validation for Food Containing Particulates," *Food Technol.,* 51(8): 60–62, 64, 66, 68.
7. Bigelow, W. D., G. S. Bohart, A. C. Richardson, and C. O. Ball. 1920. "Heat Penetration in Processing Canned Foods," *National Canners Association Bulletin.* 16-l.
8. Ball, C.O. and F. C. W. Olson. 1957. *Sterilization in Food Technology.* New York, NY: McGraw-Hill Book Co.
9. Stumbo, C. R. 1973. *Thermobacteriology in Food Processing, 2nd ed.* New York: Academic Press, Inc.
10. Pflug, I. J. 1988. "Heat Sterilization," in *Selected Papers on the Microbiology and Engineering of Sterilization Processes, 5th ed.* I. J. Pflug, ed., Minneapolis, MN: University of Minnesota Press, pp. 143–159.

11. Holdsworth, S. O. 1997. *Thermal Processing of Packaged Foods.* New York, NY: Chapman and Hall.
12. Kyereme, M., K. Swartzel, and B. E. Farkas. 1999. 'New Line Intersection Procedure for the Equivalent Point Method of Thermal Evaluation," *J. of Fd. Sci.,* 64: 565–570.
13. Damiano, D. 1997. "Issues Involved in Producing a Multiphase Food Product," *Food Technol.,* 51(10): 56–62.
14. Dignan, D. M., M. R. Berry, I. J. Pflug, and T. D. Gardine. 1989. "Safety Considerations in Establishing Aseptic Processes for Low-Acid Foods Containing Particulates," *Food Technol.,* 43(3): 118–121.
15. Burton, H., A. G. Perkin, F. L. Davies, and H. M. Underwood. 1997. "Thermal Death Kinetics of *Bacillus Stearothermophilus* Spores at Ultra High Temperatures. III. Relationship Between Data from Capillary Tube Experiments and from UHT Sterilizers," *J. Food. Technol.,* 12: 149–161.
16. Bunning, V. K., D. W. Donnelly, J. T. Peeler, E. H. Briggs, J. G. Bradshaw, R. G. Crawford, C. M. Beliveau, and J. T. Tierney. 1988. "Thermal Inactivation of *Listeria monocytogenes* within Bovine Milk Phagocytes," *Appl. Environ. Microbiol.,* 54: 364–370.
17. Mackey, B. M. and N. Bratchel. 1989. "A Review: The Heat Resistance of *Listeria monocytogenes,*" *Lett. Appl. Microbiol.,* 9: 89–94.
18. Fairchild, T. M., K. R. Swartzel, and P. M. Foegeding. 1994. "Inactivation Kinetics of *Listeria innocua* in Skim Milk in a Continuous Flow Processing System," *J. of Food. Sci.,* 59(5): 960–963.

4 Sustainability of Food Sector Development in Tropical Areas

A. L. Raoult-Wack
N. Bricas

CONTENTS

Abstract

Debates over the last few years concerning the future of the planet have seen the question of food sufficiency reappear. Developing countries in the tropics and subtropics, primarily agricultural, are currently having to adapt to unprecedented socio-economic changes (urbanization, demographic growth, globalization of trade, widening disparities) and to cope with serious problems of food security, public health, and environmental damage. How can the food sector help meet these new challenges? Can the food sector development model initiated by the countries of the North be durably applied on a global scale? Which research and development priorities face these challenges? The objective of this chapter is to try to answer such questions. The first section focuses on the possible contribution of food sector development in meeting new challenges. The second, third, and fourth sections present, respectively, an outline of the history of food research over the past 40 years, main achievements and limits of research past lines, and the concept of sustainability in food processing. This appraisal serves as a background against which to examine the new directions that need to be explored for a sustainable development of the food sector, which is the subject of the last section.

4.1 INTRODUCTION

The developing countries in the tropics and subtropics, usually primarily agricultural, are currently having to adapt to extremely rapid socio-economic change (urbanization, demographic growth, globalization of trade) and to cope with serious problems of food security, public health, poverty and environmental damage. Debates over the last few years concerning the future of the planet have seen the question of food sufficiency reappear. With the world's population growing at an ever-increasing rate and an extra 1.7 million mouths to feed each week, the Malthusian fears of a widening gap between people's needs and food production are once more coming to the fore.[1,2] The threat of medium- or long-term hardship is directing public attention to the need for a new international effort to increase food availability.

Faced with such issues, the development of these countries is generally handicapped not only by economic and political obstacles, but also by the inadequacy of their scientific and technical resources in many fields.[3,4] This is particularly true for the food sector. In fact, for over half a century, most of the efforts directed to the agri-food sector have been focused on agriculture (perfecting and popularizing improved varieties and more intensive cropping and livestock systems), whereas only a very small part of the research resources has been targeted toward the food sector (processing and trading of agricultural raw materials). It must be emphasized, however, that the situation varies considerably, from the rapidly developing countries of Asia or Latin America to the most impoverished ones.

In the North, the terms of the problem are different. The countries of the North, richly endowed with major scientific organizations and facilities, are faced not only with growing social disparities and the exclusion of some categories of their populations, but also with the disruptive effects of uncontrolled industrial and technological development on the environment and on society.[4]

As a whole, although the food sector has significantly contributed to ensuring food security over the past decades, questions are now being asked about its recent developments. The food sector is accused of being a tool used by the richer countries to establish economic and cultural domination. These countries are said to impose their own products and patterns of consumption on lesser economies, thus discouraging local production. The food industry is also alleged to expose consumers to major health risks, and the crisis associated with the danger of bovine spongiform encephalopathy (or "mad cow disease") being transmitted to man is another episode in a series of crises that include, for instance, infant formula, hormones in veal, food colorants, and ionized foodstuffs in Europe; as well as mercury in fish in Asia; and food aid cereals, flavor cubes, and mangoes treated with acetylene in Africa.

Why have developments in the food sector given rise to such concern? No doubt because food is different from other consumer products in that it passes through the body. Man is transformed by food to a greater extent than by any other product, and his well-being is more directly affected. Food contributes to growth and good health but can also cause illness or even death. It gives both sensory and social pleasure and also has a considerable effect on his sense of individual and collective identity. What is at stake in the development of the food sector, therefore, cannot be considered purely in economic terms. It must also include the question of changing patterns of consumption, examining not only their economic but also their social and cultural implications.[2]

Hence, after a century of major technical advance, essentially achieved by and for the countries of the North, it must be recognized that economic development can no longer be thought of in terms of "headlong pursuit," producing major local and global imbalances and uncontrolled effects.

This recognition has, over recent years, led both the scientific and nonscientific communities to ask (or re-ask) a certain number of questions.[3] We can no longer simply argue for "more and more" science (the experience of Eastern Europe has shown how scientific voluntarism can lead to enormous ecological and social problems). We must also ensure that science has a "social involvement," and we must correct threatening imbalances and conciliate sometimes contradictory demands.

The World Conference on Science was held in Budapest (Hungary) in June 1999 under the aegis of the United Nations Educational, Scientific and Cultural Organization (UNESCO) and the International Council for Science (ICSU). The extended title of this conference was "Science for the Twenty-First Century: a New Commitment." Presentations at this conference included "World Declaration on Science and the Use of Scientific Knowledge" and "Science Agenda: Framework for Action." Key points included science for knowledge, knowledge for progress; science in society; science for society; and science for peace and development. Presentations showed that the role of sciences and scientists in society must deeply evolve, and that ethical issues are crucial for the new commitment of science to meet new challenges in the present changing context.

How can the food sector contribute to meet the new challenges triggered by demographic growth and urbanization, globalization of trade, environmental damages, and widening disparities? Can the food sector development model initiated by the countries of the North be durably applied on a global scale? Which new research and development priorities face these challenges? How should scientists change or adapt their methods in the present context of unprecedented changes?

The objective of this contribution is to answer such questions. The first section focuses on the possible contribution of the food sector development in meeting new challenges. The second, third, and fourth sections present, respectively, an outline of the history of food research over the past 40 years, main achievements and limits of research past lines, and the concept of sustainability in food processing. This appraisal serves as a background against which to examine the new directions that need to be explored for a sustainable development of the food sector, which is the subject of the last section.

4.2 CHALLENGES FACING THE FUTURE OF THE WORLD'S FOOD SUPPLY

4.2.1 DEMOGRAPHIC GROWTH

The first challenge is demographic growth. The world's population should reach between 7.5 and 8.5 billion by the year 2020, almost four-fifths of whom will live in tropical countries. The total demand for cereals for human and animal consumption will need to have doubled by this date to about 1.7 billion tons. This challenge basically concerns agriculture. However, the food sector can help to increase the food supplies available by reducing post-harvest losses and improving the yields obtained from converting agricultural raw materials into finished foodstuffs. It is very difficult to evaluate these losses overall. They vary considerably from one country to another but are greater in hot, wet zones and are aggravated in developing countries by a lack of adequate storage and transport infrastructures. Losses also vary considerably according to the perishability of the commodities concerned. Post-harvest losses of 15–20% are often quoted for cereals in tropical areas. These may exceed 50%, and even approach 100%, for more perishable commodities like roots and tubers, fruit, or fish. Appreciable reductions in such losses appear possible not only by improving storage, conservation, and pest control techniques and processing yields but also by improving transport and marketing infrastructures and organization.

4.2.2 URBANIZATION

The second challenge is urbanization. The increase in the rate of urbanization is particularly rapid in tropical regions and exceeds that of demographic growth. It reinforces fears of a widening gap between people's needs and long-term food production. Urbanization alters dietary behavior. Town dwellers eat more meat and more processed products that have a built-in service factor (convenience foods)—in other words, calories that cost more to obtain; this is accentuated by the rising level of income. North Americans thus consume the equivalent of 800 kg of grain per annum, Italians 400 kg, and Indians 200 kg.[1] The newly industrialized and urbanized countries are moving toward an agro-nutritional model in which it appears impossible to provide a sustainable basis for the entire world, from an energetic point of view.

4.2.3 Globalization of Trade

The third challenge is the globalization of trade, which raises the question of the competitiveness of tropical foodstuffs on internal and international markets. We have seen certain foodstuffs invade the urban and then the rural markets of Africa, Latin America, and Asia. The distribution of bread, rice, chicken, dry milk solids, beer, hamburgers, and Coca-Cola® has led some authors to fear that local produce will disappear and that food will become completely standardized. However, whereas analysis of the products consumed gives the impression of convergence toward a single or dominant pattern of consumption, analysis of dietary practices reveals a considerable capacity for appropriating and reinterpreting external references through cookery and styles of consumption. The celebrated "thièbou diène" of Dakar, a rice and fish recipe that has become Senegal's national dish, is prepared using Thai rice, local fish, mainly imported vegetable oil, and vegetables introduced by the Portuguese and French. But it is far from being recognized as "international cuisine" and has become one of the symbols of African cuisine.

As a whole, as far as dietary change is concerned, the main overall trend appears to be diversification—in particular, in urban environments. Consumers enjoy a more varied diet and obtain their supplies from a wider range of sources.

In the face of this overall movement, which is very noticeable in certain countries of the South, the tropical produce and culinary preparations specific to each culture will be able to resist the globalization of trade only if their diversity is exploited and they are put to varied uses. Benin is a particularly interesting case in point: maize is processed there into about 40 different products, and this, in large part, explains the limited penetration of imported rice and wheat. The identification of products with their country of origin is becoming an important consideration in food sector development.

The competitiveness of local produce in urban markets is, nevertheless, limited by three main factors: it is often not available widely enough or over a long enough period of the year, it is sometimes expensive compared with competing imported products, and its quality does not always meet the new requirements of urban consumers.

As far as the international market is concerned, the competitiveness of tropical foodstuffs is not determined by price alone. Promotion of functional qualities is also an important factor. It is the distinctive qualities of such products as coffee, cocoa, tropical fruit flavors, herbs and spices, and acacia gum that have made them successful. Promotion of the distinctive nature of these properties is what enables such products to conquer new markets (e.g., specification of cocoa or coffee quality by geographical origin). Also, a loss of distinctiveness can make them lose markets! Coconut oil sales initially dropped on the world market as a result of the problem of aflatoxins produced by inappropriate drying processes and the cost of the refining procedures needed to remove them. Vanilla and cane sugar are likewise faced with strong competition, the former from synthetic vanillin and the latter from artificial sweeteners.

4.2.4 WIDENING DISPARITIES

The fourth challenge is widening disparities. Overall, the world currently produces enough food to meet its food requirements. Over 800 million people, however (i.e., one-seventh of the world's population) do not get enough food to lead a healthy, active life. At the same time, almost 400 million people are suffering from illnesses caused by dietary excess (e.g., obesity, diseases of the cardiovascular system). The first situation is the result of political instability, war, and poverty. Loss of identity and confusion of social and cultural reference points that previously helped maintain behavioral equilibrium are suggested as reasons for the second. In neither case does malnutrition appear as a simple question of the quantities of food available. Food security should include the notions of distribution and of sustainable access to foodstuffs for all, along with those of social and political stability, equilibrium, and consistency. Overabundance of supply may be accompanied by a demand devoid of financial resources, as is the case in Latin America; if the gap between the two widens, food security becomes a major political problem in terms of a more equitable distribution of available resources. Witness the pillaging of supermarkets in the vicinity of the shanty towns.

What has this to do with the food sector? First, it can help improve the transportation and storage of the available food resources. Second, it offers opportunities for economic activity, employment, and income in both rural and urban environments. Food sector activities today represent a major source of employment and income, particularly for women, in countries obtaining most of their resources from agriculture (most tropical countries, for instance). International comparisons show that an increasing percentage of the value added by the combined agricultural and food sectors in terms of GDP/person is attributable to the food sector and is about 10% in the poorest countries and 50% in the richest. It is, thus, of quite strategic importance in tropical countries. And third, it can play a part in reducing the cost of food production (lower losses, higher yields in terms of both materials and energy, more efficient marketing channels).

4.3 OUTLINE HISTORY OF FOOD RESEARCH IN TROPICAL AREAS

The evolution of the history of research in tropical areas can be schematically divided into four main periods, which are described below.

4.3.1 PROCESSING OF EXPORT CROPS FOR THE INTERNATIONAL MARKETS

The initial focus was on industries producing export commodities for the international market (e.g., coffee, cocoa, palm oil, coconut, groundnut, sugar cane, tinned fish, etc.) and was consistent with agricultural research in these areas.[5] The development of tropical agriculture was essentially regarded as a matter of increasing capacity for the production of cash crops in a context where trade was becoming international (product marketing was largely done locally, however; for instance, in the vegetable oil and sugar industries).

4.3.2 PROCESSING OF IMPORTED PRODUCTS TO FEED LOCAL POPULATIONS

In the second period, research was aimed at developing the food sector to feed the local population by establishing local industries to process imported products (wheat-flour mills, breweries, and soft-drink industries, powdered milk reconstitution plants, etc.) to meet a growing food demand triggered by demographic growth.

4.3.3 CREATING TROPICAL VERSIONS OF IMPORTED FOODSTUFFS

Only recently has food sector research become interested in developing food crops for local markets, mainly aimed at feeding the urban population. This research first consisted of attempts to create tropical versions of imported products. They involved, for instance, the inclusion of millet, sorghum, or maize in traditionally wheat-based foodstuffs (bread, dough, etc.). Over three-fourths of food sector research into millet and sorghum processing has been devoted to such "compound flour" programs. Research into maize and sorghum processing has also been aimed at developing products shaped like rice grains, called "maize rice" and "sorghum rice."

4.3.4 INDUSTRIALIZING THE PROCESSING OF TRADITIONAL PRODUCE

The much more recent approach to develop food crops for feeding the urban population has been to industrialize the manufacture of traditional products. This approach involves mechanizing the processing procedures and, often, marketing ready-to-cook foods that are packaged more hygienically (like industrial products) and that have a more standard quality, advantages supposedly sought after by the urban consumer. The strategy has been primarily pursued by private food-processing groups in Latin America and Asia; for example, farina and bread-making quality cassava sour starch in Brazil, panela (brown sugar from sugar cane) and patacones (plantain crisps) in Colombia, cornmeal (obtained after alkaline treatment) in Mexico, charqui (dried meat) in Brazil, tofu and tempeh (fermented soybean curd or cake) in Indonesia, and nuoc mam (fish sauce) or cana noodles in Vietnam. Similar experiments have been tried in Africa: cassava-based products like gari (grated and roasted cassava) in Togo; attiéké (a cassava product in granulated form) in the Ivory Coast; chikwangue (fermented paste) in the Congo; yam flakes in the Ivory Coast and Nigeria; millet-, sorghum-, and maize-based flour, grits, and granulated products in Senegal and Benin; baby food in Benin, Rwanda, Zaire, and Burkina Faso; fruit juice and fruit nectar and locally picked produce like mango, tamarind, guava and bissap in Senegal, Burkina Faso, Togo, Burundi, etc.

4.4 ACHIEVEMENTS AND LIMITS OF TROPICAL RESEARCH PAST LINES

Previous research priorities were certainly consistent with the development strategies adopted in tropical countries, and the targeted objectives were often achieved thanks to the results of this research. In particular, the coffee, cocoa, palm oil, and coconut subsectors could not have been so well developed if procedures had not been established to maximize the value of such products.

Also, industrialization of local produce or substitutes to imported foodstuffs have certainly contributed, to a certain extent, to filling the gap between the growing food demand and food availability in tropical countries. However, we have today a different and clearer appreciation of what is at stake in developing the food production system, and this makes it possible to analyze the limits of research past lines with respect to new challenges facing food sector development.

First, the initial focus on local industrialization of imported raw materials, although it has helped to fill the gap between urban demand and local agricultural supply in some countries, in particular Africa, has had a secondary detrimental effect of restricting outlets for local commodities.

Technology research initially focused on industrial-scale activities, either for cash crops for the international market or crops for local markets. Businesses on this scale appear more capable of rapidly supplying a suitable amount of produce of an appropriate quality to meet export market requirements or urban demand than do small-scale processing activities. In some cases, industrial technology and processing have been able to learn from other subsectors where a processing industry had been developed (for instance, industrial wheat-milling processes have been transferred or adapted to millet, maize, and sorghum mills). However, this type of business could operate at a profit over an extended period of time only under particular conditions, and satisfactory control of the supply networks was needed, which was frequently difficult to achieve because of inadequate regulation of production and marketing, especially for food crops. In addition, external technical assistance was necessary to bridge the gap between the technology employed and local technical resources.

As a whole, attempts at industrialization have, in the final analysis, had little effect on feeding the most impoverished members of society who make up most of the urban and rural population in Africa and who remain a very important part of it in Latin America and Asia. In most cases, these products have found a market, although a more limited one than expected. Only a small and largely well-to-do section of the population was prepared to pay extra for the quality advantages over domestically or traditionally produced products. Very few ersatz products have met with commercial success in Africa or Latin America for a variety of technical and economic reasons, but primarily because the products had a weak market position. Consumers were generally unwilling to buy products they considered to be of lower quality than the reference products, particularly when presented as direct substitutes.

Various observers have criticized not only the lack of success but also the ethnocentric nature of research past lines. For instance, when talking about research to develop tropical cereals, they stress that "the emphasis placed on compound flour has undeniably marginalized research into improving traditional procedures and the development of new products." Research has made only very few efforts to exploit a large number of local food crops: cereals (e.g., millet, sorghum, maize, fonio, quinoa, tef, amaranth), tropical roots and tubers (cassava, yam, taro, sweet potato, cana, aracacha, etc.), pulses (cowpea, pigeon pea, néré, etc.), vegetable oils (karite, balanites, etc.) and fruit (cupuacu, acerole, mangostan, safou, etc.) plus their role in the agricultural economy or as foodstuffs was geographically restricted.

And yet, commercial processing of such commodities has now developed with the opening up of the urban markets. As a result of past research priorities, scientists are not well equipped to meet the needs of new companies wanting to obtain information on processing procedures, improve product quality, or diversify product use. Such knowledge is rarely committed to paper, and its dissemination remains limited except in the case of a few major products. In addition, very little information on experiences and results is exchanged between Africa, Latin America, and Asia, even when they are dealing with the same subjects.

4.5 ABOUT SUSTAINABILITY OF FOOD PROCESSING

In the current renewal of the debate on global food security, there is a great danger of oversimplifying the problem by thinking that it is a question of quantity in the South and of quality in the North. Development of the industrial food sector appears essential in a context of rapid urbanization, of increased competitiveness between supply chains on a worldwide scale, and of attempts to find new forms of added value in still essentially agricultural economies.

The faith in technological progress and industrialization that typified the 1960s, 1970s, and 1980s has distracted the attention of scientists from the questions of sustainability, i.e., social control, ethics, and long-term management of changes to the food production system.

This is true in southern countries as well as in northern countries. The effects of industrialization on man's relationship to his food, on energy consumption, on the environment, and on health risks were neglected. Such concerns were too remote from the short-term requirements of business and of policy makers and often had ideological overtones. They have still not entered into fields of scientific research except to a marginal extent. Some of these concerns are given below.

4.5.1 MAN'S RELATIONSHIP TO HIS FOOD

As a whole, the history of the world's food can be seen as a process of increasing remoteness in man's relationship to his food: remoteness in space because of the internationalization of trade in agricultural products and because of urbanization; remoteness in time due to a growing offer of stabilized and out-of-season products; and remoteness resulting from the growing length and complexity of the supply chains, with the development of an intermediation sector in the food industry (storage, transportation, marketing, processing, distribution). This sector is becoming increasingly independent of agricultural production and consumption.[2]

For the consumer, the increasing remoteness in relationship to food is reflected in a certain loss and confusion of points of reference in terms of both identity and diet, contributing to certain nutritional imbalances and to loss of confidence in the industrial food sector and its lack of openness.[2]

However, a further analysis of the long history of these changes shows that man's relationship to food and to nature in general is in fact an ambivalent and simultaneous move toward greater remoteness and greater proximity.

The move toward greater proximity in the consumer's relationship to his food is shown by the development of farm produce, organically grown produce, local specialties, direct selling by the producer, home-grown garden produce, and, now, suspicion about transgenic food. This is very noticeable in the countries of the South, but it is also apparent in industrialized countries, mainly in Europe.

4.5.2 FOOD SECTOR DEVELOPMENT MODELS

The increased offer of ever more highly processed foodstuffs, particularly in the industrialized countries but now also in towns throughout the world, raises the question of the overall energy efficiency of the food-processing system. To supply individuals with enough food, more and more energy needs to be injected into the transformation system, given the increased sophistication of the technology and services involved (packaging, portioning, precooking, etc.).

An awareness of the two tendencies described above (move toward remoteness and proximity), in both the North and the South, opens the way to the elaboration of new or complementary models for food sector development. Rural, small-scale, decentralized processing is no longer regarded as the survival of archaic, outmoded activities that ought logically to make way for more "rational" industrial processing. It is only recently that the strategic importance of more decentralized enterprises, rural agro-industries, and small-scale urban workshops in the food sector has come to be appreciated.[8] As public authorities were often not officially notified of the existence of such activities, their contribution to the supply of processed food products was not usually taken into account. The view of a large number of policy makers and scientists that the food sector was an archaic symbol of underdevelopment and technological backwardness did not help its importance to be recognized, either.

It is now recognized that the consumer can maintain or re-establish reference points in his relationships with others, with himself, and with nature. In addition, long-term respect for the environment is becoming a consumer preoccupation, particularly with foods: interest in foodstuffs grown by less environmentally polluting agricultural methods and in biodegradable packaging mistrust, or even refusal of irradiated or transgenic foodstuffs, etc.[2] Over and above this, the individual as consumer is becoming a user with citizenship responsibilities.

4.5.3 MANAGEMENT OF RISKS AND TRACEABILITY

The process of increasing remoteness, already well under way in the industrialized countries, is now at work in the countries of the South as a result of their rapid urbanization and their growing involvement in international trade. This increasing remoteness means greater long-term risks for food security and the environment. Management of these risks raises the question of the sustainability of food sector development.

The greater length and complexity of the supply chains makes it more difficult to monitor the origin and the quality of foodstuffs or to react rapidly to large-scale outbreaks of food poisoning. These risks are all the greater with the growth of mass production and distribution and have to be further controlled, which is difficult in the context of developing countries.

4.6 DEFINING NEW RESEARCH PRIORITIES FOR A SUSTAINABLE DEVELOPMENT OF FOOD PROCESSING IN TROPICAL AREAS

The imbalance between the developed countries (richly endowed with scientific organizations and facilities) and the developing countries manifests itself in cooperation-oriented scientific exchanges in the following way: funding and practices remain largely dominated by the idea of the transfer of knowledge, methods, and technologies from the North to the South and are at times solely concerned with cheaply exploiting local resources. The scientific contributions of the South, potentially rich in representations of the world different from those of traditional science, have not been integrated as a real part of the cooperation model, unless they reveal themselves to be a possible object for economic exploitation.[4]

The objective of this section is to illustrate how new fields of research can be opened, taking into account ethical and long-term issues in food sector development, to meet the challenges facing the food sector development.

4.6.1 TO CONTRIBUTE TO EMPLOYMENT AND REDUCTION OF DISPARITIES

To meet the challenge of widening economic disparities, research must become involved in job creation in both rural and urban environments. Expanding food-processing activities can help to increase income levels, particularly for women, and make food more readily available to the most impoverished members of society. Research and development work already carried out in urban agriculture,[6] small-scale food production in Africa, rural agro-industries in Latin America,[7] and street food in Asia[8] show the way for investigations of this type. In addition to its major contribution to the supply of local foodstuffs to the towns, it is worth recalling its importance in terms of job and income creation and its capacity for innovation.[7]

Compared to what has so far been accomplished in the agricultural sphere (smallholder organizations, training and advice for farmers, decentralized credit facilities), efforts to give the food sector a more professional approach have been neglected. This is a new field of research for economics, the social sciences, and business studies, whose interest in this sector has up to now focused on the operation and policies of big business concerns. The diversity and complementarity of the different types of enterprise; the conditions under which they come into being and under which they operate; trade organizations; and technical, financial, and management training requirements will all need to become serious topics for research to back up their development.

4.6.2 To Make Local Products and Technologies Competitive on National and International Markets

The extremely limited interest that scientists have shown in this type of activity has led them to neglect the traditional processes used and the knowledge and skills they presuppose. A major field of research still in need of development involves identifying and characterizing the wide range of food technology know-how existing in the world. This wealth of expertise is put to good use in the North but is still relatively neglected in the South. Over and above the economic problem of providing outlets for local produce, the challenge is to exploit the diversity of man's heritage that such resources represent.

With a small number of exceptions, few countries in the South are presently familiar with their own technical resources, i.e., their areas of expertise and their own special products, while market operators are busy developing initiatives to exploit them.

Such initiatives concern the local and the international sectors. The people of Colombia and Brazil, for instance, know how to make a cassava product, so-called "sour" starch, using a combination of fermentation and sun drying. The remarkable property of sour starch is that it can be used like wheat flour to produce leavened bread (i.e., with an alveolar structure) despite an absence of gluten.[9]

This is not the case with other starches obtained from cereals, roots, or tubers, which in the current state of technology can only be used to produce flat loaves. The development of this traditional know-how has resulted in cassava being more competitive on the local markets for starch products in these two countries. Bread rolls made with sour starch are now sold in fast-food outlets in the big cities of Brazil, whereas, until recently, cassava had the image of being a poor man's food. It is still not known why this form of processing makes cassava starch suitable for bread making. Scientists are currently trying to discover how to adapt the process for use with other starch products in other countries, with the twin objectives of giving added value to cereals, roots, and tubers in the countries of the South and of manufacturing gluten-free foods essential to certain diets (baby food, food for people who are allergic to gluten).

Other examples are tropical roots and tubers (cassava, yam, sweet potato, cana, aracacha, etc.), which up to now have essentially been regarded simply as sources of starch. Better development of their distinctive properties would immediately open new markets for these products. Some of these foods have properties that are in great demand on the international market as a result of recent regulations restricting the use of modified starch. Little research has so far been carried out on properties such as the resistance of starch to heat treatment (for use in frozen foods or baby foods); the precise rheological behavior required to produce analogues of fat; its rising quality in cookery; shear strength; etc.

4.6.3 TO DESIGN ENVIRONMENT-FRIENDLY PROCESSES

Up to the present, when questions of technical aid are considered, little attention has been paid to the energy yield of the food production system or to the environmental impact of its development. There is a shortage not only of data but also of suitable methods for carrying out such assessments and providing answers to the various questions that arise. What proportion of the total energy injected into the agricultural or other sectors is used in processing, distributing, and marketing the agricultural raw materials? What are the energy requirements of the different industries according to product stability (e.g., fresh, refrigerated, frozen, dried, or sterilized products)? What is the environmental impact of these processing activities (e.g., water and wood requirements, pollutant effluents, by-products) in terms of the different processes used? Recent efforts in the industrialized countries to develop methods based on economic, ecological, and energy balance sheets may well be a path worth exploring, even though these methods still have flaws.

4.6.4 TO RETHINK MAN'S RELATIONSHIP TO FOOD AND SOCIAL CONTROL OF FOOD TECHNOLOGIES

Another field of research concerns the conditions required to give better social control of evolving food technologies. How can users, i.e., ordinary citizens, participate in the research and development process? Current research into the social control of technology represents an interesting approach.

It raises questions about long-term changes in technical systems and about man's relationship to food and to nature in general. From this point of view, further attention should be given to methods and tools to assess food risks and hazards as perceived by consumers.

It also raises questions about the role of research in the development process. Such an approach means that the issues examined and the objectives become inseparable from the way in which the research is carried out and from ethical concerns. As far as science in general is concerned, the new challenges and limits of past research results have revealed the reductive and compartmentalized representation of the world used in the traditional scientific approach.[3,10] Also, as far as scientific cooperation is concerned, former practices have been brought into question: in particular, the linear view of development based on industrialization and the "transferability" and "transportability" of technology.[4] This is particularly true for food-

sector research and cooperation, as recently analyzed in detail in another paper.[2] Over the last few years, new approaches have emerged, calling for a major reform of the practices employed by researchers, development officers, extension agents, and politicians. This new wave of thinking has four main aims: to break down the barriers between the different disciplines, to focus research procedures on the users, to adopt more flexible and interactive methods of research project management, and to develop a new science-sharing approach that goes well beyond simple knowledge transfer.[2,4]

4.7 CONCLUSION

After a century of major technological progress, one of the age-old questions facing mankind is still with us: will there be enough food for everyone tomorrow? Half a century of agronomic research has shown that the problem cannot be defined simply in terms of chasing after a constantly rising demographic curve and proposing more productive crop varieties. New priorities have emerged, and key issues in today's debate are urbanization, globalization, disparity and poverty, health risks, long-term planning, social control of technology, sustainability, and ethics.

As a consequence, food-sector development can no longer be thought of in terms of "headlong pursuit," producing local and global imbalances and uncontrolled or disruptive effects. This is not just a wish. This is a profound and explicit social demand that has expressed itself, for instance, through the rejection of scientific and technological advances such as irradiated or transgenic food in many countries, but more generally through the move toward proximity that has marked recent changes in man's relationship to food and nature. This leads to the assertion that scientists are responsible not only to society in general in their own country, but also to society in general throughout the world, as stated in the recent World Declaration on Science and the Use of Scientific Knowledge:

> Nowadays ethical implications of the use of scientific knowledge have become so profound and so much of concern to individuals and society at large, that any research or application of its results have to comply with ethical standards and principles. In this context, scientists themselves start to play an active role in defining and taking on their responsibilities.[3]

This opens the way to further thinking and rethinking of scientific priorities, methods, and responsibilities in food research and development, both in Northern and Southern countries.

REFERENCES

1. Brown, L. 1995. "Le Siècle des Limites," *Courier de la Planète*, 29: 8–10.
2. Raoult-Wack, A. L. and N. Bricas. 1998. "Controllable Development of the Food Sector in Tropical Areas: Main Challenges, Fields of Research and Research Procedures," *Outlook on Agriculture*, 27(4): 225–235.

3. UNESCO. 1999. "World Declaration on Science and the Use of Scientific Knowledge," in *General Conference, 30 C/15, Appendix III*, ed. UNESCO, 1–16.
4. Raoult-Wack, A. L., G. Toulouse, J. P. Kahane, and N. Bricas. 1999. "Sciences and Development," in *World Conference on Science, a New Commitment, French Contribution*, ed. Commission Nationale Française pour l'Unesco, 34–42
5. Asiedu, J. J. 1991. *La Transformation des Produits Agricoles en Zone Tropicale*. Paris, ed. Karthala, 335.
6. Smit, J., A. Ratta, and J. Nasr. 1993. "Urban Agriculture: Resource for Food, Jobs and Sustainable Cities." Washington, UNDP, Urban Agriculture Program.
7. Boucher, F. and J. Muchnik. 1995. *Agroindustria Rural: Recursos Técnicos y Alimentación*. San José, eds. CIRAD-CIID-IICA, 503.
8. Winarno, F. G. 1986. *Street Foods in Asia. A Proceeding of the Regional Workshop*. eds. Bogor, Food Technology Development Center and FAO, 255 pp.
9. Westby, A. and M. P. Cereda. 1994. "Production of Fermented Cassava Starch *(polvilho azedo)* in Brasil," *Tropical Science*, 34(2): 203–210.
10. Morin, E. 1995. *La Méthode, Tome 1:La nature de la nature*. Paris, ed. Le Seuil, 398.

5 International Congress on Engineering and Food: The First 25 Years

R. Jowitt

CONTENTS

Abstract

At the invitation of the president of ICEF 8, a brief and highly personal review of the first 25 years of ICEFs, with special reference to the great debt owed to their past presidents, is here presented.

5.1   ICEF 1

The first International Congress on Engineering and Food (ICEF 1) was conceived during discussions between Professor J. T. Clayton, head of the Department of Agricultural and Food Engineering at the University of Massachusetts, and the writer, who in the summer of 1972 had been invited to advise the department on the setting up there of the first U. S. undergraduate course in Food Engineering. As this is a personal view of subsequent events, it seems appropriate to use the first person. It is to my personal regret that Joe Clayton was not able to accept the invitation to be present at ICEF 8.

ICEF 1 was held on 9–13 August 1976 in the Boston, MA, Park Sheraton Hotel coincident with a hurricane, which caused widespread diversion of flights and delays

1-56676-963-9/02/$0.00+$1.50

to arriving participants. That was not the only disaster to strike the Congress. Joe Clayton had offered the publication of the Proceedings to the American Society of Agricultural Engineers, but the ASAE took so long to decide that, by the time they declined, it was too late for anyone else to undertake their publication, to the understandable widespread annoyance of many authors who then had to make private arrangements for their papers to be published piecemeal.

The task of opening the Congress fell to me as co-organizer, and I was unaware that the public address system was not functioning and that only those sitting near the platform heard what I said. So, as those historic (?) words were never heard or printed, and considering their relevance to this occasion, they are printed here below for the first time!

Ladies and Gentlemen, I didn't realize until now that the word "honorary" had different meanings on opposite sides of the Atlantic. In England an honorary office holder receives no pay; in the U. S. he does no work!

So far, my position as vice-chairman of the First International Congress on Engineering and Food has been "honorary" in both the British and the American senses, and that allows me to say some of the things I want to say *to* you and *on behalf of* you this morning.

The committee considered at length whether to hold a closing plenary session at the end of the Congress on Friday, but eventually, for diplomatic and economic reasons, adopted the ingenious plan to combine the opening and closing sessions into one! In this way, those of you who have already decided to play truant or golf for the rest of the week can now do so with a clear(er) conscience, and we can regard all attendance at the Congress sessions as sheer profit.

Seriously, I do want to say now what is usually left until later in such events but which should not go unsaid. It is because I personally have played only a small part in the work of the General Organizing Committee that I can thank them on behalf of you, and especially the large number of overseas participants, for all the work they have put in to make the Congress possible. In case you feel that such sentiments *before* the Congress are tempting Providence, as in so many endeavors—the Olympic Games, sitting exams, painting a house, or making a speech—the secret lies in the the preparation work *before* the event. As usual, the English have a phrase for it. They say that "The battle of Waterloo was won on the playing fields of Eton." In this Bicentennial Year of the United States, I hasten to change the metaphor and acknowledge that the success of this Congress has been secured in the corridors of the Department of Agricultural and Food Engineering and especially in the offices of its head, Joe Clayton. Whatever the course of the next few days, we all owe Professor Clayton and his colleagues, particularly Les Whitney and Anne Maspero, a great debt for all their work over the past two years. In your name I salute them and thank them.

This Congress, more than many, is a participatory one. Papers were only accepted on condition that the authors would be here to present them in person, which means that a large number of you are here to serve at the table and not just to eat! But we all are diners and thank the many authors who have prepared such a feast of papers for our delectation. It is a pity that so many concurrent sessions are necessary in order to get through 200-odd papers, but extending the analogy, we can regard it as a menu offering a wide choice to suit all tastes.

Next we must thank the sponsoring organization, the associate sponsors, the co operating organizations and the many liaison persons around the world without which many of us would not be here today.

I used to know a poor old man before the war who, whenever I offered him my cigarette case, would always take two, putting one between his lips and one behind his ear, saying; "And one for later, eh?" Whatever the success of the present Congress, it has deliberately been titled the *First* Congress on Engineering and Food. I know there have been conferences on food engineering before, some of them international, many of them very successful, such as that held by our Netherlands colleagues in Wageningen in 1973, but I would like to emphasize the significance of the word "First" as meaning "not the last." Like my old friend with the cigarettes, I would like to invite your consideration of not only the First but the second and subsequent Congresses on Engineering and Food.

Now is the time, when we are all together, to plan for any subsequent, similar Congresses and the secretariat invites, nay *entreats,* anyone with any views on the subject to register them at the desk, and a meeting will, if possible, be held of all interested parties before we separate after the Congress, say on Friday afternoon, to discuss future possibilities. We particularly invite suggestions and invitations as to where and when any future congresses might be held, bearing in mind that a host country will have the lion's share of the organization work and that the event will have to attract a capacity response and fit in chronologically with other competing or complementary conferences in the international calendar.

So much for the future; now back to the present. Food Engineering. What is Food Engineering? Food Engineering means *nothing, Nothing* to the vast majority of people at large and nothing to many so-called educated people. I know because of the questions that I have been asked over the past 15 years that I have been responsible for Food Engineering at the National College of Food Technology in Britain.

No doubt the experience of many of you is similar. Food Engineering means very little to most people, because they don't what it is all about. And yet to those same fellow human beings, Food Engineering is *everything!* In its widest ramifications, food engineering is essential to the well being of all human society and to the very existence of a large section of it. Without the engineering of systems for the processing, preservation, conversion, storage and distribution of food, society as we know it could not exist, and neither could many millions of its members. This truth will become increasingly important in the future as mankind struggles to balance its books of population versus food supplies.

Food Engineering has to do with the making available of adequate, wholesome, nutritive food supplies to all people at the right time and in the right place, three times a day, three billion times over and we must not forget it even if most of those three billion are oblivious of it.

To me, Food Engineering is the confluence of two of man's greatest empirical pursuits—food and engineering. *Empirical? Did he say empirical?* Yes he did, and make no mistake about that either. Science is helping us enormously to understand food and engineering, but at the end of the day, the food has to be there for the eating, and the plant, or the process or the system has to work—isn't it an engineering expression: "by guess and by God"? And no scientist *I* have met yet can guarantee either by science alone. I am not disparaging science in the least—it is our single greatest ally—but we have to keep our perspectives right, pass no bucks and carry our own responsibilities by the proper exercise of the art, skill, experience, and science, of food engineering

The significance of this First Congress on Engineering and Food is, to me, in its recognition, by food engineers and their allies, of a special kind of engineering. Not of science, not of technology, but of engineering of vital importance to society and of even greater importance to society's future and to its need to gird itself about for its future role and the great responsibilities it is certain to be called upon to shoulder. I am sure it will rise to meet and discharge with honor those responsibilities.

Thereafter, for the rest of the Congress, everyone seemed to have a good time and considered the event a great success.

5.2   ICEF 2

The meeting in Boston at the end of ICEF 1 was duly held, and the bid to host ICEF 2 by the eminent Finnish food engineer, Professor Pekka Linko, subject to his proviso of agreement by the Finnish authorities, was accepted.

That agreement was given subject to the Congress being held at the same time as the 27th IUPAC Congress in Helsinki and to the sharing with IUPAC of certain events such as the opening ceremony and concert and the congress dinner—and the corollary that it be held in 1979, three years after ICEF 1, not four as intended.

It had its own subtitle, Food Process Engineering 1979, which indicated its sharper focus on four major themes—Food Physics and Transport Phenomena; Food Processing Systems, Developments in Food Processing, and Enzyme Engineering in Food Processing—programs active in Finland at the time.

Some 400 participants from 35 countries attended; 177 papers contributed were published in 1980 by Elsevier's Applied Science Publishers (London).

It is a constant source of wonder, especially to citizens of large populations, how smaller countries like Singapore, Ireland, Belgium and Finland manage to staff all the posts that a sovereign state must maintain from such relatively small numbers of suitably qualified nationals. That they do so successfully is a tribute to those states as well as to the individuals appointed. Pekka Linko is the epitome of such, and he is well known around the world for his support and substantial contributions to almost all conceivable bodies and projects involving food, biotechnology, enzymology, instrumentation and control, extrusion cooking, computer modeling, fuzzy logic, optimization, etc.

In these circumstances, it is not surprising that ICEF 2 was a seamless success in all respects—as were the other meetings held in association with it, such as the COST 90 Physical Properties of Foods subgroup and the European Federation of Chemical Engineering Food Working Party, with its well remembered (but seldom mentioned!) sauna afterward.

If there were any problems associated with ICEF 2, they didn't show. But that's Finland. That's Pekka Linko for you. *Semper idem!*

5.3   ICEF 3

I seem to remember chairing the final session of ICEF 2, when the host and venue for ICEF were being considered and came under vigorous attack from a young

participant from an LDC for not holding such congresses in countries like his. That ICEF venues were offered, not requisitioned, had little effect on his ire, even when the offer by Brian McKenna to host ICEF 3 in Ireland was made, discussed, and accepted.

The emerging policy of ICEF supporters to try to hold ICEFs equispaced from the four-yearly IUFoST conferences was becoming increasingly difficult to apply when the latter were becoming as inconstant as, but less predictable than, the planets. By 1983, IUFoST and ICEF were not in opposition but in conjunction and, amazingly, Brian McKenna was proposing Ireland to host not only ICEF 3 in the summer of 1983 but also, a few days before it, IUFoST's International Congress on Food Science and Technology!

All that I have written under ICEF 2 about Finland and the efficiency of smaller countries applies with even greater force to ICEF 3 and to the apparently superhuman qualities called for in Brian to achieve this unprecedented and seemingly impossible task. But, he did it—without fuss and with great success. How he did it, I shall never understand. Perhaps he has a direct line to the head leprechaun! Be that as it may, some 300 participants enjoyed a smooth ride through all the by-now-familiar parts of an ICEF, and some even attended both congresses.

I had the opportunity of attending a meeting with Richard Hall and Erik von Sydow, then president and general secretary of IUFoST, at which mutual recognition of the independent roles of both IUFoST and ICEFs was agreed, on the basis of which future cooperation between the two bodies could be established.

5.4 ICEF 4

At the familiar ad hoc final meeting of ICEF 3 to decide on the venue and date for ICEF 4, the knotty problem of disengagement from coincidence with IUFoST was resolved by the next surprise—Marc LeMaguer's offer to organize ICEF 4 in two years time in Edmonton, Alberta, Canada, so that maximum separation from IUFoST's congresses could be re-established.

Marc was as good as his word, and in July 1985, we set off over the polar icecap to warm and sunny Alberta. Brian McKenna and I arrived a few days early and drove as far westward as we reasonably could to spend a few days approaching and on the Great Divide, where a small stream emerges and divides, half flowing eventually to the Pacific and the other eventually to the Atlantic.

Back in Edmonton, we found Marc, ably assisted as I am sure he would be the first to insist, by Peter Jelen, engaged in the process of not dividing but of fusing the 400 participants into a harmonious gathering for what again proved to be a successful ICEF. Not that I saw much of it. I was busy responding to the challenge posed at the Delegates' Meeting immediately preceding the Congress to produce acceptable draft Articles of Association for the proposed International Association for Engineering and Food, which would place the organization of future ICEFs on a more formal, tidy, and less ad hoc footing.

They all said it was impossible in the time, but by spending more of my time on it than on the technical program, a draft was produced that, with a little tidying

up, became the first and still extant Articles of Association for the new IAEF, and most of the "older inhabitants" formally signed up to become founder members.

5.5   ICEF 5

After the second ICEF in the New World, a return to the Old was agreed, and Walter Spiess's bid to host it in Germany, probably Cologne, in the spring of 1989 was readily accepted. As most of you will know (due sadly to the sudden demise of Lilian Marovatsonga, who was to have been IUFoST's first African president), Walter has recently assumed the presidency of IUFoST and is unique in his occupancy of the presidencies of, first, IAEF/ICEF 5 and, now, IUFoST. I am not sure which is the more fortunate, IUFoST in having an eminent food engineer for president, or Food Engineering in having such a potential influence on IUFoST. Either way, it is to be hoped that the greater recognition of each by the other will be taken a step forward by this unique event.

ICEF 5 attracted the largest number of participants so far—nearly 600, from 39 countries—and produced three volumes of proceedings with 300 papers from 580 authors. It was at a time of great political change in Europe that reflected on the atmosphere of ICEF 5 in Cologne, especially during the unscheduled periods, which for some went well into the night. A (small) team of helpers, of course, supported Walter, but none would dispute that his personal stamp characterized the whole event.

The so-called delegates' meeting of IAEF members was uncharacteristically chaotic, with more interlopers than delegates, and when the venue for ICEF 6 was debated, the bid from the United Kingdom was defeated by popular acclaim in favor of one tabled by Japan, which at the time was not even a member of IAEF and was therefore invalid! However, to have argued under the circumstances would have been pointless, and so the United Kingdom settled for a "rain check" on hosting ICEF 7, some eight years later.

Nevertheless, ICEF 5 remains in my memory as one of the great ICEFs.

5.6   ICEF 6

For many participants, ICEF 6, in Chiba, the coastal Conference City some miles from Tokyo, in May 1993, was a rare opportunity to visit legendary Japan for the first time, and nearly 500 from 39 countries attended. The strangeness to occidentals of that land and its language were largely offset by the thoughtful, detailed arrangements made by Toshimasa Yano (TY) and his team for the visitors, and only the size of the hotel rooms remained to remind burly participants where they were.

But if to us ICEF 6 ran like a well oiled machine, it was not so for those below decks. The serene swan gliding effortlessly along was in fact paddling furiously unseen underneath, and only TY and colleagues were aware of it!

TY recalls that, for the president of an international congress, one of the worst headaches is finance. The economic situation in Japan was so bad that, at one point, TY committed the whole of his personal university retirement fund to save the

Congress from the red. Fortunately, that did not prove necessary, and eventually, sufficient sponsorship was forthcoming to balance the books and even to leave a surplus, which is being used to fund scholarships for young food engineers at subsequent ICEFs.

He sees, in the progression of 25 years of ICEFs, a move away from the solution by engineers of those problems in the food industry that they could solve to a start on solving those problems that have to be solved.

For me, on this occasion, ICEFs appear as a series of events of international importance led, molded, marshaled, held together, and inspired by a succession of outstanding past presidents to whom we owe more than we realize and whom we cannot thank enough. But let us not fail to try.

5.6.1 EDITORIAL NOTE

There is, in the foregoing, no mention of ICEF 7 and its president—understandably, under the circumstances, as the author was the president. The editors have, therefore, remedied the omission by asking an eminent member of ICEF 7's organizing team for a suitable contribution. The following is provided by Donald Holdsworth, Chairman, ICEF 7 Technical Committee and member, ICEF 7 Organizing Committee.

5.7 ICEF 7

The decision to hold the 7th Congress in the United Kingdom was the result of a successful bid at ICEF 6 by Professor Ronald Jowitt on behalf of the three United Kingdom organizations in membership of IAEF. It was especially pleasing for U.K. food engineers that, after all Ron's pioneering work for "Engineering and Food," on Ron's initiative, Her Majesty, Queen Elizabeth, agreed to be its Patron. ICEF 7 was held at the Brighton Conference Centre, where the Queen was represented at various functions by her Lord Lieutenant for the East Sussex, the country in which Brighton lies. As always, Ron was a superb host, and his very polished and erudite introductory speeches showed the true mark of his personality on all occasions. Indeed, he was always at great pains to make his international guests feel at ease and most welcome.

While Ron masterminded all the events of the Congress with meticulous care, he also found time to edit, personally (but always in accord with one or more additional eminent referees), all the contributions in the two excellent volumes of the proceedings and to supervise the third volume, the *Supplement*. This was characterized by his very high and exacting standards, which he applies fearlessly to authors' manuscripts, and by his wide encyclopedic knowledge of the subject.

This was indeed a remarkable Congress, and it stands as a landmark in the annals of "Engineering and Food" and a fine tribute to Ron Jowitt's remarkable commitment to food engineering over several decades. The Presidential Chain of Office, which he presented to IAEF at ICEF 7 to be worn by his successors-in-office, will be a continuing reminder of his involvement in ICEFs from their early beginnings onward.

Part II

Physical Chemistry

6 State Transitions and Reaction Rates in Concentrated Food Systems

Y. H. Roos
S. M. Lievonen

CONTENTS

Abstract

Rates of diffusion-controlled reactions, chemical or enzymatic, have been suggested to be dependent on the physical state, whether a glassy solid or a highly concentrated liquid, of food systems. Studies on effects of glass transition, T_g, on such reactions as nonenzymatic browning, oxidation, and enzymatic changes have often provided controversial information of various effects of the T_g, temperature, and water on the

1-56676-963-9/02/$0.00+$1.50

reaction rates. It is well known that rates of several reactions in low-moisture and frozen foods are significantly reduced, providing enhanced stability. However, both food systems show rapid changes in reaction rates above some critical water content or temperature. In several cases, these critical parameters have been related to the T_g. Other changes, such as crystallization and structural transformations, may also occur in low-moisture and frozen foods, resulting in more severe changes in the rates of chemical and enzymatic reactions. For example, crystallization of amorphous lactose has been shown to accelerate nonenzymatic browning in dairy powders or to affect the rates of enzymatic changes in food models. We, among others, have shown that nonenzymatic browning may occur at temperatures below the T_g although at significantly lower rates than above the calorimetric T_g. There is, however, a change in the reaction rate constant within the glass transition temperature range. Our recent results suggest that phase separation of food components may allow reactions to occur below the T_g. In more homogeneous systems, an increase in the reaction rate correlates with the endset of the glass transition. This increase in the reaction rate seems to correlate with observed increases in molecular mobility, as derived from dielectric measurements. In both low-moisture and frozen systems, the pH, temperature, viscosity, and water content, in addition to the glass transition, are likely to affect rates of physicochemical changes.

6.1 INTRODUCTION

Food systems represent a large number of examples of nonequilibrium-state biomaterials, which exhibit complicated time-dependent changes in their physical and chemical properties during processing and prolonged storage.[1,2] Food processing often aims at controlled exposure of food materials to, for example, thermal, pressure, mechanical, or irradiative treatments. These, as a result of sufficient energy input to the food material, cause a reduction in the number of active microorganisms or enzyme activity, changes in structure and texture, or changes in chemical composition. These treatments are likely to increase the shelf life of foods or to improve their pleasantness and consumer appeal. However, foods are sensitive to high energy inputs, and adverse changes in sensory characteristics, such as texture, flavor, and taste, or nutritional value associated with processing may impair product quality.

As a food component, water often controls food temperature in processing and thereby rates of changes in microbial or enzyme activity and chemical changes. Removal of water, however, results in concentration of food solids and increases in reaction rates. It is well known that, for example, decreasing water content in the baking or processing of sugar- and protein-containing foods results in browning and formation of desired, typical flavor and odor compounds (Maillard reaction) of the particular food item. Enhanced reaction rates resulting from water removal, for example, in milk processing may also be extremely harmful, and only by proper control of time-temperature-water content relationships are high-quality products obtained.

Dehydration and freezing prevent microbial growth and allow production of foods with extended shelf life. Enzyme activity, chemical reactions, and changes in texture are also retarded in low-moisture and frozen foods. The rates of these changes may, however, be significantly affected by water content or the amount of unfrozen water.[1-3] Oxidation may occur rapidly when water is removed, while rates of other reactions are reduced. Reactions occurring between water-soluble food components, such as the Maillard reaction and several enzymatic reactions in low-moisture foods, have increasing rates above some critical water content.[1-4] It has been suggested that reaction rates at normal storage temperatures of low-moisture foods are related to water activity, a_w, defined as the ratio of the equilibrium water vapor pressure in the food, p, and the vapor pressure of pure water at the same temperature, p_0, i.e., $a_w = p/p_0$.[1] In some cases, it has been found that the maximum stability is provided by a water activity corresponding to the Brunauer-Emmett-Teller (BET) monolayer value of the food.[1]

Slade and Levine[2] have suggested the use of the "polymer science approach" in evaluating food stability. According to Slade and Levine,[2] water in foods acts as a plasticizer, depressing the glass transition of water-miscible food components such as carbohydrates and proteins. The glass transition is a well known property of amorphous materials, whether inorganic or organic. Glass transition occurs over a temperature range, which is often referred to with the *glass transition temperature,* T_g, taken as the onset or midpoint temperature of the transition determined using differential scanning calorimetry (DSC).[2,5] However, the glass transition can be observed from a number of other changes in material properties and relaxations with spectroscopic methods, such as FTIR, Raman, and NMR spectroscopies, and other thermal analytical methods including dielectric analysis (DEA) and dynamic mechanical analysis (DMA).[5-7] DEA and DMA have been found to be more sensitive in observing the glass transition than DSC, which applies especially to the thermal characterization of frozen food materials.[8]

The food polymer science approach[2] has proved to be useful in explaining temperature- and water-content-dependent changes in reaction rates of concentrated food systems, such as low-moisture and frozen foods. Low-moisture and frozen foods contain amorphous, water-miscible solids, which can exist in either a glassy or plasticized, amorphous, nonsolid, supercooled liquid state. Relationships between the physical state, plasticization (either thermal or water), and rates of possible diffusion-controlled reactions, such as nonenzymatic browning, enzymatic reactions, and oxidation, have been reported in these nonequilibrium systems. However, the reaction rates have been found to be dependent also on other factors such as temperature, pH, and water content, and not only on the diffusivity of reactants.[9-11] We will discuss state transitions of concentrated, nonequilibrium food systems and effects of phase and state transitions and other factors on diffusion, reaction rates, and stability.

6.2 CONCENTRATED SYSTEMS AND STATE TRANSITIONS

Baking, evaporation, dehydration, extrusion, and freezing are typical processes in the production of concentrated food systems. The resultant foods can be amorphous glasses or supercooled liquids, or they may be partially crystalline systems. For example, the branched starch polysaccharide, amylopectin, in baked products often exists as a partially crystalline polymer. In addition, foods may contain other compounds in separate phases, such as water and lipid phases, or, in the freeze-concentrated state, the crystalline ice and unfrozen, freeze-concentrated amorphous phases. In some cases, the amorphous components are only partially miscible, e.g., proteins and carbohydrates.[12] Obviously, the complexity of foods, their nonequilibrium nature, and the various possible states and phase separation of component compounds make characterization of foods and measurements of their state transitions difficult.

6.2.1 Low-Moisture Foods

The nonequilibrium, amorphous carbohydrates and sugars, such as lactose in dehydrated dairy foods, sugars in hard candies, and starch have been found to suffer the glass transition and affect mechanical properties and stability of, for example, freeze-dried and spray-dried foods and confectioneries.[13] The glass transition of low-moisture foods is affected by composition, molecular weight of component compounds, and extent of water plasticization.[2,5] The glass transition of foods with high sugar content (for example, dehydrated berries and fruits and candies) is detected with DSC from the change in heat capacity occurring over a temperature range of 10 to 20°C.[14,15] The glass transition temperatures of sugars are affected by molecular weight, and they decrease dramatically with water content. Glass transition temperatures for sugars typical of food systems are shown in Figure 6.1. The glass transition of bread, which contains starch and gluten, has been observed at intermediate water contents.[16] Foods containing large amounts of high-molecular-weight carbohydrates and proteins, however, have been found to exhibit glass transitions occurring over a wide temperature range with relatively low changes in heat capacity.[5] The transitions at low water contents occur at decomposition temperatures, and even in the presence of water, the transitions may be difficult to observe. The determination of such transitions with DSC often fails, and other techniques, e.g., DEA and DMA, must be used. Other factors complicating the analysis of low-moisture foods include partial crystallinity of sugars or starch, poor miscibility of carbohydrates and proteins, and the presence of lipids.

6.2.2 Frozen Foods

State transitions of frozen food systems are important in defining their freeze-drying behavior and storage stability.[5] For example, sugar solutions are known to have temperature limits in freeze-drying, referred to as *collapse* temperatures.[17] Collapse in freeze-drying occurs when the freeze-concentrated unfrozen solute phase is able to flow and cannot support the structure of the material. Collapse temperatures have been found to correspond to the onset temperatures of ice melting, T_m', in the maximally freeze-concentrated state, probably because of a rapid increase in unfrozen water content and concurrent plasticization of the unfrozen, amorphous phase.[5]

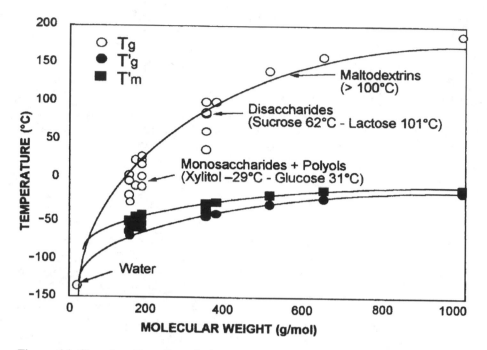

Figure 6.1 Glass transition, T_g, and glass transition in the maximally freeze-concentrated state, T_g', temperatures and onset temperature of ice melting in the maximally freeze-concentrated state, T_m', for selected monosaccharides, disaccharides, and maltodextrins.

The increase in the unfrozen water content decreases the T_g of freeze-concentrated solutes and viscosity and allows more rapid ice recrystallization.[2,18] Ice dissolution above T_m' may also result in increasing rates of diffusion-controlled reactions during storage of frozen foods.[19]

Studies of state and phase transitions in frozen food systems have used DSC, DEA, and DMA. A special disc-bending sample holder design was introduced by MacInnes[20] for DMA studies of frozen solutions and foods. The sample holder consists of two plastic membranes, and the material studied can be prepared and frozen as a disc between the membranes. DSC, DEA, and DMA studies of frozen food systems have shown that the materials suffer transitions dependent on the extent of freeze-concentration during initial freezing or dependent on thermal history. For example, rapidly cooled sugar solutions do not freeze completely even at very low temperatures.[21,22] Such partially freeze-concentrated systems exhibit during reheating a glass transition followed by devitrification and melting of ice (Figure 6.2). Devitrification refers to ice formation during reheating and concurrent freeze concentration of the unfrozen solute phase. Maximally freeze-concentrated materials are formed when freezing time at a temperature favoring maximum ice formation becomes sufficient for complete freezing. All water in food systems, however, cannot separate as ice, because ice formation is controlled by the viscosity of the unfrozen solute phase. Roos and Karel[22] found that sucrose solutions become maximally freeze

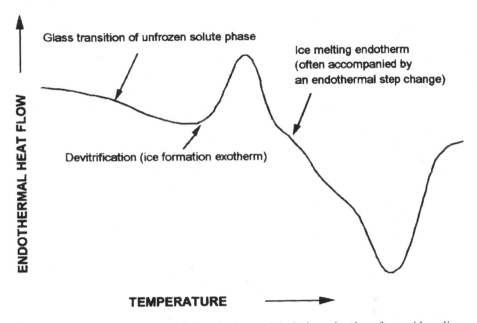

Figure 6.2 Typical transitions of frozen food materials during reheating after rapid cooling, as observed using differential scanning calorimetry. The devitrification exotherm indicates partial freeze-concentration during freezing and ice formation during rewarming.

concentrated during annealing at −35°C. The glass transition temperature of freeze-concentrated solutions and foods increases with increasing ice formation. A concurrent increase in solute concentration in the unfrozen phase decreases the equilibrium ice melting temperature. Therefore, the decreasing ice melting temperature requires that ice formation must occur at a lower temperature. However, the increasing glass transition temperature and viscosity of the unfrozen solute phase decreases the rate of ice formation. Theoretically, the T_g' and T_m' coincide at a sufficient extent of freeze concentration, and the ice formation ceases as the unfrozen solute phase is vitrified.[2] Sugar solutions often show the onset of the glass transition of the maximally freeze-concentrated unfrozen solute phase, T_g', and the onset temperature of ice melting, T_m', in two separate transitions. Increasing molecular weight increases these temperatures and, for high-molecular-weight food components, the transitions coincide.[23] Typical transitions of frozen food components are shown in Figure 6.3.

DMA and DEA as well as NMR studies of frozen solutions have provided additional information of relaxations and water mobility in freeze-concentrated systems.[8,20,24] The DMA and DEA studies have shown low-temperature relaxations below the T_g'. The temperatures of the low-temperature relaxations, as well as the α-relaxation (glass transition), are highly dependent on frequency (Figure 6.4). The lower frequency α-relaxation temperatures (<0.1 Hz) often correspond to the T_g' determined using DSC. These methods have been successful in detecting the glass transition of frozen doughs, which cannot be measured using DSC.[8]

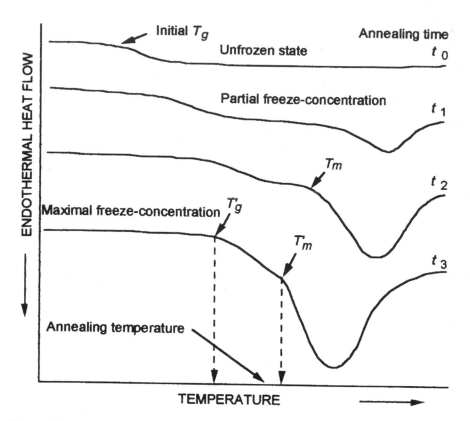

Figure 6.3 Transitions occurring in frozen food components, as observed using differential scanning calorimetry.

6.2.3 STATE DIAGRAMS

The information of phase and state transitions can be shown in state diagrams.[2] A state diagram can be established by measuring the glass transition of a material at several water contents and plotting the transition temperature against water content. Water plasticization, i.e., the decrease in T_g with water content, can be modeled using the Gordon-Taylor relationship of Equation (6.1) or other equations relating the T_g with water content.[5]

$$T_g = \frac{w_1 T_{g1} + k w_2 T_{g2}}{w_1 + k w_2} \tag{6.1}$$

where T_g is glass transition of the plasticized material, w_1 and w_2 are weight fractions of solids and water, T_{g1} and T_{g2} are glass transitions of solids and water, respectively, and k is a constant.

A full state diagram shows the water content dependence of the glass transition temperature, typical transitions of the maximally freeze-concentrated solutes, and

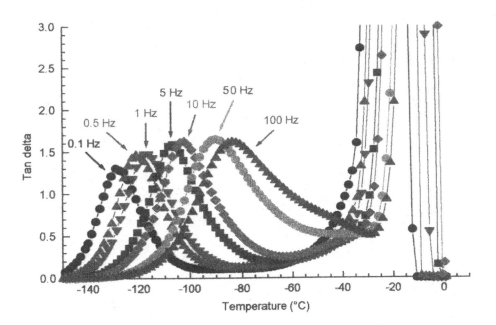

Figure 6.4 Dielectric thermal analysis of a 20% (w/w) sucrose solution showing frequency-dependent β-relaxation and α-relaxation (glass transition). The α-relaxation coincides with ice melting.

equilibrium ice melting temperatures for the partially freeze-concentrated states, as shown in Figure 6.5.[22] The state diagram may also show other important transitions or describe rates of time-dependent changes at various temperature-water content combinations.

6.3 REACTION RATE CONTROLLING FACTORS IN FOODS

Reaction rates in food materials can be controlled by a number of factors, as reviewed by Labuza and Riboh.[25] The main factors are food composition and the type of the reaction, temperature, pressure, water content, and pH. In concentrated food systems, the increasing viscosity may reduce diffusion and, as the materials at low water contents or at low temperatures vitrify, reaction rates may decrease due to diffusional limitations, and the temperature dependence of the reaction rates may deviate from Arrhenius kinetics.[25] Although water seems to be fairly mobile in food glasses, the mobility of larger food component molecules seem to be retarded in the glassy state.[24]

6.3.1 TEMPERATURE

Temperature is certainly one of the most important factors affecting reaction rates in food systems. The main approaches in relating reaction kinetics in foods to

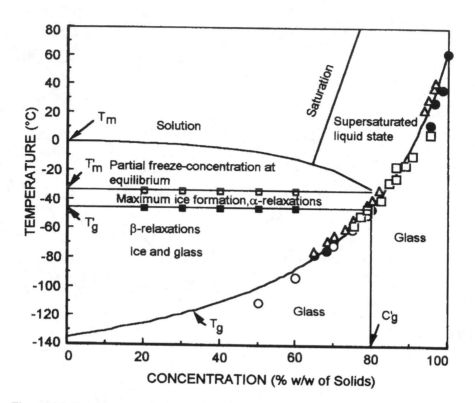

Figure 6.5 State diagram of sucrose showing the equilibrium ice melting temperature, T_m, curve, onset temperature of ice melting in the maximally freeze-concentrated state, T_m', glass transition temperature of maximally freeze-concentrated solutes, T_g', with solute concentration given by C_g', and the glass transition, T_g, curve at various solute concentration.

temperature have been the Q_{10} approach and the Arrhenius concept.[26] In several cases, the reaction rates have been shown to follow the Arrhenius relationship given in Equation (6.2).

$$\ln k = k_0 - \frac{E_a}{RT} \tag{6.2}$$

where k = rate constant
   k_0 = pre-exponential factor
   E_a = activation energy
   R = gas constant
   T = absolute temperature

It should be noticed that several other factors affecting reaction rates may also be dependent on temperature. This makes studies of the temperature dependence of reaction rates in concentrated food systems more complicated. Such interdependent factors include, for example, pH, solubility, and water activity.[25]

6.3.2 PRESSURE

Traditional food processing is based on thermal treatment, where the effect of pressure on the reaction kinetics is often negligible. The emerging high-pressure food-processing technologies are likely to introduce new aspects of the effect of pressure on reaction kinetics.[27] In thermodynamic terms, pressure has a similar effect to temperature on the rate of chemical reactions.[27]

6.3.3 WATER CONTENT

Reaction rates in foods with high water content are mainly affected by temperature and food composition. At reduced water contents, reaction rates increase as a result of increasing reactant concentration. Foods containing lipids show increasing rates of oxidation as low water activities are approached. However, lipids may also become encapsulated in carbohydrate or protein structures protecting them from oxidation. Other reactions, e.g., enzymatic reactions and the nonenzymatic browning reactions, have a rate maximum at an intermediate water activity, and the rates decrease as low water activities are approached. However, an increase in temperature may increase water activity and cause an additional increase in reaction rates.[25] In general, however, food materials with low water contents are more stable, and as water contents of foods are reduced, reaction rates even at high temperatures decrease.

6.3.4 PHASE AND STATE TRANSITIONS

The phase and state transition effects on reaction rates in low-moisture and frozen foods have been studied intensively. These studies have shown that, in low-moisture foods, diffusion-controlled reactions such as the Maillard reaction and enzymatic reactions may be affected by the physical state. Several studies have confirmed that these reactions may occur with increasing rates above the glass transition, but the reactions may also proceed below the glass transition and other factors, such as water content, pH, and temperature also affect the reaction rate. Other phase transitions, e.g., crystallization of amorphous components in low-moisture foods, have been shown to have dramatic effects on reaction rates. For example, lactose crystallization in dairy powders may cause a rapid increase in the rate of nonenzymatic browning and other deteriorative changes.[28] In frozen foods, freeze concentration increases solute concentration, and the increase in reactant concentration may enhance reactions.[25] Participation of reactants between different food phases may vary with temperature due to phase transitions and solubility.[25]

6.4   STATE TRANSITIONS AND REACTION RATES

State transitions in low-moisture and frozen foods include glass transition, low-temperature relaxations, and the ice melting transition in frozen foods. The information available on the effects of sub-T_g relaxations on reaction rates in food systems is very limited. However, an increasing amount of information of the molecular mobility and sub-T_g relaxations is being obtained.[24,29] An example of a glass transition and dielectric relaxation in systems used to study temperature and glass transition dependence of the Maillard reaction rate is shown in Figure 6.6. The effects of glass transition on reaction rates have been studied at various temperatures at constant water content as well as at constant temperatures at various water contents. Either a change in temperature or water content may result in glass transition and affect reaction rates in concentrated food systems.

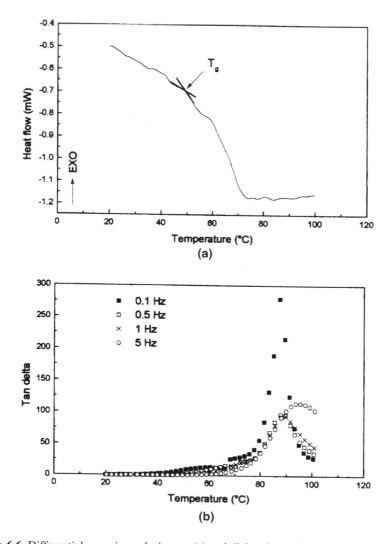

Figure 6.6 Differential scanning calorimetry (a) and dielectric analysis (b) curves at several frequencies for a polyvinylpyrrolidone model system with lysine and xylose as reactants.

6.4.1 Glass Transition, Viscosity, and Diffusion

It is generally accepted that the glass transition results in a change in the physical
state of amorphous materials, although exact glass transition temperatures cannot
be well defined, and the reported transition temperatures are dependent on the
methods and time of observation.[30] The glass transition occurs over a temperature
range with a concurrent dramatic change in viscosity. This change in viscosity results
in a decrease in relaxation times of structural transformations and apparent flow
above the transition. The changes in flow over and above the glass transition are
dependent on material properties; for example, sugar glasses become sticky syrups
not far from the glass transition, and materials containing high-molecular-weight
food polymers, such as low-moisture bread, become leathery or rubbery.

Glass Transition And Viscosity

Parks et al.[31] determined various changes occurring in glucose glasses as they were
transformed from the solid glassy state to the supercooled liquid state. Their work
included determination of viscosity among several other material properties. It has
been suggested that the viscosity data followed the Williams-Landel-Ferry (WLF)
relationship[32] of Equation (6.3) and other equations relating viscosity changes in
amorphous materials to temperature.

$$\ln a_T = \ln \frac{\tau}{\tau_s} = \ln \frac{\eta}{\eta_s} = \frac{-C_1(T - T_s)}{C_2 + (T - T_s)} \tag{6.3}$$

where a_T is the ratio of relaxation times, t and t_s, at temperatures, T and T_s, respectively.

The WLF relationship states that changes in relaxation times above the glass
transition are related to a reference temperature, T_s, which along with the constants,
C_1 and C_2, can be derived from experimental data.[5,33] Although the WLF relationship
may follow viscosity data above the glass transition, it assumes a steady viscosity
in the glassy state and downward concavity over the glass transition. Peleg,[33] however, has shown that the relationship between changes in mechanical properties over
the glass transition shows an upward concavity and follows the Fermi relationship
of Equation (6.4).

$$\frac{Y}{Y_s} = \frac{1}{1 + \exp\left[\dfrac{X - X_c}{a(X)}\right]} \tag{6.4}$$

where Y = a measure of mechanical property or "stiffness" parameter
Y_s = stiffness parameter in a reference state, e.g., in the glassy state
X = a measure of plasticization (temperature, water content, or water activity)
X_c = the value for X resulting in 50% change in Y
$a(X)$ = a measure of the steepness of the change in stiffness as a function of X

The Fermi relationship has been found to fit data on mechanical properties of foods; for example, loss of crispness as a function of water activity.[29] It seems that the relationship is also applicable to predict viscosity changes resulting from thermal or water plasticization over and above the glass transition (Figure 6.7).

Glass Transition and Diffusion

Diffusion in food systems has been related to viscosity and glass transition. Based on this assumption, the use of the WLF relationship for predicting diffusion coefficients above glass transition has been suggested.[5] Studies of diffusion in sucrose solutions,[29] however, have shown that diffusion follows the WLF relationship well above the T_g, but a decoupling between diffusion and viscosity occurs as the glass transition is approached. Therefore, in the vicinity of the glass transition, diffusivity cannot be predicted from viscosity.[29]

In polymer science, the free-volume theory is used to describe concentration and temperature dependency of solvents in amorphous polymers.[34] It is also known that diffusion coefficients of small molecules are little affected by concentration or temperature. This agrees well with the findings that water remains fairly mobile in glassy food systems.[24] In binary systems, such as food solids and water, the mutual diffusion coefficient is likely to be most sensitive to concentration and temperature in the vicinity of the T_g. This can be observed from high activation energies of

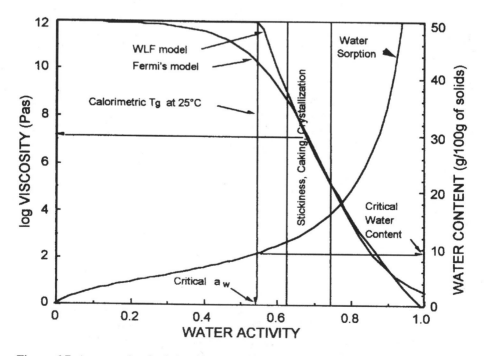

Figure 6.7 An example of relationships among water activity, water content, and viscosity as predicted using the Fermi and WLF relationships.

diffusion in polymers as the T_g is approached. Therefore, Arrhenius plots of diffusion may suffer a step change over the glass transition, although the discontinuity may not be perceptible for water.[34] Similar diffusion behavior is also likely to occur in food systems.

It is obvious that diffusion occurs below the glass transition, but the mobility of solute molecules decreases significantly below the T_g.[24] However, diffusion becomes more affected by the glass transition with increasing size of the diffusing molecules. The size of diffusing reactants may, therefore, have an effect on whether a reaction may become diffusion controlled in the vicinity of the glass transition.

6.4.2 GLASS TRANSITION AND REACTION RATES

Glass transition has been suggested to affect reaction rates in both low-moisture and frozen foods. There are fairly few systematic studies of the temperature dependence of reaction rates in diffusion-controlled and "well stirred" food systems over the same temperature range. Another option is to follow reactions over the same temperature range in food systems having different glass transition temperatures. Reaction rates in frozen foods are also affected by the extent of freeze concentration, which, being a function of composition, makes comparison of reaction rates and possible effects of glass transition on observed rates difficult.

Enzymatic Reactions

Studies of glass transition effects on enzymatic changes have included sucrose inversion by invertase,[35] rate of hydrolysis of disodium-p-nitrophenyl phosphate by alkaline phosphatase,[29] and enzymatic toughening of frozen fish.[36]

Sucrose inversion by invertase in low-moisture systems has been found to be dependent on water activity.[35] The reaction rate increases at water activities above 0.6. We found, using a freeze-dried lactose-sucrose system, that sucrose inversion occurred very slowly when the water content was not sufficient to depress the T_g to below storage temperature. In this system, the lactose crystallized above the T_g and crystallization occurred concomitantly with increasing reaction rate. Using of a maltodextrin-based system allowed observation of the reaction rate in a noncrystallizing system. However, the reaction rate was not found to increase significantly above T_g below water activity of 0.6, suggesting that a_w was a more important factor in controlling sucrose inversion than the T_g. It is possible that the enzyme mobility is not sufficient to allow the reaction to occur until the reacting molecules are plasticized by water to an appropriate extent above 0.6 a_w. Chen et al.[35] found that sucrose hydrolysis in a PVP system started at a_w 0.62, and it was more affected by a_w than the T_g of the system.

A study of the rate of hydrolysis of disodium-p-nitrophenyl phosphate by alkaline phosphatase in frozen sucrose solutions showed that the rate in the frozen state increased with dilution due to ice melting.[29] However, the solutions did not become maximally freeze concentrated, and the effect of T_g' on the reaction rate could not be estimated. In the freeze-concentrated state, the solutions had the same concentrations of sucrose, enzyme, and substrate, but, above freezing temperatures, the reaction rates were higher in less concentrated sucrose solutions.

We have conducted studies of enzymatic changes in frozen systems containing maltodextrins with different T_g' along with fish extracts containing the TMAO-demethylation enzyme system. The reaction results in formation of formaldehyde and toughening of frozen fish belonging to the cod family.[36] We also used a reaction system with glycerol, which showed that the enzymatic reaction did not occur at significant rates at temperatures below –25°C.

Maillard Reaction

The Maillard reaction as a model of a binary reaction between an amino acid and a reducing sugar has probably been given the most attention in studies of relationships between reaction rates and the glass transition. The reaction is common in dehydrated foods and, in several cases, a relationship between the occurrence of the reaction with increasing temperature difference with storage temperature and the glass transition temperature, $T - T_g$, has been observed.[37]

The effect of glass transition on the reaction rate has been related to observed discontinuities in Arrhenius plots, suggesting a high activation energy in the vicinity of the transition. It should be noticed that the reaction also seems to occur below the glass transition. This may result from heterogeneous distribution of reactants and water, phase separation of reactants from the main matrix, or the reactant mobility may not cease at the measured T_g but at some temperature lower than the measured T_g. In some food models and real food systems, the reaction rate has not increased in the immediate vicinity of the measured T_g but 20–40°C above the T_g, depending on water content.[38,39]

Studies of effects of glass transition on reaction kinetics, and the Maillard reaction in particular, are complicated because of several factors affecting plasticization and reaction rates. For example, plasticizers, i.e., water content and temperature, affect reactant concentration, glass transition, pH, and water activity. In a study of Maillard reaction kinetics in amorphous maltodextrin and PVP systems, we controlled the amount of reactants, glucose, and lysine in the water phase sorbed at 33% RH and 25°C.[38] This study showed that the reaction rate increased considerably above the glass transition (Figure 6.8). The rates in the two systems, however, differed. This was not considered to result from differences in a_w, reactant concentration, or T_g, as these parameters were comparable in the two systems. Microscopic observation of the structure of the materials suggested that the reactants were phase separated in the PVP system, which probably resulted in more rapid browning in the PVP system. In other studies using PVP systems to observe Maillard reaction kinetics, the physical state has been found to be more important in controlling browning rate than a_w.[40,41]

Other factors affecting reaction kinetics in low-moisture foods include collapse of structure and crystallization of component compounds. These factors have been found to enhance browning reactions.[37,40,41] Collapse also results in impaired diffusion, which is reflected by decreasing release of carbon dioxide[41] and glysine consumption.[40]

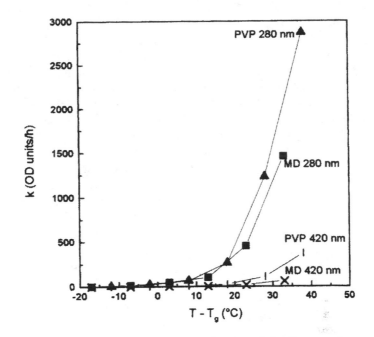

Figure 6.8 Maillard reaction rate constants as a function of the distance of the storage temperature and the glass transition temperature $(T - T_g)$ for polyvinylpyrrolidone (PVP) and maltodextrin (MD) models with lysine and glucose as reactants.

Because of the numerous factors affecting Maillard reaction kinetics in foods, the glass transition cannot be shown to have an individual effect on the reaction rate. However, the glass transition, along with material composition, water activity, temperature, and pH, provides additional information for the control of Maillard reaction in food processing and storage.

Oxidation

Oxidation is a common reaction in low-moisture and frozen foods. It may be assumed that oxidation of lipids directly exposed to atmospheric oxygen at food surfaces or surfaces of porous, dehydrated materials may occur freely. In both low-moisture and frozen foods, lipids exist phase separated from the water miscible solids, but lipids may become encapsulated in glassy matrices that often improve stability of the products.[42]

It is well known that "free fat" in, for example, dairy powders, is highly susceptible to oxidation, causing quality defects. The stability of dairy powders is related to the glassy state of lactose, and the lipids are at least partially encapsulated within the lactose-protein matrix in spray-drying. The encapsulated fat is protected from oxidation by the surrounding glassy lactose membranes. However, thermal or water plasticization resulting in the glass transition of the lactose-protein matrix allows lactose crystallization and subsequent release of the encapsulated lipids. The released lipids become accessible by atmospheric oxygen and rapid oxidation. In some systems, crystallization of the encapsulating matrix is delayed. The delayed crystallization reduces oxygen penetration even when the encapsulating matrix suffers a glass transition. It is likely that the encapsulating matrix flows above the glass transition temperature, resulting in structural collapse and partial release of the encapsulated lipids. There is, however, a possibility that some of the lipids remain encapsulated or become re-encapsulated in the matrix.

Oxidation causes quality changes in low-moisture and frozen food materials. Unfortunately, there is very little information about the relationships between oxidation kinetics and the physical state of low-moisture and frozen foods. Obviously, such studies are of utmost importance, and they will be needed for product design and optimization of storage conditions.

6.5 CONCLUSIONS

Low-moisture and frozen foods are concentrated food systems exhibiting temperature- and water-content-dependent changes in their physical state. These changes, including glass transition and ice melting, affect diffusion and, therefore, rates of quality changes in these concentrated materials. Low-moisture and frozen foods show rapid changes in reaction rates above some critical water content or temperature. Although the critical values often coincide with plasticization resulting in glass transition, several studies have shown that diffusion affecting reaction rates and component crystallization, among other quality changes in these systems, is dependent on the molecular size of the diffusing compounds, and that several reactions may occur in the glassy state. There are, however, very little experimental data available for establishing relationships between kinetics of quality changes, phase separation, temperature, water content, and food composition.

REFERENCES

1. Labuza, T. P. 1968. "Sorption Phenomena in Foods," *Food Technol.*, 22: 263–265, 268, 270, 272.
2. Slade, L. and H. Levine. 1991. "Beyond Water Activity: Recent Advances Based on an Alternative Approach to the Assessment of Food Quality and Safety," *Crit. Rev. Food Sci. Nutr.*, 30: 115–360.
3. Labuza, T. P., S. R. Tannenbaum, and M. Karel. 1970. "The Effect of Water Activity on Reaction Kinetics of Food Deterioration," *Food Technol.*, 24: 543–544, 546–548, 550.
4. Karel, M. 1985. "Effects of Water Activity and Water Content on Mobility in Food Components, and Their Effect on Phase Transitions in Food Systems," in *Properties of Water in Foods*, D. Simatos and J. L. Multon, eds. Dordrecht, The Netherlands: Martinus Nijhoff Publishers, pp. 153–169.
5. Roos, Y. H. 1995. *Phase Transitions in Foods*. San Diego, CA: Academic Press.
6. Sperling, L. H. 1986. *Introduction to Physical Polymer Science*. New York, NY: John Wiley & Sons.

7. Söderholm, S. E., Y. H. Roos, N. Meinander, and M. Hotokka. 1999. "Raman Spectra of Fructose and Glucose in the Amorphous and Crystalline States," *J. Raman Spectr.*, 30: 1009–1018.

8. Laaksonen, T. J. and Y. H. Roos. 1998. "Dielectric and Dynamic-Mechanical Properties of Frozen Doughs," in *Proceedings of the Poster Sessions, ISOPOW 7—Water Management in the Design and Distribution of Quality Foods*, Y. H. Roos, ed. Helsinki, Finland: The University of Helsinki, EKT series 1143, pp. 42–45.

9. Lim, M. H. and D. S. Reid. 1991. "Studies of Reaction Kinetics in Relation to the T_g' of Polymers in Frozen Model Systems," in *Water Relationships in Foods*, H. Levine and L. Slade, eds. New York, NY: Plenum Press, pp. 103–122.

10. Nelson, K. A. and T. P. Labuza. 1994. "Water Activity and Food Polymer Science: Implications of State on Arrhenius and WLF Models in Predicting Shelf Life," *J. Food Eng.*, 22: 271–289.

11. Bell, L. N. 1996. "Kinetics of Non-Enzymatic Browning in Amorphous Solid Systems: Distinguishing the Effects of Water Activity and the Glass Transition," *Food Res. Int.*, 28: 591–597.

12. Kalichevsky, M. T., E. M. Jaroszkiewicz, and J. M. V. Blanshard. 1993. "A Study of the Glass Transition of Amylopectin-Sugar Mixtures," *Polymer*, 34: 346–358.

13. White, G. W. and S. H. Cakebread. 1966. "The Glassy State in Certain Sugar-Containing Food Products," *J. Food Technol.*, 1: 73–82.

14. Roos, Y. H. 1987. "Effect of Moisture on the Thermal Behavior of Strawberries Studied Using Differential Scanning Calorimetry," *J. Food Sci.*, 52: 146–149.

15. Roos, Y. H. 1993. "Melting and Glass Transitions of Low Molecular Weight Carbohydrates," *Carbohydr. Res.*, 238: 39–48.

16. Laine, M. J. K. and Y. Roos. 1994. "Water Plasticization and Recrystallization of Starch in Relation to Glass Transition," in *Proceedings of the Poster Session, International Symposium on the Properties of Water, Practicum II*, A. Argaiz, A. López-Malo, E. Palou, and P. Corte, eds. Puebla, Mexico: Universidad de las Américas-Puebla, pp. 109–112.

17. Bellows, R. J. and C. J. King. 1973. "Product Collapse during Freeze Drying of Liquid Foods," *AIChE Symp. Ser.*, 69(132): 33–41.

18. Hartel, R. W. 1998. "Mechanisms and Kinetics of Recrystallization in Ice Cream," in *The Properties of Water in Foods ISOPOW 6*, D. S. Reid, ed. London: Blackie Academic & Professional, pp. 287–319.

19. Karel, M. and I. Saguy. 1991. "Effects of Water on Diffusion in Food Systems," in *Water Relationships in Foods*, H. Levine and L. Slade, eds. New York, NY: Plenum Press, pp. 157–173.

20. MacInnes, W. M. 1993. "Dynamic Mechanical Thermal Analysis of Sucrose Solutions," in *The Glassy State in Foods*, J. M. V. Blanshard and P. J. Lillford, eds. Loughborough, UK: Nottingham University Press, pp. 223–248.

21. Simatos, D. and J. M. Turc. 1975. "Fundamentals of Freezing in Biological Systems," in *Freeze Drying and Advanced Food Technology*, S. A. Goldblith, L. Rey, and W. W. Rothmayr, eds. San Diego, CA: Academic Press, pp. 17–28.

22. Roos, Y. H. and M. Karel. 1991. "Amorphous State and Delayed Ice Formation in Sucrose Solutions," *Int. J. Food Sci. Technol.*, 26: 553–566.

23. Roos, Y. H. and M. Karel. 1991. "Water and Molecular Weight Effects on Glass Transitions in Amorphous Carbohydrates and Carbohydrate Solutions," *J. Food Sci.*, 56: 1676–1681.

24. Hemminga, M. A., I. J. van den Dries, P. C. M. M. Magusin, D. van Dusschoten, and C. van den Berg. 1999. "Molecular Mobility in Food Components as Studied by Magnetic Resonance Spectroscopy," in *Water Management in the Design and Distribution of Quality Foods*, Y. H. Roos, R. B. Leslie, and P. J. Lillford, eds. Lancaster, PA: Technomic Publishing Co., Inc., pp. 255–265.

25. Labuza, T. P. and D. Riboh. 1982. "Theory and Application of Arrhenius Kinetics to the Prediction of Nutrient Losses in Foods," *Food Technol.*, 36(10): 66, 68, 70, 72, 74.

26. Villota, R. and J. G. 1992. "Reaction Kinetics in Food Systems," in *Handbook of Food Engineering*, D. R. Heldman and D. B. Lund, eds. New York, NY: Marcel Dekker, pp. 39–144.

27. Knorr, D. 1998. "Advantages, Possibilities and Challenges of High Pressure Applications in Food Processing," in *The Properties of Water in Foods ISOPOW 6*, D. S. Reid, ed. London: Blackie Academic & Professional, pp. 419–437.

28. Saltmarch, M., M. Vagnini-Ferrari, and T. P. Labuza. 1981. "Theoretical Basis and Application of Kinetics to Browning in Spray-Dried Whey Food Systems," *Prog. Food Nutr. Sci.,* 5: 331–344.

29. Le Meste, M., D. Champion, G. Roudaut, E. Contreras-López, G. Blond, and D. Simatos. 1999. "Mobility and Reactivity in Low Moisture and Frozen Foods," in *Water Management in the Design and Distribution of Quality Foods*, Y. H. Roos, R. B. Leslie, and P. J. Lillford, eds. Lancaster, PA: Technomic Publishing Co., Inc., pp. 267–284.

30. Peleg, M. 1993. "Mapping the Stiffness-Temperature-Moisture Relationship of Solid Biomaterials at and around Their Glass Transition." *Rheol. Acta*, 32: 575–580.

31. Parks, G. S. and J. D. Reagh. 1937. "Studies on Glass. XV. The Viscosity and Rigidity of Glucose Glass," *J. Chem. Phys.*, 5: 364–367.

32. Williams, M. L., R. F. Landel, and J. D. Ferry. 1955. "The Temperature Dependence of Relaxation Mechanisms in Amorphous Polymers and other Glass-Forming Liquids," *J. Am. Chem. Soc.*, 77: 3701–3707.

33. Peleg, M. 1994. "A Model of Mechanical Changes in Biomaterials at and around Their Glass Transition," *Biotechnol. Progr.*, 10: 385–388.

34. Duda, J. L. 1999. "Theoretical Aspects of Molecular Mobility," in *Water Management in the Design and Distribution of Quality Foods*, Y. H. Roos, R. B. Leslie, and P. J. Lillford, eds. Lancaster, PA: Technomic Publishing Co., Inc., pp. 237–253.

35. Chen, Y. H., J. L. Aull, and L. N. Bell. "Invertase Storage Stability and Sucrose Hydrolysis in Solids as Affected by Water Activity and Glass Transition," *J. Agric. Food Chem.*, 47: 504–509.

36. Sallinen, J. and Y. H. Roos. 1998. "Freeze-Concentration and Glass Transition Effects on Enzyme Kinetics in Frozen Food Models," in *Proceedings of the Poster Sessions. ISOPOW 7—Water Management in the Design and Distribution of Quality Foods*, Y. H. Roos, ed. Helsinki, Finland: University of Helsinki, EKT series 1143: 88–91.

37. Karmas, R., M. P. Buera, and M. Karel. 1992. "Effect of Glass Transition on Rates of Nonenzymatic Browning in Food Systems," *J. Agric. Food Chem.*, 40: 873–879.

38. Lievonen, S. M., T. J. Laaksonen, and Y. H. Roos. 1998. "Glass Transition and Reaction Rates: Nonenzymatic Browning in Glassy and Liquid Systems." *J. Agric. Food Chem.,* 46: 2778–2784.

39. Karmas, R. and M. Karel. 1994. "The Effect of Glass Transition on Maillard Browning in Food Models," in *Maillard Reactions in Chemistry, Food, and Health,* T. P. Labuza, G. A. Reineccius, V. Monnier, J. O'Brien and J. Baynes, eds. Cambridge, UK: The Royal Society of Chemistry, pp. 182–187.

40. Bell, L. N., D. E. Touma, K. L. White, and Y. H. Chen. 1998. "Glycine Loss and Maillard Browning as Related to the Glass Transition in a Model Food System," *J. Food Sci.,* 63: 625–628.
41. Buera, M. P. and M. Karel. 1995. "Effect of Physical Changes on the Rates of Nonenzymatic Browning and Related Reactions," *Food Chem.,* 52: 167–173.
42. Shimada, Y., Y. Roos, and M. Karel. 1991. "Oxidation of Methyl Linoleate Encapsulated in Amorphous Lactose-Based Food Model," *J. Agric. Food Chem.,* 39: 637–641.

7 Biopolymer-Biopolymer Interactions at Interfaces and Their Impact on Food Colloid Stability

S. Damodaran

CONTENTS

Abstract

Two-dimensional phase separation in several protein-1/protein-2/water ternary film adsorbed at the air–water interface was studied using an epifluorescence microscopy technique. The equilibrium composition of saturated mixed monolayer films of α_s-casein/β-casein, BSA/β-casein, and soy 11S/β-casein at the air–water interface at various bulk concentration ratios did not follow a Langmuir-type competitive adsorption model. This was due to a change in the binding affinity of the proteins to the interface as a result of incompatibility of mixing of the proteins in the mixed monolayer. Among the systems studied, only the α-lactalbumin/β-casein/water ternary film followed the ideal Langmuir behavior for competitive adsorption, indicating that these two proteins were thermodynamically compatible at the air–water interface. Fluorescence microscope images showed that the ternary films containing incompatible proteins underwent two-dimensional phase separation. In these phase-separated films, one of the proteins was present as dispersed phase and the other as continuous phase. It is proposed that the interface between such phase-separated regions of the protein film may act as "fault zones," promoting instability in protein stabilized foams, and possibly emulsions.

7.1 INTRODUCTION

A majority of processed foods are either foam-type (air-in-water) or emulsion-type (oil-in-water) products. The stability of these dispersed colloidal systems depends on the presence of an ampiphilic film at these interfaces. Proteins, which are amphiphilic, exhibit a high propensity to migrate and bind to air–water and oil–water interfaces and decrease the interfacial tension. In addition to lowering the interfacial tension, the adsorbed protein can form a strong viscoelastic film via intermolecular interactions that can withstand thermal and mechanical perturbations.[1,2] In addition, conformational changes in proteins at interfaces allow them to form loops that protrude from the interface into the bulk phase. The steric repulsion caused by overlapping of the layer of protruding chains of emulsion particles when they approach each other is considered to be the most important force stabilizing emulsions.[3] The latter property makes proteins more desirable than low-molecular-weight ampiphiles such as lecithin and monoglycerides as surfactants in emulsion- and foam-type food products. However, proteins differ significantly in their surface activity and in their ability to stabilize colloidal systems. This can be attributed to differences in their structural properties and their ability to undergo conformational reorientation at interfaces.[4]

Notwithstanding such structure-dependent innate differences among proteins, another important factor that might influence the stability of food emulsions and foams is the thermodynamic incompatibility among proteins in the adsorbed protein film at the interfaces. Typical food proteins, notably protein blends, used in food industries are mixtures of several proteins. As a result, the protein film formed around oil droplets of an emulsion usually consists of a mixture of proteins. The stability of the food colloid will be dependent on the nature and intensity of protein-protein interactions in the film. Generally, interactions between two dissimilar polymers are thermodynamically incompatible. This is true of proteins as well. In concentrated aqueous solutions (10–20% w/v), mixtures of two proteins exhibit thermodynamic incompatibility of mixing and, as a result, undergo phase separation.[5-8] Because the local concentration of proteins in an adsorbed protein film is equivalent to 15–30%, it is likely that two-dimensional phase separation of proteins might occur in the film. If this occurs, then the "interface" between the phase-separated regions in a mixed protein film might act as zones of high free energy. The stability and integrity of the mixed protein film might be adversely affected by these high free energy "fault zones," which, in turn, might promote coalescence of oil droplets.

The thermodynamic incompatibility between two polymers arises from energetically unfavorable interactions between them. In ternary protein solutions, even if the protein-1/protein-2 interaction parameter, χ_{12}, is negligible, thermodynamic incompatibility can still exist if there is a large difference between protein-1/solvent (χ_{1s}) and protein-2/solvent (χ_{2s}) interaction parameters.[7,9] The larger the value of $|\chi_{1s} - \chi_{2s}|$, the higher the incompatibility. Conversely, the greater the difference in hydrophilicities of proteins, the greater their incompatibility. However, one cannot correctly predict incompatibility of two proteins at an interface based on their behavior in solution, for two reasons. The thermodynamics at the asymmetrical force field of an interface is not the same as that in the bulk phase, and the structural states of proteins at an interface are not the same as that in bulk solution. Thus, proteins that exhibit compatibility or incompatibility in solution may or may not exhibit the same behavior in a mixed protein film at an interface.

Virtually no information currently exists in the literature on thermodynamic incompatibility and two-dimensional phase separation of proteins in mixed protein films at interfaces. This situation is partly due to lack of a theoretical framework and experimental approach to study phase behavior of polymer mixture at an interface. Recently, we have developed an experimental approach to study this phenomenon.[10] In this contribution, we provide experimental evidence of incompatibility and phase separation in several mixed protein films at the air–water interface.

7.2 THEORETICAL BACKGROUND

Although theoretical models to study thermodynamic incompatibility of biopolymers in solution are fairly well developed, no model currently exists for studying this phenomenon at interfaces. We propose the following model as a theoretical basis for studying thermodynamic incompatibility between biopolymers at interfaces.

Adsorption of proteins and other biopolymers at interfaces is generally assumed to follow a Langmuir adsorption model. The Langmuir adsorption model for a protein-1/solvent system is as follows:

$$\Gamma = \frac{KC}{1 + KaC} \tag{7.1}$$

where Γ = surface concentration
    K = equilibrium binding constant
    C = bulk phase protein concentration
    a = area occupied per protein molecule at saturated monolayer coverage

For a protein-1/protein-2/water ternary system, the Langmuir adsorption model for competitive adsorption is as follows:[11]

$$\Gamma_1 = \frac{K_1 C_1}{1 + K_1 a_1 C_1 + K_2 a_2 C_2} \tag{7.2}$$

and

$$\Gamma_2 = \frac{K_2 C_2}{1 + K_1 a_1 C_1 + K_2 a_2 C_2} \qquad (7.3)$$

where K_1 and K_2 are the equilibrium binding constants of protein-1 and protein-2, respectively, to the interface in protein-1/solvent and protein-2/solvent systems, C_1 and C_2 are their concentrations in bulk mixture, Γ_1 and Γ_2 are concentrations in the mixed film at the interface and a_1 and a_2 are the area occupied by protein-1 and protein-2, respectively, at saturated monolayer coverage in protein-1/solvent and protein-2/solvent systems, respectively. This Langmuir model for competitive adsorption in a protein-1/protein-2/water ternary system assumes that the surface concentrations of protein-1 and protein-2 in the mixed protein film at the interface are affected only by their relative binding affinities to the interface and their concentration ratio in the bulk phase. It assumes *a priori* that the adsorbed protein molecules do not interact with each other and that their adsorption is dictated only by their relative affinity to the interface and the availability of vacant sites at the interface. Under noninteracting conditions, from Equations (7.2) and (7.3), the Langmuir adsorption model predicts the following:

$$\frac{\Gamma_1}{\Gamma_{tot}} = \frac{K_1 C_1}{K_1 C_1 + K_2 C_2} \qquad (7.4)$$

and

$$\frac{\Gamma_2}{\Gamma_{tot}} = \frac{K_2 C_2}{K_1 C_1 + K_2 C_2} \qquad (7.5)$$

Equations (7.4) and (7.5) suggest that, when $K_1 = K_2$, the mass ratio of protein-1 to protein-2 in the mixed protein film will be exactly the same as the ratio of their concentrations in the bulk phase. That is, a plot of Γ_i/Γ_{tot} versus C_i/C_{tot} will be a straight line with a slope of 1. When $K_1 \neq K_2$, the plot will be nonlinear, and it will be a function of the ratio K_1/K_2. A family of theoretical Γ_i/Γ_{tot} versus C_i/C_{tot} curves generated for various ratios of K_1/K_2 for protein-1/protein-2/water ternary systems that obey Langmuir-type competitive adsorption is shown in Figure 7.1. Knowing the ratio of the equilibrium binding constants of the proteins to an interface, determined from their adsorption isotherms in single protein systems [Equation (7.1)], we can predict the ideal competitive adsorption behavior in a protein-1/protein-2/water ternary system.

If the proteins of a protein-1/protein-2/water ternary system exhibit thermodynamic incompatibility of mixing at an interface, then the first impact of that would be on the binding affinities of the proteins to the interface. The ratio of binding affinities of the proteins in the ternary adsorption system would not be the same as the ratio determined from single-protein adsorption systems. Thus, if thermodynamic incompatibility between two proteins exists at an interface, then the experimental Γ_i/Γ_{tot} versus C_i/C_{tot} plot would not be the same as that predicted by the Langmuir

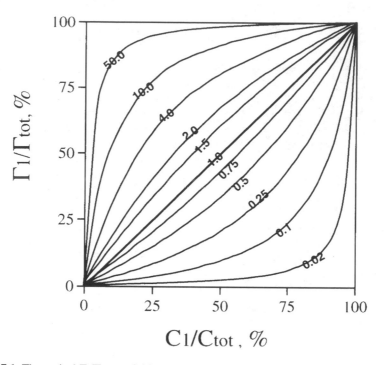

Figure 7.1 Theoretical Γ_1/Γ_{tot} vs. C_1/C_{tot} profiles at various K_1/K_2 ratios for protein-1/protein-2/water ternary systems that follow the Langmuir adsorption model according to Equation (7.4).

competitive adsorption model [Equations (7.4) and (7.5)]. The extent of deviation will be a direct measure of the extent of thermodynamic incompatibility between the proteins.

The thermodynamic incompatibility between proteins at an interface can be determined using another approach, as follows. From Equations (7.2) and (7.3),

$$\ln(\Gamma_1/\Gamma_2) = \ln(K_1/K_2) + \ln(C_1/C_2) \tag{7.6}$$

If $K_1{}'/K_2{}'$ is the ratio of the binding constants of the proteins in the protein-1/protein-2/water ternary adsorption system, then the value of $\ln(K_1{}'/K_2{}')$ for the ternary system can be determined from the intercept of a plot of $\ln(\Gamma_1/\Gamma_2)$ versus $\ln(C_1/C_2)$. If K_1 and K_2 are the binding constants of two proteins in protein-1/water and protein-2/water adsorption systems, then the absolute difference between $\ln(K_1/K_2)$ and $\ln(K_1{}'/K_2{}')$, i.e., $|\Delta \ln K|$ can be regarded as a measure of thermodynamic incompatibility among the proteins. Since $|\Delta \ln K|$ is a thermodynamic quantity representing the intensity of interaction between protein-1 and protein-2 molecules at the interface, it might in fact be a measure of the Flory–Huggins interaction parameter χ_{12}.

7.3 EXPERIMENTAL

7.3.1 ADSORPTION

The experimental parameters required to elucidate thermodynamic incompatibility of proteins at an interface are the equilibrium binding constants (K) of individual proteins in protein-solvent binary systems and equilibrium surface concentrations (Γ_1 and Γ_2) of proteins 1 and 2 in a protein-1/protein-2/solvent ternary system at various bulk concentration ratios (C_1/C_2). Adsorption of proteins from the aqueous phase to the air/water interface was studied by using the surface radiotracer method as described previously.[12–14] The proteins were radiolabeled with [^{14}C] nuclide by reductive methylation of amino groups with [^{14}C]-formaldehyde, as described elsewhere.[14]

Equilibrium adsorption of radiolabeled proteins at the air–water (20 mM phosphate buffered saline solution, pH 7.0, $I = 0.1$) interface was studied using the surface radio tracer technique, as described elsewhere.[12,15] Competitive adsorption of proteins from a protein-1/protein-2/water ternary solution was studied as follows.[10] To monitor adsorption of protein-1, stock solutions of [^{14}C]-labeled protein-1 and unlabeled protein-2 were mixed with the buffer to the required final bulk concentration ratio. The solution was poured into a Teflon® trough (19 × 5.2 × 1.27 cm) and the surface of the solution was swept using a capillary tube attached to an aspirator. The proteins were then allowed to adsorb from the bulk phase to the air–water interface. Although both protein-1 and protein-2 simultaneously adsorbed to the air–water interface, the measured surface radioactivity corresponded only to the amount of radiolabeled protein-1 at the interface. To determine the amount of protein-2 at the interface, the experiment was repeated by mixing stock solutions of [^{14}C]-labeled protein-2 and unlabeled protein-1 with the buffer. Equilibrium surface concentration (Γ_{eq}) values were obtained from surface radioactivity values after 24–30 h adsorption. Adsorption isotherms for each protein were constructed by determining Γ_{eq} at various bulk concentrations, C_b, in single-protein systems. These isotherms were used to determine the binding affinity of proteins to the air–water interface using the Langmuir adsorption model [Equation (7.1)].

7.3.2 EPIFLUORESCENCE MICROSCOPY

Phase separation in mixed protein films at the air–water interface was examined using the epifluorescence microscopy technique. In this approach, protein-1 was labeled with Fluorescin-5-EX succinimidyl ester and protein-2 was labeled with Texas Red. In both cases, the fluorescence labeling was at the amino groups of the proteins. Labeling with Fluorescin-5-EX succinimidyl ester, was carried out as follows: The protein (~1 µmol) was dissolved in 5 ml of carbonate buffer at pH 8.5 and ionic strength 0.1 M. A known amount of the fluorescent dye (~10 µmol) was dissolved in 0.5 ml anhydrous ethanol. The ethanol dye solution was added to the protein solution, and the reaction mixture was stirred for 3.5 h. After the reaction, the mixture was applied to a Sephadex G-25-50 gel-filtration column and eluted with carbonate buffer (pH 8.5 and ionic strength 0.1 M). The fractions corresponding to the first elution peak which contained the protein, as judged from absorption at 280 nm, were pooled and dialyzed extensively against phosphate-buffered saline solution (pH 7.0 and ionic strength 0.1 M) and were then lyophilized. Under these reaction conditions, about 0.85 moles of Fluorescin-5-EX succinimidyl ester was incorporated per mole of the protein.

Labeling of proteins with Texas Red was carried out by the method outlined by Sigma Chemical Co. (St. Louis, MO). Briefly, the protein (40 mg) was dissolved in 8 ml of ice-cold sodium carbonate buffer at pH 8.5 and ionic strength 0.1 M, and stirred continuously. Texas Red (4 mg) was dissolved in 1 ml of ethanol. At the onset of reaction, 0.2 ml of the Texas Red–ethanol solution were added to the protein solution and stirred under ice-cold conditions for 10 min. This procedure was repeated every 10 min by adding 0.2 ml of the dye-ethanol solution to the protein-buffer solution until all of the dye solution was consumed. After the last cycle, the reaction mixture was stirred under ice-cold conditions for an additional hour. The labeled protein was then purified as described above. Under the conditions employed, typically, about 0.132 moles of dye were incorporated per mole of protein.

For epifluorescence microscopy, first, the fluorescent-labeled proteins were allowed to adsorb to the air–water interface for a period of four days from a solution containing a mixture of the proteins in a Teflon trough. The mixed protein film formed at the air–water interface was transferred to a clean microscope slide that was precoated with 3-aminopropyl triethoxy silane (APTES).Transfer of the film was done using a horizontal lifting method in which the microscope slide held horizontally by a vacuum tube was gently lowered to make contact with the aqueous surface for about 10 s. After lifting the film, excess water was allowed to drain off for 1 min. The slide was then dipped in water to rinse loosely bound proteins from the glass slide. The slide was then air dried. A drop of SlowFade Light Antifade reagent in glycerol-water was added to the center of the protein film on the glass slide and covered thereafter with a clean cover glass. This reagent prevented photo-bleaching of the fluorophores during observation under the microscope. The slides, thus prepared, were immediately observed under a computerized Olympus epifluo-rescence microscope equipped with a wide excitation and band-pass emission Ore-gon Green optical filter cube assembly selective for green (EX: 495 ± 15 nm, EM: 545 ± 25 nm), a narrow excitation and a long-pass emission Texas Red filter cube assembly selective for red (EX: 545–550 nm, EM: 610 and above) and a wide excitation and a long-pass emission Fluorescin filter cube assembly (EX: 460–490 nm, EM: 515 nm and above) for simultaneous detection of both dyes. The fluores-cence images were digitized and analyzed with Olympus Image-Pro Analysis soft-ware.

7.4 RESULTS AND DISCUSSION

7.4.1 THERMODYNAMIC INCOMPATIBILITY

The relative affinity of various proteins to the air–water interface was determined by measuring the surface concentration (Γ) of proteins at equilibrium at various bulk concentrations (C_b) in single-protein systems. For all proteins listed in Table 7.1, the adsorption isotherms, i.e., plots of Γ versus C_b, showed a well-defined plateau in the bulk concentration range of 1.5–5.0 µg·ml^{-1}. The saturated monolayer coverage values, Γ_{sat}, for various proteins are given in Table 7.1. The equilibrium binding constant K for the proteins, calculated using Equation (7.1), is also shown in Table 7.1.

TABLE 7.1
Adsorption Isotherm Parameters of Various Proteins at the Air–Water Interface

Protein	K (cm)	Γ_{sat} (mg/m^2)
Egg lysozyme (EL)	0.41	0.87
α-Lactalbumin (α-La)	0.46	0.77
Serum albumin (BSA)	0.60	1.00
Soy 11S	0.54	1.25
α-Casein (α-CN)	1.07	1.69
β-Casein (β-CN)	1.25	1.84

To determine thermodynamic incompatibility of proteins in a mixed protein film at the air–water interface, the protein composition of the mixed protein film at equilibrium was determined at various concentration ratios in the bulk. The protein-1/protein-2/water ternary systems investigated are listed in Table 7.2. Figure 7.2 shows experimental Γ_i/Γ_{tot} versus C_i/C_{tot} curves for the ternary systems studied. It should be pointed out that similar plots for the other protein component of the ternary system would appear as inverse of the curves shown.

TABLE 7.2
Thermodynamic Incompatibility Parameters for Various Protein-1/Protein-2/Water Ternary Systems at the Air-Water Interface

Protein-1/Protein-2	$\ln(K_1/K_2)$	$\ln(K_1'/K_2')$	$|\Delta \ln K|$	X_{12}
EL/β–CN	−1.11	−2.02	0.91	0.30
α-La/β–CN	−1.00	−0.78	0.22	~0
BSA/β–CN	−0.73	−1.56	0.83	0.20
α-CN/β–CN	−0.16	−0.58	0.42	0.14
11S/β–CN	−0.84	−1.90	1.06	0.41

The theoretical curve predicted by Equation (7.4) or (7.5), based on the K_i values obtained from single-component systems for each of the ternary systems, is also shown in Figure 7.2 (dotted lines). It should be noted that, except for the α-lactalbumin/β-casein system, the experimental Γ_i/Γ_{tot} versus C_i/C_{tot} curves deviate significantly from the predicted curves. Among the ternary systems shown in Figure 7.2, only the α-lactalbumin/β-casein system showed somewhat ideal behavior at the air–water interface. This is surprising because, in physicochemical terms, these two

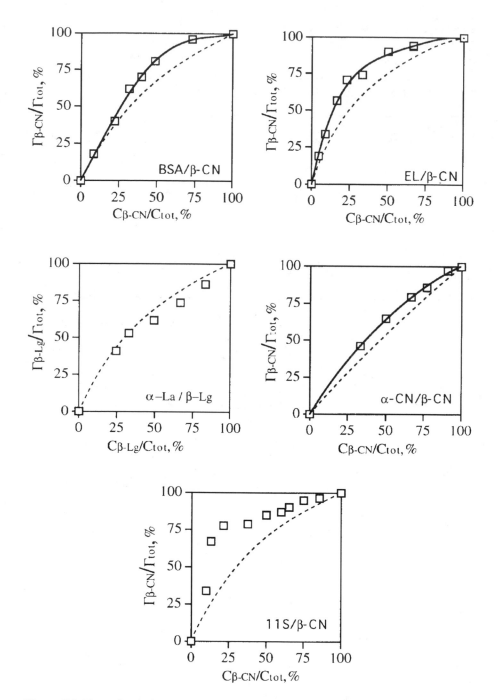

Figure 7.2 Plots of experimental Γ_i/Γ_{tot} vs. C_i/C_{tot} for various protein-1/protein-2/water ternary systems at the air-water interface. The dotted lines represent the ideal curve predicted by Equation (7.4).

proteins are very dissimilar. Whereas α-lactalbumin is a hydrophilic albumin-type protein, β-casein is a hydrophobic disordered-type protein. A mixture of such proteins is known to undergo phase separation in solution.[7] At the air–water interface, however, it is quite likely that α-lactalbumin may undergo extensive denaturation. Exposure of the interior hydrophobic groups may render it more hydrophobic in the denatured state, thus enabling it to be more compatible with the hydrophobic β-casein.

It is interesting to note that, although α-casein and β-casein belong to the same class of proteins, they exhibit some degree of incompatibility at the air–water interface. These two proteins are known to be highly compatible in solution.[7] This contradictory behavior further emphasizes the suggestion that the thermodynamics or protein-protein and protein-solvent interactions at an interface are different from those in solution.

Compatibility of two proteins in a ternary system will be dictated by the delicate interplay of four controlling parameters:

- protein-1-protein-2 interaction parameter (χ_{12})
- protein-1-solvent and protein-2-solvent interaction parameters (χ_{1s}, χ_{2s})
- molecular weights of the two protein molecules
- conformation states of the two proteins[7]

From the Flory–Huggins theory, the critical interaction parameter, χ_c, for a binary polymer-solvent system is 0.5 and for a binary mixture of polymers is 0. The significance of the critical interaction parameter is that two components will be miscible if their interaction parameter is below χ_c but will separate into two phases if the value of this parameter is above χ_c. Although intuitively two proteins in a protein-1-protein-2-water ternary system should be totally compatible when the interaction parameter χ_{12} is 0 (for instance, α-casein/β-casein), in reality, they may still exhibit incompatibility in dilute solutions due to a difference in the protein-solvent interaction parameters, $|\chi_{1s} - \chi_{2s}|$, for as little as 0.03.[9] The absolute magnitude of the polymer–solvent interaction parameter has a considerably smaller effect on polymer–polymer incompatibility than that due to the difference in the polymer–solvent interaction parameters $|\chi_{1s} - \chi_{2s}|$ for the two polymers of interest.[9] The larger the difference in hydrophilicity (i.e., $\chi_{1s} - \chi_{2s}$) of the two proteins, the greater the incompatibility and the lower the threshold for phase separation in a ternary system. On the other hand, if the solvent is equally good for each polymer (i.e., $\chi_{1s} = \chi_{2s}$), two polymers may yield a totally miscible solution in spite of a small positive value of χ_{12}.[16] It should be pointed out that because the thermodynamic state of water at the interface is different from that in bulk solution, the $|\chi_{1s} - \chi_{2s}|$ value for a ternary film at the air–water interface may not be the same as that calculated for polymer solutions. Because the density of water at the air–water interface is lower than its density in the bulk phase, a higher $|\chi_{1s} - \chi_{2s}|$ value will be expected for interactions at the air–water interface.

As discussed earlier, the deviation of the experimental Γ_i/Γ_{tot} versus C_i/C_{tot} curves from that predicted by the Langmuir equation is due primarily to incompatibility of mixing of the proteins in the mixed protein film. The degree of incompatibility between two proteins at an interface can be empirically determined from the extent of deviation of the experimental curve from the predicted one. If two proteins are totally incompatible with each other, then they cannot coexist at the interface. For this situation, the total area above the concave surface of the ideal Langmuir curve can be regarded as total incompatibility. Thus, the ratio of the area between the experimental and predicted curves to the area representing total incompatibility can be defined as the degree of incompatibility, X_{12}, between the proteins at the air–water interface. The X_{12} values for the five ternary systems are shown in Table 7.2. The data suggest that the degree of incompatibility of β-casein with other proteins at the air–water interface increases in the following order: α-lactalbumin < α-casein < BSA < egg lysozyme < soy 11S.

The X_{12} values determined by the graphical method are only empirical in nature. As discussed earlier, fundamentally, the deviation of the experimental Γ_i/Γ_{tot} versus C_i/C_{tot} curve from the predicted one is due to a change in the binding affinity of a protein to the interface in the presence of the other protein at the interface as a result of incompatibility. The relative changes in the binding affinities of proteins in a ternary system as compared to that in a protein/solvent binary system can be determined from Equation (7.6) by plotting ln (Γ_1/Γ_2) versus ln (C_1/C_2). The values of ln (K_1'/K_2') obtained from the intercept of such plots for the ternary systems at the air–water interface are given in Table 7.2. The values of ln (K_1/K_2), determined from protein/solvent binary systems at the air–water interface and the absolute difference between ln (K_1/K_2) and ln (K_1'/K_2'), i.e., $|\Delta \ln K|$, are also shown in Table 7.2. Since $|\Delta \ln K|$ represents a net change in the ratio of the binding constant between the ternary and binary systems at the interface, it represents the intensity of incompatible interactions between proteins at the interface. A linear relationship between X_{12} and $|\Delta \ln K|$ (Figure 7.3) clearly demonstrates that these two are reflections of each other.

7.4.2 Two-Dimensional Phase Separation

If $|\Delta \ln K|$ and X_{12} truly reflect the existence of thermodynamic incompatibility between proteins in a mixed film at the air–water interface, then one should be able to observe two-dimensional phase separation of proteins in the mixed film. This was investigated using the epifluorescence microscopy technique.

In initial studies, it was observed that although equilibrium adsorption of the proteins at the air–water interface took place within 24 h, the mixed protein films transferred from the air–water interface to the α_s-casein/β-casein mixed film transferred from the air–water interface to the APTES glass slide after just 24 h of adsorption did not contain distinct phase-separated regions. However, when the film at the air–water interface was aged for four days and then transferred, distinct phase-separated regions could be observed under the fluorescence microscope. This suggests that in the initial stages, the two proteins adsorb randomly to the air–water interface and then undergo two-dimensional phase separation. Phase separation of proteins at the interface seems to be kinetically limited, perhaps because of the viscosity barrier to lateral diffusion of the molecules. It should be pointed out that

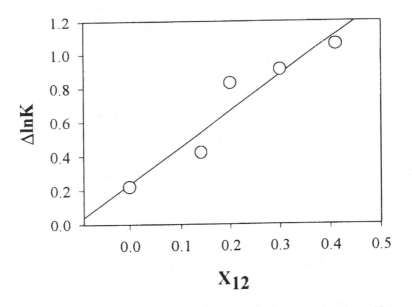

Figure 7.3 Relationship between the empirical incompatibility parameter, X_{12}, and interaction parameter, $| \Delta \ln K |$.

the local protein concentration in a saturated monolayer film at the air–water interface is almost equivalent to 20–30%, and the viscosity barrier for lateral diffusion at this high concentration can indeed be very high.

Figure 7.4 shows the fluorescence microscope image of α_s-casein and β-casein mixed film transferred from the air–water interface. In the image taken with the Texas Red filter (Figure 7.4b), there are large black patches surrounded by a red region corresponding to β-casein (light gray region in Figure 7.4b). These black patches appear as intense green patches (light gray regions in Figure 7.4a) and the red continuous region appears as green region (continuous medium-gray region in Figure 7.4a) when the image is taken with the Oregon green filter. This clearly suggests that, whereas the continuous phase contains an inhomogeneous mixture of both α_s- and β-caseins, the dispersed phase predominantly contains α_s-casein with very little or no β-casein. In all α_s-casein to β-casein concentration ratios in the mixed film, β-casein was present only in the continuous phase and α_s-casein the dispersed phase.

Figure 7.5 shows fluorescence images of BSA/β-casein and α-lactalbumin/β-casein mixed films transferred from the air–water interface. In this system, BSA was labeled with Fluorescin-5-EX succinimidyl ester, and β-casein was labeled with Texas Red. The image taken with the Texas Red filter shows a dark oval region surrounded by a red region that corresponds to β-casein (light gray region in Figure 7.5a). When the same image was taken with the green filter, the dark region, which corresponds to BSA, appeared light-green (light gray region in Figure 7.5b) surrounded by a dark-green region (dark gray region in Figure 7.5b). When the same

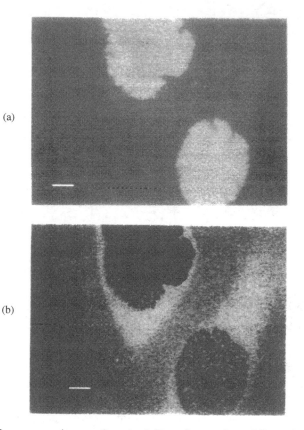

Figure 7.4 Fluorescence image of a mixed film of α-casein and β-casein at the air–water interface. (a) Light and medium gray regions correspond to green regions of α-casein. (b) Light gray regions correspond to the red regions of β-casein. Bar = 100 μm.

image was taken with the mutual filter that allowed both the red and green fluorescence, the image contained a green patch (light gray patch in Figure 7.5c) surrounded by a red continuous phase (continuous medium gray region in Figure 7.5c). It should be pointed out that if the continuous phase contained a significant amount BSA, a combination of green and red colors should have provided a yellow tinge to the continuous phase. Thus, the orange-red color of the continuous phase under the mutual filter (medium gray region in Figure 7.5c) suggests that it predominantly contained β-casein and very little BSA. It should be noted that, as in the case of the α_s-casein/β-casein system, β-casein formed the continuous phase and BSA the dispersed phase. This trend was observed at all BSA/β-casein concentration ratios at the interface.

Figure 7.6 shows a fluorescence image of a α-lactalbumin/β-casein film transferred from the air–water interface. In this system, α-lactalbumin was labeled with Fluorescin-5-EX succinimidyl ester and β-casein with Texas Red. According to

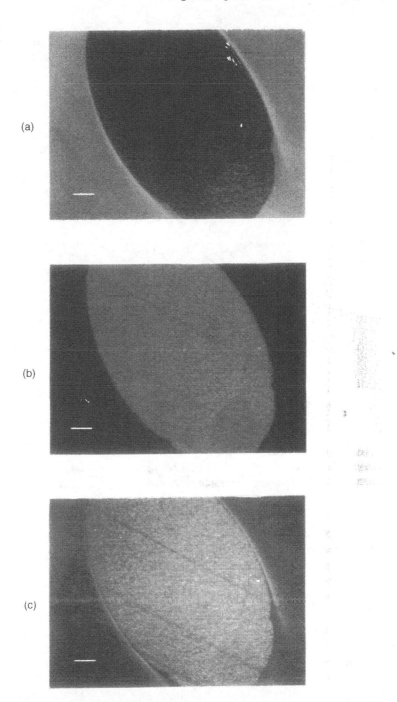

Figure 7.5 Fluorescence image of a mixed film of BSA and β-casein at the air–water interface. Light gray regions in (a) and (c) correspond to BSA, and the dark gray region in (b) and the medium gray region in (c) correspond to β-casein. Bar = 100 μm.

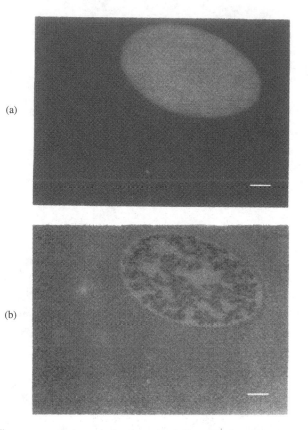

Figure 7.6 Fluorescence micrograph of a mixed film of α-lactalbumin and β-casein at the air–water interface. (a) Light gray and black regions correspond to the green regions of α-casein. (b) Light gray regions correspond to the red regions of β-casein. Bar = 100 μm.

the data presented in Figure 7.2, α-lactalbumin and β-casein appear to be thermo-dynamically compatible and, therefore, two-dimensional phase separation would not be expected in the film. However, the fluorescence image taken with the green filter shows a dense, oval, green region (light gray region in Figure 7.6a) surrounded by a dark-green continuous phase (black region in Figure 7.6a). The oval region is apparently richer in α-lactalbumin than the continuous phase. The interior of the oval region appears homogeneous with some dense internal structures. This was not observed in other systems. When the same image was taken with the Texas Red filter (Figure 7.6b), β-casein was found both in the continuous phase and in the oval region. It seems that, whereas the continuous phase contains a homoge-neous mixture of α-lactalbumin and β-casein, the dense oval-shaped region con-tains aggregated complexes of α-lactalbumin and β-casein. These aggregates might be complexes formed via hydrophobic interaction between β-casein and denatured α-lactalbumin. Formation of complexes, existence of a homogeneous phase and in the film clearly suggests that α-lactalbumin and β-casein are compatible with each other at the air–water interface. This agrees very well with the data shown in Figure 7.2.

7.5 CONCLUSIONS

The fluorescence images of several mixed protein films at the air–water interface clearly showed two-dimensional phase separation. The phase-separated regions become more prevalent during aging of the film than at the initial stages. This is probably due to hindered diffusion of the protein components in the highly viscous film. In most cases, one of the proteins in the mixed film formed a dispersed phase and the other the continuous phase. However, it can be assumed that, if sufficient time has been provided, the film may eventually phase separate into protein-1 rich and protein-2 rich phases, separated by a region of inhomogeneous mixture. The phase-separation phenomenon is due to thermodynamic incompatibility of mixing of the proteins in the two-dimensional film. The theoretical and the experimental approaches described here seem to adequately predict the intensity of thermodynamic incompatibility between proteins. The tendency of two-dimensional lateral phase-separation in mixed protein films at interfaces may adversely impact the stability of protein-stabilized foam and emulsions. Since the "interface" between the phase-separated regions of the film around oil droplets or air bubbles is at a higher free energy, they might act as zones of instability in emulsions and foam.

7.6 ACKNOWLEDGMENT

This research was supported by a grant from the National Science Foundation (Grant #BES-9712197).

REFERENCES

1. Phillips, M. C. 1981. "Protein Conformation at Liquid Interfaces and Its Role in Stabilizing Emulsions and Foams," *Food Technol.*, 35: 50–57.
2. Damodaran, S. 1997. "Protein-Stabilized Foams and Emulsions," in *Food Proteins and Their Applications*, S. Damodaran and A. Paraf, eds. New York, NY: Marcel Dekker, pp. 57–110.
3. Liu, M. and S. Damodaran. 1999. "Effect of Transglutaminase-Catalysed Polymerization of β-Casein on its Emulsifying Properties," *J. Agric. Food Chem.*, 47: 1514–1519.
4. Damodaran, S. and L. Razumovsky. 1998. "Molecular Bases for Surface Activity of Proteins," *Amer. Chem. Soc. Symp. Ser.*, 708: 2–18.
5. Polyakov, V. I., V. Y. Grinberg, Y. A. Antonov, and V. B. Tolstoguzov. 1979. "Limited Thermodynamic Compatibility of Proteins in Aqueous Solutions," *Polym. Bull.*, 1: 593–597.
6. Polyakov, V. I., I. A. Popello, V. Y. Grinberg, and V. B. Tolstoguzov. 1986. "Thermodynamic Compatibility of Proteins in Aqueous Medium," *Nahrung,* 30: 365–368.

7. Polyakov, V. I., V. Y. Grinberg, and V. B. Tolstoguzov. 1997. "Thermodynamic Incompatibility of Proteins," *Food Hydrocolloids,* 11: 171–180.

8. Ledward, D. A. 1994. "Protein/Polysaccharide Interactions," in *Protein Functionality in Food Systems*, N. S. Hettiarachchy and G. R. Ziegler, eds. New York, NY: Marcel Dekker, pp. 225–259.

9. Zeman, L. and D. Patterson. 1972. "Effect of the Solvent on Polymer Incompatibility in Solution," *Macromolecules,* 5: 513–516.

10. Razumovsky, L. and S. Damodaran. 1999. "Thermodynamic Incompatibility of Proteins at the Air- Water Interface," *Colloids Surf. B: Biointerfaces*, 13: 251–261.

11. Hill, C. G. 1977. *An Introduction to Chemical Engineering Kinetics and Reactor Design.* New York, NY: Wiley.

12. Xu, S. and S. Damodaran. 1993. "Calibration of Radiotracer Method to Study Protein Adsorption at Interfaces," *J. Colloid Interface Sci.,* 157: 485–490.

13. Xu, S. and S. Damodaran. 1993. "Comparative Adsorption of Native and Denatured Egg-White, Human, and T_4 Phage Lysozymes at the Air–Water Interface," *J. Colloid Interface Sci.*, 159: 124–133.

14. Xu, S. and S. Damodaran. 1992. "The Role of Chemical Potential in the Adsorption of Lysozyme at the Air–Water Interface," *Langmuir,* 8: 2021–2027.

15. Xu, S. and S. Damodaran. 1994. "Kinetics of Adsorption of Proteins at the Air–Water Interface from a Binary Mixture," *Langmuir,* 10: 472–480.

16. Hsu, C. C. and J. M. Prausnitz. 1974. "Thermodynamics of Polymer Compatibility in Ternary Systems," *Macromolecules,* 7: 320–324.

8 Triacylglyceride Crystallization in Vegetable Oils

J. F. Toro-Vázquez
E. Dibildox-Alvarado
V. Herrera-Coronado
M. Charó-Alonso

CONTENTS

Abstract

The phase changes of triacylglycerides (TAGS) are physical properties that determine the solid fat content in vegetable oils. In turn, the functional properties provided by oils and fats to a food system are established mainly by the proportion of TAGS in the solid phase and its dependence on temperature. Hydrogenation and interesterification of vegetable oils are used to modify the solid/liquid ratio of native vegetable oils. In addition to these chemical processes, TAGS crystallization is an alternative method to modify the solid/liquid ratio of vegetable oils and achieve the appropriate functionality in food systems.

An overview of the events involved in TAGS crystallization in vegetable oils systems investigated in this laboratory is here presented. The results obtained point out the complexity of TAGS crystallization in vegetable oils, entangling nucleation, crystal growth, and secondary crystallization. During nucleation, the involvement of local-order and sporadic nucleation depends on the cooling rate used. In contrast, secondary crystallization is involved in crystal growth. All of these events are engaged during TAGS bulk crystallization in processes occurring not sequentially but in parallel. The Avrami model does not consider all of these phenomena. Future research in TAGS crystallization must focus on developing more complex models that include the different events involved during nucleation and crystal growth.

8.1 INTRODUCTION

Triacylglycerides (TAGS) are triesters of glycerol with fatty acids. This group of nonpolar lipids is the major component of vegetable oils and animal fats and was originally named *triglycerides*. However, the use of this term is now discouraged.

The TAGS have several physical and chemical properties that determine the processing conditions of fats and oils as well as their functionalities in food systems. Thus, the phase changes of TAGS (i.e., crystallization, melting, and polymorphic transformations) are physical properties that determine the solid fat content and its temperature dependence in products such as chocolate, butter, low-fat spreads, short-enings, ice cream, and whipped cream. In turn, the functional properties provided by oils and fats to the food system, such as the visual texture, mouth feel, spread-ability, coating, and air bubble retention, are established, in great extent, by the proportion of TAGS in the solid phase (i.e., solid fat index) and its change as a function of temperature. However, vegetable oils are mixtures of different families of TAGS and consequently do not have specific physical properties. For instance, vegetable oils do not show a particular melting or crystallization temperature. Rather, vegetable oils show a melting/crystallization temperature profile. Additionally, TAGS crystallize in different polymorph states (i.e., the same TAGS observe different molecular organization corresponding to different packing arrangements of the hydrocarbon chains), which have distinct degrees of thermal stability, crystal size, and shape. These polymorph states for pure TAGS crystals are, in increasing order of thermal stability, sub-α, α, β', and β.

However, most vegetable oils do not have the adequate TAGS composition to provide the phase change useful in food systems. Thus, genetic engineering techniques have been developed and successfully utilized to alter the native TAGS composition of oilseeds.[1] Additionally, chemical processes are also used to modify the fatty acid composition of TAGS in oils, i.e., hydrogenation and interesterification.[2,3] All of these techniques have the objective of providing oils with the nutritional fatty acid profile that consumers demand and/or the required phase change properties for their use in food systems (i.e., margarine, hydrogenated soybean oil). However, research costs and regulatory issues limit faster development of genetically modified oilseed crops. On the other hand, consumer concerns associated with the atherogenic effect of *transfatty* acids constrain the future growth of hydrogenation to modify the solid-to-liquid ratio in vegetable oils. Chemical and enzymatic interesterification have economic drawbacks, mainly associated with yield efficiency and cost of the process.[4] An additional alternative involves the use of crystallization of TAGS in vegetable oils.

In the present chapter, a general overview of the results obtained in this laboratory in TAGS crystallization of vegetable oil systems will be presented. Most of these data have not been previously published. A particular emphasis will be on the complexity involved in TAGS crystallization in vegetable oils entangling TAGS nucleation, crystal growth, and secondary crystallization.

8.2 TRIACYLGLYCERIDE CRYSTALLIZATION IN VEGETABLE OILS

The industry utilizes TAGS crystallization as a process to achieve any of three objectives:[5]

1. To eliminate small quantities of high-melting compounds from an oil to improve its appearance and consumers' visual appeal at low ambient temperature (i.e., winterization)
2. To obtain TAGS fractions from oil or fat through a process known as *fractional crystallization*
3. To modify the texture of food systems

The fundamental concepts of the process of crystallization have been well addressed by Boistelle[6] and Larsson.[7] Overall, three different events are involved during crystallization; namely, the induction of crystallization (i.e., nucleation), crystallization (crystal growth), and crystal ripening (crystal perfection). These events occur nearly simultaneously at different rates, since there is a continuous variation of the conditions that produce crystallization.[6] Nevertheless, a requirement before nucleation occurs is the existence of the thermodynamic drive to develop a solid from a liquid phase, i.e., supercooling or supersaturation. From a thermodynamic point of view, there are differences between the concepts of supercooling and supersaturation. However, at the industrial scale, there is not a clear difference between the effect of supercooling or supersaturation on crystallization. The concept of supercooling is normally used when crystallization is achieved from a melt of one (i.e., a pure TAG), or often two or more components (i.e., vegetable oil or a blend of vegetable oils), to obtain either a simple solidification of a one-component system or, as in most cases, the purification of a multicomponent system (e.g., mixed crystals). In contrast, the term supersaturation is used when crystallization of a particular substance (i.e., a given family of TAGS) is achieved from a solution (i.e., a vegetable oil in hexane or acetone) to separate a pure compound. In any case, supercooling and supersaturation are associated with the development of thermodynamic conditions needed to structure the molecules in a "liquid structure," until a critical size of monomers (i.e., TAGS) aggregates and thermodynamic stable solid nuclei are formed.[6,7]

8.2.1 THE FISHER-TURNBULL EQUATION

When nucleation occurs from the liquid phase of the system, also known as the melt (i.e., a vegetable oil), the rate of nucleation (J) depends on the activation free energy to develop a stable nucleus, ΔG_c, and the activation free energy for molecular diffusion, ΔG_d. The viscosity is a physical parameter inversely proportional to molecular diffusion. Then, as supercooling increases, the viscosity in the liquid phase might become a limiting factor for nucleation or crystal growth. The Fisher-Turnbull (Equation 8.1) describes the relationship of ΔG_c and ΔG_d with J:

$$J = (N\kappa T/h)\exp(-\Delta G_c/\kappa T)\exp(-\Delta G_d/\kappa T) \tag{8.1}$$

where J = the rate of nucleation that is inversely proportional to the induction time
        of crystallization (τ_i)
       N = the number of molecules per mole
       κ = Boltzmann's constant
       T = the crystallization temperature
       h = Planck's constant

In a spherical nucleus, ΔG_c is associated with the supercooling, (ΔT), and the surface free energy at the crystal/melt interface, σ, through the following equation:

$$\Delta G_c = (16/3)\pi\sigma^3(T_M)^2/(\Delta H)^2(\Delta T)^2 \tag{8.2}$$

where $(16/3)\pi$ results from the spherical shape attributed to the nucleus
       ΔH = the heat of fusion
       T_M = the melting temperature of the crystallizing compound
       $\Delta T = T - T_M$

From the slope, s, of the linear regression of $\log[(\tau_i)(T)]$ vs. $1/T\,(\Delta T)^2$, the calculation of ΔG_c might be obtained, because $\Delta G_c = s\kappa/(\Delta T)^2$.

From a thermodynamic point of view, the supercooling (ΔT) is obtained by decreasing the temperature of the system below the point at which the smallest aggregation of molecules (i.e., unstable crystal nucleus) is in equilibrium with the molecules in the liquid phase (i.e., TAGS in lamellar "liquid organization" in the vegetable oil).[5] This parameter, known as the equilibrium melting temperature ($T_M°$), is particularly used in polymer crystallization, and a practical methodology for its determination was established[8] and later applied to oil systems.[9] $T_M°$ is a thermodynamic value independent of the concentration of the crystallizing compound in the system. When the equilibrium melting temperature is used in ΔT, an effective supercooling is utilized. This quantity is required in the analysis of J, ΔG_c, and crystal growth rate. On the other hand, relative supercooling is calculated with T_M, a value that is a function of the concentration of the crystallizing compound in the system.[10]

At temperatures below $T_M°$, the system attempts to achieve thermodynamic equilibrium through nucleation and nuclei growth, and, in the absence of foreign particles or crystals of its own type, a solid phase is developed by a process known as *homogeneous nucleation*. In the research made in the general area of crystallization, these are the conditions assumed although very difficult to achieve. When nucleation is eased through the presence of foreign particles, which work as nucleating surfaces, the process is known as *heterogeneous nucleation.* Homogeneous and heterogeneous nucleation are collectively known as *primary nucleation.* In contrast, in secondary nucleation, the presence of a solid phase, previously developed in the system or added as seed crystals, produces the development of additional solid at lower supercooling than the one needed for primary nucleation.

Figure 8.1 shows the Fisher-Turnbull plot for the crystallization of blends of palm stearin in sesame oil (26%, 42%, 60%, and 80%). This model system is a complex crystallization system. Palm stearin (PS) is a mixture of TAGS obtained through fractional crystallization from refined, bleached, and deodorized palm oil.[11] Tripalmitin is the triacylglyceride with the highest melting temperature in palm stearin,[12] and in this case, its concentration was 16.46% wt/wt (±0.17%).[9] Our previous research showed that tripalmitin mostly determines the crystallization kinetics of PS and its blends with vegetable oils (i.e., sesame oil).[9,13] In all of these experiments, the crystallization process was evaluated by DSC and involved heating the system for 30 min at 353.2 K (i.e., erasing crystallization "memory") and then cooling to the crystallization temperature at a cooling rate of 1 K/min. The induction time for tripalmitin crystallization, τ_i, was determined as the time from the start of the isothermal process to the beginning of crystallization (i.e., time where the heat capacity of the sample had a significant departure from the baseline) (Figure 8.2). The T_M° for PS in sesame oil was previously determined with a value of 344 K (70.8°C).[9]

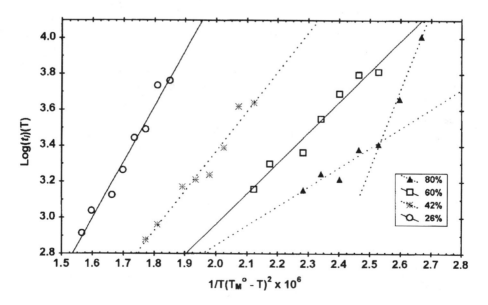

Figure 8.1 Nucleation kinetics of palm stearin blends in sesame oil according to the Fisher–Turnbull equation. Reprinted from Reference 9 with permission from AOCS Press.

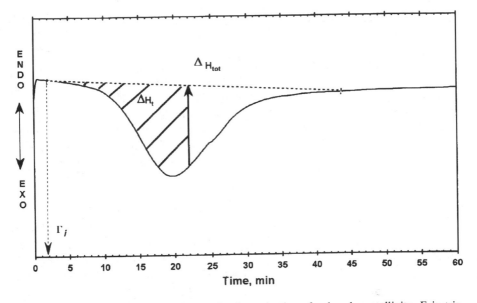

Figure 8.2 DSC isotherm that indicates the determination of reduced crystallinity, F, in triacylglyceride crystallization. Reprinted from Reference 9 with permission from AOCS Press.

A good linearity was reported within the effective supercooling interval investigated ($r > 0.98$, $P < 0.0001$) for the 26%, 42%, and 60% PS:sesame oil solutions.[9] However, the 80% PS:sesame oil solution showed a discontinuity in the plot around 307 K. This behavior, supported by additional DSC analysis to obtain the melting profile, indicated that, within the interval of effective supercooling investigated, the 26%, 42%, and 60% PS:sesame oil solutions crystallized mainly in a $\beta_1{}'$ polymorph state mixed with some α crystals, particularly at very low supercooling. In contrast, the 80% solution crystallized in two different polymorph states, i.e., $\beta_1{}'$ at $T \leq$ 307.6 K and β_1 at $T \geq$ 308.2 K.

The analysis of the ΔG_c values for the PS:sesame oil blends as a function of the crystallization temperature indicated that, in spite of the higher concentration of PS in the 80% system, crystallization in the β_1 state required more energy for its development than $\beta_1{}'$ crystallization in the 26%, 42%, and 60% PS: sesame oil blends.[9] The occurrence of the $\beta' \rightarrow \beta$ polymorphic transition is important to establish the functionality of PS in blends with sesame oil to produce value-added products like *trans*-free margarines and vegetable spreads. In general, β' form and small size confers a fine crystal network that incorporates large amounts of liquid oil and provides good spreadability and plasticity to margarines, shortenings, and spreads. In contrast, β form, larger size, and higher melting temperature than β' crystals provide a sensation of sandiness and a dull appearance. Ongoing research in our laboratory evaluates these aspects through rheological measurement during PS crystallization in vegetable oils combined with DSC and microscopic analysis.

8.2.2 THE AVRAMI EQUATION AND ITS USE IN VEGETABLE OIL CRYSTALLIZATION

The Avrami equation[14] has been used to study TAGS crystallization kinetics in vegetable oils.[13,15–17] However, its use has been mostly limited to the determination of n, the exponent in the Avrami equation associated with the nucleation mechanisms. In these reports, primary crystallization was the main basis of the discussion. Equation 8.3 shows the Avrami model for spherulite crystallization,[14] where F is the fractional crystallization. The index of the crystallization reaction, n, describes the crystal growth mechanism. In homogeneous nucleation, a crystallization process with an $n = 4$ follows a polyhedral crystal growth mechanism, a value of $n = 3$ a plate-like growth mechanism, and an $n = 2$ indicates a linear crystal growth. Non-integer n values are associated with heterogeneous nucleation. On the other hand, the rate constant of crystallization (z), depends on the magnitude of n, and it is a function of the nucleation rate and the linear crystal growth rate of the spherulite.[16] As long as conditions exist in which primary crystallization prevails (Table 8.1). The values of n and z values are usually calculated by linear regression of Equation 8.4 at values of F generally between 0.25 and 0.75. On the other hand, F is calculated as $F = (P_\infty - P_t)/(P_\infty - P_a)$, where P is a property proportional to the change in crystallinity of the material as it crystallizes as a function of time t, (P_t); P_a and P_∞ are the initial and maximum crystallinity of the system. Then, F is a reduced crystallinity, because it associates an instant crystallinity to the total one achieved under experimental conditions.

TABLE 8.1
The Constants n and z as Defined by Avrami's Original Equation

Mechanism of crystal growth	Nucleation sporadic in time (primary nucleation)	
	n	z
Polyhedral	4	$\prod G^3(I/3)$
Plate-like	3	$\prod G^2(I/3)$
Cylindrical	2	$\prod G^2(I/r)$

Note: n = index of Avrami's equation; z = rate constant in Avrami's equation; G = linear growth rate of crystal spherulite; I = sporadic nucleation rate in time; d = width of crystal fibril; r = crystal radii.

Reprinted from Reference 15, with permission from AOCS Press.

In vegetable oil crystallization, F has been determined by light transmittance,[13] DSC measurements,[9,15,16] and X-ray diffraction.[17] Recent experiments in our laboratory have been using the shear storage modulus, G', to evaluate the change of F as a function of the crystallization time (unpublished results). The instrumentation used in each methodology has particular procedures and technical limitations to achieve the thermodynamic conditions for crystallization. Then, each analytical technique has a particular effect on crystallization kinetic. Additionally, each instrument has its own experimental setup to measure a given signal associated with the development of nuclei and/or crystal growth. Consequently, the events involved in the crystallization process are evaluated at different extent, by each analytical technique.

$$1 - F = \exp(-zt^n) \tag{8.3}$$

$$\ln[-\ln(1 - F)] = \ln(z) + n[\ln(t)] \tag{8.4}$$

Figure 8.3a shows the value of F as a function of time for the isothermal crystallization of tripalmitin (0.98% wt/vol) in sesame oil. Previously, we described the experimental setup and crystallization conditions.[13] Figure 8.3b shows the logarithmic form of the Avrami equation. Figure 8.4 shows the same kind of plots for a 80% (wt/vol) PS:sesame oil blend for crystallization data obtained under the conditions described in Reference 9. These systems provide an interesting comparison between a simple crystallization system (i.e., blends of tripalmitin in sesame oil) and a complex crystallization system (i.e., blends of PS in sesame oil). As previously mentioned, our earlier investigation showed that tripalmitin mostly determines the crystallization kinetics of PS and its blends with vegetable oils (i.e., sesame oil).[9,13] In all of the experiments here discussed, the crystallization process was evaluated by DSC prior to heating of the system at 353.2 K during 30 min, and then cooling to the crystallization temperature at a rate of 1 K/min. The F value was calculated as described by Henderson[18] (Figure 8.2), with $F = \Delta H_t / \Delta H_{tot}$ where ΔH_t is the area under the DSC crystallization curve from $t = \tau_i$ to $t = t$, and ΔH_{tot} is the total area under the crystallization curve.

The sigmoidal behavior of TAGS crystallization is apparent from Figures 8.3a and 8.4a and consists of an induction period for crystallization, followed by an increase of the F value (i.e., acceleration in the rate of volume/mass production of crystals), and, finally, a crystallization plateau is reached. However, it is also evident that the plot $\ln[-\ln(1 - F)]$ vs. $\ln(t)$ does not provide a single slope associated with the value of n. Rather, at least two regions of different slopes are observed; the F value that limits these two regions varies as a function of the crystallization temperature (Figs. 8.3b and 8.4b). Figure 8.5 shows the n values obtained from each region at several crystallization temperatures with the 2.62% (wt/vol) tripalmitin blend in sesame oil (Fig. 8.5a) and the 26% (wt/vol) PS:sesame oil blend (Fig. 8.5a). Figures 8.3b and 8.4b show the n values for the second region with the 0.98% tripalmitin and the 80% PS blends in sesame oil, respectively. In this chapter, we will not discuss n and its association to the crystal growth mechanism. Thus, the n value of the first region is evidently always greater than the one obtained from the second region. A combination of heterogeneous and homogeneous nucleation along with secondary crystallization has been associated with this two-region crystallization behavior with different Avrami exponents.[19] This might also be the case with vegetable oil crystallization.

At the low cooling rate used (1 K/min), TAGS molecules have enough time to achieve local order in the liquid state (i.e., lamellar structures, References 3 and

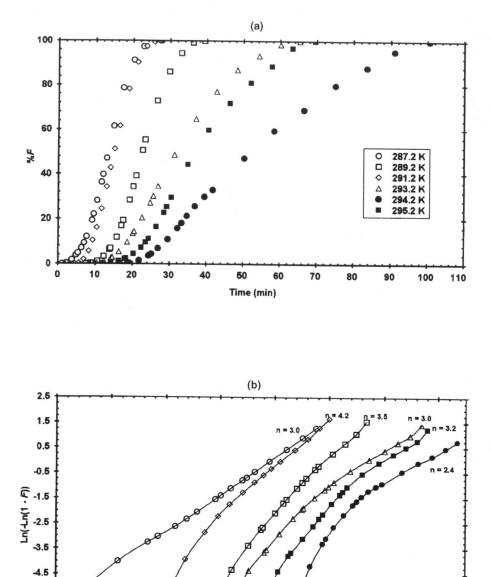

Figure 8.3 Percentage of reduced crystallinity (%F) as a function of time (a) and Avrami plots (b) for tripalmitin crystallization in sesame oil (0.98% wt/vol). The crystallization temperature and respective Avrami index (n) for the second region of the curves are indicated. See Reference 13 for additional crystallization conditions.

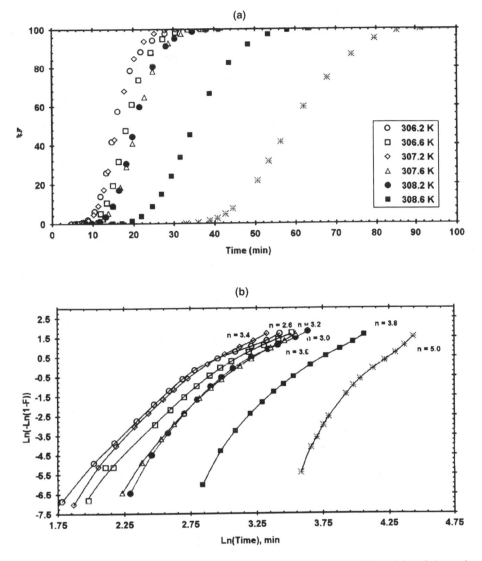

Figure 8.4 Percentage of reduced crystallinity (%F) as a function of time (a) and Avrami plots (b) for palm stearin crystallization in sesame oil (80% wt/vol). The crystallization temperature and respective Avrami index (n) for the second region of the curves are indicated. See Reference 9 for additional crystallization conditions.

5) while decreasing the temperature toward the isothermal crystallization temperature. As a result, the viscosity of the systems concomitantly increases.[5] Once the isothermal conditions have been achieved, TAGS' lamellar structures further organize and achieve a critical size to develop a stable nuclei (i.e., nucleation). This last process will take place at shorter τ_i, the longer the system takes to achieve

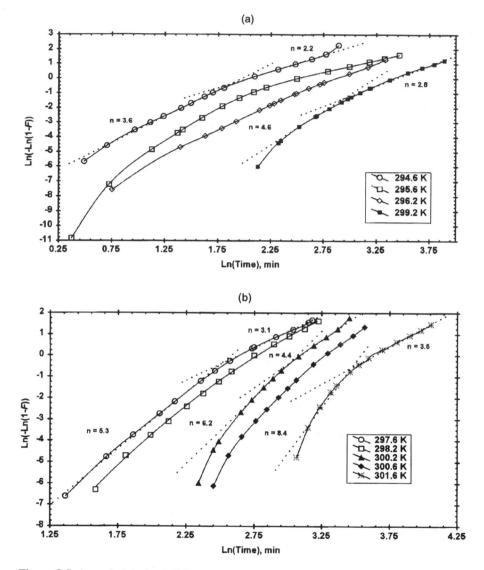

Figure 8.5 Avrami plots for 2.62% (wt/vol) tripalmitin (a) and 26% palm stearin (b) blends in sesame oil. The recrystallization temperature and the respective Avrami index (*n*) for the first and second regions of some of the curves are indicated. See References 9 and 13 for additional crystallization conditions.

isothermal conditions (i.e., the higher the effective supercooling or the lower the cooling rate). Thus, in the first region of the Avrami crystallization curve, a local order of TAGS' molecular organization is achieved during the cooling period. This process should be limited as the cooling rate is increased to values greater than 1 K/min.

In consequence, additional experiments with the PS:sesame oil blends were performed at higher cooling rates, i.e., 10 K/min and 100 K/min. These rates should limit the effect of local order during TAGS nucleation and favor the process of sporadic nucleation (i.e., in correspondence with the Avrami model) when compared with crystallization at lower cooling rates. The results of such experiments are shown in Figures 8.6 (26% PS:sesame oil) and Figure 8.7 (80% PS:sesame oil). Obviously, the isothermal conditions were achieved at shorter periods as faster cooling rates were used. Therefore, the higher the cooling rate, the less time TAGS molecules had to achieve local order in the liquid state while decreasing the temperature. As a result, as the cooling rate was increased, crystallization took place after a longer τ_i. The only exception was the 80% PS:sesame oil blend at 306.2 K (Figure 8.7), a result that we could not explain. In this case, no significant difference among the τ_i obtained at the different cooling rates was observed (i.e., τ_i between 4 and 5 min). An additional observation in all cases was that the first region in the Avrami plot was shorter and less evident as the cooling rate was increased. Then, local order must be primarily responsible for nuclei's formation in the first region of the Avrami curve. However, its effect on TAGS crystallization is reduced as a function of the increase in the cooling rate used to achieve isothermal crystallization conditions (i.e., equilibrium conditions).

Two additional phenomena must be considered to occur in this first region. The heterogeneous nucleation might play an additional role in this first region of the Avrami plot, particularly when surface irregularities of the measurement device (i.e., glass coverslips during microscopic measurements) promote the development of nuclei. However, its effect was not evaluated in this investigation. Additionally, the fractional values in the Avrami exponents obtained in this region (Figure 8.5) indicate the presence of secondary crystallization. To investigate this last issue, crystal morphology was studied in a polarized light microscope with tripalmitin and PS blends in sesame oil. Again, the experimental conditions have been reported previously[9,13] and also involved a cooling rate of 1 K/min. Figures 8.8 and 8.9 show microphotographs of the crystals developed by tripalmitin and PS as a function of time, in a 2.62% and 80% (wt/vol) blend in sesame oil, respectively. In both cases, TAGS crystal birefringence is not constant with time. Once the original solid TAGS lamellas have been obtained through the development of local order, subsequent crystallization (i.e., crystal growth) is achieved by infilling additional TAGS with the same orientation as the primary lamella. As a result, TAGS crystal birefringence increases with crystallization time, indicating that secondary lamella has the same orientation as the primary solid structure (i.e., secondary crystallization).

In contrast to the first region, in the second region of the Avrami plot, local order does not induce additional nucleation, because thermodynamic conditions predominate (i.e., isothermal conditions). Thus, sporadic nucleation gradually begins to affect the slope of the Avrami plot. However, as in the first region of the Avrami plot, secondary crystallization must also have a contribution during crystal growth (i.e., fractional n values are obtained, Figures 8.3b, 8.4b, and 8.5, and change in crystal birefringence shown in Figures 8.8 and 8.9).

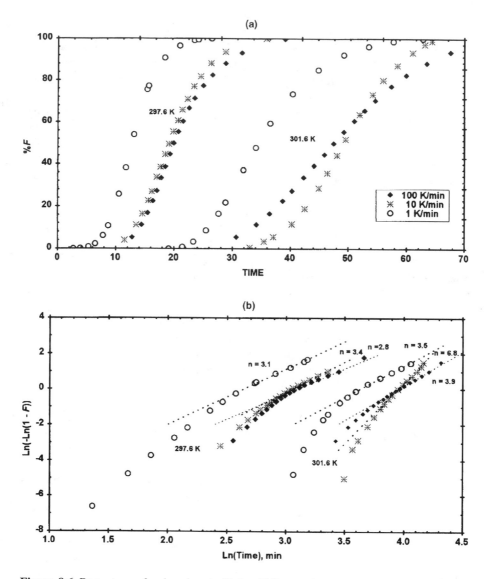

Figure 8.6 Percentage of reduced crystallinity (%F) as a function of time (a) and Avrami plots (b) for palm stearin crystallization in sesame oil (26% wt/vol). The crystallization temperature, cooling rate, and the respective Avrami index (n) for the second region of the curves are indicated.

In conclusion, TAGS crystallization from the melt involves nucleation by three mechanisms: local order, sporadic nucleation, and heterogeneous nucleation. Additionally, during crystal growth, the involvement of secondary crystallization is definite. However, all of these processes are engaged during TAGS bulk crystallization in events occurring not sequentially but in parallel. Unfortunately, the Avrami model

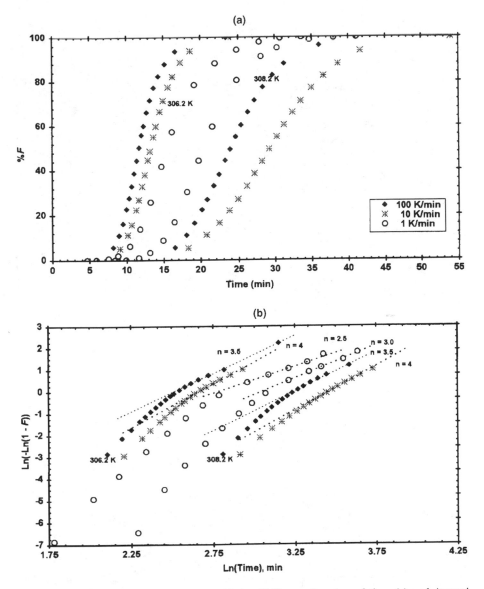

Figure 8.7 Percentage of reduced crystallinity (%*F*) as a function of time (a) and Avrami plots (b) for palm stearin crystallization in sesame oil (80% wt/vol). The crystallization temperature, cooling rate, and the respective Avrami index (*n*) for the second region of the curves are indicated.

does not consider all of these phenomena. Therefore, more complex models to describe bulk TAGS crystallization must be developed. Such models must include the nucleation mechanisms mentioned above and the effect of secondary crystallization during crystal growth to describe the whole process of TAGS crystallization.

Figure 8.8a Polarized light microphotographs of crystals obtained in a blend of tripalmitin in sesame oil (2.62% wt/vol). The crystallization time after the τ_i at 294.6 K was 5 min.

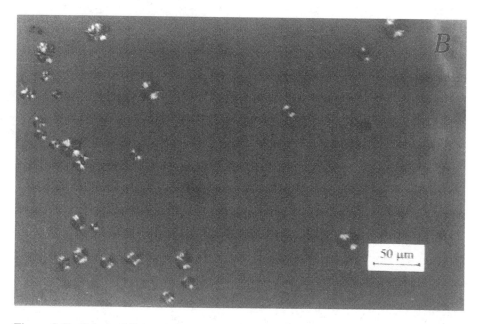

Figure 8.8b Polarized light microphotographs of crystals obtained in a blend of tripalmitin in sesame oil (2.62% wt/vol). The crystallization time after the τ_i at 294.6 K was 13.5 min.

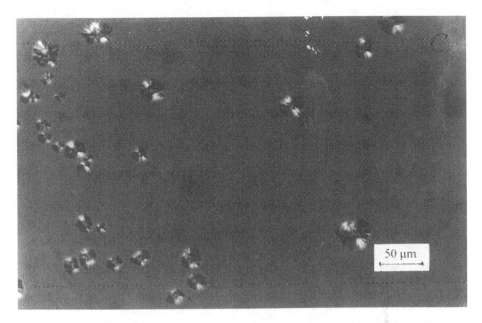

Figure 8.8c Polarized light microphotographs of crystals obtained in a blend of tripalmitin in sesame oil (2.62% wt/vol). The crystallization time after the τ_i at 294.6 K was 30 min.

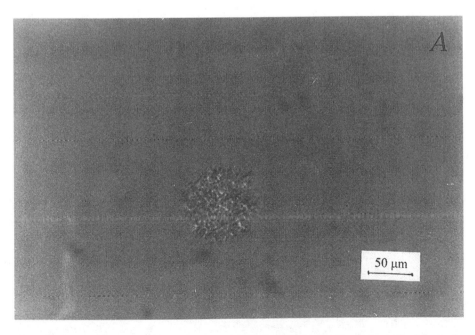

Figure 8.9a Polarized light microphotographs of crystals obtained in a blend of palm stearin in sesame oil (80% wt/vol). The crystallization time after the τ_i at 308.2 K was 3 min.

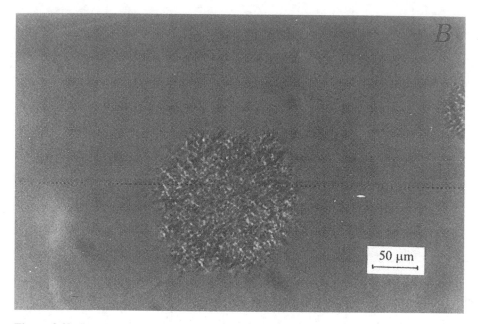

Figure 8.9b Polarized light microphotographs of crystals obtained in a blend of palm stearin in sesame oil (80% wt/vol). The crystallization time after the τ_i at 308.2 K was 12 min.

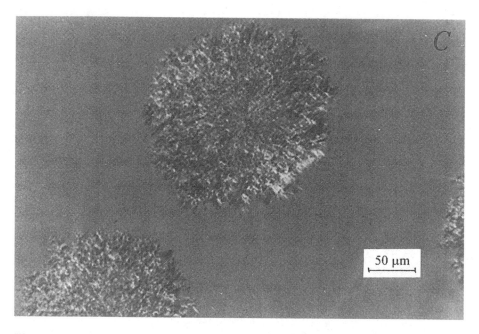

Figure 8.9c Polarized light microphotographs of crystals obtained in a blend of palm stearin in sesame oil (80% wt/vol). The crystallization time after the τ_i at 308.2 K was 20 min.

Additionally, the effect of cooling rate must also be considered, because, during nucleation, the involvement of local-order and sporadic nucleation depends on the cooling rate used. This last issue has particular relevance given the variable cooling rates achieved by industrial crystallizers used for fractional crystallization of vegetable oils or in the manufacture of margarines.

REFERENCES

1. Wilkinson, J. Q. 1997. "Biotech Plants: From Lab Bench to Supermarket Shelf," *Food Technol.*, 51(12): 37–42.
2. Fitch, H. B. 1997. "Structured Lipids Allow Fat Tailoring," *INFORM.* 8(10): 1004–1011.
3. Toro-Vázquez, J. F. and M. Charó-Alonso. 1998. "Physicochemical Aspects of Triacylglyceride and their Association to Functional Properties of Vegetable Oils," in *Functional Properties of Proteins and Lipids*, J. R. Whitaker, F. Shahidi, A. López Munguía, R. Y. Yada, and G. Fuller, eds. Washington, D. C: ACS Symposium Series 70, American Chemical Society, pp. 230–253.
4. Fitch H. B. 1994. "Tools: Hydrogenation, Interesterification," *INFORM.*, 5(6): 668–678.
5. Toro-Vázquez, J. F. and A. Gallegos-Infante. 1996. "Viscosity and Its Relationship to Crystallization in a Binary System of Saturated Triacylglycerides and Sesame Seed Oil," *J. Am. Oil Soc.*, 73: 1237–1246.
6. Boistelle, R. 1988. "Fundamentals of Nucleation and Crystal Growth," in *Crystallization and Polymorphism of Fats and Fatty Acids,* N. Garti and K. Sato, eds. New York: Marcel Dekker, Inc., pp. 189–226.
7. Larsson, K. 1994. *Lipids—Molecular Organization, Physical Functions and Technical Applications,* West Ferry, Dundee: The Oily Press, Ltd.
8. Hoffman, J. D. and J. J. Weeks. 1966. "Melting Process and the Equilibrium Melting Temperature of Polychlorotrifluoroethylene," *J. Res. Nat. Bur. Std.*, 66: 13–28.
9. Toro-Vázquez, J. F., M. Briceño-Montelongo, E. Dibildox-Alvarado, E. Charó-Alonso, and J. Reyes-Hernández. 2000. "Crystallization Kinetics of Palm Stearin in Blends with Sesame Seed Oil," *J. Am. Oil Soc.*, 77(3): 299.
10. Toro-Vázquez, J. F. and E. Dibildox-Alvarado. 1997. "Parameters that Determine Tripalmitin Crystallization in Sesame Oil," *J. Food Lipids*, 4(4): 269–282.
11. Hamm, W. 1995. "Trends in Edible Oil Fractionation," *Trends Food Sci. Technol.*, 6: 121–126.
12. Che Man, Y. B., T. Haryati, H. M. Ghazali, and B. A. Asbi. 1999. "Composition and Thermal Profile of Crude Palm Oil and its Products," *J. Am. Oil Chem. Soc.*, 76: 237–242.
13. Dibildox-Alvarado, E. and J. F. Toro-Vázquez. 1997. "Isothermal Crystallization of Tripalmitin in Sesame Oil," *J. Am. Oil Chem. Soc.*, 74: 69–76.
14. Avrami, M. 1940. "Kinetics of Phase Change. II. Transformation-Time Relations for Random Distribution of Nuclei," *J. Chem. Phys.*, 8: 212–224.
15. Metin, S. and R. W. Hartel. 1998. "Thermal Analysis of Isothermal Crystallization Kinetics in Blends of Cocoa Butter with Milk Fat or Milk Fat Fractions," *J. Am. Oil Chem. Soc.*, 75: 1617–1624.

16. Kawamura, K. 1979. "The DSC Thermal Analysis of Crystallization Behaviour in Palm Oil," *J. Am. Oil Chem. Soc.*, 56: 753–758.

17. Herrera, M. L. and F. J. Márquez-Rocha. 1996. "Effect of Sucrose Ester on the Kinetics of Polymorphic Transition in Hydrogenated Sunflower Oil," *J. Am. Oil Chem. Soc.*, 73: 321–326.

18. Henderson, D. W. 1979. "Thermal Analysis of Non-Isothermal Crystallization Kinetics in Glass Forming Liquids," *J. Non-Crystalline Solids*, 30: 301–305.

19. Medellín-Rodríguez, F. J. and P. J. Phillips. 1996. "Poly(aryl ether ketone) (PEEK) (Bulk Crystallization)," *Polym. Materials Enc.*, 75: 5513–5518.

9 The Functionality of Milkfat Fractions in Confectionery and Plastic Fats

A. G. Marangoni

CONTENTS

Abstract

Solvent fractionation of milkfat (AMF) can yield three main fractions, namely, high (HMF), medium (MMF), and low (LMF) melting fractions. To determine the suitability of incorporating AMF or one of its fractions into fat-containing food products, it is necessary to study the phase behavior of mixtures of the fats. Dropping points as a function of composition and isosolid line diagrams are useful tools for this purpose. AMF forms eutectics with cocoa butter (CB), leading to decreases in melting point and hardness. This effect is due to the extreme thermodynamic incompatibility between MMF and CB triacylglycerols (TAGs). HMF, on the other hand, is fully compatible with CB and could be added to confectionery fats as a bloom inhibitor without decreases in melting point or hardness. HMF could also be added

1-56676-963-9/02/$0.00+$1.50
© 2002 by CRC Press LLC

back to AMF to increase hardness and melting point for use in warmer climates or where butter would not be kept refrigerated for extended periods of time. Dry fractionation processes do not yield the milkfat fraction purity required for incorporation of these fractions into food products.

9.1 INTRODUCTION

The physical properties of chocolate are largely determined by the physical properties of the underlying fat phase.[1] In milk chocolate, cocrystallization of cocoa butter triacylglycerols (TAGs) and milkfat TAGs takes place; this cocrystallization is a key factor influencing the appearance and physical properties of milk chocolate.[2]

However, the amount of milkfat that can be added to chocolate is limited by the thermodynamic incompatibility between milkfat and cocoa butter TAGs in the solid state. Because of molecular geometric constraints, as well as environmental factors that influence the kinetics of crystallization, milkfat and cocoa butter TAGs do not form mixed crystals—they crystallize as separate milkfat and cocoa butter solids. This thermodynamic incompatibility results in the formation of a eutectic—a decrease in the melting point of the cocoa butter-milkfat mixture below the melting point of either of the two components. The formation of this eutectic, therefore, leads to a decrease in the hardness of the milk chocolate.[3–10] The exact amount of milkfat that can be added to chocolate before the functional properties of the material are significantly and adversely affected will depend on processing conditions such as tempering times and temperatures. Reddy et al.[8] reported the successful production of milk chocolates containing 40% (w/w of the total fat) milkfat with good gloss and demolding properties.

Besides the obvious advantages of adding milkfat to chocolate, milkfat is known to reduce the incidence of bloom formation in chocolate. Fat bloom is a defect of chocolate that results in a white-gray appearance and crumbly texture. Addition of milkfat and milkfat fractions is known to decrease the incidence of bloom in chocolate.[10–13] Bloom formation is a problem mainly associated with dark chocolate. Some manufacturers, therefore, add 2–3% milkfat to dark chocolate to control hardness and delay bloom formation.[14]

Milkfat is a complex mixture of several hundred different triacylglycerols with an extremely heterogeneous fatty acid composition.[15] Undoubtedly, it is one of the most complex fats found in nature. The physical properties of milkfat, including melting behavior, solid fat content (SFC), and polymorphism, are dependent not only on the physical and chemical properties of the constituent TAGs but also on the interaction between these constituent TAGs. For these reasons, several studies have been performed to help in understanding how TAG structure influences phase behavior and polymorphism of milkfat.[16–21] A typical melting curve of untempered native milkfat determined using differential scanning calorimetry (DSC) shows three endothermic peaks corresponding to high (>50°C), medium (35–40°C), and low (>15°C) melting fractions.[18] From DSC measurements (ratios of enthalpies),

Timms[18] determined that milkfat contains 11% HMF, 23% MMF, and 66% LMF.[18] Marangoni and Lencki[20] reported 12% (w/w) HMF, 33% (w/w) MMF, and 55% (w/w) LMF yields from solvent fractionation experiments. These fractions are chemically distinct, with HMF principally containing long-chain saturated fatty acids, MMF containing two long-chain saturated fatty acids and one short or *cis*-unsaturated fatty acid, and LMF containing one long-chain saturated fatty acid and two short-chain or *cis*-unsaturated fatty acids.[18,20]

Knowledge of the chemical composition, phase behavior, and polymorphism of these fractions and their mixtures, and how their properties influence each other, will help us to better understand, predict, and control the physical properties of milkfat and mixtures of milkfat with other fats. To obtain this understanding, milkfat first must be efficiently separated into three fractions, and the phase behavior and polymorphism of the individual fractions must be determined. Timms[18] fractionated milkfat into three fractions using acetone as a solvent and proved that HMF and MMF and LMF were, in fact, distinct fractions (at the time, MMF was believed to be a solid solution of HMF and LMF). He also studied the polymorphism of these fractions and the effects of LMF addition to HMF (50% LMF) and MMF (75% LMF). Marangoni and Lencki[20] fractionated milkfat into three major fractions using ethyl acetate as the solvent and studied the binary and ternary phase behaviors of mixtures of these three fractions.

Isosolid diagrams are useful tools in the study of the phase behavior of mixtures of natural fats.[19] These isosolid diagrams have been used in the study of the phase behavior of mixtures of confectionery fats with milkfat and milkfat fractions.[3,4,10,22,23] The type of solution behavior can usually be discerned with the aid of these diagrams. Their main use in this area has been in the identification of eutectics in mixtures of cocoa butter and cocoa butter substitutes with milkfat and milkfat fractions. This procedure constitutes a useful way to qualitatively judge the compatibility of fats.

The phase behavior of mixtures of cocoa butter and milkfat, fractionated and interesterified milkfat was originally studied by Timms.[3,4] More recently, the phase behavior of binary and ternary mixtures of confectionery fats with milkfat and milkfat fractions has also been determined.[7,10,23] A better understanding of the complex interactions between milkfat TAGs, cocoa butter TAGs, and palm kernel stearin TAGs and the resulting macroscopic properties of the blends (melting behavior, bloom formation, softening) has been obtained from these studies.

Because of the very high propensity of milkfat triglycerides to form mixed crystals,[16,20] fractionation processes based on melt crystallization are not very efficient. Milkfat fractions obtained from the melt have different properties from those obtained by solvent crystallization. It would not be possible to do justice to the large field of milkfat fractionation in this short report. Readers are directed to the comprehensive work of Kaylegian and Lindsay[24] on milkfat fractionation and fraction utilization in food products. It is the purpose of this research to study the phase behavior of solvent fractionated milkfat fractions-cocoa butter mixtures to evaluate their potential utilization as ingredients in confectionery products and plastic fats.

9.2 MATERIALS AND METHODS

9.2.1 MULTIPLE-STEP SOLVENT FRACTIONATION

Anhydrous milkfat (200 g) was melted above 80°C, cooled to 40°C, and dissolved (1:4 w/w) in room temperature ethyl acetate (Fisher Scientific, St. Louis, MO). The mixture was then transferred to a glass bottle and placed in a thermostatically controlled water bath at 5°C for one hour. The mixture was mixed by inversion every 5 min and vacuum-filtered at 5°C using a Buechner funnel. A fast-filtering Whatman #1 filter paper was used for this purpose. The collected crystal mass was immediately washed with 200 ml of 5°C ethyl acetate. The crystal mass was completely white and devoid of entrained material after this process. This fraction will be referred to as the *high melting fraction*. The filtrate plus washes were transferred to a glass bottle, which was then placed in a stainless steel bucket filled with ethylene-glycol (for improved heat transfer) in a freezer at –28°C for one hour. The mixture was mixed by inversion every 10 min and vacuum filtered at –28°C using a Buchner funnel. A fast-filtering Whatman #1 filter paper was used for this purpose. The collected crystal mass was immediately washed with 400 ml of –28°C ethyl acetate. The crystal mass was completely white and devoid of entrained material after this wash. This fraction will be referred to as the *medium melting fraction*. The wet crystal masses were spread as a thin film on stainless steel trays, and the excess solvent was allowed to evaporate overnight at room temperature in a fume hood. The excess solvent present in the –28°C filtrate and washes was removed by vacuum distillation in a rotary evaporator at 40°C. This yellow liquid was then spread as a thin film on a stainless steel tray, and the solvent was allowed to evaporate overnight at room temperature in a fume hood. This fraction will be referred to as the *low melting fraction*.

9.2.2 CHEMICAL AND PHYSICAL CHARACTERIZATION OF THE MILKFAT FRACTIONS

Solid fat content was determined by pulsed nuclear magnetic resonance using a Bruker PC20 Series NMR Analyzer (Bruker, Milton, ON, Canada) according to the AOCS Official Method Cd16-81. Dropping point and hardness index determinations were determined as described in Rousseau et al.[25] at 22°C. Differential scanning calorimetry was carried out as previously described[26] using a Dupont 2090 DSC instrument (TA Instruments, Mississauga, Ontario, Canada). Samples were melted at 60°C in DSC aluminum pans for 30 min, cooled at a rate of 5°C/min to –40°C, and immediately heated at a rate of 5°C/min until the material was completely melted again. Two to three replicate experiments were carried out simultaneously, and the averages and standard deviations were reported.

9.2.3 PHASE EQUILIBRIUM STUDIES

Mixtures (w/w) of HMF, MMF, and LMF with cocoa butter were prepared in 10% increments from 0 to 100%. The solid fat content (SFC) and dropping points of the mixtures were determined from 0 to 55°C in 5°C increments. A point-to-point spline curve was fitted to the data for interpolation purposes using the software package GraphPad Prism 2.0 (GraphPad Software, San Diego, CA). Isosolid SFC temperatures, as a function of blend composition, were derived from the data and used in the construction of the binary phase diagrams, from 100% SFC (solidus line) to 0% SFC (liquidus line) in 5% SFC intervals. Two to three replicate experiments were carried out simultaneously, and the averages and standard deviations were reported.

9.3 RESULTS AND DISCUSSION

The solid fat content (SFC) vs. temperature profiles of the high (HMF), medium (MMF), and low melting (LMF) fractions in milkfat, anhydrous native milkfat (AMF), and cocoa butter (CB) are shown in Figure 9.1. Both the HMF and MMF have narrow melting ranges, and the LMF is completely liquid above 0°C. The dropping points of the HMF, MMF, and LMF were, respectively, 51.7°C, 30.4°C, and 12.5°C, while that of native AMF was 34.3°C, and that of cocoa butter was 27.6°C. Of particular interest is the similarity between the melting profiles of cocoa butter and MMF.

The fatty acid and triacylglycerol compositions of the three milkfat fractions, anhydrous milkfat, and cocoa butter are presented in Tables 9.1 and 9.2, respectively. Table 9.3 shows the region-specific distribution of fatty acids within TAG molecules in AMF, HMF, MMF, and CB. The characteristic feature of CB is the SUS structure of its TAGs, where positions sn-1,3 are usually occupied by palmitic or stearic acids, while position sn-2 is almost exclusively occupied by oleic acid. For our purposes, HMF can be considered as a long-chain trisaturate. The conformation of TAG molecules within a crystal is reminiscent of a tuning fork, with fatty acids at positions sn-1,3 oriented parallel to the direction of the long axis of the unit cell. In this

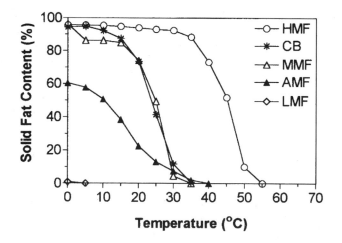

Figure 9.1 Solid fat content vs. temperature profiles for anhydrous milkfat (AMF), cocoa butter (CB), and the high (HMF), medium (MMF), and low (LMF) melting fractions of AMF.

TABLE 9.1
Fatty Acid Composition (% w/w) of Ethyl Acetate Fractionated Anhydrous Milkfat (AMF) Fractions, AMF, and Cocoa Butter (CB)

Fatty Acid	HMF	MMF	LMF	AMF	CB
4:0	—	4.78	5.24	4.51	—
6:0	0.19	3.50	3.80	3.12	—
8:0	0.34	1.45	2.03	1.64	—
10:0	1.74	3.18	4.43	3.86	—
10:1	0.09	0.47	0.67	0.80	—
12:0	3.60	3.06	4.78	4.07	—
14:0	15.40	11.82	11.14	10.99	—
14:1	0.84	1.25	2.07	1.88	—
15:0	1.77	1.51	0.88	1.46	0.1
16:0	42.50	39.46	19.84	28.73	26
16:1	1.69	1.70	3.59	3.12	0.2
17:0	0.92	0.91	0.83	0.40	—
18:0	20.55	13.63	6.00	10.45	35
18:1	9.30	12.31	29.61	20.92	34
18:2	0.54	0.44	2.17	1.86	3.5
18:3	0.39	0.23	1.38	1.65	—
20:0	0.20	0.50	1.57	0.60	0.6

conformation, the fatty acid at position sn-2 is also oriented parallel to the direction of the long axis of the unit cell, resulting in a tuning fork-like TAG structure. One could then possibly predict a strong molecular interaction along the long molecular axis formed by the sn-1 and sn-3 fatty acid chains. Using computer modeling, it is also possible to predict favorable interactions between the sn-2 oleic acid in CB TAGs and the sn-1 or sn-3 long-chain saturated fatty acids in the HMF TAGs (results not shown), namely:

In both CB and HMF TAGs, the long molecular axis is formed by long-chain saturated fatty acids. Intermolecular London dispersion forces would, therefore, be strong, and mixed crystal formation plausible. One could also envision that, if positions sn-1 or sn-3 were occupied by short-chain fatty acids or oleic acid, as in

TABLE 9.2
Triglyceride (TAG) Composition (% w/w) of Ethyl Acetate Fractionated Anhydrous Milkfat (AMF) Fractions, Native AMF, and Cocoa Butter

Carbon #	HMF	MMF	LMF	AMF	CB
22	—	0.06	—	0.24	—
24	—	0.72	1.02	0.84	—
26	—	0.24	0.6	0.37	—
28	—	0.17	1.26	0.60	—
30	—	0.18	2.37	1.04	—
32	0.20	0.62	4.74	2.64	—
34	0.64	4.94	8.16	6.40	—
36	1.05	18.18	13.37	14.04	—
38	0.73	17.59	18.72	14.73	—
40	0.96	11.14	14.73	10.78	—
42	3.23	8.90	7.66	7.69	—
44	9.23	6.98	5.66	6.91	—
46	19.13	6.10	4.80	7.42	—
48	25.67	7.84	4.34	8.63	0.1
50	24.00	9.79	5.14	9.60	16.5
52	12.25	6.04	6.28	6.42	45.8
54	1.03	0.57	1.18	1.68	36.1
56	—	—	—	—	1.5

TABLE 9.3
Fatty Acid Distribution (mol%) of HMF, MMF, AMF, and CB

Fatty acid	HMF		MMF		AMF		CB	
	sn-1 + 3	sn-2	sn-1 + 3	sn-2	sn-1 + 3	sn-2	sn-1/3	sn-2
4:0	—	—	20.0	16.1	—	18.4	—	—
6:0	0.5	1.3	0.5	1.3	—	4.2	—	—
8:0	1.7	8.5	1.1	2.6	—	2.2	—	—
10:0	0.8	5.3	1.7	3.0	4.1	3.9	—	—
12:0	3.4	7.5	3.0	5.6	8.3	3.5	—	—
14:0	11.6	15.9	9.1	16.4	19.2	18.8	—	—
16:0	38.0	30.0	30.0	45.8	35.7	20.4	34/37	2
18:0	28.2	15.3	15.3	17.3	11.5	9.4	50/53	2
18:1	15.9	16.2	20.2	8.2	21.0	19.3	12/9	87
18:2	—	—	—	—	—	—	1/–	9
18:3	—	—	—	—	—	—		

the case of MMF TAGs, intermolecular interactions would be hindered, and mixed crystal formation would be less likely to occur.

Figures 9.2, 9.3, and 9.4 show DSC crystallization and melting thermograms of HMF, MMF, and LMF, respectively. The purities of the fractions are testament to the efficiency of the ethyl acetate solvent fractionation process. This kind of fraction purity is not achievable using dry fractionation (melt crystallization) protocols.

A eutectic occurs when the melting point of a mixture is below the melting point of either of the individual components. Figure 9.5 shows the dropping points for mixtures of HMF with MMF, HMF with LMF, and MMF with LMF. On average, these mixtures displayed monotectic solution behavior. The situation for cocoa butter (CB)–AMF and milkfat fraction mixtures was, however, different. Changes in drop-

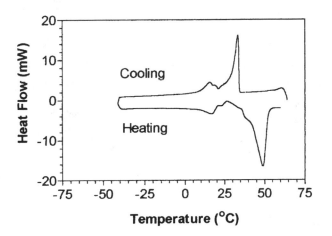

Figure 9.2 DSC thermogram for the melting and crystallization of HMF.

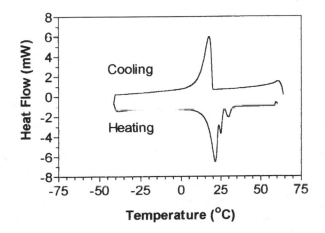

Figure 9.3 DSC thermogram for the melting and crystallization of MMF.

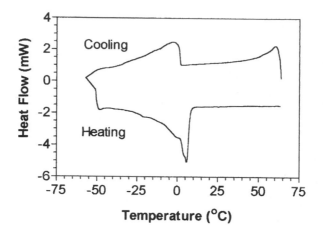

Figure 9.4 DSC thermogram for the melting and crystallization of LMF.

ping points as a function of mixture composition are shown in Figure 9.6. Evident eutectics were detected in the range of 0–30% AMF-CB and 0-60% MMF-CB. No eutectics were observed for HMF-CB mixtures, in contrast to work reported by Bystrom and Hartel.[7] These authors reported the formation of a strong eutectic between CB and HMF. This effect was probably due to the fact that fractions used in their study were obtained via melt crystallization, and significant amounts of MMF were probably present (see below). To confirm these findings, isosolid diagrams were generated for the different mixtures.

These isosolid diagrams for AMF-CB, MMF-CB, and HMF-CB mixtures, respectively, are shown in Figures 9.7, 9.8, and 9.9. A slight eutectic formation was evident in the AMF-CB system (Figure 9.7), while extreme thermodynamic incompatibility was evident in the MMF-CB mixtures (Figure 9.8). From the isosolid diagram, it is obvious that any mixture of CB and milkfat's MMF formed a eutectic. No eutectics were formed between cocoa butter and milkfat's high melting fraction (Figure 9.9). The patterns observed for the CB-HMF system are reminiscent of monotectic, partial solid solution formation,[19] where a slight amount of thermodynamic incompatibility between the two components is evident. Kaylegian et al.[22] reported similar behavior for CB-HMF and CB-MMF mixtures using milkfat fractions obtained using acetone fractionation.

The dropping point data and the isosolid diagrams agreed qualitatively. HMF and CB TAGs are thermodynamically compatible, and no eutectic is formed. MMF TAGs are, however, extremely incompatible with CB TAGs. The eutectic in the AMF-CB mixtures most probably arises from the incompatibility between MMF and CB. Our results disagree with the data reported by Bystrom and Hartel.[7] These authors reported on the thermodynamic incompatibility between HMF obtained via a dry fractionation process and cocoa butter. Most probably, their HMF fraction was contaminated with MMF, giving rise to the reported patterns. Work by the same group in the same year[13] reported that HMF obtained by solvent fractionation did

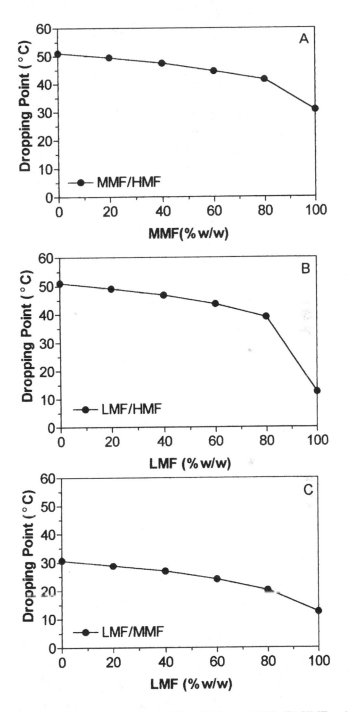

Figure 9.5 Dropping points for mixtures of (A) HMF and MMF, (B) HMF and LMF, and (C) MMF and LMF.

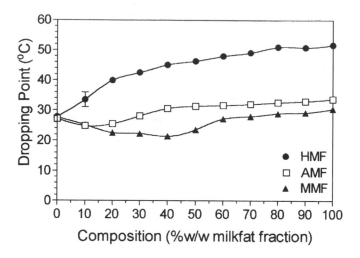

Figure 9.6 Dropping points for mixtures of cocoa butter (CB) and AMF, HMF, and MMF.

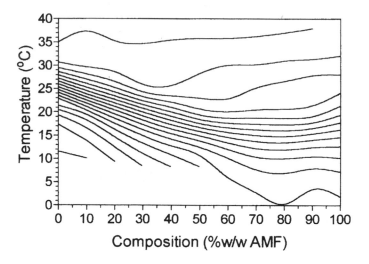

Figure 9.7 Isosolid diagram for the AMF-CB system.

not decrease hardness of cocoa butter, probably suggesting that eutectics were not formed for cocoa butter mixtures containing HMF obtained via solvent fractionation.

Figure 9.10 exemplifies the problem with the addition of milkfat and milkfat fractions to cocoa butter. After 24 h of crystallization at 22°C, addition of 10% and 40% (w/w) AMF or MMF to cocoa butter significantly decreased the hardness of the material, while HMF addition did not. After 48 h of tempering, 40% cocoa butter mixtures containing AMF or MMF were still very soft, while mixtures containing 10% AMF or MMF did not appear softer than the control CB anymore. Obviously, further crystallization, recrystallization, and/or polymorphic transformation and fat

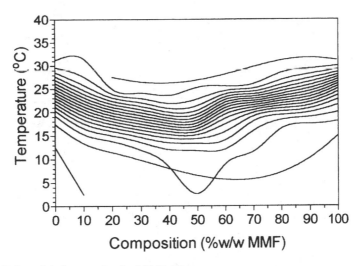

Figure 9.8 Isosolid diagram for the MMF-CB system.

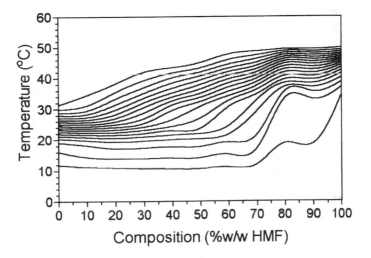

Figure 9.9 Isosolid diagram for the HMF-CB system.

crystal network setting of the mixtures occurred between 24 and 48 h, erasing any effects of AMF and MMF addition on hardness of cocoa butter. Our results disagree with those of Full et al.[9] These authors reported a decrease in hardness of chocolate containing HMF. Again, this effect is probably due to MMF contamination of the HMF fraction. Dry, melt crystallization-based fractionation of milkfat does not produce fractions that enhance the functionality of confectionery products. Lohman and Hartel[13] have clearly shown that the addition of HMF obtained via solvent fractionation does not decrease the hardness of cocoa butter. These authors also

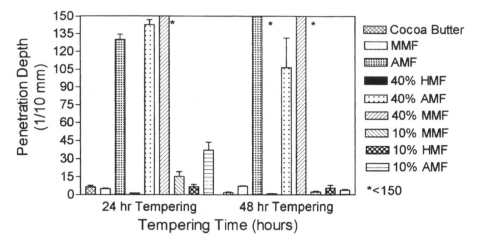

Figure 9.10 Hardness and mixtures (%w/w) of cocoa butter with milkfat and milkfat fractions measured by one penetrometry at 22°C.

clearly demonstrate the ability of HMF to delay the incidence of bloom formation in chocolate, and their equivalent to our MMF and LMF to enhance the rate of bloom formation. It would seem that the use of HMF obtained via dry fractionation (melt crystallization) is not a good idea, because it forms eutectics with cocoa butter[7] and decreases the hardness of chocolate.[9]

As far as the MMF is concerned, work in our laboratory has clearly shown that MMF can be used as a cocoa butter replacer. It is possible to manufacture high-quality confections using MMF as the sole confectionery fat.

Another possible application for milkfat fractions is in the manufacture of plastic fats. Milkfat's HMF can be added to milkfat to increase the dropping points and solid fat content (Figures 9.11 and 9.12, respectively). Increasing the SFC and dropping point of AMF would allow for the manufacture of butter for consumption in warmer countries or in situations where butter would be kept at room temperature for extended periods of time.

9.4 CONCLUSIONS

Milkfat solvent fractionation constitutes an interesting strategy to increase the utilization of milkfat and create new and innovative products with physical properties that are dependent on the presence of a milkfat fat crystal network underlying structure. HMF can be successfully used to prevent bloom formation in chocolate, while AMF is added to chocolate in the manufacture of milk chocolate and to control hardness and delay bloom formation in dark chocolate. HMF can also be utilized as a hardstock in the manufacture of high-melting butter. However, the phase behavior of mixtures of milkfat and milkfat fractions with different fats must be determined to be able to judge their suitability for inclusion in specific fat blends.

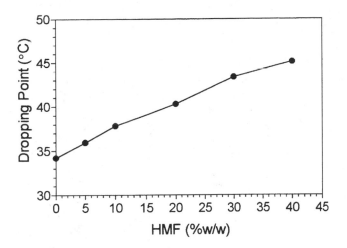

Figure 9.11 Dropping points of AMF-HMF mixtures.

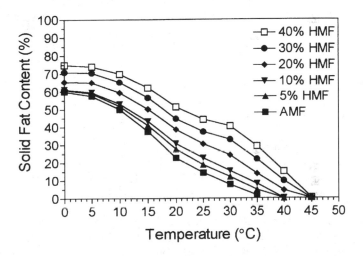

Figure 9.12 Solid fat content vs. temperature profile for mixtures of AMF and HMF.

9.5 ACKNOWLEDGMENTS

The financial assistance of the Natural Sciences and Engineering Research Council of Canada (NSERC), the Ontario Ministry of Agriculture, Food and Rural Affairs (OMAFRA), and Fractec Research and Development, Inc., are gratefully acknowledged. I would also like to thank Wendy Willis, Peter Chong, Amanda Wright, and Robert Lencki for technical assistance.

REFERENCES

1. Jeffrey, M. S. 1991. "The Effect of Cocoa Butter Origin, Milk Fat and Lecithin Levels on the Temperability of Cocoa Butter Systems," *Mfg. Confectioner*, 71: 76–82.
2. Koyano, T., J. Hachiya, and K. Sato. 1990. "Fat Polymorphism and Crystal Seeding Effects on Fat Bloom Stability of Dark Chocolate," *Food Structure*, 9: 231–240.
3. Timms, R. E. 1980. "The Phase Behavior of Mixtures of Cocoa Butter and Milk Fat," *Lebensm. Wiss. Technol.*, *13*: 61–65.
4. Timms, R. E. and J. V. Parekh. 1980. "The Possibilities for Using Hydrogenated, Fractionated or Interesterified Milk Fat in Chocolate," *Lebensm. Wiss. Technol.*, *13*: 177.
5. Barna, C. M., R. W. Hartel, and S. Metin. 1992. "Incorporation of Milk-Fat Fractions into Milk Chocolates," *Manufacturing Confectioner*, *72*: 107–116.
6. Reddy, S. Y., P. S. Dimick, and G. R. Ziegler. 1994. "Compatibility of Milkfat Fractions with Cocoa Butter Determined by Differential Scanning Calorimetry," *INFORM.*, *5*: 522.
7. Bystrom, C. E. and R. W. Hartel. 1994. "Evaluation of Milk-Fat Fractionation and Modification Techniques for Creating Cocoa Butter Replacers," *Lenbensm. Wiss. Technol.*, *27*: 142–150.
8. Reddy, S. Y., N. A. Full, P. S. Dimick, and G. R. Ziegler. 1996. "Tempering Method for Chocolate Containing Milk Fat Fractions," *J. Amer. Oil Chem. Soc.*, *73*: 723–727.
9. Full, N. A., S. Y. Reddy, P. S. Dimick, and G. R. Ziegler. 1996. "Physical and Sensory Properties of Milk Chocolate Formulated with Anhydrous Milk Fat Fractions," *J. Food Sci.*, *61*: 1068–1084.
10. Hartel, R. W. 1996. "Application of Milk-Fat Fractions in Confectionery Products," *J. Amer. Oil Chem. Soc.*, *73*: 945–953.
11. Kleinert, J. 1961. "Studies on the Formation of Fat Bloom and Methods of Delaying It," *Rev. Internat. Choco.*, *16*: 201–219.
12. Dimick, P. S., L. N. Thomas, and C. Versteeg. 1993. "Potential Use of Fractionated Anhydrous Milk Fat as a Bloom Inhibitor in Dark Chocolate," *INFORM.*, *4*: 504.
13. Lohman, M. H. and R.W. Hartel. 1994. "Effect of Milk-Fat Fractions on Fat Bloom in Dark Chocolate," *J. Amer. Oil Chem. Soc.*, *71*: 267–276.
14. Minifie, B. W. 1989. *Chocolate, Cocoa and Confectionery*, 3rd edn."New York, N. Y.: Van Nostrand Reinhold.
15. Gresti, J., M. Bugaut, C. Maniongui, and J. Bezard. 1993. "Composition of Molecular Species of Triacylglycerols in Bovine Milk Fat," *J. Dairy Sci.*, *76*: 1850–1869.
16. Mulder, H. 1953. "Melting and Solidification of Milk Fat," *Neth. Milk Dairy J.*, *7*: 149–176.
17. deMan, J. M. 1963. "Polymorphism in Milk Fat," *Dairy Science Abstracts*, *25(6)*: 219–221.
18. Timms, R. E. 1980. "The Phase Behavior and Polymorphism of Milk Fat, Milk Fat Fractions and Fully Hardened Milk Fat," *Australian J. Dairy Technol.*, *35*: 47–52.
19. Timms, R. E. 1984. "Phase Behavior of Fats and Their Mixtures," *Prog. Lipid Res.*, *23*: 1–38.
20. Marangoni, A. G. and R. W. Lencki. 1998. "Ternary Phase Behavior of Milkfat Fractions," *J. Agric. Food Chem.*, *46*: 3879–3884.
21. Grotenhuis, E. T., G. A. van Aken, K. F. Malssen, and H. Schenk. 1999. "Polymorphism of Milk Fat Studied by Differential Scanning Calorimetry and Real-Time X-Ray Powder Diffraction," *J. Amer. Oil Chem. Soc.*, *76*: 1031–1039.

22. Kaylegian, K. E., R. W. Hartel, and R. C. Lindsay. 1993. "Applications of Modified Milk Fat in Food Products," *J. Dairy Sci.*, *76*: 1782–1796.

23. Ali, A. R. and P. S. Dimick. 1994. "Melting and Solidification Characteristics of Confectionery Fats: Anhydrous Milk Fat, Cocoa Butter and Palm Kernel Stearin Blends," *J. Amer. Oil Chem. Soc.*, *71*: 803–806.

24. Kaylegian, K. E. and R. C. Lindsay. 1995. *Handbook of Milkfat Fractionation Technology.* Champaign, IL: AOCS Press.

25. Rousseau, D., A. R. Hill, and A. G. Marangoni. 1996. "Restructuring Butterfat Through Blending and Chemical Interesterification. 3. Rheology," *J. Amer. Oil Chem. Soc., 73*: 983–989.

26. Rousseau, D., K. Forestiere, A. R. Hill, and A. G. Marangoni. 1996. "Restructuring Butterfat Through Blending and Chemical Interesterification. 1. Melting Behavior and Triacylglycerol Modifications," *J. Amer. Oil Chem. Soc., 73*: 963–972.

10 Stabilization Mechanisms for Anthocyanin: The Case for Copolymerization Reactions

P. Wesche-Ebeling
A. Argaiz-Jamet

CONTENTS

Abstract

Anthocyanins belong to the flavonoid family of phenolic compounds and are widely distributed in plant tissues. They are responsible for the coloration of many fruits, flowers, and other parts of plants, and this color is a desirable attribute of many food products such as beverages, juices, nectars, marmalades, jellies, gums, and grape and fruit wines. Anthocyanins have gained recent notoriety as important nutraceuticals, mainly as natural antioxidants. The increased consumer interest in foods containing natural additives has also led to the use of anthocyanins as food colorants. The main problem with anthocyanins as additives in foods is their poor stability, because they are affected by changes in pH, the presence of ascorbic acid, sulfite, light, oxygen, and enzymatic and nonenzymatic browning reactions. Several research groups have been studying ways of stabilizing anthocyanin pigments to be used as food colorants. This contribution describes the main stabilization mechanisms under study, including structure complexity, auto-association and co-pigmentation, and, finally, the research area of our group, copolymerization reactions.

10.1 INTRODUCTION

One of the strongest tendencies in the food industry currently is the consumer's predilection toward processed foods containing no synthetic additives. This has forced researchers in the area to look for naturally occurring compounds, from plants or animals, to gradually substitute for preservatives such as benzoate, antioxidants such as BHT and BHA, and antibrowning agents such as sulfite and certified synthetic colorants. Although the advantages in the use of certified colorants are many (stability, availability in different hues, tinctoric strength, etc.), their safety has recently come into question. The alternatives to certified colors are pigments either from natural sources or those identical to the natural ones but obtained through organic synthesis (certified colorants). The main disadvantages of these pigments is that they are unstable, they cannot be combined to generate different hues, and their tinctoric power is much lower.

Several research groups have directed their attention toward anthocyanins, because this is the pigment having the widest diversity in hues in nature. Anthocyanins are present in the form of glycosides (acylated or not), but once the tissues that contain them have been altered, they are highly susceptible to degradation through factors such as light, pH, sulfite, ascorbic acid, enzymes such as glycosidases and polyphenol oxidase, and nonenzymatic browning. Not all anthocyanins are equally susceptible to degradation, and three main lines of research are currently searching for anthocyanins showing better stability.

In the case of natural red color sources, the food industry can rely exclusively on betalaines, some carotenoids, carminic acid, myoglobin/hemoglobin, and anthocyanins. Several forms of these pigments are already commercially available as certified colorants: red beet extracts, purified carminic acid, purified natural or synthetic carotenoids, and enocyanin, an anthocyanin concentrate from grape skin. Although the use of these colorants is increasing, the industry requires other sources of natural color with a wider variety of hues, higher stability, and tinctoric power.

One of the lines focuses on the search of anthocyanins having complex molecular structures (highly glycosylated and/or acylated) that confer the pigments with higher stability and tinctoric strength. Red cabbage anthocyanins belong to this group. The second line aims to find natural chemical compounds (phenyl propionic acids, flavonoids, etc.) that can form noncovalent complexes with anthocyanins, thereby protecting them from the surrounding environment, increasing their tinctoric strength through a process called co-pigmentation.

Finally, our group, among very few others, has been studying the mechanisms of polymer formation to which anthocyanins become incorporated, such as those present in red wines. The process has been named copolymerization, and some anthocyanins become part of red, low-molecular-weight polymers that are formed through the action of quinones formed through enzymatic browning reactions, and also through the polymerizing action of acetaldehyde formed during fermentation. There are several reasons why we have directed our efforts in this direction:

- The red pigments formed in red wines have shown higher stability toward most of the factors that affect monomeric anthocyanins.
- The raw materials (anthocyanins, polyphenol oxidases and their phenolic substrates, and acetaldehyde) can be obtained in large quantities from plant sources and through fermentation, and the copolymerized anthocyanins are therefore natural products that have been consumed by humans throughout history.
- It has been demonstrated through bioassays that phenolic compounds, including anthocyanins, function as antioxidants, preventing damage to membranes and other cellular components, and they have therefore been considered as nutraceutical compounds. From this, it may be concluded that anthocyanins may not only be considered as sources of color but also as natural antioxidants.

10.2　ANTHOCYANINS—NATURAL PLANT PIGMENTS

Anthocyanins belong to the group of phenolic compounds called flavonoids (C6-C3-C6). They can be found in different parts of the plant including fruits (berries, apples, grapes, etc.), flower petals (hibiscus), leaves, stems, and tubers, and they express colors from orange, pink, red, mauve, violet, blue, and black. More than 250 anthocyanins have been reported.[1]

One important function of anthocyanins is the visual attraction of animals for pollination and seed-dispersion purposes, and they have been very important in the co-evolution of plant–animal interactions.[2] Anthocyanins also appear sporadically in seedlings or new leaves with the possible function of light filtering or reception, or as a response to some type of stress.[3]

Anthocyanins are found in vacuoles in spherical bodies called *anthocyanoplasts*.[4] They are synthesized from *p*-coumaroyl-CoA and three molecules of malonyl-CoA to form an intermediate 15-carbon chalcone. This intermediate is transformed to the direct precursor of the anthocyanidins, flavan-3,4-*cis*-diol. Anthocyanidins are stabilized through hydroxylation, methoxylation, and glycosylation of some of the hydroxy groups, and acylation of the sugar moieties.[5] Acids participating in sugar acylation include *p*–coumaric, caffeic, ferulic, synaptic, acetic, oxalic, succinic, malic, and malonic.[1] According to the hydroxylation and methoxylation patterns, the following common anthocyanins can be found: pelargonidin, cyanidin, peonidin, delphinidin, petunidin, and malvinidin.

Anthocyanins are water soluble, and color is expressed by the ionized form of the pigment called *flavylium cation*. The increase in the number of hydroxy groups results in a shift toward bluer colors, while the introduction of methoxyl groups causes a shift toward red. The association of anthocyanins with other compounds also affects color expression. Four types of association exist:

1. Auto-association between anthocyanins
2. Intermolecular co-pigmentations with uncolored substances such as flavonol or flavone glycosides
3. Intramolecular co-pigmentations with the aromatic groups of hydroxycinnamic acids
4. The association of metals such as magnesium and aluminum with o-hydroxy groups of anthocyanins[6-7]

The co-pigmentation mechanism seems to be the most prevalent, and co-pigmented anthocyanins are protected from attack by water in the acidic environment of vacuoles; the anthocyanins remain in the quinonoidal base form (deprotonated flavylium cation).[7]

10.3 ANTHOCYANINS AS COLOR SOURCES

Anthocyanins are responsible for the color in many fruits and vegetables and can be found in fresh products such as apples, berries, red cabbage, cherries, grapes, red radishes, strawberries, potatoes, and onions, and in processed anthocyanin-containing fruit products such as juices, wines, marmalades, and jellies. These fresh or processed products have been continuously consumed by humans and animals, and no reports exist of health damages. This fact has allowed the use of anthocyanin extracts, mainly from grapes (enocyanin) and red cabbage, as food colorants.[6,8] Some researchers are studying the possibility of producing anthocyanins through biotechnological means from cell cultures.[9-10] It has also been recently reported that natural flavonoids may serve as antioxidants in foods without the problems associated with the synthetic antioxidants.[11-14]

The importance of relying on natural sources of red pigments is accentuated by the prohibitions or limitations recently imposed on certified red colorants such as the FD&C reds #3 and #40,[8] and the tendency of consumers is toward the ingestion of processed foods free of synthetic additives. All these events have transformed anthocyanins into one of the main options as colorants in the food industry. Nevertheless, the use of anthocyanins as food colorants has been limited due to their high susceptibility toward degradation, the difficulties in their extraction, and their low commercial availability.

10.4 ANTHOCYANIN DEGRADATION IN FOODS

Anthocyanins show great susceptibility toward pH. The red flavylium cation (FC) form exists as the dominant form only at pH < 1. The FC suffers two types of reactions with increasing pH. A very small proportion (2–4%, pH 3–5) may suffer a deprotonation and form the blue quinonoidal base, but the majority (3–87%, pH 2–5) suffer a deprotonation and hydration to form the colorless and highly unstable carbinol pseudobase that rapidly degrades through ring rupture to form a colorless chalcone (3–9%, pH 1–5).[7]

The degradation rates of anthocyanins increase when, in addition to changes in pH, there are increases in temperature or exposure to light or hydrolysis of anthocyanins to anthocyanidins due to high acidity or the presence or glycosidase.[15] In addition to the susceptibility to these factors, anthocyanins may be bleached by sulfite or degraded through enzymatic or nonenzymatic browning reactions.[15-16] In the case of red wines, color can be partially recovered through the removal of the SO_2 by the acetaldehyde generated during fermentation.

During enzymatic browning, the polyphenol oxidases (PPO) cause the oxidation of the o-phenolic compounds present into o-quinones. These react rapidly with other phenolics or with proteins in a series of polymerization reactions leading to the formation of soluble or insoluble brown pigments called melanoidins or tannins. In systems in which the pH is higher than 6.5, the o-phenolics auto-oxidize to o-quinones with the consequent browning reactions.[17]

Anthocyanins are affected through different mechanisms during the enzymatic browning reactions. Apparently, the PPO cannot use anthocyanins as substrates.[16] On the other hand, anthocyanins may become copolymerized with phenolics, proteins, or both, in most cases with the loss of color although, as has been reported for red wines, the anthocyanins may form part of the polymer and still express a red color.[18-23] In these pigments or copolymerized anthocyanins, the anthocyanin is present in the quinonoidal base form expressing red color. The copolymerized anthocyanins are less sensitive to changes in pH or to sulfite bleaching.[18] It has been reported that most probably part of the red color observed in anthocyanin-containing fruit marmalades, jellies, and juices is due to the presence of copolymerized anthocyanins.[16]

10.5 ANTHOCYANINS AS FOOD COLORANTS

Ideally, a colorant should be stable under a wide range of conditions, including changes in pH, technologically acceptable and easily available in a pure and economic form, but no colorant satisfies all these conditions, and they must be selected according to individual needs.[24] Anthocyanins have been used commercially with occasional success, because they provide a wide range of hues, but with the disadvantages of their instability and susceptibility to changes in pH.[25-26]

It has been proven that acylation of the anthocyanins results in higher stability, and there are some di- and triacylated anthocyanins from red cabbage available on the market.[8] At the same time, anthocyanins with glycosylation and acylation patterns in sites that confer them higher stability have been reported.[24,27-28] This increased stability may by explained, in part, as a result of the phenomenon of intramolecular co-pigmentation between the anthocyanin moieties and the aromatic rings of the phenolic acids, resulting in the protection of the anthocyanin against the action of water. It has been possible to study these highly complex structures thanks to the use of methodologies based on HPLC-purification of the pigments and the combined uses of diode-array detectors, nuclear magnetic resonance, and mass spectrometry.[29-30]

10.6 COPOLYMERIZED ANTHOCYANINS

The existence of copolymerized anthocyanins has been made known through studies on red wine by Somers[18] and on berries by Azar et al.[31] Several mechanisms for the incorporation of monomeric anthocyanins into the polymers have been recently proposed.[32–36] Anthocyanins present in red wines are reported to bind to flavonoids, phenyl propionic acids, and other phenolic compounds, and probably to proteins and polysaccharides. If the conditions are not appropriate, the copolymers formed may be predominantly brown and often form polymers too large to remain in solution and precipitate as tannins.[16,18–19,23,37–43]

Knowledge about the conditions that lead to the formation of stable, good-colored copolymerized pigments is necessary not only for the production of high-quality anthocyanin-containing wines and juices but also as an alternative for the production of stable pigments for use by the food industry.

Although many studies exist on the possible reactions that occur during red wine fermentation and on the characteristics of mature wines,[21–22,32,44–47] the reproducibility of these interactions with other sources of anthocyanin and plant phenolics has not been studied.

Some work has been done in our laboratory on the study of the stability of monomeric and copolymerized anthocyanins. In some of these works,[18 56] the medium-term stability of the color expressed by monomeric and copolymerized anthocyanin formed by the interactions of anthocyanin from roselle with different levels of catechin, polyphenol oxidase, and oacetaldehyde was studied using model systems. Procedures for measuring color parameters[18,57] have been used to follow color changes in the model systems. Color density (CD) expresses the sum of color from monomeric or copolymerized anthocyanin and from browning reactions; polymeric color (PC) expresses the color from copolymerized anthocyanin and from browning reactions; percent contribution of tannin (%CT) expresses the presence of color from polymers; and anthocyanin color (AC) expresses the color from monomeric anthocyanin.

Statistical analyses of the rate constant values (k) for CD have shown that PPO and catechin alone or the combinations PPO-anthocyanin and anthocyanin-catechin are significant in the decrease of CD with time. PPO, catechin, anthocyanin, or the three in combination are significant in the decrease of PC formation. All reactants and their combinations, except for the combination anthocyanin-catechin, are significant in tannin formation, while PPO, anthocyanin, catechin, and the combination anthocyanin-catechin are significant in AC loss. Between 70 and 92% of the initial monomeric anthocyanin was destroyed, but part of the red color was retained, as shown by the higher AC values. Statistical analyses show that all reactants and their combinations, except for the combination PPO-Cat, were significant in anthocyanin degradation.

Hunter L_h values did not change significantly after 35 days, and most systems became slightly lighter. The trends in color change show that the systems changed from a more saturated orange-red color to a less saturated orange-brown color.

Several conclusions came out of this preliminary study. It is important to have a PPO-substrate (catechin in this case) system for the anthocyanin to form copolymers. Anthocyanins play a key role in the copolymerization reactions; the monomeric anthocyanin is almost totally destroyed, and there is an accumulation of tannin, but red color is retained, as shown by the presence of AC. The redness remaining in some model systems seems to be due to the presence of anthocyanin copolymers and the characterization of these copolymers, and the study of the conditions leading to their formation over the formation of tannins (brown and of higher MW) is important for the production of these potential food pigments.

Work is underway to continue research on the mechanisms involved in the formation of anthocyanin copolymers as well as their stability in food systems and the search for alternative sources of anthocyanins, phenolics, and polyphenol oxidase.

10.7 ACKNOWLEDGMENTS

This work received the support of CONACyT, Mexico.

REFERENCES

1. Harborne, J. B. and R. J. Grayer 1988. In *The Flavonoids: Advances in Research Since 1980*, J. B. Harborne, edit., London: Chapman and Hall, pp. 1–20.
2. Harborne, J. B. 1988. *Introduction to Ecological Biochemistry.* 3rd. Edition. New York, NY: Academic Press, pp. 42–81.
3. Hrazdina, G. 1982. "Anthocyanins," in *The Flavonoids: Advances in Research*, J. B. Harborne and T. Mabry, eds., London: Chapman and Hall, pp. 135–188.
4. Wagner, G. J. 1982. "Cellular and Subcellular Localization in Plant Metabolism," *Recent Advances in Phytochemistry,* 16: 1–45.
5. Heller, W. and G. Forkmann. 1988. In *The Flavonoids: Advances in Research Since 1980*, J. B. Harborne, ed., London: Chapman and Hall, pp. 399–425.
6. Timberlake, C. F. and P. Bridle. 1980. In *Developments in Food Colours*, J. Walford, ed., London: Applied Science Publishers, pp. 214–266.
7. Brouillard, R. 1982. "Anthocyanins," in *Anthocyanins as Food Colors*, P. Markakis, ed., New York, NY: Academic Press, pp. 1–40.
8. LaBell, F. 1990. "Technology Diversifies Use of Natural Colors," *Food. Proc.,* 51(4): 69.
9. Seitz, H.U. and W. Hinderer. 1988. In *Cell Culture and Somatic Cell Genetics of Plants*, F. Constabel and I.K. Vasil, eds., London: Academic Press, pp. 49–76.
10. Mori, T. and M. Sakurai. 1996. "Riboflavin Affects Anthocyanin Synthesis in Nitrogen Culture Using Strawberry Suspended Cells," *J. Food Science,* 61: 698–702.
11. Shahidi, F. and P. K. J. P. D. Wanasundara. 1992. "Phenolic Antioxidants," *Crit. Rev. Food. Sci. Nutr.,* 32(1): 67–103.
12. Kanner, J., E. Frankel, R. Granit, B. German, and J. E. Kinsella. 1994. "Natural Antioxidants in Grapes and Wines," *J. Agric. Food Chem.,* 42: 64–69.
13. Pszczola, D. E. 1998. "The ABCs of Nutraceutical Ingredients," *Food Technology,* 52(3): 30–37.
14. Pekkarinen, S., I. M. Heinonen, and A. I. Hopia. 1999. "Flavonoids Quercetin, Myricetin, Kaemferol and (+)-Catechin as Antioxidants in Methyl Linoleate," *J. Sci. Food Agric.,* 79: 499–506.

15. Markakis, P. 1974. "Anthocyanins and their Stability in Foods," *Crit. Rev. Fd. Technol.,* 4(4): 437–456.

16. Wesche-Ebeling, P. and M. W. Montgomery. 1990. "Strawberry Polyphenoloxidase: Its Role in Anthocyanin Degradation," *J. Food. Sci.,* 55: 731–735, 745.

17. Mayer, A. M. and E. Harel. 1979. "Polyphenoloxidase in Plants," *Phytochem.,* 18: 193–215.

18. Somers, T. C. 1971. "The Polymeric Nature of Wine Pigments," *Phytochem.,* 10: 2175–2186.

19. Little, A C. 1977. "Colorimetry of Anthocyanin Pigmented Products: Changes in Pigment Composition with Time," *J. Food Sci.,* 42: 1570–1474.

20. Ribreau-Gayon, P., P. Pontalier, and Y. Glories. 1983. "Some Interpretations of Colour Changes in Young Red Wines During their Conservation," *J. Sci. Food. Agric.,* 34: 505–516.

21. Baranowski, E. S. and C. W. Nagel. 1983. "Kinetics of Malvinidin-3-glucoside Condensation in Wine Model Systems," *J. Food Science,* 48: 419–429.

22. Brouillard, R. and O. Dangles. 1994. "Anthocyanin Molecular Interactions: the First Step in the Formation of New Pigments During Wine Aging?," *Food Chem.,* 51: 365–371

23. Kader, F., J. P. Nicolas, and M. Metche. 1999. "Degradation of Pelargonidin-3-Glucoside in the Presence of Chlorogenic Acid and Blueberry Polyphenol Oxidase," *J. Sci. Food Agric.,* 79: 517–522.

24. Shi, Z., I. A. Bassa, S. L. Gabriel, and F. J. Francis. 1992. "Anthocyanin Pigments of Sweet Potatoes - Ipamoea Batatas," *J. Food Sci.,* 57: 755–757,770.

25. Markakis, P. 1982. *Anthocyanins as Food Colors.* New York, NY: Academic Press.

26. Francis, F. J. 1991. "Miscellaneous Food Colorants," in *Natural Food Colorants,* Chap. 7, G. Hendry, ed., Glasgow, Scotland: Blackie and Son Ltd.

27. Yoshimata, K. 1977. "An Acylated Delphinidin-3-Rutinoside-5, 3', 5'-Triglucoside form Lobelia erinus," *Phytochem.,* 16: 1857.

28. Cabrita, L., N. A. Froystein, and O. M. Andersen. 2000. "Anthocyanin Trisaccharides in Blue Berries of Vaccinium Padifolium," *Food Chemistry,* 69: 33–36.

29. Strack, D. and V. Wray 1989. "Anthocyanins," in *Methods in Plant Biochemistry, Volume 1, Plant Phenolics,* J. B. Harborne, ed., P. M. Dey and J. B. Harborne, Series eds., San Diego: Academic Press Inc., pp. 325–356.

30. Giusti, M. M., L. E. Rodriguez-Saona, D. Griffin, and R. E. Wrolstad, 1999. "Electrospray and Tandem MS as Tools for Anthocyanin Characterization," *J. Agric. Food. Chem.,* 47: 4657–4664.

31. Azar, M., E. Verette and S. Brun, 1990. "Comparative Study of Fresh and Fermented Blueberry Juices—State and Modification of the Coloring Pigments," *J. Food Sci.,* 55: 164–166.

32. Sarni, P., H. Fulcrand, V. Souillol, J. M. Souquet, and V. Cheynier. 1995. "Mechanisms of Anthocyanin Degradation in Grape Must-Like Model Solutions," *J. Sci. Fd. Agric.,* 69: 385–391.

33. Cameira-dos-Santos, P. J., J. M.Brillouet, V. Cheynier, and M. Moutounet. 1996. "Detection and Partial Characterization of New Anthocyanin-Derived Pigments in Wine," *J. Sci. Fd. Agric.,* 70: 204–208.

34. Sarni-Manchado, P., V. Cheynier, and M. Moutounet. 1997. "Reactions of Polyphenoloxidase Generated Caftaric Acid *O*-quinone with Malvinidin-3-Glucoside," *Phytochemistry,* 45: 1365–1369.

35. Vivar-Quintana, A. M., C. Santos-Buelga, E. Francia-Aricha, and J. C. Rivas-Gonzalo. 1999. "Formation of Anthocyanin-Derived Pigments in Experimental Red Wines," *Food Sci. and Technol. Intl.,* 5: 347–352.

36. Revilla, I., M. L. González-Sanjose, and M. C. Gómez-Cordoves. 1999. "Chromatic Modifications of Aged Red Wines Depending on Aging Barrel Type," *Food Sci. and Technol. Intl.,* 5: 177–181.

37. Asen, S., R. N. Steward, and K. H. Norris. 1972. "Co-Pigmentation of Anthocyanins in Plant Tissues and its Effects on Color," *Phytochem.,* 11: 1139–1144.

38. Combe, P. 1996. "Les Colorants Anthocyaniques," in *Papers—Second International Symposium on Natural Colorants.* Acapulco, México. The Hereld Organization, S.I.C. Publishing Company, pp. 85–108.

39. Frederick, F. 1985. "Pigments and Other Colorants," in *Food Chemistry.* 1st Ed., O. Fennema, ed., New York, NY: Marcel Dekker, Inc.

40. Gross, J. 1987. *Pigments in Fruits.* New York, NY: Academic Press.

41. Hoshino, T., U. Matsumoto, and T. Goto. 1980. "The Stabilizing Effect of the Acyl Group on the Co-Pigmentation of Acylated Anthocyanins with *C*-Glucosylflavones," *Phytochem.,* 19: 663–667.

42. Price, C. L. and R. E. Wrolstad. 1995. "Anthocyanin Pigments of Royal Okanogan Huckleberry Juice," *J. Food Sci.,* 60: 369–374.

43. Timberlake, C. F. and P. Bridle. 1976. "Interactions Between Anthocyanins, Phenolic Compounds, and Acetaldehyde and Their Significance in Red Wines," *Amer. J. Enol. Vitic.,* 27: 97–105.

44. Cheynier, V., J. M. Souquet, A. Kontek, and M. Moutonet. 1994. "Anthocyanin Degradation in Oxidising Grape Musts," *J. Sci. Food Agric.,* 66: 283–288.

45. Jurd, L. 1969. "Review of Polyphenol Condensation Reactions and Their Possible Occurrence in the Aging of Wines," *Am. J. Enol. Vitic.,* 20: 191–195.

46. Mayen, M., J. Mérida, and M. Medina. 1994. "Free Anthocyanins and Polymeric Pigments During the Fermentation and Post-Fermentation Standing of Musts from *Cabernet Sauvignon* and Tempranillo Grapes," *Am. J. Enol. Vitic.,* 45: 161–166.

47. Sarni-Manchado, P., H. Fulcrand, J. M. Souquet, V. Cheynier, and M. Moutounet. 1996. "Stability and Color of Unreported Wine Anthocyanin-Derived Pigments," *J. Food Sci.,* 61: 938–941.

48. Wesche-Ebeling, P., A. Argaiz-Jamet, A. López-Malo, A. Guerrero, and E. Vargas. 1995. "The Characterization and Potential as Food Colors of Polymeric Anthocyanins from Red Wines," presented at the 9th World Congress of Food Science and Technology, Budapest, Hungary.

49. Wesche-Ebeling, P., A. Argaiz-Jamet, L. G. Hernández-Porras, and A. López-Malo. 1996. "Combined Factors Preservation and Processing Effects on Anthocyanin Pigments in Plums," *Food Chemistry,* 57: 399–404.

50. Wesche-Ebeling, P., E. Ventura-Parra, and A. Argaiz-Jamet. 1998. "Effects of Water Activity and Temperature on the Stability of Polymerized Anthocyanin Fractions from Red Wine," presented at the 1998 Institute of Food Technologists Annual Meeting, Atlanta, GA.

51. Wesche-Ebeling, P., Espinosa, M. G., M. A. Meza, and A. Argaiz-Jamet. 1998. "Capulin—Black Cherry (*Prunus serotina*): Changes in Color and Composition During Maturation," presented at the 1998 Institute of Food Technologists Annual Meeting, Atlanta, GA.

52. Wesche-Ebeling, P., A. Argaiz-Jamet, and Y. Aguilar-Vásquez. 1999. "Hibiscus Anthocyanins: Effects of Interactions with Polyphenoloxidase, Acetaldehyde and Catechin on Color Stability," presented at the 1999 Institute of Food Technologists Annual Meeting, Chicago, IL.

53. Wesche-Ebeling, P., A. Argaiz-Jamet, and L. A. Saucedo-Pérez. 1999. "Red Cabbage and Strawberry Anthocyanins: Effects of pH and a_w on Degradation Kinetics," presented at the 1999 Institute of Food Technologists Annual Meeting, Chicago, IL.

54. Toledo-Flores, S., P. Wesche-Ebeling, and A. Argaiz-Jamet. 1996. "Stability of Anthocyanin Fractions Extracted from Red Wine at Different pH, a_w and SO_2 Levels," in *Papers at The Second International Symposium on Natural Pigments*, Acapulco, Guerrero, Mexico, Hamden, CT: The Hereld Organization, S.I.C. Publishing Company.

55. Rivera-López, J., C. Ordorica-Falomir, and P. Wesche-Ebeling. 1999. "Changes in Anthocyanin Concentration in Lychee (*Litchi chiensis* Sonn.) Pericarp During Maturation," *Food Chemistry,* 65(2): 195–200.

56. Ordaz-Galindo, A., P. Wesche-Ebeling, R. E. Wrolstad, L. Rodríguez-Saona, and A. Argaiz-Jamet. 1999. "Purification and Identification of Capulin (*Prunus serotina* Ehrh) Anthocyanins," *Food Chemistry,* 65(2): 201–206.

57. Wrolstad, R. E. 1993. "Color and Pigment Analyses in Fruit Products," *Oregon State University Agric. Exp. Station Bulletin* No. 624, Oregon State University, Corvallis, OR.

Part III

Mass Transport

11 Transport Properties in Food Engineering

G. D. Saravacos

CONTENTS

Abstract

The design and efficient operation of food processes and processing equipment require quantitative data on the transport properties (flow, heat, and mass transfer) of food materials. The literature contains scattered data on transport properties, usually as part of the thermophysical or engineering properties of foods. A more uniform and fundamental approach is needed, based on advances in molecular thermodynamics and food materials science.

The complex composition and structure of foods necessitate experimental measurements of transport properties, resulting in scattered values due to nonstandardized methods and the inherent variability of food materials. Processing of food significantly changes the transport properties. Food microstructure, phase transitions, density, and porosity strongly affect the transport properties of foods and can be used in semiempirical predictions of these properties. Model food systems, based mainly on starch materials, have been used experimentally to demonstrate the effect of structure on thermal and mass transport properties.

Most fluid foods are non-Newtonian, and their rheological properties are related to the concentration, particle size, and physicochemical structure of food solids. Empirical constants, obtained by regression of experimental data, are useful in predictions of the flow behavior of fluid foods.

Thermal conductivity and especially mass diffusivity are affected strongly by the physical state (rubbery, glassy, or crystalline) and the bulk porosity of the solid and semisolid foods. Simple physical models and empirical correlations can be used in predictions of these basic properties. The variation of thermal diffusivity is smaller.

The permeability of food and packaging films to moisture and other small molecules can be related to diffusivity and solubility and can be expressed in compatible fundamental units.

Interphase heat and mass transfer coefficients depend strongly on food-processing equipment. Based on limited data and semiempirical relations, representative values can be obtained for various food processes.

11.1 INTRODUCTION

The transport phenomena of momentum (flow), heat, and mass transfer have been established as the foundations of unit operations of chemical and process engineering.[1,2] The same phenomena are applicable to the food processing industry.[3] The fundamental transport properties, i.e., viscosity, thermal conductivity, and mass diffusivity of gases and liquids, have been investigated extensively, and accurate data are available in the form of tables, databases, and theoretical or empirical correlations.[4,5] Calculation and estimation of these properties are based on molecular thermodynamics, molecular simulation, corresponding states, and group contributions.

Transport properties of food systems are essential in the design of food processes and processing equipment and in the efficient operation and control of processing plants. Transport properties are also useful in physicochemical processes related to the packaging, storage, and quality of food products.

Limited information is available in the literature on transport properties of foods, especially the solid and semisolid food materials. Transport properties are usually treated as part of the thermophysical or engineering properties of foods.[6-11] The significant variation in transport properties found in the literature is due to the complex structure of foods and inaccuracy of measuring and prediction techniques. Of particular importance is the wide scattering of data on moisture diffusivity, where differences of one or more orders of magnitude have been reported for the same food.

European cooperative research projects on physical properties of foods, COST 90 and COST 90bis,[6,7] have shown that certain physical properties, for example specific heat and enthalpy, can be predicted with sufficient accuracy, based on food composition. Thermal transport properties (thermal conductivity and thermal diffusivity) can be predicted based on composition and physical structure. However, no satisfactory prediction models have been proposed for mass (moisture) diffusivity and apparent viscosity of non-Newtonian fluid foods. If experimental measurements are necessary, it is important to consider that measuring conditions and food material correspond to the actual food-processing operation (context).

If reliable data on the transport properties of a food system are not available, and if experimental measurements are not feasible, it may be possible to adapt published values, carefully considering the composition and physical structure of the food material and the context of food application. For purposes of design of food processes and equipment, an accuracy of about 10% is satisfactory. However, higher accuracy of transport properties may be required in analyzing safety and quality of food products during food processing and storage.

Recent advances in computer-aided engineering require reliable transport property data for food applications in process and equipment design and control.[12] The engineering properties of foods, particularly transport properties, are important in the computer simulation of food processes. Because of their complex structure, recent work has focused on the transport properties of fruit and vegetable products.[13,14] Databases for food properties are limited,[15] and an effort is underway in the European Union to compile a detailed database of physical properties of foods (EU FAIR project DOPPOF).

The present review of transport properties of foods is an effort to treat transport properties in a uniform manner, based on the fundamental transport mechanisms of flow, heat, and mass, and the physical structure of foods.

11.2 FOOD STRUCTURE AND TRANSPORT PROPERTIES

Most foods are solid or semisolid materials, and their transport properties are strongly affected by physical structure. The transport properties of fluid foods are affected by their composition and structure, e.g., aqueous solutions, colloidal dispersions, emulsions, and particle suspensions.

Food structure has a profound effect on mass (e.g., water) diffusivity and permeability and on thermal conductivity, but less on thermal diffusivity. It is, therefore, important to know the physical structure of a food material if literature data or prediction models are to be used. Food structure is of fundamental importance in the developing field of food materials science.

11.2.1 POLYMER STRUCTURE

Advances in polymer science demonstrate the importance of physical structure on the transport (diffusion) of small molecules in polymer matrices.[16] Polymer materials can exist in any of three states: crystalline, which is practically impermeable to water and other small molecules; glassy, in which mass transport is very slow; and rubbery, which facilitates mass transport. In the rubbery state, the polymer chains are flexible, creating free volumes through which water and other small molecules can be transported.

The effect of polymer structure on mass transport of small molecules is utilized in the design of various membranes/films, which are used as packaging materials, protective coatings, and selective membranes for ultrafiltration, reverse osmosis, and gas separation.

The diffusion of water molecules in synthetic polymers is related to the transport of water in food biopolymers, which form the basic physical matrix of complex food materials.[17] Swelling of amorphous polymers by absorption of water and relaxation phenomena strongly affect the mechanism of mass transport (Fickian or non-Fickian diffusion). Relaxation is related to the transition from glassy to rubbery state, or the glass transition temperature (T_g), which is decreased by the absorption of water.

11.2.2 MODEL FOOD MATERIALS

Model food materials are useful in measurements and investigations of physical and engineering properties of foods, because samples can be prepared with controlled composition, structure, and shape. The conclusions reached with model materials must be validated using real foods.

Starch materials constitute the basic matrix of several vegetable and cereal foods, and they are convenient for preparing model structures for experimental measurements of physical properties. Starch materials have been used in several studies of mass (moisture) diffusivity and thermal conductivity. Pectin, cellulose, and other biopolymers have been used to a smaller extent for the same purpose. Starch (usually corn or potato) can be used in the granular or gelatinized form. Granular starch resembles porous foods, while gelatinized material represents a more homogeneous food structure. Sugars and other food components can be incorporated in the starch structure.

The transport properties of model foods are strongly affected by their bulk density and bulk porosity.[18] The bulk porosity (ε) of a porous material is the fraction of the empty volume (void fraction) and is usually estimated from the bulk density (ρ_b) and the solids density (ρ_s) of the material, according to Equation (11.1). Bulk density is measured by volumetric displacement method, while the solids density is measured using a stereopycnometer, which measures the volume of the solids, excluding the void space.

$$\varepsilon = 1 - \rho_b/\rho_s \qquad (11.1)$$

Solids density (ρ_s) decreases linearly as moisture content (X) is increased, which means that starch swells linearly as water is absorbed in the biopolymer matrix. A maximum of (ρ_s) has been observed at very low moisture content, presumably due to strong water-starch interaction.[18]

The bulk porosity of starch materials decreases, in general, when the moisture content is decreased (Figure 11.1). The highest porosities are obtained in granular starches, while gelatinized material gives the lowest porosity. Incorporation of water-soluble sugars in granular starch reduces the porosity but has little effect on the gelatinized material.

Pore size distribution of model materials and food products is difficult to estimate, and the literature on this subject is very limited. The main reason is that pore size and food structure are not fixed, but they change substantially during processing and storage and during actual measurement. High-pressure mercury porosimetry, used to determine small pores, may result in a collapse of the food structure. When low-pressure mercury porosimetry (less than 100 bar) was used in the estimation of pore size distribution of starch materials,[19] it was found that most of the porosity was due to large pores and cracks and that the contribution of very small pores was very low. Thus, bulk porosity, estimated from simple density measurements, is a good measure of the porosity of model foods and food products.

Crack formation during drying affects transport properties and may cause considerable damage to the quality of some food products, like pasta. The mechanism of crack formation in cylindrical shape (extending from the surface inward) was simulated using heat/mass transfer and mechanical stress models and was validated on model starch materials.[20,21,22] Hygrostresses are formed at the initial stage of drying by high moisture gradients (uneven moisture distribution) at the surface of

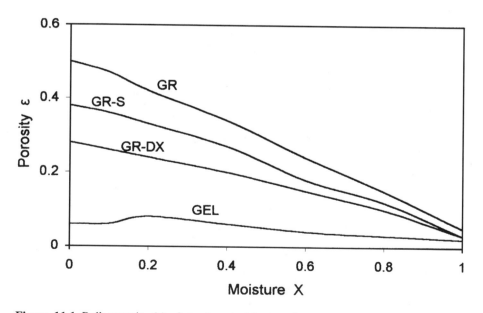

Figure 11.1 Bulk porosity (ε) of starch materials at moisture contents X (kg/kg d.m.). GR = granular starch, GEL = gelatinized starch, S = sucrose, and DX = dextrin.

the product. Crack formation and propagation (e.g., in pasta) can be prevented by controlling relative humidity in the early stage of drying.

Crack formation and propagation during drying of spherical samples of granular starch have been observed visually and microscopically.[18] Radial cracks and channels, extending from the surface to the center of the sample, are formed during air drying, evidently due to the capillary and hydrodynamic flow of liquid water and water vapor as drying progresses inward.

11.2.3 FOOD MATERIALS

The characterization of the physical structure of model (starch) materials can be extended to various food products, providing important information on the transport properties. Measurements of bulk density and bulk porosity of fruits and vegetables at various moisture contents have provided useful data on the drying, rehydration behavior, and mechanical properties of these food materials. Simple mathematical models, based on physical principles, have been proposed, which are applicable to other foods as well.[23–25]

The basic physical properties of a porous solid food are particle (solids) density (ρ_p) and bulk density (ρ_b), defined by Equations (11.2) and (11.3), respectively. The derived property of bulk porosity (ε) is estimated from Equation (11.1).

$$\rho_p = (1 + X)/(1/\rho_s + X/\rho_w) \tag{11.2}$$

$$\rho_b = (1 + X)/(1/\rho_{bo} + \beta X/\rho_w) \tag{11.3}$$

Both properties are functions of the moisture of material X (kg/kg d.m.). The parameters of Equations (11.2) and (11.3) are: ρ_s = dry solids density, ρ_w = water density, ρ_{bo} = bulk density of dry solids, and β = shrinkage coefficient.

The parameters ρ_s, ρ_{bo}, ρ_w, and β are estimated by regression analysis of several experimental data of the basic properties ρ_p and ρ_b. The density of water in the food materials (ρ_w) was estimated to be slightly higher than the density of pure water (1030 kg/m^3), presumably due to the strong water/food component interaction. Typical values of the parameters ρ_s, ρ_{bo}, and β for potato and apple are shown in Table 11.1.

Figure 11.2 shows the changes in bulk porosity (ε) during the drying of potato and apple as a function of the moisture content (X), using various drying methods. In general, the porosity increases as water is removed from the material by evaporation. Freeze drying produces the highest porosity, whereas air drying results in lower porosity. Vacuum drying produced higher porosity in apple than in potato, evidently due to the different cellular structure of the two materials. Microwave drying resulted in higher porosity in potato than in apple, due to the collapse of the sugar-containing apple product. Incorporation of sugar in osmotic dehydration of apple resulted in lower porosity, due to the precipitation of sugar in the cellular structure.

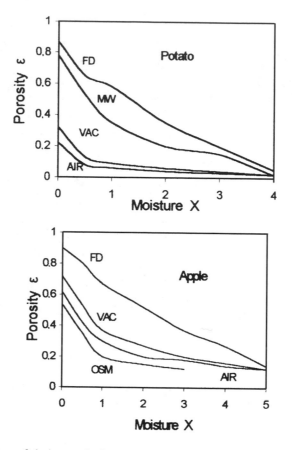

Figure 11.2 Effect of drying method on bulk porosity (ε) of potato and apple at moisture contents X (kg/kg d.m.). FD = freeze, MW = microwave, VAC = vacuum, OSM = osmotic, AIR = air drying.

TABLE 11.1
Estimated Parameters for Potato and Apple

Material	Drying Method	ρ_s	ρ_{bo}	β
Potato	Air drying	1.60	1.50	1.03
	Vacuum drying	1.58	1.29	1.03
	Microwave drying	1.63	0.44	0.81
	Freeze drying	1.57	0.18	0.29
Apple	Air drying	1.69	0.56	0.99
	Vacuum drying	1.61	0.39	0.96
	Microwave drying	1.66	0.56	1.01
	Freeze drying	1.67	0.34	0.34
	Osmotic dehydration	1.67	0.73	1.10

The differences in drying behavior between potato and apple are caused by the quite different cellular structure of the two plant materials. Apple tissue has higher porosity (more intercellular void space) than potato tissue, resulting in higher mass transport rates, as in drying, rehydration, and osmotic impregnation. The vacuum impregnation in the osmotic dehydration process of apple can be based on the hydrodynamic mass transport in the porous structure of the material.[26]

The macrostructural properties of apple tissue during osmotic dehydration (density and porosity) are related to the microstructural properties (cellular arrangement) of the samples, as observed by SEM.[27]

The transport properties of foods are affected strongly by the phase transitions (crystallized, amorphous states) of their components. Phase transitions in solid and semisolid foods are similar to changes in polymers, and the principles of polymer science can be adapted to complex food materials.[28] The glass transition temperature (T_g) from glassy to rubbery state is very important, especially in mass transport. Water and other small molecules are transported (diffused) faster in the rubbery state.

The structure of freeze-dried and other porous foods can collapse, resulting in significant changes of quality and transport properties.[29] Sugars considerably decrease the (T_g) of fruit/vegetable materials, and the difference of the drying temperature $(T - T_g)$ should be kept small to prevent collapse of the product.

The porosity of dehydrated plant foods can be controlled by combined drying processes, e.g., freeze-drying followed by air drying.[30] The pore size distribution of freeze-dried carrots showed three maxima at 21, 1, and 0.2 μm, which shifted to lower values by air drying, reducing the bulk porosity.

Tailor-made porous solid foods can be prepared by controlling the food matrix, incorporating sugars and/or edible oils.[31,32] Cellular food structures of a wide range of bulk porosity (0.04–0.85) were prepared by freeze-drying sodium alginate/starch gels impregnated with sugars in sucrose solutions of 10–60° Brix.

11.3 VISCOSITY OF FLUID FOODS

Viscosity is one of the three fundamental transport properties of liquids and gases, the other two being thermal conductivity and mass diffusivity. In fluid foods, viscosity is important in the design of food processes and food processing equipment, in product development, and in quality control of foods.

In most applications, and in this review, the dynamic or shear viscosity (η) of fluid foods is used, which for a Newtonian fluid is defined by the following equation:

$$\tau = \eta\gamma \tag{11.4}$$

where τ = the shear stress
　　　　γ = the shear rate

The units of these quantities in the SI system are τ (Pa), γ (1/s), and η (Pa s); 1 cP = 0.001 Pa s =1 mPa s.

Other rheological properties of fluid and semisolid foods, such as extensional viscosity (orifice flow, extrusion) and viscoelastic properties (gels, etc.), are treated in specialized publications.[33,34]

Shear viscosity is usually determined by measuring the shear stress at various shear rates in a controlled flow system (capillary or rotational viscometer).

Composition and structure strongly affect the viscosity of fluid foods. Water, clear aqueous solutions, and clear food liquids like edible oils are Newtonian fluids. Most fluid foods are complex dispersions of various components and do not obey the basic Equation (11.4), i.e., they are non-Newtonian fluids.

11.3.1 NEWTONIAN FLUIDS

Gases and vapors of importance to food processing and storage (such as air, oxygen, nitrogen, ethylene, carbon dioxide, and water vapor) are Newtonian fluids. The viscosity of simple gases and liquids at various temperatures and pressures can be predicted using thermodynamic or empirical models or can be obtained from tables and databases in the literature.[4,5] Typical viscosity values at atmospheric pressure are air at 20°C, 0.018 mPa s, and water vapor at 100°C, 0.013 mPa s.

The viscosity of gases increases at higher temperatures and pressures. The viscosity of liquid decreases with temperature and more sharply at higher viscosities.

Liquid viscosities are very sensitive to the structure of the constituent molecules. The viscosity of liquid mixtures can be estimated by interpolation of the viscosities of the constituents using empirical equations.[4] The viscosity of water decreases from 1.792 mPa s (0°C) to 0.284 mPa s (100°C), and that of ethanol from 1.77 mPa s (0°C) to 0.458 mPa s (75°C).

Aqueous sugar solutions, clarified (depectinized) juices, honey, vegetable oils, and other clear fluid foods are Newtonian fluids, and their viscosity strongly depends on their composition and concentration. The effect of temperature is negative and can be expressed by the Arrhenius equation.[6,33,36]

$$\eta = \eta_0 \exp(\Delta E_a / RT) \tag{11.5}$$

where η_0 = pre-exponential factor
R = 8.13 kJ/mol K
T = temperature, K

The energy of activation for shear flow (ΔE_a) increases exponentially as the solute concentration is increased and takes values for clarified fruit juices of 20 kJ/mol (15°Brix) to 70 kJ/mol (70°Brix). Properties of cooking oil are viscosity 80 mPa s (25°C) and $\Delta E_a = 28.3$ kJ/mol.

11.3.2 NON-NEWTONIAN FLUIDS

The viscometric behavior of non-Newtonian fluids is usually determined from experimental shear stress (τ) – shear rate (γ) diagrams. The ($\tau - \gamma$) relationship for time-independent fluids can be expressed by one of the following empirical models:

$$\tau = K\gamma^n \tag{11.6}$$

$$\tau = \tau_o + K\gamma^n \tag{11.7}$$

$$\tau^{1/2} = \tau_o + K\gamma^{1/2} \tag{11.8}$$

The rheological constants in these equations are K = consistency index (Pa s^n), n = flow behavior index, and τ_o = yield stress (Pa).

Equation (11.6) refers to the power law model, which is a simplification of the Herschel–Bulkley model, Equation (11.7), for the case $\tau_o = 0$. The Casson model, Equation (11.8), has been used mostly with chocolate. Most fluid foods are shear-thinning, i.e., $n < 1$.

Time-dependent non-Newtonian (e.g., thixotropic) fluids are characterized by loops of the rheological diagram; e.g., in thixotropic fluids, the descending curve lies below the ascending path.

The apparent viscosity (η_a) of a non-Newtonian fluid is defined by Equation (11.9):

$$\eta_a = K\gamma^{n-1} \tag{11.9}$$

In shear-thinning fluids, the apparent viscosity decreases at higher shear rates.

Temperature has a significant negative effect on the consistency index K (Arrhenius equation), while it slightly affects the flow behavior index (n).

The viscosity of fluid suspensions reflects the hydrodynamic flow forces and the particle/solvent (aqueous solution) and particle/particle interactions. The particle size, shape, and concentration strongly affect the rheological properties of fluid foods. In a simple suspension of spherical particles, the relative viscosity of the suspension to the viscosity of continuous phase (η_r) is related to the volume fraction of the particles (φ) by the Einstein equation:

$$\eta_r = 1 + 2.5\varphi \tag{11.10}$$

The Einstein equation is not applicable in suspensions of spherical particles at high concentrations.[6] Relative viscosities higher than those predicted by Equation (11.8) are obtained with nonspherical particles, especially when the particles become more elongated (e.g., food fibers).

Fluid food suspensions are complex materials of hydrophilic and interacting particles of a wide particle size distribution, deviating considerably from Equation (11.10).

Particle size is affected by processing,[33] for example by the screen size of tomato pulpers/finishers (mean particle sizes from 120–400 μm). The mean particle size of cornstarch increases from about 15–30 μm during heating before gelatinization, sharply increasing the apparent viscosity of the suspension. Table 11.2 shows typical rheological constants for food products.[6,33,35,36,37,38]

TABLE 11.2
Rheological Constants of Fluid Food Products (25°C)

Material		K, Pa s^n	n	ΔE_a, kJ/mol
Apple juice	65° Brix	0.12	1.00	50
Orange juice	44° Brix	0.60	0.65	25
Tomato paste	28° Brix	27.50	0.34	21
Apple sauce	12° Brix	12.00	0.30	10
Conc. milk	30% solids	0.016	0.91	27

11.4 THERMAL CONDUCTIVITY AND DIFFUSIVITY

Thermal conductivity (λ) and the thermal diffusivity (α) are basic transport properties of a material and are defined by the Fourier equations.

$$dQ/dt = -\lambda(dT/dz) \tag{11.11}$$

$$\delta T/\delta t = \alpha(\delta^2 T/\delta z^2) \tag{11.12}$$

$$\alpha = \lambda/\rho C_p \tag{11.13}$$

Thermal diffusivity is related to thermal conductivity by Equation. 11.13 The SI units for (λ) are W/m K, and for (α) m^2/s, with ρ = density and C_p = specific heat.

In simple gases and liquids, thermal conductivity (like the two other fundamental transport properties, viscosity and mass diffusivity) can be estimated from thermodynamic and molecular quantities or can be obtained from literature tables and data banks.[4,5] Typical values of (λ) are (at 25°C) water 0.609, ethanol 0.169, water vapor 0.0179, and air 0.0258 W/m K. It is interesting to note that λ (water) is about 20 times higher than λ (air).

For food materials, experimental data are necessary, and empirical models are useful for approximate estimations.

Experimental methods of measurement have been used and reported in the literature.[10,11,39,40] Experimental values of several foods at various compositions and temperatures have been published in the form of tables, data banks, and empirical models.[8,10,11,15,39]

11.4.1 FLUID FOODS

The thermal conductivity of aqueous liquid foods is close to that of pure water, decreasing almost linearly with increasing soluble and insoluble solids content. Thus, the λ of sugar solutions or fruit juices decreases from 0.540 W/m K (10° Brix) to 0.400 W/m K (60° Brix). Temperature has a small positive effect on λ. Vegetable (cooking) oils have considerably lower λ values than aqueous fluid foods and are not affected by temperature, e.g., 0.170 W/m K.

The thermal diffusivity (α) of fluid foods does not change much with the composition and temperature, with a typical value of 1.3×10^{-7} m²/s. According to Equation (11.13), any changes of density (ρ) are compensated with changes of thermal conductivity (λ) in the same direction.

11.4.2 SOLID FOODS

Thermal conductivity (λ) and diffusivity (α) of solid foods are affected considerably by the physical structure of the material. Composition and temperature influence both properties, and pressure has a significant effect on porous foods. Moisture content and temperature have weak positive effects in most foods. A linear effect of moisture content has been observed at moistures higher than 10%. However, at lower moistures, α changes nonlinearly, with a maximum near 5%.[41] This anomaly may be caused by the strong water-biopolymer interaction, which is manifested by the anomalous change of solids density in this region.[18]

Thermal diffusivity can be estimated indirectly from λ, using Equation (11.13).[42] Typical values of α for nonfrozen solid foods are $1.0–2.0 \times 10^{-7}$ m²/s.[11]

Changes in the bulk porosity (ε) of solid food materials at low moisture contents [Figures (11.1) and (11.2)] have a strong effect on thermal conductivity (λ), evidently due to the large differences between the λ values of the solids and the air in the pores.

For granular starches, the effective thermal conductivity (λ) can be described by the parallel model:[43, 44]

$$\lambda = (1 - \varepsilon)\lambda_s + \varepsilon\lambda_a \tag{11.14}$$

For gelatinized starches, the perpendicular model has been found to be more suitable:

$$1/\lambda = (1 - \varepsilon)/\lambda_s + \varepsilon/\lambda_a \tag{11.15}$$

In these equations, λ_a is the thermal conductivity of air, and λ_s is the thermal conductivity of the solids, estimated by regression analysis of experimental data.

Figure 11.3 shows the estimated values of λ_s of granular and gelatinized starch, which represent a porous and a homogeneous food material, respectively. The lower values of λ_s of granular starch may be due to the presence of small pores within the starch granules, which disappear upon gelatinization.

The thermal conductivity of fibrous food is higher parallel to the fiber than in the perpendicular direction. The λ of freeze-dried foods is very low, due to the high

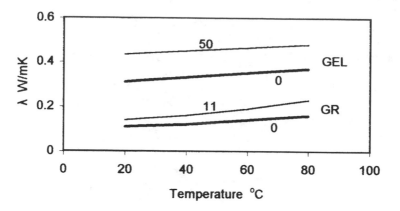

Figure 11.3 Thermal conductivity (λ) of granular (GR) and gelatinized (GEL) starch at 0, 11, and 50% moisture content. Reprinted from Reference 43 with permission from Elsevier Science.

bulk porosity of these materials ($\varepsilon > 0.9$), and it decreases considerably as the (air) pressure is reduced below the atmospheric.[45]

11.4.3 FROZEN FOODS

Frozen foods have higher thermal conductivity than nonfrozen materials, due to the high λ value of ice ($\lambda = 2.0$ W/m K). Thermal conductivity remains nearly constant, about 1.5 W/m K, from very low temperatures to about $-5°C$, and it drops to about 0.5 W/m K near the freezing point. The thermal diffusivity of frozen foods is about three to five times higher than for nonfrozen materials, ranging from $3.0–10.0 \times 10^{-7}$ m^2/s.[10,11]

11.5 MASS DIFFUSIVITY

Mass diffusivity is a fundamental transport property of food systems, which is very useful in the quantitative analysis of several processes: storage, product development, and quality control of food products. Most research and development is related to the diffusivity of water (moisture), but the transport of other food components, such as salt and aroma compounds, is also important.

Diffusion is the transport of mass in a medium by a concentration gradient, in analogy to the heat transport by conduction Equations (11.11) and (11.12). By analogy to thermal conductivity (λ), mass diffusivity (D) is defined by the Fick equations for steady or unsteady transport.

$$dm/dt = -D(dC/dz) \qquad (11.16)$$

$$\delta C/\delta z = (\delta/\delta z)(D\delta C/\delta z) \qquad (11.17)$$

In moisture transport processes, the concentration (C) is usually replaced by the moisture content (X, kg water/kg dry solids). The units of D are m^2/s.

The rigorous thermodynamic analysis of diffusion is based on the assumption that mass is transported by the chemical potential of the diffusing substance. However, the simpler approach of concentration gradient is preferred in food engineering.

Most solid and semisolid foods are complex heterogeneous materials in which mass may be transported, in addition to molecular diffusion, by other mechanisms such as hydrodynamic flow, Knudsen flow, and capillary flow. The term *effective diffusivity* (D_{eff} or simply D) is used as an overall transport property, assuming that all mass is transported in the system by a concentration gradient. This simplification is convenient for engineering calculations involving mass transport, but the underlying transport mechanisms should be kept in mind when analyzing the physics of transport processes in food systems.

11.5.1 DIFFUSIVITY IN FLUIDS

The mass diffusivity of a compound/component in pure gases and liquids can be predicted by molecular or empirical correlations, as a function of molecular weight, temperature, pressure, or some empirical molecular constants.[4,5] The diffusivity (D) of some simple fluids of interest to food engineering are given in Table 11.3.

TABLE 11.3
Diffusivity (D) in Simple Gases and Liquids (25°C)

Diffusant	Medium	D, m^2/s
Oxygen	Air	1.6×10^{-5}
Carbon dioxide	Air	1.5×10^{-5}
Water vapor	Air	2.8×10^{-5}
Ethylene	Air	1.4×10^{-5}
Oxygen	Water	2.4×10^{-9}
Carbon dioxide	Water	2.0×10^{-9}
Water	Water	1.0×10^{-9}
Ethanol	Water	1.2×10^{-9}

The diffusivity of small molecules in liquids is about 10,000 times lower than in gases, and this has major implications in various mass transport processes.

The diffusivity (D) of gases is inversely proportional to the absolute pressure (P), according to the empirical equation:

$$DP = \text{constant} \tag{11.18}$$

The diffusivity (D_{AB}) of a molecule or particle (A) of radius (r_A) in a liquid (B) of viscosity (η_B) is given by the hydrodynamic Einstein–Stokes equation,

$$D_{AB} = (RT)/6\pi\eta_B r_A \tag{11.19}$$

Thus, at a given temperature, T, the diffusivity, D_{AB}, is inversely proportional to the viscosity of the liquid medium; i.e., $D_{AB}\, \eta_B$ = constant.

High pressure has a negative effect on diffusivity, while the effect of temperature in liquid diffusion follows the Arrhenius equation.

The diffusivity of ions in electrolytic solutions (e.g., salts) is estimated from empirical equations, which include molecular and thermodynamic quantities and the viscosity of the solution. For sodium chloride, the D in dilute aqueous solution is 1.5×10^{-9} m^2/s, and it goes through a minimum as the molality increases.

11.5.2 DIFFUSIVITY IN POLYMERS

The diffusion of small molecules in polymer systems is important in several industrial applications, such as membrane separations, packaging materials, and protective coatings.[16] The transport mechanism through polymers is very important in food engineering, because the physical structure of most solid and semisolid foods is a biopolymer matrix. As explained earlier, the structure of the polymer decisively affects the transport mechanism and the transport rate.

In food engineering, the transport of moisture (water) in biopolymers, in the liquid or vapor state, is of particular importance. A simple and widely used system for studying diffusivity is the sorption/diffusion of small molecules in a film (thin plate), applying the empirical equation

$$M_t/M_0 = kt^n \tag{11.20}$$

The following mass transport mechanisms may take place:[17]

$$n = 0.5 \text{ (Fickian diffusion)}$$
$$0.5 < n < 1.0 \text{ (non-Fickian diffusion)}$$
$$n = 1.0 \text{ (case 2 diffusion)}$$

In non-Fickian and case 2 diffusion, the transport of water is accompanied by swelling and stresses, leading to relaxation of the macromolecules and transition from the glassy to the rubbery state.

The diffusion type in polymers can be predicted by the Deborah number, De, which is defined by the following equation:

$$De = \chi/\theta = \chi D/l^2 \tag{11.21}$$

where χ and θ = the characteristic relaxation and diffusion times, respectively
l = the (film) thickness of the material[17]

For $De \gg 1$ or $De \ll 1$ Fickian diffusion, and $0 < De < 1$ non-Fickian diffusion.

For Fickian diffusion/sorption in a polymer film of thickness (l), the diffusivity (D) can be estimated from the simplified solution of Equation (11.17).

$$D = 0.049/(t/l^2)_{1/2} \tag{11.22}$$

where $(t/l^2)_{1/2}$ corresponds to half equilibrium sorption, i.e., $M_t/M_0 = 0.5$.

Molecular simulation, based on molecular dynamics, has been used recently to describe and predict the sorption/diffusion of small molecules in polymers.[46]

11.5.3 MOISTURE DIFFUSIVITY IN FOODS

The effective moisture diffusivity in solid and semisolid foods is determined by various experimental methods appropriate to the system.[47,48] The sorption technique is used in homogeneous materials and films, like gelatinized starch. The drying method is applied widely, utilizing the simplified (method of slopes) or numerical solutions of the diffusion Equation (11.17). The moisture distribution and permeability methods are experimentally difficult for food materials, but they are useful for special applications. Numerical solutions are necessary when D is a strong function of concentration. The effect of temperature can be described by the Arrhenius equation, and the energy of activation for diffusion (E_a) is affected by the physical structure of the food material and the type of moisture transport (liquid or gaseous diffusion).

Moisture diffusivity of foods varies over a wide range, due primarily to the physical structure of the food material. Table 11.4 shows typical values of moisture diffusivity and energy of activation for diffusion of some food materials.[3,47,49,50]

TABLE 11.4
Typical Moisture Diffusivities of Foods (60°C)

Material	$D, \times 10^{10}$ m²/s	E_a, kJ/mol
Porous granular starch	10.0	16.7
Starch gel	1.0	33.4
Starch/sugar	0.1	41.8
Pasta	0.5	50.0
Wheat	0.5	54.0
Potato	1.0	54.3
Fruit	0.6	62.7

Moisture diffusivity in porous foods at low moistures is strongly affected by the physical structure of the material, particularly the bulk porosity. Figure 11.4 shows the effective moisture diffusivity (D) in starch materials of different structures, produced by air-drying processes. The highest diffusivities were observed in highly porous extrusion-cooked starch and the lowest in gelatinized material. In gelatinized starch, D increases continuously when the moisture content is increased, as expected

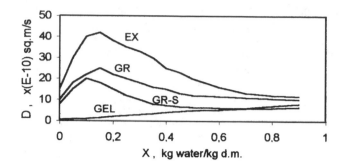

Figure 11.4 Effective moisture diffusivity in starch materials during air drying at 60°C. (EX = extruded, GR = granular, GR-S = granular/sucrose, GEL = gelatinized.)

for a polymer gel, swelling linearly with the sorption of water. Porous materials, like granular and extrusion-cooked starch, showed an anomalous behavior at moistures below 30%, with a maximum D near 10%. The shape of the diffusivity curve indicates that liquid diffusion of water may be controlling at high moistures, while water vapor diffusion may be important at low moisture, when a porous structure was developed by the rapid evaporation of water within the material. At moistures below 5%, D decreased sharply, presumably due to the lower mobility of the biopolymer-bound water.

Values of moisture diffusivity lower than those in pure biopolymers (starch) are obtained when soluble sugars, proteins, or lipids are incorporated in the polymer matrix (Table 11.4). The reduction in D is analogous to the reduction in bulk porosity ε, as shown in Figures 11.1 and 11.2. A nonlinear model for prediction of D in starch materials, as a function of porosity, moisture content, and temperature, has shown the prevailing effect of bulk porosity.[49]

Moisture diffusivity during the drying of food materials can be estimated from an overall heat and mass transfer analysis of the whole system.[51,52] The resulting equations are solved numerically, and the diffusivity is expressed by an exponential function of moisture content and temperature.

Moisture diffusivity of cellular foods, like fruits and vegetables, can be estimated by a mass transfer analysis at the cellular level.[3,53] The proposed model applies the chemical potential as the driving force and requires determination of the mass conductivity in the different phases of the cell. Recent advances in measuring the chemical potential of water in cellular food systems may facilitate the application of this new technique.[54]

11.5.4 MASS DIFFUSIVITY OF SOLUTES IN FOODS

The transport of small molecules (solutes and volatiles) is important in several food processes and in the storage and quality of food products. The transport of solutes is assumed to be controlled by molecular diffusion, and the (effective) diffusivity (D) can be estimated from the solution of Equation (11.17) for a semi-infinite solid (usually a cylinder).[47] The diffusivity of salt (sodium chloride) in food gels, cheese, meat, fish, and pickles is close to the D of salt in water: 16.1×10^{-10} m²/s. Lower values of D were estimated in green olives (3.2×10^{-10} m²/s), presumably due to the resistance to mass transfer of the olive skin and oil.[55] Low D values were obtained in food gels for glucose (3.3×10^{-10} m²/s), sucrose (3.2×10^{-10} m²/s), and tripalmitin (fat) (0.35×10^{-10} m²/s.)[47]

The transport of volatiles in foods is strongly affected by sugar content, which was found to decrease from 10×10^{-10} m²/s (water) to 0.1×10^{-10} m²/s (60° Brix).[47]

Mass transport during osmotic dehydration is a complex process involving mainly liquid diffusion and hydrodynamic flow. The cellular structure of the food material (e.g., apple) plays a decisive role in the transport of water from the food to the osmotic solution and the transport of sugar in the opposite direction.[26,27] Measurements of water removal and solute (sugar) pickup in apple slices in a sucrose solution of 65° Brix at 30°C yielded effective diffusivities of 3×10^{-10} m²/s for water and 1×10^{-10} m²/s for sugar.[56]

11.6 PERMEABILITY IN FOOD SYSTEMS

The transport of small molecules, and especially water, in polymer films and membranes is of great interest to separation processes,[16] to food packaging,[58] and, in general, to food processing and engineering.[50,57] The same principles apply to packaging films and protective coatings of foods. In addition to water, the permeability of films to gases such as oxygen, carbon dioxide, and ethylene are of interest to food packaging.

The permeability (P) of a film or thin layer of thickness (z) is related to the diffusivity (D) and the solubility (S) of the penetrant in the material, according to the following equation:

$$J = P(\Delta p/z) = DS(\Delta p/z) \tag{11.23}$$

where J is the mass transfer rate (kg/m² s) and ($\Delta p/z$) is the pressure gradient (Pa/m). The solubility (S) is the gas/liquid equilibrium constant ($S = C/p$), where C is the concentration (kg/m³) and p the pressure (Pa). The solubility (S) is usually determined as the slope of the sorption isotherm (C vs. p).

The SI units of permeability are (kg/m s Pa) or (g/m s Pa), but various other units are used in packaging, reflecting the measuring technique or the particular food/package application.[58]

The permeability (P) is related to the permeance (PM) and the water vapor transfer rate ($WVTR$) by the following equation:

$$P = (PM)z = (WVTR)/\Delta p \tag{11.24}$$

The units of permeance are (kg/m² s Pa), which are identical to the units of the mass transfer coefficient (k_p). The units of ($WVTR$) are (kg/m s).

The SI units are useful in relating and comparing the literature data on P and $WVTR$ to the fundamental mass transport property of diffusivity (D, m²/s). Table 11.5 shows typical literature data of P and D.[47,50,57,58]

TABLE 11.5
Permeability (P) and Diffusivity (D) of Water (25°C)

Film or coating	P, 10^{10} g/m s Pa	D, × 10^{10} m²/s
HDPE	0.002	0.005
LDPE	0.014	0.01
PP	0.010	0.01
PVC	0.041	0.05
Cellophane	3.7	1.0
Protein films	0.1–10.0	0.1
Polysaccharide films	0.1–1.0	0.1
Lipid films	0.003–0.1	0.01
Gluten	5.0	1.0
Corn pericarp	1.6	1.0

Note: LPDE = low-density polyethylene, HDPE = high-density polyethylene, PP = polypropylene.

The permeability of films and food coatings is affected by the chemical and physical structure (morphology) of the material. Moisture content (water activity) and temperature have similar effects with diffusivity, and the energy of activation for permeation varies in the range of 20–70 kJ/mol.[58] Typical values of the permeability of oxygen at 25°C are LDPE: 4.4×10^{-14} and HDPE: 6.0×10^{-15} g/m s Pa. For carbon dioxide, P values at 25°C are typically LDPE: 1.6×10^{-13} and HDPE: 5.3×10^{-15} g/m s Pa.

11.7 HEAT AND MASS TRANSFER COEFFICIENTS

Interface heat and mass transfer are important in the design of food processes and processing equipment and in the control of food packaging and storage. Heat transfer coefficients are essential in thermal processing and in cooling/freezing and storage operations. Mass transfer coefficients are important in drying and storing foods and in separation processes such as solvent extraction, gas absorption, and reverse osmosis. One basic feature of both coefficients is that they are strongly affected by the characteristics of the processing equipment.

11.7.1 HEAT TRANSFER COEFFICIENTS

The surface heat transfer coefficient (h, W/m² K) and the overall heat transfer coefficient (U, W/m² K) are defined by the following equation:

$$q/A = h\Delta T = U(\Delta T)_{LM} \qquad\qquad (11.25)$$

where q/A = the heat flux (W/m^2)

ΔT = temperature difference (K)

ΔT_{LM} = log-mean temperature difference between the two media

The overall heat transfer coefficient (U) is of major importance to the design and operation of food-processing equipment as heat exchangers, coolers/freezers, and dryers. It is a function of the surface heat transfer coefficients and the thermal conductivity and fouling of the heat transfer wall.

The values of h can be determined experimentally or calculated from empirical equations, which involve the following dimensionless numbers:

Nusselt, $Nu = hd/\lambda$;   Reynolds, $Re = u\rho d/\eta$;   Prandtl, $Pr = C_p\eta/\lambda$ (11.26)

where d = diameter-dimension

u = velocity

ρ = density

In engineering design and applications, experimental and empirical values of U are preferred, because they include the effects of the product and the fouling of the heat transfer surface. Typical values of U are given in Table 11.6.[13,52]

TABLE 11.6
Typical Heat Transfer Coefficients (U)

Equipment/product	U, W/m^2 K
Agitated kettle, sugar soln. 40° Brix	1200
Agitated kettle, fruit puree	800
Falling film evaporator, clear juice	1700
Falling film evaporator, cloudy juice	1000
Agitated film evaporator, juice 40° Brix	2000
Air dryers	10–200

Since most fluid foods are non-Newtonian, modified heat transfer correlations of the heat transfer coefficient are applied in which the apparent viscosity is used as a function of the flow velocity and the rheological constants of the power-law or the Herschel-Bulkley models.[33] In shear-thinning food fluids, the heat transfer coefficients (h, U) increase considerably at higher velocities or agitation rates.

11.7.2 Mass Transfer Coefficients

Mass transfer between phases, e.g., gas/solid (drying), is usually analyzed by the two-film theory.[47] The mass transfer rate J (kg/m^{2}s) is defined by the following equation in which the driving force can be the partial pressure difference (Δp, Pa), the concentration difference (ΔC, kg/m³) or the mol fraction difference (Δy, –):

$$J = k_p(\Delta p) = k_c(\Delta C) = k_y(\Delta y) \tag{11.27}$$

The three mass transfer coefficients are related as follows:

$$k_y = k_c C = k_p p \tag{11.28}$$

The SI units of the coefficients are k_p (kg/m² s Pa), k_c (m/s), and k_y (–).

In food engineering, the ideal gas law is usually obeyed (atmospheric pressure and not high temperatures), and then $k_c = k_p(RT)$, where T is the absolute temperature (K), and $R = 8.31$ Pa m³/(MW) K; for water vapor (MW) = 18.

The mass transfer coefficients can be determined experimentally or estimated from empirical equations, analogous to the heat transfer correlations. These equations involve the following dimensionless numbers:

$$\text{Sherwood, } Sh = k_c d/D; \text{ Schmidt, } Sc = \eta/\rho D; \text{ Reynolds, } Re = u\rho/\eta \tag{11.29}$$

In moisture transfer operations, some important simplifications can be made to make the calculations easier. The heat/mass transfer analogies are assumed to be applicable, and the mass transfer coefficient (k_c) can be estimated from the heat transfer coefficient (h) by the Lewis equation:

$$h/k_c = \rho C_p \tag{11.30}$$

The density and specific heat of air (atmospheric pressure) can be taken as $\rho = C = 1$ kg/m³ and $C_p = 1000$ J/kg K, and then, according to Equations (11.28) and (11.30)

$$k_c \text{ (m/s)} = h/1000, \text{ or } k_c \text{ (mm/s)} = h \text{ (W/m}^2 \text{ K)} = k_y \text{ (g/m}^2 \text{ s)}$$

Table 11.7 shows some typical values of the mass transfer coefficient (k_c).[47,50,52]

TABLE 11.7
Typical Mass Transfer Coefficients

Air/food system	k_c, mm/s
Drying wet solids, 2 m/s	35
Baking oven	15–30
Freezing food	10–30
Thawing food	20–40
Desiccation of frozen food	0.3–5

The mass transfer coefficients shown in Table 11.7 are numerically similar to the heat transfer coefficients (h, W/m^2 K) of similar air/food systems. In drying and desiccation processes, the values of (k_c) decrease sharply as the food surface forms a dried layer, and the drying rate is controlled by internal diffusion.[50]

11.8 ACKNOWLEDGMENT

The author wishes to acknowledge the contributions of his associates and graduate students at the National Technical University of Athens, Rutgers University, and CAFT, and Cornell University, Geneva.

REFERENCES

1. Bird, R. B., W. F. Stewart, and E. N. Lightfoot. 1960. *Transport Phenomena*. New York, NY: Wiley.
2. Geankoplis, C. J. 1993. *Transport Processes and Unit Operations*. 3rd ed. New Jersey: Prentice Hall.
3. Gekas, V. 1992. *Transport Phenomena of Foods and Biological Materials*. New York: CRC Press.
4. Reid, R.C., J. M. Prausnitz, and B.E. Poling. 1987. *The Properties of Gases and Liquids*. New York: McGraw-Hill.
5. DIPPR. 1998. *Handbook and Databases of Transport Properties*. New York: AIChE.
6. Jowitt, R., F. Escher, H. Hallstrom, H. F. Th. Meffert, W. E. L. Spiess, and G. Vos., eds. 1983. *Physical Properties of Foods*. London: Applied Science Publ.
7. Jowitt, R., F. Escher, M. Kent, B. McKenna, and M. Roques, eds. 1987. *Physical Properties of Foods—2*. London: Elsevier Applied Science.
8. Okos, M. R. ed. 1986. *Physical and Chemical Properties of Foods*. St. Joseph, MI: ASAE.
9. Lewis, M. A. 1987. *Physical Properties of Foods and Food Processing Systems*. London: Ellis Horwood.
10. Rao, M. A. and S. S. H. Rizvi, eds. 1995. *Engineering Properties of Foods*. New York: Marcel Dekker.
11. Rahman, S. 1995. *Food Properties Handbook*. Boca Raton, FL: CRC Press.
12. Datta, A. K. 1998. "Computer-Aided Engineering in Food Process and Product Design," Food Technol., 52(10): 44–52.
13. Saravacos, G. D. and A. E. Kostaropoulos. 1995. "Transport Properties in Processing of Fruits and Vegetables," *Food Technol.*, 49(9): 99.
14. Saravacos, G. D. and A. E. Kostaropoulos. 1996. "Engineering Properties in Food Processing Simulation," *Computers Chem. Engng.*, 20: S461–S466.
15. Singh, R. P. 1993. *Food Properties Database*. Boca Raton, FL: CRC Press.
16. Vieth, W. R. 1991. *Diffusion in and through Polymers*. Munich: Hauser Publ.
17. Peppas, N. A. and L. Brannon-Peppas. 1994. "Water Diffusion and Sorption in Amorphous Macromolecular Systems and Foods," *J. Food Eng.*, 22: 189–210.
18. Marousis, S. N. and G. D. Saravacos. 1990. "Density and Porosity in Drying Starch Materials," *J. Food Sci.*, 55: 1367–1372.
19. Karathanos, V.T. and G. D. Saravacos, 1993. "Porosity and Pore Size Distribution of Starch Materials," *J. Food Eng.*, 18: 259–280.

20. Izumi, M. and K.-I. Hayakawa. 1995. "Heat and Mass Transfer in Hygrostress Crack Formation and Propagation in Cylindrical Elastoplastic Food," *Intl. J. Heat Mass Transfer,* 38(6): 1033–41.

21. Akiyama, T., H. Liu, and K.-I. Hayakawa. 1997. "Hygrostress—Multicrack Formation and Propagation in Cylindrical Viscoelastic Food Undergoing Heat and Mass Transfer Processes," *Intl. J. Heat Mass Transfer,* 40(7): 1601–1607.

22. Liu, H., L. Zhou, and K.-I. Hayakawa. 1997. "Sensitivity Analysis for Hygrostress Crack Formation in Cylindrical Food during Drying," *J. Food Sci.,* 62: 447–450.

23. Zogzas, N. P., Z. B. Maroulis, and D. Marinos-Kouris. 1994. "Densities, Shrinkage and Porosity of Some Vegetables during Air Drying," *Drying Technol.,* 12(7): 1653–1666.

24. Krokida, N. K. and Z. B. Maroulis. 1997. "Effect of the Drying Method on Shrinkage and Porosity," *Drying Technol.,* 15(10): 1145–1155.

25. Maroulis, Z. B., M. K. Krokida, and G. D. Saravacos. 1999. "Structural and Mechanical Properties of Dehydrated Products during Rehydration," presented at the Annual Meeting of IFT, paper No 8-8, Chicago, July 24–28.

26. Fito, P. 1994. "Modeling of Vacuum Osmotic Dehydration of Food," *J. Food Eng.,* 22: 313–325.

27. Barat, J. M., A. Albors, A. Chiralt, and P. Fito. 1998. "Equilibrium of Apple Tissue in Osmotic Dehydration. Microstructural Changes," Proceedings of the 11th International Drying Symposium, Halkidiki, Greece, August 19–22, vol. A, pp. 827–835.

28. Roos, Y. 1992. "Phase Transitions and Transformations in Food Systems," in *Handbook of Food Engineering.* New York: Marcel Dekker, pp. 145–197.

29. Karathanos, V. T., S. A. Anglea, and M. Karel. 1996. "Structure and Collapse of Plant Materials during Freeze Drying," *J. Thermal Analysis,* 46(5): 1541–1555.

30. Karathanos, V. T., N. K. Kanellopoulos, and V. G. Belessiotis. 1996. "Development of Porous Structure during Drying of Agricultural Plant Products," *J. Food Eng.,* 29(2): 167–183.

31. Aguilera, J. M. and D. W. Stanley. 1999. *Microstructural Principles of Food Processing and Engineering,* 2nd ed., Gaithersburg, MD: Aspen Publ.

32. Rassis, D., A. Nussinovitch, and I. S. Saguy. 1997. "Tailor-Made Porous Solid Foods," *Intl. J. Food Sci. Technol.,* 32: 271–278.

33. Rao, M. A. 1999. *Rheology of Fluid and Semisolid Foods,* Gaithersburg, MD: Aspen Publ.

34. Rao, M. A. 1995. "Rheological Properties of Fluid Foods," in *Engineering Properties of Foods,* M. A. Rao and S. S. H. Rizvi, eds. New York: Marcel Decker, pp. 1–53.

35. Saravacos, G. D. 1968. "Tube Viscometry of Fruit Purees and Juices," *Food Technol.,* 22:1585–88.

36. Saravacos, G. D. 1970. "Effect of Temperature on the Viscosity of Fruit Juices and Purees," *J. Food Sci.,* 35: 122–125.

37. Ibarz, A., C. González, and S. Esplugas. 1994. "Rheology of Clarified Fruit Juices. III. Orange Juices," *J. Food Eng.,* 21: 485–494.

38. Velez-Ruiz, J. F. and G. V. Barbosa-Cánovas. 1998. "Rheological Properties of Concentrated Milk as a Function of Concentration, Temperature and Storage Time," *J. Food Eng.,* 35: 177–190.

39. Kostaropoulos, A. E. 1971. *Warmeleitzahlen von Lebensmitteln und Methoden zu deren Bestimmung,* VDMA No. 6, Frankfurt, Germany: Maschinenbau-Verlag GmbH.

40. Maschinenbau Verlag, GmbH and P. Nesvadba. 1982. "Methods for Measurement of Thermal Conductivity and Diffusivity of Foodstuffs," *J. Food Eng.,* 1: 93–113.

41. Kostaropoulos, A. E. and G. D. Saravacos. 1997. "Thermal Diffusivity of Granular and Porous Foods at Low Moisture Content," *J. Food Eng.*, 33: 101–109.

42. Drouzas, A. E., Z. B. Maroulis, V. T. Karathanos, and G. D. Saravacos. 1991. "Direct and Indirect Determination of the Effective Thermal Diffusivity of Granular Starch," *J. Food Eng.*, 13: 91–101.

43. Maroulis, Z. B., A. E. Drouzas, and G. D. Saravacos. 1990. "Modeling of Thermal Conductivity of Granular Starches," *J. Food Eng.*, 11: 226.

44. Maroulis, Z. B., K. K. Shah, and G. D. Saravacos. 1991. "Thermal Conductivity of Gelatinized Starch," *J. Food Sci.*, 56: 773–776.

45. Saravacos, G. D. and M. N. Pilsworth. 1965. "Thermal Conductivity of Freeze-Dried Model Food Gels," *J. Food Sci.*, 30: 773–778.

46. Theodorou, D. N. 1996. "Molecular Simulations of Sorption and Diffusion in Amorphous Polymers," in *Diffusion in Polymers*, P. Neogi, ed., New York: Marcel Dekker., pp. 67–142.

47. Saravacos, G. D. 1995. "Mass Transfer Properties of Foods," in *Engineering Properties of Foods,* M. A. Rao and S. S. H. Rizvi, eds. New York: Marcel Dekker, pp. 169–221.

48. Zogzas, N. P., Z. B. Maroulis, and D. Marinos-Kouris. 1994. "Moisture Diffusivity Methods of Experimental Determination. A Review," *Drying Technol.*, 12: 483–515.

49. Marousis, S. N., V. T. Karathanos, and G. D. Saravacos. 1991. "Effect of Physical Structure of Starch Materials on Water Diffusivity," *J. Food Proc. Preserv.*, 15: 183–195.

50. Saravacos, G. D. 1996. "Moisture Transport Properties of Foods," in *Advances in Food Engineering,* CoFE 4, G. Narsimham, M. R. Okos, and S. Lombardo, eds. Purdue Univ.

51. Kiranoudis, C. T., Z. B. Maroulis, and D. Marinos-Kouris. 1995. "Heat and Mass Transfer Model Building in the Drying with Multidisperse Data," *Intl. J. Heat and Mass Transfer*, 38: 463–480.

52. Marinos-Kouris, D. and Z. B. Maroulis. 1994. "Thermophysical Properties in the Air-Drying of Solids," in *Handbook of Industrial Drying*, A. S. Mujumdar, ed. New York: Marcel Dekker.

53. Rotstein, E. 1987. "The Prediction of Diffusivity and Diffusion-Related Transport Properties in the Drying of Cellular Foods," in *Physical Properties of Foods—2*, R. Jowitt, F. Escher, M. Kent, B. McKenna, and M. Roques, eds. London: Elsevier Applied Science.

54. Doulia, D., K. Tzia, and V. Gekas. 1999. "Knowledge Base for the Apparent Mass Diffusion Coefficient (D_{eff}) of Foods," *Intl. J. of Food Properties* (in press).

55. Drusas, A. E., G. K. Vagenas, and G. D. Saravacos. 1988. "Diffusion of Sodium Chloride in Green Olives," *J. Food Eng.*, 7: 211–222.

56. Lazarides, H. N., V. Gekas, and N. Mavroudis. 1997. "Apparent Mass Diffusivities in Fruit and Vegetable Tissues Undergoing Osmotic Processing," *J. Food Eng.*, 31: 315–324.

57. Krochta, J. M., E. A. Baldwin, and M. Nisperos-Cacieda, eds. 1994. *Edible Films and Coatings to Improve Food Quality,* Lancaster, PA: Technomic Publishing Co., Inc.

58. Hernández, R. J. 1997. "Food Packaging Materials, Barrier Properties, and Selection," in *Handbook of Food Engineering Practice*, K. J. Valentas, E. Rotstein, and R. P. Singh, eds. New York: CRC Press, pp. 291–360.

12 Mass Transfer and Related Phenomena in Plant Tissue during Heat Treatment and Osmotic Stress

W. E. L. Spiess
D. Behsnilian
M. Ferrando
M. Gaiser
U. Gärtner
M. Hasch
E. Mayer-Miebach
A. Rathjen
E. Walz

CONTENTS

Introduction

In classical engineering, consideration of the process as such and the unit operations involved were of prime interest; heat and mass transfer were mainly considered with respect to the phenomena in the interfaces between the product and the surrounding medium. The product played a role only insofar as its geometry and other physical factors had an impact on the process. In this chapter, emphasis is given to the product and to possible changes that affect mass transfer from the product into the surrounding media, and vice versa, during heating and exposure to osmotic solutions. The considerations are based on work that has been carried out in the last six years at the Federal Research Centre of Nutrition, Institute of Process Engineering, in Karlsruhe. The main interest in the work arose from the fact that, in food science and nutrition, more and more emphasis is put on the role of vegetables in a healthy diet. In this context, the rather controversial discussion of the degree to which vegetables should be heat treated also plays a role. On the other hand, there is a school of thought that asks for non-denatured produce, which allows the food to be as close to nature as possible; on the other hand, there is the understanding that through heat treatment, the bioavailability of relevant components can be considerably improved.[1] Despite the fact that sensory sensations of raw vegetables and fruits received during consumption are highly appreciated by most consumers, the consumption of raw vegetables incurs the danger of hygienic risks. Heat treatment in aqueous environments to reduce microbial contamination not only improves digestibility, it leads to the loss of crunchiness and other desired quality attributes, and the leaching of water-soluble substances. This leaching can be considered an undesired effect in the case of minerals or vitamins; it is, however, an advantage in the case of substances considered anti-nutritive, such as nitrate or oxalate.

The desire for biologically intact and hygienic material, especially fruits and fruit products, has led to the development of processes like "high-pressure treatment," "pulsed electric fields," "high-intensity light pulses," and so on, but also to processes like the "sous-vide treatment" or "extremely high temperature short time processing."[2] At the Karlsruhe Nutrition Research Centre, projects have been completed that allowed the microbial contamination of green salads to be reduced without affecting their biological integrity and, to give another example, that allowed the nitrate concentration of processed spinach and similar products to be reduced with a minimized loss of other water-soluble substances.[3,4]

12.1 THE PLANT CELL

To understand some of the phenomena involved, the plant cell must be considered. In a very simplified view (Figure 12.1), the cell consists of the cell wall, the plasmalemma or cell membrane surrounding the protoplast, the nucleus and other cell-organs embedded in the cytoplasm, and the cell vacuole separated from the cytoplasm by the vascular membrane, the tonoplast. In the case of a living cell, the cell wall, the plasmalemma, as well as other cellular membranes as the tonoplast, are semipermeable, allowing, besides the apoplasmatic transport in the cell, only transmembrane transport processes through the membrane which are osmotically controlled. Through the plasmodesma (200 nm dia.), the symplasmatic transport takes place so that normal diffusional or bulk flow processes are also possible within a tissue Figure 12.2.

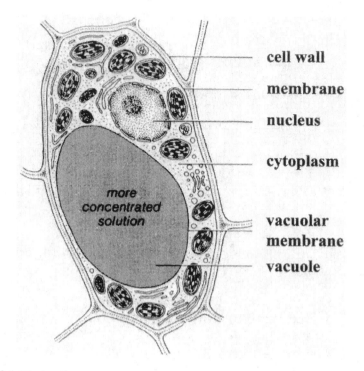

Figure 12.1 Plant cell.

The various biological transport processes within the plant cell and across the membranes are not yet entirely understood. Under normal environmental conditions, plant material is protected against any type of loss other than the loss of water (Figure 12.3). To make the cell content more easily available during digestion, or to extract material from a cell or a complex tissue, either an osmotic stress must be exercised or the relevant membranes have to be denatured in a way that the cell content is released. The denaturation can be done by heat treatment or by chemical treatments, e.g., acidification, salination, etc.

12.2 PROCESSING OF PLANT MATERIAL IN AQUEOUS ENVIRONMENTS

To produce hygienically satisfactory fruit and vegetable products, the raw materials are subjected to a sequence of processes including washing, cutting, trimming, and blanching, if required. Blanching may be followed by freezing, drying, or canning. The washing process is usually carried out with tap water; in certain cases, the addition of safe detergents, salts, or organic acids is observed to improve the cleaning effect and increase the reduction of microorganisms adhering to the product. The latter can also be supported by elevated temperatures of the washing media. An upper temperature limit, however, is given by the loss of the semipermeability of

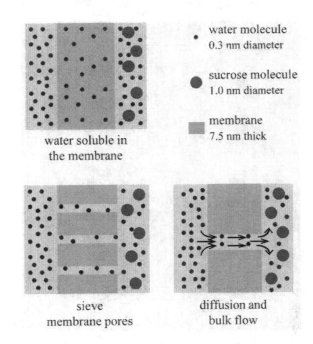

Figure 12.2 Movement of water across membranes.

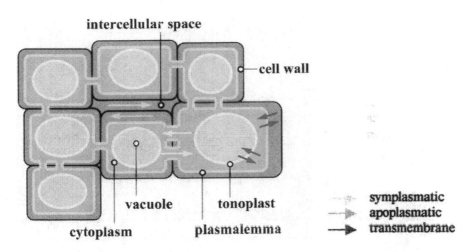

Figure 12.3 Transport mechanisms in plant tissue.

the cell membrane and related properties. The temperature limit at which the permeability is lost must be considered as the minimum temperature for the extraction of components of the cell content. In the following text, two stress situations that may lead to transport process across the cell membrane and the cell wall will be considered.

12.2.1 Exposure of Plant Cells to Osmotic Stress

When living plant cells are exposed to an environment with a higher osmotic pressure, water is drained from the cell into the system with the higher pressure so as to reduce the pressure. The loss of water is associated with a shrinkage process (Figures 12.4, 12.5, and 12.6).

Because the kinetics of the water transport across the system of semipermeable membranes are obviously related to the number of membranes that pass, the rate is higher in the case of protoplasts than for a tissue system with complete cells. It is also obvious that the loss of water does not necessarily lead to a loss of viability.

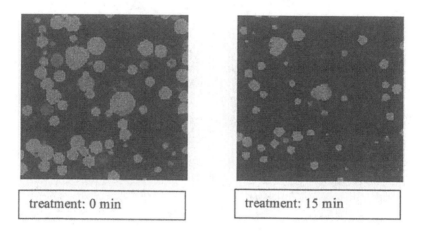

| treatment: 0 min | treatment: 15 min |

Figure 12.4 Carrot protoplast after 40% sucrose treatment.

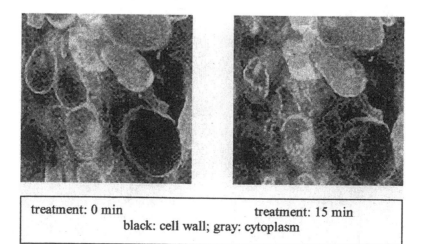

treatment: 0 min   treatment: 15 min
black: cell wall; gray: cytoplasm

Figure 12.5 Strawberry cortex tissue after 40% sucrose treatment.

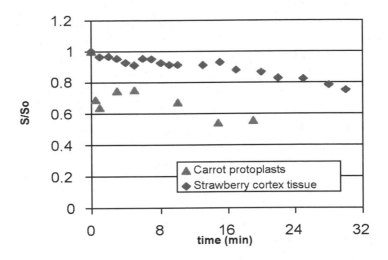

Figure 12.6 Shrinkage (S/So) of protoplast and cortex tissue in 40% sucrose solution.

A precise definition of lethal osmotic pressures is, however, rather difficult, because different types of cells and cells of different origin react in different ways (Figure 12.7).[5]

12.2.2 Exposure of Plant Cells to Elevated Temperatures

When plant cells or tissues are exposed to an isotonic aqueous environment with a temperature above ambient, a gradual loss of the semipermeability of the cell mem-

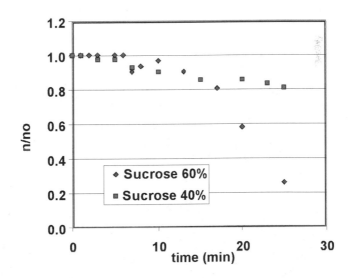

Figure 12.7 Living cell ratio (n/no) of strawberry tissue.

brane system may be observed. This loss of semipermeability can be experienced, e.g., through the release of minerals and organic substances into the surrounding media. Most sensitive is the loss of minerals, which increases the electrical conductivity of the aqueous heating media; the critical temperature for most produce is between 50 and 60°C. The effect as such is time dependent (Figures 12.8 and 12.9).

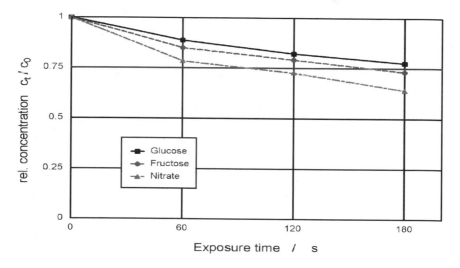

Figure 12.8 Extraction of water-soluble substances (spinach) at 90°C.

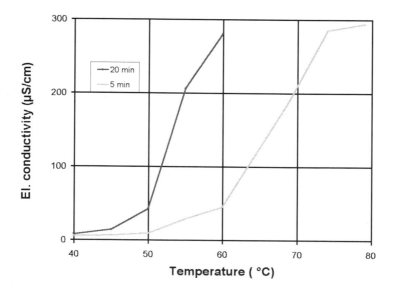

Figure 12.9 Electrical conductivity of blanching media.

Initial observations revealed that the minerals are leached first, followed by water-soluble carbohydrates (sugar). As experiments with spinach have demonstrated, the electrical conductivity starts to rise at water bath temperature of 55°C after 5 min, whereas the organic carbon content in the water bath rises only after 15 min. It must be acknowledged, however, that the extraction rate is dependent on the physiological state of the produce in question, its mechanical integrity, the organs of the plant involved (leaf, stalk, root), and the species under investigation (Figures 12.10 through 12.12). Associated with the drain of water-soluble substances are

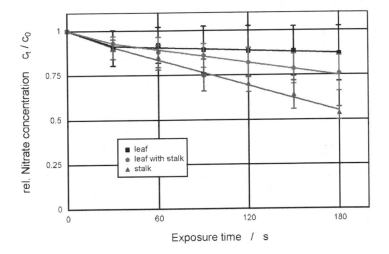

Figure 12.10 Nitrate extraction of cut vegetable material (80°C).

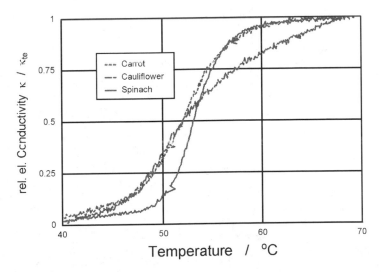

Figure 12.11 Thermal stability of vegetables.

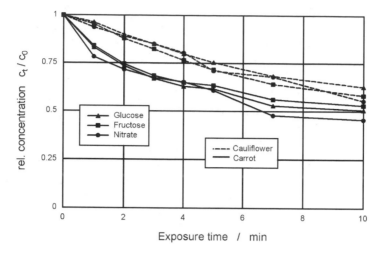

Figure 12.12 Extraction of water-soluble substances.

reactions in the plant material itself, such as structural changes in the cell wall that may lead to an increase in firmness immediately after heat exposure and to oxidative reactions expressed in a loss of vitamin C, so that the decrease in vitamin C and the loss of the antioxidant power of the processed materials is not only caused by leaching effects (Figures 12.13 and 12.14).

The loss of quality becomes more obvious after a storage period of two days (Figure 12.15). Besides the improvement of the hygienic status of the material, which

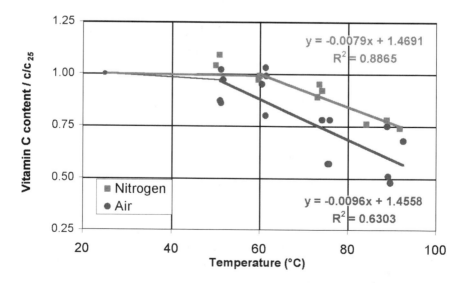

Figure 12.13 Vitamin C concentration in spinach after blanching (exp. time 1.5 min).

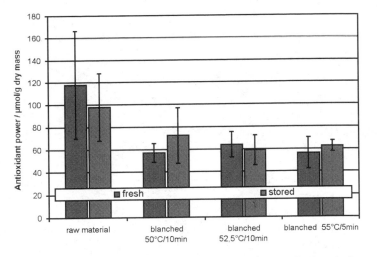

Figure 12.14 Antioxidant power of processed green salad (endive), fresh and after two days.

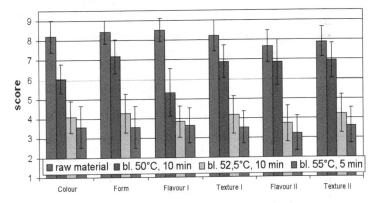

Figure 12.15 Sensory quality of blanched green salad after two days of storage.

has not been discussed in the course of this chapter, changes in the product micro-structure caused by the heat treatment led to changes associated with the loss of semipermeability. This loss of semipermeability also leads through the uptake of oxygen to oxidative reactions and through the internal mass transfer process to reactions between the constituents of the cell.

12.2.3 MODELING THE EXPOSURE OF PLANT CELLS TO ELEVATED TEMPERATURE

The mass-transfer phenomena taking place in plant systems on exposure to elevated temperature in an aqueous environment are rather pronounced in the case of blanching, where the water temperature is, in general, above 90°C. In a blanching unit operated at 95°C, an extraction of all water-soluble substances contained in a plant cell can be observed. This extraction is concentration controlled so that, in principle, in a freshwater operation, almost all soluble substances can be extracted (Figures 12.16 and 12.17). If the blanching media is not renewed or is recycled after a cleaning operation, the concentration of soluble substances runs into an equilibrium situation, the level of which is dependent on the amount of blanching media and throughput. The extractions of the various substances at elevated temperatures follow Fick's law of diffusion after a convection phase so that the extraction mechanism and the concentration of the extracted material in the blanching can be described through physical parameters (Figures 12.18 and 12.19).

A mathematical model (Figures 12.20 through 12.22) of the situation developed on the basis of empirical results and a rigorous application of Fick's law for an unsteady state situation allows calculation of the concentration built up for all water-soluble substances involved. The model calculation for vitamin C demonstrates that the actual concentration in the spinach is slightly lower than that calculated due to oxidative effects.

12.3 FINAL CONSIDERATIONS

The comprehensive description of various projects carried out to study the possibilities of improving the quality of plant material through osmotic and heat treatment has revealed that osmotic and thermal treatments of green plants produce a variety of product conditions of technical interest. In the case of osmotic treatment, it was observed that exposure to environments with a moderate osmotic pressure for cleaning or other technical purposes does not necessarily lead to a loss of viability, only

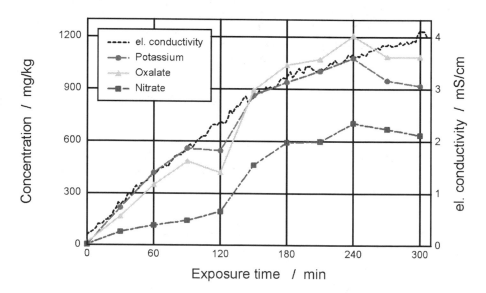

Figure 12.16 Extraction of water-soluble substances (pilot plant, freshwater operation).

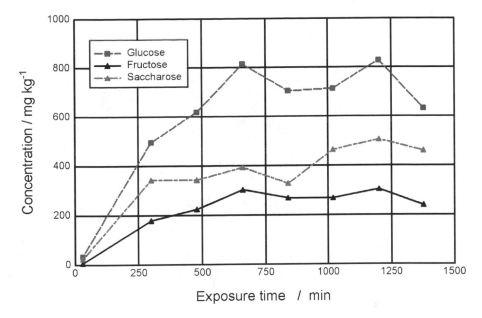

Figure 12.17 Extraction of sugars in blanching medium.

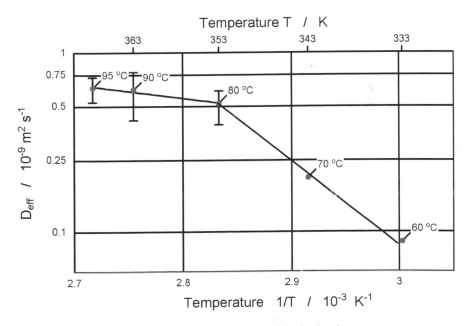

Figure 12.18 Diffusion coefficient of extracted material of spinach.

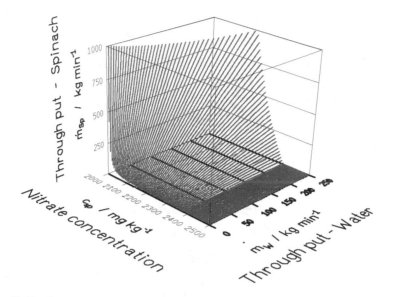

Figure 12.19 Nitrate extraction in a continuous blanching unit (1 min, 90°C).

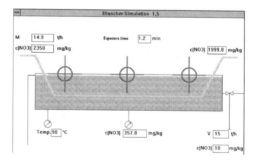

Figure 12.20 Simulation of blanching process.

to an exposure to very high osmotic pressures (corresponding to $a_w \sim 0{,}9$), and a long exposure time causes a total loss of viability.

With respect to thermal treatment, two temperature regimes can be recognized:

- *Temperatures below 50 to 60°C.* In this temperature range, the semipermeability of plant membranes of different natures is hardly affected. The plant materials exposed to this temperature do not allow a leaching of cell content; extraction of water is possible only through osmotic processes. The exact critical temperature at which denaturation of membranes is initiated depends very much on the species, on the organs involved, and on the physiological states of the raw materials. The temperatures immediately below the critical temperature are of special interest for the production of hygienically safe raw vegetable products.

$$c_{Sp,out} = \frac{c_{Sp,in} \cdot Z_1 + c_{W,in} \cdot Z_2}{N}$$

$$Z_1 = (1-f) \cdot \exp\{-\beta a t_b\} \cdot (1 - \frac{\dot{m}_{Sp}}{\dot{m}_W}) + \frac{\dot{m}_{Sp}}{\dot{m}_W}$$

$$Z_2 = 1 - (1-f) \cdot \exp\{-\beta a t_b\}$$

$$N = 1 - (1-f) \cdot \exp\{-\beta a t_b\} \cdot (\frac{\dot{m}_{Sp}}{\dot{m}_W}) + \frac{\dot{m}_{Sp}}{\dot{m}_W}$$

$$c_{W,out} = \frac{\dot{m}_{Sp}}{\dot{m}_W} \cdot (c_{Sp,in} - c_{Sp,out}) + c_{W,in}$$

Figure 12.21 Mathematical model to evaluate the extraction of components during blanching. In this model, c (mg/kg) is concentration (mass related), f is washout factor (dimensionless) β (m/s) is a mass transfer coefficient, a (m²/m³) is a surface/volume quotient, tb (s) is blanching time, m (kg/s) is mass flux, Z is a numerator, and N is a denominator. Sp refers to spinach, W to water, in to the beginning, and out to the exit.

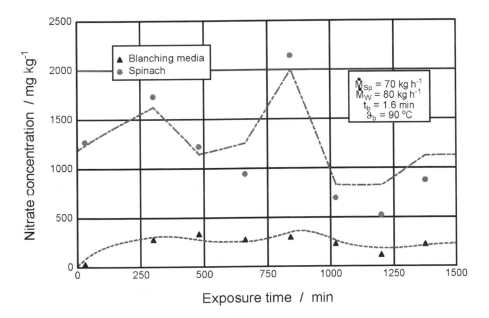

Figure 12.22 Nitrate concentration in blanching and in spinach.

- *Temperatures above 60°C.* In the temperature range above 60°C, the various membranes in a plant cell will be denatured in any case, and this leads to leaching processes primarily to a flux of cell content into its immediate environment, and, in certain cases, foreign materials penetrate into the cell as oxygen or salts. The leaching processes can be described with rigorous physical models and can be modeled for industrial application such as the operation of a blanching unit. This temperature range is important with respect to improving the biological availability of physiologically relevant substances and the extraction of anti-nutritive substances.

REFERENCES

1. Gärtner, C., W. Stahl, and H. Sies. 1997. "Lycopene is More Bioavailable from Tomato Paste than from Fresh Tomatoes," *Am. Journal Clin. Nutrition*, 66: 116–122.
2. Walz, E. and W. E. L. Spieß. Moderne Verfahren der Lebensmittelbe- und verarbeitung Verbraucherdienst 42–2/97.
3. Weisbrodt, Th. 1999. Untersuchungen zur Verminderung des Nitrat- und Oxalatgehaltes in Gemüseprodukten und Blanchiermedien durch selektiven Ionenaustausch; Diss. Universität Hohenheim.
4. Gärtner, U. 2006 Modifizierung konventioneller Blanchierverfahren zur Reduktion der mikrobiellen Oberflächenkontamination sowie zur gleichzeitigen Verminderung der Verluste ernährungsphysiologisch wertvoller Inhaltsstoffe, Diss. Universität Hohenheim—In preparation.
5. Ferrando, M. and W. E. L. Spiess. 1998. "Use of Cellular Systems to Characterise Plant Tissue Behaviour," *in Proceedings of the 3rd Karlsruher Nutrition Conference*, 18–20 October, 2, 337, ISDN 0933–5463.

13 Non-Fickian Mass Transfer in Fruit Tissue

V. Gekas
F. A. R. Oliveira
G. H. Crapiste

CONTENTS

Abstract

Many fruit processing unit operations involve the application of heat or immersion of the fruit in different types of solutions, causing significant mass transfer of various food components as well as of exogen solutes. The mass transfer process may have important consequences on food texture, nutritional value, and quality in general. Application of minimal processing to obtain convenience fresh-like foods with extended shelf life is gaining interest worldwide, and this type of process often involves mass transfer in living tissues. A better understanding of mass transfer phenomena is crucial for controlling and optimizing food processing. The increasing awareness of this importance is revealed by the large amount of research in this field in recent years. The current state of the art is still dominated by the Fickian approach (implying use of lumped effective diffusion coefficients).

1-56676-963-9/02/$0.00+$1.50

However, transport in living tissues requires basic knowledge of biology, as the cell structure plays a major role in the transport mechanisms. On the other hand, evidence supplied from the food-processing area shows changes in the pattern of mass transfer in some species with temperature and environmental conditions that may be intimately related to phenomena at the cell level. Therefore, the function, the mechanisms of transport of the biological membrane, and its fate during various processing conditions are most important for the food process engineer. This review covers cases of non-Fickian behavior, including convection phenomena due to turgor and to buoyancy, and agitation effects, passive membrane transport, and active membrane transport. The possibility of application of the thermodynamic approach as a basis for mass transfer modeling when mechanisms other than Fickian diffusion are present is covered as well.

13.1 INTRODUCTION

Food processing implies a wide spectrum of operating temperatures, from freezing up to cooking and frying temperatures, and sometimes extreme conditions of other environmental factors such as ambient humidity, osmotic pressure, or water chemical potential, which are quite different from physiological conditions. These conditions induce mass transfer of water and other components that is quite important in food-processing unit operations such as drying, blanching, osmotic treatment (dehydration or impregnation), cooking, and frying, because of its influence on texture and overall quality. The material is divided into three parts concerning the effects of low-, intermediate-, and high-temperature conditions, respectively.

Mass transfer elucidation, in general, requires the knowledge of structure aspects and driving force aspects. Before addressing the processing aspects, we anticipate two parts on structure and driving force, starting with cell walls, which are the first internal constraints of the system, hindering the free diffusion of certain components—namely, macromolecules. Living cell membranes present special mechanisms of mediated and active transport that enable the metabolic activities of the cells and of the whole plant organism. A number of textbooks cover these subjects thoroughly; therefore, in the first part of this review, we intend to provide only an engineering-oriented short description. We also propose a classification scheme for transport processes to clarify terminology that sometimes leads to confusion.

We discuss the driving force aspects following the thermodynamic approach as being the most plausibly fundamental. In the thermodynamic ensembles of Gibbs and Callen, the very important concept of chemical potential is derived from either G, the free enthalpy function, or from S, the entropy function. The Gibbs ensemble, using measurable parameters such as temperature T, pressure P, and number of moles n_i of any component i to be the thermodynamic coordinates, leads to the definition of measurable thermodynamic quantities, e.g., of the chemical potential. The entropy ensemble, taking extensive properties as thermodynamic coordinates, leads to the deduction of all important driving forces and provides the basis for the theory of irreversible thermodynamics. The coupling of fluxes and driving forces could potentially explain qualitatively and quantitatively the facilitated and active transport phenomena. Finally, needs for further research are identified.

13.2 STRUCTURE ASPECTS

13.2.1 THE CELL WALL AS AN ULTRAFILTRATION MEMBRANE

From the mass transfer point of view, the cell wall behaves as an ultrafiltration (UF) membrane that hinders the entrance of foreign particulate matter, viruses, etc., into the cell. It also controls the transport of macromolecules, such as proteins, from and to the cell. Like a UF membrane, the cell wall allows water, sugars, and salts to pass freely, while macromolecules are hindered,[1,2] except in the case of the symplastic mode of direct cell-to-cell transfer of macromolecules through the plasmodesmata. In the latter case, common in mass transfer from a mother cell to the daughter cells, the hindering effect of the cellular membranes is also absent.[3] The cell wall, therefore, acts as a UF membrane in the apoplastic mode, i.e., cell wall to cell wall, eventually involving the plasmalemma membrane, with the singularity of symplastic transport of macromolecules through the plasmodesmata. Plasmodesmata are tortuous channels with a diameter of 30–60 nm, lined by plasma membrane at the periphery, and containing the desmotubule (tubular plasma membrane). Cell walls typically have 1–15 of these channels per square micrometer.

As with any UF membrane, data and properties of interest are thickness, constituting material (charge and hydrophilic/hydrophobic properties), pore size, and cutoff.[1,2] Thickness is basically determined by the primary wall that ranges from 100–1000 nm, because the other two parts of the intermediate lamella (less than 30 nm) and the second wall (which is found in specialized cells only, with the main function of cell type recognition) are much thinner. Because of the cellulosic nature of the constituting material, cell walls are normally hydrophilic, but they may become hydrophobic in some cases (lignification, encrustation on the epidermis/endodermis). Lignification may also affect (reduce) water and dissolved solute permeability. The (net) charge of the cell wall is negative, due to uronic acid residues of pectin and xylans.

The pore size (3.5–5.5 nm) is of the same order of magnitude of typical pore sizes of a "good," i.e., sharp, cutoff UF membrane with cutoff values between 10 and 50 kDa, depending on the size and shape of the macromolecule. Similar values are shown by synthetic polymeric or inorganic UF membranes. Using the concept of hydrodynamic volume value for polymers, however, it is possible to obtain a unique cutoff value for macromolecules such as dextrans, PEGs, and proteins.[4] For the sake of simplicity, some cells specialized in secreting larger macromolecules are not considered in this review. Our main interest is "normal" cells of the parenchymatic tissue, like the ones in potato tubers or in fruits such as apples.

Mass transfer modeling takes into account both the structure of the medium across which the mass transfer occurs and the applied driving force. In commercial ultrafiltration or microfiltration membranes, the applied driving force is a pressure difference. However, in almost all cases, secondary concentration differences also play an important role, and the chemical potential of the component studied becomes the driving force for the transfer of that component. It is known that, in the extended form, the chemical potential includes both pressure and concentration gradients, and even electrical potential gradients, in the case of ions or charged molecules (known in the last case also as *electrochemical potential*).[5] This is the universal driving force for mass transfer and is also valid in the living cell, except in the case of active transport.[6]

Because the cell wall acts as a UF membrane separating macromolecules that are hindered from passing and microsolutes (ions, sugars, organic acids) that move freely both to and from the cell, the resistance of the cell wall to the transport of small solutes is negligible when compared with cell membrane transport. An exception occurs when specific interactions are established between a small solute and the cell wall (as, for example, calcium, participating in reactions with pectin molecules in the middle lamella). On the other hand, macromolecules represent the other extreme case of infinite cell wall resistance (or zero permeability). Although the cutoff values reported for the cell wall indicate the possibility of movement of macromolecules with low molecular mass, we will not consider these cases, as they touch applications outside the interest of this review. Water should also be mentioned, as it plays an interesting role due to the *matric* potential created by bound water in the cell wall.

13.2.2 CELL MEMBRANES

Description

Every living cell is surrounded by a cell membrane, the plasma membrane, which defines the periphery of the cell, separating its content from its surroundings. Even organelles inside of the cell are surrounded by membranes. The membranes are composed of a bilayer of polar lipids, mostly phospholipids, but also include other components such as cholesterol and triacyl glycerol. Both the internal and the external surfaces of a membrane are hydrophilic (because of the polar phospholipid heads). The fatty acetyl chains in the interior of the membrane form a fluid, hydrophobic region. Integral membrane proteins float in this sea of lipid, held by hydrophobic interactions in their nonpolar amino acid side chains. Both proteins and lipids are free to move laterally in the plane of the bilayer. In a plant cell, there is a also a vacuole, which is surrounded by a membrane, the tonoplast. The vacuole represents, especially in mature cell, a big part of the cell volume, sometimes as much as 90%. It contains ions, metabolites, and digestive enzymes that degrade and recycle macromolecular components no longer useful to the cell. Because of the high solute concentration in the vacuole, water passes into it by osmosis. This mechanism creates a pressure on the cytosol and on the cell wall. This turgor pressure within the cells stiffens the plant tissue, so the vacuole provides a physical support to the plant cell.

Data reported on turgor pressure (estimated or modeled values) for roots, in general, fall in the range of 0.432–0.534 MPa; barley roots (measured values) show a narrower range of 0.31–0.50 MPa. In general, 0.5 MPa or 5 bar seems to be a typical order of magnitude for full turgor values. Several procedures may be used to measure tissue osmotic pressure, such as pressure volume (PV) method (theoretically more sound, but time consuming), methods based on sap expression (rapid but with errors due to cell disrupture), and other methods based on measurements of water chemical potential.

Classification of Transport Mechanisms

Some proteins are constructed as channels for ion transport. Their amino acid sequences span the membrane several times, forming a transmembrane channel. The basal permeability, i.e., the permeability of the phospholipidic phase, is usually expressed in the same units as synthetic polymeric or inorganic membranes, i.e., as fluxes divided by the driving force. On the other hand, the permeability of channels is given as conductance or as numbers of ions transported through a channel per second. The same format is used for mediating protein transporters, both passive and active. Active transport uses energy to transport species against the electrochemical potential gradient. Most primary active transporters use ATP, but some use light or substrate oxidation. Secondary transport occurs when uphill transport of one solute is coupled to the downhill flow of a different solute (the lactose symporter from *Escherichia coli* is a typical example). This transporter utilizes the proton electrochemical gradient generated by the respiratory electron transport chain to drive the uptake of lactose into the cell. In this case, two different solutes (protons and lactose) are simultaneously transferred across the membrane; therefore, the term symporter is applied. Antiporters, on the other hand, couple the transport of solutes in opposite directions, for example, the Band 3 from erythrocyte, responsible for the transport of Cl^- and HCO_3^- in opposite directions across the red blood cell membrane.[7] Permeases, translocases, and carriers are other terms found in the literature applied to protein transporters other than primary active transporters. Engineers should also have in mind the terms high-affinity and low-affinity transporters. This level of affinity is quantified by the Km values and may be described in analogy to enzymes, as the affinity of the protein transporters to the substrates they transport.

Water passes through channels and pores in the membrane and also diffuses, to some extent, through the phospholipidic phase (simple diffusion). Some dissolved gases (e.g., O_2, N_2, and CH_4) are transported by simple diffusion through the phospholipidic phase. This transport could be explained by the hydrophobic nature of the phospholipidic phase, once gas solubilization is favored in this phase. However, for the transport of oxygen, mediated transport has been reported as well. Some small organic molecules, e.g., urea, and some organic acids (e.g., acetic acid, lactic acid, salicylic acid) also pass through the phospholipidic phase. Amino acids and sugars are transported by active secondary symport transport in the intestine by coupling to the flow of sodium. In bacteria, lactose is transported in the same way, using protons instead of sodium. Glucose enters the erythrocyte by facilitated diffusion via a specific glucose permease, which allows glucose to enter into the cell at a rate about 50,000 times greater than its unaided diffusion through a lipid layer.

Glucose also has been found to be transported through facilitated transport in plant cells.[8] Chloride (Cl^-) and bicarbonate (HCO_3^-) are cotransported across the erythrocyte membrane in an antiport process. The chloride-bicarbonate exchanger increases the permeability of the membrane to bicarbonate by a factor of more than a million. Calcium (Ca^{2+}) is transported into the cells in vertebrate animals to keep the calcium concentration low inside of the cell. This is an antiport process with three sodium ions for each calcium ion, and it uses the transmembrane Na^+ gradient. The free Ca^{2+} of the cytoplasm in animal cells has been reported to be very low, between 10^{-6} and 10^{-8} M depending on the cell type. This low level of free Ca^{2+} has to be maintained against a Ca^{2+} concentration in the extracellular fluid, which is a few orders of magnitude higher.

Studies with intact red blood cells have shown that Ca^{2+} is expelled from the cytoplasm by an active energy consuming transport mechanism through the plasma membrane. Similar experiments with plant cells (tobacco protoplasts, carrot cells, pea epicotyls, onion roots) have provided evidence that, similar to animal cells, plant cells maintain a low cytoplasmatic concentration of free Ca^{2+}, mainly by an active efflux mechanism in the plasma membrane. Intracellular organelles and compartments also take part in the regulation of the cytoplasmatic concentration of free Ca^{2+}.[9] Potassium is the most abundant cation, in higher plants, and is crucial for nutrition growth tropism and osmoregulation. K^+ accumulation can be rate limiting for agricultural production. K^+ is also a key factor for nutrition value of foods.[10] Whereas Ca^{2+} is maintained at low concentrations in the cytoplasm, the uptake of K^+ by the cell guarantees high intracellular concentrations. This implies active transport, as first studied in animal cells. Potassium (K^+) is transported into cells of animals by an active antiport cotransport with sodium (Na^+). The enzyme $Na^+K^+ATPase$ couples breakdown of one ATP to the simultaneous movement of both three Na^+ and two K^+ ions, against their concentration gradients. This mechanism creates a higher potassium concentration inside the cell.[11]

The cell's membrane potential, the difference between the electrical potential inside and outside the cell, is partly determined by the $Na^+K^+ATPase$ and potassium channels. These channels regulate a wide range of functions, including salt and water flow from kidney cells and guard cells, insulin release from pancreatic β-cells, electrical excitability and synaptic plasticity in neurons, and even the rate at which the heart beats. Potassium channels of known sequence fall into two distantly related families, the voltage-gated and the so-called *inwardly rectifying channels*. These two classes of potassium channels share the same basic pore design. The activity of voltage-gated channels, which have six putative membrane-spanning domains in each subunit, is insensitive to changes in the membrane potential. These channels are typically activated at membrane potentials above the resting potential (the potential that causes no net flow of ions across the membrane, typically around −60 mV). In animals, this is slightly above the potassium equilibrium potential, while in plants, it is slightly below it. Inwardly rectifying potassium channels, by contrast, have only two putative transmembrane spanning domains in each subunit and are sensitive to changes in the potassium concentration. They are effectively unidirectional, as they allow a much larger potassium influx than efflux, allowing them to control the resting potential without causing massive potassium loss. All potassium channels are at least 100 times more permeable to potassium ions than to sodium ions and allow a flux greater than a million potassium ions per second. Because sodium has a smaller radius than potassium, it is unlikely that selectivity is achieved by physical occlusion (sieving effect). Other features, such as multiple binding sites arranged in a single file along each pore channel, may explain this selectivity.[12] K^+ channels were first studied in animal cellular tissues, but evidence of the existence of such channels in plants has also been reported. Analogies between the mechanisms of animal and plant K^+ channels have also been found. For example, it was shown that the blockade of the K^+ channel of *Chara contraria* by Cs^+ and tetraethylammonium resembles that of K^+ channels in animal cells. In a recent publication concerning potassium uptake by roots of higher plants, it was found that the transport mechanism was based on a K^+–H^+ cotransporter. This way, when a plant is deprived of potassium, a high-affinity K^+ uptake takes place in the roots. Sodium (Na^+) transport is important for halophytic plants. Na^+ transport was studied in the marine euryhaline alga, *Enteromorpha intestinalis* in seawater (465 mM Na^+) and in low salinity medium [Artificial Cape Bank Spring Water (ACBSW), 25.5 mM Cl^-, 20.4 mM Na^+, 0.5 mM K^+]. Most of the Na^+ of the *Enteromorpha* tissue was bound to the fixed negative charges of the cell wall, and this binding has, in previous studies, led to great overestimates of the intracellular Na^+ of this plant. The Na^+ flux in *Enteromorpha* plants in seawater was about 3 nmol $m^{-2} s^{-1}$ and in low-salinity plants was about 0.2 nmol $m^{-2} s^{-1}$. Sodium in *Enteromorpha* is far from electrochemical equilibrium (more than -100 mV) in plants in both seawater and ACBSW medium so that Na^+ is actively excluded from the cells. The plasmalemma has a very low Na^+ permeability (seawater, 3 pm s^{-1}; ACBSW plants, either 3 or 100 pm s^{-1}, depending on the compartmentation model used).

Typical Transport Rates

Typical transport rates are found in textbooks, such as Reference 7, both for basal and specialized permeability. While basal permeability is expressed in units of fluxes divided by molar concentrations in ms^{-1}, specialized permeabilities are expressed, in literature, in number of ions or molecules transported per unit time and per transporter. To enable comparison, the units of basal permeabilities should be expressed in numbers of transported molecules or ions by dividing the former units by membrane surface for a typical apple plasmalemma cell membrane. In the case of ion transport, the permeabilities are sometimes expressed in conductance. Significant differences are apparent in the permeabilities of different species between the two main categories (basal and special) but also within each category. For the components allowed to pass the phospholipidic phase, the permeabilities are comparable to values of hindered diffusion in solids. The basal permeability of the ions and the sugars is two to three orders of magnitude lower than water or urea; that is, the phospholipidic phase can be considered almost impermeable to ions and sugars. Ion permeabilities in channels demonstrate comparable transport rates as water and small organic compounds in the phospholipidic phase (ca 10^8 ions per second per transporter). It should not be forgotten, however, that channel transport, as happens with all forms of protein-mediated transport, occurs only at fractions of time when the channels are open, a parameter that is difficult to know in practice. It is especially difficult to know this parameter for a plant tissue after harvesting and under processing conditions. Therefore, a strict comparison between the two types of permeability is not possible. A comparison within the special permeability between the transport rates of channels and transporters that change their conformation shows differences of many orders of magnitude (100–1000 molecules per second per transporter), the active transport rates being by far the slower ones (ca 30 per second per transporter).

13.3 DRIVING FORCE ASPECTS

13.3.1 CONCENTRATION APPROACHES: MICHAELIS–MENTEN VS. FICKIAN

Physiologists have modeled mediated (channeled, facilitated, and active) transport with regard to the influence of the concentration on the rate of transport. In these kinetic studies, it was shown that the protein transporter or carrier follows a typical Michaelis–Menten behavior similar to enzymatically catalyzed reactions.[8,13] The concentration of the transferred component (substrate) is noted by S, and the flux, known as transport velocity, by V. V_{max} and K_m have analogous meanings, V_{max} being, in this context, the maximum transport rate and K_m the concentration of the transferred component when V equals $V_{max}/2$. The kinetics of mediated transport may be described by

$$V = \frac{V_{max}S}{S + K_m} \tag{13.1}$$

and, if competitive inhibition occurs, by

$$V = \frac{V_{max}S}{S + K_m(1 + IK_i)} \tag{13.2}$$

where I and K_i stand for the concentration and the inhibition constant of the inhibitor. Using similar terminology, simple Fickian diffusion may be described by

$$V = K_d S \tag{13.3}$$

where K_d is a mass transfer coefficient corresponding to the ratio between the diffusion coefficient D and the membrane thickness L. Equations (13.1) through (13.3) imply steady state conditions. For steady state simultaneous diffusion and mediated transport (without inhibition), the kinetics equation becomes

$$V = \frac{V_{max}S}{S + K_m} + K_d S \qquad (13.4)$$

if uptake is considered through a cell membrane, as, for example, uptake of water through both the phospholipidic phase and through a protein carrier. Transport away from the cell could be similarly considered, and generalization of Equation (13.4) to this situation is straightforward. Low K_m values mean a high affinity of the protein transporter for the component being transported (the substrate), while high K_m values imply the opposite (low-affinity transport). When considering unsteady coupled diffusional and mediated transport, modeling may be based on the generalized transport equation using a source term, R, to account for the flux due to the mediated transport, in a way analogous to the approach followed for enzymatic reactions.[5]

13.3.2 IRREVERSIBLE THERMODYNAMICS

Wherever a membrane is present in a system, trials to model water and solute transport using a single diffusivity or permeability coefficient for each species have proved to be inadequate, because the coupling between fluxes and driving forces previewed by irreversible thermodynamics was not considered. There is an interactive effect of the driving forces for water and solutes on their movement. Thermodynamic driving forces are based on chemical potentials rather than on concentrations, thus including both composition forces (activities or concentrations) and pressure or even voltage potentials. Based on thermodynamic driving forces, the transport equations for biological membranes were developed for a typical system containing a solvent (water) and a solute.[14] The driving force was expressed in terms of chemical potentials of water and the solute, and using the entropy production theorem of irreversible thermodynamics (IT) and the Onsagers law of reciprocity,[15] the following equations and inequalities for the volumetric flux, J_V (conjugate to a pressure-driving force) and the diffusive flux, J_D, conjugate to the osmotic pressure or composition driving force, were proposed.

$$J_v = L_p \Delta P + L_{PD} \Delta \Pi$$

$$J_D = L_{DP} \Delta P + L_D \Delta \Pi \qquad (13.5)$$

For the phenomenological coefficients, entropy considerations require that

$$L_p > 0, \quad L_D > 0 \text{ and } L_p L_D > L_{pD}^2$$

(The cross-phenomenological coefficients L_{pD} and L_{Dp} are equal, by Onsagers law.) After a number of simplifications for dilute solutions, i.e., use of concentration instead of activity in the composition term, neglecting the solute contribution in the volumetric flux and using an average concentration C_{av} within the membrane, the two transport equations[5] yield the following equations:

$$J_v = L_p(\Delta P - \sigma \Delta \Pi)$$

$$J_s = L_s \Delta C_{av} + (1 - \sigma)J_v C_{av} \qquad (13.6)$$

If $\Delta \Pi$ instead of ΔC_{av} is used in the expression for solute flux, the latter becomes

$$J_s = \omega \, \Delta \Pi + (1 - \sigma) \, J_v C_{av}$$

L_p and L_s or ω have to be positive, whereas σ has to vary between 0 and 1, as a consequence of the entropy production term (or dissipation factor). In the above equations, three parameters are needed to phenomenologically describe a membrane: L_p is the water (volume) permeability of the membrane when a pressure is used as the driving force, L_s (or ω taking into account different units) is the diffusive permeability of the membrane toward the solute, and sigma (σ) is the reflection coefficient showing the deviation of the membrane behavior from ideal semipermeability. (If $\sigma = 1$, then the membrane is ideally semipermeable—it passes the water through, but the solute is totally reflected; if $\sigma = 0$, the membrane does not discriminate between water and solute.)

The IT model has had success and has also been extended to polymeric reverse osmosis membranes and even to ultrafiltration membranes. It has been shown that the application of the IT model in UF yields results similar to those of Fick's law and the Maxwell–Stefan model.[16] An example of applications of the IT model in plant physiology is given[17] for organic solute uptake by plants. In this interesting work, the fate of an exogen organic substance being uptaken by the roots of a plant is modeled, considering division of the plant in various compartments (xylem, phloem, symplast, and apoplast found in the roots, the stem, and leaves), each one surrounded by a membrane with typical values toward water and the exogen solute. When solving the IT equations, they simulated the compartmentalization of the exogen substance in the plant as a function of time. The IT is also useful in the case of membranes being selective to ions, where electrical potential driving forces also appear. Some authors use the term *electrochemical potential* to denote the chemical potential concept extended by the addition of an electric potential term.[5] If a membrane is subjected to two driving forces, pressure and electric potential, then the application of IT equations would be given by

$$J_V = L_{11}\Delta P + L_{12}\Delta V$$

$$I = L_{21}\Delta P + L_{22}\Delta V \qquad (13.7)$$

If the conjugate flux of the electric potential is a current, I, and the conjugate flux of pressure is the volumetric flux, as above in the Kedem–Katchalsky approach, then the phenomenological coefficients L_{11} and L_{22} are the hydraulic permeability and the reciprocal electric resistance, respectively.[18] How could it be possible to include the mediated transport in the IT approach? In general, as stated above, one possibility might be to use the generalized transport equation (from the second law of Fick) and include the mediated transport rate in the source term as if it were a reaction term.

Using the IT approach, the action of the protein carrier or transporter could be coupled to the transport of the solute. Acting driving forces should be the driving force of diffusion, which is the chemical potential of the solute and the affinity, A, of the protein carrier's reaction.

Let us define affinity. In the entropy ensemble, Prigogine gives the following expression for the entropy production term, per unit time, s, for a system into which heat transfer, mass transfer, and chemical reaction occur:[15]

$$s = 1/T \cdot d\Phi/dt + \mu_\gamma/T \cdot dn_\gamma/dt + A/T \cdot d\xi/dt \qquad (13.8)$$

In this equation, the generalized driving forces are $1/T$ for heat transfer, μ_γ/T for mass transfer, and A/T for chemical reaction, where T, μ_γ, and A stand for temperature, chemical potential of a given component γ, and chemical affinity, respectively. The fluxes are $d\Phi/dt$ for heat transfer, dn_γ/dt for mass transfer (n number of moles of the component γ), and $V = d\xi/dt$ for the chemical reaction, ξ being the extent of the reaction. The affinity A of a reaction in general is related to the stoichiometric coefficients and the chemical potentials of the reaction components as follows:

$$A = \sum v_l \mu_l \qquad (13.9)$$

Ideas also exist for coping with active transport, where coupling with the reaction of ATP hydrolysis could be considered. The solute's flux against its own chemical potential gradient can be understood as an effect of major affinity (the driving force of ATP reaction) on that flux rather than the effect of the conjugative driving force (chemical potential of the solute). Mathematically, this implies a negative cross-coefficient combined with high affinity values, which can easily be fulfilled in those cases.

13.4 EFFECTS OF PROCESSING CONDITIONS

13.4.1 LOW-TEMPERATURE EFFECTS

The food engineer interested in low-temperature processes (cooling, freezing, thawing) can take advantage of literature data on responses of plant cells (cell walls and cell membranes) to low temperatures. Plant physiologists have studied those effects during cold acclimation studies of plants cultivated in cold climates and frozen environments. Data on effects of low-temperature conditions on the integrity of cell membranes function is of great interest. Physical disruption of tissue during freezing is often attributed to ice crystal formation, but this mechanism is insufficient to explain the wide, observed variations in the response of different species to chilling or freezing. Other mechanisms suggested to explain these differences are disulphide bond formation between membrane-bound protein molecules (upon thawing, those bonds remain intact, affecting protein mobility and activity);[19,20] phase transition;[21,22] [this suggests a primary (reversible) event associated with a threshold temperature, and secondary effects (such as increase of membrane permeability as a result of membrane protein disfunction or overall membrane disfunction) that may follow until general cellular degradation]; and, although there is no full agreement between physiologists, the importance of membrane phospholipid composition (which can explain acclimation phenomena: certain plants show the ability to fight chilling injury by inducing alterations in fatty acids—the more the latter become unsaturated, the more successful the cold hardening). Membrane composition alteration is also known as *membrane turnover.*

Susceptibility to chilling injury implies an increased risk to freezing injury. Both in freezing and in chilling, membrane integrity is lost, and the metabolic activity is altered, and it has been suggested that the injury mechanisms are the same for the two cases. Nevertheless, important differences are noticeable and should be stressed. Both the electrolytic leakage and the electrical impedance methods have been reported to be applicable for detection of average membrane damage in complex tissues, whereas DTA can be applied to evaluate major and minor freezing points of water. Work with juvenile and mature apple trees (cv. Spur Mac) has shown that the normalized electrical impedance test appears to measure changes in the cell wall fluids within the tissues, while electrolytic determination is based on average changes of the diffusion-limited efflux of electrolytes. These differences between the tests' basic principles appears to account for the lack of a complete linear fit between Znf and electrolytic leakage. Znf measurements were found to predict hardness in juvenile and mature apple trees throughout the winter period.[23]

Chilling has different effects, depending on light conditions. Under dark conditions, ion leakage indicates that the plasmalemma membrane plays a major role, as well as the tonoplast and mitochondrial membranes. This crucial importance of the plasmalemma is explained by its function as ion transporter and its role in keeping cell wall integrity. Light induces restricting photosynthetic ability, and, in this case, plastids are the sites of major importance. The plasmalemma has been considered as the primary site of freezing injury, the decisive step being freeze-induced water transport from the interior of the cell to the intercellular spaces. The cell membrane regulates the level of supercooling of the cytoplasm by controlling the water flux through an alteration of membrane permeability, although details of this mechanism are not yet fully understood. This prevents ice formation from extending from intercellular to intracellular spaces. Maintenance of membrane integrity upon thawing is critical for the survival of the cell. The altered water status at the membrane interface, due to dehydration, is said to be of particular interest in freezing injury. Compositional changes, phase transitions, lipid peroxidation within plant membranes, and their effects on membrane fluidity have also been considered important. Needs for further studies are based on the following:

1. Similarities and differences between chilling and freezing injury must be further explored.
2. A single mechanism is insufficient to account for all low-temperature and acclimation responses, where subtle rather than dramatic changes may play an important role. At the cell membrane level, lipid-protein interactions and interfacial phenomena need more investigation.
3. Biophysical studies must go beyond phase transitions and fluidity. Diffusion and mass transport related aspects, for example, how cell membrane permeability is affected by temperature, constitute challenges for future studies. Ion transport is of particular importance—the role of Ca^{2+} on cold acclimation mechanisms involving the cell membrane is likely to be quite relevant.
4. Although focusing on the cell membrane is important, we believe that the role of the cell wall has been almost totally neglected. The cell wall constitutes the first barrier of the cell to the extracellular environment, both mechanically and functionally. It is possible, for example, that freezing injury starts with ice nucleation in the intercellular capillaries, and this may cause mechanical damage to the plant cell walls. Those could serve as additional nucleation sites for both extracellular and intracellular ice formation. Furthermore, damage of the cell walls would leave the cell membrane unprotected, with accumulation of various species on its surface as a consequence. This could accelerate cell membrane responses and deteriorative mechanisms.

13.4.2 Intermediate-Temperature Effects

Between low-temperature and high-temperature conditions, there is a wide spectrum of the quasi-ambient temperature conditions, common in postharvest, minimal processing, storage, and distribution. Minimal processing involves a large number of different operations such as grading, cutting, mild heat treatments, etc. Minimally processed products are living, damaged tissues that require refrigeration and adequate packaging. Therefore, understanding the involved phenomena at the cell level is most important in this case. The boundary between minimal processing and conventional processing is not very clear, and some processes, such as osmotic dehydration, lie exactly in this boundary. Osmotically treated products do not show a fresh-like appearance, but the final product has a water activity well within the range of minimally processed products, and processing temperatures are below the critical temperature at which membrane protein denaturation occurs (this temperature will be further referred to as T_d). Nonthermal processing methods such as high pressure, electrical or magnetic fields, ohmic heating, microwaves, etc., are also milder treatments, although already out of the range of minimal processing. These applications are at an early development stage, and their effect on cell structure has only been incipiently studied, although there is evidence that they increase cell membrane permeability.[24–25]

Effects on Cell Walls

Cell walls are susceptible to minimal processing unit operations such as cutting, sizing, slicing, or grinding. The consequences are that cut cells will show less resistance to oxidative browning and to the entrance of bacteria as compared with intact cells. The effects of divalent ions on the cell wall strengthening (through the interaction with the pectin methyl esterase -PME- enzyme action and pectin molecule bridging) is well documented, not only for low and ambient temperature conditions but also in blanching and osmotic dehydration.[23,26–29]

The effects of divalent cations on the response of apple fruit tissue to accelerated aging, temperature, or osmotic stress have been evaluated[29] by measuring ethylene evolution, electrolyte leakage, and membrane microviscosity. Apple tissue slices, incubated in an isotonic sorbitol solution at 25°C for 24 h, underwent rapid aging, as expressed by a sharp drop in ethylene production (70–90%) and leakage of potassium from the tissue. At a higher temperature and in hypotonic medium, these symptoms were accelerated and enhanced. Addition of Ca^{2+} to the isotonic incubation medium inhibited ethylene production during the first 6 h, but thereafter partially prevented the drop in ethylene production that took place in the aged tissue slices and completely prevented K^+ leakage beyond 4 h of incubation. The influence of Ca^{2+} on the physical state of the membranes was observed by an increase in membrane microviscosity relative to control, as determined by the fluorescence depolarization technique. Using fluorescent probes that bind to different sites in the membrane, it was found that the effect of Ca^{2+} was more pronounced at the membrane surface and diminished toward the hydrophobic region of the membrane bilayer. The effect of Mg^{2+} on ethylene production and membrane microviscosity was similar though weaker. However, membrane permeability, as expressed by K^+ leakage, was unaffected by Mg^{2+} at similar concentrations. Ca^{2+} is suggested to inhibit stress-induced senescence by maintaining membrane integrity.

Glenn et al. investigated the effect of Ca^{2+} on various parameters of apple fruit senescence. Distinct and specific changes in polypeptide and phosphoprotein patterns were observed in Ca^{2+}-treated fruit, when compared to control fruits. Transmission electron micrographs of the cell showed Ca^{2+} to be effective in maintaining the cell wall structure, particularly the middle lamella. Furthermore, increase in fruit Ca^{2+} reduced CO_2 and C_2H_4 evolution and altered chlorophyll content, ascorbic acid level and hydraulic permeability.[27]

Postharvest calcium treatment is, however, sometimes insufficient to accomplish cold storage preservation. Peaches harvested in mature-green stage were dipped in a $CaCl_2$ solution at 49°C for 2.5 min and wrapped in PVC film of 15 µm thickness and placed in constant cold storage (0°C and RH 90–95%). Half of the samples were then submitted to intermittent warming for 48 h at ambient temperature (21°C). In spite of the high level of calcium, the cold-stored apples showed ripening and cold injury symptoms. Intermittent warming to ambient temperature gave more satisfactory results.[28]

Other effects related to cell walls are linked to the enzymatic hydrolysis of cell wall components caused by slicing. Pectinolytic and proteolytic enzymes liberated from cells damaged by slicing could diffuse into inner tissues. High migration rates of macromolecules through, for example, kiwi fruit tissue, determined with labeled enzymes, were observed (penetration rate, 1 mm/h). Compared with normal maturation conditions, a difference in cell wall hydrolysis was noticed, protopectin solubilization predominating in the former case.

Effects on Cell Membranes

When heating or treating a plant cellular tissue with osmotic solutions, loss of membrane functions, including changes in transport and permeability characteristics of membranes, may occur. In both cases (heating, immersion in osmotic solutions) the role of divalent ions such as Ca^{2+} on the strengthening of the structure has been identified.[26,27,29]

Millimolar concentrations of extracellular Ca^{2+} have been reported to be necessary for proper membrane function and to protect the cell against adverse conditions of low pH, toxic ions, and nutrient imbalance. Ca^{2+} may influence membrane structure due to its ability to induce asymmetrical distribution of negative phospholipids in membrane bilayers and alter membrane fluidity. Ca^{2+} has been found to be effective in preventing senescence-related increases in membrane microviscosity.[27] Under modified atmosphere packaging (MAP) conditions, the transport rates of the respiratory (or respiration-related) gases O_2 and CO_2 are very important. Slicing of tuber or root tissue (for example, potato) could invoke an immediate two- to fourfold increase in respiration.

Taking into account that gas permeability in the cell is controlled by the cell membranes, slicing obviously causes damage at the cell membrane level. Study of the cell membrane structure changes with temperature could explain the anomalous Arrhenius patterns sometimes observed.[21] In a number of tissues, low temperatures may induce a rise in respiration, probably due to phase transitions of the phospholipidic phase.[30] A typical example is potato tubers, where storage at 1°C evokes a respiration rise above that observed at 10°C.[31] Special care should, therefore, be taken when applying the Arrhenius equation to predict the respiration of a fresh product at a given temperature.

13.4.3 HIGH-TEMPERATURE EFFECTS

Physical Changes

Studies on biological membrane behavior are mainly done by physiologists under physiological conditions, and the high temperatures in physiologist's language are restricted to the 35–45°C interval. Although this temperature range is important to the food engineer, knowledge of the behavior of cell membranes at temperatures higher than the denaturation temperature ($T > Td$) is of utmost importance, because many food-processing operations are carried out at these temperatures (canning, aseptic processing, drying, etc.). A few studies have been reported on the behavior of membranes at these temperatures, both by physiologists and engineers.

Paszewski and Spiewla, for instance, studied the changes in the electrical resistance of internodal giant cells of *Characeae* in a range of temperatures from 4 to 48°C.[32] Cells of *Characeae,* because they are large, have often been used as models for plant cellular tissues. In the whole temperature range examined, the electrical resistance of the cell membranes underwent from five to nine distinct changes; the overall changes were approximately hyperbolic while, in certain narrow temperature ranges, they were nonmonotonous and, in others, linear. Arrhenius plots of these dependencies provided data that suggest a specific, stepwise change in permeability to ions, produced by temperature changes and probably connected with corresponding modifications of cell membrane fluidity. This seems to support the view that the cell membranes of the examined plant species may exist in different, discrete, physiological states, characteristic of certain temperature ranges. Values of the unitary resistance r_m (kΩ cm^2) were found in the region of 10–50 and 20–90 for two different species of *Characeae;* i.e., for *Nitellopsis obtusa* and *Lychnothamnus barbatus,* respectively.

The higher resistance values refer to 4°C and the lower ones to 48°C. In an earlier study, the cytoplasmic membranes of sugar beet and sugar parenchyma tissues fresh and also during heat treatment (78°C, 30 min) and mechanical stressing (pressures of 350.105 Pa, for cane only) were observed by electron microscopy. Whereas, in the intact cell membrane, the pores were 0.3 μm and round, after treatment, they became oval, creating channels doubled in size, of a diameter much greater than needed by sucrose and other sugar molecules to pass through. The interpretation given by the authors is disputable, because evidence suggests that the transport of sugar molecules in fresh, untreated tissues is protein mediated and thus not dependent on pore size. Nevertheless, their results show evident alterations suffered by cell membranes at processing temperatures and pressures. Concerning the effects of high temperatures on cell walls and the relation to texture properties of plant tissues, we have recently published a critical review focusing on potato.[26] Cell walls are known to be thermal resistant. On heating, they may become thinner and mechanically weaker, but they can keep the cell structure even at cooking conditions.[26]

Determination of T_d

There is considerable evidence of a discontinuity in the temperature dependence of mass transfer rates of various solutes from the cells of plant tissues (leaching) immersed in treatment solutions (brine, blanching, osmotic). Discontinuities are also sometimes found when solutes are transferred in the opposite direction, from the solution to the tissue (uptake). Thus, above a critical temperature (T_d, ≈50°C[10,33–41]), apparent diffusivities seem to increase one and two orders of magnitude for solutes such as sugars (glucose, fructose) and for ions (K$^+$, Ca^{2+}), respectively, compared with the respective values at $T < T_d$. Because the above solutes present a very low basal permeability and are transferred by mediated transport, this discontinuity clearly reflects an abrupt change of the integrity of the protein phase of the cell membranes that is responsible for the mediated transport.

For nonmediated solute transport but high basal permeability solutes (for example, acetic acid[7]), experiments done at ESB, Porto, have shown that the temperature dependence of this solute in plant tissue was as smooth as in a gel; i.e., a medium lacking cell structure.[39,42] In another work[38] on batch pickling of carrots at temperatures below T_d, increase of the apparent diffusivity of electrolytes with concentration may suggest clear evidence of the mediated transport by membrane proteins being intact. In general, discontinuities in the Arrhenius dependence of apparent diffusivities offer a diagnosis possibility for the detection of the transport mechanism of a given component, either leaching from or uptaken from a plant tissue material in a given temperature interval.[7–8,12,14,18,24–25,27–28,30–31,33–34,43–48]

In the case of potatoes, however, another phenomenon occurs at an interval of temperatures close to 55°C, namely the gelatinization of starch granules,[10,26] making the determination of T_d more difficult. Until a collaborative study was undertaken by Bahia Blanca and Lund Universities, there was no way to differentiate the effects due to membrane denaturation or starch gelatinization. The aim of this study was twofold, to discriminate between the two mechanisms and to more accurately determine T_d. Starch (potato) and nonstarch (apple) commodities were used, and the experiments were carried out in isotonic media to limit the transport to the solute only (in this case, potassium). In those experiments, the tissue was first preheated to a given temperature in the range of 25–75°C, and then the loss rate of the solute was measured at ambient temperature so as to avoid the interference of the Arrhenius temperature dependence of the mass transfer rates. The findings of this study were, in our opinion, very interesting. Under similar experimental conditions, apples and potatoes have both shown a sharp increase of the potassium loss from the tissue to the isotonic solution at temperatures between 45 and 50°C. The pattern of the sharp increase was, however, different for the two commodities tested. Apples exhibited a single linear sharp increase, whereas potatoes showed a two-phase sharp increase, one from 45°C up to approximately 50–52°C, and the other, less sharp, between 50–52°C and 57–60°C. Most probably, the second period corresponds to the gelatinization of starch in the potato, whereas the T_d for both tissues appears to be a temperature between 45 and 50°C.[69–70]

A third interesting finding was that, for temperatures beyond 60°C, a slight decrease in the potassium loss rate, rather than an increase, was observed. A possible explanation of this behavior is related to interactions with the cell wall and the thermostability of the PME enzyme; phase transition of the cell membrane may also be associated, but more investigation is needed to clarify this point.

A fourth finding provides strong evidence that ATPase activity shows likewise patterns; i.e., with sharp changes, as above, in the temperature intervals, where the sharp changes in observed mass transfer rates occur.

Modeling of T_d

By far, the Fickian approach is the dominating basis for modeling mass transfer in food processing (unit operations such as blanching, drying, frying, etc.). This approach is based on the use of an overall, apparent or effective, diffusion coefficient that accounts for a number of important phenomena and anomalies, such as shrink-age, simultaneous heat transfer, concentration effects (reflecting the difference between activity of a component and its concentration), the anisotropic structure of the media, and sometimes even multicomponent effects. Application of irreversible thermodynamics (IT) in food processing is uncommon, and the Maxwell–Stefan's approach (well suited for multicomponent transport) is even more rare. To our knowledge, the only study based on Maxwell–Stefan's approach is a Ph.D. thesis concerning mass transfer in a model food ternary system composed of water, protein, and sugar.[51] IT has also been applied in air convective drying.[45,52]

The need for a more fundamental approach considering both true driving forces (activities or chemical potentials) and the cellular structure is being increasingly considered to be of great importance. This approach would allow better insight on the mechanisms of mass transfer at cell level and their relationship to food product quality.[5–6,35,37–38,41,53–56]

To accomplish this objective, the food engineer needs a mathematical model that may be chosen from among the Fickian approach modified to include other effects such as source terms, multicomponent effects, and shrinkage; irreversible thermodynamics; Maxwell–Stefan's approach[5]; and a physical model to simulate the plant or vegetable tissue cellular structure, both under physiological conditions, and its changes upon processing (cell size, intercellular spaces, shrinkage, crust formation in some cases acting as a dynamic membrane, etc.). Convection phenom-ena in relation to osmotic treatment of fruits and vegetables has been studied by the team of Prof. Fito, in Valencia, who has considered the so-called hydrodynamic mechanism in the mass transfer inside intercellular spaces. This mechanism is enhanced when vacuum conditions are applied to the tissue with a twofold aim, increasing the driving force per se between the intracellular phase and the outer phase through the increase of the pressure term of the chemical potential and increasing the driving force by replacing the air contained initially in the intercellular spaces by the osmotic solution, where the chemical potential of water is lower than it is in the air.[55,56] The chemical potential as the driving force has also been considered for convective air drying[6,53–54] and also by Le Maguer et al. for osmotic dehydra-tion.[35,37] Characterization of the osmotic solution in terms of its chemical potential for both water and solutes has also been studied.[43] The equation for the water chemical potential of the cell is given by

$$\mu - v_m P + RT \ln a_w + V r n \psi_m \qquad (13.10)$$

where P is the hydrostatic pressure (due to the turgor pressure) of the water solution in the vacuole, V_m is the molar volume of water, a_w is the water activity corresponding to the effect of soluble solids in the cell, and ψ_m is the so-called matric potential term, corresponding to the effect of the insoluble solids on the water chemical potential.[6] If μ_0, the reference potential of the water in its pure state, is set equal to zero, and all the terms are divided by the molar volume, the units of each term of the chemical potential are those of pressure; the components of the water chemical potential are, therefore, the pressure potential (ψ_p), the matric potential (ψ_m), and the solute potential (ψ_s). Attention should be paid to avoid confusion with the more rigorous term of thermodynamics solute chemical potential, which refers to the chemical potential of the solute, whereas the term solute potential as used in biology is the effect of the solutes on the water chemical potential. Then, the solute term is negative, because the presence of soluble solids in a solution lowers the water activity, and the pressure term is positive, because it increases the tendency of water to leave the vacuole, and it therefore increases the chemical potential. The matric term is also negative, but it can be neglected in many cases (for example, in the case of relatively high-porosity commodities such as apples), since its contribution to the total water potential becomes very small.[6]

Whereas, for the mathematical modeling, the food engineer has access to a reasonably large body of literature, information on how to proceed with the second problem, i.e., the cellular structure simulation, is scarce. Therefore, we shall focus the discussion in this section on cellular structure modeling. Starting from the simple Fickian approximation, a biphasic model of a porous medium, expressing the effective diffusivity as the diffusivity in an aqueous solution corrected by porosity and tortuosity, would be a clear improvement. This has already been attempted by some authors. More sophisticated methods imply the use of the Voronoi tessellation[57] or the averaging volume method introduced by Whitaker and applied to food tissues by References 6, 53–54, and 58.

The Voronoi tessellation is based on a random division of a plane (or a volume) in a number of polygons (or polyhedra). It is based on the observation that the plant cells seem to have a polyhedral shape (polygonal plane cross section) with a variable number of faces, from 9–20, averaging 13.8.[57] After the division of a plane (or volume) into a number of polygons (or polyhedra), according to appropriate constraints, some of them are randomly chosen to represent the intercellular spaces or pores, respecting the real porosity experimentally observed.

Mattea et al.[57] used the Voronoi tessellation to simulate the shrinkage and deformation of cellular tissue during dehydration. The same method was also applied to model penetrometric texture studies in apples. A further step in this structure simulation would be to couple it with the study of mass transfer properties of the processed tissues.

The averaging volume method applied by Rotstein and Crapiste in foods aimed at a more detailed and realistic picture of the cellular structure in a way that the cell compartments, such as the vacuole, the cytoplasm, the cell wall, and the intercellular spaces, constitute the phases over which the volume averaging of the mass transfer properties was made. In this model, the cellular membranes were taken into account as interfaces between the phases. Modeling of the drying of apples and potatoes was performed.

13.5 CONCLUSION: NEED FOR FUTURE RESEARCH

There is an obvious need to focus research on mass transfer at the cell level on biological tissues undergoing processing and to develop more fundamental mathematical models, taking into account all the driving forces acting in the systems studied. Knowledge of the thermodynamics of the system cell and its environment is a prerequisite.

More research is needed for the determination of the membrane deterioration temperature (T_d) in various systems, covering not only thermal processing conditions but also different types of environments and application of pressure, electrical, and magnetic fields.

Thermodynamic modeling of solutions containing sugars and ions, typically found in the vacuole of the plant cells, as well as the processing solutions (blanching, osmotic) also have to be studied. There is an obvious gap compared with similar data reported in classical chemical engineering (hydrocarbon compounds, petrochemical organic chemistry). Modeling of the cellular structure is also in its infancy, only a little having been done until now, but emerging studies are providing clear paths for future research. At the cell level, studies have to be clearly accompanied by adaptation of powerful analytical tools [e.g., particle-induced X-ray emission (PIXE) or fluorescence microscopy] to the special features of plant cells.

REFERENCES

1. Gekas, V., G. Traegardh, and B. Halistroem, eds. 1993. *Ultrafiltration Membrane Performance Fundamentals*, Lund, Sweden: The Swedish Membrane Foundation.
2. Gekas, V. 1988. "Terminology for Pressure Driven Membrane Operations," *Desalination*, 68, 77–92.
3. Brett, C. and K. Waidron. 1961. "Physiology and Biochemistry of Plant Cell Walls," in *Electrolytes and Plant Cells,* U. Hyman, G. E. LondonBriggs, A. B. Hope, and R. N. Robertson, eds. Oxford: W. O. James, Blackwell Scientific Publications.
4. Rogissart, I. 1991. "Etude de Selectivite' d'une membrane d' Ultrafiltration," Ph.D. thesis, Universite Paul Sabatier, Toulouse, France.
5. Gekas, V. 1992. *Transport Phenomena of Foods and Biological Materials,* R. P. Singh and D. Heldman, eds. Boca Raton, FL: CRC Press.
6. Crapiste, G. H. 1985. *Fundamentals of Drying of Foodstuffs,* Ph.D. Thesis, Planta Piloto de Ingeniería Química, Bahia Blanca, Argentina.
7. Gennis, R. B. 1989. *Biomembranes,* chapters 7 and 8, New York: Springer Verlag.
8. Komor, E. 1982. "Transport of Sugar," chapter 17, in *Plant Carbohydrates,* F. A. Loewus and W. Tunner, eds. New York, Berlin, Heidelberg: Springer Verlag.
9. Marme, D. 1982. "Calcium Transport and Function," chapter 1, in *Plant Carbohydrates V,* F. A. Loewus and W. Tunner, eds. Berlin Heidelberg, New York: Springer Verlag.
10. Gekas, V., R. Öste, and I. Lamberg. 1983. "Diffusion in Heated Potato Tissue," *J. Food Science*, 58: 4.
11. Lehninger, A. L., D. L. Nelson, and M. M. Cox. 1993. *Principles of Biochemistry,* 2nd edition, New York: Worth Publishers.
12. Jan, L. Y. and Y. N. Jan. 1994. "Potassium Channels and Their Involving Gates," *Nature,* 371, 8, Sept.
13. Neame, K. D. and T. G. Richards. 1976. *Elementary Kinetics of Membrane Carrier Transport*, Madrid: Herrman Blume Ediciones.

14. Kedem, 0. and A. Katchalsky. 1958. "Thermodynamic Analysis of the Permeability of Biological Membranes to Non-electrolytes," *Biochemica and Biophysica Acta,* 27: 229.

15. Prigogine, I. 1968. *Introduction to Thermodynamics of Irreversible Processes,* 3rd edition, New York: John Wiley & Sons, Inc.

16. Gekas, V., P. Aimar, J. P. Lafaille, and V. Sanchez. 1993. "Diffusive Flows in Ultra-filtration," *J. Membrane Sci.*

17. Boersma, L., F. T. Lindstrom, C. MeFarlane, and E. L. McCoy. 1988. "Uptake of Organic Chemicals by Plants. A Theoretical Model," *Soil Science,* 146(6), 403–417.

18. Hernández, A., J. I. Calvo, L. Martínez, J. Ibañez, and F. Tejerina, 1991. "Measurement of the Hydraulic Permeability of Microporous Membranes from the Streaming Potential Decay," *Separation Science and Technology,* 26(12): 1507–1517.

19. Levitt, J. 1962. "A Sulfhydry–Disulphide Hypothesis of Frost Injury and Resistance in Plants," *J. Theor. Biol.,* 3: 335–391.

20. Levitt, J. 1973. *Responses of Plants to Environmental Stresses,* New York: Academic Press.

21. Lyons, J. M. 1973. "Chilling Injury in Plants," *Ann. Rev. Plant Physiol.,* 24: 445–466.

22. Raison, J. K. and J. M. Lyons. 1986. "Chilling Injury: A Plea for Uniform Terminology," *Plant Cell Environm.,* 9: 685–686.

23. Leshem, Y. Y. 1992. *Plant Membranes,* chapter 8, Dordrecht, The Netherlands: Kluwer Academic Publishers, pp. 155–170.

24. Knorr, D. 1993. "Effects of High Hydrostatic Pressure Processes on Food Safety and Quality," *Food Technology,* 47(6): 156.

25. Knorr, D. 1994. "Non-Thermal Processes for Food Preservation," in *Minimal Processing of Foods and Process Optimization,* R. P. Singh and F. A. Oliveira, eds. Boca Raton, FL: CRC Press, pp. 3–15.

26. Andersson, A., V. Gekas, I. Lind, F. Oliveira, and R. Öste, 1994. "Effect of Preheating on Potato Tissue," *Critical Reviews in Food Science and Nutrition,* 34(3): 229–251.

27. Glenn, G. M., A. S. N. Reddy, and B. W. Poovaiah. 1988. "Effects of Calcium on Cell Wall Structure, Protein Phosphorylation and Protein Profile in Senescing Apples," *Plant Cell Physiol.,* 29(4): 565–572.

28. Holland, N., M. I. F. Chitarra, and A. B. Chitarra. 1994. "Potential Preservation of Peach Fruits cv biuti, Effect of Calcium and Intermittent Warming during Cold Storage under Modified Atmospheres," in *Minimal Processing of Foods and Process Optimization,* R. P. Slugh and F. A. Oliveira, eds. Boca Raton, FL: CRC Press, pp. 467–473.

29. Nur, T., R. Ben-Arie, S. Lurie, A. and Altman. 1986. "Involvement of Divalent Cations in Maintaining Cell Membrane Integrity in Stressed Apple Fruit Tissues," J. *Plant Physiol.,* 125: 47–60.

30. Larsson, K. 1996. Food Technology at Lund University, personal communication.

31. Isherwood, A. C. 1973. "Starch-Sugar Interconversion in *Solanum tuberosum,"* *Phytochemistry,* 12: 2579–2591.

32. Paszewski, A. and E. Spiewla. 1986. "Temperature Dependence of the Membrane Resistance in *Characeae* Cells," *Physiol. Plant.,* 66: 134–138.

33. Lazarides, H. and N. Mavroudis. 1995. "Freeze/Thaw Effects on Mass Transfer Rates During Osmotic Dehydration," *J. Food Science,* 60(4): 826–829.

34. Lazarides, H., V. Gekas, and N. Mavroudis. 1997. "Mass Diffusivities in Fruit and Vegetable Tissues Undergoing Osmotic Drying Processing," J. *Food Engineering,* 31(3): 315–324.

35. Le Maguer, M. 1988. "Osmotic Dehydration, Review and Future Directions," in *Proceedings Symposium on Progress in Food Preservation Processes*, pp. 183–309.

36. Lenart, A. 1994. "Osmotic Dehydration of Fruits Prior to Drying," in *Minimal Processing of Foods and Process Optimization,"* R P. Singh and F. A. Oliveira, eds. Boca Raton, FL: CRC Press, pp. 87–106.

37. Marcotte, M. and M. Le Maguer. 1992. "Mass Transfer in Cellular Tissues, Part 2. Computer Simulations vs. Experimental Data," *J. Food Engineering*, 17: 177.

38. Menezes, R. M. 1995. "Mathematical Modelling of Batch Pickling of Carrots via Lactic Acid Fermentation," Ph.D. thesis, ESBIUCP, Porto, Portugal.

39. Moreira, L., F. Oliveira, and T. R. Silva. 199. "Prediction of pH Change in Processed Acidified Turnips," *J. Food Science*, 57(4): 928–931.

40. Moreira, L. 1994. "Analyses of Mass Transfer, Texture and Microstructure During Acidification of Vegetables," Ph.D. thesis, ESB, Porto, Portugal.

41. Oliveira, F. A. R. and C. L. M. Silva. 1992. "Freezing Influences Diffusion of Reducing Sugars in Carrot Cortex*," Int. J. Food Science and Technology*, 57: 932.

42. Azevedo, I. and F. Oliveira. 1995. "A Model Food System for Mass Transfer in the Acidification of Cut Root Vegetables," *International Journal of Food Science and Technology*, 30: 473–483.

43. Gekas, V., H. Lazarides, and N. Mavroudis. 1995. "Mass Diffusivities in Plant Tissues Undergoing Osmotic Drying Processing," presented at the Annual COPERNICUS Meeting, Porto, 7–9 December.

44. González, C., V. Gekas, P. Fito, H. Lazarides, and I. Sjoeholm. 1995. "Osmotic Solution Characterization," presented at the Annual COPERNICUS Meeting, Porto, Portugal, 7–9 December.

45. Honda, Sota and Nozu. 1987. "An Ultrastructural Study of Photo-Induced Conidiogenesis and Dedifferentiation in *Alternaria solani," Bull. Fac. Agr. Schimane Univ.*, 21: 141–154.

46. Husain, A., C. S. Chen, and J. T. Clayton. 1973. "Simultaneous Heat and Mass Diffusion in Biological Materials," *J. Agr. Eng. Res.,* 18: 343.

47. Jowitt, R., F. Escher, M. Kent, B. MeKenna, and M. Roques, eds. 1983. "Physical Properties of Foods (COST—project results)," in *Thermophysical Properties of Foods,* chapters 15–17, Applied Science Publishers, pp. 229–328.

48. Lafta A. and C. B. Rajashekar. 1990. "Cell Wall Pore Size and Tensile Strength of Apple Fruit Cultured Cells in Relation to Treatment with Calcium, Ethylene, pH and Low Temperature," *Plant Physiology,* Abstract no. 498, 93: 86.

49. Tufvesson, F. 1995. "Preheating Effects on Potassium Leakage from Potato Tissue with Emphasis on Cellular Level Understanding," M.Sc. thesis, Food Engineering Div., Lund University.

50. Wenz, J., V. Gekas, and C. Crapiste. 2000. "Effects of Preheating on Potassium Leakage Potato and Apple Tissue," *Food Engineering*, In press.

51. Meerdink, G. 1993. "Drying of Liquid Food Droplets, Enzymatic Inactivation and Multicomponent Diffusion," Ph.D. thesis, AUW, Wageningen, NL.

52. Fortes, M. and M. R. Okos. 1981. "A NonEquilibrium Thermodynamics Approach to Transport Phenomena in Capillary Porous Media," *Transactions of the ASAE*, pp. 756–760.

53. Crapiste, G. H., S. Whitaker, and E. Rotstein. 1988. "Drying of Cellular Material, I. A Mass Transfer Theory," *Chem. Eng. Sci.*, 43: 2919–2928.

54. Crapiste, G. H., S. Whitaker, and E. Rotstein. 1988. "Drying of Cellular Material, I. B. Experimental and Numerical Results," *Chem. Eng., Sci.,* 43: 2929.

55. Fito, P. and R. Pastor. 1995. "On Some Non-Diffusional Mechanisms Occurring during Vacuum Osmotic Dehydration," *J. Food Engineering*.
56. Fito, P., A. Andrés, R. Pastor, and A. Chiralt. 1994. "Vacuum Osmotic Dehydration of Fruits," in *Minimal Processing of Foods and Process Optimization,* R. P. Singh and F. A. Oliveira, eds. Boca Raton, FL: CRC Press, pp. 107–122.
57. Mattea, M., M. J. Urbicain, and E. Rotstein. 1989. "Computer Model of Shrinkage and Deformation of Cellular Tissue during Dehydration," *Chemical Engineering Sci.*, 44(12): 2853–2859.
58. Ochoa, J. A., P. Stroeve, and S. Whitaker. 1988. "Diffusion and Reaction in Cellular Media," *Chem. Eng., Sd.,* 14(12): 2999–3013.

14 The Cellular Approach in Modeling Mass Transfer in Fruit Tissues

M. Le Maguer
G. Mazzanti
C. Fernandez

CONTENTS

Abstract

Osmotic dehydration processes in plant materials involve complex phenomena in the tissues. The parameters for the cellular properties of the material (e.g. diffusivity, tortuosity and porosity), the properties of the solution (e.g. viscosity, diffusivity and density) and processing conditions (e.g. temperature and shape of the material) are described.

1-56676-963-9/02/$0.00+$1.50
© 2002 by CRC Press LLC

A mass transfer model was developed based on the above parameters that can compute for the water flow, the advance of the solute front as well as estimate the average concentration and equilibrium conditions inside the cells. The model also considered the effects of the boundary layer of solution to compute for the fluxes using overall mass transfer coefficients.

The overall mass transfer coefficient increases as the flow of water slows down with respect to the flow of sucrose. The model developed has been able to predict mass transfer behavior by considering the cellular properties of the material and its interaction with the external solution.

14.1 SYMBOLS

α	Coefficient for the extracellular concentration
τ	Tortuosity of the active zone of the tissue
Δv_p	Change in volume of the piece of material, m^3
ε	Diffusive porosity of the active zone of the tissue
ρ_o	Density of the fresh tissue, kg/m^3
ρ_l	Density of the osmotic solution, kg/m^3
μ	Viscosity, $Pa \cdot s$
θ	Correction factor for molar mass transfer coefficient
ξ	Elastic modulus of the tissue, Pa
A_m	Surface area of cells per unit volume, m^2/m^3
A	Interface area of a single piece of tissue, m^2
$\overline{C}_f$	Average concentration of solute in extracellular space, $kmol/m^3$
C_f	Molar concentration of solute in extracellular space, $kmol/m^3$
C_i	Molar concentration of solute at interface, $kmol/m^3$
C_T	Total molar concentration of osmotic solution, $kmol/m^3$
C_w	Molar concentration of water in extracellular space, $kmol/m^3$
D	Binary diffusivity, m^2/s
d_i	Internal diameter of a ring, m
d_o	External diameter of a ring, m
i_p	Constant parameter for turgor pressure, Pa
J_w^c	Flux of water through cell membrane, $kmol/m^2 \cdot s$
J_w	Flux of water at the interface, $kmol/m^2 \cdot s$
J_s	Flux of sucrose at the interface, $kmol/m^2 \cdot s$
k_c'	Mass transfer coefficient for the solution, m/s
k_M	Molar mass transfer coefficient for the active zone of material, $kmol/m^2 \cdot s$
K_x	Overall molar mass transfer coefficient, $kmol/m^2 \cdot s$
k_x	Molar mass transfer coefficient for the solution, $kmol/m^2 \cdot s$
k_x^*	Corrected molar mass transfer coefficient for the solution, $kmol/m^2 \cdot s$

L	Superficial length of the piece of material, m
L_e	Equivalent permeability coefficient, $L_e = A_m R T L_p$ kmol/m^3·s
L_p	Permeability coefficient of cell membrane, kmol^2/Pa·m^5·s
M_s	Molecular weight of water, kg/kmol
M_{ss}	Apparent molecular weight of cellular solutes, kg/kmol
M_w	Molecular weight of water, kg/kmol
P	Turgor pressure, Pa
PR	Excess plasmolysis ratio
R	Universal gas constant, 8314 Pa·m^3/K·kmol
Re	Reynolds number
R_f	Flux ratio for correction factor
Sc	Schmidt number
Sh	Sherwood number
sg_a	Absolute solids gain, kg
s_p	Parameter for the turgor pressure, Pa
T	Temperature, K
t	Time, s
u_z	Velocity of liquid flowing out in the extracellular space, m/s
u_{sup}	Superficial velocity of liquid flowing past surface of tissue, m/s
V_c	Volume of a cell, m^3
v	Volume, m^3
v_{oz}	Theoretical volume of active zone at time zero, m^3
v_{avg}	Average molar specific volume at interface, m^3/kmole
v_{az}	Total volume of active zone, m^3
v_s	Molar-specific volume of osmotic solute, m^3/kmole
V_w	Molar-specific volume of water, m^3/kmole
wl_a	Absolute water loss, kg
wl_{ac}	Absolute water lost by the cells of active zone, kg
W_{sso}	Mass fraction of soluble solids in fresh tissue
W_{wo}	Mass fraction of water in fresh tissue
$\bar{x}_c$	Average molar fraction of solute in cellular space
x_c	Molar fraction of solute in cellular space
$\bar{x}_c^*$	Average equivalent molar fraction of solutes in cells
x_c^*	Equivalent molar fraction of solutes in cells
x_{co}	Initial molar fraction of solute in cellular space
x_{co}^*	Initial equivalent molar fraction of solute in cellular space
x_{cp}	Molar fraction of solute in cellular space at plasmolysis
x_f	Molar fraction of solute in extracellular space
$\bar{x}_f$	Average molar fraction of solutes in extracellular space

x_i	Molar fraction of solute at the interface
x_1	Molar fraction of solute in osmotic solution
z	Direction perpendicular to the surface, m
z_p	Depth of penetration of the front of solute in the tissue, m

14.2 INTRODUCTION

Each phase of the solid-liquid contact operation requires its own treatment, according to the basic engineering principles of mass transfer. The liquid-phase mass transfer has been extensively studied and documented already by classic authors.[1]

Several different approaches have been used to model the internal mass transfer in the tissue, mostly following the diffusion in solids assuming independent effective diffusivity for water and the osmotic solute. The application of traditional solutions[2] gives a reliable computational tool to estimate the diffusivity. Some empirical approaches seem to be quite useful as well, e.g., those that rely on the square root of time[3] or on an asymptotic behavior.[4]

The most difficult challenge for modeling the operation is introduced by the complexity of the solid. It is, in fact, a composite of several phases: the insoluble solids, the cellular solution, the extracellular solution, and the extracellular air. The pathways for the movement of water in the tissue include the extracellular space (apoplast) and sometimes the intercellular plasmodesmata (symplast), according to Nobel.[5] The fraction of extracellular space filled with air is not available for the mass transfer. Therefore, the tortuous route that the substances travel inside the tissue constitutes an important parameter for the modeling. The way the structure changes when it loses water will affect the porosity and the tortuosity, as well as the pressure inside the cells. For plant tissues, the turgor pressure can be several atmospheres, significantly contributing to the chemical potential of cellular water. As extensively studied by Salvatori,[6] the dehydration occurs from the surface inward, thus creating a variable set of conditions for mass transfer both in space and time.

Discrete units have been used to represent a typical cell and its surroundings as building blocks of the tissue. This naturally leads to discrete element numerical methods.[7–8]

Another methodology is to average properties in a reasonable way and use simplified analytical procedures.[9–10] The presence of a nondiffusional water flux creates special conditions for the opposite diffusing solutes that can be related by the basic laws of mass transfer.

When the superficial cells of the piece of tissue come in contact with the osmotic solution, they start dehydrating, generating a water flow into the solution. It takes less than 3 min for a typical vegetable cell to reach equilibrium.[11] Simultaneously, the diffusive penetration of the osmotic solute into the extracellular capillary network creates an active dehydration zone.[6,12] This is the zone from the surface inward in which the cells are in contact with a solution that produces an osmotic flow of water. From there, the water has to find its way out of the tissue through the capillary spaces. Cells that are located deeper down in the tissue keep their natural state (Figure 14.1).

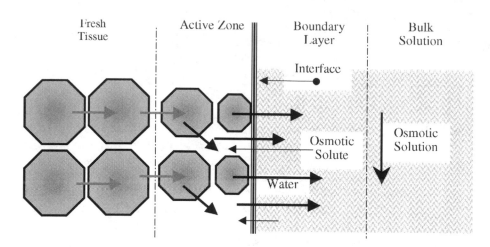

Figure 14.1 Different mass transfer mechanisms in osmotic dehydration: osmotic flow of water from the cells to the pores or the solution, convective flow of water out of the tissue through the pores, diffusion of solute against the outflowing water, and occasional symplastic flow of sap between cells.

14.3 PARAMETERS INVOLVED IN THE MODEL

14.3.1 TISSUE PARAMETERS

Parameters for the Water Flux through the Cell Membrane

The flux will be a function of the difference in chemical potential between the water in the cell and in the solution surrounding the cell. Only water is considered to be moving through the membrane during the dehydration. For solutions with small *mole fraction* of solutes, as is the case for the cells and concentrated sugar solutions, it may be simplified to the following:

$$J_w^c = RTL_p(x_f - x_c^*)$$ (14.1)

The pressure, though, will modify the *effective mole fraction* of solutes in the cell.

$$x_c^* = x_c - \left(\frac{v_w}{RT}\right)P$$ (14.2)

The *effective membrane permeability*, L_p, of plant cell membrane varies between 5×10^{-11} and 3×10^{-9} kmol²/j·m²·s.[5,13] The actual driving force for the dehydration of the cells is the difference between the molar fractions of the cell and its surroundings.

The *turgor pressure* inside the cells of a fresh fruit or vegetable is usually around 400 or 500 kPa. The turgor pressure also has a considerable effect on the chemical potential of the water. The pressure will drop to zero at the onset of *plasmolysis*.

A common expression used to describe the relationship between turgor pressure and volume is[5,14]

$$\frac{dP}{dV_c} = \frac{\xi}{V_c} \tag{14.3}$$

In practice, this can be approximated as a function of the mole fraction in the cells as

$$P = s_p x_{ss} + i_p \tag{14.4}$$

Parameters for Water Flow and Solute Diffusion in the Extracellular Space

The average *diffusivity* of the osmotic solute in water is a function of temperature and concentration.

The *tortuosity* is the ratio between the distance measured from the surface of the piece and the actual length of the path that water and solutes have to travel in the tissue, due to the entangled matrix of capillary pores and cell walls.

Porosity, as far as mass transfer is concerned, is the volume fraction available for diffusion of the species in the extracellular space. It is different from the void fraction of the tissue, which is filled with air. The space has to be a liquid medium to allow the diffusion of the osmotic solute to reach the inner cells. The total extracellular volume of the active zone as well as the total volume of the zone is changing during the dehydration. In a very rigid structure, the cells shrink, leaving behind a space that has a volume equal to the water released from the cell. In a very flexible structure, the cell walls shrink together with the cells, and the total extracellular space does not change.

Most tissues have air in the extracellular space. The *void fraction* is the part of the volume of tissue that is filled with air, and hence, is not available for the mass transfer process. During the osmotic dehydration, however, part of this air may be displaced, either by shrinkage or by being replaced by solution.

14.3.2 OSMOTIC SOLUTION PARAMETERS

The estimation of the interface concentration of the solution, as well as the variation of the binary *diffusivity* with concentration, has to be taken into account for the modeling, along with the *viscosity* and the *density* of the solution. These define the boundary layer around the piece of tissue. Values for these parameters for osmotic solutions can be found in several publications.[15–17]

14.3.3 PROCESS PARAMETERS

The *shape* of the piece determines the surface area exposed to the solution, as well as the volume enclosed by that surface.

The *temperatures* for this process usually range between room temperature and 50°C. Higher temperatures produce a disruption of the cell membranes, causing the phenomenon to be more diffusive and less osmotic. On the other hand, the higher the temperature, the lower the viscosity and the higher the diffusivity. Thus, a compromise has to be reached when developing a process. The interaction between the *flow regime* and the mass transfer within the tissue will determine the value for the interface concentration.

The *ratio of liquid to solids* and the *countercurrent or parallel* contacting will produce different gradients of external concentration. These are reflected in the way the solid reacts to the solution and must be taken into account when developing a cellular model of the osmotic treatments.

14.4 MASS TRANSFER MODELING

14.4.1 ACTIVE ZONE OF THE TISSUE

A more detailed deduction of the equations presented in this section can be found in Mazzanti et al.[18]

Computation of the Water Flow

The flux of water at the surface of the tissue comes from the integral of the transmembrane flux minus the water left in the extracellular space. Both terms are included in the following expression:

$$wl_a = M_w \left[\int_0^t L_e \int_0^{v_{az}} (x_f - x_c^*) dv dt - C_w v_{az} \varepsilon \right]$$ (14.5)

The flux of water coming out of the cells is a function of the difference between the average concentrations of the cellular and extracellular spaces.

$$\int_0^{v_{az}} (x_f - x_c^*) dv = v_{az}(\bar{x}_f - \bar{x}_c^*)$$ (14.6)

When determining the volume integrals, the fact that the function that relates v_{az} with z_p depends on the shape of the particle must be considered.

Only experimental observation of the shrinkage can give an appropriate idea of the kind of functions that better describe the relationship between water loss from the cells and shrinkage. Macroscopic observations should be linked with microscopic images to provide a clear structural picture of the shrinking patterns.

Advance of the Solute Front and Extracellular Concentration

In the extracellular free space, the solute is diffusing against an advection flow caused by the transmembrane flow. Assuming an average diffusivity, the concentration of solute, C_f, at any point in the capillary changes with time, t, following Equation (14.7):[1]

$$\frac{1}{\tau^2}\frac{\partial C_f}{\partial t} = D\frac{\partial^2 C_f}{\partial z^2} - u_z\frac{\partial C_f}{\partial z} \qquad (14.7)$$

The steep profiles yielded by this equation are consistent with the fact that the osmotically dehydrated products always show only a small superficial layer of solute,[12] even after several hours of treatment.

If the penetration front of solute, z_p, is considered to be where $C_f/C_i = 0.001$, solving Equation (14.7) for the depth of the active zone is defined as

$$z_p = \frac{4.7}{\tau}\sqrt{Dt} \qquad (14.8)$$

Since the solute is diffusing only in the active zone, the total mass of solids gained, sg_a, is computed as

$$sg_a = \overline{C}_f M_s \varepsilon v_{az} \qquad (14.9)$$

with the average extracellular concentration being

$$\overline{C}_f = \frac{1}{v_{az}}\int_0^{v_{az}} C_f dv \qquad (14.10)$$

For the diffusion happening against the advection flow,

$$\overline{C}_f = \alpha C_i \quad \alpha < 0.5 \qquad (14.11)$$

Average Concentration Inside the Cells

The volume that the active zone would have had before the dehydration is its present volume plus the volume lost, $-\Delta v_p$.

$$v_{oz} = v_{az} - \Delta v_p \qquad (14.12)$$

Assuming no solutes are leached from the cells, the average molar fraction of the cells in the active zone is

$$\overline{x}_c = \frac{M_w W_{sso}}{M_w W_{sso} + M_{ss} W_{wo} - \dfrac{w l_{ac} M_{ss}}{v_{oz}\rho_o}} \qquad (14.13)$$

The term

$$x_c^* = \left(x_c - \frac{v_w}{RT}\Delta P\right)$$

is zero for a cell in its fully turgid state. As explained before, the value of ΔP may change linearly with x_c until plasmolysis. PR is a plasmolysis excess ratio, ranging from 0.1 to 0.5. Consequently, the term x_c^* can vary linearly from zero to the plasmolysis value of x_c and afterward will be equal to x_c alone. The value x_{cp} is the concentration at which plasmolysis will happen, and x_{co} is the initial concentration of the cell. Thus,

$$\left.\begin{array}{l} x_{cp} \approx (PR + 1) \cdot x_{co} \\[2mm] \bar{x}_c < x_{cp} \Rightarrow \bar{x}_c^* = x_{cp}\left(\dfrac{\bar{x}_c - x_{co}}{x_{cp} - x_{co}}\right) \\[2mm] \bar{x}_c \geq x_{cp} \Rightarrow \bar{x}_c^* = \bar{x}_c \end{array}\right\} \tag{14.14}$$

Quasi-Equilibrium Conditions in the Tissue

Equilibrium is established when the chemical potential of water inside the cell equals the chemical potential of its surroundings. As mentioned before, this can be simplified to equal effective mole fraction in both phases.

At any time during the dehydration, the mass balance (in kg) for the water *in the active zone* of the tissue is

Initial water = Water left in cells + Extracellular water + Water lost

And the mass balance for the osmotic solute is

Solute in extracellular space = Solute gained

Since the response time of the cells is quite short as compared with the velocity at which the osmotic solute advances, it can be assumed that they are quite close to equilibrium with the extracellular space, i.e., that their mole fraction is the same as the extracellular space x_f. The balance then yields

$$v_{oz}\rho_o W_{wo} = \left(\frac{v_{oz}\rho_o W_{sso}}{M_{ss}} + \frac{sg_a}{M_s}\right)\left(\frac{1 - \bar{x}_f}{\bar{x}_f}\right)M_w + wl_a \tag{14.15}$$

Therefore, the diffusive porosity can be estimated as

$$\varepsilon = \frac{sg_a}{M_s}\left[\hat{v}_s + \hat{v}_w\left(\frac{1 - \bar{x}_f}{\bar{x}_f}\right)\right]\frac{1}{v_{az}} \tag{14.16}$$

If the extracellular mole fraction is not required, then

$$\varepsilon = \frac{sg_a}{M_s}\left[v_s + \frac{v_w}{M_w}\frac{v_{oz}\rho_o W_{wo} - wl_a}{\left(\frac{v_{oz}\rho_o W_{sso}}{M_{ss}} + \frac{sg_a}{M_s}\right)}\frac{1}{v_{az}}\right] \tag{14.17}$$

The volumes of the active zone before (v_{oz}) and after (v_{az}) dehydration are computed from the depth of penetration and the shrinkage of the tissue.

14.4.2 BOUNDARY LAYER OF THE SOLUTION

The interface flux of water or solids, J, is defined as

$$J_w = \frac{1}{A}\frac{dwl_a}{dt} \qquad J_s = \frac{1}{A}\frac{dsg_a}{dt} \tag{14.18}$$

According to Geankoplis (1983), for a laminar flow past a submerged flat surface,

$$Sh = \frac{k'_c L}{D} = 0.664 Re^{1/2} Sc^{1/3} (\pm 25\%) \tag{14.19}$$

$$Sc = \frac{\mu}{\rho_l D} \tag{14.20}$$

$$Re = \frac{Lu_{sup}\rho_l}{\mu} \tag{14.21}$$

To express the driving forces as differences in mole fraction, k_c' should be converted into k_x, according to the equation below:

$$k_x = k_c'\left(\frac{1}{C_T}\right) \tag{14.22}$$

C_T is defined as

$$C_T = \left(\frac{\text{kmoles}_{\text{sucrose}}}{\text{m}^3}\right) + \left(\frac{\text{kmoles}_{\text{water}}}{\text{m}^3}\right) \tag{14.23}$$

The independent generation of an osmotic bulk flow of water perpendicular to the external flow requires the use of a corrected mass transfer coefficient, according to the following expression:[1]

$$J_s - x_i(J_s + J_w) = k_x^*(x_i - x_l) \tag{14.24}$$

The effect of the water flux on the mass transfer coefficient is then corrected by computing for the flux ratio, R_f.[1]

$$R_f = \frac{(x_i - x_l)}{\left(\dfrac{J_s}{J_s - J_w} - x_1\right)} \tag{14.25}$$

The correction factor, θ, is defined as

$$\theta = \frac{\ln(R_f + 1)}{R_f} \tag{14.26}$$

The corrected mass transfer coefficient expressed in $kmol/m^2 \cdot s$ is

$$k_x^* = \theta k_x \tag{14.27}$$

The procedure requires the simultaneous determination of the mole fraction at the interface x_i.

14.4.3 OVERALL AND TISSUE MASS TRANSFER COEFFICIENTS

It is convenient for design purposes to compute the fluxes using overall mass transfer coefficients. It is as desirable to express those fluxes as functions of the difference in concentration between the cells and the bulk solution. Because the tissue at the deeper end of the active zone is in its fresh state, it seems quite natural to use the initial cellular concentration of the tissue to compute this driving force.

The overall driving force is the difference between the solution and the effective mole fraction inside the cells.

$$Js - x_i(Js + Jw) = K_x(x_c^* - x_l) \tag{14.28}$$

For cells that are fully turgid, the effective concentration is zero, due to the pressure effect, as described in Equation 14.2. Therefore,

$$Js - x_i(Js + Jw) = -K_x x_l \tag{14.29}$$

This overall mass transfer coefficient results from the combined resistance of the material and the solution.

$$\frac{1}{K_x} = \frac{1}{k_x^\bullet} + \frac{1}{k_M} \tag{14.30}$$

Therefore, the mass transfer coefficient for the active zone of the tissue is

$$k_M = \frac{1}{\left(\dfrac{1}{K_x}\right) - \left(\dfrac{1}{k_x^{\bullet}}\right)} \qquad (14.31)$$

This mass transfer coefficient should be proportional to the conditions in the active zone of the tissue:

$$k_M \propto \frac{\varepsilon D}{\tau z_p v_{avg}} \qquad (14.32)$$

14.5 APPLICATION TO EXPERIMENTAL RESULTS

14.5.1 MATERIALS AND METHODS

Rings of Granny Smith apples (d_i = 20 mm, d_o = 75 mm) of 10 mm and 15 mm thickness were submerged in a circulating bath. A 60% w/w sucrose solution at 40°C was circulated at a superficial velocity of 0.11 m/s. The internal and external edges of the rings were sealed with a rubber solution. The rings were sampled after 5, 10, 20, 30, 60, and 90 min. The weight and dimensions of each ring were recorded before and after treatment. Cylinders were bored from the ring to analyze moisture (vacuum-drying oven) and sucrose contents (HPLC). At 5, 20, and 90 min, extra cylinders were bored and sliced to follow the penetration of the front. These methods are similar to those used by Barat[19] and Salvatori.[6]

14.5.2 WATER LOSS, SOLIDS GAIN, AND VOLUME REDUCTION

By combining the expressions presented in the Section 14.4.1, "Active Zone of the Tissue," it was possible to model the absolute water loss wl_a and solids gain sg_a in time. There was no significant difference observed between the 10-mm and 15-mm ring behaviors, since the solute did not reach the mid-plane of the rings. The average water loss for the rings and the correspondent sucrose gain values are shown in Figure 14.2. The lines show how the model fits the experimental results, by applying Equations (14.5) and (14.10).

The model used the volume reduction correlated to the absolute water loss, as shown in Figure 14.3. The volume reduction was linear with the thickness reduction of the pieces, since the area was reduced during the treatment. The data on penetration are shown in Figure 14.4. These four sets of data were smoothed before going on to the subsequent analysis. The size of the active zone before and after dehydration was determined using these data.

Application of Equation (14.15) yielded the values of the extracellular concentration. The tortuosity was computed following Equation (14.8) using the penetration data, and the porosity was computed following Equation (14.17). The values of tortuosity, porosity, and shrinkage ratio are plotted as a function of the extracellular concentration shown in Figure 14.5.

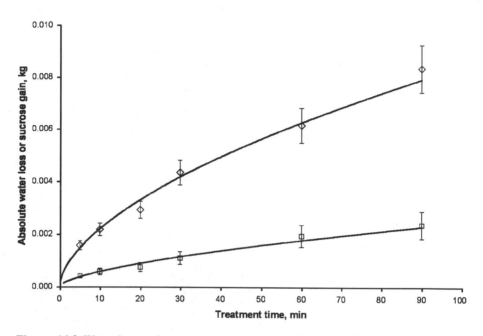

Figure 14.2 Water loss and sucrose gain are expressed here in absolute terms, i.e., in kilograms, without referring them to the initial mass of material, as is customary in many publications. The reason for this is that the mass transfer depends on the surface and on the depth of penetration in the material rather than its initial mass.

14.5.3 FLUXES AND MASS TRANSFER COEFFICIENTS

After smoothing the original data, it was possible to compute the fluxes crossing the interface, following Equation (14.18). They are plotted in Figure 14.6. Application of Equations (14.19) to (14.27) allowed the determination of the interface mass transfer coefficient for the solution, as well as the interface concentration.

Their relationship is linear, as can be seen in Figure 14.7. Through the equations presented, the overall and tissue mass transfer coefficients were determined. The mass transfer coefficient for the tissue, plotted in Figure 14.8, does in fact follow the proportionality suggested in Equation (14.32). Therefore, from the surface fluxes and the composition of the tissue, it is possible to estimate the internal mass transfer coefficient.

14.6 CONCLUSIONS

The proposed model offers new insights into the interpretation of the mass transfer phenomena in osmotic dehydration. By accounting for the cellular nature of the materials and the interaction with the surrounding solution, it is possible to estimate the main parameters of the tissue so as to predict its behavior. A relationship between the parameters of the tissue and the traditional definitions of mass transfer coeffi-

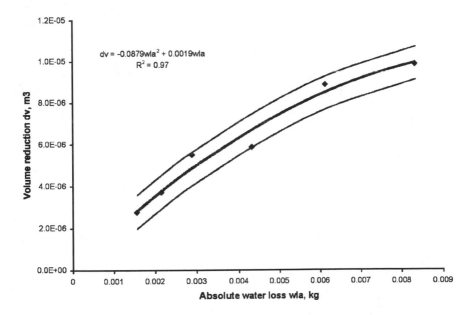

Figure 14.3 Volume reduction is related to the amount of water lost by the cells. As an approximation, it can be correlated to the water lost by the tissue.

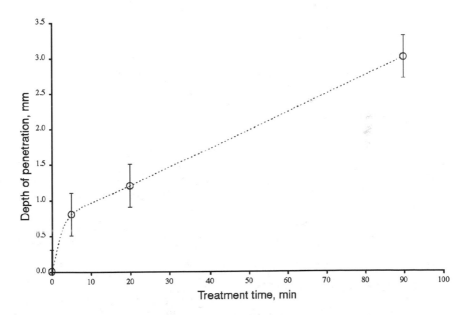

Figure 14.4 The penetration of the solute into the tissue determines the depth of the active zone, i.e., the fraction of the tissue that is involved in the dehydration. This concept is important for short time treatments, in which the front of penetration does not reach the mid-plane of the tissue. It also allows estimation of the internal parameters of the tissue.

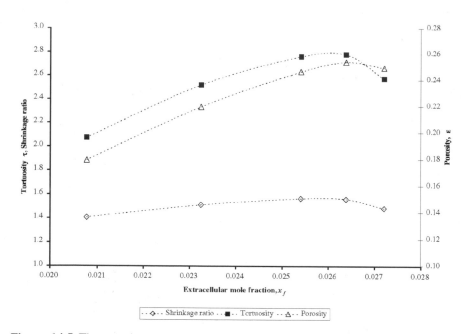

Figure 14.5 The porosity and tortuosity of the tissue are functions of the extracellular concentration of the tissue, as is the shrinkage ratio. The penetration of the front does not happen at a constant concentration, though. Therefore, the extracellular mole fraction presented here is an averaged value, as are the parameters and the shrinkage ratio.

cients was also proposed. This opens new perspectives for the integration of the experimental work with the engineering design of industrial equipment.

REFERENCES

1. Bird, R., W. Stewart, and E. Lightfoot. 1965. *Transport Phenomena.* New York: Wiley.
2. Crank, J. 1975. *The Mathematics of Diffusion.* Oxford: Clarendon Press.
3. Magee, T. R. A., A. A. Hassaballah, and W. R. Murphy. 1983. "Internal Mass Transfer During Osmotic Dehydration of Apple Slices in Sugar Solutions," *Ir. J. Food. Sci. Technol.* Dublin: An Foras Taluntais. 7(2): 147–155.
4. Azuara, E., C. I. Beristain, and H. S. García. 1992. "Development of a Mathematical Model to Predict Kinetics of Osmotic Dehydration," *J. Food Sci. Technol.*, 29(4): 239–242.
5. Nobel, P. S. 1983. *Biophysical Plant Physiology and Ecology.* New York: W. H. Freeman and Company.
6. Salvatori, D. M. 1997. "Deshidratación Osmótica de Frutas: Cambios Composicionales y Estructurales a Temperaturas Moderadas," Ph.D. thesis, Universidad Politécnica de Valencia, Valencia, Spain.
7. Toupin, C. J., M. Marcotte, and M. Le Maguer. 1989. "Osmotically-Induced Mass Transfer in Plant Storage Tissues: A Mathematical Model. Part I," *Journal of Food Engineering*, 10(2): 13–38.

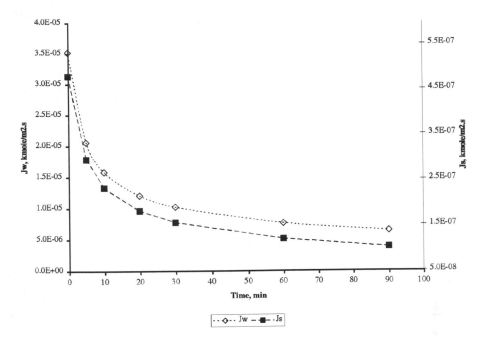

Figure 14.6 The molar fluxes are the result of superimposed diffusive and bulk flows. This makes necessary the correction factor for the presence of bulk flow perpendicular to the interface.

8. Spiazzi, E. and R. Mascheroni. 1997. "Mass Transfer Model for Osmotic Dehydration of Fruits and Vegetables—I. Development of the Simulation Model," *Journal of Food Engineering*, 34: 387–410.
9. Le Maguer, M. and Z. M. Yao. 1995. "Mass Transfer during Osmotic Dehydration at the Cellular Level," in *Food Preservation by Moisture Control. Fundamentals and Applications*. G. V. Barbosa-Cánovas and J. Welti-Chanes, eds. Lancaster, PA: Technomic Publishing Co., Inc.
10. Mazzanti, G. and M. Le Maguer 1998. "Sensitivity Analysis for a CounterCurrent Osmotic Dehydration Unit," in *Proceedings of the IV AcoFoP Conference*, Göteborg, Sweden.
11. Chu, H. C. 1996. "Three-Dimensional True Colour Optical Sectioning Microscopy: Development and Application to Plant Cells," Ph.D. thesis, University of Guelph, Guelph, ON, Canada.
12. Lazarides, H. N. 1994. "Osmotic Preconcentration: Developments and Prospects," in *Minimal Processing for Foods and Process Optimization*, R. P. Singh and F. A. R. Oliveira, eds., pp. 73–84.
13. Stadelmann, E. J. 1977. "Passive Transport Parameters of Plant Cell Membranes," in *Regulation of Cell Membrane Activities in Plants*, E. Marrè and O. Ciferri, eds. North-Holland.
14. Dainty, J. 1976. "Water Relations of Plant Cells," in *Transport in Plants II. Part A. Cells*, U. Luttge and M. G. Pitman, eds., New York: Springer-Verlag.

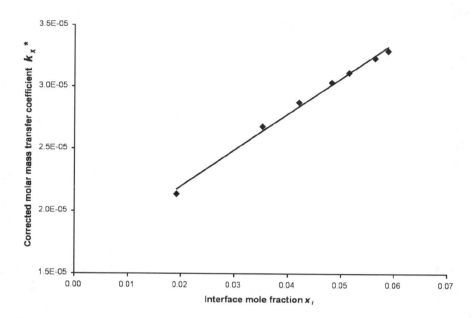

Figure 14.7 The concentration at the interface is lower at the first stages of the dehydration, when the flux of water is at its higher values and, therefore the gradient necessary to transport the sugar into the tissue becomes steeper. The mass transfer coefficient increases as the flow of water becomes less significant with respect to the diffusive flow of sucrose.

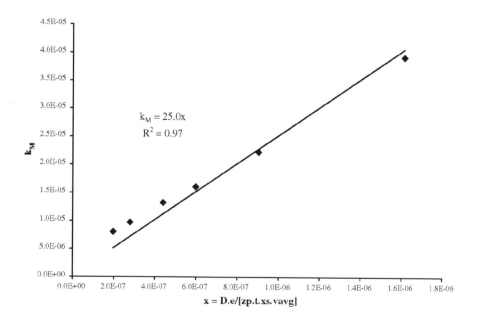

Figure 14.8 The mass transfer coefficient for the active zone of the tissue, k_M, is proportional to the penetration depth and other parameters of the tissue.

15. Gekas, V., C. González, A. Sereno, A. Chiralt, and P. Fito. 1998. "Mass Transfer Properties of Osmotic Solutions. I. Water Activity and Osmotic Pressure," *International Journal of Food Properties*, 1(2): 95.

16. Gekas, V. and N. Mavroudis. 1998. "Mass Transfer Properties of Osmotic Solutions. II. Diffusivities," *International Journal of Food Properties*, 1(2): 181.

17. Bouhon, P., M. Le Maguer, and A. L. Raoult-Wack. 1997. "Densities and Viscosities of Ternary Systems NaCl-Sucrose-Water from 283.15 to 303.15 K," *Journal of Chemical Engineering Data*, 42: 266–269.

18. Mazzanti, G., J. Shi, and M. Le Maguer (in preparation). "The Cellular Approach to Mass Transfer in Osmotic Treatments," in *Vacuum Impregnation and Osmotic Dehydration in Food Systems*, P. Fito, et al., eds., Aspen Pub. Co. (in press).

19. Barat, J. M. 1998. "Desarrollo de un Modelo de la Deshidratación Osmótica como Operación Básica," Ph.D. thesis, Universidad Politécnica de Valencia, Valencia, Spain.

15 Mass Transport and Deformation Relaxation Phenomena in Plant Tissues

P. Fito
A. Chiralt
J. M. Barat
J. Martínez-Monzó

CONTENTS

Abstract

Mass transfer in the dehydration process of plant tissues promotes a great sample volume reduction due to cell water loss. As a consequence, cells deform greatly while linked through bonding zones, giving a shrunken cell network. This network may store a great amount of mechanical stress, depending on the elastic/viscous properties of cell walls and bonding zones. In osmotic dehydration, water loss and solute gain define the total volume of product liquid phase (LP) lost throughout the process and the final ratio between LP and solid matrix phase (MP); the higher the ratio, the lower the sample volume loss. In highly porous tissues, the gas phase (GP) volume collapse will also affect the sample volume development. LP and GP volume changes are greatly affected by sample shape.

After sample deformation as a result of water loss, the free energy stored in the sample as mechanical stress can be liberated, leading the system to true equilibrium. Then, the sample volume relaxes, and the product recovers volume and, if it is immersed in an external liquid phase, mass. Hydrodynamic mechanisms promoted by internal pressure gradients associated with volume relaxation imply a gain in the external liquid phase. Kinetics of volume relaxation and volume-mass recovery degree are greatly affected by particular characteristics of the plant tissues and by the osmotic dehydration process conditions. The control of these phenomena may be a useful tool to reformulate fruit and vegetable products in processes focused on improving their quality, stability, and nutritional or health properties.

15.1 INTRODUCTION

In the processing of fruits and vegetables, the solid-fluid systems (SFSs) are very frequent in different fields of food technology, such as osmotic dehydration, rehydration, candy processing, boiling and cooking, etc. The heat and mass transfer processes in such systems have usually been modeled considering the food solid as a continuous phase. Nevertheless, the tissue cellular structure (intercellular spaces and cell compartmentation) plays an important role in the definition of mechanisms involved in the process and, therefore, in process kinetics.[1,2] Recently, several works have been carried out in determining the influence of product porosity in the response of the fruit tissue to solid-liquid operations.[3-6] In this sense, hydrodynamic mechanisms (HDMs) have been described as fast mass transfer phenomena occurring in process operations in SFS when changes in temperature or pressure take place. During HDM action, the occluded gas inside the product pores is compressed or expanded according to pressure or temperature changes, while the external liquid is pumped into the pores in line with the gas compression. An effective exchange in the product (internal gas for the external liquid) can be promoted by vacuum impregnation (VI). In such an operation, a vacuum pressure ($p_1 \sim 50\text{--}100$ mbar) is imposed on the system for a short time (t_1) and afterward, the atmospheric pressure (p_2) is reestablished, while the product remains immersed in the liquid for a time t_2.[5]

The pressure changes may cause deformation of the sample volume, and, consequently, HDM usually occurs coupled with deformation-relaxation phenomena (DRP) of the sample structure. The role of DRP is relevant when the sample behaves as a viscoelastic structure, and characteristic impregnation and relaxation times of the product are in the same order.[5] The volume fraction of the initial sample impregnated by the external liquid when mechanical equilibrium is achieved in the sample (X) has been modeled as a function of the compression ratio (r), sample effective porosity (ε_e), and sample volume deformations (γ) at each step of the process.[5] Vacuum impregnation implies a significant structural change of the product: volume deformation and great reduction of porosity with the introduction of external liquid in the pores. So, different mass transfer behavior has been observed for VI products.[6-8]

Mass transfer in many processes (such as hydration or dehydration operations) implies important changes in the sample volume and shape, mainly in regard to the gain or loss of the major component (water). Cellular structures such as plant tissue deform to a great extent after osmotic or air dehydration (shrinkage) as well as after hydration (swelling). In such systems, the occurrence of pressure gradients also concerns both mass transfer and deformation-relaxation phenomena. Deformations will be greatly dependent on the cell network arrangement, the gas-phase volume trapped or contained in their porous structure and, in osmotic dehydration, on the facility of the tissue to allow the contradiffusion of external solutes to replace the water lost. After shrinkage, the elastic component of mechanical energy involved in the volume changes and the cellular deformations may be liberated, thus promoting volume recovery of the samples.[8-11] The generated changes in the internal pressure associated with volume recovery will promote hydrodynamic fluxes and, consequently, mass gain.

In this chapter, the volume changes promoted by osmotic dehydration as affected by sample porosity and shape are analyzed. Likewise, the volume recovery capability of some osmosed fruits and its implication in promoting hydrodynamic flow after chemical equilibrium is discussed. Finally, the influence of the replacement of the sample gas phase by VI with an external liquid on sample volume changes and deformation-relaxation phenomena is analyzed.

15.2 FRUIT VOLUME CHANGES IN OSMOTIC TREATMENTS

Changes of the volume of a fruit or vegetable sample throughout an osmotic dehydration process may be considered in a simplified way as being made up of three phases:[7] the fruit liquid phase (LP), the gas phase occupying pores (GP) and the insoluble solid matrix (MP). The first is constituted by water plus soluble solids (native or incorporated from the external solution). Volume or mass loss during an osmotic process may be explained in terms of the partial losses of each one of the phases. In samples where there is a gas phase, either losses or compression-expansion of gas will contribute to volume change. Equation (15.1) reflects the total volume variation in terms of the changes of each phase, all referring to the sample's initial volume. In practical terms, it is possible to assume that $\Delta V_{MP} = 0$, due to its very small initial value; thus, only changes in the liquid and gas phase will describe the volume development. The values of ΔV_{LP} can be calculated, if sample composition (water and soluble solid content) and weight loss are known, by applying Equation (15.2), where the density of the LP (ρ_{LP}) can be estimated from its empirical relationship with the LP solute mass fraction (z_s). Equation (15.3) shows an example of a previously established equation for some fruits by Barat et al.[10] At a determined level of LP concentrations, ΔV_{LP} values will be lower when the solute gain is higher and the water loss lower.

$$\Delta V = \Delta V_{LP} + \Delta V_{GP} + \Delta V_{MP} \tag{15.1}$$

$$\Delta V_{LP} = \left(\frac{\rho^0}{M^0}\right)\left[\frac{M^t(x_w^t + x_s^t)}{\rho_{LP}^t} - \frac{M^0(x_w^0 + x_s^0)}{\rho_{LP}^0}\right] \tag{15.2}$$

$$\rho_{LP} = 141z_s^2 + 376z_s + 1000 \qquad (15.3)$$

In Figure 15.1, differences in sample volume development, in line with concentration of the LP, can be observed for the same product (Granny Smith apple) in osmotic treatments with 65° Brix sugar solutions. Total volume change of the samples is plotted as a function of the solid mass fraction of the LP (z_s) reached at different times. Volume pathway differences appear due to sample shape (2 × 2 cm cylinders, C, and 1 cm thick slices, S) and sample porosity reduction by previous VI treatment with isotonic solutions of different viscosity (with and without 2% HM pectin: samples C-I-2 C-I-1, respectively). Comparison of cylinder and slice behavior throughout osmotic treatment shows that at a determined LP concentration, greater volume reduction was promoted in cylinders. This is because the greater solid gain-water loss ratio was reached in slices, as shown by the development of ΔV_{LP}, also plotted in Figure 15.1 for both samples. Sugar gain:water loss ratio is improved in slices because of their higher surface-volume ratio and their lower mass transfer characteristic dimension, both making the solid penetration into the tissue through the pores more effective.

In porous samples, the change in the gas-phase volume will also contribute to volume loss. Figures 15.2a and 15.2b show the comparison between ΔV and ΔV_{LP} for cylinders and slices, respectively. Total volume losses are greater than the corresponding ΔV_{LP}, although a linear correlation was observed for both magnitudes. In cylinders, ΔV is 1.07 times the ΔV_{LP}, which, taking into account Equation (15.1), implies a gas volume loss in line with process progression of $0.065\Delta V$. In slices, gas

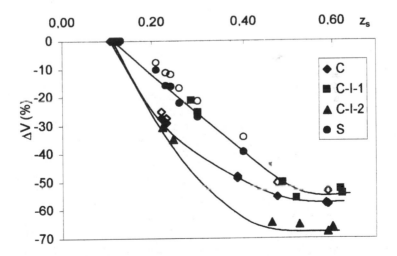

Figure 15.1 Relative volume change of apple samples (2 × 2 cm cylinders, C, and 1 cm thick slices, S) throughout osmotic dehydration as a function of the fruit liquid phase concentration. Samples C-I-2 and C-I-1 were previously impregnated with a sugar isotonic solution, with and without 2% of HM pectin, respectively. (Open symbols concern the respective values of ΔV_{LP}).

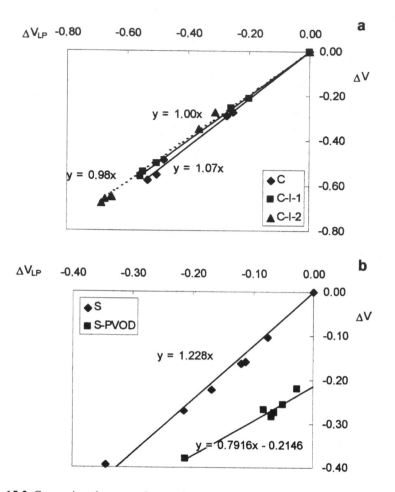

Figure 15.2 Comparison between changes in sample volume and changes in the volume of fruit LP for apple samples throughout osmotic dehydration. (a) Cylindrical (2×2 cm) samples. Samples C-I-2 and C-I-1 were previously impregnated with a sugar isotonic solution, with and without 2% of HM pectin, respectively. (b) Apple slices (1 cm tick). S-PVOD samples were submitted to a vacuum pulse (50 mbar for 5 min) at the beginning of the osmotic process.

volume losses contribute to the total volume loss to a greater extent, on the order of 0.19 ΔV. From these results, it can be deduced that sample shape plays an important role, not only in the relative water-solute transport rate, but also in the cell network collapse during dehydration and thus in porosity changes of the sample. Figure 15.3 shows the development of apple sample porosity throughout osmotic dehydration as a function of the reached z_s values. In both cases, porosity increases in line with sample concentration, as has been previously reported.[10,11] Nevertheless, the greater gas phase collapse in slices is reflected in Figure 15.3, as the porosity increase is less marked than in cylinders.

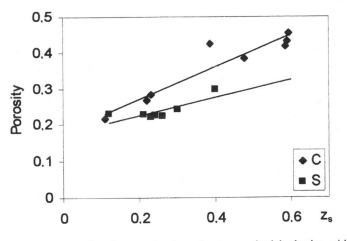

Figure 15.3 Development of apple porosity throughout osmotic dehydration with 65° Brix sugar solution at 30°C for 2 × 2 cm cylindrical samples (C) and 1 cm thick slices (S).

Reduction of sample porosity by VI with isotonic solutions modifies the pathway of sample volume change depending on the viscosity of the impregnating liquid. Apple cylinders impregnated with the low-viscosity solution (samples C-I-1 in Figure 15.1) show smaller volume losses, at a determined concentration level, than nonimpregnated cylinders or than those impregnated with the higher viscosity solution (samples C-I-2 in Figure 15.1). For impregnated samples where practically no residual gas remains, total volume change will be equal to the ΔV_{LP} value, as shown in Figure 15.2a for samples C-I-1 and C-I-2. Therefore, differences in sample volume change behavior are explained in terms of the differences in sugar gain-water loss ratio reached in the tissue. In Figure 15.2a, the close agreement between ΔV and ΔV_{LP} values for C-I-1 and C-I-2 samples can be observed. The presence of a low-viscosity liquid in the intercellular spaces promotes solute diffusion in these non-compartmented (and open at the interface) spaces, whereas this is more limited in samples containing a very viscous liquid in the pores. So, in low-porosity products, the compositional equilibrium may be reached in each case with different relative levels of sugar gain and water loss, defining the total volume change of the samples. The greater the LP/MP mass ratio, the higher the sample final volume.

In osmotic dehydration processes, sample porosity reduction occurs when a vacuum pulse is applied at the beginning of the process in the dehydration tank. This process has been called pulsed vacuum osmotic dehydration (PVOD),[12] and differences in the sample volume change pathway may be obtained in these cases. In this operation, an osmotic solution is introduced into the fruit pores, replacing the initial gas, but sample volume compression may occur when atmospheric pressure is reestablished, if characteristic times of liquid penetration and sample deformation range in the same order. This takes place when sample pores are very narrow or liquid viscosity is high. In VI with highly concentrated osmotic solutions, both factors may be present: highly concentrated sugar solutions are viscous, and the surface pore entrance may be narrowed because of the fast water loss by surface cells.[12] Sample volume changes vs. ΔV_{LP} for apple slices (1 cm thick) in PVOD treatments with 65° Brix sucrose solutions are plotted in Figure 15.2b. A linear relationship is observed for both variables, but a significant value of the straight-line intercept was obtained. The slope of the straight line is not near 1, as in previously impregnated samples with isotonic solutions, but less than 1. In this case, Equation (15.4) shows the volume balances, where two additional terms appear: a possible sample volume change due to compression-relaxation by the vacuum pulse ΔV_{C-R}, and a liquid phase volume impregnating the pores (ΔV_{LP-VI}) that does not contribute to sample volume [Equation (15.4)], both terms replacing the initial ΔV_{GP} in Equation (15.1). These terms explain the straight-line intercept value. On the other hand, a sample volume relaxation when taken out of the viscous osmotic solution (and the subsequent GP recovery) may explain the slope deviation from 1.

$$\Delta V = \Delta V_{LP-VI} + \Delta V_{C-R} + \Delta V_{LP} \tag{15.4}$$

Table 15.1 shows the changes in mass and volume of cylindrical samples of some osmotically treated fruits, with different porosities, at time of chemical equilibrium with the osmotic solution (t_c). The effects of previous impregnation of the sample with an isotonic solution of different viscosity on sample volume and mass, commented on above, can be observed at this time. On the other hand, values of change in volume and mass of some fruits equilibrated with 55° Brix sucrose solution ($z_s = 0.55$) can be observed in Table 15.1. Fruits with low porosity show very close values of ΔV and ΔV_{LP}, since the reduction of the gas volume phase is not appreciable due to its low initial value. Different feasibility of the fruit tissue to promote solid gain was observed through the DVLP values. In this sense, apple and kiwi fruit with very different initial porosities show the smallest LP loss and, thus, the highest sugar gain-water loss ratio. In all fruits, PVOD process implied lower values of ΔV_{LP}, and, as expected, the higher the fruit porosity, the greater the difference, according to the LP gain in the sample due to the vacuum pulse.

15.3 CHANGES AT THE CELLULAR LEVEL

Sample volume loss in the osmotic process is the result of a great cell volume reduction because of water loss. Cell water loss implies the generation of an internal void and subsequent internal pressure reduction, which promotes hydrodynamic mechanisms. Internal pressure gradients also contribute to a different cell micro-structural development depending on the kind of fluid (gas or liquid) in the tissue pores.[9] Figures 15.4a and 15.4b show cryoSEM micrographs of osmosed apple tissue for two kinds of osmotic treatments, OD (carried out at atmospheric pressure) and PVOD, where a very different aspect can be observed. In this kind of micrograph, an insoluble matrix (MP) can be observed as the cell wall network and membranes. The fruit LP appears in the intracellular or extracellular volumes with a dentritic aspect, according to the description of Bomben and King,[13] for aqueous phases in

TABLE 15.1
Volume and Mass Changes at Chemical Equilibrium with the External Solution (between Brackets) at 30°C of Different Cylindrical (2 × 2 CM) Samples Osmosed in OD and PVOD Processes

Fruit	ε_e	Treatment	t_c (h)	ΔM	ΔV	ΔV_{LP}	ΔV_{GP}
Apple	0.23	OD	120	−0.63	−0.58	−0.54	−0.04
(RCGM)		C-I-1-OD	96	−0.45	−0.54	−0.54	0.00
		C-I-2-OD	120	−0.62	−0.68	−0.68	0.00
Apple	0.23	OD	72	−0.52	−0.50	−0.45	−0.05
(SS)		PVOD	72	−0.24	−0.48	−0.27*	−0.21**
Pear	0.03	OD	96	−0.50	−0.56	−0.58	−0.02
(SS)		PVOD	96	−0.42	−0.52	−0.50*	−0.04**
Mango	0.06	OD	48	−0.47	−0.53	−0.55	−0.02
(SS)		PVOD	48	−0.20	−0.36	−0.32*	−0.04**
Kiwi	0.01	OD	48	−0.38	−0.46	−0.47	−0.01
(SS)		PVOD	48	−0.30	−0.41	−0.43*	−0.02**

Note: Fruit varieties: apple *Granny Smith*, pear *Conference*, mango *Kent*, kiwi *Hayward*. RCGM: Rectified concentrated grape must (65° Brix); SS: 55° Brix sucrose solution.

*Include the LP gain in the initial vacuum pulse.

**Include the GP loss in the initial vacuum pulse.

the tissue containing soluble solids. GP volume can also be observed occupying the intercellular spaces (*is*) in the OD, treated samples. No GP is observed in the micrograph of the PVOD, treated sample, as this was quantitatively exchanged by external liquid during the vacuum pulse. Another difference between the microstructures of OD and PVOD samples concerns the behavior of the cell wall-plasmalemma ensemble throughout dehydration. When GP is present in the *is*, such as in OD treated apple, no cell wall-plasmalemma separation is observed,[14] and the latter deforms together with the cell wall throughout the cell shrinkage. In the PVOD-treated sample, plasmalemma separates from the cell wall at the very beginning of the osmotic process in the cell, and the LP in the *is* flows through the permeable cell wall, flooding the volume between plasmalemma and wall.[15] This feature can be observed in Figure 15.4b for apple PVOD treated with rectified-concentrated grape must. The well-preserved cell volume surrounded by the cell membrane during sample fracture after cryofixation is remarkable. This suggests a great firmness of the apparently undisrupted membrane.

The differences in behavior of the cell wall plasmalemma ensemble have been explained in terms of the different pressure drops of gas or liquid phases in the *is* during their flux toward the generated volumes when the cell shrinks.[9] A force balance on both sides of the cell-cell wall layer during the cell dehydration can be created (Figure 15.5). [The reaction forces to the shrinkage-associated force (F_S) are

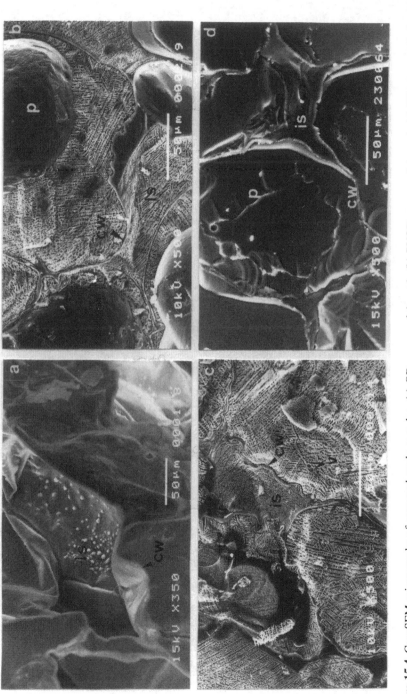

Figure 15.4 CryoSEM micrographs of osmosed apple samples. (a) OD processed ($t < t_c$), (b) PVOD processed ($t < t_c$), (c) OD processed after VI of the sample with an isotonic solution containing 2% HM pectin ($t < t_c$), and (d) PVOD processed after volume relaxation ($t >> t_c$). (cw = cell wall, p = plasmalemma, v = vesicles, is = intercellular space)

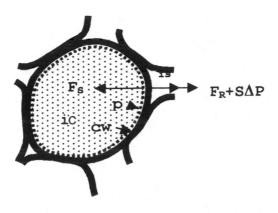

Figure 15.5 Force balance on cell wall (*cw*)-plasmalemma (*p*) ensemble throughout osmotic dehydration (*ic* = intracellular content, *is* = intercellular space).

the deformation resistance of the layers (F_R), plus the resistance associated with the fluid pressure drop ($S\Delta P$) when it flows in the *is*.] The module of action-reaction forces on the elastic double layer in dynamic equilibrium increases in line with the overall water loss (degree of layer deformation), and with the water loss rate (deformation rate) as well. When the *is* is occupied by gas (e.g., in OD), the force component $S\Delta P$ is negligible, and *cw* deforms bonded to *p* while the gas phase flows into the *is*. On the contrary, when a liquid phase occupies the *is* (e.g., in PVOD), the pressure drop of liquid is much greater, and the F_S value overcomes the cell wall-plasmalemma adherence force F_A quickly. In this situation, cell wall-plasmalemma separate, thus promoting the liquid flow through the permeable *cw*, whereas this remains less deformed. The water loss rate, which defines the cell deformation rate, will also affect the critical value of water loss at which *cw-p* separation occurs, due to the viscoelastic response of the bonding zones.[9,11]

Figure 15.4c shows a cryoSEM micrograph of an osmodehydrated apple sample previously impregnated with an isotonic solution containing 2% HM pectin. The dentritic aspect of the intercellular liquid appears more compact due to the vitrifying effect of polymer in the impregnated liquid. Cell walls are more folded than in the sample impregnated with a less viscous solution (Figure 15.4b). Another notable difference with respect to this sample is that sample fracture after cryofixation does not show the entire cell volume surrounded by plasmalemma, but it shows a non-continuous membrane and vesicles. The different response of cell wall plasmalemma ensemble in this impregnated sample can be understood on the basis of the model of force balance explained above. When F_S increases in line with dehydration, the *is* liquid will pass through the cell wall, but pectin macromolecules will be retained in the cell wall as in a filtration membrane, as the process occurs with a very high pressure drop. Cells lose water (and volume) faster than the external liquid passes through the cell wall, and so this deforms to a greater extent than in the sample of Figure 15.4b. These great forces acting on the *cw-p* layer seem to break membranes that appear, in some cases, as small vesicles.

15.4 CELL NETWORK RELAXATION AND HYDRODYNAMIC FLOW

In long-term osmotic processes, sample mass and volume decrease in line with osmotic dehydration until a minimum value is reached at compositional equilibrium time t_c. This is when the fruit LP and the osmotic solution have the same value of water chemical potential, which, in practical terms, implies the similar mass fraction of soluble solids of both liquid phases.[10] From this point, sample mass and volume begin a slowly increasing pathway until the initial values are almost recovered in some cases. This was observed initially in apple samples[10,16] and has been confirmed for other fruits.[17] Sample mass and volume recovery has been explained in terms of the relaxation of the shrunken cell wall network, greatly deformed in the initial dehydration step. A great amount of free energy was stored in the structure as mechanical energy during cell dehydration-shrinkage. The true equilibrium in the system implies the free energy reduction by release of stored mechanical stress. If the product remains immersed in the external solution, volume recovery will be coupled with liquid suction, and, therefore, with mass gain by hydrodynamic mechanism (HDM). The volume relaxation rate will be affected by the process driving force or stored mechanical energy in the sample, and also by the pressure drop of in-flowing liquid, because the system (fruit plus in-flowing liquid) behaves as a viscoelastic structure. The total recovery of volume or mass will be affected by the elastic character of the tissue cellular matrix, which will be greatly dependent on the integrity of bonding zones or cell middle lamella after chemical equilibrium in the osmotic process, as well as on the preservation of the laminar cell wall ultra-structure.

Figures 15.6 and 15.7 show the mass and volume recovery of several fruit cylinders after their chemical equilibrium with the osmotic solution at t_c. Time scale was plotted in a reduced way $(t - t_c)$, and mass and volume recovery was calculated as the variation between t and t_c, divided by the sample initial value, according to Equations (15.5) and (15.6). In Figure 15.6, the influence of previous impregnation of the sample with an isotonic solution on the relaxation pathway of apple samples can be observed, as well as the influence of the liquid viscosity of the impregnating solution. The impregnated sample volume relaxes faster than that of nonimpregnated samples. Nevertheless, when the impregnated solution has HM pectin (sample C-I-2), the relaxation rate slows down. Differences in volume relaxation rate may be due to different cell network microstructure reached in each case, as commented on above, which will imply a different level of stored mechanical energy. The different viscosity of the fruit LP will also affect relaxation rate, because internal LP will flow inside the sample throughout its volume recovery. As expected, mass and volume recovery undergoes parallel variations, although mass gain of a nonimpregnated sample is faster than volume gain, because a part of the external liquid flows into the empty *is*, reducing sample porosity at the same time as the cell network relaxes.[10,11]

$$\Delta M'(t) = \frac{M^t - M^{tc}}{M^0} \tag{15.5}$$

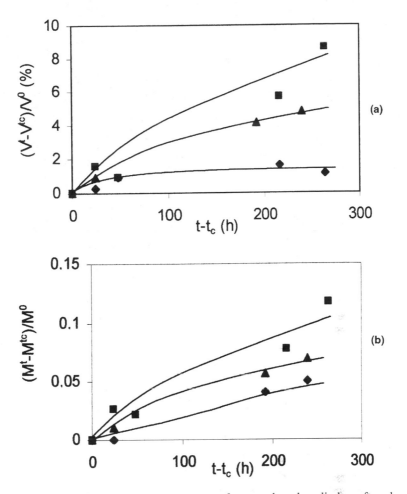

Figure 15.6 (a) Volume and (b) mass recovery of osmosed apple cylinders after chemical equilibrium at t_c with 65° Brix rectified concentrated grape must at 30°C. ◆ = non-impregnated samples, ▲ and ■ = samples previously impregnated with an isotonic solution, with and without 2% HM pectin, respectively.

$$\Delta V'(t) = \frac{V^t - V^{tc}}{V^0} \qquad (15.6)$$

On the other hand, Figure 15.7 shows the pathway of initial volume relaxation and mass gain of some fruits OD and PVOD treated with 55° Brix sucrose solution. The volume of pear and kiwi fruit relaxes to a smaller degree than that of mango and apple, which relaxes quickly. The kind of treatment, OD or PVOD, appears to have no notable influence on volume relaxation of pear and apple, although the volume of PVOD-treated mango (and kiwi to a much lesser extent) relaxes faster

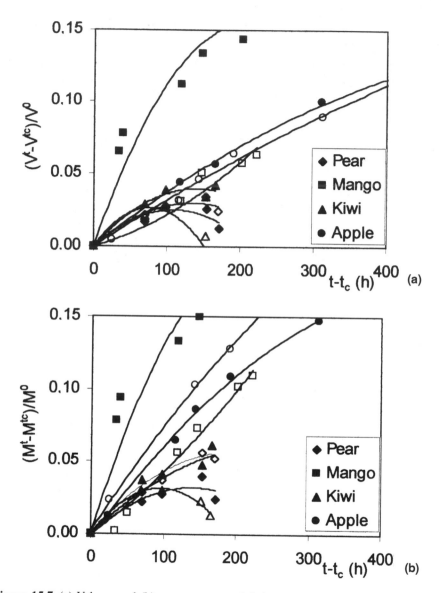

Figure 15.7 (a) Volume and (b) mass recovery of fruit cylindrical samples after chemical equilibrium at t_c with 55° Brix sucrose solution at 30°C. Open symbols = OD processed samples, closed symbols = PVOD processed samples.

than OD samples. Mass recovery is faster than volume recovery, especially for the more porous OD-treated fruit, due to the pore impregnation in line with cell network relaxation commented on above, as it has been described for apple in previous works.[11] The roundness of osmosed cells at long-term relaxation time can be appreciated in Figure 15.4d for apple tissue PVOD treated with 65° Brix sucrose solution. The reduced volume of the cell membrane can be observed in the round cell cavity.

The effect of sample vacuum impregnation with a sugar solution on the cell network relaxation rate could be explained in terms of a reinforcement of cell walls during passing through of the external liquid in line with cell wall-plasmalemma separation. Solids of external liquid may positively interact with the cell wall polymeric arrangement giving a more elastic structure. Several works have reported the beneficial effect of PVOD treatments on the preservation of cell wall ultrastructure and its implications in fruit texture improvement as compared with OD-treated products with the same water activity reduction.[18-20]

The HDM flux in the matrix relaxation process has been modeled on the basis of the Peleg model[21] for relaxation of viscoelastic structures [Equation (15.7)]. In Equation (15.7), F_0 and $F(t)$ are, respectively, the initial force acting on the sample at a given deformation, and the force at a determined relaxation time t. The constants A and B represent, respectively, the total relative relaxation level and the relaxation rate. To fit Equation (15.7) to the experimental mass recovery data ($\Delta M'$ vs. $t - t_c$), the next hypotheses were considered: the mechanical stress on the matrix is released as pressure gradients on the external liquid, which promote a determined flow pressure drop; likewise, a laminar flow was assumed. From these hypotheses, $F(t)$ can be obtained from the relationship given in Equation (15.8), where L is the pore characteristic dimension, D_p the pore diameter, and μ and ρ_{IS} the viscosity and density, respectively, of the impregnating solution and m' the liquid flux. The latter can be obtained from the slope of the curve $\Delta M'$ vs. t', according to Equation (15.9), t' being equal to $t - t_c$.

$$\frac{F_0 t}{F_0 - F(t)} = \frac{1}{AB} + \frac{t}{A} \tag{15.7}$$

$$-\Delta p = \frac{32 \mu L m'}{\rho_{IS} D_p^2} = \frac{4F}{\pi D_p^2} \tag{15.8}$$

$$m' = \left(\frac{\partial \Delta M'(t')}{\partial t'} \right) \tag{15.9}$$

From Equations (15.8) and (15.9), the relationship between $F(t)$ and data $\Delta M'(t)$ can be obtained [Equation (15.10)]. By substituting Equation (15.10) in Equation (15.7) at time t and 0, assuming constant L, the volume relaxation equation is obtained in terms of the slopes of the curve $\Delta M'(t)$ and parameters A and B [Equation (15.11)].

$$F(t) = \frac{8 \pi \mu L}{\rho_{IS}} \left(\frac{\partial \Delta M'(t')}{\partial t'} \right) \tag{15.10}$$

$$\frac{(\partial \Delta M'(t')/\partial t')_0 t'}{(\partial \Delta M'(t')/\partial t')_0 - (\partial \Delta M'(t')/\partial t')_{t'}} = \frac{t'}{Y_F} = \frac{1}{AB} + \frac{t'}{A} \qquad (15.11)$$

This model has been fitted to mass recovery data, throughout approximately two months, of apple cylinders OD and PVOD treated with sucrose solutions of different concentrations (ranging from 35 to 65° Brix) at 30, 40, and 50°C.[17] To fit Equation (15.11) to the experimental data the function $m'(t')$ was obtained by fitting a biexponential function to the experimental curve $\Delta M'(t')$, and calculating the derivative equation and its values at each time.

Figure 15.8 shows the linear relationship between experimental points plotted as defined by Equation (15.11) for OD and PVOD treatments of apple, with 55° Brix sucrose solution, at each temperature. As can be observed in Figure 15.8, a slight influence of temperature and kind of treatment on the kinetics of HDM mass flux during mechanical relaxation of apple cylinders was detected. Table 15.2 shows the mean values obtained for A and B parameters for OD and PVOD treatments in the range 30–50°C. The values of A are near 1 in all cases, which indicates that the sample recovers about 100% of initial mass throughout the examined time, the relaxation rate being affected by the syrup concentration. Until 25° Brix, the lower the sucrose concentration, the faster the relaxation, in agreement with the lower viscosity values of the solutions.

15.5 FINAL REMARKS

Deformation-relaxation phenomena of plant tissue occur in line with osmotic dehydration processes, these contributing to define mechanisms involved in mass transfer.

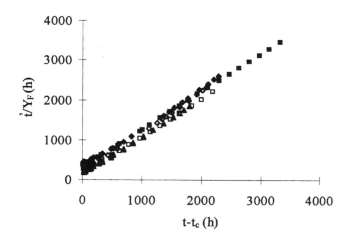

Figure 15.8 Linearization of mass recovery data of osmosed apple cylinders in 55° Brix sucrose solution. Treatment temperatures: ◆ = 30°C, ■ =40°C, ▲ = 50°C. Closed symbols = OD treatments, open symbols = PVOD treatments.

TABLE 15.2
Parameters A and B [Equation (15.11)] for Mass Recovery of OD and PVOD
Treated Apple Cylinders in Sucrose Solutions of Different Concentration

	OD		PVOD	
Osmotic solution, °Brix	A	B	A	B
65	1.02	0.002	0.98	0.009
55	1.05	0.004	1.07	0.005
45	1.02	0.012	1.05	0.037
35	1.01	0.024	1.00	0.094
25	0.98	0.162	0.96	0.082
20	1.01	0.024	1.01	0.080

Plant tissue microstructure (cell network arrangement) and porosity, as well as sample shape, affect the sample volume pathway throughout the dehydration-deformation step. They also affect sample volume recovery in line with the release of stored mechanical energy. By understanding and modeling the action of all these mechanisms, better process control will be possible.

15.6 NOTATION

μ	Solution viscosity
ε_e	Sample effective porosity
L	Sample pore length
D_p	Pore radius
ρ^0	Density of the initial product (kg/m³)
ρ_{IS}	Density of the impregnating solution (kg/m³)
$\rho_{LP}{}^t$	Density of the product liquid phase at time t (kg/m³)
x_i^t	Mass fraction in the product of component i at time t of treatment
z_i^t	Mass fraction of component i in the food liquid phase at time t of treatment
M^t	Mass of the sample at time t
V^t	Sample volume at time t
ΔV	Relative volume change with respect to the product initial value, occurred between $t = 0$ and t
$\Delta V'$	Relative volume change with respect to the product initial value, occurred between t_c and t
$\Delta M'$	Relative mass change with respect to the product initial value occurred between t_c and t
m'	Mass flux of external liquid during the sample volume relaxation period
t	Process time
t_c	Time of chemical equilibrium (equal solute concentration in product liquid phase and osmotic solution)
t'	$t - t_c$
F	Force
A	Parameter of volume relaxation equation (dimensionless)
B	Parameter of volume relaxation equation (time⁻¹ dimension)

15.7 ACKNOWLEDGEMENTS

The authors wish to thank to Comisión Interministerial de Ciencia y Tecnología (Spain), E.U (DGXII) and CYTED program for their financial support.

REFERENCES

1. Le Maguer, M. 1997. "Mass Transfer Modeling in Structured Foods," in *Food Engineering 2000*, P. Fito, E. Ortega-Rodríguez, and G.V. Barbosa-Cánovas, eds. New York: Chapman & Hall, pp. 253–270.
2. Aguilera, J. M. and D. W. Stanley. 1999. *Microstructural Principles of Food Processing and Engineering.* Gaithersburg, Maryland: Aspen Publishers, Inc.
3. Fito, P. 1994. "Modelling of Vacuum Osmotic Dehydration of Food," *J. Food Eng.,* 22: 313–328.
4. Fito, P. and R. Pastor. 1994. "Non-diffusional Mechanism Occurring during Vacuum Osmotic Dehydration," *J. Food Eng.,* 21: 513–519.
5. Fito, P., A. Andrés, A. Chiralt, and P. Pardo. 1996. "Coupling Of Hydrodynamic Mechanism and Deformation Relaxation Phenomena During Vacuum Treatments In Solid Porous Food-Liquid Systems," *J. Food Eng.,* 27: 229–240.
6. Fito, P. and A. Chiralt. 1997. "Osmotic Dehydration: An Approach to the Modelling Of Solid Food-Liquid Operations," in *Food Engineering 2000*, P. Fito, E. Ortega-Rodríguez, and G.V. Barbosa-Cánovas, eds. New York: Chapman & Hall, pp. 231–252.
7. Lazárides, H. N., P. Fito, A. Chiralt, V. Gekas, and A. Lenart. 1999. "Advances in Osmotic Dehydration," in *Processing of Foods: Quality Optimization and Process Assessment*, F. A. R. Oliveira and J. C. Oliveira, eds. Boca Raton: CRC Press, pp. 175–199.
8. Fito, P., A. Chiralt, J. Martínez-Monzó, and J. M. Barat. 1999. "Hydrodynamic Transport in Plant Tissues. A Tool in Matrix Engineering," in *Proceedings of the 6th Conference of Food Engineering (CoFE '99)*, G. V. Barbosa-Cánovas and S. P. Lombardo, eds. New York: American Institute of Chemical Engineers, pp. 59-66.
9. Fito, P., A. Chiralt, J. M. Barat, and J. Martínez-Monzó. 1999. "Vacuum Impregnation in Fruit Processing," in *Trends in Food Engineering*, J. E. Lozano, G. V. Barbosa-Cánovas, E. Parada–Arias, M. C. Añón, eds. Gaithersburg, Maryland: Aspen Publishers, Inc. (in press).
10. Barat, J. M., A. Chiralt, and P. Fito. 1998. "Equilibrium in Cellular Food Osmotic Solution Systems as Related to Structure," *J. Food Sci.,* 63: 1–5.
11. Barat, J. M., A. Albors, A. Chiralt, P. Fito. 1999. "Equilibration of Apple Tissue in Osmotic Dehydration: Microstructural Changes," *Drying Technol.,* 17 (7 and 8): 1375–1386.
12. Fito P. and A. Chiralt. 1995. "An Update on Vacuum Osmotic Dehydration," in *Food Preservation by Moisture Control: Fundamentals and Applications*, G. V. Barbosa-Cánovas and J. Welti-Chanes eds. Lancaster: Technomic Pub. Co., pp. 351–372.
13. Bomben J. L. and C. J. King. 1982. "Heat and Mass Transport in the Freezing of Apple Tissue," *J. Food Technol.,* 17: 615–632.
14. Salvatori, D., A. Andrés, A. Albors, A. Chiralt, and P. Fito. 1998. "Structural and Compositional Profiles in Osmotically Dehydrated Apple," *J. Food Sci.,* 63: 606–610.

15. Martínez-Monzó, J., N. Martínez-Navarrete, P. Fito, and A. Chiralt. 1998. "Mechanical and Structural Changes in Apple (var. Granny Smith) due to Vacuum Impregnation with Cryoprotectants," *J. Food Sci.,* 63: 499–503.

16. Fito P., A. Chiralt, J. M. Barat, D. Salvatori, and A. Andrés. 1998. "Some Advances in Osmotic Dehydration of Fruits," *Food Sci. and Technol. Int.*, 4: 329–338.

17. Barat, J. M. 1998. "Desarrollo de un Modelo de la Deshidratación Osmótica como Operación Básica," *Ph.D. thesis.* Universidad Politécnica. Valencia, Spain.

18. Alzamora, S. M., L. N. Gerschenson, S. L. Vidales, and A. Nieto. 1997. "Structural Changes in Minimal Processing of Fruits: Some Effects of Blanching and Sugar Impregnation," in *Food Engineering 2000*, P. Fito, E. Ortega-Rodríguez, and G.V. Barbosa-Cánovas, eds. New York: Chapman & Hall, pp. 117–139.

19. Muntada, V., L. N. Gerschenson, S. M. Alzamora, and M. A. Castro. 1998. "Solute Infusion Effects on Texture of Minimally Processed Kiwifruit," *J. Food Sci.*, 63: 616–620.

20. Nieto A., M. A. Castro, D. Salvatori, and S. M. Alzamora. 1998. "Structural Effects of Vacuum Solute Infusion in Mango and Apple Tissues," in *Drying'98*, C. B. Akritidis, D. Marinos–Kouris and G. D. Saravacos, eds. Thessaloniki: Ziti Editions, vol. C, pp. 2134–2141.

21. Peleg, M. 1980. "Linearization of Relaxation and Creep Curves of Solid Biological Materials," *J. Rheol.*, 24: 451–463.

16 The Relation between Sublimation Rate and Volatile Retention during the Freeze-Drying of Coffee

K. Niranjan
J. M. Pardo
D. S. Mottram

CONTENTS

Abstract

The relationship between mass transfer of water and the retention of selected volatile compounds has been studied in coffee solutions undergoing freeze-drying. A statistical experimental design enabled the evaluation of the influence of five processing parameters (pressure, heating temperature, freezing rate, initial solid contents, and thickness of the sample) on ice sublimation rate and volatile retention. Five marker volatile compounds were added to coffee extract, which was previously stripped of all volatile compounds. The solution was then freeze-dried in trays, and the amount of each volatile retained was measured using gas chromatography (GC). Models that contained process variables in linear, quadratic, and interactive terms described the results.

Temperature, pressure, and initial solid content were found to have a strong influence on sublimation rate and volatile retention. Temperature showed a direct and significant influence on the sublimation rate and volatile retention. On the other hand, an increase in solid content and pressure resulted in a decrease in sublimation rate as well as volatile retention. Thickness and freezing rate did not show a significant influence on either parameter. A more fundamental explanation to account for these trends is being developed using theories of heat and mass transfer.

16.1 INTRODUCTION

The use of freeze-drying in the food industry started after the second World War, and key developments took place in the 1950s and 1960s. Harper and Tapel[1] have reviewed its practice in this period. It is noteworthy that the reference to freeze-dried coffee in this article is very brief, and the application of the process in the coffee industry has been described as an idea "with little possibility of commercialization." The current situation is evidently in stark contrast with the predictions of this paper; soluble coffee is one of the most successful freeze-dried products available on the market, and the retained aroma is the main reason for its success.

Freeze-drying basically involves freezing a product and removing the ice by sublimation under reduced pressure (below 100 Pa) and at low temperature (below 60°C). Two different stages can be identified during drying. In the primary stage, the ice sublimes, leaving behind a porous layer. During this period, the two layers (frozen and porous) are separated by a moving interface known as the *ice front*. Ice sublimation and the re-treatment of the ice front initially dominate the drying rate, although some unfrozen water also evaporates from the porous layer left after ice sublimation. The movement of the ice front can be modeled by making use of uniform retreating ice front (URIF) equations developed by Karel.[2] Throughout the process, the removal of ice and unfrozen water is accompanied by a concomitant evaporation of volatile components.

Coffee extract has a large number of organic compounds that give it its characteristic flavor and odor. Recent studies on impact odorants of coffee listed 22 compounds as potent odorants.[3] It is known that all the volatile compounds are not perceived by human senses. It is therefore important to study the retention of volatile compounds individually rather than lumping their effects together. Fosbol[4] studied the retention of volatile compounds using different industrial freeze drying methods and found no significant difference in the average area of all peaks in the chromatogram but found variations in the retention of individual compounds.

Two basic mechanisms have been proposed to explain the retention of volatile compounds during the drying of materials that are initially in a solid state.

1. *Microregion entrapment.* This name comes from the observed fact that the volatile compounds are trapped inside localized regions with small areas.[5]

2. *Selective diffusion.* This theory, with a more mathematical approach than microregion entrapment, is built on the observed variation of diffusion coefficients (water and volatile compounds) with changes in water content.[6]

The former theory can be summarized as follows. During freezing of the aqueous solution, crystallization of water results in continuous concentration of the unfrozen solution (water, solids and volatile compounds), which remains between the growing crystals. The lowering of the water content of the unfrozen solution leads to a molecular association of carbohydrate molecules due to hydrogen-hydrogen bonding. This association produces a structure which entraps volatile compounds. This structure is stabilized after the ice has sublimed, and if moisture and temperature conditions in the sample are maintained, the volatile compounds will be retained.[7] The influence of processing parameters on volatile retention during freeze-drying can be explained by their effect on the integrity of the microregion structure.

Thijssen et al., on the other hand, developed the so-called selective diffusion theory and applied it to both spray drying and freeze-drying.

Volatile compounds in general are those having higher volatility than water at a given temperature (see, for example, Reference 8). This implies that the compounds are likely evaporate faster than water from the surface of a drying sample, decreasing to zero as soon as drying commences. Afterward, rate of loss is entirely controlled by their movement through the sample to the surface layer. This movement can be convective or molecular. Due to the characteristics of the sample during freeze-drying (high concentration of solids in the unfrozen solution) and the operating condition (low temperature), the convective movement is virtually absent and molecular diffusion governs the transport of both water and volatile.

Selective diffusion theory is based on the difference between the diffusion rates of water and volatile compounds through concentrated solutions. The diffusion coefficients of water and volatile compounds are both known to decrease with increasing concentration, but the latter has been reported to decrease more sharply.[8] Menting et al.[9] reported measurements of the diffusion coefficients of water, D_w, and acetone, D_a, in a maltodextrin-water-acetone system. The values of D_a showed a steeper decrease with the change of water content than the values of D_w. The data were correlated as follows:

$$D = A \cdot e^{-\frac{B}{C_w}} \qquad (16.1)$$

where A, B = constants
C_w = concentration of water in the sample
D = diffusion coefficient

This equation has the same form as the one proposed by Crank et al.[10] to explain the "hole" theory for the diffusion of molecules in amorphous polymer matrices. Basically, the theory describes how molecules move to a free neighboring site if they have sufficient energy to surmount the barrier separating its position from the possible new location. As the solids concentration increases, the energy necessary for one molecule to move from one free space to another increases, making the diffusion slower and, consequently, the diffusion coefficient smaller, as can be seen from Equation (16.2).

$$D = D_0 \cdot e^{-\frac{E}{RT}} \tag{16.2}$$

where D_0 = the diffusion coefficient at a reference temperature

E = energy necessary to overcome the activation barrier as well as the energy necessary to create a mol of vacancies

R = gas law constant

T = absolute temperature

It is important to highlight the positive effect of temperature on the diffusion coefficient. This influence was pointed out by Clarke[11] while reviewing published data on diffusion coefficients. He expressed this with an equation containing the ratio of the diffusion coefficient of volatile to water.

$$\log \frac{D_v}{D_w} = A - \frac{B}{T} \tag{16.3}$$

As the ratio D_v/D_w is always less than one at freeze-drying conditions,[8] the water molecules will reach the surface and evaporate faster than the volatile molecules. At some stage during drying, the water concentration at the interface will be so low that $D_v/D_w \rightarrow 0$, according to Equation (16.1), which in practice means that volatile compounds will be immobilized inside the dried sample. The moisture content at which volatile compounds are immobilized is known as critical moisture content, and it varies from 7 to 30%, depending on the compound and the nature of the solid.

The desorption rate of water and volatile compounds can be modeled making use of Fick's second law.

$$\frac{\partial C}{\partial t} = \frac{\partial}{\partial x}\left(D\frac{\partial C}{\partial x}\right) \tag{16.4}$$

This equation is applicable to calculate the desorption of water and volatile compounds in every layer of the frozen slab. As evident from Equations (16.2) and (16.3), the diffusion coefficients decrease with decreasing absolute temperature. This suggests that the volatile compounds are trapped in the frozen sample and only move when the ice has sublimed. Menting et al.[9] observed some loss of volatile compounds during freezing, but it is suggested to be ignored. Therefore, it can be concluded that aromas escape only above the ice front and below the region in which the water concentration has achieved its critical value. Above this layer, the volatile compounds that cannot escape will be retained in the final product.

To predict aroma retention, heat and mass transfer equations can be used to determine the velocity at which the ice front retreats. This rate can be coupled with diffusion movements of water and aroma from the inner part of the solid matrix. The dependence of both D_v and D_w on water content can be used to link water loss with that of volatile compounds and to build a complete mathematical description of the process. The solution for the partial differential Equation (16.4) is not simple, and the dependence of diffusion coefficients on water concentration makes it even more difficult. Bruin[12] has suggested a simplification of this approach by correlating drying time with aroma retention. A similar approach has been adopted successfully for the spray drying of coffee.[11] The aim of this paper is to explore further the relation between drying rate during sublimation and volatile retention.

16.2 MATERIALS AND METHODS

The experimental method involved four steps:

1. Preparation of coffee solution stripped of all volatile compounds
2. Adding marker volatile compounds
3. Freeze-drying
4. Analysis of the marker compound

16.2.1 PREPARATION OF THE COFFEE SOLUTION

Nescafé® granules were dissolved in deionized water and stripped of volatile compounds using a pilot plant climbing film evaporator under vacuum. The temperature of evaporation was maintained between 40 and 48°C, and the solids content increased from 10 to 35% w/w. To remove the maximum amount of volatile compounds, the concentrated solution was rediluted to 10% and concentrated once again using the same procedure. A comparison between the initial and final amount of volatile compounds can be seen in Figure 16.1.

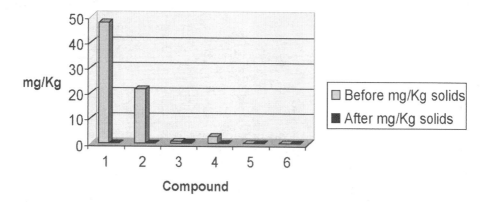

Figure 16.1 Levels of volatile compound before and after stripping of Nescafé.

Stripped coffee solutions were stored in a cold room at −18°C. One day prior to freeze drying, the coffee sample was thawed and diluted with distilled water to the desired solid content for each experiment.

16.2.2 Addition of Marker Volatile Compounds

Six volatile compounds were selected and added to the volatile-depleted coffee solutions before freeze-drying. These were methylpyrazine, dimethylpyrazine, benzaldehyde, methoxyphenol, 4-ethylbenzaldehyde, and 2-methoxy-4-methylphenol.

Dissolving 520 mg of each compound in ethanol made up a 100 ml solution of the volatile; 1 ml of this solution was diluted to 100 ml with distilled water. This stock was kept in a cold room at 5°C in glass containers.

The coffee and aroma stock solutions were mixed in a room at 10°C for 5 min using a magnetic stirrer. The volume of stock solution added to the different coffee solutions was, in all cases, calculated so as to maintain the ratio of the mass of volatile compounds to that of total solids, constant at 2.60 mg/kg solids. This concentration was selected on the basis of the amount of guaiacol present in the coffee, which has been described as an impact odorant in brews.[3]

16.2.3 Freeze-Drying

Freeze-drying was carried out in aluminum trays having an area of 0.09 m². The thickness of the slab depended on the mass of coffee solution. A pilot plant Stokes freeze dryer was used. The sample was heated by radiation from two plates, one above and the other, below the tray. The pressure in the chamber was controlled manually by a needle valve. The weight of the sample was monitored using a mechanical balance, and thermocouples were used to measure the temperature of the heating plates, the surface of the slab and the center of the sample.

A statistical experimental design enabled the evaluation of the effects of different process variables by selecting a model containing linear, quadratic, and interactive terms. A set of 30 experiments was planned, and the variables were allowed to vary in the range described in Table 16.1.

16.2.4 Analysis of Volatile Compounds

A sample of 100 ml was separated from each batch of liquid coffee before freeze-drying. These samples were stored at −35°C until they were analyzed. Fifty g of freeze-dried samples were also stored at −35°C. Volatile compounds were extracted from the samples (5 g of freeze-dried coffee or 25 g of coffee solution) using Licker–Nickerson continuous distillation with pentane (3 ml) and ether (27 ml). The extract was refrigerated to solidify the water. The unfrozen part was concentrated using a Vigreaux column. The marker volatile compounds were identified and quantified using gas chromatography mass spectroscopy (GC-MS). A dilute solution of dichlorobenzene was used as an internal standard. Details relating to the sample preparation for analysis are described by Shultz et al.[13]

TABLE 16.1
Experimental Variables and Their Units

Parameter	Symbol	Range
Initial solid contents (%)	S	[10–30]
Thickness of the slab (m)	L	[0.01–0.02]
Pressure in drying chamber (Pa)	P	[30–70]
Heating temperature (K)	T	[298.15–318.15]
Freezing rate	F	[three levels: slow, medium, fast*]

*Slow = 18 h @ –18°C + 6 h @–35°C; medium = 6 h @ –18°C + 18 h @ –35°C; fast = 24 h @ –35°C.

16.3 RESULTS AND DISCUSSION

Genstat version 4.1 for Windows was used to analyze the effect of process variables (pressure, temperature, freezing rate, initial solids content, and thickness of the slab) on the drying rate and the volatile retention of each compound. The result of the analysis was an expression similar to Equation (16.5) and included the effect of each parameter using linear, quadratic and interactive terms.

$$R = \left(\sum_{1}^{n} K_j \cdot V_j \right) + \left(\sum_{1}^{n} L_j \cdot V_j^2 \right) + \left[\sum_{1}^{n} M_j \cdot (X \cdot Y)_j \right] \tag{16.5}$$

where R = the response variable (drying rate or retention)
K, L, M = constants

V_j, V_j^2 = process variables

$(X \cdot Y)_j$ = pairs of interactive variables

16.3.1 DRYING RATE

By applying Equation (16.5) to the experimental data, the following equation was obtained for the drying rate (correlation coefficient = 0.87):

$$dr = (11.36 - 0.55\overline{P} + 1.5\overline{T} - 1.06\overline{S} - 0.47\overline{L} + 0.79\overline{F} + 0.66\overline{PT}$$
$$- 0.62\overline{PS} + 0.26\overline{PL} + 0.22\overline{TL} - 1.04\overline{PF} + 0.49\overline{LF}) \times 10^{-4} \tag{16.6}$$

It is important to note that all parameters were normalized so that their values lie in the range [−1,0,1]. The normalization was done using Equation (16.7).

$$\overline{V} = \frac{U - U_0}{abs(U - U_0)} \tag{16.7}$$

where $\bar{V}$ is the standardized variable, U_0 is the variable to standardize, and U is the mean value of the variable. Equation (16.6) was used to generate the plots shown in Figures 16.2 through 16.6.

Drying rate consistently increased with temperature under all conditions examined. This trend is expected, due to an increase in the temperature driving force for heat conduction. Equation (16.8) shows the effect of temperature on the rate at which heat enters a slab by conduction from top and bottom.

$$q_I + q_{II} = q_T = \frac{AK_s(T_s - T_i)}{L(t)} + \frac{K_iA(T_{bot} - T_i)}{L - L(t)} \tag{16.8}$$

where K_s and K_i = thermal conductivities of porous and frozen layers
        T_i, T_{bot}, and T_s = the ice front, bottom, and surface temperatures
        L = thickness of the slab
        $L(t)$ = position of the ice front measured from the top

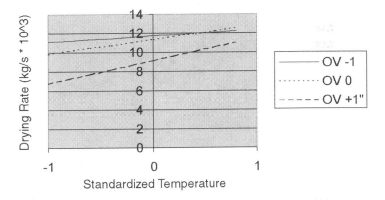

Figure 16.2 Influence of temperature on drying rate. OV = other variables.

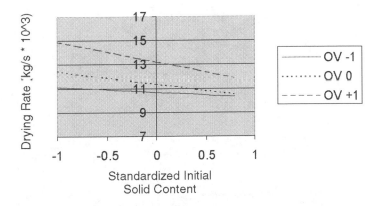

Figure 16.3 Influence of initial solid content on drying rate. OV = other variables.

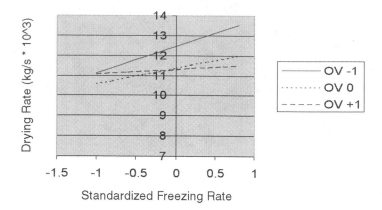

Figure 16.4 Influence of freezing rate on drying rate. OV = other variables.

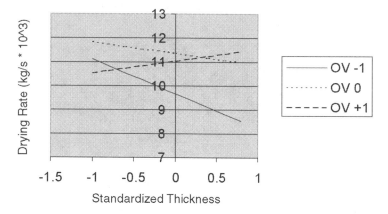

Figure 16.5 Influence of thickness on drying rate. OV = other variables.

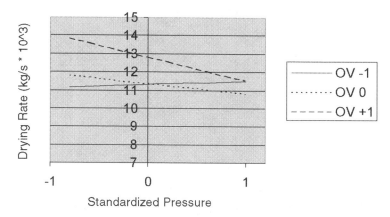

Figure 16.6 Influence of pressure on drying rate. OV = other variables.

In Figure 16.2, the influence of the other process parameters can be observed. Each line represents one of the three arbitrary levels assumed by the other parameters (−1,0,1). At constant temperature, the overall influence of increasing other parameters is to reduce the drying rate. Consider, for instance, the effect of increasing solid contents at a given temperature. Equation (16.9) shows the influence of solid contents on the thermal conductivity of coffee, which has been taken from Costherm.[14]

$$K_s = \left(-0.547\frac{\rho_t S}{\rho_s}\right) + 0.59 \qquad (16.9)$$

where K_s = thermal conductivity of the porous layer
ρ_s, ρ_t = densities of the porous layer and of the whole slab
S = initial solid content

It is evident that increasing the solids content would increase the thermal conductivity of the dry layer and therefore improve heat transfer rates. However, the higher solid content also reduces pore size according to the following equation:

$$S = \frac{AL - \pi d^2 n L \rho_s}{4W_t} \qquad (16.10)$$

where   the pores are assumed to have the volume of a cylinder
d = pore diameter
n = number of pores
L = thickness of the slab
W_t = weight of the slab
S = solids content

The smaller pore will reduce the drying rate, as evident from the following equation, which is a simplification of the mass transfer equation for a slab, where vapor is escaping only from the top.

$$\frac{dW}{dt} = \frac{ACd(p_i - p_s)}{L(t)}\sqrt{\frac{M}{T}} \qquad (16.11)$$

where A = area of the tray
d = pore diameter
L = position of the ice front
p_i, p_s = partial pressures of vapor at the ice front and surface
T = temperature
M = molecular weight of water vapor
C = a constant depending on the structure of the porous matrix

Thus, increasing initial solid content has two conflicting effects.

1. It increases the heat transfer rate due to higher thermal conductivity.
2. It reduces the mass diffusion rate due to a reduction in pore size.

In our experiments, the latter appears to dominate, as shown in Figure 16.3.

The freezing rate shows a slight influence on the drying rate (Figure 16.4, OV0). From our experiments, it is difficult to confirm the trend observed by Thijssen and Rulkens,[6] who reported a decrease in drying rate with increasing freezing rate, based on the well-known fact that faster freezing results in smaller ice crystals and therefore smaller pores. It is likely that the difference between the three freezing levels used in this work was not high enough to detect any significant changes.

Thickness by itself shows small influence on the drying rate (Figure 16.5, OV0). This is an expected trend and is in agreement with observations made by Brutini et al.[15] They explained this trend as a consequence of the dominance of heat transfer on the process; their conclusion neglects the influence of mass transfer on the drying rate during the sublimation stage.

Thickness shows importance as an interactive variable as evident from Equation (16.6) and in Figure 16.5. The interaction of the thickness with the temperature has its basis in the collapse of the structure, as explained by Mackenzie.[16] This collapse is caused by the frozen sample attaining the glass transition temperature.[7] At the start of drying, before the working vacuum has been achieved (5 to 10 min), the sample is subjected to a high temperature gradient ($-30°C$ in the sample and $+20°C$ in the drying chamber), which will inevitably affect the sample: several layers of the slab will reach the critical temperature range of -24 to $-20°C$, the range in which the glass transition temperature is found for coffee solutions.[11] At this range of temperatures, the frozen sample will be transformed from a solid material to a viscous liquid, leading to a collapse in the structure, which will reduce the drying rate. The thinner the sample, the higher the ratio of collapsed layer to total thickness; that means a higher influence of the collapsed layer on the drying rate.

Pressure is expected to have influence on both heat and mass transfer. Several researchers[17-19] reported significant variations of thermal conductivity with the variation of pressures. The following empirical equation, developed by Harper, indicates how pressure influences thermal conductivity within a transition region between high vacuum (below 0.1 mm Hg) and low vacuum (>10 mm Hg). Increasing the pressure will increase the thermal conductivity.

$$K_d = (1 - \varepsilon)*K_s + \frac{K_{go}}{1 + \dfrac{c}{p}} \qquad (16.12)$$

where K_s, K_{go} = the thermal conductivities of the solids and the gas

$\qquad\qquad p$ = absolute pressure

$\qquad\qquad \varepsilon$ = porosity of the material

$\qquad\qquad c$ = a constant relative to the nature of the solids

As seen in Equation (16.11), pressure has a direct influence on mass transfer. Increase in total pressure in the cabinet will increase the partial pressure (*pi*) and consequently will decrease the driving force (*ps-pi*) and the drying rate. Figure 16.6 shows the negative influence of pressure on the drying rate, which further confirms that the drying is being governed by mass transfer.

Thus, lower pressures, higher temperatures, and lower initial solid content seem to result in higher drying rates in our work.

16.3.2 VOLATILE RETENTION

Given that the mass of each marker volatile added to a sample was exactly the same, the volatile retention between the various samples was determined by comparing the concentration of the compounds retained after drying (milligrams final volatile/2.6 mg added).

Due to the minute quantities of compounds used and the high volatility of two of the compounds (methylpyrazine and dimethylpyrazine), it was difficult to separate the area under the peaks in the chromatogram from the background noise due to residual volatile compounds that remained in the coffee after stripping.

No significant difference was detected in the retention of benzaldehyde between the different experimental conditions. It is important to note that a small amount of benzaldehyde remained in the coffee after stripping (Figure 16.1); this shows that it is difficult to remove this compound completely. Further experiments with higher concentrations of Benzaldehyde are necessary to confirm this observation.

A model was fitted using the data for the retention of 4-ethylbenzaldehyde and 2-methoxy-4-methylphenol. The correlation factor R^2 was low for the ethylbenzaldehyde (46%). In the case of 2-methoxy-4-methylphenol, the correlation factor was much greater at 85%; the equation relating the retention with process parameters is as follows:

$$\text{Re } t = (580 - 93.8\bar{P} + 47.4\bar{T} + 3.46\bar{S} + 17.5\bar{L} + 0.85\bar{F} + \qquad (16.13)$$
$$100\bar{L}^2 + 160\overline{TS} - 24\overline{TL} + 42.4\overline{TF} - 14.8\overline{LF})10^{-3}$$

A positive influence of increasing temperature on the volatile compound retention can be observed in Figure 16.7 (OV0). From Equation (16.3), it is seen that an increase in temperature will change the ratio D_v/D_w. Depending on the volatile substance, especially its polarity, this ratio can either increase or decrease. In the case of methyl-guaiacol, the ratio of diffusion coefficients appeared to decrease; this is evident from the greater retention at higher temperature.

Experimental data shows a positive influence of increasing the initial solid content on volatile retention Figure 16.8. From the phase diagram reported by Clarke[11] and by Sivetz and Desrosier,[20] it is possible to see that the concentration of unfrozen water in the solid matrix is independent of the initial solid contents of the solution; it depends only on the freezing temperature. Thus, the initial solid contents of the solution will only affect porosity; the lower the initial solid concen-

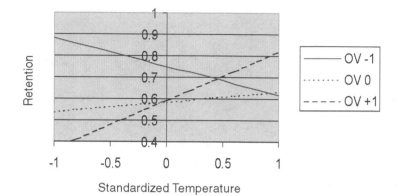

Figure 16.7 Influence of temperature on retention. OV = other variables.

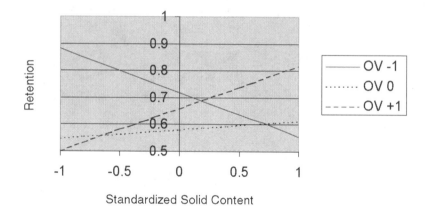

Figure 16.8 Influence of solid content on volatile retention. OV = other variables.

tration, the higher the porosity. Consequently, the diffusion path length for the volatile compounds will be lower, causing more aroma to evaporate.

It is important to highlight the considerable significance of the interactive term in Equation (16.13); both high levels of temperature and high levels of solid content lead to high volatile retention. However, no explanation has been found for the adverse trend at low values of temperature and solids due to this interactive term.

It can also be seen that the interactive terms containing freezing rate ($\overline{F}$) affect volatile retention to a greater extent than the linear term in $\overline{F}$. Regardless, it is evident from Figure 16.9 that freezing rate has a weak influence on volatile retention. This, coupled with the earlier observation on drying rate, confirms that the variation between the three freezing levels was not significant enough to observe differences.

From Equation (16.13) and Figure 16.10, it can be seen that there is an inverse relationship between volatile retention and pressure. The absolute pressure in the drying cabinet has direct influence on the partial pressure of vapor at the ice front

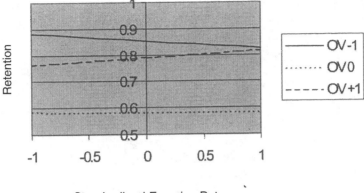

Figure 16.9 Influence of the freezing rate on volatile retention. OV = other variables.

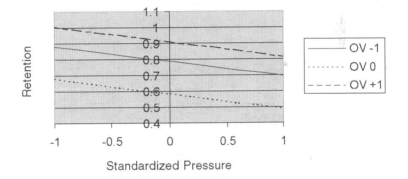

Figure 16.10 Influence of pressure on volatile retention. OV = other variables.

and, consequently, on the temperature of the ice front. Lowering the temperature of
the ice front has two important consequences directly related to volatile retention.
The first one is the lowering of the D_v and D_w due to the influence of temperature
on these values, as can be seen from Equation (16.2). The second and perhaps most
important consequence is the lowering of the concentration of unfrozen water in the
solid matrix. As stated above, the amount of unfrozen material decreases with
decreasing temperature. The change in water concentration has a positive effect on
the diffusion coefficients, as stated in Equation (16.1), and therefore, decreasing the
amount of unfrozen water will result in an increase in volatile retention. It is
important to highlight that this lowering of water concentration in the solid matrix
of coffee occurs until the value moisture content drops to around 0.30 kg water/kg,
when a thermodynamic equilibrium is achieved.[11] Lowering the temperature further
will have no effect on water concentration.

The effect of thickness (Figure 16.11) is evident in the quadratic term. Its influence on volatile retention is not clear. No influence of this parameter on volatile retention was expected. Its interactive character $\overline{TL}$ can have the same explanation as the one given for the influence of this parameter on drying rate; i.e., structure collapse will liberate the entrapped volatile compound and lower retention.

16.4 CONCLUDING REMARKS

Temperature, pressure, and initial solid content showed a strong influence on the drying rate and in the volatile retention demonstrated by the marker 2-methoxy-4-methylphenol. The interactive effect of the variables on volatile retention is very strong—indeed, stronger in some cases than their effect on drying rate.

Freezing rate did not show a significant effect on the drying rate or on the retention of volatile compounds over the range of conditions investigated. The possibility of using a wider range of freezing levels should be explored. Thickness did not show a significant effect on the drying rate, but showed a strong favorable influence on volatile retention.

Microregion entrapment and selective diffusion theory provide a suitable framework to explain the observed trends experimentally. A more in-depth knowledge of the diffusion coefficients of the marker volatile compounds would be useful to apply Thijssen's theory in a more quantitative manner.

It would be desirable to add higher levels of volatile markers in order to confirm the observed trends.

16.5 ACKNOWLEDGEMENTS

The support of America Latina Formación Académica (ALFA) programme, between The University of Reading (U.K.) and La Universidad de La Sabana (Colombia), is gratefully acknowledged.

J. M. Pardo would like to acknowledge the support of Dr. Stephen Elmore and Mr. Andrew Dodson in the GC-MS analysis, and Dr. Steven Gilmour in the statistical design of the experiments.

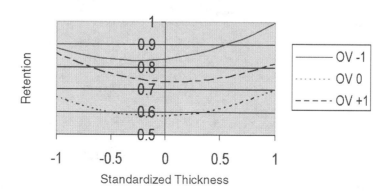

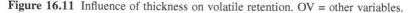

Figure 16.11 Influence of thickness on volatile retention. OV = other variables.

REFERENCES

1. Harper, J. and Tapel. 1957. "Freeze-Drying of Food Products," in *Adv. In Fd. Research*, E. Mark and Stewart, eds., New York: A.C. Press.
2. Karel, M. 1975. "Heat and Mass Transfer in Freeze-Drying," in *Freeze-Drying and Advances in Food Technol.*, S. A. Goldblith, L. R. Rey, and W. W. Rothmayer, eds. London: Academic Press.
3. Semmerlcoh, P. and W. Grosch. 1996. "Studies on Character Impact Odourants of Coffee Brews," *J. Agric. Fd. Chem.* 44: 537–543.
4. Fosbol, P. and E. Petersen. 1980. *Flavour Retention During Freeze Drying of Coffee.* Proceedings, London: ASIC Colloquium, pp. 251–263.
5. Flink, J. 1975. "The Retention of Volatile Components during Freeze: A Structurally Based Mechanism," in *Freeze-Drying and Advances in Food Technol.*, S. A. Goldblith, L. R. Rey, and W. W. Rothmaye, eds. London: Academic Press.
6. Thijssen, H. and W. Rulkens. 1969. "Effect of Freezing Rate on the Rate of Sublimation and Flavour Retention in Freeze-Drying," *in Recent Developments in Freeze-drying*, International Institute of Refrigeration Commission X, ed., Laussane.
7. Ross, Y. 1995. "Glass Transition–Relate Physicochemical Changes in Food," *Food Technology*, Oct. 97–102.
8. Thijssen, H. A. C. 1975. "Effect of Process Conditions in Freeze-Drying on Retention of Volatiles Components," *in Freeze Drying and Advances in Food Technol.*, S. A. Goldblith, L. R. Rey, and W. W. Rothmayer, eds. London: Academic Press.
9. Menting, L., B. Hoogstad, and H. A. C. Thijssen. 1970. *J. Fd. Technol.* 5: 11–126.
10. Crank, J., N. R. McFarlane, J. C. Newby, G. D. Paterson, and P. Pedey. 1981. *Diffusion Processes in Environmental Systems.* V2 Technology, Hong Kong: Macmillan Press Ltd.
11. Clarke. 1987. "Drying," in *Coffee*, R. J. Clarke and R. Macrae, eds. London: Elsevier Applied Science.
12. Bruin, S. 1992. "Flavour Retention in Dehydration Processes," in *Advances in Food Engineering*, P. Singh and M. Wirakartakusumah, eds. U.S.A.: CRC.
13. Schultz, T. H., R. A. Flath, T. R Mon, S. B. Eggling, and R. Terranishi. 1977. *J. Agric Food Chem.* 25: 446-449.
14. Jowitt, R., F. Escher, B. Hallstrom, H. F. Th. Meffert, W. E. L. Spiess, and G. Vos. 1983. *Physical Properties of Foods,* vol. 1. London: Elsevier.
15. Mackenzie, A. P. 1975. "Collapse during Freeze-Drying—Qualitative and Quantitative Aspects," in *Freeze-Drying and Advances in Food Technol.* S. A. Goldblith, L. R. Rey, and W. W. Rothmayer, eds. London: Academic Press.
16. Brutini, R., G. Rovero, and Baldi. 1991. "Experimentation and Modeling of Pharmaceutical Lyophylization Using a Pilot Plant," *The Chemical Engineering Journal*, 45: 367–377.
17. Harper, J. 1962. "Transport Properties of Gases in Porous Media at Reduced Pressure with Reference to Freeze-Drying," *A.I.Ch.E. Journal*, 298–303.
18. Mellor, J. D. 1978. *Fundamentals of Freeze-Drying.* London: Academic Press.
19. Liapis, A. 1987. "Freeze Drying," in *Handbook of Industrial Drying*, A. Mujmdar, ed. New York: Marcel Decker.
20. Sivetz, M. and N. Desrosier. 1979. *Coffee Technology.* U.S.A.: The AVI Publishing Company.

17 A Proposal of Analysis of the Drying Phenomena by Means of Fractal Theory

G. F. Gutiérrez-López
J. Chanona-Pérez
L. Alamilla-Beltrán
A. Hernández-Parada
A. Jiménez-Aparicio
R. Farrera-Rebollo
C. Ordorica-Vargas

CONTENTS

1-56676-963-9/02/$0.00+$1.50
© 2002 by CRC Press LLC

Abstract

A proposal for studying the kinetic aspects of convective drying is presented herein, based on the fractal evaluation of the drying curve, taking slabs and spheres as working geometric systems. Drying kinetics of these models were determined and the fractal dimensions of the surface temperature distribution (STD) of the slabs and of the mass and size of spheres were obtained. It was possible to correlate the different stages of the drying curve with the fractal dimension of the STD of the slabs along this curve. The STD presents a Sierpinski carpet pattern suggesting the absence of a constant rate period of drying. Based on the determinations of the STD and of the range of invariance of the fractal dimensions found, a falling rate period of drying governed by randomness was identified. Euclidian (linear) behavior was observed toward the end of this period and, along the period of adjustment of conditions, a chaotic region characterized by the presence of a strange attractor was found. When comparing surface temperature kinetics at different operating conditions, self-similarity behavior is presented within the points of measurement, which could make possible the prediction of the surface temperature of these systems. Also, fractal dimensions of mass and fractal dimensions of the drying process for the dehydration of spheres were evaluated, and correlations between the rate of drying and the homogeneity of the process were found. This study may constitute a proposal toward the systematic study of drying kinetics with the aid of the fractal theory and can be applied to predict the degree of homogeneity of the process and its consequence in product quality.

17.1 INTRODUCTION

The word *fractal* was created by Mandelbrot.[1] A fractal object presents similar structure at different observation scales. One of the first uses of fractal numbers was to determine tortuosity of coastlines and of irregular objects.[1] Values of fractal dimensions (D) are between 0 and 3. The most widely used methods to determine D are Richardson's graph, box-counting dimension, surface area, and fractal mass.[2–5] Also, fractal interpolation and experimental data treatment like the one by Barnsley[6] are used. Application of fractal analysis has spread and food engineering is not an exception.[7] One example in this area is the characterization of rugosity of particles such as instant coffee, milk powder, casein aggregates, and others.[8–12] Also, fractal isotherms have been obtained from GAB and BET models.[13–15] Kinetics of decomposition and generation of toxic substances may be irregular or chaotic and thus described by fractal models. Within food dehydration, kinetics of moisture, surface area, and the shape of the product, as well as superficial temperature distribution, may be random along the process, given the complexity of the structures and variation of process conditions. Fractal analysis may help to characterize the process and the product. This may be useful in the setting and appraisal of drying conditions and evaluation of the product and may also contribute to the better understanding of drying operations. In this work, fractal analysis of the drying curve and of the superficial temperatures of the product is presented, which may lead to the evaluation of properties of the drying surface during drying.

17.2 OBJECTIVE

The objective is to contribute to the description of the drying operation of food-related products by means of the fractal theory.

17.3 EXPERIMENT

17.3.1 MODEL FOOD

Plates of a carbohydrate-based food model were prepared by mixing a 6% w/w agar solution with glucose syrup 40% w/w in a 1:3 ratio. Solution so prepared was heated at 80°C and cast into a perspex tray $10 \times 10 \times 0.15$ cm (Figure 17.1). Also, jellified spheres containing 1% agar solution and skimmed milk 20% total solids (ts) were prepared by injecting the solution into an aluminium mold with a 0.9 cm internal diameter. Two plates and two spheres were prepared for each experiment. With one plate, the drying kinetics were followed, while the second one was used to measure surface temperatures (plates) and changes in size (spheres). In the case of the spheres, three measurements of diameter were carried out along the same axis each time and perpendicular to each other using a calliper (Mitutoyo no. 39652-20). Food model plates were tested for shrinkage during drying by measuring decrease in length and thickness with the above-mentioned caliper; less than 0.5% of decrement in size was detected. Plates were dehydrated in an experimental drying tunnel with airflow parallel to the surface of the samples (Figure 17.2). Spheres were dried in an experimental tunnel with air flowing around the samples (Figure 17.3).

17.3.2 EXPERIMENTAL DESIGN

Drying conditions for plates are shown in Table 17.1 and for spheres in Table 17.2. A complete factorial design 2^3 and 2^4 was applied for the experiments and included three repetitions for each point.

TABLE 17.1
Drying Conditions for Plates

T (°C)	Velocity (m/s)		
45	1	2	3
55	1	2	3
65	1	2	3

17.3.3 SURFACE TEMPERATURE (ST) EVALUATION

Twenty-five surface temperatures were evaluated, homogeneously covering the plate. An infrared thermometer (Raytek RAYST6LXE) with a laser-pointing beam was used (Figures 17.1 and 17.2). ST kinetics were determined until the plates reached a constant temperature. (Figures 17.5 and 17.7).

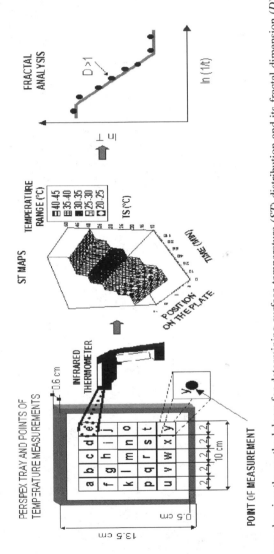

Figure 17.1 Diagram showing the methodology for determining surface temperature (*ST*) distribution and its fractal dimension (*D*).

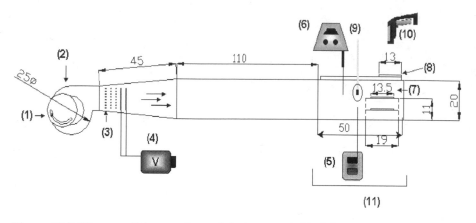

Figure 17.2 Diagram of the experimental drying tunnel (1) airflow controller, (2) fan, (3) heating elements, (4) variable resistance, (5) thermoanemometer, (6) temperature recorder, (7) support for test sample, (8) rail for thermometer, (9) wet bulb temperature recorder, (10) infrared thermometer, (11) test section. Dimensions are annotated in centimeters.

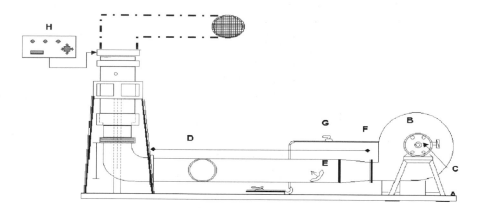

Figure 17.3 Experimental drier for spheres. (A) support, (B) fan, (C) motor, (D) air conduct, (E) airflow regulator, (F) heater, (G) gas regulating valve, and (H) thermoanemometer.

17.3.4 DRYING KINETICS

The weight of a plate undergoing drying was registered at different times by means of an analytical balance (Ohaus, Analytical Plus), and moisture vs. time curves were thus constructed.

17.3.5 FRACTAL ANALYSIS

Fractal kinetics of ST distribution were determined according to methodology reported elsewhere.[6,16] In Richardson's graph, ln(TS) vs. ln($1/t$), where t is drying

TABLE 17.2
Drying Conditions for Spheres

T (°C)	Velocity (m/s)			
50	0.5	1.5	6.8	8
70	0.5	1.5	6.8	8
80	0.5	1.5	6.8	8
90	0.5	1.5	6.8	8

time, was obtained. To find the fractal dimension D, the theorem of the box was used,[1] which may be represented by

$$D = 1 + |\alpha| \qquad (17.1)$$

where $|\alpha|$ = absolute value of Richardson's plot
D = the fractal dimension

For spheres undergoing drying, the fractal dimension of mass (D_{me}) was obtained along drying kinetics by relating the ln of the weight (W) of the sphere to its diameter according to Equation (17.2). The fractal dimension of the process (D_{mp}) was calculated according to Reference 4 using Equation (17.3).

$$W = R^{D_{me}} \qquad (17.2)$$

$$W(R) = R^{D_{mp} - D_c} \qquad (17.3)$$

In these equations, R is the average diameter of the sphere; D_c is an indicative factor of the geometry of spheroids particles, which for this case adopts a value of 1, given that one dimension (diameter) is being measured.[17] A Richardson's graph was constructed by plotting $\ln(R)$ vs. $\ln(W)$. D_{mp} characterizes the structure of particles along the process and D_{me} the structure of an individual particle.[4,18]

17.4 RESULTS AND DISCUSSION

17.4.1 ANALYSIS OF DRYING KINETICS

In Figure 17.4, the effect of air velocity is shown. A slight increase in drying rate is observed with air velocity, probably due to the increased heat transfer film coefficient. Values of effective diffusion coefficient D_{eff} evaluated by Arrhenius equation are shown in Table 17.3. It is possible to note that the fractal dimension D, its period of invariance, and the value of D_{eff} may characterize the falling rate period and may indicate an oscillating trend of surface temperature distribution, like those described in this chapter, and the uneven distribution of water inside the solid. These parameters may also be correlated among each other to help in the characterization of the drying curve.

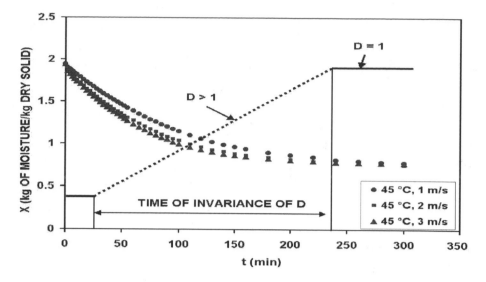

Figure 17.4 Drying kinetics of slabs at different air velocities.

TABLE 17.3
Fractal Dimension of Surface Temperatures, Time Invariance of D, and Effective Diffusion Coefficient at Drying Conditions

Drying conditions	Fractal dimension	Time of invariance of D	Effective diffusion coefficient $\times 10^5$ [m²/h]
45°C, 1 m/s	1.290	22–200 min	2.07
45°C, 2 m/s	1.255	16–160 min	2.36
45°C, 3 m/s	1.237	16–160 min	2.76
55°C, 1 m/s	1.268	19–220 min	2.44
55°C, 2 m/s	1.253	19–200 min	2.36
55°C, 3 m/s	1.212	16–140 min	2.69
65°C, 1 m/s	1.261	19–160 min	2.85
65°C, 2 m/s	1.226	16–160 min	2.90
65°C, 3 m/s	1.177	13–110 min	2.96

The absence of a constant rate period is noticeable, since the superficial temperature of the slab is higher than the wet bulb temperature of the air (Figure 17.5). In Figure 17.6, the effect of temperature on drying rate is observed. From these graphs, the effective diffusion coefficients were obtained and this parameter correlated with the fractal dimension and its period of invariance (Table 17.3). Over the major part of the falling rate period, a constant value of D can be used and this value correlated with the fractal dimension, which in all cases was higher than 1, indicating

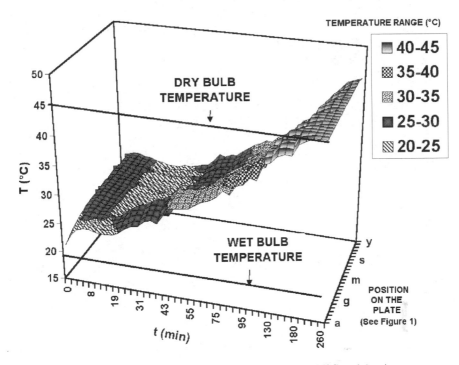

Figure 17.5 ST distribution of slabs when drying with air at 45°C and 1 m/s.

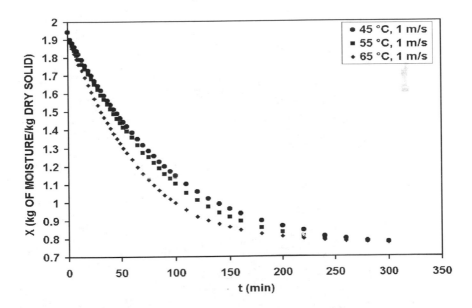

Figure 17.6 Drying kinetics of a slab at different air temperatures.

a nonlinear behavior. These findings may also have a correlation with the structure of the material undergoing drying, since structures through which water moves may have geometric self-similarity as the value of D is above 1. Similar findings were obtained for the dehydration of spheres (Figures 17.12 and 17.13).

17.4.2 KINETICS OF SURFACE TEMPERATURE DISTRIBUTION AND FRACTAL ANALYSIS

In Figures 17.5 and 17.7, it is possible to observe that ST distribution does not follow a uniform pattern. However, when analyzing Figure 17.8, a tendency toward a periodic behavior is observed that is typical of nonlinear functions.[19,20] The shape of the lines in Figure 17.8 suggests that the surface of the plate has a self-similarity pattern that repeats along drying time and that is perpendicular to the direction of evaporation of water.

This suggests that the drying zones appear randomly on the surface, along the falling rate period (Figure 17.8, Figure 17.11, and Table 17.3). The fractal dimension of the ST distribution kinetics for all experiments showed values lower than 2, which corresponds to a Sierpinski carpet[1] and agrees with the postulates of Kopelman,[16] who reported catalytic models for chemical kinetics of mono-dispersed isles on the surface of radioactive materials, characterized by fractal analysis.

These findings may support the postulates of the classical theory of drying in which there is a period when zones of lower moisture content (mono-dispersed isles),

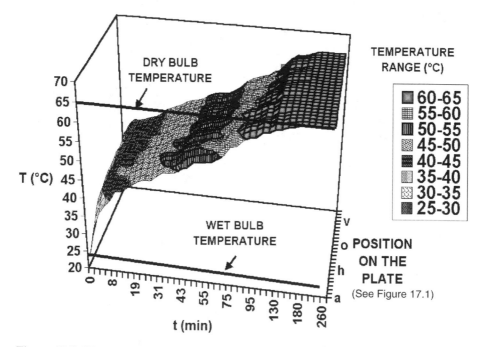

Figure 17.7 ST distribution of a slab when drying with air at 65°C and 2 m/s.

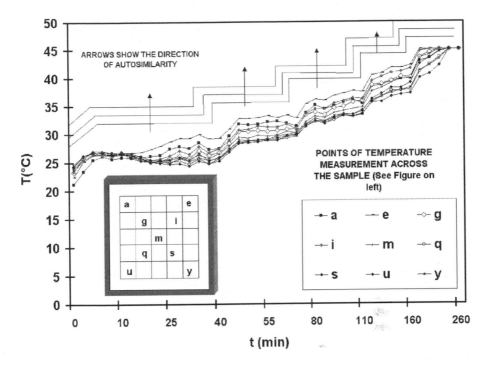

Figure 17.8 Self-similarity behavior of ST kinetics for plates dried at 45°C and 1 m/s.

and thus of lower temperature appear, randomly on the surface of the solid.[21-23] Different values of fractal dimensions are observed in Figures 17.9 and 17.10, which are the Richardson's graphs applied to the drying kinetics of the plates. This diversity in data is also found in systems with finite self-similarity, with invariance on the scale and which obey mathematical power laws.[24] The different slopes found may characterize the different periods of the drying phenomena. The first stage corresponds to a period of adjustment of drying conditions and represents around 3% of the total drying time, as reported in the literature.[25] A second stage with a D higher than 1 is followed by a third stage with a D equal to 1 (Euclidian behavior), which corresponds to the end of the process. In consequence, it is possible to postulate that the fractal dimension of ST distribution may be used as a criterion to determine stages of drying and zones of transition between them.[26]

For the falling rate period, different values of D are found for the various conditions tested; these are reported in Table 17.3 and in Figure 17.11. The influence of temperature and air velocity on D can be observed. D is lower at higher values of temperature and air velocity. This is due to the fact that ST distributions have a frequency of variation more homogeneous (less fractal) than when conditions are less drastic at lower temperatures and air velocities. However, the time at which the zone of invariance of D (Figures 17.4, 17.9, 17.10, and 17.11, and in Table 17.3) starts and finishes, and its duration, varies. It has been found that fractal dimension

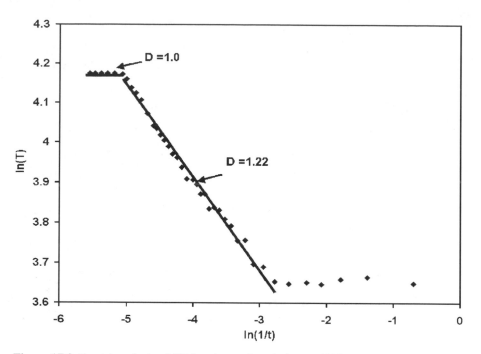

Figure 17.9 Fractal analysis of ST for plates when drying at 65°C and 2 m/s.

of surface temperature distribution is characteristic for each drying condition. These findings may be useful for the analysis of the drying operation and to help to predict quality and economics of the process. Also, stages of the operation may be differentiated, and the drying time can be estimated.

Prediction of ST distribution is useful in quality assessment, given the fact that zones exposed to longer drying times are more prone to deteriorate. Information on the structure of materials and its impact on drying is obtained from the considerations of self-similarity. In another work,[27] when drying the same material under similar conditions as those used in this work, a microstructural pattern of superimposed layers of similar structure was found.[27] This may partially explain the self-similarity pattern of ST distribution found in this work, since these layers would receive heat one after the other starting from the top. Consequently, water will evaporate from each layer starting from the top one, which will dehydrate first and therefore increase its temperature prior to those layers beneath it. This proposal could have consequences on relating drying conditions and product architecture, and it may give information on the homogeneity of the drying conditions when treating the material. The importance of this has been widely discussed in the literature.[28,29]

Distribution of surface temperatures is governed by randomness along the plate and may be characterized by the value of D. In an analogous way, fractal analysis of the thermal history of spheres and of their drying curves may be useful when considering droplets subjected to dehydration in spray-drying operations.

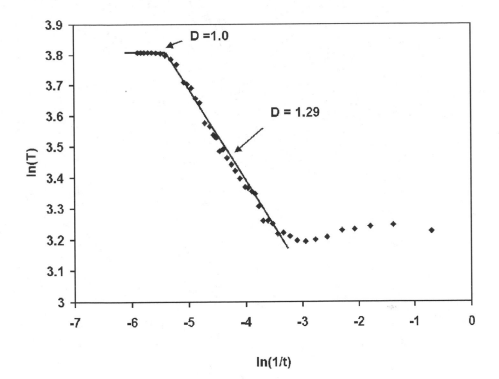

Figure 17.10 Fractal analysis of ST for plates when drying at 45°C and 1 m/s.

17.4.3 FRACTAL DIMENSION OF SPHERES

Using Equations (17.2) and (17.3), the fractal dimension of the mass of spheres during drying and the fractal dimension of mass during the process were evaluated. Results showing the maximum value of (D_{me}) and the various (D_{mp}) are presented in Tables 17.4 and 17.5 and in Figures 17.12 and 17.14. In Figure 17.12, the corresponding drying kinetics are also found.

Analysis of curves D_{me} vs. time (Figure 17.12), suggest that values of D_{me} may be correlated with stages of drying. Values of maximum mass fractal dimension $(D_{me})_M$ may be characteristic for each drying condition (Table 17.4 and Figure 17.12). It is possible to observe that at higher airflows and temperatures, $(D_{me})_M$ is also higher, indicating a more random behavior at higher mass transfer rates. During spray drying, it has been shown[27] that the major part of the drying process takes place in the outer radial half of the chamber, in which the values of the relative Reynolds number of air droplets approaches zero. In this region, changes in the size, shape, and nature of the particles are more pronounced due to the high drying rate, and thus higher values of $(D_{me})_M$ are expected, as has been found in this study. The results presented here may thus be useful when studying spray-drying operations.

TABLE 17.4
Maximum Mass Fractal Dimensions $(D_{me})_M$ for spheres and Time to Reach Them, for Different Drying Conditions

Drying conditions	50°C		70°C		80°C		90°C	
	t (min)*	$(D_{me})_M$	t (min)*	$(D_{me})_M$	t (min)*	$(D_{me})_M$	t (min)*	$(D_{me})_M$
0.5 m/s	340	1.79	280	1.81	220	1.84	200	1.81
1.5 m/s	260	1.79	255	1.80	220	1.81	170	1.82
6.0 m/s	240	1.71	240	1.69	180	1.81	160	1.73
8.0 m/s	230	1.81	200	1.75	180	1.81	120	1.82

*Time to reach maximum D_{me}

TABLE 17.5
Process Mass Fractal Dimension (D_{mp}) of Spheres under Different Drying Conditions

Drying conditions	D_{mp} 50°C	D_{mp} 70°C	D_{mp} 80°C	D_{mp} 90°C
0.5 m/s	3.4325	2.9260	3.0682	3.2078
1.5 m/s	3.1808	2.4562	2.6237	3.1618
6.0 m/s	2.8214	1.9647	2.4785	2.7970
8.0 m/s	1.4667	1.1634	2.4986	1.8149

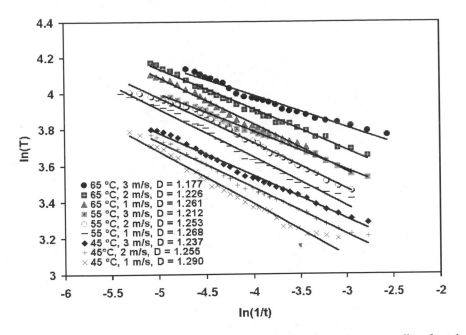

Figure 17.11 Richardson's graph for ST distribution of slabs and corresponding fractal dimension for the various combinations of air velocity and temperatures used.

17.4.4 THE CHAOTIC PATTERN OF ST DISTRIBUTION OF PLATES DURING THE FIRST PERIOD OF DRYING

The ST distribution of plates during the period of adjustment of conditions suggests a chaotic behavior. An analysis of this period by nonlinear dynamics suggests the presence of a strange attractor when plotting ST distribution trajectories in a bidimensional phase space (Figures 17.16 and 17.17). This analysis consists of taking a time series of ST distribution and moving out of phase along the y-axis. A plot of these values vs. temperature is known as bidimensional phase space, in which each point corresponds to one in the drying kinetics.[30] The presence of this attractor is characteristic of deterministic chaotic systems.[30] As drying proceeds into the falling rate period, this attractor tends to disappear. This may be due to internal temperature and moisture gradients being established. This finding corresponds to those reported in the fractal analysis and period of invariance of D, in which, when the attractor disappears, a periodic behavior with a characteristic value of D prevails.

Figures 17.15 and 17.16 show that, in spite of having uncertain initial TS distribution, the process seems to be nonsensitive to these variations, since TS trajectories during the falling rate period, as well as equilibrium conditions, are subjected to open centers of attraction that converge in the final equilibrium temperature.

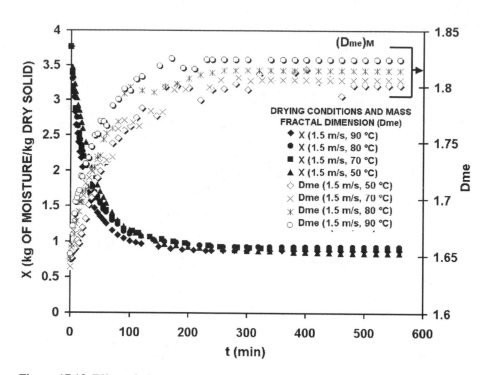

Figure 17.12 Effect of air temperature for drying of spheres and evolution of D_{me} along drying kinetics.

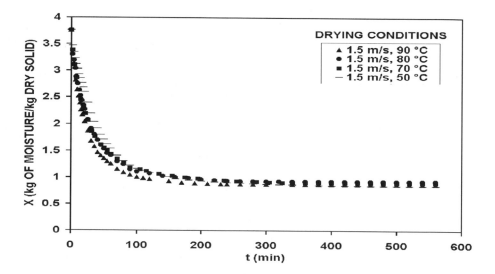

Figure 17.13 Effect of air temperature on drying kinetics of spheres.

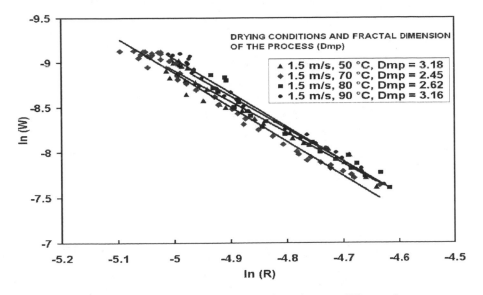

Figure 17.14 Richardson graph for obtaining D_{mp} for spheres at different air temperatures and an air velocity of 1.5 m/s.

From the results described in this work, it is possible to interpret global drying kinetics with the fractal theory in terms of zones along drying and structure of the material. The findings may also lead to inclusion of the concept of chaos in the drying curve.

REFERENCES

1. Mandelbrot, B. B. 1977. *The Fractal Geometry of Nature.* New York: W. H. Freeman, pp. 30–41.
2. Harrison, A. 1995. *Fractals in Chemistry.* U.K.: Oxford Science Publications, pp. 4–20.
3. Jones, C. L. and G. T. Lonergan. 1997. "Prediction of Phenol-Oxidase Expression in a Fungus using the Fractal Dimension," *Biotechnology Letters*, 19(1): 65–69.
4. Ben Ohoud, M., F. Obrecht, L. Gatineau, P. Levitz, and H. Van Damme. 1988. "Surface Area, Mass Fractal Dimension, and Apparent Density of Powders," *Journal of Colloid and Interface Science*, 124(1): 156–161.
5. Belloutio, M., M. M. Alves, J. M. Novais, and M. Mota. 1997. "Flocs vs. Granules: Differentiation by Fractal Dimension," *Water Research*, 31(5): 1227–1231.
6. Barnsley, M. F. 1993. *Fractals Everywhere.* Boston: Academic Press, 180–187, 205–238.
7. Peleg, M. 1993. "Fractals and Foods," *Food Science and Nutrition*, 33(2): 149–165.
8. Peleg, M., and M. D. Normand. 1985. "Characterization of the Ruggedness of Instant Coffee Particles by Natural Fractals," *Journal of Food Science*, 50(3): 829–831.
9. Barletta, B. J. And G. V. Barbosa-Cánovas. 1993. "Fractal Analysis to Characterize Ruggedness Changes in Tapped Agglomerated. Food Powders," *Journal of Food Science*, 58(5): 1030–1035, 1046.

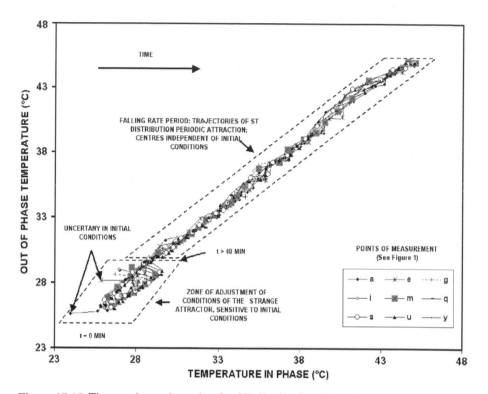

Figure 17.15 Time series trajectories for ST distribution along the drying of the slab shaped model food at 45°C and 1 m/s in a bidimensional space of phases, showing the presence of a strange attractor in the period of adjustment of conditions and a periodic behavior for the falling rate period.

10. Yano, T. And T. Nagai. 1989. "Fractal Surface of Starchy Materials Transformed with Hydrophilic Alcohols," *Journal of Food Engineering*, 10: 123–133.

11. Yano, T. 1996. "Fractal Nature of Food Materials," *Bioscience Biotechnology Biochemistry*, 60 (5): 739–744.

12. Vétier, N. S., Desobry-Banon., E. M. M. Ould, and J. Hardy. 1997. "Effect of Temperature and Acidification Rate on the Fractal Dimension of Acidified Casein Aggregates," *Journal of Dairy Science*, 80: 3161–3166.

13. Aguerre, R. J., P. E. Viollaz, and C. Suárez. 1996. "A Fractal Isotherm for Multilayer Adsorption in Foods," *Journal of Food Engineering*, 30: 227–238.

14. Aguerre, R. J., M. P. Tolabada, and C. Suárez. 1999. "Multilayer Adsorption Surfaces: Effect of the Adsorbate-Adsorbate Interactions on the Sorptional Equilibrium," *Drying Technology*, 17(4–5): 869–881.

15. Nagai, T. and T. Yano. 1990. "Fractal Structure of Deformed Potato Starch and Its Sorption Characteristics," *Journal of Food Science*, 55(5): 1334–1337.

16. Kopelman, R. 1988. "Fractal Reaction Kinetics," *Science*, 241: 1620–1626.

17. Van Damme, H., P. Levitz, L. Gatineau, J. F. Alcover, and J. J. Fripiat. 1988. "On the Determination of the Surface Fractal Dimension of Powders by Granulometric Analysis," *Journal of Colloid and Interface Science*, 22(1): 1–8.

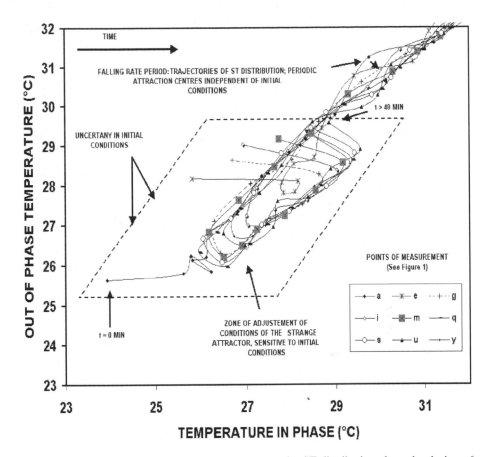

Figure 17.16 Zooming in of time series trajectories for ST distribution along the drying of the slab shaped model food at 45°C and 1 m/s in a bidimensional space of phases, showing the presence of a strange attractor in the period of adjustment of conditions and a periodic behavior for the falling rate period.

18. Rahman, S. M. 1997. "Physical Meaning and Interpretation of Fractal Dimensions of Fine Particles Measured by Different Methods," *Journal of Food Engineering*, 32: 447–456.

19. Braun, E. 1996. *Caos, Fractales y Cosas Raras*. México D.F.: Fondo de Cultura Económica, pp. 33–36.

20. Gerbogi, C., E. Ott, and J. A. Yorke. 1987. "Chaos, Strange Attractors, and Fractal Basin Boundaries in Nonlinear Dynamics," *Science*, 238: 632–638.

21. Treybal, R. E. 1990. *Operaciones de Transferencia de Masa*. 2nd edición. México D.F.: McGraw-Hill, pp. 655–716.

22. Foust, A. S., L. A. Wenzel, C. W. Clump, L. Maus, and L. B. Andersen. 1979. *Principios de Operaciones Unitarias*. New York: John Wiley & Sons, Inc., pp. 411–417.

23. Brennan, J.G., J. R. Butters, N. D. Cowell, and A. E. Lilly. 1979. *Food Engineering Operations*. 2nd ed. London, U.K.: Applied Science Publishers Limited, pp. 323–325.

24. Brown, A.F., M. E. Yépez, G. I. Ramos, J. A. Peralta, C. G. Pavia, and Miller. 1997. "Análisis de Fractalidad en Anomalías de Autopotenciales Eléctricos," *Acta Mexicana de Ciencia y Tecnología*, 44(13): 107–111.
25. Sano, Y. and B. R. Keey. 1981. "The Drying of a Spherical Particle Containing Colloidal Material into a Hollow Sphere," *Chemical Engineering Science*, 36(6): 881–889.
26. Derbour, L., H. Desmorieux, and J. Andrieu. 1998. "Determination and Interpretation of the Critical Moisture Content (C. M. C.) and the Internal Moisture Content Profile During the Constant Drying Rate Period," *Drying Technology*, 16(3–5): 813–824.
27. Gutiérrez-López, G. F. 1989. Convective Drying and Solid-Moisture Interactions. Ph. D. Thesis. University of Reading, U.K. pp. 72–77, 237–256.
28. Cronin, K. and S. Kearney. 1998. "Monte Carlo Modelling of Vegetable Tray Dryer," *Journal of Food Engineering*, 35: 233–250.
29. Strumillo, C. and J. Ademiec. 1996. "Energy and Quality Aspects of Food Drying," *Drying Technology*, 14(2): 423–448.
30. Monroy, O. C. 1997. *Teoría del Caos*. México D. F.: Alfaomega, pp. 60–116.

Part IV

Food Rheology

18 Relationship between Rheology and Food Texture

M. C. Bourne

CONTENTS

Abstract

There is considerable overlap between rheology and food texture. The deformation of a food item squeezed in the hand is both a textural property and a rheological property. The flow of the bolus of chewed food in the mouth, and the flow of fluid and semifluid foods, are both a textural property and a rheological property. However, the fracturing and grinding of solid foods that occurs during mastication is not a rheological phenomenon, and neither are the textural perceptions of particles, or the release or absorption of moisture or oil. Therefore, we conclude that food texture is partly rheology and partly non-rheology. Some rheological properties are probably not detected by the tactile sense, which means that some rheological properties are related to texture and some are not.

18.1 DEFINITIONS

Rheology is defined as the study of the deformation and flow of matter. *Food rheology* is defined as the study of the deformation of the raw materials, the intermediate products, and the final products of the food industry.[1]

The International Standards Organization defines *texture* as "all the rheological and structure (geometrical and surface) attributes of a food product perceptible by means of mechanical, tactile, and, where appropriate, visual and auditory receptors" (International Organization for Standardization, Standard 5492/3).

My definition of texture, which is a little more elaborate, defines textural properties as "the textural properties of a food are that group of physical characteristics that arise from the structural elements of the food; are sensed by the feeling of touch; are related to the deformation, disintegration, and flow of the food under a force; and are measured objectively by functions of mass, time, and distance."[2]

18.2 FOOD RHEOLOGY

There are several reasons for performing rheological tests:

1. Engineering process design. The flow properties and deformation properties of foods need to be understood so as to design equipment for handling foods, including conveyor belts, storage bins, pumps, pipelines, spray devices, etc.
2. To obtain information on the structure of the food or on the conformation of the molecular constituents of the foods, especially the macromolecular constituents.
3. To make measurements that will predict the sensory assessment of the textural attributes of the product. Based on these measurements, the process or the formula for a given product will be changed to produce a finished product that falls within the range of textural parameters that experience has shown is desirable to the consumer.

The science of rheology has many applications in the field of food processing and handling. For example, the agricultural engineer is interested in the creep and recovery of agricultural products that are subjected to stresses, particularly long-term stresses resulting from storage under confined conditions such as the bottom of a bulk bin. The food engineer is interested in the ability to pump and mix liquid and semiliquid foods. A number of food processing operations depend heavily on rheological properties of their products at an intermediate stage of manufacture, because this has a profound effect on the quality of the finished product. For example, the rheology of dough, milk curd, and meat emulsions are important aspects in the manufacture of high-quality bread, cheese, and sausage products.

Rheology also has many applications in the measurement of food acceptability, which may be divided into three major categories:

1. *Appearance* (color, shape, size, etc.), which is based on optical properties. There is a small component of rheology in appearance, because certain structural and mechanical properties of some foods can be determined by appearance; e.g., we can see how runny the food is on the plate.
2. *Flavor* (taste and odor), which is the response of the receptors in the oral cavity to chemical stimuli. Rheology has no direct part in this category, although the manner of food breakdown in the mouth can affect the rate of release of flavor compounds.
3. *Touch,* which is the response of the tactile senses to physical stimuli that result from contact between some part of the body and the food. This has been called "kinesthetics" by Kramer[3] and "haptaesthesis" by Muller.[4] Rheology is a major contributor to the sense of touch in the evaluation of food quality. Foods are squeezed in the hand and the sense of deformability and recovery after squeezing gives an impression of their textural quality. For example, fresh bread is highly deformable, while stale bread is not; the flesh of fresh fish recovers quickly after squeezing, while that of stale fish does not. During the process of mastication, a number of rheological properties are sensed in the mouth such as the deformation that occurs on the first bite and the flow properties of the bolus (the mass of chewed food mixed with saliva).

Non-Newtonian viscometry is an important component of the quality of most fluid and semifluid foods. Plasticity, pseudoplasticity, and the property of shear thinning are important quality factors in many foods. A wide variety of foods (e.g., butter, margarine, applesauce, catsup, mayonnaise, and starch pastes) are either plastic or pseudoplastic in nature. They are required to spread and flow easily under a small force but to hold their shape when not subjected to any external force other than gravity. Shear thinning is a desirable property of most fluid and gel-like foods. Szczesniak and Farkas[5] found that solutions of gums that exhibited little shear thinning were considered slimy by sensory panels and were rated as undesirable. This finding was confirmed by Stone and Oliver.[6]

18.2.1 Deformation by Machine

The deformation of a food under the influence of a force is frequently used as a measure of quality. Foods that deform to a large extent are classified as soft, flaccid, or spongy, while foods that deform to a small extent are classified as firm, hard, or rigid. Softness may be associated with good or poor quality, depending on the food. For example, a soft marshmallow is of a higher quality than a firm marshmallow; a soft loaf of bread is fresher than a firm loaf; a soft head of lettuce has fewer leaves than a firm head; and a soft peach, tomato, or banana is riper than a firm one. This test is probably the most widely used sensory test for measuring the firmness of foods. It is usually performed by gently squeezing the food between the thumb and fingers, and it is a nondestructive test.

Bourne[7] described a technique for measuring deformation of foods with a high degree of precision. Briefly, the method consists of compressing the food between two extensive rigid horizontal metal platens in a universal testing machine and recording the force-distance curve produced. It is an official method for measuring the firmness of bread.[8] It has been shown that, for many foods, a low deforming force gives the best resolution between samples (see Table 18.1).

TABLE 18.1
Deformation of Jumbo Marshmallows

Deforming force	Deformation, mm		Ratio, deformation of A/ deformation of B
	A. Soft	B. Firm	
5,000 g	18.7	14.9	1.25
1,000 g	10.7	6.9	1.55
100 g	2.32	0.92	2.52
20 g	0.80	0.022	3.64

Reprinted from Reference 7 with permission from the *Journal of Food Science.*

The geometry of the test sample needs to be controlled to obtain reproducible results in the deformation test. Figure 18.1 shows deformation curves of four upright cylinders cut from a single frankfurter. The first curve was obtained on a test piece with plane, parallel ends. The next two curves were obtained on test pieces with plane ends that were not parallel. The last curve was obtained on a test piece with one end curved. The effect of these irregularities in shape is to produce a "tail" at the low force end of the curve. These tails represent that portion of the deformation when less than the complete cross-sectional area of the test piece is being compressed, and it can introduce large errors into the measurement. Test pieces A, B, C, and D, all cut from the same frankfurter, show deformations of 1.6, 2.4, 3.4, and 2.5 mm under 1 kg force. This source of error can be overcome by measuring the deformation from some reference force level that is sufficiently above zero to eliminate the effect of the tails. In this case, the deformations of these four test pieces measured between 0.05 kg and 1.05 kg are 1.6, 1.6, 1.5, and 1.5 mm, which is as uniform as can be expected. While every care should be taken to cut the ends plane and parallel, corrections can be made in this way for small deviations from ideal shapes.

In the engineering field, it is customary to use specimens of standard dimensions for testing purposes. The force per unit area is called the *stress,* and the deformation per unit dimension is called the *strain.* With many foods, it is possible to prepare pieces with standard dimensions for the purpose of the deformation test. The most common technique used is to cut out cylinders of the food with a power-driven cork borer and trim the end to give a standard length; however, this procedure is destructive. If the deformation could be measured on intact units of the food, it would become nondestructive, and the samples could be tested again or reused for some other purpose.

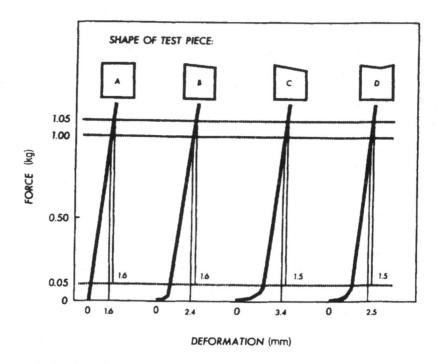

Figure 18.1 Effect of irregularities in surface of test piece on deformation test. Note how the errors are overcome by starting the deformation a little above zero force. Reprinted from Reference 7 with permission from *Journal of Food Science*.

Some foods are not suited for cutting into standard-size pieces. For example, it is impossible to cut a standard-size piece of tissue from a whole tomato that is representative of the intact fruit. There would be great difficulty in cutting standard-size pieces from citrus fruits, nuts, onions, lettuce, bell peppers, whole eggs, or hard candies, to name just a few foods. In these cases, we are forced to measure whole units of the food, with all their natural variations in size and shape. It is desirable that different readings reflect differences in the deformability of the food only, and not those caused by variations in size and shape. It is possible, for example, to have two peaches that are identical in deformability (i.e., of equal textural quality) but that give different deformation measurements because one peach is larger than the other. There is, therefore, a need to understand how size and shape affect the deformability of a food. When this is known, it will be possible to separate the deformability measurement into differences due to changes in textural quality of the food, and differences due to changes in size.

Brinton and Bourne[9] made agar gels, which can be assumed to have uniform textural quality, and cut them into rectangular solids, spheres, and cylinders. Then they measured the deformation by gentle compression in a universal testing machine using a force small enough to stay within the Hookean elastic range. They obtained the results described below.

Rectangular Solids

The deformation is directly proportional to height and inversely proportional to the cross-sectional area of the specimens.

Young's modulus of elasticity in compression for an elastic material that obeys Hooke's law is defined as the ratio of the compressive stress to the compressive strain in a specimen of a uniform cross section.

$$E = \frac{\text{stress}}{\text{strain}} = \frac{\text{force per unit area}}{\text{deformation per unit length}} = \frac{F/A}{\Delta L/L_0} \tag{18.1}$$

where E = Young's modulus
   F = compression force
   A = cross-sectional area of specimen
   L_0 = unstressed specimen height
   ΔL = deformation, which is change of height under influence of F

This equation can be rearranged as follows:

$$\Delta L = \frac{FL}{EA} \tag{18.2}$$

Equation (18.2) shows that the deformability of a substance with a constant Young's modulus E under a standard force F should be directly proportional to the height of the test sample and inversely proportional to the cross-sectional area. Brinton and Bourne's experimental data with shaped agar gels show excellent agreement with Equation (18.2), which leads to the conclusion that, in the form of regular rectangular solids, agar gels subjected to small stress behave as an elastic material and obey the equation defining Young's modulus. Under these conditions, it is valid to make corrections for changes in dimensions so as to compare deformabilities of different size samples.

Cylinders with Vertical Axis

These show the same type of deformation as rectangular solids, that is, the deformation is directly proportional to height and inversely proportional to the cross-sectional area. Equations (18.1) and (18.2) are applicable to cylinders with a vertical axis.

Cylinders with Horizontal Axis

Brinton and Bourne[9] showed that

- Deformation is linearly and inversely related to the length of the cylinder.
- Deformation is directly proportional to the diameter.

According to Roark,[10] the change in diameter of a horizontal cylinder under compression is given by the equation

$$\Delta D = \frac{4}{3}P\left(\frac{1-v^2}{\pi E}\right) + 4P\left(\frac{1-v^2}{\pi E}\right)\log_e\frac{(DE)^{1/2}}{1.075P^{1/2}} \qquad (18.3)$$

where D = diameter
$\quad\Delta D$ = deformation
$\quad P$ = force per unit length
$\quad v$ = Poisson's ratio
$\quad E$ = Young's modulus

According to this equation, the deformation of a horizontal cylinder (ΔD) is the sum of two terms. The first term $[4/3\ P(1-v^2)/\pi E]$ is independent of the diameter and represents the intercept on the ordinate. The second term is a complex function of the diameter. Brinton and Bourne's data for agar gels[9] is in general agreement with this equation.

Spheres

The deformation vs. diameter plots are U-shaped. As the diameter increases, the deformation first increases and then decreases. The peak deformation is found at smaller diameters as the stiffness of the gel (concentration of agar) increases. The peak also occurs at smaller diameters as the deforming force is decreased.

Figure 18.2 shows that changing the diameter of a sphere can cause the deformation to increase under some circumstances and to decrease under other circumstances, depending on whether increasing the diameter causes the deformation to approach the peak or to move away from it. The peak deformation appears to be a complex function, depending on the rigidity of the gel and the force of deformation. If a 5% gel deformed between 50–200 g force is considered typical of the type of test that is commonly applied to whole fruits, it can be seen that there is a comparatively slight change in deformability over a wide range of diameters. From 1 cm up to 5 cm diameter, the minimum deformation is 1.1 mm, and the maximum deformation is 1.4 mm. A 500% change in diameter causes a maximum change of about 30% in deformation. Except with small-diameter 3% soft gel spheres, the rate of change in deformation is comparatively small as compared with the change in diameter. From this evidence, we draw the tentative conclusion that small changes in diameter will have only a slight effect on the deformation of a food that is approximately spherical in shape.

The Boussingesq theory for engineering materials predicts that the deformation is directly related to the deforming force.

$$P = F/2\pi a(a^2 - r^2)^{1/2} \qquad (18.4)$$

where P = the pressure at any point under the punch
$\quad F$ = total force applied to punch
$\quad a$ = radius of punch
$\quad r$ = distance from center of punch to stressed area

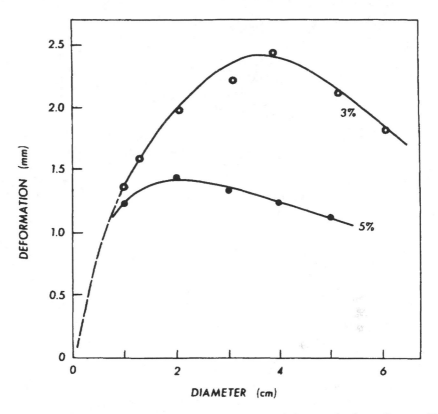

Figure 18.2 Effect of diameter on deformation of 5% and 3% agar gel spheres illustrated by ● and ○, respectively. Reprinted from Reference 9 with permission from Food & Nutrition Press, Inc.

For agar gels, the theory holds up to the peak deformation but not afterward; that is, the theory agrees for small diameter spheres, but not for larger diameter spheres.

According to Hertz's theory, the deformation of a rounded specimen compressed between parallel plates is given by the equation

$$E = \frac{0.338F(1-\mu^2)}{D^{3/2}}\left[K_1\left(\frac{1}{R_1}+\frac{1}{R_1'}\right)^{1/3}+K_2\left(\frac{1}{R_2}+\frac{1}{R_2'}\right)^{1/2}\right]^{3/2} \qquad (18.5)$$

where
E = modulus of elasticity
F = force
μ = Poisson's ratio
D = deformation
K_1, K_2 = constants
R_1, R_1', R_2, R_2' = radii of curvature at contact points

Equation (18.5) can be rearranged and simplified for the case of flat plates compressing a sphere to give

$$D = \left[\frac{0.338F(1-\mu^2)}{E}\right]^{2/3} \frac{k}{R^{1/3}} \tag{18.6}$$

According to Equation (18.6), there is an inverse relationship between deformation and radius (deformation $\alpha R^{-1/3}$). The Hertz theory agrees qualitatively with the experimental data for large spheres, but it fails to agree for small-diameter agar spheres.

Roark[10] gives the following formula for deformation of a sphere under a flat plate:

$$y = 1.55\left(\frac{P^2}{E^2 D}\right)^{1/3} \tag{18.7}$$

where y = combined deformation of sphere and compressing plate
        P = total load
        E = Young's modulus
        D = diameter of sphere

According to this formula, the deformation should be proportional to $(D)^{-1/3}$. The data obtained in this study are not in accord with this equation. It should be pointed out that Roark's equation is intended to be applied to ball bearings and similar objects in which the compressing plate and the test sphere presumably have similar mechanical properties. The modulus of an agar gel is different from that of the metal compression plate by several orders of magnitude. However, Roark's equation may apply over a limited range of large diameters.

Real foods of approximate spherical shapes (e.g., cherries, apples, onions, tomatoes, and papayas) have a far more complex structure than agar gels. As Morrow and Mohsenin[11] have pointed out, "an intact product such as fruit violates all fundamental assumptions of homogeneity, isotropy, and continuity that are normally required in solving elementary materials science problems." When we add to the complex function between deformation and diameter of a uniform material such as an agar gel the additional complexities of the structure of intact foods, the need for additional research, both practical and theoretical, is indicated to understand how the diameter affects deformation. It may be that there is a different relationship for each commodity.

18.2.2 Deformation by Hand

Peleg[12] studied the sensitivity of the human tissue in squeeze tests and pointed out that, in these types of tests, there can be significant deformation of the human tissues (e.g., the balls of the fingers) in addition to the deformation of the specimen. He pointed out that the combined mechanical resistance in a squeezing test is given by the equation

$$M_c = \frac{M_1 M_x}{M_1 + M_x}$$ (18.8)

where M_c = combined mechanical resistance of the sample and the fingers
    M_1 = resistance of the human tissue
    M_x = resistance to deformation of the test specimen

This equation provides a simple explanation of why there are differences in the sensing range between the fingers and the jaws and why the human senses are practically insensitive to hardness beyond certain levels.

Three different types of responses can be drawn from this equation:

Case No. 1: $M_1 \gg M_x$. This case occurs when a soft material is deformed between hard contact surfaces (e.g., a soft food is deformed between the teeth). Under these conditions, Equation (18.8) becomes $M_c = M_x$ (since $M_1 + M_x \approx M_1$). In this situation, the sensory response is primarily determined by the properties of the test specimen.

Case No. 2: M_x and M_1. These are of comparable magnitude. In this case, the response is regulated by both the properties of the test material and the tissue applying the stress, as given in Equation (18.8).

Case No. 3: $M_x \gg M_1$. This occurs when a very firm product is compressed between soft tissues; for example, pressing a nut in the shell between the fingers. Under these conditions, the equation becomes $M_c = M_1$ (because $M_1 + M_x \simeq M_x$). In this situation, the response is due to the deformation of the tissue and is insensitive to the hardness of the specimen. This appears to be interpreted as "too hard to detect" or "out of range."

18.2.3 FLOW

The flow of fluids has been the subject of intense study for many fluid and semifluid products. Generally speaking, the equations developed for non-food products are also applicable to foods. Figure 18.3 shows the major types of flow.

Newtonian

This type of flow is characterized by a linear relationship between shear stress and shear rate, and it passes through the origin. Water, oil, milk, and sugar syrups, including most honeys, exhibit Newtonian flow. The equation for dynamic viscosity of Newtonian liquid is

$$\eta = \sigma/\dot{\gamma}$$ (18.9)

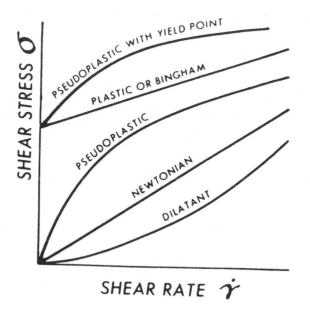

Figure 18.3 Shear stress vs. shear rate plots for various types of flow. Reprinted from Reference 2 with permission from Academic Press, Inc.

where η = viscosity
  σ = shear stress
  $\dot{\gamma}$ = shear stress

Non-Newtonian

These fluids show either a nonlinear relationship between shear stress and shear rate, or the flow does not begin until a measurable shear stress is applied.

Pseudoplastic Flow

An increasing shear force gives a more-than-proportional increase in shear rate, but the curve begins at the origin. Salad dressings are a good example of this kind of flow. (See Figure 18.3.)

Dilatant Flow

The shear stress-shear rate plot for this type of flow begins at the origin and is characterized by equal increments in shear stress, giving less than equal increments in shear rate. (See Figure 18.3.) High solids suspensions of raw starch and some chocolate syrups are examples of this kind of flow.

The power equation is usually effective in describing pseudoplastic and dilatant flow.

$$\sigma = k\dot{\gamma}^{n} \tag{18.10}$$

where σ = shear stress

$\dot{\gamma}$ = shear rate

k = consistency index

n = dimensionless number that indicates closeness to Newtonian flow

For a Newtonian liquid, $n = 1$; for a dilatant liquid, n is greater than 1; and for a pseudoplastic liquid, n is less than 1. Taking logarithms reduces this equation to the linear form.

$$\log \sigma = \log k + n \log \dot{\gamma} \tag{18.11}$$

Plastic Flow

Plastic flow requires that a minimum shear stress, known as yield stress, be exceeded before flow begins. The shear stress-shear rate plot does not begin at the origin but, rather, at some point up the y axis. It is exhibited by many foods, including tomato paste, mayonnaise, margarine, and fruit jelly. These foods are usually consistent with the Herschel–Bulkley equation.

$$\sigma = \sigma_0 + k\dot{\gamma}^n \tag{18.12}$$

where σ_0 = yield stress

Conclusion

The above discussion clearly establishes that rheology is a major component of food quality and food texture. The numerous research papers published every year demonstrate that food rheology is an active field of research and that steady progress is being made year by year.

18.3 FOOD TEXTURE

Texture comprises all the tactile sensations felt by the human body from the moment food is picked up in the store until mastication is complete and the bolus is swallowed. The sense of deformation and flow properties is present in many of these experiences. However, there are many other sensations that have no connection to rheology.

The texture notes that are perceived sensorially appear all through the masticatory cycle, from the first bite until the final swallow. Some of the texture notes are detected on the initial bite, others during the masticatory period, and some during the final swallowing. Mastication is a dynamic process in which food is crushed and diluted with saliva, and the temperature is brought close to that of the body. Rheological measurements can measure only the quality of the food as it is placed in the mouth; they cannot measure the changes that occur during mastication. The texture notes that are sensed during mastication may be divided into three classes.[13]

1. *Mechanical characteristics,* which are the reaction of the food to stress. Most of these can be measured by instrumental tests.
2. Characteristics that are related to the arrangement of the physical constituents of the food, such as size, shape, arrangement of particles within a food, and surface roughness. These properties have been called *geometrical properties* by Brandt et al.,[14] and *particulate properties* by Hutchings and Lillford.[15] These texture notes are not rheological properties.
3. Another group of texture properties that are not rheological are those classed as *chemical properties.*[14] This term refers to properties such as the sensation of moistness, oiliness, or greasiness as perceived in the mouth. These sensations may or may not be correlated with moisture and fat content as determined by conventional chemical analysis, but they are an important component of the textural sensation experienced during mastication.

18.3.1 MASTICATION

The deformation texture test described above is a nondestructive test. In contrast, the process of mastication is a highly destructive procedure.

Size Reduction

Mastication is a process in which pieces of food are ground into a fine state, mixed with saliva, and brought to approximately body temperature in readiness for transfer to the stomach, where digestion, absorption, and utilization begin. Table 18.2 summarizes the amount of size reduction that occurs before food can be absorbed and utilized. The process of mastication is the beginning, or an early part, of this process of size reduction down to small molecules. A mouth-size portion of food is usually around 5 g. During mastication, this will usually be reduced by 2 to 3 orders of magnitude before going to the stomach. In the stomach, approximately another 20 orders of magnitude of size reduction is accomplished by biochemical and chemical means before the food can be absorbed. If the food cannot be reduced to particles of the order of a few multiples of 10^{-22} g, it is not absorbed or utilized but is excreted.

This process of mastication imparts pleasurable sensations that seem to fill a very basic human need. Chewing is a sensual pleasure that we enjoy throughout life, from the cradle to the grave. The existence of a large dental industry is due primarily to the fact that people enjoy chewing their food and do not want to be deprived of these gratifying sensations. From a nutritional standpoint, it is possible to have a completely adequate diet in the form of fluid foods that require no mastication. The mechanics of size reduction (or comminution, disintegration, pulverization, and trituration) are an important part of the spectrum of food texture that do not belong in the field of rheology.

Dynamics of Mastication

The rheological properties of the food change markedly during the process of mastication, and food technologists are just as interested in the dynamics of these

TABLE 18.2
Steps in the Comminution of Food Before Absorption by the Body

State	Approx. particle mass, g	Process	Location	Implement
Large Cookie				
Whole cookie	20	Biting off	Mouth	Incisors
Mouth-size portion	5	Grinding, crushing	Mouth	Molars
Swallowable past (bolus)	1×10^{-2}	Biochemical attack	Stomach, intestine	Acid, enzymes
Hexose sugar molecules	3×10^{-22}	Absorption	Intestines	—
Whole dressed steer				
Whole carcass	3×10^{5}	Sawing and cutting	Butcher shop	Saw, knives
Cooked steak	3×10^{2}	Cutting	Plate	Knife and fork
Mouth-size portion	5	Shearing, grinding	Mouth	Teeth
Swallowable paste	3×10^{-2}	Biochemical attack	Stomach, intestines	Acid, enzymes
Amino acid molecules	3×10^{-22}	Absorption	Intestines	—

Reprinted from Reference 2, with permission from Academic Press, Inc.

changes as in the initial rheological properties of the food. Conventional rheological tests on fluids usually assume that the fluid will maintain its physical properties unchanged during the course of the test. From a human standpoint, a fluid food is manipulated with the tongue or teeth, diluted with saliva, and brought to near body temperature. Conventional rheological tests on solids are usually one-event tests; as soon as the material has broken into two pieces, the test is terminated. In mastication, a piece is broken during the first chew, then broken further into smaller pieces on the second chew, and so on until the food is swallowed. Thus, with food, one deals with the situation of sequential tests on material that has been mechanically damaged and reduced in size by preceding tests, and has also been altered by moisture and sometimes by thermal effects and enzymes.

Particulates

During mastication, the size and shape of food particles and their surface roughness are sensed. These are important attributes of the overall textural sensation. Szcz-csniak's group[14] described these surface properties of food particles in sensory terms such as powdery, chalky, grainy, gritty, coarse, lumpy, beady, flaky, fibrous, pulpy, cellular, aerated, puffy, and crystallized. A person can be trained to identify and quantify on a semiquantitative scale each of these properties, but these are not rheological properties. Bourne[16] suggested *rugosity* (surface roughness) as a term that should be familiar to those who work with the textural properties of foods, but rugosity is not a rheological property.

Saliva Effects

The ability of many foods to wet with saliva is another important textural property that is not a rheological property. An essential factor in the quality of most snack foods, such as potato chips and similar products, is that they hydrate rapidly with saliva and lose their brittleness.

Other important non-rheological factors in the overall texture sensation are the amount of saliva absorbed by the bolus and the rate at which it is absorbed, the sensation of oiliness or fattiness that results from release of lipids, and the amount and rate of release of moisture from the food during mastication.[14,17] Most varieties of fresh apples release a considerable amount of juice into the mouth during mastication, whereas apples that have been in storage for a long period of time release little juice. Differences in the rate of release of moisture are exemplified by foods manufactured from animal protein, which usually give a uniform sensation of moistness for the duration of mastication, while their analogs, manufactured from texturized vegetable protein, usually exhibit excessive moisture release during the first few chews, followed by no release of moisture, which gives a sensation of dryness just prior to swallowing.

Temperature Effects

In some foods, phase changes that result from temperature changes occur in the mouth. Ice cream and chocolate melt, while the oil in hot soup may solidify into fat in the mouth. Foods based on gelatin gels melt in the mouth, while foods based on agar gels do not melt; this is a basic difference between the mouth feel of a gelatin gel and an agar gel. The food technologist is as interested in the dynamics of the temperature-induced rheological changes as in the original rheological properties.

18.4 CONCLUSION

We conclude, therefore, that food texture measurements fall partly within and partly outside the field of conventional rheology. The food technologist needs to define and measure certain rheological properties of foods, but there are large areas of interest where the classical science of rheology is of no help in the study of some textural properties of foods. Therefore, we are forced to develop non-rheological techniques to measure these textural properties.

Food rheologists have borrowed the concepts, theories, and instrumentation developed by the thousands of rheologists who work with products other than food. There is no similar large body of knowledge from nonfood products that texture technologists can apply to their problems. The result is that the theory and practice of non-rheological texture measurements is not nearly as well developed as it is for the rheological texture measurements.

REFERENCES

1. White, G. W. 1970. "Rheology in Food Research," *J. Food Technol.*, 5: 1–32.
2. Bourne, M. C. 1982. *Food Texture and Viscosity. Concept and Measurement.* New York, NY: Academic Press.
3. Kramer, A. 1973. "Food Texture—Definition, Measurement and Relation to Other Food Quality Attributes," in *Texture Measurements in Foods*, A. Kramer and A. S. Szczesniak, eds. Dordrecht: D. Reidel Publishing Co., Chap. 1, pp. 1–9.
4. Muller, H. G. 1969. "Mechanical Properties, Rheology and Haptaesthesis of Food," *J. Texture Studies,* 1: 38–42.
5. Szczesniak, A. S. and E. Farkas. 1962. "Objective Characterization of the Mouthfeel of Gum Solutions," *J. Food Science*, 27: 381–385.
6. Stone, H. and S. Oliver. 1966. "Effect of Viscosity on the Detection of Relative Sweetness Intensity of Sucrose Solutions," *J. Food Science*, 31: 129–134.
7. Bourne, M. C. 1967. "Deformation Testing of Foods I. A Precise Technique for Performing the Deformation Test," *J. Food Science*, 32: 601–605.
8. American Association of Cereal Chemists. Method 74–09.
9. Brinton, R. H. and M. C. Bourne. 1972. "Deformation Testing of Foods. 3. Effect of Size and Shape of the Test Piece on the Magnitude of Deformation," *J. Texture Studies,* 3: 284–297.
10. Roark, R. J. 1965. *Formulas for Stress and Strain.* New York: McGraw-Hill.
11. Morrow, C. T. and N. N. Mohsenin. 1966. "Consideration of Selected Agricultural Products as Viscoelastic Materials," *J. Food Science,* 31: 686–698.
12. Peleg, M. 1980. "A Note on the Sensitivity of Fingers, Tongue and Jaws as Mechanical Testing Instruments," *J. Texture Studies,* 10: 245–251.
13. Szczesniak, A. S. 1963. "Classification of Textural Characteristics," *J. Food Science*, 28: 385–389.
14. Brandt, M. A., E. Z. Skinner, and A. S. Szczesniak. 1963. "Texture Profile Method," *J. Food Science,* 28: 404–409.
15. Hutchings, J. B. and P. J. Lillford. 1988. "The Perception of Food Texture—The Philosophy of the Breakdown Path," *J. Texture Studies*, 19: 103–115.
16. Bourne, M. C. 1975. "Size Reduction of Food Before Swallowing."
17. Civille, G. V. and I. H. Liska. 1975. "Modifications and Applications to Foods of the General Foods Sensory Texture Profile Technique," *J. Texture Studies,* 6: 19–32.

19 Relevance of Rheological Properties in Food Process Engineering

J. Vélez-Ruíz

CONTENTS

Abstract

Basic aspects related to the rheology of foods are introduced, including the definition of rheology, stress and strain concepts, as well as the classification of food materials. Physical relationships between force, deformation, and material properties or rheological properties, as well as the most known rheological models, are mentioned. More commonly used rheological techniques are briefly commented on regarding their main features: rotational rheometry, tube rheometry, back extrusion rheometry, extensional viscometry, and squeezing flow rheometry, as well as mixer viscometry. Mainly, flow properties and their relation to food process operations are discussed; studies in food process engineering, momentum transfer operations, heat transfer processes, mass transfer unit operations, and physical-structural changes are presented. Momentum transfer operations and their relation to flow properties of liquid materials are established; and the relationships between rheological parameters and food transport systems, mechanical separations, mixing, and pumping are emphasized. For heat transfer operations, the influence of rheological properties such as apparent viscosity, consistency coefficient, and flow behavior index are related to the heat transfer coefficient,

1-56676-963-9/02/$0.00+$1.50
© 2002 by CRC Press LLC

presenting four representative equations developed and proposed by recognized authors. Similarly, in mass transfer operations, exemplified by spray drying and fermentation, the role of rheological behavior and its influence on process performance are mentioned. Finally, the relationship of structural changes and/or physical changes to the rheology of food products and components are briefly discussed.

19.1   INTRODUCTION TO FOOD RHEOLOGY

The goal of this chapter is to present a general overview of the rheology applied to food materials, with emphasis on the relationship between rheological behavior and food process engineering.

Rheology, defined as the science of flow and deformation of materials, is a fundamental interdisciplinary science that has been gaining importance in the field of foods. According to Rao,[1] Steffe,[2] Holdsworth,[3] Vélez-Ruiz and Barbosa-Cánovas,[4] Bhattacharya et al.,[5] and Vélez-Ruiz,[6] among others, there are numerous topics of interest to the food industry related to rheology, such as

- process engineering applications involved in equipment and process design
- physical characterization of liquid and solid foods
- development of new products or reformulation
- quality control of intermediate and final products
- correlation with sensorial evaluation
- understanding of food structure

Rheology can be used to characterize not only flow behavior of biological and inorganic materials, but also structural characteristics. Flow properties, such as viscosity, yield stress, thickness, pourability, softness, spreadability, and firmness, contribute substantially to facilitate transport and commercial processing as well as to promote consumer acceptance. Insight into structural arrangement helps to predict behavior or stability of a given material with storage, change in humidity and temperature, and handling.[7] Consequently, basic rheological information on materials is important not only to engineers but also to food scientists, processors, and others who might utilize this knowledge and find new applications.

Although food exists in a variety of forms, solids and liquids are of primary importance in food rheology. Many foods are neither solid nor liquid but exist in an intermediate state of aggregation known as semi-solid. As a consequence of the complex nature and lack of a precise boundary among solids and semi-solids, many foods may exhibit more than one rheological behavior, depending on their specific characteristics and the measuring conditions used during physical characterization. Deformation may be conservative or dissipative, depending on whether it is related to solids or liquids. Flow is a time-dependent form of deformation and, consequently, is more related to fluids.[8]

Two mechanical parameters (stress and strain) are the basis for material classification, from a rheological viewpoint, into three recognized groups: elastic, plastic, and viscous. Mohsenin,[9] based on the physical response of biological material to a given stress or strain, expressed a paramount visualization of rheology using three fundamental parameters: force, deformation, and time. This general classification is outlined in Figure 19.1. Accordingly, the specific relationship developed between an applied stress and the resulting deformation of the material is known as a rheological property.[10]

A wide range of models are available for the rheological characterization of foods. According to Holdsworth,[3] the rheological models may be divided into three main groups: *time-independent models*, including Bingham, Power Law, and Herschel–Bulkley; *time-dependent models,* such as Carreau, Hahn, Powell–Eyring, and Weltmann; and *viscoelastic* models, with Kelvin–Voigt element and Maxwell body being the best known. Certainly, there exist other, less popular models in food products, and others specifically applied to fit the effect of concentration and temperature.

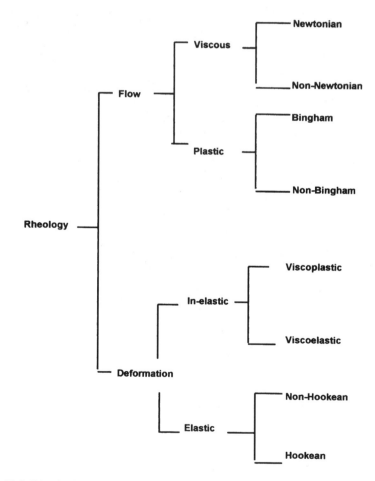

Figure 19.1 Rheological classification of food materials.

In rheology, *strain* and *stress* are two relevant physical variables that need to be considered when a material deforms in response to applied forces. Strain represents a relation of a change in length with respect to the original dimension. This parameter is essentially a relative displacement, and there are many definitions associated with this concept. For instance, the engineering strain is expressed by the following equation:[2,11,12]

$$\varepsilon_c = \frac{\Delta L}{L_0} = \frac{L - L_0}{L_0} \tag{19.1}$$

where ε_c = engineering strain, also called Cauchy strain
   ΔL = change in length
   L = final length after deformation of the material
   L_0 = original length before deformation

This definition is expressed in terms of a simple shear. Strain is determined by displacement gradients, and strain rate by velocity gradients. Strain and stress are tensor quantities and are represented by nine components.[10,13]

Stress that relates the magnitude of the force over the surface of application can be compressive, tensile, or shear, depending on how the force is applied. Nine separate components are required to adequately describe the state of stress in a material.[2,10,14] The stress at any point in a body may be represented by the following matrix:

$$\tau_{ij} = \begin{bmatrix} \tau_{11} & \tau_{12} & \tau_{13} \\ \tau_{21} & \tau_{22} & \tau_{23} \\ \tau_{31} & \tau_{32} & \tau_{33} \end{bmatrix} \tag{19.2}$$

where τ_{ij} is the stress tensor; the first subscript indicates the orientation of the face upon which the force is acting, and the second subscript refers to the direction of the force. This matrix may be simplified, depending on each specific physical system. For instance, in steady-simple shear flow, also known as *viscometric* flow, the matrix is reduced to only five components.[2,10]

A deformed body, as a result of an applied force, will develop internal stresses and strains that may be of various types.[15] The relationship shown by any food material between applied stress and resulting strain defines the *rheological properties of the material*. These relationships can be expressed either empirically or in terms of a rheological equation of state.[10] Figure 19.2 illustrates the correspondence between the physical forces and stress (τ), as well as the functional relation between deformation and strain (γ) and shear rate (strain/time).

There exists a particular approach in which the rheological behavior of a material is analyzed on a simplified deformation called *single shear* or *uniaxial* deformation. This approach is the basis for many rheological measurement techniques and permits the characterization of many food materials.[10,16]

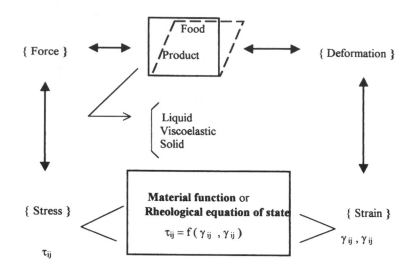

Figure 19.2 Physical relationships between force, deformation, and material properties.

19.2 MEASURING TECHNIQUES

Measuring the rheological properties and identifying the rheological behavior of food materials is necessary when employing any of the available commercial instruments that allow objective characterizations. The accuracy of these rheometers has been improved due to the incorporation of microcomputer technology and frictionless devices.

Based on their elemental geometry, the existing *common instruments* are divided into two major groups (rotational class and tube class[2,7]), but there are other rheometric approaches that have not been completely developed and therefore are not commercially available. These include back extrusion flow,[17,18] extensional viscometry,[19] mixing devices of various kinds,[20] and squeezing flow.[21] Each group of rheometers has advantages and disadvantages that must be considered whenever a rheological characterization of food systems is carried out. The selection of a rheometer for a particular task is a matter of great importance.

In food rheology, the *rotational type* is more commonly used than tube fixtures.[22] These rheometers are most suited for low-viscosity fluids,[23] and they have three main performance features:

1. A uniform shear rate may be applied.
2. The effect of time on the flow behavior can be followed.
3. Different geometries may be utilized: concentric cylinders, cone and plate, and parallel plates.

Most oscillatory or dynamic experiments are carried out in these rheometers. On the other hand, *tube viscometers* have been broadly used in flow characterization of different viscous foods such as fruit purees and juices, vegetable concentrates, gum solutions, and food dispersions. Tube viscometers that apply higher shear rates than rotational instruments can be easily manufactured following basic recommendations; therefore, they are simple and low-cost equipment, with several limitations.[24-28]

In *back extrusion*, the technique is simple and requires a texture meter, such as the Instron Universal Testing Machine or a similar instrument. Back extrusion requires graduated cylinders to hold the samples and plungers. Osorio and Steffe[18] developed mathematical relationships for the rheological characterization of Newtonian and power law food fluids. This approach is also known as *annular pumping*,[18] or the compression-extrusion test,[29] and it offers a good measuring alternative.

In many industrial processes (e.g., blow molding, dough sheeting, extrudate expansion, fiber spinning, flow through porous beds, mouth feel and swallowing of beverages, food spreading, and vacuum forming), it has been recognized that elongational deformation is the most important type of deformation.[23,30] Therefore, *elongational* or *extensional deformation* has been developed as a proper technique for rheological characterization of some food materials.[30] Data and applications of extensional deformation in foods are scarce. This technique is adequate for foods with proteins and polysaccharides and requires a generation of controlled extensional flows for rheological measurements.[19,23,30]

Squeezing flow viscometry, as a variation of the extensional flow measuring, is based on compression of a fluid material between two parallel plates. This type of measurement has been mathematically analyzed by Campanella and Peleg,[31] who generated the respective equations for lubricated squeezing flow. Fluids with high viscosity, such as cheese,[32] peanut butter,[31] mustard, mayonnaise, and tomato ketchup,[33] have been tested to evaluate their rheological properties.

The application of *mixing impellers* for the estimation of rheological parameters has been proposed by various authors.[34-40] Mixers have often been applied in rheological studies for fermentation processes. Particularly, the helical ribbon impeller has been very useful for complex fluids and suspensions, generating stable process viscosity values and avoiding phase separation,[20] a very common flow problem in food dispersions. Unfortunately, the adoption of mixer viscometry techniques has been limited because of the development of complex flow and the high costs of the equipment.

Rao[41] established that, due to the complex composition and particular structure of foods, approaches to characterizing a food's rheological behavior may be classified into five methods: empirical, phenomenological (including fundamental and imitative), linear viscoelastic, nonlinear viscoelastic, and microrheological. In the case of foods, empirical and phenomenological approaches have been predominant in those studies associated with flow properties.[41,42] With respect to viscoelastic nature, the approaches mainly include fundamental and linear viscoelasticity; most food systems are considered to be linear viscoelastic materials at small strains. Nonlinear viscoelastic materials exhibit physical properties that are a function not only of time but also of the magnitude of applied stress.[43]

On analyzing the published studies, one may observe that there are four main directions in rheological research.[23]

1. Description of the macroscopic phenomena developed during the deformation of materials
2. Explanation of the phenomena from a molecular point of view
3. Experimental characterization of parameters and functional relationships describing these phenomena
4. Practical application of the aforementioned directions

19.3 RHEOLOGICAL STUDIES IN FOOD PROCESS ENGINEERING

After commenting on some aspects of food rheology, it is interesting and important to identify some of the food process engineering operations in which flow and viscoelastic properties play an outstanding role. Those unit operations may be grouped in four areas:

1. Pipeline transport, mixing, pumping, and mechanical separations
2. Heat transfer operations such as heating, cooling, and evaporation
3. Mass transfer processes such as air drying, fermentation, osmotic concentration, and membrane separations
4. Physical changes during processing and tailoring the consistency and textural characteristics of foods

19.3.1 MOMENTUM TRANSFER OPERATIONS

Rheological properties of fluid foods are intimately related to power consumption, possibly the most important design parameter for momentum transfer operations involved in food processing. Although there have been advances in correlating the rheological behavior of non-Newtonian foods and power requirements or pressure drop, it is still a research field that needs more studies, mainly when some large-size particles are suspended in the fluid.[44]

Mechanical energy losses due to friction may be expressed by Equation (19.3),[45] which includes the energy losses due to fittings.

$$E_f = \frac{2f\,u^2 L}{g_c D} + \sum_{i=1}^{n} \frac{k_f u^2}{2g_c} \tag{19.3}$$

where E_f = energy loss per unit mass due to friction
  f = Fanning friction factor
  u = mass average or bulk velocity
  L = total pipeline length
  g_c = proportionality factor or constant (1.0 in SI, 32.2 in FPS)

D = internal diameter
k_f = friction loss coefficient
i = fitting

Other works have analyzed the influence of rheological properties on less important engineering design parameters. For instance, Li[46] studied the effect of the flow behavior index on the hydrodynamic entrance lengths in laminar falling films.

Some representative works related to momentum transfer in food process engineering are summarized in Table 19.1.

TABLE 19.1
Rheological Studies Related to Pipeline Transport and Pumping Needs

Ref./year	Material/experimental	Purpose/hypothesis	Important remarks
47/1971	–Dilatant starch system –Laminar flow	Study of flow through pipelines and fittings	Non-Newtonian fluid (NN) influences frictional losses.
48/1973	–CMC and starch solution –Mixing	Power consumption of non-Newtonian fluids	Development of empirical equations for power needs.
28/1980	–Pipe transportation of minced fish paste	Evaluation of pressure losses of a Bingham plastic food.	Pressure losses were related to rheological nature of minced fish
49/1984	–Non-Newtonian fluids –Theoretical approach	Expression to calculate kinetic energy	Mathematical and graphical solutions for kinetic energy evaluation.
50/1984	–Applesauce –Evaluation of pressure loss	Expression to evaluate frictional losses through fittings	Friction loss coefficients correlated to Reynolds number
51/1986	–Herschel–Bulkley fluids –Theoretical development	Determine optimum economic diameter	Optimal diameter estimated from mass, economic, and flow parameters.
52/1986	–Agitated systems –Newtonian (N) and non-Newtonian fluids	Effect of rheological behavior on power consumption	Develop empirical correlations for pseudoplastic fluids.
53/1987	–Friction factors for non-Newtonian fluids –Laminar and turbulent flows	Comparison and development of relationships for friction factors	Friction factor depends on the flow behavior index and yield stress.
54/1991	–Sodium CMC solutions –Two phase tube flow	Modeling two phase, non-Newtonian flow	Drag forces of suspended particles and N-N fluids.
55/1993	–Newtonian and non-Newtonian materials –Mixing fluids	Effect of rheological properties on power consumption	Elasticity of fluid materials increased the power.
56/1994	–Sodium CMC solutions –Horizontal pipe	Critical Reynolds number for pseudoplastic fluids	An equation for critical Reynolds number as a function of low index.

TABLE 19.1
Rheological Studies Related to Pipeline Transport and Pumping Needs

Ref./year	Material/experimental	Purpose/hypothesis	Important remarks
57/1994	–Non-Newtonian fluid –Two-phase tube flow	Influence of upstream and downstream on the drag force	An equation was developed to compute the drag correction factor for Stokes's equation.
58/1995	–H-B fluids –Theoretical approach	Develop expressions to estimate the kinetic energy correction factor	Two simple and useful equations were developed.
59/1996	–Wheat flour with low moisture content –Twin-screw extruder	Effect of rheological properties on die design.	The extrudate performance was related to the flow parameters of the dough.
60/1997	–Newtonian and non-Newtonian fluids –Ducts of complex cross-sectional shape	Friction curves for laminar and turbulent regimes Re: 10^{-2} to 10^5	Single friction curve for all types of flow, Newtonian, pseudoplastic and Herschel–Bulkley fluids.

19.3.2 Heat Transfer Operations

Flow properties are also closely related with those unit operations involving heat transfer phenomena. Cooling, heating, pasteurization, and sterilization are the most common heat transfer operations utilized in food processing, and the main purpose implies flow of food fluids inside of the exchanger.[61] Evaporation is also extensively used in food processing of food liquids, but there are few data on the effect of concentration and heat transfer on rheological properties, or the influence of flow properties on evaporation phenomena, due to very complex flow patterns developed during this unit operation.

Since the flow profile controls mixing and thermal-time effects in the heat equipment, it is essential to know how rheological properties influence the heat transfer mechanisms. Numerous empirical correlations have been developed for the heat transfer of Newtonian fluids, but the existing correlations for non-Newtonian foods are scarce.

Peeples[62,63] established the constant values for the forced convection, heating, and cooling of fluid milk products; this was accomplished by inclusion of the Reynolds number and parameters related to product composition. In a most recent study, Dodeja et al.[64] obtained a correlation to express the heat transfer coefficient of concentrated milk (19 to 70% of solids content) in a thin-film scraped-surface heat exchanger, generating the following equation:

$$Nu = 6615.06 GRe^{0.1331} \, GPr^{0.0764} \left[\frac{\Delta T}{T_s}\right]^{0.2843} \tag{19.4}$$

where Nu = Nusselt number

GRe = generalized Reynolds number
GPr = generalized Prandtl number
ΔT = temperature difference
T_s = temperature of condensing steam

Fortunately, the number of papers related to heat transfer coefficients and rheological properties has increased.[65–69] The proposed empirical equations express the Nusselt number as a function of the generalized Reynolds and generalized Prandtl numbers, including in both of them the flow behavior index and the consistency coefficient. For instance, Alhamdan and Sastry[66] proposed empirical correlations for heating and cooling of aqueous solutions of sodium carboxymethylcellulose and an irregular shaped particle. Chandarana et al.[67] studied a similar system with a starch solution. Bhamidipati and Singh[69] also determined the fluid-to-particle convective heat transfer coefficient and evaluated magnitudes in the range of 108.66 to 195.58 W/m²°F for carboxymethylcellulose at different concentrations.

In relation to thermal processing of non-Newtonian fluids, Palmer and Jones[70] reported a study for the prediction of holding times in continuous thermal processing of power-law fluids. They generated a plot of the velocity ratio in laminar and turbulent flow as a function of the generalized Reynolds number, for different values of the flow index (0.20–2.0). The holding time can then be evaluated as the length over the maximum velocity, the last being obtained from the velocity ratio. More recently, Sandeep et al.[71] and Moyano et al.[72] obtained an experimental equation to compute the residence time of non-Newtonian suspensions and fruit pulp, respectively, including the rheological properties in the following proposed equations:

$$t_{mean} = t_{min}1.004Re_p^{\,0.293-0.288/n}PC^{0.044}n^{0.091} \tag{19.5}$$

$$t_{min} = t_{mean}0.698GRe^{-0.26} \tag{19.6}$$

where t_{mean} = mean residence time
t_{min} = minimum residence time
Re_p = particle Reynolds number
n = flow behavior index
PC = particle concentration
GRe = generalized Reynolds number

For the evaporation process, several studies have been conducted to determine the influence of rheological properties on evaporator performance. Chen et al.[73] developed an expression for the heat transfer coefficient in a rotary steam-coil vacuum evaporator, for concentration of tomato paste, involving the consistency of the fluid as apparent viscosity. Stankiewicz and Rao[74] established a correlation for the heat transfer of sugar solutions in a thin-film wiped-surface evaporator. Similarly, they utilized the apparent viscosity as part of the equation. Bouman et al.[75] did the same for falling films evaporators in the dairy industry. Table 19.2 includes relevant studies to emphasize the importance of rheological properties in food process operations based on heat transfer.

There is more information related to the concentration process, which is widely utilized in the food industry for production of fruit juices, evaporated milk, and vegetable purees. Many studies have focused on the effect of concentration and temperature on rheological properties.[81–93] Most of these studies have incorporated solids content, temperature, or both as part of the equations for the rheological parameters. The relationships express the flow behavior index, consistency coeffi-

TABLE 19.2
Rheological Studies Related to Heat Transport in Food Process Operations

Ref./year	Material/experimental	Purpose/hypothesis	Important remarks
76/1979	–Continuous sterilization –Bingham plastic fluids	Modeling continuous sterilization for a non-Newtonian fluid	The effect of T over the rheological properties is necessary for a good sterilization process.
74/1988	–Thin film evaporation –Water and sugar solutions	Simulation of fruit juice evaporation	Evaporation regimes were related to flow properties.
77/1989	–Newtonian fluids –Numerical applications	Modeling of heat and mass transfer in a falling film	Pseudo-stationary evaporation was related to Reynolds number.
78/1989	–Double tube heat exchanger –Guar gum and CMC solutions	Predicting heat exchange during laminar flow	Correlations were developed for a non-Newtonian fluid.
66/1990	–Model fluid system –Heating/cooling	Measurement of free convective heat transfer coeff. for particle/fluid	h was higher for heating and lower with increasing viscosity.
67/1990	–Starch solutions and water –UHT process	Evaluations of the particle/fluid interface convective heat transfer coefficient	h for particles heated in N fluid were 20% higher than in NN.
79/1993	–Whey protein solution –Tubular heat exchanger	Effect of viscosity on whey protein fouling	Fouling from whey was related to Reynolds number.
69/1995	–Particle in CMC sols. –Aseptic treatment	Determination of fluid convective heat transfer coefficient	Fluid viscosity significantly affected h.
80/1994	–Water and starch suspensions –Particulate flow	Calculate incipient fluid velocity for particulate flow in holding tubes	Incipient velocity was affected by porosity, density, and viscosity.
61/1999	–Non-Newtonian and Newtonian fluids –Scraped surface heat –Numerical simulation	Effect of heat transfer on flow profiles	Figures of velocity profiles for three fluids ($n = 0.28$, 0.77 and 1.0) were obtained.

cient, yield stress, or apparent viscosity as a function of concentration and/or temperature.

19.3.3 MASS TRANSFER OPERATIONS

There are mass transfer operations in which fluid foods are handled; therefore, air and spray drying, fermentation processes, membrane separations, and vacuum drying may be influenced by the rheological behavior of the fluid materials. Other unit operations have less dependency on flow properties, such as extraction, frying, leaching, and osmotic dehydration.

Filková[94] conducted a study of spray drying and concluded that the size distribution of the final dry product is related to the consistency of the sprayed liquid and that the viscosity of the sprayed slurry was necessary to predict the droplet diameter. Later, Weberschinke and Filková[95] obtained the apparent viscosity from the rheological properties to evaluate the particle diameter. Hayashi and Kudo[96] concluded that the particle size of powdered skim milk increased with the viscosity of the concentrate. Recently, Hayashi[97] observed that a viscosity of less than 500 mPa-s must be kept to obtain constant good spraying of soy milk.

In fermentation processes, large quantities of substances are involved; diffusion and convective mass transfer then necessarily occur. In formulating mass transfer coefficients, either gases or solids need to pass through the liquids, showing very different rates. The comprehensive correlation of mass transfer coefficients and interfacial areas involves the rheological properties of the liquid phase. Most of the systems take into account only the viscosity of the continuous phase that, in agitated vessels, is usually part of the Reynolds number.[98] Tucker and Thomas[99] investigated the relationship between biomass concentration, mycelial morphology, and the rheological properties of broths, and they proposed that the rheology of fungal fermentation broths should be related to clump properties rather than to the morphology of the freely dispersed mycelia.

Table 19.3 summarizes some relevant research works related to mass transfer operations in the food industry.

19.3.4 STRUCTURAL CHARACTERISTICS AND PHYSICAL CHANGES

Fluid and semisolid food materials exhibit flow behavior that is both strain- and rate-of-strain dependent; therefore, food components play an important role in rheological behavior and sensory characteristics. Each component (mainly carbohydrates and proteins) responds differently to industrial processes in which the raw food material is performed to manufacture food products.

The structural features of plant food dispersions and their rheological parameters have been studied and related to physical properties of the fluid. Solids content, particle size distribution of solids, and serum viscosity play an important role in the rheological behavior of plant food dispersions, whereas, in the case of milk and milk products, the physicochemical variables such as solids content, type of protein, degree of denaturation, and presence of fat globules, just to mention a few, are very important.

TABLE 19.3
Rheological Studies Related to Mass Transfer Processes.

Ref./year	Material/experimental	Purpose/hypothesis	Important remarks
100/1976	–Skim milk concentrates –Ultrafiltration	Study the relationship between separation and flow properties	Diffusion was related to viscosity.
101/1977	–Sweet potato puree –Dehydration	Flow characteristics of potato puree as indicators of flake quality	Flow properties were significantly related to mouth feel-sensory descriptors.
94/1980	–Water and CMC solutions –Spray drying	Influence of viscosity on drop size of dried products, nf	Comparison between droplet size of N and NN fluids.
95/1982	–Theoretical derivation –Spray drying	Droplet diameter of power law liquid	An equation for drop diameter of non-Newtonian fluids was developed.
102/1989	–Power law fluids –Drum drying –Theoretical approach dryer	Film thickness of drying material for a drum	Thickness of the film depended on velocity ratio of cylinders and flow behavior index.
103/1989	–Roasted peanuts –Drying	Effects of drying conditions on textural properties	The higher T, the lower the shear-compression force.
104/1995	–Potato, apple, and carrot –Dehydration	Determination of heat transfer during drying	Coefficients for cylinders and slices by a dimensionless expression.
97/1996	–Soy protein milk –Spray drying	Effect of viscosity on drying of soy protein milk	The viscosity must be controlled to get good atomization and milk quality.

On the other hand, processing variables such as applied stresses, cooking time, concentration, holding time, pasteurization time, pH, previous treatments, rate of heating, salts types, temperature, water/solids ratio, and other variables play a transcendental role on the rheological behavior, depending on each product and its manufacturing process.

Important advances in rheological, structural, and textural characterization have been reached by taking advantage of the recent advances in computer technology, microscopy, and rheometric instrumentation. Thus, this part of food process engineering is very abundant in papers and studies in relation to physical changes and structural characteristics related to rheology. Table 19.4 includes some studies regarding this field.

TABLE 19.4
Rheological Studies Related to Structural Characteristics of Foods.

Ref./year	Material/experimental	Purpose/hypothesis	Important remarks
105/1981	–Egg custard –Mayonnaise	Characterizing of stress decay due to structural changes	Steady shear provided information about structure.
106/1983	–Ketchup, butter, margarine, cream, cheese, and peanut butter	Relationship between viscoelastic properties of food materials.	Viscosity and normal stress followed a power-law behavior.
107/1984	–Coagulating milk –Dynamic tests	Changes of viscoelastic properties	Dynamic techniques gave information on milk gel.
108/1987	–Food gels –Thermal gelation of proteins	Rheological changes during gelation phenomena	Rheological and textural methods to follow the gelation of proteins.
109/1990	–Processed cheese analogs –Sensorial tests	Relation between structure, rheology, and sensory texture	Stress and work were related to sensory and structure characteristics.
110/1993	–Whey protein concentrates –Thermal gelation –Dynamic tests	Study of gel structure formation	Higher protein concentrations produced higher storage modulus.
111/1993	–Whey protein concentrates –Dynamic tests	Relation between molecular and rheological properties	The rheology of WPC changed from time-independent to time-dependent shear thinning.
112/1994	–β-lactoglobulins A and B –Thermal gelation/denaturation	Following of gelation process by a dynamic rheological technique	Increasing protein conc. increased the storage modulus being A >> B.
113/1995	–Concentrated skim milk –Membrane concentration	Gelation related to viscoelastic behavior	G' and G'' were related to the gelation temperature.
114/1996	–Surimi-like material from beef and pork	Examine the functional properties of surimi	Gel hardness of samples affected by protein and cooking temperature.
115/1999	–Dynamic tests –Wheat dough –Microscopy studies	Mixing parameters affect microscopy and rheology	Microscopy helped to interpret rheometric data.

19.4 FINAL REMARKS

The role of the flow properties of materials on different food process operations and those related to physical changes has been emphasized. All unit operations involving food liquids as part of the performance will have a more or less strong influence on the rheological properties.

Important progress has been made in characterizing the rheological behavior and structural features of food products, but this is not the case for the rheological influence on food process engineering. However, the use of rheometers and application of rheological models for many food fluids is quite generalized; there are many food materials that still have not been rheologically characterized.

Although rheological and engineering data for food process operations are available, and there are more studies focused on this field, additional research is needed. Studies based on advanced instrumentation, computer technology, digital image processing, microscopy techniques, numerical analysis, and textural measuring devices, among others, will provide a better understanding and modeling of the relationships between the rheological behavior and unit food process operations.

REFERENCES

1. Rao, M. A. 1986. "Rheological Properties of Fluid Foods," in *Engineering Properties of Foods*, M. A. Rao, and S. S. H. Rizvi, eds. New York: Marcel Dekker, Inc., pp. 1–47.
2. Steffe, J. F. 1992. *Rheological Methods in Food Process Engineering*. East Lansing, MI: Freeman Press.
3. Holdsworth, S. D. 1993. "Rheological Models Used for the Prediction of the Flow Properties of Food Products: A Literature Review," *Trans. I Chem. E.*, 71 (part C): 139–179.
4. Vélez-Ruiz, J. F. and G. V. Barbosa-Cánovas. 1997. "Rheological Properties of Selected Dairy Products," *Crit. Rev. Fd. Sci. & Nutr.,* 37(4): 311–359.
5. Bhattacharya, S., N. Vasudha, and K. S. K. Murthy. 1999. "Rheology of Mustard Paste: A Controlled Stress Measurement," *J. Fd. Eng.*, 41: 187–191.
6. Vélez-Ruiz, J. F. October, 1999. "Rheology of Dairy Products," Conference presented in Symposium of Food Rheology at the XXX Mexican Congress of Food Science and Technology. Veracruz, Ver., Mexico.
7. Motyka, A. L. 1996. "An Introduction to Rheology with Emphasis on Application to Dispersions," *J. of Chem. Education.,* 73(4): 347–380.
8. Finney, E. E. 1973. "Elementary Concepts of Rheology Relevant to Food Texture Studies," in *Texture Measurements of Foods*, A. Kramer and A. S. Szczesniak, eds. Boston: D. Reidel Publishing Company, pp. 33–51.
9. Mohsenin, N. N. 1986. *Physical Properties of Plant and Animal Materials.* New York: Gordon and Breach Science Publishers, pp. 129–347.
10. Darby, R. 1976. *Viscoelastic Fluids. An Introduction to Their Properties and Behavior.* New York: Marcel Dekker, Inc., pp. 1–70.
11. Prentice, J. H. 1992. *Dairy Rheology. A Concise Guide.* New York: VCH Publishers, Inc.
12. Stroshine, R. and D. D. Hamann. 1994. *Physical Properties of Agricultural Materials and Food Products*, R. Strohine ed. North Carolina, pp. 82–142.
13. Bird, R. B., R. C. Armstrong, and O. Hassager. 1977. *Dynamic of Polymer Liquids. Volume 1. Fluid Mechanics.* New York: John Wiley and Sons.
14. Whorlow, R. W. 1980. *Rheological Techniques.* Chichester, West Sussex, England: Ellis Horwood Ltd.
15. ASTM. 1968. *Physical and Mechanical Testing of Metals; Nondestructive Tests.* Part 31, ASTM Standards, pp. 150–159.
16. Kokini, J. L. 1992. "Rheological Properties of Foods," in *Handbook of Food Engineering*, D.R. Heldman and D. B. Lund, eds. New York: Marcel Dekker, Inc., pp. 1–38.
17. Osorio, F. A. 1985. *Back Extrusion of Power Law, Bingham Plastic and Herschel-Bulkley Fluids.* M. Sci. Thesis. Michigan State University. East Lansing, MI.

18. Osorio, F. A. and J. F. Steffe. 1987. "Back Extrusion of Power Law Fluids," *J. Texture Stud.*, 18: 43–63.

19. Kokini, J. L. 1992. "Measurement and Simulation of Shear and Shear Free (Extensional) Flows in Food Rheology," in *Advances in Food Engineering*, R. P. Singh and M. A. Wirakartakusumah, eds. Boca Raton, FL: CRC Press, pp. 439–477.

20. Brito-De la Fuente, E., J. A. Nava, L. M. López, L. Medina, G. Ascanio, and P. A. Tanguy. 1998. "Process Viscometry of Complex Fluids and Suspensions with Helical Ribbon Agitators," *Can. J. Chem. Eng.*, 76(Aug.): pp. 689–695.

21. Campanella, O. H. and M. Peleg. 1987. "Determination of the Yield Stress of Semi-Liquid Foods from Squeezing Flow Data," *J. Fd. Sci.*, 52(1): pp. 214–217.

22. Shoemaker, C. F., J. I. Lewis, and M. S. Tamura. 1987. "Instrumentation for Rheological Measurements of Food," *Fd. Technol.*, (March): pp. 80–84.

23. Ferguson, J. and Z. Kembloswski. 1991. *Applied Fluid Rheology*. Cambridge, England: Elsevier Applied Science, pp. 85–134.

24. Rao, M. A. 1977. "Rheology of Liquid Foods-A Review," *J. Texture Stud.*, 8: pp. 135–168.

25. Rao, M. A. 1977. "Measurement of Flow Properties of Fluid Foods-Developments, Limitations, and Interpretation of Phenomena. Review Paper," *J. Texture Stud.*, 8: pp. 257–282.

26. Bhamidipati, S. and R. K. Singh, 1990. "Flow Behavior of Tomato Sauce With or Without Particulates in Tube Flow," *J. Fd. Proc, Eng.*, 12: pp. 275–293.

27. Griskey, R. G., D. G. Nechrebecki, P. J. Nothies, and R. T. Balmer. 1985. "Rheological and Pipeline Flow Behavior of Corn Starch Dispersions," *J. of Rheolog.*, 29 (3): 349–360.

28. Nayakama, T., E. Niwa, and I. Hamada. 1980. "Pipe Transportation of Minced Fish Paste," *J. Fd. Sci.*, 49: 844–847.

29. Bourne, M. C. 1982. *Food Texture and Viscosity: Concept and Measurement*. New York: Academic Press., pp. 135–144.

30. Padmanabhan, M. 1995. "Measurement of Extensional Viscosity of Viscoelastic Liquid Foods," *J. Fd. Eng.*, 25: 311–327.

31. Campanella, O. H. and M. Peleg, 1987. "Squeezing Flow Viscometry of Peanut Butter," *J. Fd. Sci.*, 52 (1):180–184.

32. Campanella, O. H. and M. Peleg, 1987. "Determination of the Yield Stress of Semi-Liquid Foods from Squeezing Flow Data," *J. Fd. Sci.*, 52 (1): 214–217.

33. Campanella, O. H., L. M. Popplewell, J. R. Rosenau, and M. Peleg. 1987. "Elongational Viscosity Measurements of Melting American Process Cheese," *J. Fd. Sci.*, 52(5): 1249–1251.

34. Rao, M. A. 1975. "Measurement of Flow Properties of Food Suspensions with a Mixer," *J. Texture Stud.*, 6: 533–539.

35. Griffith, D. L., and V. N. M. Rao. 1978. "Flow Characteristics of Non-Newtonian Foods Utilizing a Low-Cost-Rotational Viscometer," *J. Fd. Sci.*, 43: 1876–1877.

36. Rao, M. A. and H. J. Cooley, 1984. "Determination of Effective Shear Rates in Rotational Viscometers with Complex Geometries," *J. Texture Stud.*, 15: 327–335.

37. Ford, E. W. and J. F. Steffe. 1986. "Quantifying Thixotropy in Starch-Thickened, Strained Apricots Using Mixer Viscometry Techniques," *J. Texture Stud.*, 17: 71–85.

38. Nienow A. W. and T. P. Elson. 1988. "Aspects of Mixing in Rheologically Complex Fluids," *Chem. Eng. Res. Des.*, 66(Jan.): 5–15.

39. Steffe, J. F., M. Castell-Pérez, K. J. Rose, and M. E. Zabik. 1989. "Rapid Testing Method for Characterizing the Rheological Behavior of Gelatinizing Corn Starch Slurries," *Cereal Chem.* 66(1): 65–68.

40. Castell-Pérez, M. and J. F. Steffe, 1992. "Using Mixing to Evaluate Rheological Properties," in *Viscoelastic Properties of Foods*, M. A. Rao and J. F. Steffe, eds. New York: Elsevier Applied Science, pp. 247–279.

41. Rao, M. A. 1992. "Measurement of Viscoelastic Properties of Fluid and Semisolid Foods," in *Viscoelastic Properties of Foods*, M. A. Rao and J. F. Steffe, eds. London, England: Elsevier Applied Science, pp. 207–231.

42. Barbosa-Cánovas, G. V., A. Ibarz, and M. Peleg. 1993. "Rheological Properties of Fluid Foods, Review," *Alimentaria,* April: 39–89.

43. Rao, V. N. M. 1992b. "Classification, Description and Measurement of Viscoelastic Properties of Solid Foods," in *Viscoelastic Properties of Foods*, Rao, M. A. and J. F. Steffe, eds. London, England: Elsevier Applied Science, pp. 3–48.

44. Martínez-Padilla, L. P., L. Cornejo-Romero, C. M. Cruz-Cruz, C. C. Jáquez-Huacuja, and G. B. Barbosa-Cánovas. 1999. "Rheological Characterization of a Model Food Suspension Containing Discs Using Three Different Geometries," *J. Fd. Process Eng.*, 22: 55–79.

45. Steffe, J. F. and R. G. Morgan. 1986. "Pipeline Design and Pump Selection for Non-Newtonian Fluid Foods," *Fd. Technol.,* 40(12): 78–85.

46. Li, R. 1991. "Hydrodynamic Entrance Lengths of Non-Newtonian Laminar Falling Films," *Can. J. Chem. Eng.*, 69(Feb.): 383–385.

47. Griskey, R. G. and R. G. Green. 1971. "Flow of Dilatant (Shear Thickening) Fluids," *AICHE J.* 17(3): 725–728.

48. Rieger, F. and V. Novák, 1973. "Power Consumption of Agitators in Highly Viscous Non-Newtonian Liquids," *Trans. Instn. Chem. Engrs.* 51: 105–111.

49. Osorio, F. A. and J. F. Steffe. 1984. "Kinetic Energy Calculations for Non-Newtonian Fluids in Circular Tubes," *J. Fd. Sci.,* 49: 1295–1296, 1315.

50. Steffe, J. F., I. O. Mohamed, and E. W. Ford, 1984. "Pressure Drop Across Valves and Fittings for Pseudoplastic Fluids in Laminar Flow," *Transactions of the ASAE*, 27: 616–619.

51. García, E. J. and J. F. Steffe. 1986. "Optimum Economic Pipe Diameter for Pumping Herschel-Bulkley Fluids in Laminar Flow," *J. Fd. Process Eng.*, 8: 117–136.

52. Sesták, J., R. Zitny, and M. Houska. 1986. "Anchor-Agitated Systems: Power Input Correlation for Pseudoplastic and Thixotropic Fluids in Equilibrium," *AICHE J.,* 32(1): 155–158.

53. García, E. J. and J. F. Steffe. 1987. "Comparison of Friction Factor Equations for Non-Newtonian Fluids in Pipe Flow," *J. Fd. Process Eng.*, 9:93–120.

54. Subramaniam, G., C. A. Zuritz, and J. S. Ultman. 1991. "A Drag Correlation for Single Spheres in Pseudoplastic Tube Flow," *Transactions of the ASAE,* 34 (5): 2073–2078.

55. Carreau, P. J., R. P. Chhabra, and J. Cheng. 1993. "Effect of Rheological Properties on Power Consumption with Helical Ribbon Agitators," *AIChE J.,* 39(9): 1421–1428.

56. Campos, D. T., J. F. Steffe, and R. Y. Ofoli. 1994. "Statistical Method to Evaluate the Critical Reynolds Number for Pseudoplastic Fluids in Tubes," *J. Fd. Eng.,* 23: 21–32.

57. Sandeep, K. P. and C. A. Zuritz. 1994. "The Drag Force on Sphere Assemblies Fixed in the Flow Regime of a Non-Newtonian Fluid Flowing Through Holding Tubes," Paper 94–6508, ASAE Meeting, Atlanta, Dec. 13–16.

58. Briggs, J. L. and J. F. Steffe. 1995. "Kinetic Energy Correction Factor of a Herschel-Bulkley Fluid," *J. Fd. Process Eng.*, 18: 115–118.

59. Arhalias, A., J. P. Pain, J. Bouzaza, and J. M. Bouvier. 1996. "The Role of Rheological Properties and Die Design on Extrudate Expansion," in Proceedings of the 5th *World Congress of Chemical Engineering,* Vol. II:156–162, San Diego, CA.

60. Delplace, F., G. Delplace, S. Lefebvre, and J. C. Leuliet. 1997. "Friction Curves for the Flow of Newtonian and Non-Newtonian Liquids in Ducts of Complex Cross-Sectional Shape," in *Engineering and Food* at ICEF 7, R. Jowitt ed. Food Rheology Section: E36-E39.

61. Wang, W., J. H. Walton, and K. L. McCarthy. 1999. "Flow Profiles of Power Law Fluids in Scraped Surface Heat Exchanger Geometry Using MRI," *J. Fd. Process Eng.,* 22: 11–27.

62. Peeples, M. L., I. A. Gould, C. D. Jones, and W. J. Harper. 1962. "Forced Convection Heat Transfer Characteristics of Fluid Milk Products," *J. Dairy Sci.,* 45: 303–310.

63. Peeples, M. L. 1962. "Forced Convection Heat Transfer Characteristics of Fluid Milk Products During Cooling," *J. Dairy Sci,.* 45: 1456–1458.

64. Dodeja A. K., S. C. Sarma, and H. Abichandani. 1990. "Heat Transfer During Evaporation of Milk to High Solids in Thin Film Scraped Surface Heat Exchanger," *J. Fd. Process Eng.,* 12: 211–225.

65. Anantheswaran, R. C. and M. A. Rao. 1985. "Heat Transfer to Model Non-Newtonian Liquid Foods in Cans During End-over-end Rotation," *J. Fd. Eng.,* 4: 21–35.

66. Alhamban, A. and S. K. Sastry. 1990. "Natural Convection Heat Transfer Between Non- Newtonian Fluids and an Irregular Shaped Particle," *J. Fd. Process Eng.,* 13: 113–142.

67. Chandarana, D. I., A. Gavin, and F. W. Wheaton. 1990. "Particle/Fluid Interface Heat Transfer Under UHT Conditions at Low Particle/Fluid Relative Velocities." *J. Fd. Process Eng.* 13: 191–206.

68. Awuah, G. B., H. S. Ramaswamy, and B. K. Simpsom. 1993. "Surface Heat Transfer Coefficients Associated with Heating of Food Particles in CMC Solutions," *J. Fd. Process Eng.,* 16: 39–57.

69. Bhamidipati, S. and R. K. Singh. 1995. "Determination of Fluid-Particle Convective Heat Transfer Coefficient," *Transactions of the ASAE.,* 38(3): 857–862.

70. Palmer, J. A. and V. A. Jones. 1976. "Prediction of Holding Times for Continuous Thermal Processing of Power-Law Fluids," *J. Fd. Sci.,* 41: 422 427.

71. Sandeep K. P., C. A. Zuritz, and V.M. Puri. 1997. "Mathematical Modeling and Experimental Studies on RTD and Heat Transfer During Aseptic Processing of Non-Newtonian Suspensions," in *Engineering and Food* at ICEF 7, R. Jowitt ed. Food Rheology Section: E52-E55.

72. Moyano, P. C., C. A. Esquerre, and F. A. Osorio. 1997. "Temperature and Generalized-Reynolds Number Effects on Minimum Residence Time of Fruit Pulp," in *Engineering and Food* at ICEF 7, R. Jowitt ed. Food Rheology Section: E56-E59.

73. Chen, C. S., V. M. Lima, and A. Marsaioli. 1979. "A Heat Transfer Correlation for Rotatory Steam-Coil Vacuum Evaporation of Tomato Paste," *J. Fd. Sci.,* 44(1): 200–203.

74. Stankiewics, K. and M. A. Rao. 1988. "Heat Transfer in Thin-Film Wiped-Surface Evaporation of Model Liquid Foods," *J. Fd. Process Eng.,* 10: 113–131.

75. Bouman, S., R. Waalewijn, P. De Jong, and H. J. L. J. Van Der Linden. 1993. "Design of Falling Film Evaporators in the Dairy Industry," *J. of the Soc. of Dairy Technol.,* 46(3): 100–106.

76. Guariguata, C., J. A. Barreiro, and G. Guariguata. 1979. "Analysis of Continuous Sterilization Processes for Bingham Plastic Fluids in Laminar Flow," *J. Fd. Sci.,* 44: 905–910.

77. Barkaoui, M., A. Ramadane, Z. Aoufoussi, G. Louis, and P. Le Goff. 1989. "Modeling of Coupled Heat and Mass Transfer in a Falling Film Evaporator," in *Drying '89*, A. S. Mujumdar, ed. Hemisphere, pp. 246–253.

78. Pereira, E. C., M. Bhattacharya, and R.V. Morey. 1989. "Modeling Heat Transfer to Non-Newtonian Fluids in a Double Tube Heat Exchanger," *Transactions of the ASAE.*, 32(1): 256–262.

79. Belmar-Beiny, M. T., S. M. Gotham, W. R. Paterson, P. J. Fryer, and A. M. Pritchard. 1993. "The Effect of Reynolds Number and Fluid Temperature in Whey Protein Fouling," *J. Fd. Eng.*, 19: 119–139.

80. Grabowski, S. and H. S. Ramaswamy. 1995. "Incipient Carrier Fluid Velocity for Particulate Flow in a Holding Tube," *J. Fd. Eng.*, 24(1): 123–136.

81. Rao, M. A., M. C. Bourne, and H. J. Cooley. 1981. "Flow Properties of Tomato Concentrates," *J. Texture Stud.*, 12: 521–538.

82. Crandall P. G., C. S. Chen, and R. D. Carter. 1982. "Model for Predicting Viscosity of Orange Juice Concentrate," *Fd. Technol.*, May: 245–252.

83. Vitali, A. A. and M. A. Rao. 1982. "Flow Behavior of Guava Puree as a Function of Temperature and Concentration," *J. Texture Stud.*, 13: 275–289.

84. Vitali, A. A. and M. A. Rao. 1984. "Flow Properties of Low-Pulp Concentrated Orange Juice: Serum Viscosity and Effect of Pulp Content," *J. Fd. Sci.*, 49: 876–881.

85. Costell, E., E. Carbonell, and L. Duran. 1987. "Chemical Composition and Rheological Behaviour of Strawberry Jams," *Acta Alimentaria.*, 16(4): 319–330.

86. Qiu, C. G. and M. A. Rao. 1988. "Role of Pulp Content and Particle Size in Yield Stress of Apple Sauce," *J. Fd. Sci.*, 53(4): 1165–1170.

87. Rao, M. A. and H. J. Cooley. 1992. "Rheological Behavior of Tomato Paste in Steady and Dynamic Shear," *J. Texture Stud.*, 23: 415–425.

88. Tang, Q., P. A. Munro, and O. W. McCarthy. 1993. "Rheology of Whey Protein Concentrate Solutions as a Function of Concentration, Temperature, pH and Salt Concentration," *J. Dairy Res.*, 60: 349–361.

89. Hernández, E., C. S. Chen, J. Johnson, and R. D. Carter. 1995. "Viscosity Changes in Orange Juice After Ultrafiltration and Evaporation," *J. Fd. Eng.*, 25: 387–392.

90. Vélez-Ruiz, J. F. and G. V. Barbosa-Cánovas. 1997. "Effects of Concentration, Temperature on the Rheology of Concentrated Milk," *Transactions of the ASAE.*, 40(4): 1113–1118.

91. Trinh, K. T. and J. A. Yoo. 1997. "On Modeling Rheology of Concentrated Whole Milk Solutions," in *Engineering and Food* at ICEF 7, R. Jowitt ed. Food Rheology Section: E44–E47.

92. Aguiar, G. I. L., T. Gehrke, F. E. X. Murr, and D. Cantú-Lozano. 1997. "Effect of Temperature and Concentration on Viscosity of Acerola Pulp," in *Engineering and Food* at ICEF 7, R. Jowitt ed. Food Rheology Section: E56–E59.

93. Vélez-Ruiz, J. F. and G. V. Barbosa-Cánovas. 1998. "Rheological Properties of Concentrated Milk as a Function of Concentration, Temperature and Storage Time," *J. Fd. Eng.*, 35: 177–190.

94. Filková, S. 1980. "Dropsize Distribution of Non-Newtonian Slurries," in *Drying '80*, A. S. Mujumdar, ed. New York: Hemisphere, Pub. Corporation, pp. 346–350.

95. Weberschinke, J. and I. Filková. 1982. "Apparent Viscosity of Non-Newtonian Droplet on the outlet of Wheel Atomizer," in *Drying '82*, A. S. Mujumdar, ed. New York: Hemisphere Pub. Corporation, pp. 165–170.

96. Hayashi, H. and N. Kudo. 1989. "Effect of Viscosity on Spray Drying of Milk," *Drying '89,* ed. New York: Hemisphere Pub. Corporation, pp. 365–369.

97. Hayashi, H. 1996. "Effect of Viscosity on Spray Drying of Soy Protein Milk," in *Proceedings of the 5th World Congress of Chemical Engineering* at San Diego, CA. Vol. II: 98–102.

98. Skelland, A. H. P. and J. M. Lee. 1981. "Drop Size and Continuous-Phase Mass Transfer in Agitated Vessels," *AIChE J.,* 27(1): 99–109.

99. Tucker, K. G. and C. R. Thomas. 1993. "Effect of Biomass Concentration and Morphology on the Rheological Parameters of *Penicillium chrysogenum* Fermentation Broths," *Trans. IchemE.,* 71(Part C): 111–117.

100. Randhahn, H. 1976. "The Flow properties of Skim Milk Concentrates Obtained by Ultrafiltration," *J. Texture Stud.,* 7: 205–217.

101. Gross, M. O. and V. N. M. Rao. 1977. "Flow Characteristics of Sweet Potato Puree as Indicators of Dehydrated Flake Quality," *J. Fd. Sci.,* 42(4): 924–926.

102. Daud, W. R. B. W. 1989. "Theoretical Determination of the Thickness of Film of Drying Material in a Top Loading Drum Dryer," in *Drying'89,* A. S. Mujumdar, ed. New York: Hemisphere Publishing Corporation, pp. 496–500.

103. Hung, Y. C. 1989. "Predicting the Effect of Drying Conditions on the Textural Properties of Roasted Peanuts," *Transactions of the ASAE.,* 32(3): 968–972.

104. Ratti, C. and G. H. Crapiste. 1995. "Determination of Heat Transfer Coefficient During Drying of Foodstuffs," *J. Fd. Process Eng.,* 18: 41–53.

105. Figoni, P. I. and C. F. Shoemaker. 1981. "Characterization of Structure Breakdown of Foods from Their Flow Properties," *J. Texture Stud.,* 12: 287–305.

106. Bistany, K. L. and J. L. Kokini. 1983. "Dynamic Viscoelastic Properties of Foods in Texture Control," *J. of Rheology,* 27(6): 605–619.

107. Bohlin, L., P. O. Hegg, and H. Ljusberg-Wahren. 1984. "Viscoelastic Properties of Coagulating Milk," *J. Dairy Sci.,* 67: 729–734.

108. Hamann, D. D. 1987. "Methods of Measurement of Rheological Changes During Thermally Induced Gelation of Proteins," *Fd. Technol.,* March: 100–108.

109. Marshall, R. J. 1990. "Composition, Structure, Rheological Properties, and Sensory Texture of Processed Cheese Analogues," *J. Sci. Fd. Agric.,* 50: 237–252.

110. Tang, Q., O. J. McCarthy, and P. A. Munro. 1993. "Oscillatory Rheological Study of the Gelation Mechanism of Whey Protein Concentrate Solutions: Effects of Physicochemical Variables on Gel Formation," *J. Dairy Res.,* 62: 257–267.

111. Tang, Q., P. A. Munro, and O. J. McCarthy. 1993. "Rheology of Whey Protein Concentrate Solutions as a Function of Concentration, Temperature, pH and Salt Concentration," *J. Dairy Res.,* 62: 349–361.

112. McSwiney, M., H. Singh, O. Campanella, and L. Creamer. 1994. "Thermal Gelation and Denaturation of Bovine *B*-Lactoglobulins A and B," *J. Dairy Res.,* 61: 221–232.

113. Tobitani, A., H. Yamamoto, T. Shioya, and S. B. Ross-Murphy. 1995. "Rheological and Structural Studies on Heat-Induced Gelation of Concentrated Skim Milk," *J. Dairy Res.,* 62: 257–267.

114. Park, S., M. S. Brewer, J. Novakofski, P. J. Bechtel, and F. K. McKeith. 1996. "Process and Characteristics for a Surimi-Like Material Made from Beef or Pork," *J. Fd. Sci.,* 61(2): 422–427.

115. Létang C., M. Piau, and C. Verdier. 1999. "Characterization of Wheat Flour-Water Doughs. Part I: Rheometry and Microstructure," *J. Fd. Eng.,* 41: 121–132.

20 Rheological Properties of Concentrated Suspensions: Applications to Foodstuffs

P. J. Carreau
F. Cotton
G. P. Citerne
M. Moan

CONTENTS

Abstract

Foodstuffs are in many instances concentrated suspensions that exhibit a very complex rheological behavior. In this chapter, we review basic concepts on the rheology of suspensions. Problems and difficulties encountered in rheometry are discussed. Examples of the complex behavior of model suspensions and of a few foodstuffs are presented. The rheological properties of peanut butter are typical of many food products and our recent results are used to illustrate rheological complexity. In particular, we discuss wall slip effects, yield stress measurement, thixotropic behavior, strain-induced hardening, and nonlinear effects.

1-56676-963-9/02/$0.00+$1.50
© 2002 by CRC Press LLC

20.1 INTRODUCTION

Concentrated suspensions are widely encountered in the food industry and include everyday items such as chocolate,[1] fruit juices, purees,[2] ketchup, wheat flour dough,[3] and dairy products such as yogurts.[4] Rheological properties are needed to understand the phenomena encountered and changes occurring during processing and can also be related to their consistency and texture.[5] Rheological methods could be powerful tools to establish relationships between structure, formulation, processing, and properties of food products that are concentrated suspensions.

20.2 BASIC CONCEPTS

Suspensions are defined as mixtures of fluid and solid particles. The suspending fluid has its own rheological behavior, which could already be complex. It is mixed with particles that could have different shapes, sizes, size distributions, and properties. What makes suspensions even more complex are the interactions between the fluid and the particles and, above all, the interactions between the particles themselves.

Various forces affect the rheological behavior of concentrated suspensions, especially in the case of colloidal particles (submicron particles, which is often the case with food suspensions) for which Brownian forces and particle–particle interactions play a major role. These interactions can be subdivided into hydrodynamic and non-hydrodynamic interactions.

Hydrodynamic interactions result from the relative motion of the suspending medium with respect to the particles. Hydrodynamic interactions are responsible for migration of particles, alignment or orientation in the case of non-spherical particles, and, finally, structure breakdown in the cases of flocs and aggregates.

Non-hydrodynamic interactions consist mainly of the Brownian forces responsible for the internal motion of the particles and diffusion, and forces arising from physical and chemical interactions. These interactions are effective at a short particle–particle distance and are driven by forces like Van der Waals or electrostatic forces. With the help of Brownian motion, which can bring particles closer together, low-range attractive forces promote floc building as well as network and weak gel formation. On the other hand, the forces could also be repulsive, with a tendency to distribute the particles in the suspension. This is the case of steric stabilization using a stabilizer that absorbs on the surface of the particles.

The *volume fraction of particles* is an important parameter when comparing one suspension to another. Generally, suspensions are considered to be concentrated at volume fraction over 0.20. But this general rule depends on the maximum packing of the suspension, i.e., the maximum volume fraction of particles that could be added in a suspension. The maximum packing is a function of the particle geometry. It can vary from 0.18 for fibers ($L/D = 27$, L being the length and R the radius of fiber)[6] to 0.96 for elastic particles that could be deformed.[7] The value of the maximum packing fraction also depends on the arrangement of particles. The maximum packing value for monodispersed particles ranges from 0.52 for cubic arrangement to

0.74 for a hexagonal close packed arrangement. The maximum packing organization can also vary with flow conditions.

Shear thinning is defined as the decrease of viscosity with the increase of shear rate. This behavior is encountered with concentrated suspensions in Newtonian fluids when the solids have particle–particle interactions. As the flow rate increases, the bonds between particles are broken, the mobility of the particles is increased, and the viscosity of the suspension decreases. Other shear thinning effects are observed for nonspherical particles that could orient during flow and/or when the suspending fluid is non-Newtonian. The interparticle bonds can be rebuilt, and the rheological properties of the suspensions are, in general, time dependent or thixotropic. *Thixotropy* is defined as a reversible decrease in viscosity with time and shear rate.

The rheological behavior of concentrated suspensions (with strong particle–particle interactions) is governed by a competition between the microstructure breakdown due to stresses and the build-up due to collisions between particles induced by Brownian motion and flow. As the microstructure breakdown and build-up are time and shear dependent, there is an equilibrium viscosity value that depends on the shear rate. This equilibrium viscosity could be attained after a short or very long shearing time, depending on the interactions between the particles. In the definition of thixotropy, "reversible" means that the particle microstructure can be rebuilt to its "original" state after the suspension has been left for a period of rest. It may not be true for most foodstuffs, in which irreversible changes are usually observed.

Classically, the *yield stress* (σ_o) is defined as the stress value below which a material cannot deform or flow. This definition has been the subject of controversy among rheologists but, for the purpose of this chapter, we will use this basic definition. A good example of yield-stress material is ketchup. Everyone has experienced that ketchup will not flow until sufficient stress in applied. In the case of concentrated suspensions with attractive particle–particle interactions, the yield stress is the stress that is enough to partially break down the particle microstructure to allow the suspension to flow. Yield stress could be a useful rheological parameter to characterize particle–particle interactions in suspensions. In principle, the stronger the interactions, the larger the yield stress. Yield stress is usually hard to measure, especially when it is very small. Phenomena like wall slip and sample fracture can also lead to erroneous results. When facing these problems, vane geometry[8,9] could be used to measure yield stress with success. With concentrated suspensions of neutral particles or with repulsive particle–particle interactions, an apparent yield stress can be measured. But this yield stress behavior is caused by the inhomogeneous distribution of the particles in the suspension sample. In a zone of high concentration, the motion of the particles may be restricted by the surrounding particles. This problem can usually be solved by appropriate conditioning of the sample (see the next section).

Shear thickening is defined as the increase of viscosity with shear rate and/or time. Most concentrated suspensions exhibit shear thickening at high shear rates, but the onset and the importance of shear thickening effects depend on the volume fraction, particle size distribution, and viscosity of the suspending fluid. The shape of the particles is also quite important, and particle anisotropy tends to produce shear thickening effects at a lower shear rate and for a smaller solids fraction. The increase of viscosity has been attributed to a transition from a two-dimensional layered arrangement of particles to a (random) three-dimensional form. The layered arrangement of particles allows the suspension to flow easily as each layer moves without obstruction by other particles. The hydrodynamic forces created by the motion of the particles can dismantle this arrangement into a more distributed one. So, the transition from the easiest flow arrangement to another one creates an increase in viscosity. Shear thickening in suspensions of aggregated particles depends strongly on the shear history.

Suspensions are usually considered to be homogeneous materials (a continuum), and the solid particles are assumed to move with the surrounding fluid (the so-called *affine assumption*). Close to maximum packing, the continuum hypothesis is no longer valid, as the motion of a particle is restricted by the surrounding particles. We have shown, using many different concentrated suspensions, that such steric effects lead to *strain-hardening* phenomena. This strain hardening also yields an increase in viscosity, but it depends on the deformation the fluid has sustained.[10] It is observed for strain values close to unity or less depending on the distance between surrounding particles. Strain hardening has been observed in different deformation modes such as oscillatory, creep, or shear stress growth experiments. Some typical results are reported in the next section. Strain hardening is responsible for the nonlinear behavior observed at very low strains, apparent yield stresses, and primary normal stress difference detected for concentrated suspensions of so-called non-interactive particles in inelastic fluids.

20.3 DIFFICULTIES IN RHEOMETRY OF CONCENTRATED SUSPENSIONS

When carrying rheological measurements of suspensions, basic precautions must be taken to obtain reliable data. The experimenter must always keep in mind that he is working with very complex fluids and that it takes time and effort to get a good characterization of the behavior of a concentrated suspension.

The first criterion that must be respected is related to the size of the particles vs. the geometry of the measuring element. In principle, the largest particle of the suspension must be at least 10 times smaller than the gap of the measuring element to avoid artificial wall effects. This is why cone and plate geometry, which requires a very small gap, is rarely used with suspensions.

Consideration should be given to the buoyancy effects that become important when the particle and the fluid densities are significantly different. These effects are responsible for sedimentation or flotation of the particles, affecting the stability of the suspensions and resulting in concentration profiles changing with time and apparent thixotropic or anti-thixotropic effects. To obtain significant rheological results, buoyancy effects should be negligible compared with the rheological effects. Suspensions in transient experiments may be subjected to very large inertia effects, responsible unrealistic transient stress responses, and possible concentration profiles. The following criteria[10] could be used as guidelines to avoid these two effects.

$$\frac{t_{\exp} (d_p)^2 |\rho_p - \rho_f| g}{h \quad \eta_f} << 1 \tag{20.1}$$

$$\frac{d_p V \rho}{\eta_s} << 1 \tag{20.2}$$

where d_p = characteristic dimension of the particles
h = gap of the measuring device
ρ_p = density of the particles
ρ_f = density of the suspending fluid
η_f = viscosity of the suspending fluid
g = gravitational acceleration
V = average velocity of the fluid
ρ = average density of the suspension
η_s = viscosity of the suspension

The first criterion relates the experimental time, $t_{\exp}$, to a characteristic time for the gravitational or buoyancy effects. The second criterion expresses that the Reynolds number based on the particle and the average velocity of the fluid has to be very small (creeping flow regime).

Special care should be taken to characterize or prevent slip at the walls of the measuring element. With suspensions, there is always a slight decrease of particle concentration near the walls of the measuring element, due to steric effects. If the suspending fluid is not viscous enough, it may not transfer enough momentum in transient experiments from the walls to the bulk of the suspension. A small lubricating layer of fluid could then be formed and create an apparent slip at the walls. Slippage effects can be avoided by using serrated or rough surfaced measuring elements. Common practice is to use rough sandpaper glued to the surfaces of the parallel plates.

As with most fluids, evaporation during rheological measurements must be avoided (especially with suspensions, where evaporation increases the particle concentration). Air bubbles, introduced during the suspension preparation, must be removed.

With concentrated suspensions, the way the measured sample is installed in the rheometer can have some influence on the results. When installing a sample, inevitably there is some deformation and flow of the suspension, which can cause the breakdown or the build-up of a structure and even the migration of particles. The rheological results will then be affected and the suspension needs to be conditioned to remove the effects of the installation history. Conditioning may be different for different suspensions. We suggest carrying out strain amplitude ($\cong 0.05$) oscillatory experiments at high frequency ($\cong 10$ Hz) for concentrated suspensions of solid non-colloidal particles. With concentrated suspensions of hollow glass beads ($\phi = 0.5$) in a viscous Newtonian fluid (low-molecular-weight polybutene), we have observed that it could take up to one hour to obtain a good conditioning (steady state data) for these experimental conditions. Figure 20.1 presents the results of such a condi-

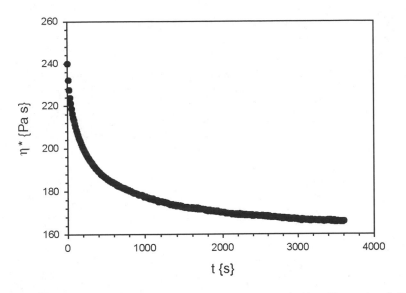

Figure 20.1 Conditioning of a suspension of hollow glass beads (d_p = 10 µm, ϕ = 0.5) in a Newtonian polybutene. Complex viscosity vs. time at ω = 62.8 rad/s and γ^0 = 0.05.

tioning. We always observe a marked decrease in the complex viscosity with time as the particles are redistributed by hydrodynamic interactions due the oscillatory motion. The hydrodynamic forces tend to homogenize the suspension and break down agglomerates or the structure of particles created by installing the sample in the rheometer. For colloidal suspensions, we suggest high rate shearing experiments (>100 s^{-1}) followed by a rest period, from 10 min up to 1 h, depending on the strength of the colloidal interactions.

Also, the preparation (mixing) of the suspension can strongly affect its rheological properties. To obtain reproducible data, the suspension has to be homogeneous, and it is not always trivial to prepare homogeneous suspensions. For highly concentrated suspensions, the solids are not, in general, evenly distributed, and aggregates and flocs may be present. There could be a strong influence of the mixing history on the size of aggregates and breakdown or build up of the microstructure that could yield different results from one preparation batch to another. During measurement or processing, the degree of dispersion of the solids or the structure due to particle–particle interactions may be considerably modified with time. Adequate experimental characterization and modeling of the rheological properties of suspensions, especially for interactive particles, remain difficult.

20.4 RHEOLOGICAL PROPERTIES OF MODEL SUSPENSIONS

It is well known that the viscosity of suspensions increases exponentially with solids concentration. But they could also exhibit a complex rheological behavior that might be important for processing and characterization. In this section, we briefly review and present some of the key results obtained for model suspensions made of low interactive (non-colloidal) particles.

The viscosity of non-interactive spherical particles in Newtonian fluids is illustrated in Figure 20.2 in terms of the relative or reduced viscosity, $\mu_r = \mu_s/\mu_m$, where μ_s is the viscosity of the suspension and μ_m is the viscosity of the matrix or suspension fluid. These data, compiled by Thomas,[11] show that the relative viscosity is approximately a unique function of the volumetric fraction of solids. The Maron and Pierce[12] empirical equation can describe the whole curve.

$$\mu_r = \frac{\mu}{\mu_m} = \left[\frac{1 - \phi}{\phi_m}\right]^{-2} \tag{20.3}$$

where ϕ_m is the solid fraction at maximum packing, theoretically equal to 0.68 for solid spheres of narrow size distribution. The dashed line in Figure 20.2 represents the Einstein classical relation, clearly valid for very low concentration ($\phi < 0.01$) only.

The non-Newtonian behavior of polymers filled with non-interactive spheres is very similar to that of non-filled polymers, at least up to a solid fraction close to maximum packing. Figure 20.3 shows the data obtained by Poslinski et al.[13] for filled thermoplastics and the description of the data using the Carreau equation:[14]

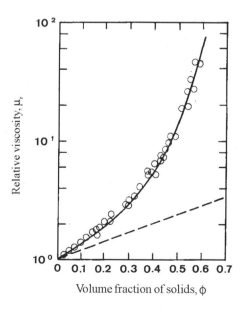

Figure 20.2 Relative viscosity vs. concentration for suspensions of spheres of low interaction, narrow size distributions—particle diameters in the range of 0.1 to 440 μm [adapted from Thomas,[11] Equation (3), Einstein's theory].

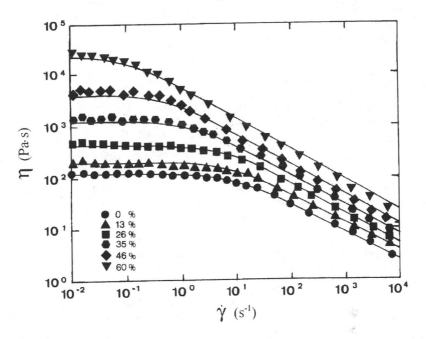

Figure 20.3 Shear viscosity of glass spheres of various volume fractions, dispersed in a thermoplastic polymer at 150°C. The glass spheres' average diameter was 15 μm with a narrow size distribution. The solid lines are the predictions using the Carreau equation. (Replotted from Poslinski et al.[13] with permission from the *Journal of Rheology*.)

$$\eta_s = \eta_{so}[1 + (t_s\dot{\gamma})^2]^{(n-1)/2} \tag{20.4}$$

where the subscript s refers to the suspension, and η_{so} is the zero-shear viscosity. We note that the data of Poslinski et al.[13] were obtained by using a cone-and-plate rheometer (Rheometrics RMS) for the lower shear rates and an Instron capillary viscometer for high shear rates. The Mooney correction was found to be negligible, and the excellent agreement between both sets of data lends support to the idea that these filled systems behave like homogeneous fluids. The time constant for the suspension in Equation (20.4), could be described by a Maron–Pierce equation [Equation (20.3)].

Another very important finding of Poslinski et al.[13] is that Equation (20.4) can be applied to non-Newtonian polymeric systems, provided that the viscosity values of the composite and of the matrix are compared at the same shear stress; i.e., Equation (20.4) has to be rewritten in the following form:

$$\eta_r = \left.\frac{\eta_s}{\eta_m}\right|_{\sigma_{21}} = \left[1 - \frac{\phi}{\phi_m}\right]^{-2} \tag{20.5}$$

where η_r = reduced viscosity
η_s = non-Newtonian viscosity of the suspension
η_m = non-Newtonian viscosity of the matrix
σ_{21} = shear stress

This means that the viscosity of thermoplastics filled with narrow size distribution and low interacting spheres can be predicted from the viscosity of the matrix.

Even the simplest concentrated suspensions have a complex rheological behavior, as illustrated by Cotton,[15] who observed strain hardening phenomena for suspensions of neutral spherical particles in a Newtonian fluid. Spherical hollow glass beads of about 10 μm in diameter and a narrow size distribution were mixed with a low-molecular-weight polybutene to obtain the simplest possible noncolloidal suspensions. The particles and the fluid had comparable density to avoid sedimentation or buoyancy problems. The viscosity of the polybutene was high enough to ensure accurate measurements. Strain hardening has been observed with these suspensions in creep as well as in oscillatory experiments.

Figure 20.4 presents the results of a creep test for an applied shear stress of 10 Pa, and Figure 20.5 presents the corresponding results for oscillatory measurements at a frequency of 0.628 rad/s, both for a suspension with a volume fraction of 0.5 of solid particles. Before these tests, the suspension samples were conditioned (see Figure 20.1) and were assumed to be almost homogeneous after conditioning. In both cases, at low strain the suspension behaves like a Newtonian fluid, indicating low or negligible interactions between the particles. The suspending fluid dictates the suspension behavior. At a strain of about 0.4, two neighboring particles have

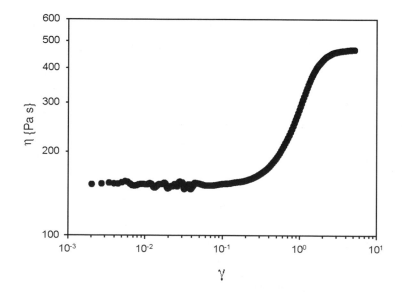

Figure 20.4 Strain hardening in a creep test with a suspension of hollow glass beads (10 μm, ϕ = 0.5) in a Newtonian polybutene. Viscosity vs. strain at σ = 10 Pa.

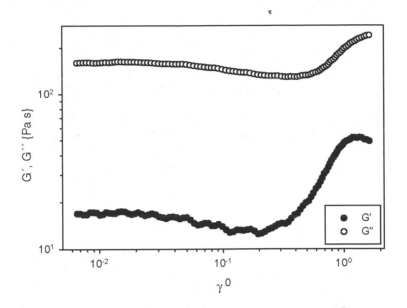

Figure 20.5 Strain hardening in a creep test with a suspension of hollow glass beads (10 μm, φ = 0.5) in a Newtonian polybutene in oscillatory measurements. Elastic and loss moduli as a function of strain at ω = 0.628.

traveled the relative distance initially separating them. They then start to block each other's motion and build up a shear induced structure. The resulting increase in viscosity is quite large, as illustrated by the figures (about three times for the creep viscosity). Note in Figure 20.5 that the initial decreases for the loss and elastic moduli are probably due to further homogenization of the suspension. The increases at the critical strain are spectacular and the strain-induced structure is responsible for the elastic character of the suspension of glass beads in a Newtonian fluid.

20.5 RHEOLOGICAL EXAMPLES OF FOODSTUFFS*****

In this section, we briefly present some results of recent studies and then discuss extensively our recent results on peanut butter.

Figure 20.6 illustrates the typical nonlinear behavior for foodstuff suspensions in dynamic measurements. The loss and elastic moduli as functions of the strain amplitude have been observed in this case for a 40% moisture hard wheat flour dough.[16] It is interesting to note that the moduli start to decrease at strain values around 0.001, the typical order of magnitude for concentrated suspensions of inter-active particles. Also worth noting is the magnitude of the elastic modulus, which is about three times larger at low strains than the loss (viscous) modulus. This suggests strong particle–particle interactions and/or network structure.

Another interesting example is presented in Figure 20.7, which illustrates how the shear viscosity vs. shear rate of a 25% starch suspension in water changes with

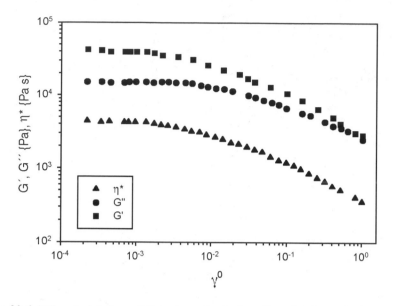

Figure 20.6 Dynamic data for a 40% moisture hard wheat flour dough suspension: complex viscosity, loss, and elastic moduli vs. strain amplitude at w = 10 rad/s. (Adapted from Reference 16 with permission of *Journal of Rheology* and J.L. Kokini.)

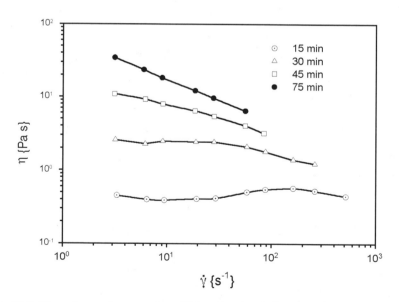

Figure 20.7 Viscosity vs. shear rate for a 25% maize starch dispersion in water after various cooking times at 65°C. The data were obtained at 60°C. (Adapted from Reference 17 with permission of Elsevier Science.)

cooking time at 65°C.[17] Initially, the viscosity of the maize starch suspension was too low to be measurable. During cooking, the starch particles were hydrolyzed, and the water swollen particles increased the effective volume fraction. The slight shear thickening effect observed at a shear rate around 100 s[-1] after 15 min was attributed to the rigidity of the swollen particles. For longer times, the effective volume fraction increased, and the rheological behavior became highly shear thinning—typical of the behavior of some polymer solutions or gels.

20.5.1 Rheological Properties of Peanut Butter

Peanut butter has been the subject of few studies. For example, Bistany and Kokini[18] showed that the Cox–Merz rule did not hold for peanut butter and reported outward migration of peanut butter samples. In another study, Campanella and Peleg[19] determined the rheological characteristics of peanut butter using lubricated squeezing flows. Nevertheless, as far as we are aware, no comprehensive study has been carried out on the subject. Thus, the aim of our study was to gain a better understanding of the rheology of peanut butter.

Peanut butter is a concentrated suspension of peanut particles in peanut oil. It is generally stabilized with vegetable oil and contains other ingredients such as salt and sugar in very small quantities. We have observed that commercial peanut butter contains approximately 60% volume of peanut particles, which have about 6 μm in diameter with a quite narrow particle size distribution. The rheological behavior of peanut butter is governed by the strong interactions between particles. So, peanut butter exhibits shear thinning, thixotropy, and yield stress.

The detailed methodology and results of this work are discussed by Citerne et al.[20] Two types of commercial peanut butter have been studied. Both products were concentrated suspensions and differed by the presence of additives. The first type, sold as "100% peanuts," was an unstabilized suspension consisting only of solid peanut particles in peanut oil (Newtonian fluid, η = 70 mPa·s at 24°C). The second type, sold as "smooth," consisted of the same suspension stabilized with a vegetable oil and contained other ingredients such as salt and sugar in very small quantities. Strain controlled (Scientific Rheometric ARES) as well as stress-controlled rheometers (Scientific Rheometric SR-5000 and Bohlin CSM) were used in this work. The geometry most often used consisted of parallel plates of diameter 25 mm on which sandpaper was glued to reduce slip. At one point, a double Couette geometry was used (ARES), with inside and outside gaps of 2 mm and a depth of 50 mm.

To determine the particle size and the particle distribution, "smooth" samples were diluted (1/20) in a commercial peanut oil and analyzed through optical microscopy. The number, area, and volume average diameters were found to be 3.9, 5.8, and 6.6 μm, respectively. The polydispersity, as defined by the ratio of the volume average diameter to the arithmetic average diameter, was equal to 1.7—characteristic of a narrow particle distribution. These results were corroborated by electronic photomicrographs of solid particles.

The volume fraction (unknown) proved more difficult to determine, and a new method was developed for this purpose. Samples of "100% peanuts" were diluted with a known amount of peanut oil until a Newtonian behavior was observed. The volume fraction of the diluted samples was calculated, assuming that well known rheological models, such as the Maron–Pierce,[12] Krieger–Dougherty[21] and Thomas,[11] were applied. Knowing the dilution of the sample for which the volume fraction had been calculated, the initial volume fraction for the solid particles could then be estimated to be around 0.6 for the "100% peanuts."

All peanut samples were conditioned when placed in the rheometer.[20] The major problem was due to slip effects evidenced by the dependence of the results on gap size, as shown in Figure 20.8 for stress growth experiments. One common method used to eliminate wall slip is the use of rough plates.[22] With the use of rough plates, made by gluing sandpaper to the parallel plates of the rheometers, the results obtained during stress growth experiments were no longer dependent on gap size, demonstrating that slip had been greatly reduced (if not eliminated), as shown by Figure 20.9. Approximate reproducible data have thereby been obtained.

The shear thinning behavior of peanut butter is illustrated in Figure 20.10, both as functions of the shear rate, for steady shear measurements, and frequency at small strain amplitude ($\gamma = 0.01$), for oscillatory measurements, respectively. Strangely enough, the shear viscosity and the complex viscosity fall approximately on the same curve (the so-called Cox–Mere rule). This contradicts the findings of Bistany and Kokini.[18] The highly shear thinning properties of this smooth sample of peanut butter are typical of a plastic behavior, which is evidenced by the data of the shear stress or stress amplitude in dynamic mode, which are nearly independent of shear rate or frequency. Note that the yield stress in simple shear (plateau at low shear rates) is considerably larger than the limiting value observed in dynamic measurements (approximately 300 Pa compared with about 40 Pa). The difference is attrib-

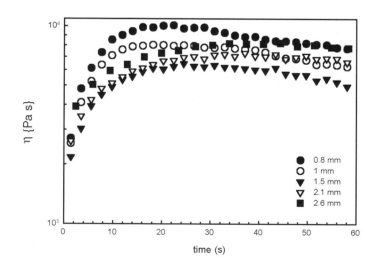

Figure 20.8 Evidence of slip for a smooth sample of peanut butter stress—growth experiments for different gap sizes at a shear rate of 10^{-2} s^{-1}.

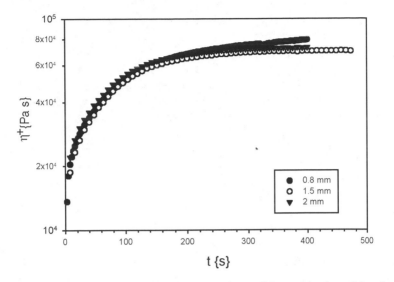

Figure 20.9 Shear viscosity of peanut butter (smooth sample) as a function of time in stress growth experiment for an applied shear rate of 5×10^{-3} s^{-1} for different gaps, using rough plates.

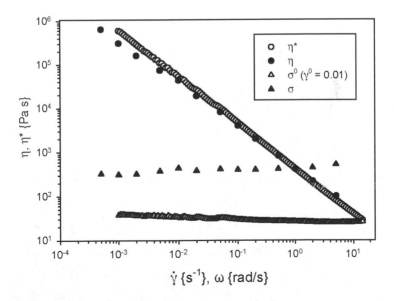

Figure 20.10 Steady shear and dynamic data for peanut butter (smooth sample) as functions of the shear rate or frequency.

uted to the strain induced structure, discussed in the previous section, that is undeveloped in low stain flows (see also Reference 10).

In this study, the yield stresses of the suspensions were calculated in three ways, with good agreement between the various methods used. First, the yield stress can be interpreted as the low shear rate limit of the stress on the flow curves (shear stress as a function of shear rate, as reported in Figure 20.10). Second, the yield stress can be taken as a parameter of the Bingham[23] and Casson[24] models, thus obtained by curve fitting the data. Last, the yield stress can be seen as the stress for which the behavior changes from solid-like to liquid-like behavior during creep experiments.[25] In all cases, the yield stresses were found to be around 15 Pa and 300 Pa for the unstabilized and stabilized suspensions, respectively. The difference of magnitude (on the order of one decade) between the yield stresses of the unstabilized ("100% peanuts" sample) and the stabilized suspensions ("smooth" sample) was attributed to the presence of the stabilizing agent.

Both suspensions were found to be highly nonlinear in oscillatory shear experiments. The first evidence of nonlinearity was given by strain sweeps performed on both suspensions. No linear domain was found for frequencies in the range of 6.28 10^{-3} rad/s and 6.28 rad/s. The results for a strain sweep performed on a "smooth" sample at frequency $\omega = 6.28 \ 10^{-3}$ rad/s are shown in Figure 20.11. At higher frequencies (e.g., $\omega = 6.28$ rad/s), the viscosity curves for both types pass through a minimum, for a strain amplitude around 0.3. The increases of the dynamic moduli and shear viscosity with strain above a critical value have been shown in the previous section and reported in previous works.[10,26,27] The decreases in the moduli shown in Figure 20.11 denote a structure breakdown under strain in the material. Note that the elastic modulus is considerably larger than the loss (viscous) modulus. This is

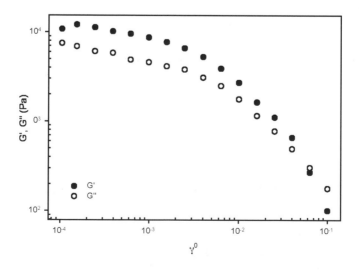

Figure 20.11 Dynamic moduli as a function of strain amplitude for a sample of smooth peanut butter, frequency $\omega = 6.28 \ 10^{-3}$ rad/s.

another indication that peanut butter is more of a solid-like than liquid-like material, and the large elastic modulus reflects strong particle–particle interactions and/or the network-type structure in stabilized peanut butter.

To characterize the nonlinearity of the materials, the strain time signals were recorded and analyzed during oscillatory measurements at constant stress amplitude. These experiments were carried out on a CSM rheometer equipped with rough plates. Lissajous curves were obtained in which the strain was plotted as a function of the applied stress (sinusoidal of frequency $\omega = 6.28$ rad/s). For a recorded strain amplitude of $1.2 \ 10^{-3}$ the Lissajous curve was elliptical whereas, for a higher strain amplitude of $4 \ 10^{-2}$, the Lissajous curve was no longer elliptical, indicating a strong departure from a linear response.[20]

Time-dependent (thixotropic) effects were highlighted via oscillatory measurements. For example, the moduli, the viscosity, and the recorded torque of both suspensions were shown to decrease when high enough strain amplitudes were applied. Under small strain (of the order of $5 \ 10^{-3}$) or at rest, the moduli were found to increase with time. Such a complex thixotropic behavior is illustrated in Figure 20.12 for a smooth sample taken at the production line. The initial decreases observed during oscillatory experiments at very low strain in both moduli are probably due to effects of conditioning or homogenization occurring after placing the sample between the plates of the rheometer. The following large increases with time are evidence of a structure build-up in presence of the stabilizer.

Last, frequency sweeps were performed on the stabilized suspension. The moduli of a "smooth" sample showed almost no dependence on the frequency. Such behavior is a characteristic of a solid or weak gel, as reported in previous studies on stabilized suspensions. The existence of a G'-plateau at small frequencies has been linked by

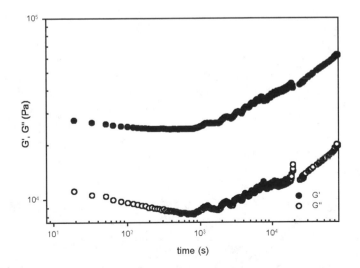

Figure 20.12 Variations of the dynamic moduli with time for a sample of smooth peanut butter taken at the production line. The initial time was 40 min after production.

some authors[28] to the existence of a yield stress, as shown in Figure 20.10 for the stress amplitude at low frequencies.

20.6 CONCLUDING REMARKS

In this chapter, we have presented and discussed rheological properties of concentrated suspensions in relation to the rheology of foodstuffs. Foodstuffs are very complex materials and, largely because of this complexity, little progress has been achieved over the last decades in understanding the phenomena. High-performance rheometers are now commercially available, and there is no doubt that, within a few years, rheometry will prove to be a powerful method for the characterization of polymers, suspensions, and foodstuffs.

Food products, in many instances, behave similarly to other industrial suspensions such as latex paints, coating colors for paper, molten filled polymers, etc. The rheology of suspensions of colloidal particles is quite complex because of the strong particle–particle interactions, which are responsible for yield stress and thixotropic behavior. Thixotropy is the result of a kinetics of structure build-up and that of a breakdown under strain or strain rate. We have shown that careful experimental procedures have to be carried out if one wishes to obtain relevant and significant rheological data. Thus, we have undertaken a major research program on model suspensions, starting with the simplest suspensions of noncolloidal, neutrally buoyant spheres. Yet, these reveal a very complex rheological behavior at high concentrations (above 40 vol%); strain hardening is observed in simple shear as well as in oscillatory shear flow for strain or strain amplitude around 0.3 to 1. This is attributed to a strain induced structure buildup responsible for the nonhomogeneity of the suspensions. Other interesting rheological effects are the presence of normal stresses and nonlinear behavior in oscillatory shear flow at large enough strains. Similar strain induced structure effects have been observed for colloidal suspensions,[10,27] as well as for peanut butter.[20] However, with colloidal suspensions, particle–particle interactions dominate, and yield stress and thixotropy are more interesting phenomena.

Peanut butter is, in our point of view, a good example of a foodstuff that exhibits most of the rheological characteristics and complexity encountered in suspensions of colloidal particles. Major experimental difficulties have been related to wall slip effects and irreproducibility. These have been largely solved by using rough plates, obtained by gluing gross sandpaper to the plate surface. The rheological behavior of stabilized peanut butter was found to be that of a plastic (Bingham-type) material with a large yield stress (about 300 Pa) and a viscosity close to that of the suspending oil. However, there is much more complexity in stabilized peanut butter. The elastic and loss moduli are strong functions of the strain, and no linear viscoelastic regime is observed for strain values down to 0.0001. The behavior is highly thixotropic, with long characteristic times and highly complex kinetic evolutions of the microstructure.

What are the challenges now? We badly need more and accurate (reproducible) data in transient flows for such suspensions of colloidal particles. These transient data are required to better characterize thixotropy, to determine key parameters that control thixotropy, and to assess constitutive equations proposed to describe and predict their behavior. Stress growth and stress relaxation experiments could yield useful information. Also, a promising experimental method is based on large amplitude oscillatory shear experiments, for which the nonlinear (deformed sinewave) response of the material can be analyzed in light of higher harmonics.[15]

REFERENCES

1. Tscheuschner, H. D. 1997. "Rheological Properties of Liquid Chocolate and Their Influencing Factors," presented at the 1st International Symposium on Food Rheology and Structure, March, 16–21, 1997.
2. Rao, M. A. 1977. "Rheology of Liquid Foods—A Review," *J. Texture Studies,* 8: 135–168.
3. Berland, S. and B. Launay. 1995. "Shear Softening and Thixotropic Properties of Wheat Flour Doughs in Dynamic Testing at High Shear Strain," *Rheol. Acta,* 34: 622–625.
4. de Kruif, C. G., D. van den Ende, M. E. van Marle, and J. Mellema. 1997. "A Microrheological Model for the Viscosity of Stirred Yoghurt," presented at the 1st International Symposium on Food Rheology and Structure, March, 16–21, 1997.
5. Sone, T. 1972. *Consistency of Foodstuffs.* Dordrecht: Reidel.
6. Kitano, T., T. Kataoka, and T. Shirota. 1981. "An Empirical Equation of the Relative Viscosity of Polymer Melts Filled with Various Inorganic Fillers," *Rheol. Acta,* 20, 207.
7. Mewis, J., J. W. Frith, T. A. Strivens, and W. B. Russel. 1989. "The Rheology of Suspensions Containing Polymerically Stabilized Particles," *AIChE J.,* 35(3): 415.
8. Liddell, P. V. and D. V. Boger. 1996. "Yield Stress Measurements with the Vane," *J. of Non-Newtonian Fluid Mech.,* 63: 235–261.
9. Nguyen, Q. D. and D. V. Boger. 1983. "Yield Stress Measurement for Concentrated Suspension," *J. Rheol.,* 27(4): 321–349.
10. Carreau, P. J., P. A. Lavoie, and F. Yziquel. 1999. "Rheological Properties of Concentrated Suspensions," in *Advances in the Flow and Rheology of non-Newtonian Fluids,* D. Siginer, D. De Kee, and R. P. Chhabra, eds. Amsterdam: Elsevier, pp. 1299–1345.
11. Thomas, D. G. 1965. "Transport Characteristics of Suspensions: VIII. A Note on the Viscosity of Newtonian Suspensions of Uniform Spherical Particles," *J. Colloid Sci.,* 20: 267–277.
12. Maron, S. H. and P. E. Pierce. 1956. "Application of Ree-Eyring Generalized Flow Theory to Suspensions of Spherical Particles," *J. Colloid Sci.,* 11: 80–95.
13. Polinski, A. J., M. E. Ryan, R. K. Gupta, S. G. Seshadri, and F. J. Frechette, 1988. "Rheological Behavior of Filled Polymeric Systems I. Yield Stress and Shear-Thinning Effects," *J. Rheol.,* 32: 703.
14. Carreau, P. J., d. De Kee, and R. P. Chhabra. 1997. *Rheology of Polymeric Systems: Principles and Applications,* Munich: Hanser.
15. Cotton, F. 1998. "Comportement Rhéologique Non-linéaire des Suspensions Concentrées de Particules Non-Colloïdales," M. Sc. A. Thesis, École Polytechnique de Montréal.

16. Dus S. J. and J. L. Kokini. 1990. "Prediction of the Nonlinear Viscoelastic Properties of a Hard Wheat Flour Dough using the Bird-Carreau Constitutive Model," *J. Rheol.*, 34(7): 1069–1084.
17. Bagley E. B. and F. R. Dintzis. 1999. *Shear Thickening and Flow Induced Structures in Foods and Biopolymer Systems, Advances in the Flow and Rheology of Non-Newtonian Fluids* Part A. Amsterdam: Elsevier, pp. 63–86.
18. Bistany, K. L., J. L. Kokini. 1983. "Dynamic Viscoelastic Properties of Foods in Texture Control," *J. Rheol.*, 27(6): 605–620.
19. Campanella, O. H., M. Peleg. 1987. "Squeezing Viscosimetry of Peanut Butter," *J. Food Sci.*, 52(1): 180–184.
20. Citerne, G. P., P. J. Carreau, M. Moan. 1999. "Rheological Properties of Peanut Butter," *submitted to Rheol. Acta*.
21. Krieger, I. M., T. J. Dougherty. 1959. "A Mechanism for Non-Newtonian Flow in Suspensions of Rigid Spheres," *Trans. Soc. Rheol.*, III: 137–152.
22. Barnes, H. A. 1995. "A Review of the Slip (wall depletion) of Polymer Solutions, Emulsions and Particle Suspensions in Viscometers: its Cause, Character, and Cure," *J. Non-Newtonian Fluid Mech.*, 56: 221–251.
23. Bingham, E. C. 1922. *Fluidity and Plasticity.* New York: McGraw-Hill.
24. Casson, N. 1959. *Rheology of Dispersed Systems.* London: Pergamon Press.
25. Cheng, D. C.-H. 1985. "Yield Stress: a Time-dependent Property and How to Measure it," *Rheol. Acta*, 25: 542–554.
26. Miller, R.R., E. Lee, R.L. Powell. 1991. "Rheology of Solid Propellant Dispersions" *J. Rheol.*, 35(5): 901–920.
27. Yziquel, F., P. J. Carreau, P. A. Tanguy. 1999. "Non-linear Viscoelastic Behavior of Fumed Silica Suspensions," *Rheol. Acta*, 38: 14–25.
28. Nguyen, Q. D., D. V. Boger. 1992. "Measuring the Flow Properties of Yield Stress Fluids," *Annu. Rev. Fluid Mech.*, 24: 47–88.

21 Mixing of Viscoelastic Fluids: A New Strategy to Elucidate the Influence of Elasticity on Power Consumption

E. Brito-De La Fuente
F. Bertrand
P. Carreau
P. A. Tanguy

CONTENTS

Abstract

In this chapter, power consumption with elastic second order and Boger fluids is studied in the laminar mixing regime (Re $\leq$ 10). Elasticity significantly increases power input for Re > 1.0. The classical representation of power input data (N_p vs. Re) is presented and discussed. By following this classical representation, the role of elasticity on power input is not clear, and different representations are suggested. On the other hand, using dimensional analysis and a generalization of the Metzner–Otto concept, a master power input curve for both Newtonian and elastic fluids is obtained.

21.1 INTRODUCTION

Polymer-based products such as foods are rheologically complex materials. Very often, these materials show shear dependent viscosity and viscoelastic properties. Although viscosity has been considered as the most important rheological function for industrial applications, it is now known that other rheological properties, e.g., elasticity, may also play an important role on flow processes. As viscoelastic materials are becoming more employed in industrial processes, it is important to understand the role played by elasticity on flow processes such as mixing in stirred tanks. This will lead to improved efficiency in mixing processes and to providing materials of higher quality.

Flow processes with viscoelastic materials can be improved by a more profound knowledge and understanding of fluid viscoelastic properties, in particular those properties associated with deformation conditions closer to the actual flow process. It should be pointed out that the nonhomogeneous nature of food materials (i.e., presence of several phases) adds complexity to the experimental determination of viscoelastic properties.

It is well known that the normal stress functions (e.g., the primary normal stress coefficients and difference) are closely related to fluid elasticity.[1-3] However, recently, scattered reports on normal stresses of foods have appeared. Primary normal stress measurements have been reported for food systems such as peanut butter, whipped butter, whipped dessert topping, marshmallow cream, whipped cream cheese, squeeze and tube margarine, apple butter, ketchup, mustard, mayonnaise, margarine, and canned frosting, among others.[4] Normal stresses for model fluids simulating fluid food systems have also been reported recently.[2,5] Most of the food systems mentioned above are prepared in agitated tanks; thus, knowledge about the role of elasticity on mixing is important.

In continuation with our previous work on mixing with helical ribbon impellers,[1,6,7] it is now our purpose to show that dimensional analysis can elucidate the influence of elasticity on the increase of power consumption. The case studied here is the use of helical ribbon impellers and second-order fluids. The results shown in this work are for the laminar mixing regime.

21.2 LITERATURE REVIEW

Mixing of viscoelastic fluids is typical in many industrial operations including food processing, fermentation, polymerization processes, and paper coating color impregnation. A review of the literature on the mixing of these rheologically complex materials in stirred tanks reveals that the flow pattern created by a given impeller geometry is strongly dependent on elasticity. Thus, elastic properties are responsible for significant changes in power consumption, a macro-mixing parameter closely related to the economy of the process.[1,2,6] However, reports on the quantitative effect of elasticity on mixing are not consistent from one study to another, although similar impeller geometry and elastic fluid properties are kept similar.

Early reports agree on the fact that elasticity does not increase power consumption in the laminar mixing regime.[8,9] Since 1983, many authors have consistently noted a marked increase of power consumption with viscoelasticity beyond a Reynolds number threshold.[1,3,5,10,11] A review of studies on mixing of elastic fluids is given elsewhere.[4] In all these papers, elastic effects on power input have been analyzed using dimensionless numbers. It must be pointed out that studies on elasticity effects on other macro-mixing parameters such as mixing and circulation times are scarce.

Several definitions of the Weissenberg number, Wi, are found in the literature to characterize the fluid elasticity.[6] The definition used here is[3]

$$\mathrm{Wi} = \frac{\Psi_1 N}{\mu} = \frac{N_1 N}{\gamma^2 \mu} \tag{21.1}$$

where Ψ_1 = primary normal stress coefficient
N_1 = primary normal stress difference
N = impeller rotational speed
μ = Newtonian viscosity
γ = shear rate

The elasticity number, which is the result of a balance between inertial and elastic forces, has also been used[1,3] and is given by

$$\mathrm{El} = \frac{\mathrm{Wi}}{\mathrm{Re}} \tag{21.2}$$

It must be noted that of the groups Wi, El, and Re, only El and Re are independent. On the other hand, for second-order fluids, El is constant.[1]

21.3 MATERIALS AND METHODS

The helical ribbon agitators used have been fully described previously.[1] The impeller diameter, d, is 0.185 m and equal to its height, P. The ratio of the height of the liquid to the height of the impeller was kept constant and equal to 1.14. It must be noted that other geometric characteristics of the impellers were kept according to standards. The mixing system and its main geometric characteristics are described in Figure 21.1.

Newtonian second-order fluids having constant viscosity and a quadratic dependence on N_1 (here also named HECV fluids) and Boger-type fluids are analyzed in this chapter. A more detailed description on the composition, preparation procedures, and rheological measurements has been reported previously.[1] As an example of typical elastic properties, Figure 21.2 shows the normal stress difference, N_1, as a function of shear rate for several second order fluids with constant viscosity at 25°C.

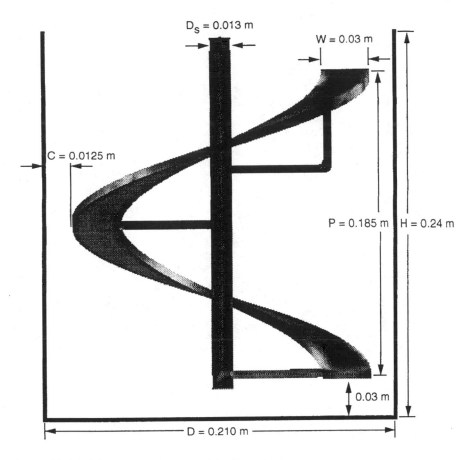

Figure 21.1 Mixing system and geometric dimensions.

As the results suggest, N_1 has a quadratic dependence on shear rate. Therefore, the elasticity number as defined in Equation (21.2) is a constant for these fluids. This property, and the fact that these fluids have a constant viscosity, allow separation of the effects of shear dependent viscosity from the effects caused by the elasticity. Symbol α in Figure 21.2 represents an Arrhenius-type function of the temperature[1] viscosity at 25°C.

Steady shear viscosity and normal stress differences as a function of shear rate for a typical Boger fluid are shown in Figure 21.3. In the same figure, the experimental values of Collias and Prud'homme[3] are represented. From Figure 21.3, a slight shear thinning behavior (n from the power law model is equal to 0.97) can be observed. A similar behavior has been observed for fluids with similar composition.[3] On the other hand, as the results of Figure 21.3 suggest, the Boger fluid is not a second-order fluid, because N_1 is not quadratic in shear rate. Consequently, the elasticity number defined in Equation (21.2) is not a constant as it was for the second-order fluids.

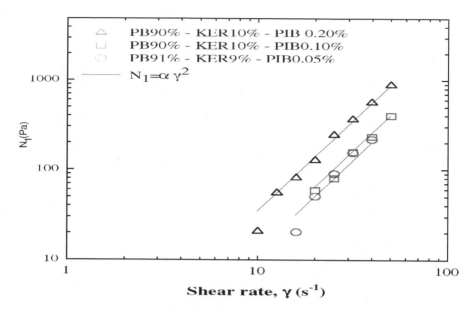

Figure 21.2 Normal stress difference for second-order fluids (HECV) with constant viscosity at 25°C. PB = polybutene, KER = kerosene, PIB = polyisobutylene.

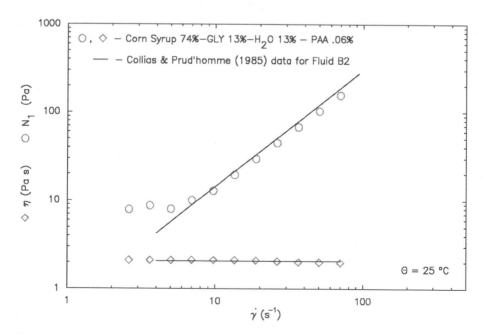

Figure 21.3 Steady shear viscosity and primary normal stress difference data for the Boger fluid (corn syrup 74%, glycerol 13%, water 13%, polyacrylamide 0.06%) at 25°C.

21.4 POWER CONSUMPTION RESULTS AND DISCUSSION

Power input results are typically expressed by using dimensionless numbers like power number, Np, and Reynolds number, Re. The definitions of these numbers are, respectively,

$$N_p = \frac{P}{\rho N^3 d^5} \tag{21.3}$$

and for Newtonian fluids,

$$Re = \frac{\rho N d^2}{\mu} \tag{21.4}$$

Then, for the laminar mixing regime (Re $\leq$ 10),

$$N_p Re = K_p = \frac{P}{\mu N^2 d^3} \tag{21.5}$$

where K_p is known as the power constant and d is the impeller diameter.

Now, a dimensional analysis on power input with second-order fluids is as follows.

• A second-order fluid is represented by

$$\tau = \mu\gamma + \Psi_1\gamma^2 \tag{21.6}$$

• Power input in an agitated vessel is estimated by

$$P = \int_\Omega \tau : \nabla V dv = \int_\Omega \tau : \gamma d\gamma \tag{21.7}$$

Then, for second-order fluids,

$$P \alpha \mu N^2 d^3 + \Psi_1 N^3 d^5 \tag{21.8}$$

where α = a proportionality constant

The power constant K_p now becomes

$$K_p(Wi) = \frac{P}{\mu N^2 d^3}(1 + Wi)^{-1} \tag{21.9}$$

Note that, for nonelastic Newtonian fluids (Wi = 0), Equation (21.9) reduces to Equation (21.5).

Using an analogy of the Metzner–Otto concept[6] for second-order fluids, an effective shear rate is obtained as

$$\gamma_{eff} = K_e N \tag{21.10}$$

Using Equation (21.10), a generalized Re number is now defined as

$$Re_g = \frac{\rho N d^2}{\mu(1 + K_e Wi)} \tag{21.11}$$

and the power number N_p and constant K_p now become

$$N_p = N_{p\,Newt}(1 + K_e Wi) \tag{21.12}$$

$$K_e = \frac{1}{Wi}\left[\frac{N_p}{N_{p\,Newt}} - 1\right] \tag{21.13}$$

where N_{pNewt} is Newtonian power number and K_e is the generalized Metzner–Otto constant.

Equation (21.12) is comparable to the one proposed previously.[6]

$$N_p = N_{p\,Newt}(1 + a Wi^b) \tag{21.14}$$

where a and b are fitting constants.

The analysis of experimental data will first follow the classical representation of power input results (N_p vs. Re). Then, the new model [Equation (21.12)] will be tested with Newtonian and second-order (HECV) fluids.

Figure 21.4 shows power input data for the different fluids studied here. The Newtonian results are well represented by

$$N_p = 135.2 Re^{-1} \tag{21.15}$$

The Newtonian data expressed by Equation (21.15) compares quite well with the predictions from the Couette flow analogy.[1,8]

Regarding the experimental data for the second-order fluids (HECV), as seen in Figure 21.4, elasticity significantly increases power input for Re > 1.0. With respect to Newtonian curve, this increase can be as high as 75%. The effect of different levels of elasticity on power input has been discussed previously.[1] It was shown that the use of raw experimental data (i.e., torque, impeller rotational speed, and viscosity) as the ratio torque/viscosity versus N gives a better representation and under-

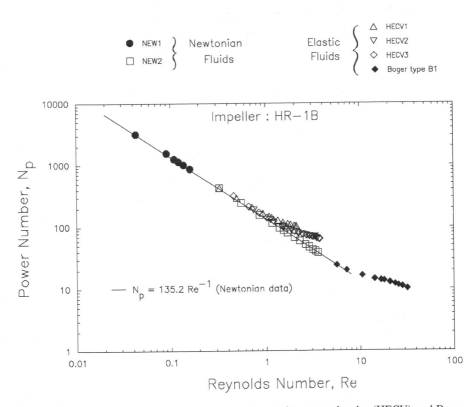

Figure 21.4 Power consumption curve for the Newtonian, second-order (HECV) and Boger fluids.

standing of the role of elasticity on power input. Unfortunately, this type of representation is not dimensionless.[1]

The classic power consumption curve for the Boger fluid is also shown in Figure 21.4. As the results shown suggest, elasticity also increases power input. According to these results, it may be concluded that power input falls mainly in the transition regime. In addition, for this fluid, the role played by elasticity is unclear. Nevertheless, for Re < 10, the experimental points are well predicted by the Newtonian power correlation [Equation (21.15)].

To determine whether the power input results for the Boger fluid fall in the laminar or the transition region, the experimental power consumption data were analyzed by using

$$P = aN^b \tag{21.16}$$

where P is the power consumption, and a and b are fitting constants.

Figure 21.5 shows the representation of power consumption as a function of impeller rotational speed for the Boger fluid and two Newtonian fluids. From this

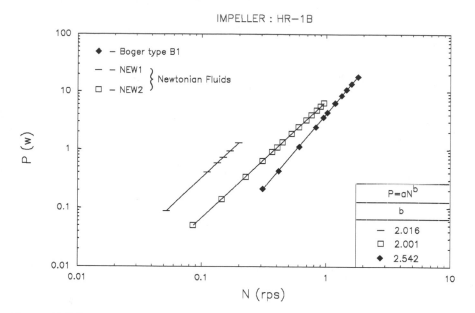

Figure 21.5 Power as a function of the impeller rotational speed for Newtonian and Boger-type fluids.

figure, the least-mean-squares fit yields an exponent b, which is not equal to 2, the resulting exponent when the dimensionless power number is correlated with the inverse of the Reynolds number in the laminar regime.

The value of the exponent b for the Boger fluid is equal to 2.542, indicating that the viscous regime has its upper limit at less than Re = 10, this last the upper limit of the laminar regime. A similar behavior was reported by Collias and Prud'homme[3] for Boger-type fluids agitated with a Rushton turbine. It should be noted that the average b value for Newtonian fluids is 2.009.

Because, for the Boger fluid studied here, N_1 is not quadratic in shear rate as it was for the second-order fluids, then it is possible to find a function showing the elasticity number, El, as a function of Re. Figure 21.6 shows this function as well as the results found when fitting the experimental results to the following polynomial expression:

$$\text{El} = \frac{A}{Re} + B + CRe + ERe^2 \tag{21.17}$$

where A, B, C, and E are fitting constants.

The values of the coefficients in Equation (21.17) are given in Figure 21.6. In the same figure, the data for the B2 fluid by Collias and Prud'homme[3] are also shown for comparison.

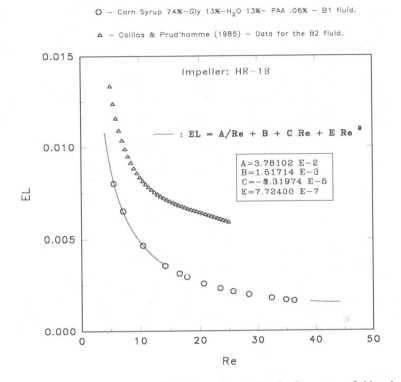

○ — Corn Syrup 74%–Gly 13%–H₂O 13%– PAA .06% – B1 fluid.

△ — Collias & Prud'homme (1985) – Data for the B2 fluid.

Figure 21.6 Elasticity as a function of the Reynolds number for Boger-type fluid and representation of Equation (21.17).

Data displayed in Figure 21.6 clearly show that the elasticity number decreases as the Reynolds number increases. It must be pointed out that, for the comparison made in Figure 21.6 between the results of this work and those reported by Collias and Prud'homme,[3] one may have rheological similarity (same fluid), dynamic similarity (same Re), but not elastic similarity (different El numbers), nor geometric similarity (Rushton turbine vs. Helical ribbon impeller). Nevertheless, the trend followed by the function El = f(Re) is the same for both impellers.

Finally, in Figure 21.7, the new model [i.e., Equation (21.12)] with a proposed K_e value of 30 is shown. In this case, the analysis is performed using Equations (21.11) through (21.13) As the results suggest, a unique master representation is obtained for the Newtonian and the second order fluids studied here.

The results shown in this chapter suggest that rise in power input owing to elastic properties can be modeled using dimensional analysis. In this sense, there is no need to use data regression analysis in viscoelastic mixing. On the other hand, a generalized Metzner–Otto constant (K_e) can be calculated and used to generate a master curve, as for non-Newtonian inelastic fluids.

This new model is also tested with experimental data obtained with a helical ribbon impeller and elastic fluids by Carreau et al.[5] Figure 21.8 shows the raw

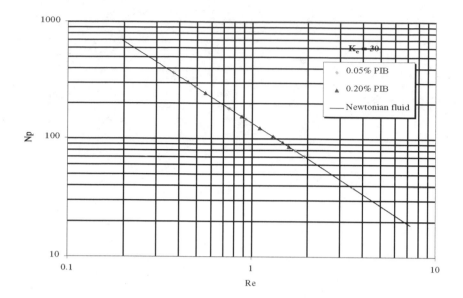

Figure 21.7 Rise in power consumption owing to elasticity. Data analysis using Equations (21.11) through (21.13). PIB = polyisobutylene.

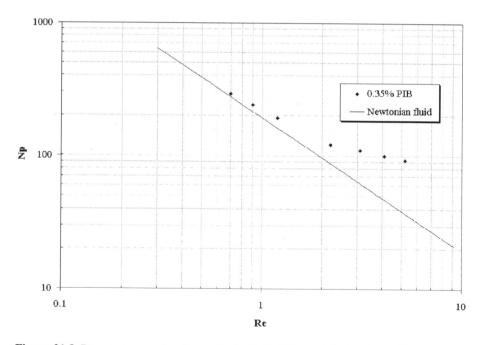

Figure 21.8 Power consumption for an elastic fluid. PIB = polyisobutylene. (Adapted from Carreau, et al.)[5]

experimental data using the classical representation. As can be seen again in this case, elasticity increases power consumption. These results are in agreement with the results shown in Figure 21.4, suggesting that the increase in power consumption due to elasticity shift the upper limit in the laminar regime towards lower values of Re.

Figure 21.9 shows another validation of the proposed model [Equation (21.12)] using the experimental data of Carreau et al.[5] and a proposed K_e value of 15. As the results suggest again, a master curve can be obtained.

21.5 CONCLUSIONS

This work investigates the role of elasticity on power input using second-order and Boger-type fluids. The main conclusions are:

- Elasticity significantly increases power input when mixing with helical ribbon impellers. This increase may be as high as 75% with respect to Newtonian fluids.
- Viscoelastic power consumption data can be represented with a simple analytical relation using the Newtonian power number and the Weissenberg number.
- An elastic Metzner–Otto constant, K_e, can be used to generate a master curve, as in the case of non-Newtonian shear thinning in elastic fluids. The value of K_e is a function of the impeller geometry.

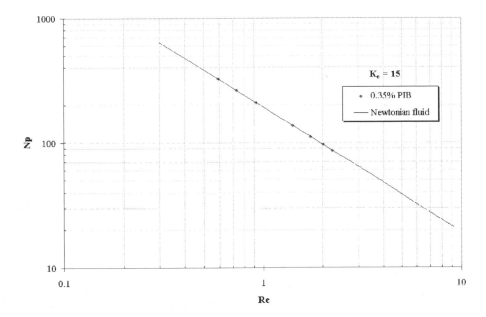

Figure 21.9 Rise in power consumption owing to elasticity. Experimental data from Carreau et al.[5] Analysis using Equations (21.11) through (21.13) with a proposed K_e value of 15.

REFERENCES

1. Brito-De La Fuente, E., J. C. Leulliet, L. Choplin, and P. A. Tanguy. 1991. "On the Role of Elasticity on Mixing with a Helical Ribbon Impeller," *Chem. Eng. Res. & Dev.*, 69(A4): 324–331.
2. Ulbrecht, J. J. and P. Carreau. 1985. "Mixing of Viscous Non-Newtonian Liquids," *in Mixing of Liquids by Mechanical Agitation,* J. J. Ulbrecht and G. R. Patterson, eds. New York: Gordon and Breach Science Pub., 1, pp. 93–97.
3. Collias, D. J. and R. K. Prud'homme. 1985. "The Effect of Fluid Elasticity on Power Consumption and Mixing Times in Stirred Tanks," *Chem. Eng. Sci.,* 40: 1495–1505.
4. Rao, M. A. and J. F. Steffe. 1992. *Viscoelastic Properties of Foods*, New York: Elsevier Applied Food Science Series.
5. Carreau, P. J., R. P. Chabra and J. Cheng. 1993. "Effects of Rheological Properties on Power Consumption with Helical Ribbon Agitators," *A.I.Ch.E.J.,* 39(9): 1421–1430.
6. Brito-De La Fuente, E., L. Choplin, and P. A. Tanguy. 1990. "Mixing and Circulation Times in Rheologically Complex Fluids," *I. Chem. Eng. Symposium Series*, 121: 75–96
7. Brito-De La Fuente, E., L. M. López, L. M. Medina, G. Ascanio, and P. A. Tanguy. 1998. "Estimation of Viscous Food Properties Using a Helical Ribbon Impeller," in *Functional Properties of Proteins and Lipids,* J. R. Whitaker, F. Shahidi, A. López-Munguía, R.Y. Yada, and G. Fuller, eds. ACS Symposium Series 708. ACS, pp. 52–64.
8. Chavan, V. V. and J. J. Ulbrecht. 1973. "Power Correlations for Close-Clearance Helical Impellers in Non-Newtonian Liquids," *Ind. Eng. Chem. Process Des. Develop.,* 12(4): 472–476.
9. Kelkar, J. V., R. A. Mashelkar, and J. J. Ulbrecht. 1972. "On the Rotational Viscoelastic Flows Around Simple Bodies and Agitators," *Trans. I. Chem. Eng.*, 50: 343–352.
10. Nienow, A. W., D. J. Wisdon, T. J. Solomon, V. Machon, and J. Vleck. 1983. "The Effect of Rheological Complexities on Power Consumption in an Aerated Agitated Vessel," *Chem. Eng. Commun.,* 19: 273–293.
11. Youcefi, A., D. Anne-Archard, H. C. Boisson, and M. Sengelin. 1997. "On the Influence of Liquid Elasticity on Mixing in a Vessel Agitated by a Two-Bladed Impeller," *J. Fluids Engineering, Trans. A.S.M.E.,* 119: 616–622.

22 Microstructuring of Multiphase Food Systems in Shear and Elongational Process Flows

E. J. Windhab
P. Fischer
M. Stranzinger
S. Kaufmann

CONTENTS

Abstract

A wide variety of food systems are treated in dispersing operations to generate a homogeneous disperse microstructure that allows adjustment of structure-related quality characteristics of the final products. For fine dispersing of disperse structures like solid particles/biological cells, emulsion drops, gas cells, and aggregates thereof, the volumetric power and energy inputs are the crucial integral processing parameters but are not sufficient to optimize dispersing efficiency. The flow field characteristics are of additional importance. For most concentrated food systems containing a disperse phase, the dispersing flow field in any apparatus is laminar. However, the ratio of shear and elongational laminar flow contributions influences the dispersing efficiency strongly. Beside disperse components, most multiphase food systems also contain macromolecular stabilizing components in the continuous phase as well as surfactants at the interfaces. The various disperse and macromolecular components have, in general, different sensitivities to mechanical stresses acting in dispersing flows.

Consequently, there is a need to optimize dispersing flow apparatus and adapt the stresses acting during processing in order to adjust the resulting microstructure. The most important structural characteristics for the disperse phase(s) are size, size distribution, and shape, all of which have a strong impact on the rheology of the fluid system and which are preferably focussed within the work reported here.

This chapter discusses well defined shear and/or elongational dispersing flow fields; the disperse particle size distribution and shape can be adjusted so as to generate specific rheological and quality properties of related food systems. The synergistic application of dispersing flow experiments and computational fluid dynamics (CFD), which allow optimization of dispersing flow processes, and related product properties are also demonstrated.

22.1 LIST OF SYMBOLS

η_0	zero shear viscosity	λ	time constant
η_∞	upper qu. Newt. viscosity	n	power law index
G_{G1}	interfacial tension modulus	el	elongation deformation
G_{G2}	interfacial skin elasticity	α	shear/elong. flow ratio
τ_0	yield value	$\eta^*(\dot\gamma)$	equil. viscosity function
τ	shear stress	$\eta(\dot\gamma^*, \dot\gamma)$	structure viscosity fct.
σ	interfacial tension	$\dot\gamma^*$	shear structure index
x	particle diameter	A_0	spherical droplet surface
λ	viscosity ratio (droplet)	A	deformed droplet surface
η_d	disperse phase viscosity	γ	shear deformation
η_c	continuous phase viscosity	$\dot\gamma$	shear rate

22.2 INTRODUCTION

Micro-structuring in shear and elongational dispersing flow fields links the areas of process engineering and material engineering. The latter includes structure-rheology relationships. Rheologically, the viscous behavior of multiphase food systems is generally non-Newtonian, mostly shear-thinning, and time-dependent. In many cases, elastic properties are also not negligible. Dispersing as a flow structuring unit operation, as regarded here, generates the product microstructure due to shear and normal stresses that act in the dispersing flow velocity field. A detailed description of the velocity field can be received from flow visualization experiments and from computational fluid dynamics (CFD). The coupling of local velocity field information and rheological material functions received from rheometric shear and elongation experiments allow determination of the related local shear and normal stress distributions. Microstructuring is triggered by the stresses acting locally in the flow field and generally occurs if critical stresses of the structuring units like networks, droplets, and particle aggregates are exceeded. To generate a flow-induced structure, the structuring units have to experience along their flow tracks a certain total deformation that requires sufficient shear/elongational rates and residence times.

22.3 STRUCTURE RHEOLOGY RELATIONSHIPS

Laminar flow fields are of uni- or multiaxial shear or elongational nature or a mixture of both. Within such flows, the rheological behavior mirrors the dynamic behavior of structural units, which can change by orientation, deformation, and aggregation or deaggregation (i.e., structuring mechanisms). The latter is synonymous with dispersing.

Figure 22.1 gives a schematic overview on such structure rheology relationships related to the viscous behavior in shear flow.[1] The flow induced structuring generates non-Newtonian behavior, even for dispersions with Newtonian continuous fluid phases. The shear rate-disperse phase concentration- and time-dependent viscous behavior can be described by the structure-related fluid immobilization behavior, which depends strongly on the structuring mechanisms described before and investigated in previous works.[2]

Elongational flow fields more efficiently induce microstructural changes due to the fact that there is no rotational contribution of the velocity field, which, as in shear flows, causes structuring units to rotate. Rotation prevents structuring units from being strongly deformed, because the acting stresses may not act long enough to exceed a critical deformation. This is more the case if the structuring units are highly viscous or of solid-like nature. This is demonstrated in Figure 22.2 for the breakup of droplets with a diameter x in uniaxial shear or elongation flows by the well known critical Weber number, We [see Equation (22.1)] dependency from the viscosity ratio λ ($= \eta_{\text{disperse}}/\eta_{\text{continuous}}$).[3]

$$\text{We} = 0.25\tau\frac{x}{\sigma} \tag{22.1}$$

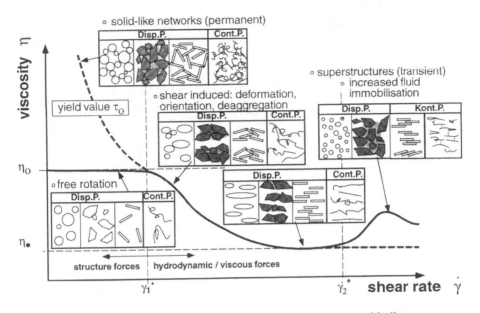

Figure 22.1 Structure–rheology relationships for multiphase systems with disperse components (droplets, particles, fibers, macromolecules).

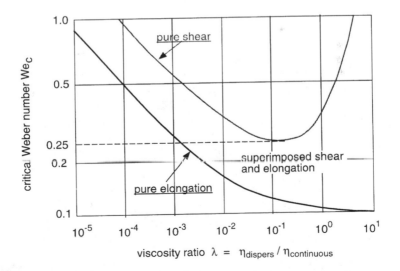

Figure 22.2 Critical Weber number as a function of the viscosity ratio $\lambda = \eta_{disperse}/\eta_{continuous}$ for uniaxial shear and elongational flows.

Droplets are preferably used as model structuring units due to their isotropy, homogeneity, and symmetry. In uniaxial shear, the drop breakup is limited to the domain $\lambda \leq 4$; in uniaxial elongational flow fields, this limitation does not exist (Figure 22.2).

22.4 EXPERIMENTS IN DISPERSING FLOW FIELDS

22.4.1 MODEL FLOW FIELDS

To get experimental results from laminar dispersing flow fields that are close to real flow fields in dispersing apparatus, the experimental setup should allow investigation of pure shear and elongational flows, but also their well defined superposition. These conditions are locally fulfilled in a Four-Roller Apparatus (FRA) and Cross-Slot Channels (CSC). These flow apparatus are constructed in such a way that laser optics for optical rheometry, small angle light scattering (SALS) and microscopic/image analysis tools (e.g., CSLM) can easily be adapted. Within the FRA, the velocity ratio α of the two diagonal roller pairs determines whether pure shear ($\alpha = 0$), pure elongational ($\alpha = 1$) or mixed flows of defined elongation to shear ratio ($0 < \alpha \leq 0$) are generated (see Figure 22.3).

Figure 22.3 shows the FRA built in a laser optic arrangement for optical rheometry. Figure 22.4 demonstrates drop deformation and breakup in the stagnation flow domain of the FRA for two different viscosity ratios λ under pure planar elongational flow conditions. Either drop breakup at the tips or the elongated drop filament breakup allow generation of very narrowly size distributed droplets.

22.4.2 SINGLE-DROP RHEOLOGY

From the dynamic drop behavior during deformation and relaxation, interfacial elasticity and viscosity (as well as interfacial tension) can be estimated under small drop deformation conditions (interface ratio $\xi = A/A_0 \leq 1.05$) as investigated in References 22.4 and 22.5. Equation (22.3) is derived for the drop relaxation from a mechanical spring/dashpot model of the drop interface.[4]

$$\sigma = \frac{\gamma Y \times \eta_c}{2D} \frac{19\lambda + 16}{16\lambda + 16} \tag{22.2}$$

$$A/A_0 = \exp\left\{ \frac{\tau^2 (G_{G1} + G_{G2})^2}{2G_{G1}^2 G_{G2}^2} \exp\left(-\frac{2G_{G1}G_{G2}}{(G_{G1} + G_{G2})((\eta_d + \eta_c)} t \right) \right\} \tag{22.3}$$

22.4.3 EMULSION RHEOLOGY

Within multidrop systems (emulsions), the disperse structure is determined by a dynamic equilibrium between drop breakup and recoalescence. Experimentally, equilibrium drop size distributions were measured in pure shear flow fields (con-

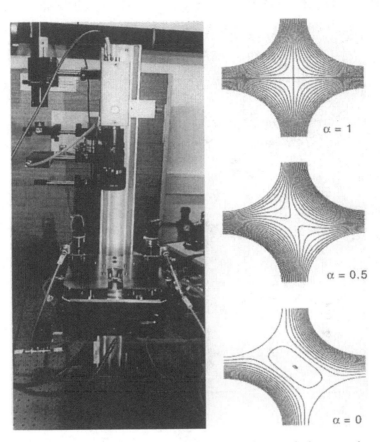

Figure 22.3 Four Roller Apparatus (FRA) connected to a laser optical setup and streamline patterns for various diagonal roller pair speed ratios.

centric cylinder gaps) as well as in superimposed shear and elongational flow fields within eccentric cylinders for diluted ($\phi V \leq 0.05$) and concentrated ($\phi_V = 0.3$) model O/W emulsions.

Figure 22.5 shows equilibrium drop size distributions from pure shear flow experiments.[6] The related equilibrium viscosity function $\eta^*\dot{\gamma}$, as well as so-called structure viscosity functions $\eta(\dot{\gamma}^*,\dot{\gamma})$ (= viscosity functions at constant drop size, measured by step-shear experiments), are described elsewhere.[7]

22.5 MODELING AND SIMULATIONS

22.5.1 FLOW SIMULATION

To get information about the locally acting shear and normal stresses, as well as the related shear and elongational rates, computational fluid dynamics (CFD) was applied to model flows in the geometries described above. Using a partly modified

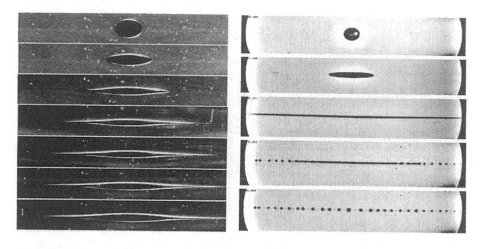

Figure 22.4 Droplet breakup under pure elongational flow conditions in a Four Roller Apparatus (FRA). Left: tip-stream breakup for $\varepsilon \approx 5$ s^{-1}, $\lambda \approx 10^{-4}$ (water/silicon oil). Right: filament breakup for $\varepsilon \approx 5$ s^{-1}, $\lambda \approx 1$ (water + PEG/silicon oil).

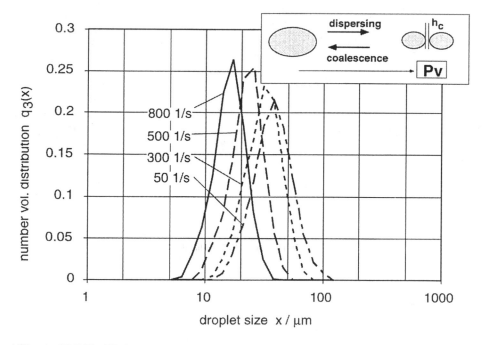

Figure 22.5 Equilibrium drop size distributions after shear at various constant shear rates (pure uniaxial shear in concentric cylinder gap).

finite element code (SEPRAN), the shear and elongation rate fields were calculated by solving the typical equations for mass, energy, and momentum conservation. Results are shown for the FRA in Figure 22.6.[8] Figure 22.7 additionally demonstrates elongation rate fields in three different experimentally investigated cross-slot flow channels. From such calculations, the stress/strain history (SSH) of droplets or other structuring elements along flow tracks (particle tracks) within the apparatus flow can be derived. The application of such a particle tracking calculation is demonstrated in the chapter where real rotor/stator-dispersing flow geometries are described.

22.5.2 SIMULATION OF DROP DEFORMATION AND BREAKUP

Using a boundary integral element method (BIM) developed by Löwenberg et al.,[9] the drop deformation in superimposed shear/elongation flows was calculated, assuming clean interfaces (no surfactants). As the only interfacial property a constant equilibrium interfacial tension s was taken into account. Some results from such drop deformation simulations are demonstrated in Figure 22.8.[10]

If critical capillary numbers Ca (= Weber numbers We) are derived from such calculations, the expected increase from pure elongation to pure shear is found, as shown in Figure 22.9.

22.6 APPLICATIONS TO FOOD SYSTEMS

As derived from the results of fundamental model dispersing flow investigations as described above, laminar dispersing flows to be applied in dispersing apparatus should preferably be of elongational nature. It is obvious that close to apparatus walls, at which the continuous phase adheres, shear always dominates elongation. However, from practical experience, even a minor elongational contribution is expected to have a relevant impact on the dispersing efficiency, in particular at higher viscosity ratio ($\lambda > 2$), which shall be quantified.

22.6.1 PROCESSING IN CONTINUOUS NARROW ANNULAR GAP REACTORS (NAGRs)

In a narrow annular gap reactor (NAGR), which consists basically of a concentric cylinder gap geometry with rotating inner cylinder, the treated fluid systems experience a rather narrow range of shear stresses. Commonly, dispersing/mixing elements are adapted to the rotating inner and/or outer cylinder wall. In the case of in-built wall scraping rotor or stator blades or pins, which are in general inclined relative to a tangential plane to the cylinder wall (see Figure 22.10), high elongation rates act locally. This has been proved experimentally using particle imaging velocimetry (PIV) as well as by CFD calculations which apply a particle tracking method.[11] For the investigated flow gap between a rotating dispersing blade that is mounted to the inner cylinder, scraping the surface of the outer cylinder, and the inner cylinder surface, three particle tracks are shown in Figure 22.11. For various flow gap ratios of $r_S = 0.25, 0.4, 0.6$ and a constant blade angle of $\beta = 30°$, the calculated shear and

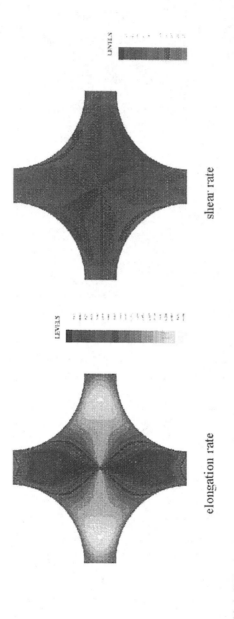

elongation rate   shear rate

Figure 22.6 CFD simulation of elongation (left) and shear rate fields (right) for a Four Roller Apparatus (FRA). Program = SEPRAN.

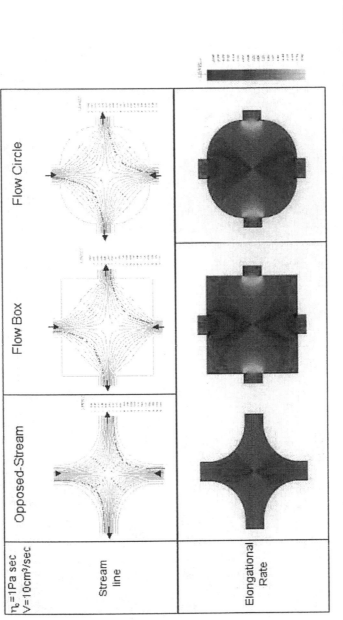

Figure 22.7 CFD simulation of streamlines (top) and elongation rate fields (bottom) for different cross-slot flow channels. Program = SEPRAN.

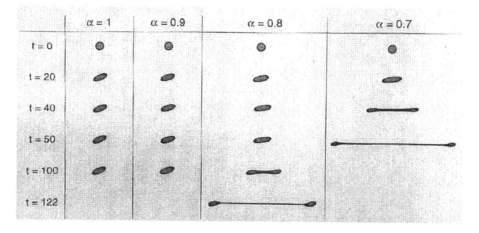

Figure 22.8 Calculated droplet deformations under mixed shear-elongational flow conditions ($\alpha = 1$: simple shear, $\alpha = 0$: elongation); simulation with boundary BIM; We = 0.35, $\lambda = 1$.

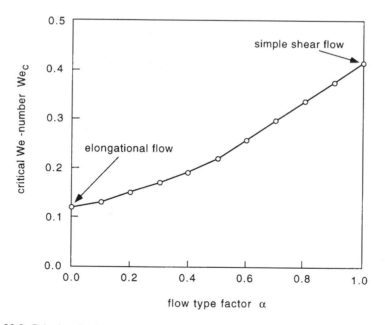

Figure 22.9 Calculated critical Weber numbers (from BIM simulation); $\lambda = 1$.

elongation rates ($\dot{\gamma}, \varepsilon_1$) along the three selected particle tracks are shown in Figure 22.12.[11]

In addition to maximum shear and elongation rates, the shearing and elongation times (and consequently the total shear and elongational deformations) are considered so as to generate a flow-induced (i.e., dispersed) microstructure.

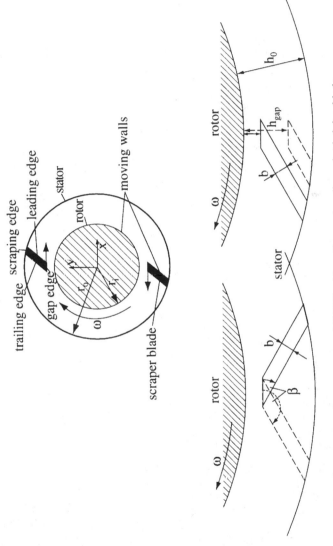

Figure 22.10 Cross section of narrow annular gap reactor (NAGR) with wall scraping dispersing/mixing blade.

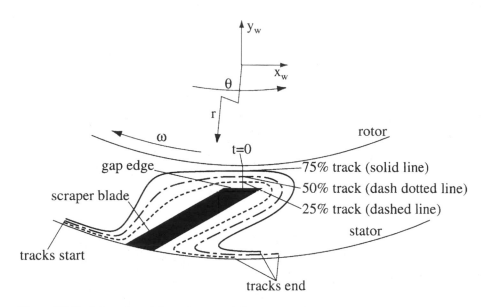

Figure 22.11 Selected particle tracks around dispersing/mixing blade in Narrow Annular Gap Reactor (NAGR).

Elongational and shear rate integrated over time provide the elongational and shear deformation (ε, γ). Thus, the calculated elongational rate $\dot{\varepsilon}_1$ and the shear rate $\dot{\gamma}$ with respect to the local (e_1, e_2) coordinate system are integrated between $t = t_{\text{start}}$ and $t = t_{\text{end}}$, as defined in Figure 22.13.[12] Equations (22.4) and (22.5) describe the time integrals for ε_1 and γ, which are approximated numerically applying a trapezoidal rule according to Burden and Faires,[13] using a time step Δt, according to the time increments used for numerical particle tracking (NPT).

$$\varepsilon_1 = \int_{t_{\text{start}}}^{t_{\text{end}}} \dot{\varepsilon}_1(t)\,dt \qquad (22.4)$$

$$\gamma = \int_{t_{\text{start}}}^{t_{\text{end}}} \dot{\gamma}(t)\,dt \qquad (22.5)$$

The total particle deformations ε_1 and γ, shown in Figure 22.14, are compared for the three representative particle tracks, positioned at 25%, 50%, and 75% for a shear-thinning fluid ($n = 0.65$), with flow incidences of $\beta = 150°$, at Re = 80.

For optimization of a structuring process, the total energy density rates separately for the elongational and for the shear component are of further interest. The time integral of the elongational dissipation component in primary flow direction e_1 describes the elongational energy density e_E (energy per volume), given in Equation

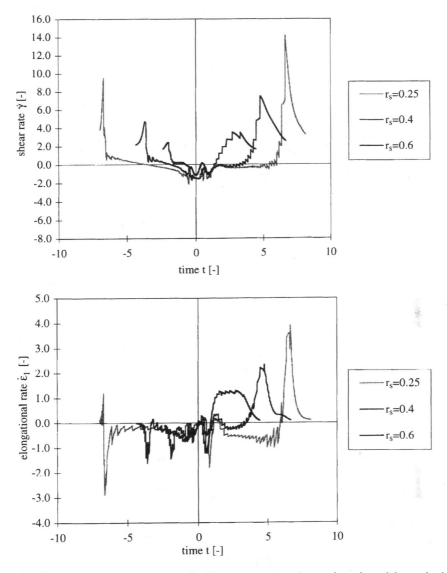

Figure 22.12 Shear (top) and elongational (bottom) rates along selected particle tracks in Narrow Annular Gap Reactor (NAGR) with blade angle $\beta = 30°$; Re = 10 and various scraper blade gap size ratio r_s.

(22.6), whereas the shear dissipation component contributes to the integral of the shear energy density e_S, as given in Equation (22.7). The velocity vector is formulated in the polar coordinate system (r, θ). Applying the rotational coordinate transformation to the velocity gradient, $\nabla \underline{v}$, e_E and e_S are found in the local coordinate system (e_1, e_2).

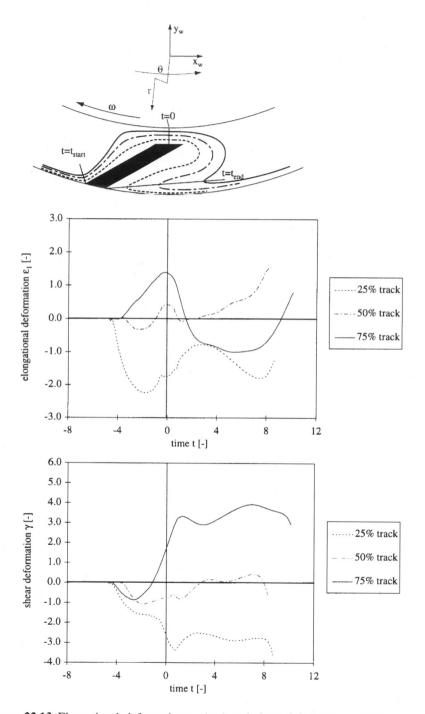

Figure 22.13 Elongational deformation ε_1 (top) and shear deformation γ (bottom) along selected particle tracks from $t = t_{start}$ to $t = t_{end}$. $Re = 80$, $n = 0.65$, $\beta = 150°$.

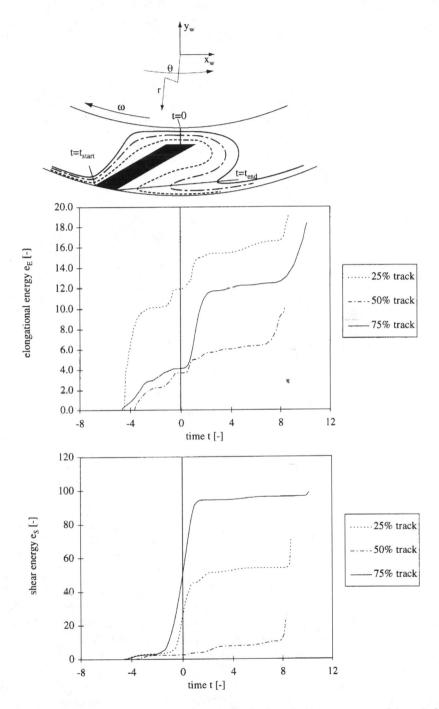

Figure 22.14 Volumetric elongational energy e_E (top) and shear energy e_S (bottom) for selected particle tracks from $t = t_{start}$ to $t = t_{end}$. Re = 80, $n = 0.65$, β = 150°.

$$e_E = \int\limits_{t_{start}}^{t_{end}} 2\eta_s(\dot{\gamma})\left[\left(\frac{\partial v_r}{\partial r}\right)^2 + \left(\frac{1}{r}\frac{\partial v_\theta}{\partial \theta} + \frac{v_r}{r}\right)^2\right]dt \qquad (22.6)$$

$$e_S = \int\limits_{t_{start}}^{t_{end}} \eta_s(\dot{\gamma})\left[\frac{\partial v_\theta}{\partial r} + \frac{1}{r}\frac{\partial v_r}{\partial \theta} - \frac{v_\theta}{r}\right]^2 dt \qquad (22.7)$$

Both Equations (22.5) and (22.6) are formulated in terms of the shear viscosity $\eta_s\dot{\gamma}$ as a function of the shear rate $\dot{\gamma}$, which is transformed onto the local coordinate system (e_1, e_2), where $\dot{\gamma}$ is the magnitude of the rate-of-strain tensor $\underline{\dot{\gamma}}$. For the investigated non-Newtonian model fluid system with a watery CMC solution as the continuous fluid phase, $\eta_s(\dot{\gamma})$ was described by a Carreau–Yasuda (CY) model defined by Equation (22.8).

$$\frac{\eta - \eta_\infty}{\eta_0 - \eta_\infty} = [1 + (\lambda *\dot{\gamma})^a]^{(n-1)/a} \qquad (22.8)$$

Approximating the time integrals between $t = t_{start}$ and $t = t_{end}$ applying a trapezoidal rule, the expressions e_E and e_S can be found numerically. The flow structuring shear and elongation related volumetric energy inputs according to Equations (22.5) and (22.6) are also compared for the three representative particle tracks, positioned at 25%, 50%, and 75% of the scraper blade gap width gap (hgap). Figure 22.14 shows the increase of e_E and e_S as a function of time between $t = t_{start}$ and $t = t_{end}$ along the three representative particle tracks for a shear-thinning fluid ($n = 0.65$) with $\beta = 150°$ at $Re = 80$.[12] All elongational and shear energies commonly develop in a step function manner. This behavior is caused by the pronounced turning sections of the tracks around the scraper blade.

Summarizing the CFD results, it can be stated that the main influence on the deformation energy depicts the flow incidence, which varies the total average elongational energy by about 95% between blade angles of $\beta = 30°$ and $\beta = 110°$. The maximum change for the total average shear energy at $\beta = 90°$ and $\beta = 110°$ is calculated as 33%.

Comparing the impact of the rotor velocity with the influence of the flow incidence, variations of the scraper blade angle cause higher structuring anisotropy with respect to elongational and shear strain than increasing the Reynolds number between 10 and 80. All flow cases show larger total shear energies than total elongational energies. The minimum difference is found for $Re = 10$, $\beta = 30°$, $n = 0.65$ with a factor of 3.5

To express the flow structuring energy contributions of the two distinguished flow types (shear, elongation), Equations (22.9) and (22.10) define the (local) flow type contribution factors χ_E, χ_S with regard to the investigated scraper blades, for elongational flow χ_E and shear flow χ_S, respectively.

$$\chi_E = \frac{e_E}{e_{tot}} \tag{22.9}$$

$$\chi_S = \frac{e_S}{e_{tot}} \tag{22.10}$$

$\chi_E \geq 0.14$, which was the lowest contribution factor for the elongational flow type, showed a qualitative improvement of the dispersing efficiency (i.e., drop size reduction) at constant energy input, compared with a pure shear flow field (concentric cylinder gap). To quantify the improvement of dispersing efficiency as a function of χ_E experimentally in more detail is the subject of our ongoing research work in this field. Within the work reported here the maximum value of $\chi_E = 0.35$ was received for Re = 10, $\beta = 150°$, $r_S = 0.6$; $n = 0.65$.

22.6.2 DEVELOPMENT OF MULTIPHASE FOOD PRODUCTS WITH SHAPED PARTICLES

As demonstrated in model flow experiments described in previous sections, the application of well defined shear, and in particular elongational flow fields, allows the shaping of disperse components like immiscible fluid droplets. If such disperse fluid components reach a certain shape during the flow structuring procedure (i.e., a certain equilibrium shape by balancing flow stresses and Laplace pressure, which is proportional to the interfacial tension), shape fixing was applied. This was done by changing temperature or adding skin building agents.[14,15]

For different two-phase fluid/biopolymer systems [i.e., gelatin (w) /oil; gellan/κ-carraghenan], drop shaping experiments were carried out within pure shear (concentric cylinder gap) and mixed shear/elongation flow fields (FRA, eccentric cylinder gap). Resulting shape fixed particle structures are shown in Figure 22.15. If the shaped particles are separated by centrifugation and then mixed with a fluid at constant volume fraction, a shape-related rheological behavior is detected.

If ellipsoidally or fiber-like shaped particles with a large aspect ratio ($L/D \geq 10$) are mixed into a continuous fluid with a solids fraction of about $\phi V \geq 0.15$, a strong network results for the randomly oriented particles. Rheologically, a yield value is exhibited.

If such systems are sheared and/or elongated, particle orientation in the shear/elongation flow direction is induced. This reduces the viscosity strongly. In the concentrated system, a further particle rotation is hindered by the oriented neighbor particles.

For a model suspension with fiber-like particles, Figure 22.16 demonstrates the impact of particle orientation on the viscosity function investigated by up/down curve shear flow measurements.[15] If the particles are small enough (<1 μm), Brownian motion related diffusion can restructure the fluid system randomly. If this principle is applied to food products, gel-type systems can be created that become liquid if they are sheared between tongue and plate or if they are stirred, and that solidify reversibly by heating.

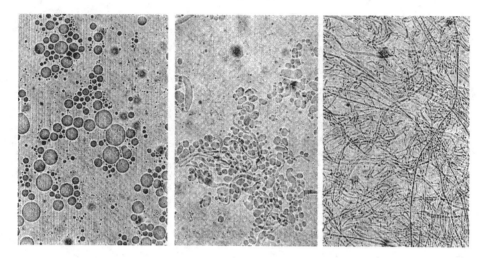

Figure 22.15 Shape-adjusted biopolymer particles (gellan) in another continuous watery biopolymer solution (κ-carraghenan); shear/elongation induced structuring in excentric cylinder gap at increasing shear and elongation rates (from left to right).

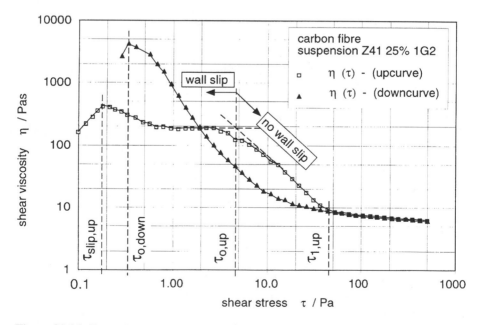

Figure 22.16 Shear viscosity functions (up/down curves) for a concentrated fiber suspension system; influence of fiber orientation.

REFERENCES

1. Windhab, E. 2000. "Fluid Immobilization-A Structure Related Key Mechanism for the Viscous Flow Behaviour of Concentrated Suspension Systems," in *Proc. of 2nd Int. Symp. Of Food Rheology and Structure,* P. Fischer, I. Marti, and E. Windhab, eds. Zürich, pp. 13–24.

2. Grace, H. P. 1982. "Dispersion Phenomena in High Viscosity Immiscible Fluid Systems," *Eng. Found. Res. Conference Mixing,* 3rd, Andover, N.H. Republished 1982 in *Chem. Eng. Commun.* 14, 225–277.

3. Windhab, E. 1996. "The Influence of Structure/Rheology Relationships on the Processing of Food Suspensions," in *Annual Transactions of the Nordic Rheological Society,* vol. 4: 3–15.

4. Wolf, B. 1995. Untersuchungen zum Formverhalten Mikroskopisch Kleiner Fluidtropfen in Stationären und Instationären Scherströmungen, Ph.D. Thesis, ETH Zürich, Nr. 11068.

5. Cox, R.G. 1969. *Journal of Fluid Mechanics*, 37: 601–623.

6. Windhab, E. and B. Wolf. 1992. "Influence of Deformation and Break-Up for Emulsified Droplets on the Rheological Emulsion Properties," in *Theoretical and Applied Rheology,* P. Moldenaers and R. Keunings, eds. Brussels, Belgium, vol. 2, Seite 681–683.

7. Windhab, E. 1988. "A New Method for Describing the Time Dependent Rheological Behaviour of Concentrated Suspension," Suppl. to *Rheologica Acta;* Progress and Trends in Rheology II, Steinkopf Verlag Darmstadt.

8. Kaufmann, S., P. Fischer, and E. Windhab. 2000. "Investigation of Droplet Dispersing Processes in Shear and Elongational Flow," in *Proc. of 2nd Int. Symp. of Food Rheology and Structure,* P. Fischer, I. Marti, and E. Windhab, eds. Zürich, pp. 404–406.

9. Christini, V., J. Blawzdziewicz, and M. Löwenberg. 1998. "Drop Break-Up in Three-Dimensional Viscous Flow," *Physics of Fluids,* 10(8): 1781–1783.

10. Feigl, K. 2000. "Use of Numerical Simulation in Food Processing," in *Proc. of 2nd Int. Symp. of Food Rhreology and Structure*, P. Fischer, I. Marti, and E. Windhab, eds. Zürich, pp. 69–76.

11. Stranzinger, M. 1999. Numerical and Experimental Investigations of Newtonian and Non-Newtonian Flow in Annular Gaps with Scraper Blades, Ph. D. Thesis, ETH Zürich, Nr. 13369.

12. Bieder, A., M. Stranzinger, and E. Windhab. 2000. "Einfluss der Prozessgeometrie auf das Strukturierungsverhalten in Ringspaltreaktoren," Semesterwor ETH Zürich.

13. Burden, R. L. and J. D. Faires. 1993. *Numerical Analysis,* 5th Edition. PWS Publishing Company

14. R. Scirocco, B. Wolf, P. Fischer, and E. Windhab. 2000. Diploma thesis, ETH-Zürich.

15. Wolf, B. 2000. "Biopolymer Suspensions with Spheroidal and Cylindrical Particle Shapes: Generation and Flow Behaviour," in *Proc. of 2nd Int. Symp. of Food Rheology and Structure,* P. Fischer, I. Marti, and E.Windhab, eds. Zürich, pp. 404–406.

16. Eischen, J. C. 1999. Bildanalytische und Rheologische Untersuchungen zum Orientierungs—und Strukturierungsverhalten von Faserförmigen Partikeln in Laminaren Scherströmungen, Ph. D. Thesis, ETH-Zürich, Nr. 13217.

23 Modeling Viscosity of Starch Dispersion and Dough during Heating: Master Curves of Complex Viscosity

J. Tattiyakul
H.-J. Liao
M. A. Rao

CONTENTS

1-56676-963-9/02/$0.00+$1.50
© 2002 by CRC Press LLC

Abstract

Reduced complex viscosity (η_R^*) vs. temperature master curves were derived for complex viscosity η^* data on:

- An 8% tapioca starch dispersion (STD) at heating rates of 1.5 to 6.0°C and frequency (ω) values of 3.14 to 62.83
- An 8% waxy rice STD at a heating rate 2°C min⁻¹ and ω values 1.27 to 125.6 rad s⁻¹
- A 45.5% waxy rice dough at a heating rate 4°C min⁻¹ and ω values 1.27 to 94.2 rad s⁻¹

Because the influence of oscillatory frequency (ω, rad s⁻¹) was scaled by a frequency shift factor; the master curves describe the η^* vs. temperature data during gelatinization of the STDs and the dough.

23.1 NOMENCLATURE

c	starch concentration, w/w
G'	storage modulus, Pa
G''	loss modulus, Pa
G^*	complex modulus
	$G^* = \sqrt{(G')^2 + (G'')^2}$ (Pa)
R	gas constant (8.314 J/mol K); radius of can
STD	starch dispersion
T	temperature of fluid (°C)
β	exponent in the frequency shift factor, $\left(\dfrac{\omega}{\omega r}\right)^{\beta}$
η_a	apparent viscosity (Pa s)
η^*	complex viscosity (Pa s), $\eta^* = (G^*/\omega)$
η_R^*	reduced complex viscosity, $\eta_R^* = \eta^*\left(\dfrac{\omega}{\omega_r}\right)^{\beta}$
ω	dynamic frequency, rad s⁻¹
ω_r	the reference frequency, rad s⁻¹

23.2 INTRODUCTION

When a starch dispersion (STD) is heated in excess water, above a certain temperature, the hydrogen bonding is broken, and the granules swell as they absorb water. Because of the swelling of the granules, the viscosity of the dispersion increases and reaches a maximum value. Continued heating of a starch dispersion, especially an unmodified starch, results in the rupture of the granules and a concomitant decrease in viscosity.[1]

For practical application, one should obtain rheological data while continuously heating a dispersion over the range of temperature of interest at specific heating rates.[2,3] Because complex viscosity (η^*) data can be obtained by means of dynamic rheological tests at low strains with minimal alteration of the STD structure, they provide unique opportunities for studying applicable models during starch gelatinization. It will be recalled that from a dynamic rheological test conducted in the linear viscoelastic range, the storage modulus (G') and the loss modulus (G'') can be obtained as a function of oscillatory frequency (ω). In addition, the complex modulus, G^*, and η^* can be obtained as follows:

$$G^* = \sqrt{(G')^2 + (G'')^2} \tag{23.1}$$

$$\eta^* = G^*/\omega \tag{23.2}$$

Viscosity vs. temperature profiles of dispersions of a specific starch (e.g., corn or bean) are usually similar over narrow ranges of starch concentrations.[4,5] Shift factors have been used successfully in time-temperature superposition of viscoelastic properties (e.g., η^*) obtained using frequency sweeps at several temperatures of many synthetic polymers[6,7] and a few food polymer dispersions.[8] Therefore, superposition of dynamic viscoelastic properties of starch dispersions should be useful in quantitative comparison of the properties of different starches obtained under different experimental conditions.

During gelatinization of starch, η^* data can be obtained as a function of either the heating time, t, or temperature, T, at different heating rates and dynamic frequencies. By using temperature instead of time as the independent variable and a frequency shift factor $(\omega/\omega_r)^\beta$, where ω is the frequency at which the test was performed and ω_r is a reference frequency, Yang and Rao[9] developed reduced η^*, designated as η_R^*, vs. temperature (60–95°C) master curve for an 8% corn STD over the heating rates 1.6–6.0°C min^{-1} and ω (0.63–78.5 rad s^{-1}). Therefore, the effects of different heating and shear rates on data were taken into account quantitatively. Starches from different botanical sources and dispersions with different moisture content exhibit different sol-gel transition temperatures and rheological properties. Applicability of the η_R^* vs. temperature master curve concept described in Reference 9 to η^* data obtained during the heating (gelatinization) of 8% tapioca and 8% waxy rice STDs, and a 45.5% waxy rice dough will be described here. The effect of starch concentration on η^* vs. temperature data was also discussed in an earlier paper.[10]

23.3 MATERIALS AND METHODS

23.3.1 MATERIALS

Tapioca starch (National Starch and Chemical Co.) was used. The moisture content of the starch, based on 11 replicates (5 g sample dried to a constant weight at 105°C), was 11.0%; the amylose content determined using a rapid colorimetric method[11] was 19.3% (dry weight basis). The amylose content of the waxy rice starch (California Natural Products) was found to be less than 1% using the method for rice starch based on the strong iodine affinity of amylose.[12]

Weighed amounts of starch and water used in high-pressure liquid chromatography were mixed manually and allowed to hydrate at room temperature (~23°C) prior to rheological testing; the 8% w/w waxy rice STDs were held six hours, the 8% w/w tapioca STDs for two hours, and the low-moisture 45.5% w/w waxy rice starch dough samples for six hours in sealed plastic bags.

23.3.2 RHEOLOGICAL PROCEDURES

Dynamic rheological data were obtained (temperature sweeps) with a Carri-Med CSL[2] 100 rheometer (TA Instruments) using a 4-cm stainless steel parallel plate system with 500-μm gap as the tapioca STDs were heated from 58 to 88°C, and the waxy rice STDs and dough were heated from 55 to 95°C. A cone-and-plate geometry was not used, because the relatively narrow gap contributed to capillary suction into the STD[9] of the paraffin oil placed at the edge of the geometry to prevent moisture loss. A concentric cylinder geometry was ruled out, because of the potential for settling of raw starch granules in the early stages of heating of the STDs.

To obtain reliable dynamic rheological data on the STDs, the following procedures were carried out:[3,13]

1. A fixed amount of sample (0.75 ml) was carefully loaded on the rheometer plate, avoiding splashing, since excessive amount of sample contributed to measurement errors due to edge effects.
2. Evaporation of water from the starch sample during the tests was minimized by placing paraffin oil on the open edge of the parallel plate geometry just after the storage modulus (G') reached values that were measurable. With this procedure, the paraffin oil did not penetrate into the STD.

In comparison with experiments on the STDs, obtaining dynamic rheological data on waxy rice starch dough was not as difficult with regard to placing a sample on the rheometer plate and minimizing moisture loss during an experiment. The paraffin oil applied to minimize moisture loss was suitable, because the starch did not contain a significant amount of lipids.

In a temperature sweep test, the dynamic frequency (ω), strain, and heating rate are the independent variables. A 3% strain, determined to be in the linear viscoelastic range in isothermal experiments on several gelatinized STDs,[14] was used in the experiments on the STDs, and a 0.5% strain that was also in the linear viscoelastic range was used in the experiments on the dough. Having selected the value of strain, rheological data were obtained either at a fixed ω as the test samples were heated at different heating rates, or at a fixed heating rate as the test samples were subjected to different ω values. All rheological measurements were conducted in triplicate.

23.3.3 RHEOLOGICAL EXPERIMENTS

The heating rates and dynamic frequencies employed in obtaining data on η^* of 8% tapioca STDs, 8% waxy rice STDs, and a 45.5% waxy rice dough are summarized in Table 23.1.

TABLE 23.1
Temperature Sweep Experiments

Experiment	Controlled variables	Changed variables
Tapioca 8% STDs		
Effect of heating rates	$\gamma = 3\%$, $\omega = 6.28$ rad s^{-1}	HR: 1.5 to 6.0 °C min^{-1}
Effect of frequency	$\gamma = 3\%$, HR = 3.5°C min^{-1}	$\omega = 3.14$, to 62.83 rad s^{-1}
Waxy Rice 8% STDs		
Effect of heating rates	$\gamma = 3\%$, $\omega = 1.26$ rad s^{-1}	HR: 1 to 4 °C min^{-1}
Effect of frequency	$\gamma = 3\%$, HR = 2°C min^{-1}	$\omega = 1.27$ to 125.6 rad s^{-1}
Waxy Rice 45.5% Dough		
Effect of heating rates	$\gamma = 0.5\%$, $\omega = 1.26$ rad s^{-1}	HR: 1 to 4 °C min^{-1}
Effect of frequency	$\gamma = 0.5\%$, HR = 2°C min^{-1}	$\omega = 1.27$ to 125.6 rad s^{-1}

γ = strain, ω = dynamic frequency, and HR = heating rate.

23.4 RESULTS AND DISCUSSION

Results on the 8% tapioca STDs will be discussed in detail and, because similar trends were observed with the waxy rice STD and dough, they will be discussed in less detail.

23.4.1 DYNAMIC RHEOLOGICAL PROPERTIES OF TAPIOCA STARCH DISPERSION

The influence of heating rate (1.5 to 6.0°C min^{-1} on η^* data of 8% tapioca STDs at 6.28 rad s^{-1} and 3% strain) is shown as a function of heating time in Figure 23.1 and of temperature in Figure 23.2. While, in Figure 23.1, the η^* data obtained at different heating rates followed separate curves, in Figure 23.2, most of the data collapsed to a single curve, reflecting the important role of temperature in describing the data. In 8% corn STDs[9] heated from 60 to 92°C at 1.6 to 6.0°C min^{-1}, 1.26 rad s^{-1} and 1% strain, the higher heating rates resulted in slightly lower peak values of η^*, probably due to limited diffusion of water into the starch granule at the higher heating rates. For the 8% tapioca STD, it seems that water diffusion into the starch granules and viscosity development occurred rapidly so that high heating rates did not significantly affect the peak values of η^*.

The η^* vs. temperature profile of 8% tapioca STD (Figure 23.2) had maximum complex viscosity values in the range 67 to 74°C that represented the peak of starch gelatinization and the strongest starch gel network. After peak viscosity was reached, further increase in temperature brought about a decrease in complex viscosity that

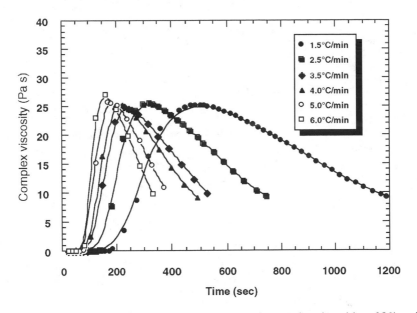

Figure 23.1 The influence of different heating rates on the complex viscosities of 8% tapioca starch dispersions at 6.28 rad s^{-1} and 3% strain as a function of heating time is seen as separate curves.

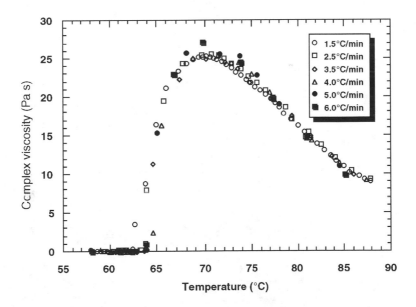

Figure 23.2 Complex viscosities of 8% tapioca starch dispersions at 6.28 rad s^{-1} and 3% strain as a function of heating temperature follow a single curve.

may be attributed to the disruption of starch granule structure at high temperatures. Similar viscosity vs. temperature behavior was observed with dispersions of other starches, albeit at a specific dynamic frequency.[9,15–17]

Effect of ω on η^* vs. Temperature Profile

The η^* vs. temperature profiles of 8% tapioca STDs obtained at values of ω between 3.14 rad s^{-1} to 62.83 rad s^{-1}, 3% strain and a heating rate 3.5°C min^{-1} are shown in Figure 23.3. As expected, an increase in ω resulted in a decrease in values of η^*. Because η^* vs. temperature curves had similar profiles, it was possible to reduce them to a single curve in terms of the reduced complex viscosity η_R^*, defined as:

$$\eta_R^* = \eta^* \left(\frac{\omega}{\omega_r} \right)^\beta \qquad (23.3)$$

where $(\omega/\omega_r)^\beta$ is the frequency shift or scaling factor, ω is the frequency at which the test was performed, and ω_r is the reference frequency.[9] In this study, a frequency of 1 Hz (6.28 rad s^{-1}) was chosen as the ω_r. The exponent of the shift factor β was determined by graphical shift of the experimental data obtained at different frequencies. When η_R^* was used, values of complex viscosities determined at different dynamic frequencies collapsed to a single master curve (Figure 23.4). Values of exponent β ranged from 1.0 for data at the lower frequencies (3.14 and 6.28 rad s^{-1}) to 0.70–0.77 for data at the higher frequencies (12.57 to 62.83 rad s^{-1}).

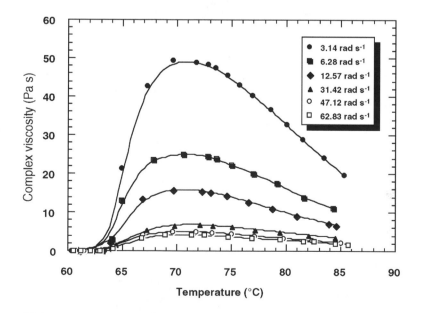

Figure 23.3 Complex viscosities vs. temperature profiles of 8% tapioca starch dispersion at different dynamic frequencies, at 3.5°C min^{-1} and 3% strain.

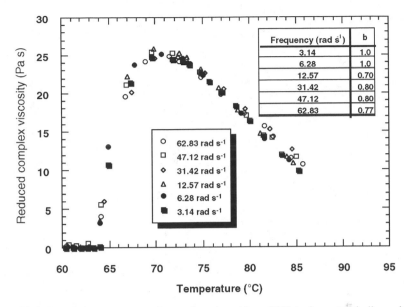

The inset table in the figure reads:

Frequency (rad s⁻¹)	b
3.14	1.0
6.28	1.0
12.57	0.70
31.42	0.80
47.12	0.80
62.83	0.77

Legend:
○ 62.83 rad s⁻¹
□ 47.12 rad s⁻¹
◇ 31.42 rad s⁻¹
△ 12.57 rad s⁻¹
● 6.28 rad s⁻¹
■ 3.14 rad s⁻¹

Figure 23.4 Sample master curve of complex viscosities of 8% tapioca starch dispersions at different heating rates and dynamic frequencies. Values of the frequency shift factor β are shown in Table 23.1.

23.4.2 DYNAMIC RHEOLOGICAL PROPERTIES OF 8% WAXY RICE STARCH DISPERSION

In a manner similar to the tapioca STD, a η_R^* vs. temperature master curve was derived (Figure 23.5) for η* data of 8% waxy rice starch dispersion obtained at 2°C min⁻¹ and ten different dynamic frequencies. The scaling factor β was determined to be 0.93 for data at the lower frequencies: 1.27 to 47.1 rad s⁻¹, and 0.84 for data at the higher frequencies: 62.8 to 125.6 rad s⁻¹.

23.4.3 DYNAMIC RHEOLOGICAL PROPERTIES OF 45.5% WAXY RICE STARCH DOUGH

For the 45.5% waxy rice dough, the effect of temperature on η* and the effect of ω on η* vs. temperature profiles were examined as described for the 8% tapioca and waxy rice STDs. The η_R^* vs. temperature master curve (Figure 23.6) of the 45.5% waxy rice dough was also derived as described for the STDs. The exponent β = 0.93 was suitable for data at all the dynamic frequencies employed.

23.5 CONCLUSIONS

The superposition of the η_R^* vs. temperature data suggests that η* of tapioca and waxy rice STDs and 45.5% waxy rice dough at various dynamic frequencies had

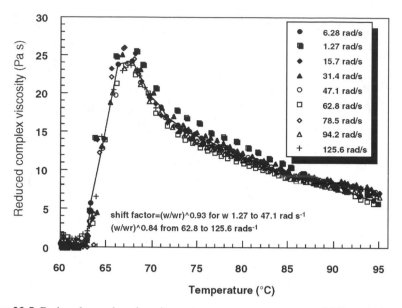

Figure 23.5 Reduced complex viscosity vs. temperature master curve of 8% waxy rice starch dispersion; data obtained at 2°C min^{-1} and several dynamic frequencies.

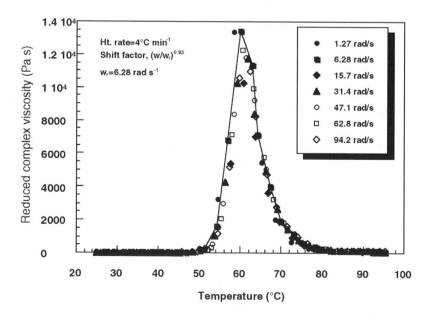

Figure 23.6 Reduced complex viscosity vs. temperature master curve of 45.5% waxy rice dough; data obtained at several dynamic frequencies and at 4°C min^{-1}.

the same temperature dependence. Time-temperature superposition was shown to be applicable to melt of polyisobutylene (PIB), a linear amorphous polymer, and to other synthetic polymers,[6,7] as well as to locust bean gum dispersions.[8] The term *thermorheologically simple* was used to describe such behavior of PIB[7] where molecular mechanisms within related groups of polymers tend to have the same temperature dependence. Furthermore, as Plazek[7] suggested for synthetic polymers, when other food polymers exhibit different behavior in the reduction of viscosity vs. temperature data, it may be possible to identify different kinds of η^* vs. temperature mechanisms.

The complex viscosity vs. temperature master curves of a 3.5% corn starch and a 4% waxy rice starch dispersion were used in studies on simulation of heat transfer to the dispersions in a stationary can and flowing in a tube by Yang[3] and Liao,[13] respectively. Therefore, the results of studies on superposition of rheological data of starch dispersions are also useful in studies on thermal processing. Lastly, the superposition technique is also applicable to apparent viscosity vs. temperature data on starch dispersions obtained over a wide range of shear rates.[13,18]

23.6 ACKNOWLEDGEMENTS

The authors thank the USDA for NRI Grant #97-35503-4493; the Royal Thai Government for a scholarship to JT; and the National Starch and Chemical Co., Bridgewater, NJ, and the California Natural Products, Lathrop, CA, for donation of starch samples.

REFERENCES

1. Okechukwu, P. E. and M. A. Rao. 1995. "Influence of Granule Size on Viscosity of Corn Starch Suspension," *J. Texture Studies*, 26: 501–516.
2. Dolan, K. D., J. F. Steffe, and R. G. Morgan. 1989. "Back Extrusion and Simulation of Viscosity Development During Starch Gelatinization," *J. Food Process Eng.*, 11: 79–101.
3. Yang, W. H. 1997. *Rheological Behavior and Heat Transfer to a Canned Starch Dispersion: Computer Simulation and Experiment*, Ph.D. Thesis, Cornell University, Ithaca, NY.
4. Launay, B., J. L. Doublier, and G. Cuvelier. 1986. "Flow Properties of Aqueous Solutions and Dispersions of Polysaccharides," in *Functional Properties of Food Macromolecules*, J. R. Mitchell and D. A. Ledward, eds. London: Elsevier Applied Sci. Publishers, pp. 1–78.
5. Champenois, Y. F., M. A. Rao, and L. P. Walker. 1998. "Influence of Gluten on the Viscoelastic Properties of Starch Pastes and Gels," *J. Sci. Food Agric.*, 78: 119–126.
6. Ferry, J. D. 1980. *Viscoelastic Properties of Polymers*, 3d. ed., New York: John Wiley & Sons.
7. Plazek, D. J. 1996. "1995. Bingham Medal Address: Oh, Thermorheological Simplicity, Wherefore Art Thou?" *J. Rheol.*, 40: 987–1014.

8. Lopes Da Silva, J. A., M. P. Gonçalves, and M. A. Rao. 1994. "Influence of Temperature on Dynamic and Steady Shear Rheology of Pectin Dispersions," *Carbohydrate Polymers*, 23: 77–87.

9. Yang, W. H. and M. A. Rao. 1998. "Complex Viscosity-Temperature Master Curve of Cornstarch Dispersion During Gelatinization," *J. Food Process Eng.*, 21: 191–207.

10. Liao, H-J., J. Tattiyakul, and M. A. Rao. 1999. "Superposition of Complex Viscosity Curves During Gelatinization of Starch Dispersion and Dough," *J. Food Process Eng.*, 22: 215–234.

11. Williams, P. C., F. D. Kuzina, and I. Hlynka. 1970. "A Rapid Colorimetric Procedure for Estimating the Amylose Content of Starches and Flours," *Cereal Chem.*, 47: 411–420.

12. Juliano, B. O., C. M. Perez, A. B. Blakeney, T. Castillo, N. Kongseree, B. Laignelet, E. T. Lapis, V. V. S. Murty, C. M. Paule, and B. D. Webb. 1981. "International Cooperative Testing of the Amylose Content of Milled Rice," *Starch/Stärke*, 33: 157–162.

13. Liao, H-J. 1998. *Simulation of Continuous Sterilization of Fluid Food Products: the Role of Thermorheological Behavior of Starch Dispersion and Process Optimization*, Ph. D. Thesis, Cornell University, Ithaca, NY.

14. Tattiyakul, J. 1997. *Studies on Granule Growth Kinetics and Characteristics of Tapioca Starch Dispersion During Gelatinization Using Particle Size Analysis and Rheological Methods*, M. S. Thesis, Cornell University, Ithaca, NY.

15. Eliasson, A. C. 1986. "Viscoelastic Behaviour During the Gelatinization of Starch. I: Comparison of Wheat, Maize, Potato and Waxy Barley Starches," *J. Texture Studies*, 17: 253–265.

16. Doublier, J. L. 1987. "A Rheological Comparison of Wheat, Maize, Faba Bean and Smooth Pea Starches," *J. Cereal Sci.*, 5: 247–262.

17. Kokini, J. L., L-S. Lai, and L. L. Chedid. 1992. "Effect of Starch Structure on Starch Rheological Properties," *Food Technol.*, 46 (6): 124–139.

18. Tattiyakul, J. 2000. *Thermorheology and Heat Transfer to a Canned Starch Dispersion Under Agitation: Numerical Simulation and Experiment*, Ph.D. Thesis, Cornell University, Ithaca, NY.

24 The Role of Rheology in Extrusion

O. H. Campanella
P. X. Li
K. A. Ross
M. R. Okos

CONTENTS

Abstract

The study of the expansion phenomenon of extrudates relies on knowledge of melt rheological properties. Viscosities of highly viscous materials have commonly been performed using a piston-driven capillary rheometer. However, when a food material is turned into a melt under conditions of high temperature, pressure, and shear, complex physicochemical changes take place. These changes are highly sensitive to the processing history of the melt and therefore influence the measurement of the rheological properties. Thus, it is important to obtain rheological properties of melts that have experienced the complete history in the extruder. Rheological measurements indicative of the complete processing history of food melts cannot be directly measured on standard rheometers. In this chapter, the use of on-line viscometry for extrusion is reviewed, and a new modified on-line slit viscometer is presented. Unlike many other viscometers used on-line, the measurements obtained with the new on-line viscometer do not interfere with the operation of the extruder. The effects of moisture content, temperature, and composition of ingredients on the viscosity measured with the modified on-line viscometer are discussed.

Melt viscosity is an important parameter in determining the expansion of extrudates. Models that predict the expansion of extrudates are reviewed. The role of the melt glass transition temperature during expansion, notably shrinkage, is also discussed. A new model to predict expansion of extrudates, taking into account the role of glass transition and the diffusion of moisture through the melt during expansion, is discussed.

24.1 INTRODUCTION

Food extrusion is a reactive process in which many biochemical reactions such as starch gelatinization, protein denaturation, and starch dextrinization, to name a few, take place. To study the kinetics of these reactions and how they are affected by the extrusion processing variables is a formidable and complex, if not impossible, task. It has been recognized, however, that during extrusion, and due to these reactions, drastic changes in the properties (e.g., rheological, chemical) of the material occurs. Thus, when properly determined, these changes can be used to monitor the extrusion process.

The quality of expanded extruded products is commonly determined by properties such as bulk density, cell structure, expansion ratio, mechanical strength, and sensory characteristics such as crunchiness and crispness. Good expansion of the extrudate melt is one of the main characteristics required from the extrusion process, as the quality of expanded extruded products is closely related to and controlled by this process.

The production of extrusion-processed polymeric foams is dictated primarily by the rheology of the mixture of the polymeric matrix along with a blowing/expansion agent.[1] In food extrusion, the relationship between the melt rheology and the product properties plays a fundamental role in the control of the expansion process.[2,3] The food industry, owing the idea to polymer and plastic processors, is now starting to consider the possibility of using on-line measurements of the extrudate melt properties.

Figure 24.1 schematically shows a list of variables that can affect the extrusion process and illustrates the interrelationship between these variables and their possible effects on the extrusion process. Some of these variables, such as the raw material characteristics and the extruder operational conditions, can be regarded as *input variables*. The interactions among these input variables result in operational conditions that yield another group of variables such as specific mechanical energy (SME), residence time distribution (RTD), temperature, pressure, and viscosity of the melt. These can be regarded as *process variables*. These process variables, in turn, influence the product quality as defined by properties such as expansion, bulk density,

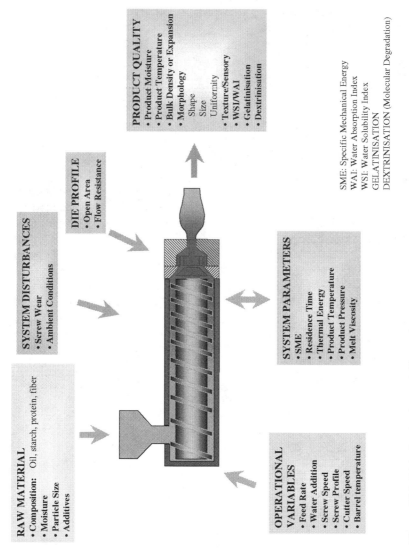

Figure 24.1 List of extrusion variables and parameters.

texture, and product morphology. Variables used to define the extruded product quality can be regarded as *output variables*. It is also important to note that extrusion can be affected by external disturbances such as ambient conditions and equipment wear, notably the screws.

The many variables involved in extrusion and the complexity of the interactions occurring during the process make it very difficult to establish relationships that could help us to understand the process and the effect of variables on product quality. Because of this difficulty, a simplified approach for research in food extrusion has been adopted in the past. In many studies, the effect of only one variable at a time has been investigated. This approach generally diverts the study into different directions, focusing on only a part of the whole extrusion process. If the whole process were to be investigated, the approach would involve such a large number of experiments as to make the study impractical. Although the approach has its disadvantages, it has been widely used and accepted in extrusion research.

By defining key variables and establishing the basis of the interactions between these and product quality, progress toward modeling the extruder performance, as well as the development of relationship between input and output variables, can be made. As discussed above, one crucial measurement defining product characteristics is the viscosity of the melt within the extruder barrel. If the melt viscosity could be properly measured, it would help not only in the understanding of the extrusion process but also aid in the design of an extrusion control system.

24.2   THE ROLE OF RHEOLOGY IN EXTRUSION

A stable sample is an important prerequisite for meaningful rheological measurements. This requirement excludes materials with rapidly changing structures.[4] This concept is important in extrusion, wherein processes with different time scales can occur. For instance, the transformation of raw materials into a melt due to thermal and mechanical input occur in a time that could be considered of the same order of magnitude as the extruder residence time (less than one minute). Conversely, melt expansion is a process occurring in very short times (milliseconds), wherein a homogeneous viscoelastic melt is converted into a semisolid foam due to rapid changes of temperature and moisture contents. Thus, in the extrusion process, the times of two characteristics could be considered for meaningful rheological measurements. Furthermore, the extrudate melt is a viscoelastic material, and the variable time has a large influence on the rheological measurements and transformations associated with the extrusion process. This is further illustrated by defining a dimensionless number known as the Deborah number.[5]

$$N_{De} = \frac{t_R}{t_P} \tag{24.1}$$

where t_R is a material or relaxation time that characterizes the rheological behavior of the melt, whereas t_P is a characteristic time associated with the process. By using

Equation (24.1), it can be readily noted that Deborah numbers for the raw material-melt conversion process ($t_p > t_R$) are small, whereas those for the expansion process ($t_R > t_P$) are large.

Values of the Deborah number associated with processes in the areas of food and polymers are discussed in the literature.[6,7] In general, when the Deborah number increases, the material's solid-like (elastic) behavior becomes more important. Thus, it is expected that, if present, elastic behavior would be important during the expansion process, whereas the viscous or liquid-like behavior of the melt would govern the transformation of raw materials into a melt inside the extruder.

To describe the role of rheology in the extrusion process, the following will be discussed:

1. On-line measurements to determine melt viscosity during food extrusion; advantages and disadvantages of the adopted methods
2. Melt rheology and how it is influenced by the extrusion process
3. The effect of rheological properties on product expansion and the importance of melt rheology

24.2.1 On-Line Measurements

To develop relationships between fundamental rheological properties of the extrudate melt, final product characteristics, and the extruder operation, it is necessary to develop suitable tools for on-line measuring of melt viscosity with minimal interference from and to the extrusion process. The determination of the melt properties and how they are affected by operational and system variables has two main advantages. First, the knowledge of these properties is an excellent tool to fully monitor the extrusion process. Second, on-line determination of these properties can provide a means for controlling the process.

During extrusion, the raw material is subjected to a combination of shear, temperature, and pressure conditions that are very difficult to achieve in any commercial viscometer. Furthermore, the rheology of extrudates changes rapidly. Thus, it is very difficult to collect material so that its rheology can be determined off-line. On-line measurements are therefore needed to characterize the rheology of extrudates.

On-line rheological determination of melt extrudates has received considerable attention in the food processing area. Extrudate melts are viscoelastic materials and exhibit shear-dependent rheological properties. Thus, it is essential to study the rheology of the melt at different shear rates. Two major on-line viscometers have been used in food extrusion: capillary and slit die viscometers. Both viscometers are based on the same measuring principle; that is, the determination of the melt pressure profile (used for calculation of shear stresses) at different volumetric flow rates (used for calculation of shear rates). To minimize end effects, many of these viscometers are commonly designed with a length much longer than that of standard extruder dies. This design, when used as an on-line attachment, unfortunately affects the extruder pressure, product residence time, and product shear and thermal history.

Unlike plastic materials, foods are very sensitive to shear and thermal history, and different shear and thermal treatments result in products with different levels of molecular modification and, therefore, different rheological properties. This indicates that the use of long-tube on-line capillary or slit viscometers may not be appropriate to determine the melt rheology of food products. This problem has recently been recognized, and, in the last few years, new designs or measurement techniques have appeared in the literature, taking into account the inherent error associated with on-line measurements. Conventional and some new designs and procedures have been described elsewhere.[8]

A simplified approach to avoid the errors associated with on-line measurement has been adopted.[3] By using very short capillary tubes and two pressure transducers, it is possible to determine melt rheological properties with minimal interference to the extrusion process, but with appreciable end effects. Although the proposed method does not provide fundamental rheological properties, it is useful in determining relationships between rheological properties (semi-empirical) and the expansion properties of the final product.

Capillary Viscometer

A capillary rheometer is a device that can create a well defined and simple shear flow. The apparent viscosity of the fluid can be calculated by using the following equation:

$$\mu_{app} = \frac{\tau_w}{\dot{\gamma}_w} \tag{24.2}$$

where τ_w and $\dot{\gamma}_w$ are the shear stress and the shear rate at the capillary wall, respectively. They are calculated as

$$\tau_w = \frac{R}{2}\frac{\partial P}{\partial x} \cong \frac{R}{2}\frac{\Delta P}{L} \tag{24.3}$$

$$\dot{\gamma}_w = \left(\frac{3n+1}{4n}\right)\frac{4Q}{\pi R^3} \tag{24.4}$$

where P = pressure
Q = volumetric flow rate
R = capillary radius
x = direction parallel to the capillary axis
L = capillary length and n the index flow of the melt

It is important to note that, for the calculation of shear stress using the approximation given on the right side of Equation (24.3), a linear pressure profile must exist along the capillary tube. It has been observed that a linear pressure profile can be achieved a few diameters away from the entrance and exit regions. Furthermore, the shear rate is not uniform along the tube radius and therefore is calculated at the capillary wall. The calculation of the shear rate at the wall, as given by Equation (24.4), involves the knowledge of melt rheology (flow index n); thus, an iterative algorithm is needed. Details of the algorithm for calculation of the shear rate are given in the literature.[6,9] The requirement for meaningful measurements of linear pressure profiles makes the design of on-line capillary viscometers long and impractical. Long capillaries at the extruder exit can create excessive pressures affecting the extruder operation, residence time, and measurements. Entrance effects can be corrected,[10] but it is necessary to use at least two capillaries with different L/D diameters. Changing the capillary for entrance effects corrections is time consuming, thereby precluding the use of on-line capillary viscometers in commercial operations.

Slit Viscometer

The working principle of the slit viscometer is similar to that of the capillary viscometer, using a thin rectangular channel rather than a capillary to produce the viscometric flow. One advantage in using slit viscometry is related to the measurement of the pressure. Pressure transducers can be flush mounted on the flat sides of the slit, reducing errors occurring in capillary viscometers due to tube curvature.

Shear stress τ_w and shear rate $\dot{\gamma}_w$ at the slit wall can be calculated as

$$\tau_w = -\frac{H}{2}\left(\frac{\partial P}{\partial x}\right) \cong \frac{H}{2}\frac{\Delta P}{L} \tag{24.5}$$

$$\dot{\gamma}_w = \frac{(2n+1)}{3n}\frac{6Q}{WH^2} \tag{24.6}$$

where W is the slit width and H the slit height.

Similar to capillary viscometry, the calculation of the shear stress at the wall of a slit viscometer requires a linear pressure profile and calculation of the shear rate, an iterative algorithm.[6,9] In general, for melts exhibiting shear thinning behavior the flow index n can be calculated by the slope of a log $(\Delta P/L)$ versus log $(6Q/WH^2)$ plot.

Once the shear rate and the shear stress are determined, the apparent viscosity can be determined using Equation (24.2).

Melt Rheology

A power-law rheological model has been used to describe the behavior of the melt. The model, although empirical, fits the data reasonably well and predicts the fluid flow adequately.[6] The model can be expressed as

$$\tau_w = K\dot{\gamma}_w^n \tag{24.7}$$

The apparent viscosity μ_{app} defined by Equation (24.2) can now be expressed as

$$\mu_{app} = K\dot{\gamma}_w^{n-1} \qquad\qquad (24.8)$$

where K and n are named consistency and flow indexes, respectively. Values of K are related to the fluid viscosity. The value of n defines three types of fluids. Liquids with a value of $n = 1$ are known as Newtonian fluids. Their viscosity is independent of the shear rate and is only a function of temperature. Fluids with $n < 1$ are known as pseudoplastics, or shear thinning fluids. Their viscosity is a function of shear rate and decreases when shear rate increases. Materials with $n > 1$ are known as dilatant or shear thickening fluids. Their viscosity increases when shear rate increases.

On-Line Measurements: Interference with the Extrusion Process

To measure the viscosity of the extrudate melt on-line, the viscometer is mounted as a die block at the extruder exit. However, due to the shear thinning behavior of the melt, its rheological characterization requires that measurements of viscosity be performed at different shear rates. In classical on-line viscometry, different shear rates are achieved by varying the throughput of the extruder,[11-14] which, depending on the type of extruder used, can vary in two different manners. For single-screw extruders, changes in throughput can be achieved by varying the screw speed, whereas, for twin-screw extruders, changing the extruder feed rate modifies the throughput. It is important to note that, in both cases, varying screw speed and feed rate alters the conditions under which the extruder is operating, and the extruded material is subjected to different thermomechanical treatment of each operating condition fixed to reach the desired shear rate. Thus, the flow curves (τ vs. $\dot{\gamma}$ curves) obtained using the classical on-line viscometer are affected by uncontrolled variations in the operating conditions of the extruder and may not represent the flow behavior of a single product. Therefore, the resulting flow curves obtained by using the classical approach should be treated with caution. A typical flow curve for cornmeal using the classical on-line viscometer and a twin screw extruder is illustrated in Figure 24.2. In this experiment, shear rates were changed by varying the extruder feed rate. The experimental data points depicted in the figure clearly show that the different processing conditions used to achieve the selected shear rates result in material with different thermomechanical treatment. It appears that increases in shear rates produced by increases in feed rate yielded materials with lower viscosity due to the larger molecular degradation produced at the higher feed rate conditions. Similar results have been obtained by other researchers[11] who compared on-line rheological data obtained with linear polyethylene, corn grits, and potato starch. Rheological measurements on polyethylene were not affected by the processing history, while data for corn grits and potato starch showed a strong dependence on the thermomechanical history.

The effect of thermomechanical history during on-line measurements is better illustrated by the negative flow indexes n as reported in the literature.[15]

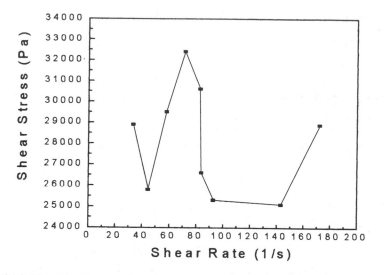

Figure 24.2 Rheological properties of cornmeal obtained with a classical on-line slit viscometer attached to a twin-screw extruder.

Experiments using both polymeric and food materials clearly show that extreme care must be taken when testing food materials due to their sensitivity to thermomechanical treatment. The potential problem of using a classical on-line viscometer with food materials is also illustrated in Figure 24.2. However, it is has been claimed[16] that, by keeping the degree of fill (defined as the ratio between extruder feed rate and screw speed) constant, the shear rate of the on-line slit viscometer can be varied without strongly affecting the material thermomechanical treatment.

Improved On-Line Slit Viscometer Design

With clear evidence of the interaction between the extrusion process and on-line rheological measurements, researchers have begun to design new on-line viscometers wherein these interactions are minimized. One of the first designs included a side stream valve to vary the flow rate in the viscometer while keeping the same thermomechanical treatment.[17,18] With the new designs, interactions between measurement and the extrusion operation were largely minimized. This is clearly evidenced by more meaningful flow curves obtained from the experiments (flow indexes n larger than 0.3). Other designs followed,[19] in most cases aimed at minimizing interactions between the measurements and the extrusion operation.

Using an on-line computerized data acquisition and control system and a modified slit die viscometer, flow curves of extrudate melts were obtained with minimum changes in the thermomechanical treatment of the sample.[20] A schematic of the modified on-line slit viscometer is illustrated in Figure 24.3.

The viscometer had an adjustable slit height, and pressure and temperature were measured in five locations along the slit. Heating elements controlled by a computer were employed to maintain a selected temperature in the viscometer. The flow rate,

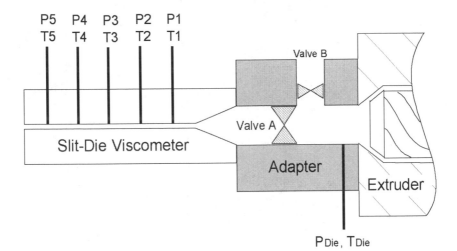

Figure 24.3 Schematic diagram of the modified slit viscometer.

and thus the shear rate, was varied by carefully adjusting the openings of valves A and B. The condition of a constant thermomechanical treatment during the measurements was monitored by the die pressure and specific mechanical energy (SME), measured and/or calculated in real time by computerized data acquisition and control system. Conditions of constant pressure and SME were used to ensure that the extruder operating conditions remained unchanged.

24.2.2   MELT RHEOLOGY AND HOW IT IS INFLUENCED BY THE EXTRUSION PROCESS

Moisture Content

The effect of moisture content on the rheological properties of extruded grits was studied with the modified slit viscometer at 120°C. Typical results for a moisture content of 31.7% are illustrated in Figure 24.4.

As indicated in the figure, pressure decreased linearly with the distance measured from the exit of the viscometer. This indicates that a fully developed flow was established in the slit channel and that the necessary conditions for the calculations of shear rate and shear stress given by Equations (24.5) and (24.6) were met.

Plots of apparent viscosity as a function of shear rate at different moisture contents are shown in Figure 24.5. The figure illustrates that the behavior of the melts at all moisture contents were non-Newtonian and that the viscosity decreased with the shear rate. It is worth noting that the slope of the plots for each moisture content are similar, indicating that the power law index n did not change with moisture content. Values of the consistency index K, however, were highly dependent on moisture content.

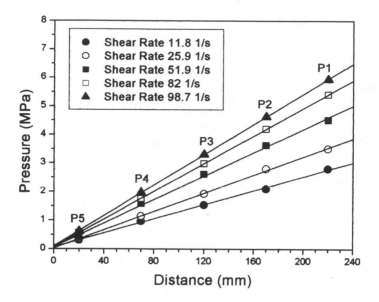

Figure 24.4 Pressure profile along the on-line viscometer for various shear rates at 31.7% moisture content and 120°C.

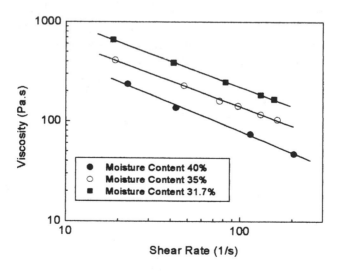

Figure 24.5 Melt viscosities at different shear rates and moisture contents.

Measured rheological values and parameters measured during the extrusion process are given in Table 24.1. The table shows that the lower values of viscosity obtained at high moisture contents provided lower resistance for the extruder rotation and flow through the extruder die, resulting in lower values of torque, SME, and die pressure. It is also worth noting that at each moisture content values

TABLE 24.1
Effect of Moisture Content on Extruder Operating Conditions

Moisture content (%)	Test #	Apparent shear rate (1/s)	SME (Wh/kg)	Torque (Nm)	P_{thrust} (bar)	P_{die} (Psi)
	1	11.8	50.0	18.0	72.7	1217.7
	2	25.9	51.3	18.3	73.7	1248.9
31.7	3	51.9	50.3	18.1	73.0	1222.0
	4	82.0	52.2	18.5	74.8	1261.2
	5	98.7	50.8	18.2	72.7	1228.9
	1	12.2	29.9	13.8	49.2	843.5
	2	30.3	29.9	13.8	48.9	832.3
35.0	3	48.9	28.6	13.5	47.3	810.1
	4	61.7	29.3	13.6	47.7	810.2
	5	83.4	29.6	13.7	47.6	810.2
	6	103.2	30.7	14.0	48.5	830.7
	1	12.7	13.0	10.3	27.0	481.8
40.0	2	23.8	11.9	10.0	26.0	463.5
	3	64.1	12.7	10.2	26.4	471.1
	4	117.6	12.0	10.0	26.4	468.5

of die pressure and SME were approximately constant regardless of whether the shear rate was modified.

The approximate constant values of the extruder operating conditions at each moisture content reported in Table 24.1 indicate that the properties of the melt were not modified when the shear rate was varied. Thus, for each moisture content, it can be assumed that the raw material was subjected to the same thermomechanical treatment and that the measurements did not interfere with the extruder operation.

The process variables—die pressure, SME, and torque—decreased with moisture content as well as the measured apparent viscosity of the melt, indicating a strong correlation between the rheology of the melt and the extrusion process variables.

Barrel Temperature

Extrusion temperature has an important influence on the extrusion process, and it is expected that changes in extrusion temperatures will result in changes in the melt rheology. The transformation of raw materials, notably starch, during extrusion is very different from what occurs in a low or shear free environment. During extrusion, shear force physically tears starch granules apart, contributing to water penetration, starch gelatinization, and fragmentation. Shear forces have a large influence on the gelatinization process during extrusion. It would be almost impossible for starch to undergo complete gelatinization under the conditions of short residence times, low heat transfer, and low moisture contents existing in the extruder without these forces.

Apparent viscosities of melts produced at different extrusion temperatures are given in Figure 24.6. The shear thinning behavior of the melts is observed at all temperatures. In general, viscosities increase with temperature until a temperature of 140°C and after decrease. Fitting the rheological data to the power model yielded values of consistency index K that increase with temperature up to a temperature of approximately 140°C. This is clearly illustrated in Figure 24.6, wherein maximum values of apparent viscosity for different shear rates are observed for temperatures near 140°C. Values of the flow index n did not change significantly with barrel temperature, ranging between 0.40 and 0.50. However, a trend was observed in which n increased with increases in barrel temperatures. Similar results have been obtained by other researchers.[14] It is thought that the rheological properties of the melt are closely related to its molecular structure (i.e., average molecular weight), and thus changes in the shear thinning behavior could be attributed to a decrease in the size of starch molecules due to degradation at high temperatures. To study the effect of temperature on the modification of the raw material, the degree of gelatinization of the extruded samples were measured; the results are illustrated in Figure 24.7. This figure shows that the degree of gelatinization of the starch increases linearly with temperature up to approximately 120°C. After 120°C, the rate of increase in degree of gelatinization is reduced, and it approximates to almost 100% of temperatures near 140°C.

These results seem to follow the same trend as apparent viscosity (Figure 24.6); thus, the initial increases in viscosity could be attributed to starch gelatinization, and the decrease for temperatures higher than 140°C, to starch fragmentation created by the combination of high shear and high temperatures.

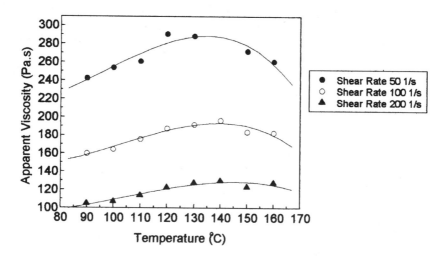

Figure 24.6 Effect of barrel temperature on apparent viscosity.

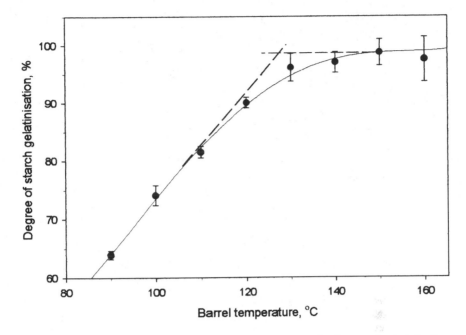

Figure 24.7 The effect of the extruder barrel temperature on the degree of gelatinization.

Starch Composition

Manufacturers of extruded snack foods carefully specify the properties of their raw materials. As indicated in Figure 24.1, the list of raw material properties that may affect the extrusion process is extensive, and their interactions could be complex. Thus, the effect of raw material properties on the extrusion process and the rheology of the melt need to be known. To test the suitability of the newly designed slit die viscometer, and given the importance of the amylose/amylopectin ratio in the processing of snacks, an experiment was designed to determine the rheology of melts produced with starch containing different contents of amylose and amylopectin. Results are illustrated in Figure 24.8. Starches utilized were a commercial high-amylose starch (Colflo 67™) with 45% amylose and 55% amylopectin, a commercial high-amylopectin starch with 98% amylopectin (Crispfilm™), and a 50/50 mixture of these two starches. The extrusion moisture content was 35%wb, and the extrusion and measurement temperatures were 120°C.

Results in Figure 24.8 show that the viscosity of melts produced from starch containing a high fraction of amylopectin is considerably lower than the viscosity of starch having a high amylose content. Scission of the amylopectin branched molecules, which are more prone to breakage by the action of high shear and temperature, could be the cause of the decreased viscosity of the melt. Other researchers have reported similar results.[13,21] Results illustrated in Figure 24.8 show that starch composition, notably amylose/amylopectin ratio, has a large influence

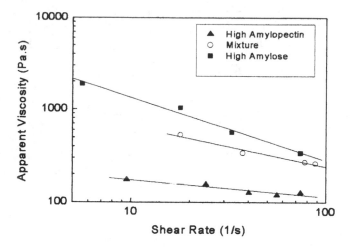

Figure 24.8 Effect of starch composition on the melt rheology.

on the rheological properties of the melt. It is also probable that SME affects the different polymeric forms of the starch molecules in a different manner. Therefore, on-line viscosity of the melt could provide a better control of the extrusion process than SME.

24.2.3 THE EFFECT OF RHEOLOGICAL PROPERTIES ON THE PRODUCT EXPANSION

The structural and textural quality of expanded food products depends largely on the melt expansion mechanism. Extrudate expansion is a complex problem primarily affected by the rheological properties of the melt before leaving the die. It is considered that the extrudate expands for two different mechanisms: (1) die swell due to viscoelastic effects and (2) vapor driven bubble growth (puffing). Die viscoelastic swell is produced by the elastic component of the melt and the presence of normal stresses (radial stresses). It has been stated that to describe the elastic extrudate swell phenomenon, it would be necessary to carry out a complete rheological analysis involving the elastic recoil mechanism and the melt viscoelastic properties. The die swell phenomenon has been described qualitatively in the polymer literature as the phenomenon under which a fluid with elasticity flowing through a die is subjected to extra tension along the streamlines due to the presence of normal stresses. These normal stresses or tensions are maintained as long as the tube walls are present. Once the fluid flows out of the die, the walls no longer constrain the flow, thereby allowing the fluid to expand.[17] The equation used to predict die swell is given as

$$\frac{D_e}{D} = \left[1 + \frac{1}{8}\left(\frac{\tau_{11} - \tau_{22}}{\tau_{12}}\right)\right]^{1/6}$$

(24.9)

where D = the diameter of the die

         D_e = final diameter of the extrudate

$\dfrac{\tau_{11} - \tau_{22}}{\tau_{12}}$ = the ratio between the normal stress difference (a.k.a. N_1) and shear stress

Equation (24.9) is adequate only for estimation purposes, as it excludes some very important factors, such as extensional viscosity and vaporization of moisture, which may strongly affect die swell of food materials.[6] The on-line slit and capillary viscometers described above are not suitable for determination of normal stresses and other viscoelastic effects, and this is an area that needs further research.

The expansion of the extrudate due to vapor-driven air cell growth is also related to the rheology of the melt, but the expansion mechanism is appreciably different from that of the die swell phenomenon. When the melt exits the die, the sudden pressure drop causes an extensive flashing off of superheated vapor. This process forms bubbles in the molten extrudate that grow due to the pressure difference between the melt and the atmospheric pressure, resulting in product expansion.[3]

The effect of melt rheology on expansion has been investigated in the polymer[22] and food literature.[23] It has been was shown that melt expansion dynamics are controlled by the ratio $\Delta P/\mu_a$. ΔP, the difference between die pressure and atmospheric pressure, represents the disequilibrium that induces puffing, whereas the apparent viscosity μ_a measured at the die provides an indication of the resistance of the material to expansion. It has also been shown that radial and longitudinal expansion of extrudates is influenced by melt rheological properties.[3] Although the proposed models provide a good qualitative approximation to the expansion phenomenon, it is important to note that they are far from being complete and do not include other effects taking place during expansion of extrudates, notably extrudate shrinkage and cell collapse.

In food extrusion, water induces plasticization of the extruded material, which in turn causes extrusion temperature to surpass the glass transition and melting temperature of the extrudate, thus turning it into a melt. When the melt exits the die, moisture flashes out of its liquid state due to sudden drop in pressure. The melt then experiences a sudden viscosity increase due to the combined effect of evaporative cooling and decrease of moisture content, which helps to freeze in place the extrudate structure prior to extrudate solidification. A similar phenomenon is described for the expansion of cellular polymers.[1] Using a modified on-line viscometer, it has been demonstrated that the viscosity of the melt is highly dependent on moisture content and temperature, and knowledge of variations is needed if the expansion process is to be modeled.

24.2.4 THE ROLE OF GLASS TRANSITION ON EXPANSION

Starch is a semicrystalline polymer with an amorphous component (amylose) and a partially crystalline component (branched amylopectin).[18] When amorphous polymers are subjected to low temperatures, the molecular motion of the polymer chain is permanently maintained in a fixed conformation, rendering it glassy. Cohesive forces along the polymer chain hinder the molecular movement. Thus, the molecular motion and chain mobility of the polymer are integral in defining the glass transition of materials.[18] The application of thermal energy will induce molecular motion, and, when the molecules gain sufficient energy to slide past one another, the macroscopic properties observed in the material are those of a material that is viscous, rubbery, and flexible. At the glass transition temperature, there is a discontinuous change in the material properties, notably heat capacity and thermal expansion coefficient. As stated above, rheological and/or mechanical properties also exhibit large changes. These changes are used to determine glass transition temperatures of materials by continuously measuring these properties to detect the temperature at which the transition occurs.

Properties of raw materials are highly dependent on several operational variables (listed in Figure 24.1) such as moisture content and temperature. Glass transition and melting properties of the extrudate are also affected by the polymeric structure of the raw materials. It has been reported that the specific mechanical energy (SME) applied during the extrusion process has a large effect on the polymeric structure and average molecular weight of extrudates.[23] When SME increases, the extent of fragmentation also increases, affecting the rheological properties of the melt and the glass transition of the extrudate, which largely affect its final properties—notably crispness. It is also thought that lower-weight fragments in the rubbery state may act as plasticizers for higher-molecular-weight fragments, thus decreasing the glass transition temperature of the larger fragments.

The effect of the melt glass transition temperature on the expansion process has been described in the polymer literature related to the expansion of thermoplastics.[24] The primary reason for bubble growth cessation has been attributed to the solidification of material that occurs below the glass transition temperature, i.e., as viscosity reaches a maximum value, above which bubble growth stops. Conversely, it has been stated that a minimum viscosity must be maintained in the extrudate to prevent the bubbles formed in an expanded extrudate from collapsing.[25] Therefore, it can be explicitly stated that the same factors that affect the glass transition of materials (namely, moisture content, temperature, and macromolecular composition) will affect melt viscosity and will ultimately influence the mechanism of bubble growth and cellular expansion of the extrudate.

A theoretical model in which a spherical vapor bubble surrounded by a shell of fluid with known rheological properties was developed to qualitatively describe the expansion process of foods, notably extrusion.[2] The model assumes that, during the growth of a bubble, the melt temperature decreases as a result of heat loss to the surroundings and evaporative cooling of water at the interface between the melt and the expanding bubble. Theoretical results obtained with this model showed that bubble-wall movement ceased when the extrudate temperature fell below T_g + 30 K. The glass transition concept was also used to describe shrinkage of the cellular structure, due to the fact that a partial vacuum can be created within the bubble.

Although the above model can give a good qualitative account of the expansion process and a sound explanation about the role melt viscosity and glass transition may play on expansion and shrinkage of the extrudate, it is far from reality. The model assumes that, during the expansion process, the temperature of the extrudate is uniform. However, during the later stages of the process, i.e., when the foam structure starts to form, the thermal conductivity of the product is reduced. The reduction of the extrudate conductivity creates a nonuniform temperature distribution, with lower temperatures at the product surface and higher temperatures at the center, because the rate of heat transfer from the surface to the surroundings is always higher than from the center to the surface. Thus, the extrudate surface, which is subjected to both rapid water evaporation and fast cooling, could experience a structure collapse and solidification with the formation of a skin or crust. The formation of a crust at the product surface could reduce further the water loss at the product surface and the water diffusion in the extrudate matrix, thus creating a moisture gradient in the product and increasing the residual moisture in the product. The formation of an extrudate skin and shrinkage can be visually observed during high-temperature and relatively high-moisture extrusion. The formation of crust, structure collapse, and the production of high-density materials have also been observed during drying of biomaterials at high temperature.[26] It has been observed that the viscosity of the matrix governs the rate at which it can deform and is dependent on the glass transition temperature of the matrix. A similar model to that of drying is necessary to study the expansion/shrinkage of extrudate materials.

24.3 MODIFIED EXTRUDATE EXPANSION MODEL

A modified expansion model based on that applied to the drying of biomaterials was elaborated and is schematically illustrated in Figure 24.9.[27] The different stages of the process are as follows:

STAGE 1.   At the extruder die, the melt is a homogeneous material having a temperature higher than the boiling point of water at atmospheric pressure.

STAGE 2.   Bubble nucleation starts; the product is exiting the extruder die. The product temperature is still higher than the boiling temperature of water at atmospheric pressure.

STAGE 3.   Bubble expansion starts; a crust is formed at the product surface. Melt temperature is still higher than boiling temperature, but crust prevents

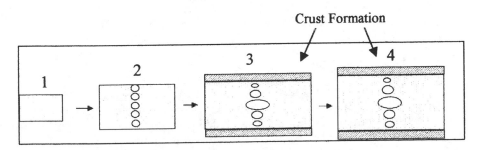

Figure 24.9 Schematic of the various stages during extrudate expansion.

expansion of the bubbles at the surface. Bubbles in the center expand more than those at the surface.

STAGE 4.   Final product. Temperature of the product is lower than the boiling point of water, so bubble expansion stops. Temperature of the product is lower than the glass transition temperature; foam structure sets.

Equations needed to solve this model are given elsewhere[27] and can be summarized as described below.

24.3.1 PRODUCT TEMPERATURE

$$\rho C_p \frac{dT_p}{dt} = Ah(T_p - T_a) \tag{24.10}$$

where ρ = the product density

C_p = its heat capacity

T_p and T_a = the average product temperature and the ambient temperature, respectively

A = the heat transfer area of the product and h is the heat transfer coefficient

24.3.2 BUBBLE EXPANSION

$$\frac{P^v - P_{atm}}{\Delta V} = K \tag{24.11}$$

where P^v and P_{atm} are the vapor pressure of the water and the atmospheric pressure respectively. K is the bulk modulus. Both P^v and K are functions of temperature and moisture content. Although temperature varies with position, Equation (24.10) assumes that no significant internal gradients exist when compared with moisture gradients. The moisture content gradient in the product is obtained using a moisture diffusion equation that accounts for shrinkage and crust formation.[26] The solution of that equation implies knowledge of the rheological properties of the melt (viscosity and elasticity) as a function of temperature and moisture.

24.4 CONCLUSIONS

Rheological properties of extrudate melts could be directly linked to a number of critical variables in the extrusion process, and they are important not only to monitor and control the process, but they are also needed to model the expansion of extrudates. Rheological characterization of extrudate melts cannot be carried out using off-line viscometers, because their properties change rapidly once melts exit the extruder die. Thus, rheological characterization of melts can be performed only using appropriate on-line methods. When food ingredients are turned into a melt under conditions of high temperature, pressure, and shear, complex physicochemical changes take place (e.g., starch gelatinization, protein denaturation). These changes are highly sensitive to the processing history of the melt and therefore may influence the on-line measurement of rheological properties. An on-line rheological method to determine flow curves (τ vs. $\dot{\gamma}$) of extrudates that are independent of the extruder processing condition is presented in this chapter.

The rheological properties of the melt and, subsequently, the extrudate are affected by processing conditions (e.g., shear rate, SME, moisture, temperature) and raw material characteristics, notably composition. There are other factors (namely, nucleation, water diffusion, heat transfer, and glass transition) that need to be considered to completely understand the expansion phenomenon.

REFERENCES

1. Gendron, R. and A. Correa. 1998. "The Use of On-line Rheomtery to Characterize Polymer Melts Containing Physical Blowing Agents," *Cellular Polymers,* 17(2): 93–113.
2. Fan, J., J. R. Mitchell, and J. M. Blanshard. 1994. "A Computer Simulation of the Dynamics of Bubble Growth and Shrinkage During Extrudate Expansion," *J. Food Eng.,* 32: 239–262.
3. Bouzaza, D., A. Arhaliass, and J. M. Bouvier. 1996. "Die Design and Dough Expansion in Low Moisture Extrusion-Cooking Process," *J. Food Eng.,* 29: 139–152.
4. Mours, M. and H. H. Winter. 1994. "Time-Resolved Rheometry," *Rheol. Acta,* 33: 385–397.
5. Reinier, M. 1964. "The Deborah Number," *Physics Today,* January, 62.
6. Steffe, J. F. 1992. *Rheological Methods in Food Process Engineering,* 2nd ed. East Lansing: Freeman Press.
7. Bird, R. B., R. C. Armstrong, and O. Hassager. 1987. *Dynamics of Polymeric Liquids,* vol. 1, 2nd ed., New York: John Wiley and Sons.
8. Kokini, J. L., C. T. Ho, and M. V. Karwe, eds. 1992. *Food Extrusion Science and Technology.* Marcel Dekker, Inc.
9. Li, P. X.-P., O. H. Campanella, A. K. Hardacre, and K. J. Kirkpatrick. 1997. "A Design of a New On-line Slit-Die-Viscometer to Use in Food Extrusion," in *Engineering and Food at ICEF '97,* R. Jowitt, ed. UK: Sheffield Academic Press.
10. Bagley, E. B. 1957. "End Corrections in the Capillary Flow of Polyethylene," *J. App. Phys.,* 28: 624–627.
11. Senouci, A. and A. C. Smith. 1988. "An Experimental Study of Food Melt Rheology. I. Shear Viscosity Using a Slit Die Viscometer and a Capillary Rheometer," *Rheologica Acta,* 27: 649–655.
12. Lai, L. S. and J. L. Kokini. 1990. "The Effect of Extrusion Operating Conditions on the On-Line Apparent Viscosity of 98% Amylopectin (Amioca) and 70% Amylose (Hylon 7) Corn Starches during Extrusion," *J. of Rheol.,* 34: 1245–1266.
13. Lai, L. S. and J. L. Kokini. 1991. "Physicochemical Changes and Rheological Properties of Starch During Extrusion. A Review," *Biotechnology Progress,* 7: 251–266.
14. Padmanabhan, M. and M. Bhattacharya. 1993. "Effect of Extrusion Process History on the Rheology of Corn Meal," *J. of Food Eng.,* 18: 335–349.

15. Padmanabhan, M. and M. Bhattacharya. 1991. "Flow Behavior and Exit Pressures of Corn Meal under High-Shear-High-Temperature Extrusion Conditions Using a Slit Die," *J. of Rheol.*, 35: 315–343.
16. van Lengerich. 1990. "Influence of Extrusion Processing on In-line Rheological Behavior, Structure and Function of Wheat Starch," in *Dough Rheology and Baked Product Texture*, H. Faridi and J. M Faubion, eds. Van Nostrand Reinhold Ltd.
17. Padmanabhan, M. and M. Bhattacharya. 1989. "Extrudate Expansion During Extrusion Cooking of Foods," *Cereal Foods World,* 34: 945–949.
18. Wu, J. P. C. 1991. "Study of Mechanisms for Vapour-Induced Puffing of Starch-Rich Materials," Ph.D. Thesis, University of Massachusetts.
19. Vergnes, B., G. Della Valle, and J. Tayeb. 1993. "A Specific In-Line Rheometer for Extruded Starchy Products. Design, Validation and Application to Maize Starch," *Rheologica Acta*, 32: 465–476.
20. Li, P. X. 1998. "Milling and Extrusion Characteristics of New Zealand Corn. Development of a Hardness Test and an On-Line Extruder Viscometer," Ph.D. Thesis, Massey University, New Zealand.
21. Chinnaswamy, R. and M. A. Hanna. 1988. "Relationship Between Amylose Content and Extrusion-Expansion Properties of Corn Starches," *Cereal Chemistry*, 65(2): 138–143.
22. Amon, M. and C. D. Denson. 1984. "A Study of the Dynamics of Foam Growth: Analysis of the Growth of Closely Spaced Spherical Bubbles," *Polym Eng. Sci.*, 24: 1026–1034.
23. Kokini, J. L., C. N. Chang, and L. Lai. 1992. "The Role of Rheological Properties on Extrudate Expansion," in *Food Extrusion, Science and Technology*, J. L. Kokini, C. T. Ho, and M. V. Karwe, eds. New York: M. Dekker.
24. Kaletunc, G. and K. L. Breslauer. 1996. "Construction of a Wheat-Flour State Diagram. Application to Extrusion Processing," *Journal of Thermal Analysis*, 47: 1267–1288.
25. Burt, J.G. 1978. "The Elements of Expansion of Thermoplastics Part II," *J. of Cellular Plastics*, Nov./Dec., 341–345.
26. Achanta, S., T. Nakamura, and M. R. Okos. 1998. "Stress Development in Shrinking Slabs During Drying," in *The Properties of Water in Foods ISOPOW 6*, D. S. Reid, ed. Blackie Academic and Professional, pp. 253–271.
27. Achanta, S. 1995. "Moisture Transport in Shrinking Gels during Drying," Ph.D. Thesis Purdue University.

25 Nonlinear Viscoelasticity Modeling of Vegetable Protein-Stabilized Emulsions

C. Gallegos
J. M. Franco
J. M. Madiedo
A. Raymundo
I. Sousa

CONTENTS

Abstract

This chapter reviews the work carried out on the rheological characterization of lupin protein-stabilized oil-in-water emulsions. With this aim, the influences that processing variables and emulsifier composition exert on the viscous and linear viscoelasticity functions have been studied. Special attention has been given to the nonlinear viscoelasticity modeling of these emulsions, using a factorable nonlinear constitutive equation.

1-56676-963-9/02/$0.00+$1.50
© 2002 by CRC Press LLC

25.1   INTRODUCTION

Food emulsions such as mayonnaise and salad dressings are typically stabilized by egg yolk. However, many authors have studied the emulsifying capacity of vegetable proteins to replace the traditional egg yolk.[1,2] The main advantage of this substitution results from the growing interest in low-cholesterol and salmonella-free food emulsions. In this sense, protein isolates from *Lupinus albus* seeds may be another important alternative to egg products, encouraging in this way the use of a food additive ingredient from a legume plant with a positive agricultural impact.

As has been shown by some authors,[3] the research focused on the use of new surface active agents to manufacture oil-in-water emulsions must include the optimization of both emulsion formulation and emulsification process. This chapter reviews the work we have carried out on the influences that processing variables and emulsifier composition exert on the bulk rheology and droplet size distribution of concentrated oil-in-water emulsions stabilized by lupin protein. Special attention has been given to the nonlinear viscoelasticity modeling of these emulsions, using a factorable nonlinear constitutive equation.

25.2   EXPERIMENTAL

Oil-in-water emulsions containing a sunflower oil concentration of 65% wt were prepared. A lupin protein isolate, L9020, from Mittex Alangenbau GmbH (Germany) was used as received. In addition, different low-molecular-weight emulsifiers were used: two sucrose stearates of HLB 7 (SS7) and HLB 15 (SS15), respectively, and a sucrose laurate of HLB 15 (SL15), all from Mitsubishi Food Corporation (Japan); and a polyoxyethylene (20) sorbitan monolaurate, Tween 20, of HLB 17 (PSOL) from Aldrich (UK). The total emulsifier concentration was fixed at 6% wt. The protein/surfactant ratios studied were 6/0, 4.5/1.5, 3/3, 1.5/4.5, and 0/6. Both macromolecular and low-molecular-weight emulsifiers were dispersed in distilled water at 60°C so as to favor both the solubility of the sucrose ester and protein denaturation, which increases its hydrophobicity. Emulsification was performed using an Ultra Turrax T-25 homogenizer from Ika (Germany) at different agitation speeds (8000–20500 rpm) and emulsification times (3–10 min).

Droplet size distribution (DSD) measurements were performed in a Malvern Mastersizer-X (Malvern, UK). Values of the Sauter mean diameter,[4] which is inversely proportional to the specific surface area of droplets, were obtained as follows:

$$d_{SV} = \frac{\sum_i n_i d_i^3}{\sum_i n_i d_i^2} \tag{25.1}$$

where n_i = the number of droplets having a diameter d_i

Dynamic viscoelasticity and steady-state flow measurements were carried out in a controlled stress rheometer (RS-75) from Haake (Germany). Oscillatory tests were performed inside the linear viscoelastic region using a cone and plate sensor system (35 mm, 2°) in a frequency range of 0.05–200 rad/s. Steady state flow curves were obtained with a serrated plate-plate sensor system (20 mm) to prevent wall slip phenomena.[5]

25.3 RESULTS AND DISCUSSION

25.3.1 LUPIN PROTEIN-STABILIZED EMULSIONS: INFLUENCE OF PROCESSING

As has been previously reported,[4,6] processing variables strongly influence emulsion structural parameters such as droplet size distribution (DSD), interdroplet interactions, and continuous phase rheology. Consequently, they affect the bulk rheology of the emulsion. Thus, an increase in the energy input during the emulsification process usually reduces the mean droplet size of the emulsion.[7] Figures 25.1 and 25.2 show the effect of agitation speed and emulsification time on the DSD curves for selected lupin protein-stabilized emulsions. As can be observed, all DSD curves show a bimodal shape with a secondary maximum at relatively low values of the droplet diameter and the maximum distribution value at higher sizes. An initial increase in the agitation speed (8000–14250 rpm) yielded a large reduction in droplet size, the maximum of the distribution appeared at lower diameters, and the secondary

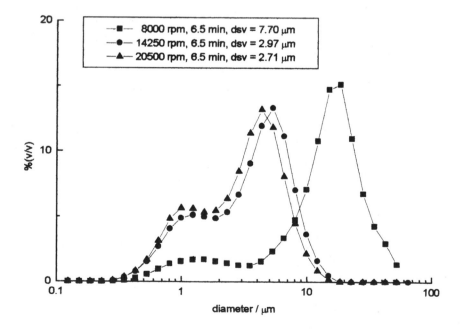

Figure 25.1 Evolution of the DSD with the agitation speed for lupin protein-stabilized emulsions. Reproduced from Reference 7 with permission of the American Chemical Society.

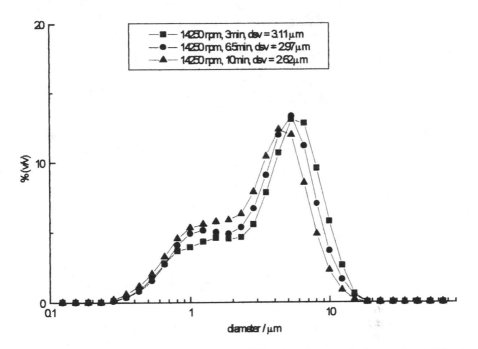

Figure 25.2 Evolution of the DSD with emulsification time for lupin protein-stabilized emulsions. Reproduced from Reference 7 with permission of the American Chemical Society.

maximum became more pronounced. On the contrary, a further increase in the agitation speed produced only a slight decrease in the mean droplet diameter (d_{sv}). The emulsification time also reduced the droplet diameter, although this influence is less important, especially at higher agitation speeds, as previously reported by Sánchez et al.,[8] for surfactant-stabilized emulsions.

The evolution of the storage (G') and loss (G'') moduli with frequency shown by these types of emulsions is similar to that found for flocculated emulsions.[9–11] Therefore, the storage modulus is always higher than the loss modulus within the experimental frequency range, with a tendency to the development of a plateau region in G', accompanied by a minimum in the loss modulus as energy input during the emulsification process increases. Thus, as can be observed in Figures 25.3 and 25.4, a clear tendency to a crossover of both viscoelastic functions at low frequencies is found when the emulsion is processed at low agitation speeds and/or during short emulsification times. On the contrary, an extended plateau region can be observed as the values of both variables increase. The presence of a pseudo-terminal region at low frequencies has been previously reported for emulsions stabilized by a mixture of egg yolk and a high HLB sucrose stearate,[6] although, in this case, this region was always shown within the range of processing variables chosen (5000–8000 rpm; 3–10 min). In any case, for both types of emulsions, the plateau region in the relaxation spectrum spreads as energy input increases.

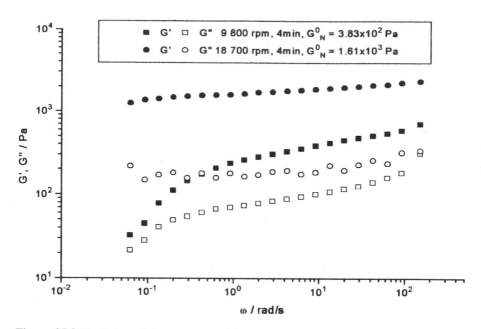

Figure 25.3 Evolution of the storage and loss moduli with frequency for lupin protein-stabilized emulsions as a function of the agitation speed. Reproduced from Reference 7 with permission of the American Chemical Society.

Steady-state flow curves always show a shear-thinning behavior with a clear tendency to a zero-shear-rate-limiting viscosity, η_o, at very low shear rates. Thus, Figures 25.5 and 25.6 show selected steady-state flow curves as a function of agitation speed and emulsification time, respectively. As may be clearly observed, the steady-state viscosity increases with both variables within the whole shear rate range studied, although the influence is more relevant at low shear rates.

The above-mentioned results may be explained on the basis of the development of an entanglement network, due to bridging flocculation of oil droplets as a result of increased physical interactions among proteins chains, as energy input during the emulsification process increases. The rise in temperature (up to 50–55°C) during the emulsification process, induced by an intense agitation or long emulsification times, may significantly affect the structure of the protein, increasing its hydrophobicity. This would explain why the values of the rheological parameters are much larger at high agitation speeds or long emulsification times, although the droplet size of the emulsions may not be altered. In other words, the rheological results obtained cannot be explained taking into account only the variation of the mean droplet size and polydispersity of these concentrated emulsions with the above mentioned processing variables.

Consequently, these results suggest that another important processing variable may be the previous thermal treatment applied to the lupin protein aqueous dispersion. Thus, protein heat treatment induces protein unfolding, with a subsequent

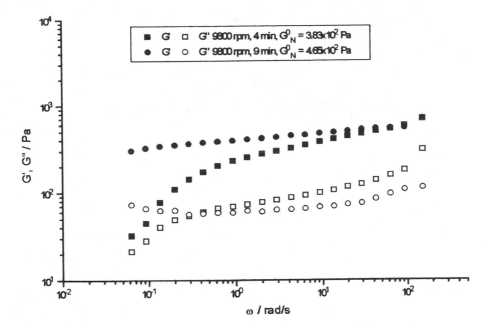

Figure 25.4 Evolution of the storage and loss moduli with frequency for lupin protein-stabilized emulsions as a function of the emulsification time. Reproduced from Reference 7 with permission of the American Chemical Society.

increase in surface hydrophobicity,[12] which may increase the film thickness and its mechanical resistance.[13] This results in a reinforcement of the entanglement network, denoted by the appearance of a plateau region in the mechanical spectrum of the emulsion and/or a significant increase in the values of the viscoelastic functions and emulsion viscosity. This can be observed, for instance, in Figure 25.7, where frequency sweep curves are shown for selected emulsions containing lupin protein submitted to different thermal treatments. Thus, the pseudo-terminal region appears at low frequencies for the emulsion containing lupin protein preheated at 50°C, whereas a well developed plateau region was obtained when emulsions were prepared with lupin protein previously submitted at higher treatment temperatures (70 or 90°C).

In the same way, steady-state viscosity increases with protein treatment time and temperature. However, protein thermal treatment induces only slightly lower values of emulsion Sauter's diameters.

25.3.2 Lupin Protein-Stabilized Emulsions: Influence of Emulsifier Composition

As has been previously mentioned, a crucial factor influencing the rheology of emulsions is the strength and nature of interparticle interactions. If the attractive forces exceed the repulsive electrostatic and/or steric interactions, an aggregation of

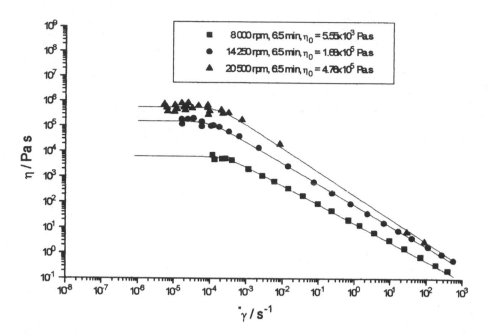

Figure 25.5 Steady-state flow behavior of lupin protein-stabilized emulsions as a function of the agitation speed. Reproduced from Reference 7 with permission of the American Chemical Society.

oil droplets occurs, and coalescence may take place, depending on the viscoelasticity of the surface layer. The strength and nature of the interactions among droplets depend on the nature and concentration of the emulsifier used, which determine if the flocculation is reversible or irreversible. The extension of this aggregation process influences the bulk rheology of emulsions.[14] The use of mixtures of macromolecular and low-molecular-weight emulsifiers in concentrated emulsions may increase stability. Thus, positive surfactant-protein interactions influencing rheological parameters and emulsion stability have been found in different systems.[10,11,15] However, a complex behavior may also be found, depending on the nature of the emulsifiers and the surfactant/protein ratio.[16]

With the aim of studying the influence of the surfactant/protein ratio on the bulk rheology of emulsions stabilized by mixtures of lupin protein and a low-molecular-weight emulsifier, a constant emulsifier concentration (6% wt) was used. The experimental results obtained demonstrate that the rheological behavior found is dependent on the type of surfactant used.[17] Substitution of 1.5% wt protein by an identical amount of a low-molecular-weight emulsifier always produces a decrease in the linear viscoelasticity functions (Figure 25.8). This decrease should be related to a loss in surface viscoelasticity of the adsorbed emulsifier layer at the oil-water interface, in spite of an apparent decrease in Sauter's diameter. Furthermore, as can be observed in Figure 25.9, the values of the plateau modulus for the emulsions

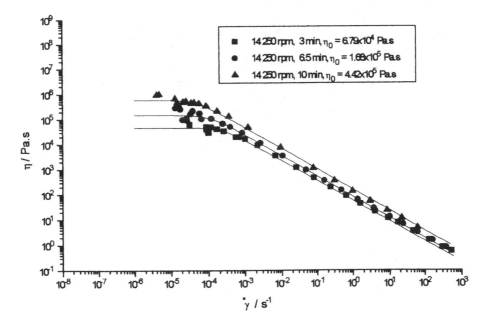

Figure 25.6 Steady-state flow behavior of lupin protein-stabilized emulsions as a function of the emulsification time. Reproduced from Reference 7 with permission of the American Chemical Society.

containing PSOL, SL15 and SS15 nonionic surfactants are practically the same (G_N^0 ≈ 700 Pa), although the values of the Sauter diameter are slightly different (2.0–2.3 μm). A further increase in the surfactant/protein ratio (3/3) yields a similar influence on the three types of emulsions previously mentioned, indicating the predominance of the protein on the viscoelastic characteristics of the emulsions. However, if the surfactant/protein ratio increases further (4.5/1.5), the plateau modulus for the PSOL and SL15-stabilized emulsions still decreases, but an almost unchanged value for the SS15-stabilized emulsion is noticed (Figure 25.9), which may be related to an increased influence of its continuous phase viscosity. Thus, if only the low-molecular-weight emulsifier is used to stabilize these emulsions, a further decrease in G_N^0 is noticed for the PSOL and SL15-stabilized emulsions, while a dramatic increase is observed for the SS15-stabilized emulsion, a fact that is closely related to the development of a gel-like continuous phase in these emulsions.[18]

On the other hand, the emulsion stabilized by the SS7 emulsifier shows a different behavior. An initial increase in the surfactant/protein ratio (1.5/4.5) also produces an important decrease in the plateau modulus, due to a decrease in surface viscoelasticity of the adsorbed emulsifier at the oil-water interface, but less pronounced than for the emulsions stabilized by emulsifier blends containing any of the other three low-molecular-weight emulsifiers. These differences are even more important after a further increase in the surfactant/protein ratio (3/3). For this emul-

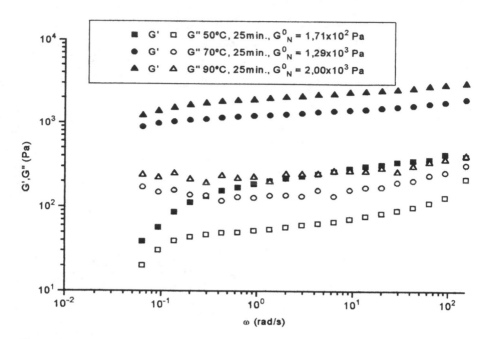

Figure 25.7 Influence of the previous thermal treatment on the evolution of the storage and loss moduli with frequency for lupin protein-stabilized emulsions. Reproduced from Reference 12 with permission from Wiley-VCH.

sion, a significant increase in the plateau modulus is observed, reaching a value similar to that shown by the surfactant-free emulsion. An additional increase in the surfactant/protein ratio (4.5/1.5) produces, however, a dramatic decrease in the plateau modulus, down to values relatively close to those shown by the emulsions stabilized by other emulsifier blends. This peculiar behavior may be related to the development of interactions between unadsorbed sucrose stearate and protein molecules via hydrophobic interactions.[16] These interactions would explain the maximum in G_N^0 at an intermediate surfactant/protein ratio and the significantly higher values of this parameter in relation to emulsions stabilized by other emulsifier blends. The specific behavior of this surfactant may be related to its hydrophobicity, the largest of the four low-molecular-weight emulsifiers used. Finally, the protein-free emulsion shows maximum values of the plateau modulus, a fact that should be related to the phase behavior of this sucrose stearate, which forms a gel-like phase with water in a very wide range of concentrations.[19]

25.3.3 LUPIN PROTEIN-STABILIZED EMULSIONS: NONLINEAR VISCOELASTICITY MODELING

Within the linear viscoelasticity range, the results obtained from a given rheological test can be used to predict the values of other material functions.[20] This, for instance, may be achieved by means of the continuous linear relaxation spectrum,[21,22] which

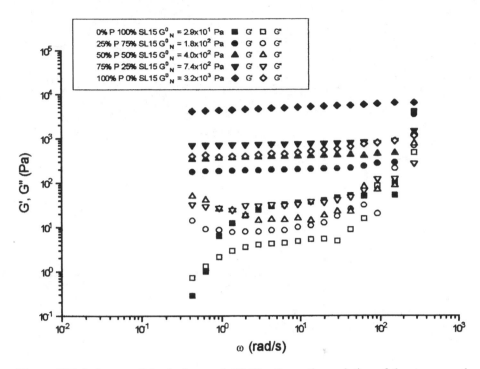

Figure 25.8 Influence of the lupin protein/SL15 ratio on the evolution of the storage and loss moduli with frequency for emulsions stabilized by a mixture of emulsifiers. Reproduced from Reference 17 with permission from AOCS.

is related to linear viscoelasticity functions by means of Fredholm integral equations of the first kind.[23,24] Thus, for instance, the dynamic moduli, $G'(\omega)$ and $G''(\omega)$, and the linear relaxation modulus, $G(t)$, are related to the continuous relaxation spectrum, $H(\lambda)$, by means of the following equations:

$$G'(\omega) = G_e + \int_{-\infty}^{\infty} H(\lambda)\frac{\omega^2\lambda^2}{1 + \omega^2\lambda^2}d[\ln(\lambda)] \qquad (25.2)$$

$$G''(\omega) = \int_{-\infty}^{\infty} H(\lambda)\frac{\omega\lambda}{1 + \omega^2\lambda^2}d[\ln(\lambda)] \qquad (25.3)$$

$$G(t - t') = G_e + \int_{-\infty}^{\infty} H(\lambda)\exp[-(t - t')/\lambda]d\ln\lambda \qquad (25.4)$$

As this spectrum cannot be obtained from direct experimentation, the problem that usually arises is its calculation from the experimental data of a given linear viscoelasticity function. Once the spectrum is known, it can be used to predict any linear viscoelasticity function by inserting it into the corresponding integral equation.[23,24]

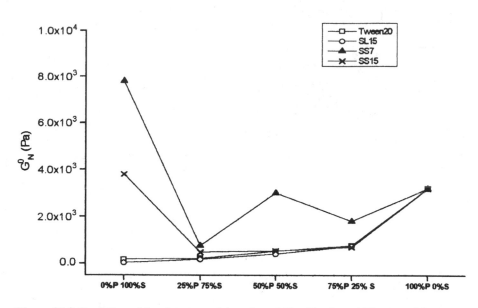

Figure 25.9 Evolution of the plateau modulus of emulsifier blends–stabilized emulsions with lupin protein/surfactant ratio (wt/wt). Reproduced from Reference 17 with permission from AOCS.

On the other hand, linear responses cannot, in principle, be used to predict the values of material functions within the nonlinear range; nor can nonlinear data obtained in a given test be used to make, from basic principles, predictions of rheological responses in other nonlinear tests. To make these predictions, it is necessary to use constitutive equations that allow modeling of the nonlinear behavior. One of these equations is the Wagner model, which derives from the K-BKZ equation[25,26] by considering the time-strain separability and dropping the term that involves the Cauchy tensor.[27]

$$\tilde{\tilde{\sigma}}(t) = -\int_{-\infty}^{t} m(t-t')h(I_1,I_2)\tilde{\tilde{C}}^{-1}dt'$$

(25.5)

where C^{-1} is the Finger tensor, $h(I_1, I_2)$ is the damping function that quantifies the magnitude of the nonlinearity, I_1 and I_2 are the first and second invariants of the Finger tensor, and $m(t - t')$ is the memory function related to the linear relaxation modulus by differentiation.

$$m(t-t') = \frac{dG(t-t')}{dt'}$$

(25.6)

Models that involve the time-strain separability, as, for instance, the Wagner model, have been successfully applied to predict the nonlinear rheological response of a great variety of materials.[28–33]

Considering only the simple shear component of the stress tensor, the Wagner model yields

$$\tau(t,\dot{\gamma}) = \int_{-\infty}^{t} \frac{dG(t-t')}{dt'} h(\gamma)\gamma(t,t')dt' \tag{25.7}$$

where $\tau(t,\dot{\gamma})$ is the transient shear stress and $h(\gamma)$ is the damping function, which can be easily calculated from the ratio between the nonlinear relaxation modulus, $G(\gamma, t - t')$, and the linear relaxation modulus, $G(t - t')$[27]

$$h(\gamma) = \frac{G(\gamma,t-t')}{G(t-t')} \tag{25.8}$$

On the other hand, the relaxation modulus is related to the continuous linear relaxation spectrum by the Fredholm integral Equation (25.4). Substituting Equation (25.4) in Equation (25.7), and taking into account the relationship between shear stress and apparent viscosity, the Wagner model in simple shear gives for the transient viscosity

$$\eta(t,\dot{\gamma}) = -\frac{1}{\dot{\gamma}} \int_{-\infty}^{t} \int_{-\infty}^{\infty} H(\lambda)\exp(-(t-t')/\lambda)h(\gamma)\gamma(t,t')d\ln\lambda dt' \tag{25.9}$$

The calculation of the linear relaxation spectrum, $H(\lambda)$, is, however, much more complex. It implies inverting any of the Fredholm integral equations, which relate it to the values of a given linear viscoelasticity function, but small changes in the rheological functions give rise to strong oscillations in the spectra[23,34,35] so that unstable and even physically senseless solutions may be obtained. For that reason, many approximation methods have been developed to perform such calculations by simplifying the original problem.[20,23] On the other hand, the Tikhonov regularization method has been proposed as a powerful tool for solving the difficulties that arise with Fredholm equations, allowing one to obtain relaxation and retardation spectra.[19,24] This method is based on the minimization of functionals that relate the spectrum to the experimental values of a given linear viscoelasticity function. Thus, for instance, if the relaxation spectrum is calculated from the storage modulus $G'(\omega)$, from the loss modulus $G''(\omega)$ or from the relaxation modulus $G(t)$, the following functionals must be minimized:

$$\Gamma(v) = \sum_{i=1}^{n}\left[\left(G'_{i\ \exp} - G_e - \int_{-\infty}^{\infty} H(\lambda)\frac{\omega_i^2\lambda^2}{1+\omega_i^2\lambda^2}d\ln\lambda\right)^2\right] + v\|\Xi(H(\lambda))\|^2 \tag{25.10}$$

for the calculation of $H(\lambda)$ from $G'(\omega)$,

$$\Gamma(v) = \sum_{i=1}^{n}\left[\left(G''_{i\ \exp} - \int_{-\infty}^{\infty} H(\lambda)\frac{\omega_i\lambda}{1+\omega_i^2\lambda^2}d\ln\lambda\right)^2\right] + v\|\Xi(H(\lambda))\|^2 \qquad (25.11)$$

for the calculation of $H(\lambda)$ from $G''(\omega)$, and,

$$\Gamma(v) = \sum_{i=1}^{n}\left[\left(G_{i\ \exp} - G_e - \int_{-\infty}^{\infty} H(\lambda)\exp((-t_i)/\lambda)d\ln\lambda\right)^2\right] + v\|\Xi(H(\lambda))\|^2 \quad (25.12)$$

for the calculation of $H(\lambda)$ from $G(t)$.

In Equations (25.10) through (25.12), the factor v is called the regularization factor, and Ξ is an operator for which, usually, the second derivative with respect to relaxation time is used.[19,24,35]

The rheological data for a selected lupin protein-stabilized emulsion containing 3% wt emulsifier are shown in Figure 25.10. The experimental data shown are typical of the oil-in-water emulsions studied in this chapter. Thus, Figure 25.10A shows data for a frequency sweep carried out in the linear viscoelasticity range fitted with a continuous spectrum of relaxation times. For the continuous spectrum fit, a time domain equivalent to the experimental frequency domain was chosen. Recalculated values from the relaxation spectrum fit fairly well to experimental ones. Figure 25.10B shows the continuous spectrum of relaxation times. The results of step-strain experiments are shown in Figure 25.10C. The nonlinear behavior of the material is apparent in this figure, where an increase in the applied strain from 1 to 200% results in a marked decrease in the relaxation modulus. Step-strain data have been used to determine the damping function to characterize the degree of nonlinearity. The Wagner, Soskey–Winter, and Papanastasiou damping functions have been used Equations (25.13), (25.14), and (25.15), respectively.[28,36,37]

$$h(\gamma) = \exp(-k\gamma) \qquad (25.13)$$

$$h(\gamma) = \frac{1}{1+a\gamma^b} \qquad (25.14)$$

$$h(\gamma) = \frac{\alpha}{\alpha+\gamma^2} \qquad (25.15)$$

The parameter values of the three models for three emulsions containing different lupin protein concentrations are shown in Table 25.1. As can be seen, an increase in protein content results in a decrease in the values of both Wagner's damping factor *(k)* and Soskey–Winter's parameter *(a)*. This marked strain softening confirms the experimental observation that emulsion viscoelasticity is more delicate and more limited to smaller strain deformations than polymer systems.[10]

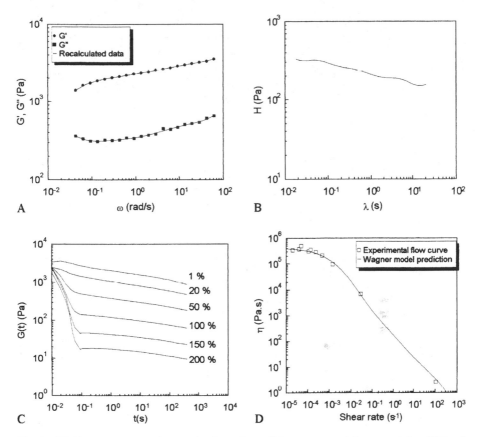

Figure 25.10 Rheological characterization of an oil-in-water emulsion containing 65% oil and 3% lupin protein. (A) Experimental values of the dynamic moduli and recalculated data from the relaxation spectrum, (B) relaxation spectrum, (C) experimental data of the relaxation modulus at different deformation values, (D) experimental and predicted values of viscosity vs. shear rate.

TABLE 25.1
Damping Parameters Corresponding to Models given by Equations (25.13), (25.14), and (25.15)

Lupin protein concentration (% wt)	5%	3%	2%
k	2.13	2.65	3.93
a	12.36	17.25	28.16
b	1.79	1.86	1.10
α	0.08	0.05	0.03

These damping models have been used to predict the flow behavior of these emulsions. Figure 25.10D shows the experimental flow curve along with the predicted values from the Wagner constitutive equation. The three damping models considered provided similar values of the steady-state viscosity in a wide range of shear rates.

REFERENCES

1. Tornberg, E. 1978 "Functional Characteristics of Protein Stabilized Emulsions: Emulsifying Behaviour of Proteins in a Valve Homogenizer," *J. Food Sci.*, 29: 867–879.
2. Elizalde, B. E., G. B. Bartholomai, and A. M. R. Pilosof. 1996. "The Effect of pH on the Relationship between Hydrophilic/Lipophilic Characteristics and Emulsification Properties of Soy Proteins," *Lebens. Wiss. u Technol.*, 29: 334–339.
3. Gallegos, C. and J. M. Franco. 1999. "Rheology of Food Emulsions," in *Advances in the Flow and Rheology of Non-Newtonian Fluids,* D. A. Siginer, D. De Kee, and R. P. Chhabra, eds., Amsterdam: Elsevier, 87–118.
4. Sprow, F. B. 1967. "Distribution of Drop Sizes Produced in Turbulent Liquid–Liquid Dispersion," *Chem. Eng. Sci.*, 22: 435–442.
5. Franco, J. M., C. Gallegos and H. A. Barnes. 1998. "On Slip Effects in Steady-State Flow Measurements of Oil-In-Water Food Emulsions," *J. Food Eng.*, 36: 89–92.
6. Franco, J. M., A. Guerrero, and C. Gallegos. 1995. "Rheology and Processing of Salad Dressing Emulsions," *Rheol. Acta,* 34: 513–524.
7. Franco, J. M., A. Raymundo, I. Sousa, and C. Gallegos. 1998. "Influence of Processing Variables on the Rheological and Textural Properties of Lupin Protein-Stabilized Emulsions," *J. Agric. Food Chem.*, 46: 3109–3115.
8. Sanchez, M. C., M. Berjano, A. Guerrero, E. Brito, and C. Gallegos. 1998. "Evolution of the Microstructure and Rheology of O/W Emulsions During the Emulsification Process," *Can. J. Chem. Eng.*, 76: 479–485.
9. Pal, R. 1995. "Oscillatory, Creep and Steady-State Flow Behaviour of Xanthan-Thickened Oil-in-Water Emulsions," *AIChE J.*, 41: 783–795.
10. Dickinson, E. and Y. Yamamoto. 1996. "Rheology of Milk Protein Gels and Protein-Stabilized Emulsion Gels Cross-Linked with Transglutaminase," *J. Agric. Food Chem.*, 44: 1371–1377.
11. Franco, J. M., M. Berjano, and C. Gallegos. 1997. "Linear Viscoelasticiy of Salad Dressing Emulsions," *J. Agric. Food Chem.*, 45: 713–719.
12. Raymundo, A., J. M. Franco, C. Gallegos, J. Empis, and I. Sousa. 1998. "Effect of Thermal Denaturation of Lupin Protein on its Emulsifying Properties," *Nahrung*, 42: 220–224.
13. Dickinson, E. and G. Stainsby. 1988. "Emulsion Stability," in *Advances in Food Emulsions and Foams.* Ed. Elsevier Applied Science, 1–41.
14. Pal, R. and E. Rhodes. 1989. "Viscosity/Concentration Relationships for Emulsions," *J. Rheol.*, 33: 1021–1045.
15. Clark, D. C., P. J. Wilde, D. R. Wilson, and R. C. Wunsteck. 1992. "The Interaction of Sucrose Esters with β-Casein from Bovine Milk," *Food Hydrocoll.*, 6:173–186.
16. Dickinson, E. and S. T. Hong. 1995. "Influence of Water-Soluble Non-ionic Emulsifier on the Rheology of Heat-Set Protein Stabilized Emulsion Gels," *J. Agric. Food Chem.*, 43: 2560–2566.

17. Raymundo, A., J. M. Franco, P. Partal, I. Sousa, and C. Gallegos. 1999. "Effect of the Lupin Protein/Surfactant Ratio on Linear Viscoelasticity Properties of Oil-in-Water Emulsions," *J. Surf. Deterg.,* 2: (in press).

18. Bower, C., C. Gallegos, M. R. Mackley, and J. M. Madiedo. 1999. "The Rheological and Microstructural Characterisation of the Nonlinear Flow Behaviour of Concentrated Oil-In-Water Emulsions," *Rheol Acta*, 38: 145–159.

19. Madiedo, J. M. 1996. Ph.D. Thesis. University of Seville, Seville.

20. Ferry, J. D. 1980. *Viscoelastic Properties of Polymers,* New York: Wiley.

21. Madiedo, J. M. and C. Gallegos. 1997. "Rheological Characterisation of Oil-In-Water Emulsions by Means of Relaxation and Retardation Spectra," *Recent Res. Devel. Oil Chemistry*, 1: 79–90.

22. Madiedo, J. M. and C. Gallegos. 1997. "Use of the Tikhonov Regularization Method for the Linear Viscoelastic Characterization of Emulsifying Systems," *CIT-Información Tecnológica*, 8: 333–340.

23. Tschoegl, N. W. 1989. *The Phenomenological Theory of Linear Viscoelastic Behavior,* Berlin: Springer Verlag.

24. Honerkamp, J. and J. Weese. 1993. "A Non-Linear Regularization Method for the Calculation of Relaxation Spectra," *Rheol. Acta*, 32: 65–73.

25. Kaye, A. 1962. "Non-Newtonian Flow in Incompressible Fluids," College of Aeronautics, Cranfield, CoA Note No. 134.

26. Bernstein, B., E. A. Kearsley, and L. J. Zapas. 1964. "Thermodynamics of Perfect Elastic Fluids," *J. Res. Natl. Bur. Stand.*, 68B: 103–113.

27. Larson, R. G. 1988. *Constitutive Equations for Polymer Melts and Solutions*, Boston: Butterworth-Heinemann.

28. Wagner, M. H. 1976. "Analysis of Time-Dependent Non-Linear Stress-Growth Data for Shear and Elongational Flow of a Low-Density Branched Polyethylene Melt," *Rheol. Acta*, 5: 136–142.

29. Campanella, O. H. and M. Peleg. 1987. "Analysis of the Transient Flow of Mayonnaise with a Coaxial Viscometer, *J. Rheol.*, 31: 439–452.

30. Laun, H. M. 1978. "Description of the Non-Linear Shear Behavior of a Low-Density Polyethylene Melt by Means of an Experimentally Determined Strain Dependent Memory Function," *Rheol. Acta*, 17: 1–15.

31. Mackley, M. R., R. T. J. Marshall, J. B. Smeulders, and F. D. Zhao. 1994. "The Rheological Characterization of Polymeric and Colloidal Fluids," *Chem. Eng. Sci.*, 49: 2551–2565.

32. Liang, R. F., M. R. Mackley. 1994. "Rheological Characterization of the Time and Strain Dependence for Isobutylene Solutions," *J. Non-Newtonian Fluid Mech.*, 52: 387–405.

33. Wang, C. F. and J. L. Kokini. 1995. "Simulation of the Non-Linear Rheological Properties of Gluten Dough using the Wagner Constitutive Model," *J. Rheol.*, 39: 1465–1482.

34. Groetsch, C. W. 1984. *The Theory of Tikhonov Regularization for Fredholm Equations of the First Kind*, London: Pitman.

35. Press, W. H., S. A. Teukosky, W. T. Vetterlingm, and B. P. Flannery. 1992. *Numerical Recipes in C*, Cambridge: Cambridge University Press.

36. Soskey, P. R. and H. H. Winter. 1984. "Large Step Shear Strain Experiments with Parallel-Disk Rotational Rheometers," *J. Rheol.*, 28: 625–645.

37. Papanastasiou, A. C., L. E. Scriven, and C. W. Macosko. 1983. "An Integral Constitutive Equation for Mixed Flows: Viscoelastic Characterization," *J. Rheol.*, 27: 387–410.

26 Rheological Properties of Food Materials under Standard and High Pressure and Temperature Conditions

A. Gaspar-Rosas

CONTENTS

Abstract

The fundamental definition of rheology indicates that the flow of any material cannot occur before its original structural composition exceeds a critical deformation.

When heat and pressure are applied, the rheology of many materials can change significantly. While rheological testing is easily and commonly performed at standard temperatures and pressures, the information obtained should be carefully considered because of the nonlinear rheological behavior of materials against temperature and pressure. The goal is to determine characteristic parameters under conditions similar to processing, application, storage, packaging, etc.

A new pressure cell device enables the performance of rheological tests under actual processing, storage, and/or application conditions. Extrapolating traditional

1-56676-963-9/02/$0.00+$1.50
© 2002 by CRC Press LLC

data to predict high-temperature and -pressure performance is no longer necessary. Sample boiling and evaporation no longer occur. Testing can now be performed at temperatures up to 300°C and at pressure up to 150 atm.

Technological advancements incorporated in current testing rheometers now allow determination of the strength of even the weakest structures under a wide number of controlled conditions (temperature, pressure, flow, UV light, humidity, electromagnetic field, etc.). In all cases, testing protocols must consider the rheological history imposed on the sample during handling before a final viscoelastic behavior is assigned to the sample under evaluation. In addition, it is now clear that the viscoelastic properties of a material depend on the relationship that exists between its characteristic response time and the experimental time allowed for its observation. Even fluids can be made to appear as viscoelastic solids.

The characteristic parameters (τ_y, p, η, G', G'', tan δ, λ, γ_c, etc.) obtained from each rheological test are interwoven and complement each other.

This paper describes procedures for reliable characterization of food products under normal and high conditions of pressure and temperature. Tomato ketchup was used as the representative food product.

26.1 BACKGROUND

In 1929, the American Society of Rheology accepted the definition of rheology as the science that studies how materials deform and flow under the influence of external forces. Based on the equations (basic linear concepts) of Robert Hooke,

$$\tau = G^* \bullet \gamma \tag{26.1}$$

and Isaac Newton,

$$\tau = \eta^* \bullet \gamma' \tag{26.2}$$

where τ = the applied shear stress (external force per unit of area)
  γ = the resulting shear strain (deformation)
  γ' = the resulting shear rate (flow)

The rheological properties G^* (shear modulus) and η^* (viscosity) of materials result from the form the shear stress and the shear strain or shear rate relate to each other. Without the asterisk, G and η represent properties determined when the shear stress applied produces a steady flow. With the asterisk, G^* and η^* represent properties determined when the shear stress is of an oscillatory nature and, consequently, the type of flow imposed to the sample. Because most materials do not behave according to the linear concepts of Hooke (ideal solid, $\eta \sim 0$) or Newton (ideal fluid, $G \sim 0$) due to their fundamental composition, their properties are considered viscoelastic.

With oscillatory flow testing, the magnitude of the oscillatory deformation is kept small and carefully controlled. Under such conditions, the strength of the original internal structure of the material can be accurately determined. When the properties of the structure are graphically presented vs. strain, a linear viscoelastic region reveals the properties of the structure, and the point where the behavior becomes nonlinear indicates the initiation of flow and structural rearrangement.

Steady flow testing allows for significant structural changes, and the resulting behavior can be labeled as Newtonian, pseudoplastic, and/or dilatant. Because a material cannot flow before its internal structure exceeds a critical deformation, rheological characterization should include oscillatory and steady flow testing.

26.1.1 OSCILLATORY FLOW

The functional representation for the oscillatory shear stress applied to the sample is

$$\tau(t) = \tau_0 \bullet \sin \omega t \tag{26.3}$$

The resulting oscillatory shear strain or shear rates are

$$\gamma(t) = \gamma_0 \bullet \sin \omega t \tag{26.4}$$

$$\gamma'(t) = \gamma'_0 \bullet \cos \omega t \tag{26.5}$$

A general arrangement of Equations (26.4) and (26.5) can lead to:

$$\gamma(t) = \gamma_0 \bullet \sin(\omega t + \delta) \tag{26.6}$$

and

$$\gamma'(t) = \gamma'_0 \bullet \sin(\omega t + \delta) \tag{26.7}$$

where δ is the phase angle between the oscillatory shear stress and the resulting deformation or flow. When a material behaves as a viscoelastic solid, $\delta = 0$, and when it behaves as a viscoelastic fluid, then $\delta = \pi/2$. Mathematical arrangement of the ratio of the periodic functions of Equations (26.3) and (26.6) yield the complex shear modulus,

$$G^* = \tau(t)/\gamma(t) \tag{26.8}$$

where

$$G^* = G' + iG'' \tag{26.9}$$

and the ratio of Equations (26.3) and (26.7) produce the complex viscosity,

$$\eta^* = \tau(t)/\gamma'(t) \tag{26.10}$$

where

$$\eta^* = \eta' - i\eta'' \tag{26.11}$$

The interrelation between the complex expressions is given by

$$G^* = -(i\omega\eta^*) \tag{26.12}$$

The components of the complex shear modulus known as storage modulus (G' or elastic component) and loss modulus (G'' or viscous component),[2] when presented as a ratio are:

$$\tan(\delta) = G''/G' \tag{26.13}$$

describing the way the sample dissipates ($G'' > G'$) or stores energy ($G' > G''$).

Thus, the rheological properties of the structure of a material at apparent conditions of experimental time (ω_o), external force (τ_o); and shear deformation (γ_o) are basically defined by G', G'', and $\tan(\delta)$.

26.1.2 Steady Shear Flow

Steady shear flow results from increasing the shear stress applied to the sample vs. time,

$$\tau(t) = \tau_o \bullet t \tag{26.14}$$

in such a way that enough time is allowed for the sample to reach steady-state flow. Otherwise, the information obtained should be considered as transient responses. Steady flow can also be produced by adjusting the shear stress until the specified shear rate is achieved. In both cases, controlling the shear stress or demanding specific shear rates, the relationship between the characteristic response time of the material (λ) and the experimental time (t_o), known as the Deborah number,

$$De = \lambda/t_o \tag{26.15}$$

should definitively be smaller than unity. However, when the De is larger than unity, additional and revealing information can also be obtained, but one must be aware of the testing conditions.

26.1.3 Oscillatory Shear Flow

Oscillatory shear flow results from controlling either the amplitude of the shear stress at a constant frequency over time,

$$\tau(t) = \tau_0 \bullet \sin(\omega_o t) \tag{26.16}$$

or the amplitude of the shear strain or deformation,

$$\gamma(t) = \gamma_0 \bullet \sin(w_o t) \tag{26.17}$$

When the shear stress is controlled, the resulting shear deformation is measured; when the shear deformation is controlled, the required shear stress is measured.

In general, oscillatory shear testing produces information about the structural properties of the material, and steady shear testing reveals how the structure reorganizes to make the material flow.

26.2 NORMAL CONDITIONS RHEOLOGY

Rheological characterization of a commercial tomato ketchup sample was determined with a Paar Physica MCR300 rheometer (torque range of 0.0005 to 150 mNm and angular resolution of < 0.3 μrad) using a cone-plate measuring system (50 mm diameter, 2° angle), controlling the temperature of the sample (22°C) with a Peltier thermal unit TEK150P and ambient pressure conditions. To protect the sample against dehydration and air currents that could introduce temporary temperature gradients, a special hood was placed around the sample and measuring system. Sample volume, air bubbles, and the cleanliness of the surfaces in contact with the sample were methodically controlled. To standardize the handling procedure of the sample, and as a way to impose the same rheological history on the structure, the sample was pre-sheared at 50 s^{-1} for 60 s, allowed to rest for 90 s for thixotropic recovery, and then the desired rheological test was performed.

The rheological response of the sample to increasing levels of steady shear is shown in Figure 26.1. The viscosity curve clearly shows a Newtonian region that defines the consistency of the original structure, followed by a pseudoplastic or shear thinning region that defines the consistency of a continuously changing structure. Although not shown, the shear thinning reaches a maximum level and a shear induced Newtonian region develops. The shear stress curve shows the point (τ_y) where the force of the internal structure is exceeded by the external force, causing restructuring, which yields flow.

The viscous response can be accurately characterized by a Carreau mathematical model to reduce the information to a few significant variables (η_0, η_∞, p, a), where η_0 represents the initial viscosity of the original structure of the sample (also known as the zero shear viscosity], η_∞ represents the terminal viscosity at very high shear rate (also known as the limiting viscosity), p represents the flow index and is associated with the pseudoplastic region, and a represents a material consistency parameter. At specific conditions of temperature and pressure those variables, along with the yield stress τ_y and the critical maximum deformation γ_c, can be used for quality control, formulation, and product development, and other industrial purposes. For materials like ketchup, the Carreau mathematical model should be modified to include the yield stress property of the material.

The shear stress value that produces the critical maximum deformation characteristic of the original structure of the sample is presented in Figure 26.2. The change

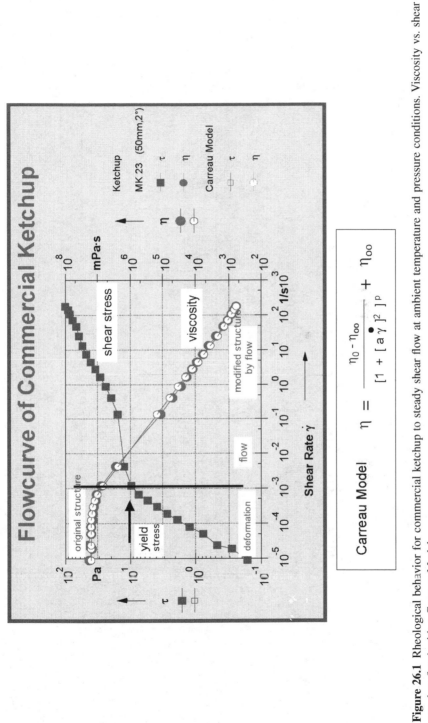

Figure 26.1 Rheological behavior for commercial ketchup to steady shear flow at ambient temperature and pressure conditions. Viscosity *vs.* shear rate data fitted with a Carreau Model.

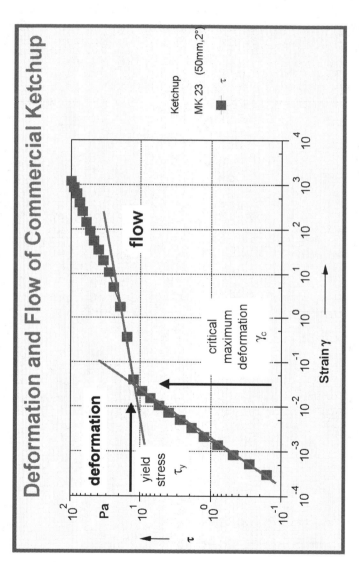

Figure 26.2 Determination of critical maximum deformation (γ_c) and yield stress (τ_y) for commercial ketchup at ambient temperature and pressure conditions.

in slope shows a transition point where shear deformation turns into shear flow. This point is also known as the yield point of the material.

The storage (G') and loss (G'') modules resulting from oscillatory testing describe the viscoelastic properties of the structure of the sample. Even though, for the evaluated ketchup, using the selected MCR300 rheometer, steady shear testing can define an initial Newtonian region, the viscoelastic properties of the structure are accurately described by the oscillatory amplitude sweep test.

The region where the properties G' and G'' show a constant behavior defines the linear viscoelastic region (LVER), characteristic of the original structure of the material. When in the LVER $G' > G''$, the structure of the material is defined as a viscoelastic solid, and when $G'' > G'$, the structure is defined as a viscoelastic fluid.

The response of ketchup to the amplitude sweep test at constant angular frequency of 6.28 rad/s (1 Hz) is shown in Figure 26.3. The curves clearly show the LVER and the point where the original structure stops deforming and starts flowing. At low deformations, the loss factor, tan (δ) is smaller than 1, and the original structure presents properties of a viscoelastic solid.

Figure 26.3 also shows that the critical deformation (γ_c) from oscillatory testing matches the value obtained from steady shear (rotational) flow testing. The curves also show that when the material flows, the elastic component G' drastically decreases as the internal structure reorganizes to present less and less resistance to flow. As the value of G' decreases with the increase of deformation, and the difference between G'' and G' becomes more than one order of magnitude, the contribution of the elastic component (G') is less and less significant.

The response of the structure of a material to different experimental times can be obtained from the frequency sweep test. The test is typically performed at a constant shear strain. Magnitude is a value that can be located inside the linear viscoelastic region of the material. The frequency sweep response for the ketchup sample investigated is presented in Figure 26.4. The shear strain selected for the test was 1%, and the allocated testing time over the frequency range (1–300 rad/s) was logarithmically controlled to ensure steady-state conditions and sufficient signal integration time. Thus, at low frequency, the testing time was long, while, at high frequency, the testing time was short to prevent shear heating in the structure. The curves show that as the frequency increases, the experimental time decreases, and the rheological properties G' and G'' increase because of a conflict that develops between the characteristic time of the material (λ) and the experimental time ($1/\omega$) applied to each testing apparent condition. Figure 26.4 also shows the values of G', G'', and tan (δ) from the selected point from the LVER of the amplitude sweep test of Figure 26.3.

26.3 THE PRESSURE CELL

When heat and pressure are applied, the rheology of many materials can change significantly. Rheological testing performed at standard temperature and pressure should be carefully considered because of the nonlinear rheological behavior of materials against temperature and pressure. To determine the behavior of materials

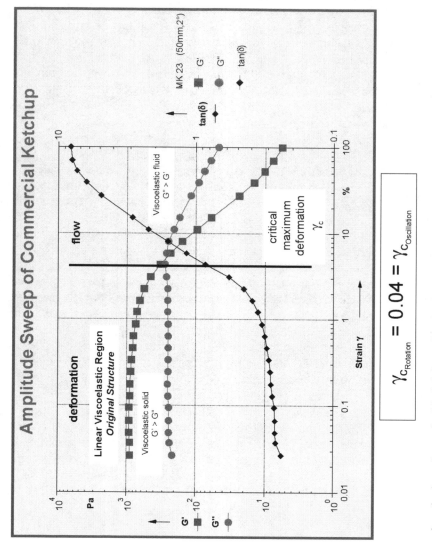

Figure 26.3 Linear viscoelastic region and critical deformation (γ_c) for ketchup at 22°C and normal pressure from the amplitude sweep test.

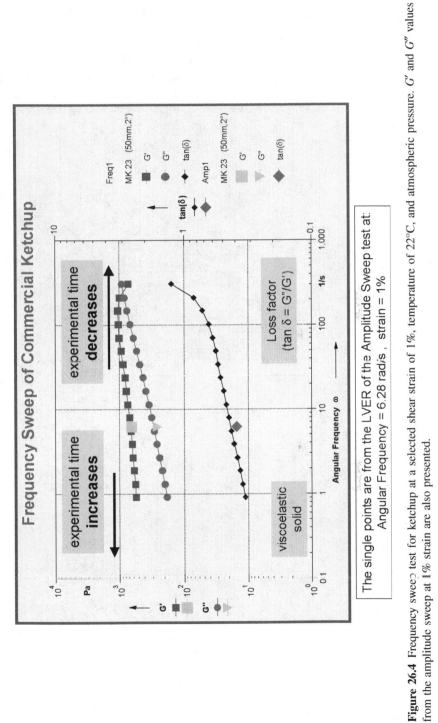

Figure 26.4 Frequency sweep test for ketchup at a selected shear strain of 1%, temperature of 22°C, and atmospheric pressure. G' and G'' values from the amplitude sweep at 1% strain are also presented.

at elevated conditions of temperature and pressure to simulate processing, applications, storage, and packaging conditions, a pressure cell device can be used. Testing can be performed at temperatures up to 300°C and pressure up to 150 atm. Extrapolating traditional data to predict high-temperature and -pressure performance is no longer necessary. Sample boiling and evaporation no longer occur. When temperature and pressure are added as variables, the properties of the sample are determined using the same fundamental principles of Hooke and Newton, and the information obtained at specific conditions is still considered as apparent. Figure 26.5 shows the schematic of the pressure cell used to measure rheological properties of ketchup at elevated temperature and pressure.

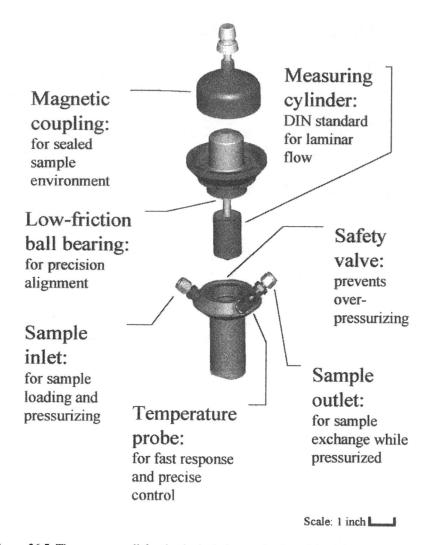

Magnetic coupling: for sealed sample environment

Measuring cylinder: DIN standard for laminar flow

Low-friction ball bearing: for precision alignment

Safety valve: prevents over-pressurizing

Sample inlet: for sample loading and pressurizing

Temperature probe: for fast response and precise control

Sample outlet: for sample exchange while pressurized

Scale: 1 inch

Figure 26.5 The pressure cell for rheological characterization of food products.

The pressure cell allows performance of rheological tests on samples at high pressures and temperatures. The technical specifications of the pressure cell used for testing ketchup samples are

- Pressures up to 150 bars (2200 psi)
- Temperatures up to 300°C
- Stresses from 0.1 Pa to 3400 Pa
- Shear rates up to 1300 s⁻¹

A low-pressure version P_{max} = 6 bar (88 psi) and T_{max} = 140°C at all pressures can also be used. The performance of the pressure cell is presented in Figure 26.6.

26.4 HIGH-TEMPERATURE AND -PRESSURE RHEOLOGY

Food products are often processed at temperatures above 100°C in sealed chambers to preclude boiling and drying (ketchup is processed at 125°C), stored at sub-ambient temperature (5°C) in sealed containers to extend shelf life, and stored in households at room temperature (25°C), in sealed containers, and applied at room temperature out of the containers. The rheology of ketchup in the pressure cell was determined at the above-indicated three temperatures.

To identify the effects of temperature on stress required for the structure of the material to yield into flow, the shear stress was logarithmically increased. The response of the material at each testing temperature is described in Figure 26.7 in terms of strain and stress. The curves of the figure show that, at storage and application temperatures, elastic properties of the structure are quite similar, but, at elevated temperature, the initial structure is weaker and flows more readily. The viscosity curves of Figure 26.8 show similar flow properties at storage and application temperatures but significantly lower viscous properties at the processing temperature.

High pressure and high temperature are often used to reduce the cooking time of some food products. Some complex fluids (foams and polymer melts) are often

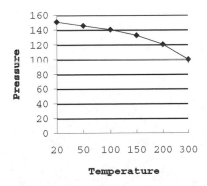

Figure 26.6 Pressure and temperature performance of the pressure cell for rheological testing.

Tomato Paste: Elasticity and Yield Stress

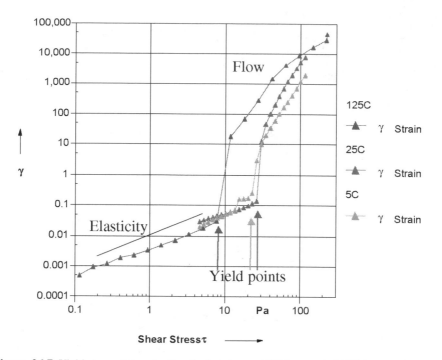

Figure 26.7 Yield stress determination for ketchup at 5°C (storage), 25°C (storage/application), and 125°C (processing). Sample contained in pressure cell. Reprinted from Reference 1 with permission from Physica Messtechnik GmbH.

applied at pressures well above atmospheric conditions. To demonstrate the effect of pressure on the rheology of materials, an aqueous foam commonly used in oil-well pumping were tested at two equivalent pressure conditions: surface (1 bar) and at 1 km below the surface inside a well (10 bar). The temperature of the sample was kept constant at 20°C.

The viscosity curves of Figure 26.9 show that, when pumping the foam at low shear rates (low flow), the resistance to flow presented by the foam is approximately the same regardless of the significant difference in pressure. However, as the shear rate increases (flow increases), the viscosity becomes significantly different. The pressurized foam sample flows more easily, making the pumping easier, too.

The main advantages of using a pressure cell to perform rheological testing of materials can focus on two aspects: temperature and pressure. At high temperature, testing can be performed without boiling the material, and, at low temperature, testing can be performed without moisture condensation. At high pressure, testing can mimic actual environmental conditions. Other advantages include testing in a sealed environment to prevent evaporation, chemical reaction with air and the escape of toxic fumes, monitoring of gas saturation such as CO_2 into aqueous food, mon-

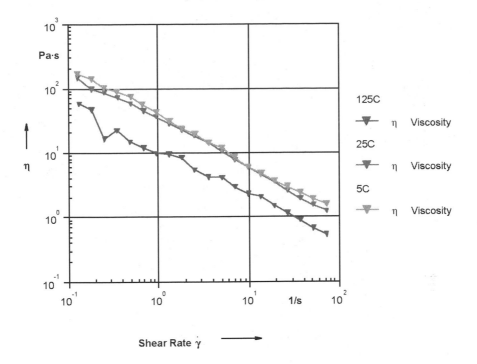

Tomato Paste: Flow Curve

Figure 26.8 Viscosity curves for ketchup at 5°C (storage), 25°C (storage/application), and 125°C (processing). Sample contained in pressure cell. Reprinted from Reference 1 with permission from Physica Messtechnik GmbH.

itoring of chemical reactions between sample and added gas, and monitoring of fast pressure changes such as during the generation of solid foam.

26.5 CONCLUSIONS

Rheology testing is a very powerful tool to describe the structural nature and properties of food products. Materials are viscoelastic by nature. Even for the weakest structure, flow cannot occur before the structure exceeds a critical maximum deformation. Rheological parameters must carefully consider the relationship between experimental time and the material characteristic response and recovery time. The rheological tests must be selected according to the material properties to be investigated or the process to be simulated. Tests can be performed with a pressure cell at controlled pressure and temperature conditions.

Parameters from rheological tests are interwoven, complement each other, and describe properties of formulations during processing, application, storage, and they can even be translated into economic value.

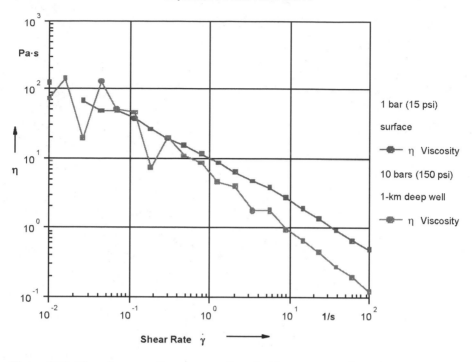

Figure 26.9 Viscosity curves from aqueous foam inside a pressure cell at 1 bar (surface) and 10 bar (1 km below surface inside a well) conditions. Temperature was kept constant at 20°C. Reproduced from Reference 1, with permission from Physica Messtechnik GmbH.

REFERENCES

1. Mezger, T. and G. Maier. 1998. *A Little Course in Rheology*, Sttutgart: PHYSICA Messtechnik GmbH.
2. Barnes, H. A., J. F. Hutton, and K. Walters. 1991. *An Introduction to Rheology*, New York: Elsevier.

27 Rheo-Reactor for Studying Food Processes: Specific Cases of Foaming and Freezing

L. Choplin

B. E. Chavez-Montes

E. Schaer

CONTENTS

Abstract

Food processing often implies important modifications of the product's rheology throughout each processing step or unit operation. We have developed a new tool, called the *rheo-reactor,* in which this rheological evolution can be followed up *in situ* under operating conditions similar to the real ones. Ice cream is a good example of a complex food that results from a process in which the raw material (mix) undergoes great rheological and structural modifications, mainly in the aerating and freezing steps. We have carried out and studied these two consecutive steps in a specially configured rheo-reactor, and the effect of relevant operating conditions on the product's rheology is analyzed in this chapter.

27.1 INTRODUCTION

Food processing usually involves significant rheological modifications of the product that are closely related to structure and quality. Industry and research constantly seek better operating conditions to obtain the best quality products by means of simple and traditional ways. These consist of observing the changes that have occurred between the inlet and the outlet of a processing unit at different controlled operating conditions and considering the processing unit as a "black box." A better understanding of what happens in the processing unit (heat exchangers, extruders, mixers, etc.) is needed, but difficulties are often confronted in performing any measurements inside the processing equipment, where the product is unreachable, or in obtaining samples for off-line examination.

Rheological follow-up of dynamic systems is possible with the use of what we have called rheo-reactors, i.e., mini-reactors installed on rheometers. In these rheo-reactors, with the help of some specific accessories, we can impose some processing conditions that are similar to real ones and obtain real time rheological information during processing. As the rheo-reactor also performs conventional rheometry, it is also possible to characterize the product at the inlet and at the outlet of an operation, *in situ* and without the need of sampling.

This experimental tool can be useful for a wide variety of applications, not necessarily food related, where there is a manifest interest in studying the rheological evolution of a medium under controlled conditions. In food processing, this is the case, for instance, in situations where dispersing a phase into another one takes place; e.g., emulsification or foaming, or where a physical or chemical change occurs such as gelation, thickening, phase transitions, etc. We have focused in this chapter on the particular cases of foaming and freezing, the two key unit operations in the fabrication process of ice cream, and we describe them from a rheological point of view.

27.1.1 THE RHEO-REACTOR

Theory

A rheo-reactor consists simply of a vessel and an impeller adapted to a rheometer. An analytical method based on a Couette analogy allows quantitative analysis of torque-rotational speed data so as to extract absolute viscosity-shear rate data using such non-conventional geometries.[1]

The Couette analogy consists first of determining a radius, R_i, of an equivalent Couette inner cylinder having the same height, $L,$ as the impeller (e.g., a ribbon), which, for a constant rotational speed, N, gives the same torque, C, in a cylindrical vessel of radius R_e. Solving the equations of change in this virtual Couette geometry, and supposing steady state, laminar regime, and isothermal conditions, for a power-law fluid,

$$\eta = k(\dot{\gamma})^{n-1} \tag{27.1}$$

where k and n are the consistency and the flow index, respectively, the following expression for R_i is obtained:

$$R_i = \frac{R_e}{\left[1 + \left(\dfrac{4\pi N}{n}\right)\left(\dfrac{2\pi k L R_e^2}{C}\right)^{1/n}\right]^{n/2}} \tag{27.2}$$

For a given set of (N, C) values, it is found that R_i is a weak function of n; therefore, the determination of R_i can be done in the particular case of $n = 1$; that is, with a Newtonian fluid of known viscosity. Once R_i has been determined in such a virtual Couette geometry, we can calculate the shear stress and the shear rate profiles, from which it is easy to deduce the viscosity at a given position in the virtual gap. The shear stress is given by

$$\tau = \frac{C}{2\pi L r^2} \tag{27.3}$$

and the shear rate, for a power-law fluid, by

$$\dot{\gamma} = \left[\frac{\dfrac{4\pi}{n}\left(\dfrac{R_i}{r}\right)^{2/n}}{1 - \left(\dfrac{R_i}{R_e}\right)^{2/n}}\right] N \tag{27.4}$$

Even for a large virtual gap, we found that a specific position $r = r^*$ exists in the gap at which the term in brackets in Equation (27.4) is essentially independent of the power-law index n—in other words, of the rheology of the fluid. This r^* value can be calculated using the Equation (27.4), taking the case of $n = 1$, and then both shear stress and shear rate are evaluated at this specific r^* value.

$$\tau_{r*} = \frac{C}{2\pi L (r^*)^2} \tag{27.5}$$

and

$$\dot{\gamma}_{r*} = \frac{4\pi N\left(\dfrac{R_i}{r*}\right)^2}{1 - \left(\dfrac{R_i}{R_e}\right)^2} \qquad (27.6)$$

The viscosity is then obtained by the ratio between τ_{r*} and $\dot{\gamma}_{r*}$, and corresponds to a shear rate equal to $\dot{\gamma}_{r*}$.

This method has been tested for several impeller combinations and for a great number of rheologically complex systems, such as food emulsions like mayonnaise, salad dressing,[2] and ice cream mix ("Results and Discussion," p. 454). The experimental results were found to fit fairly well with those obtained using conventional geometries over a wide range of shear rates, within the experimental error estimated to be no more than 5%.

This analytical method has also been transposed to small-amplitude oscillatory strain tests. In this case, the complex modulus is expressed as follows:

$$G* = G' + iG'' = \frac{\tau}{\gamma_0} e^{i\delta} \qquad (27.7)$$

Storage (G') and loss (G'') moduli can be calculated knowing the expressions of stress τ and strain γ_0, in our virtual Couette geometry, for $r = r*$ and $n = 1$. Stress is obtained directly from Equation (27.5) and strain from Equation (27.6).

$$\gamma_0 = \frac{2\theta}{1 - \left(\dfrac{R_i}{R_e}\right)^2}\left(\frac{R_i}{r*}\right)^2 \qquad (27.8)$$

where θ is the deformation angle. In the linear viscoelasticity domain, the mechanical spectra obtained with a standard geometry and with a rheo-reactor configuration compare fairly well.[2]

27.1.2 APPLICATION TO FOAMING AND FREEZING FOR ICE-CREAM PROCESSING

Ice cream is a complex frozen foam in which different dispersed phases (e.g., air bubbles, ice crystals, and partially coalesced fat globules) coexist in a freeze-concentrated solution containing mainly sugars, salts, and proteins. This particular structure is classically obtained by simultaneous and vigorous whipping and cooling of ice cream mix in a freezer. Soft ice cream at $-6°C$ is collected at the outlet before hardening and storage. The freezer is a scraped-surface heat exchanger in which it is very difficult to observe the product's behavior for many reasons, but mainly because the quickly rotating scraping blades impede the placement of any measurement device, and because residence time is short (about 30 s). This is why it has traditionally been treated like a "black box," and the freezing or foaming processes themselves, inside such exchangers, have rarely been treated. Our approach consists of separately studying foaming and freezing of ice cream mix in a particular rheo-reactor configuration that mimics the real process. This sequence has proven to be useful for controlling process parameters and product quality.[3] With this strategy, we will be able to follow in detail, for a given formulation and under controlled conditions, the product's structure buildup, first when air is dispersed into it and second when ice crystal nucleation and growth occurs. Thus, we will attempt to establish some connections between operating conditions and the product's rheological quality.

27.2 MATERIALS AND METHODS

27.2.1 Mix Fabrication and Characterization

Plain ice cream mix was prepared in the laboratory according to a simple typical recipe shown in Table 27.1. Mix processing was carried out as follows (see Figure 27.1). Ingredients were blended in a jacketed agitated vessel, batch-pasteurized for 30 min at 60°C, homogenized in two steps with a Microfluidizer M-110L processor (Microfluidics, U.S.A.) at 30 and 10 MPa, cooled in a freezer, and stored at 4°C for at least 8 h for aging. Storage duration before use of the mix never exceeded seven days.

Mix rheology was characterized by steady deformation rate and frequency sweep (1% strain, in the linear viscoelasticity domain) tests at 4°C, in the rheo-reactor and also with a double Couette geometry. Shear rates ranged from 0.01 and 100 s⁻¹, and frequency from 0.01 and 100 rad/s.

27.2.2 Foaming in the Rheo-Reactor

Mix foaming was carried out in a rheo-reactor coupled with the accessories illustrated in Figure 27.2. The rheo-reactor consisted of a cylindrical dish-bottomed 100 ml vessel equipped with a close-clearance impeller to ensure good macromixing, especially as the product became highly viscous. Three impeller geometries

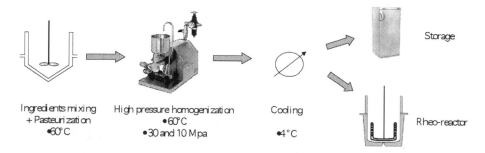

Figure 27.1 Mix fabrication process.

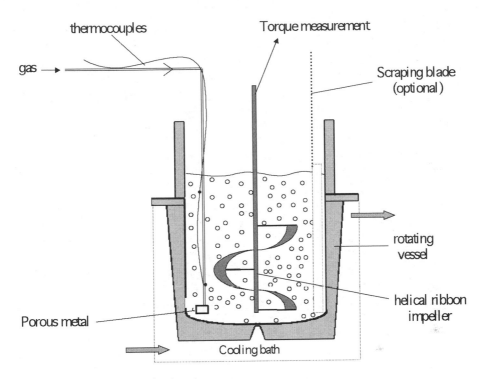

Figure 27.2 Rheo-reactor for foaming.

TABLE 27.1
Standard Mix Formulation

	% weight
Milk solids, nonfat	11
Milk fat	12
Sucrose	15
Guar gum stabilizer	0.15
Tween 80 emulsifier	0.15
Water	61.7
Total solids	38.3

can be used: a helical ribbon (L = 42 mm, D = 43 mm, noted HR1), a small helical ribbon (L = 30 mm, D = 27 mm, noted HR2) or an anchor (L = 42 mm, D = 47 mm). The vessel was installed on an RFS II Mechanical Spectrometer (Rheometric Scientific, U.S.A.), and connected to the rheometer motor. On its top, a Plexiglass® prolongation was adapted to avoid ejection of the foam formed when the vessel turned at high speed.

The accessories were a gas injection system, four flexible thermocouples, and a conductivity probe. Temperature control was ensured by a cool fluid circulating bath. Foaming was realized by injecting nitrogen gas (in replacement of air) into the mix through a small porous metal cylinder at the bottom of the rheo-reactor (see Figure 27.2) while the vessel rotated at high speed to ensure dispersion and retention of air bubbles. The pore diameter of the metallic porous cylinder had a specified value of 25 μm. Constant gas flow rate is ensured by a mass flow controller for N_2 (Aalborg GFC17, U.S.A.).

The mix was aerated at 2°C with an N_2 flow rate of 60 ml/min until foam reached approximately 100% overrun [overrun = (foam volume – initial volume)/initial volume × 100], corresponding to 50% volume of gas in the foam. Viscosity and local temperatures were meanwhile recorded, and viscosity or temperature vs. time graphs were drawn. With a simple volume calibration using the thermocouples as volume level sensors, we could also determine an approximate gas volume fraction (ϕ) at some points of the curves; thus, we also suggest interesting relationships between apparent viscosity and gas volume fraction for a non-concentrated air bubble dispersion as is the aerated mix.

During mix aeration, the following parameters were maintained constant: temperature, gas flow rate, and shear rate. The effect of shear rate, rotation speed, and impeller geometry on the foaming kinetics and on the rheology of the foam were studied. The experimental conditions for foaming experiences are shown in Table 27.2, each shear rate corresponding to a rotational speed.

TABLE 27.2
List of Operating Conditions for Foaming

	HR1				HR2		Anchor	
$\dot{\gamma}$ (s^{-1})	70	100	140	220	85	100	220	100
N (rpm)	200	290	400	600	400	480	400	175

The resulting foamed mix was characterized by means of small amplitude oscillatory strain tests. As for the mix, frequency sweeps were performed between 0.01 and 100 rad/s at 1% strain in the linear viscoelasticity domain.

27.2.3 FREEZING IN THE RHEO-REACTOR

Freezing was induced by cooling the fluid circulating in the bath to −12°C, with the use of a cryo-thermostat (CC240 Huber, 1.2 kW cryo-powered). When the mix reached its initial freezing point, ice crystals were created and grew along the vessel's wall; a scraper (see Figure 27.2, tool in dashed lines) was then used to homogenize the system, as in scraped surface exchangers. By intervals, dynamic rheological measurements were performed. Successive cycles of scraping/oscillatory testing were done. Structural development of ice cream was examined in terms of G' and G'' moduli vs. time or temperature. The complete experimental protocol is summarized in Figure 27.3.

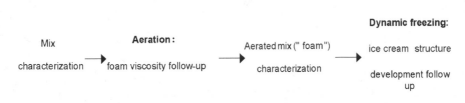

Figure 27.3 Experimental sequence.

27.3 RESULTS AND DISCUSSION

27.3.1 MIX CHARACTERIZATION

Ice cream mix is an oil-in-water emulsion of 12% fat globules with a typical diameter of 1 μm.[4] It is a non-Newtonian shear-thinning fluid (see Figure 27.4). Flow curves can be fitted to the power-law model shown in Equation (27.1) or, more accurately, to the Sisko model shown in Equation (27.9).[5]

$$\eta_{mix} = \eta_\infty + k\dot{\gamma}_{eff}^{n-1} \tag{27.9}$$

The mechanical spectrum of the mix emulsion is presented in Figure 27.5. Ice cream mix is a highly structured viscoelastic system, since G' and G'' exhibit low dependence on frequency. As mentioned in the theory section, experimental mea-

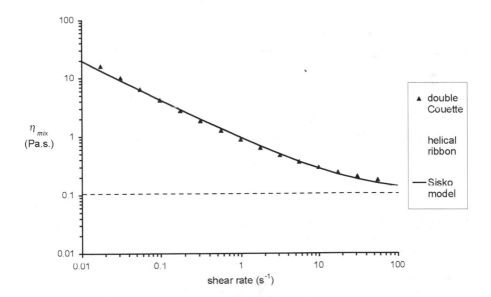

Figure 27.4 Steady rate sweep of laboratory-made ice cream mix at 4°C. Sisko model parameters are $\eta_\infty = 0.10$ Pa · s, $k = 0.851$ Pa · s^n and $n = 0.32$.

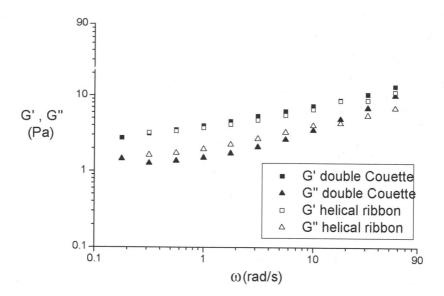

Figure 27.5 Mechanical spectra for ice cream mix at 4°C (1% strain).

surements with the rheo-reactor fit reasonably well with those obtained with a conventional double Couette geometry.

27.3.2 FOAMING

Aerated mix is not a foam in the common sense, because its gas volume fraction is lower than in common foams. Gas volume fraction in aerated mix (thus, in ice cream) does not exceed 0.6 and is typically 0.5, a value corresponding to 100% overrun.[4] Therefore, it is reasonable to assume that air bubbles should be spherical and separated enough for interactions between bubbles to be negligible. Viscosity dependence on dispersed phase volume fraction is, in the dilute regime, an extension of the Einstein's result to slightly deformable particles proposed by Taylor.[6]

$$\eta_{foam} = \eta_{mix}\left[1 + \frac{1 + \frac{5\gamma}{2}}{1 + \lambda}\phi\right] \tag{27.10}$$

where η_{mix} = continuous phase viscosity

λ = ratio of viscosity of dispersed phase to that of the continuous phase

ϕ = gas volume fraction

In the so-called foam limit case (or in the case of gas bubbles), λ is negligible, and Equation (27.10) reduces to its simpler form,

$$\eta_{foam} = \eta_{mix}(1 + \phi) \tag{27.11}$$

This result correctly predicts that the presence of tiny gas bubbles increases the viscosity of a liquid; however, as in the case of rigid spheres, this result is valid only for very low concentrations (Figure 27.6). The data for η_{mix} are evaluated at shear rates ranging from 70 s^{-1} up to 230 s^{-1} (shear rates prevailing during foaming); in other words, considering the viscous behavior of the mix (see Figure 27.4) at the high shear rate Newtonian plateau $\eta_\infty = 0.1$ Pa.s. Considering the interfacial tension (gas-mix) $\Gamma = 0.055$ N/m, and an average size of the gas bubbles $d_b \approx 100$ μm, the characteristic time for the deformation process is

$$\lambda_d = \frac{d_b \eta_\infty}{\Gamma} = 1.8 \times 10^{-4} \text{s} \tag{27.12}$$

This means that, under the aerating conditions and viscous measurement conditions, the gas bubbles will remain spherical.

The Maron–Pierce empirical equation [Equation (27.7)] for rigid spheres was also plotted for comparison purposes.

$$\eta_{rel} = \frac{\eta_{foam}}{\eta_{mix}} = \left(1 - \frac{\phi}{\phi_m}\right)^{-2} \tag{27.13}$$

In this equation, ϕ_m is the solid fraction at maximum packing, equal to 0.68 for narrow size distribution and 0.72 for broad size distribution.

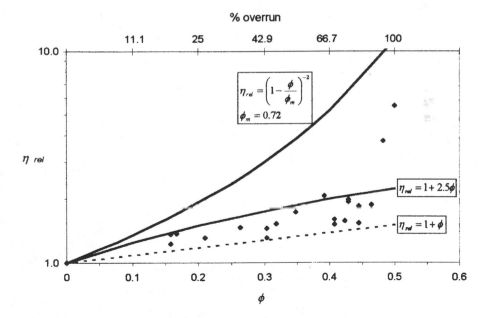

Figure 27.6 Relative shear viscosity η_{rel} vs. gas volume fraction ϕ from foaming experiments with HR1 at shear rates ranging from 70 to 220 s^{-1}.

Our results show only a slight deviation from Equation 27.11, up to a value of ϕ on the order of 0.45, where the relative viscosity seems to exhibit a sharp increase.

The mechanical spectrum of the mix is significantly modified by the presence of gas bubbles, particularly when we approach 100% overrun. Figure 27.7 illustrates the mechanical spectra of both foamed and non-foamed mix. If the frequency dependence does not seem to be modified, the mechanical spectra are markedly different. A more significant increase in loss modulus than in storage modulus can be noted. Certainly, interfacial interactions, as well as structural rearrangements of proteins and emulsifiers, should participate in these modifications of the mechanical spectrum of the mix, and further investigations are needed for understanding them.

Effect of Impeller Geometry and Macromixing Conditions

For a given impeller geometry, the rotational speed influences both the effective shear rate in the whole vessel and the macromixing conditions, especially in terms of power consumption as well as pumping capacity. The shear rate in the immediate vicinity of the gas sparger, which can be very different from the above-mentioned effective shear rate (depending on both the relative positioning of the sparger and the geometry and rotational speed of the impeller), influences the size of the gas bubbles: the higher the shear rate close to the sparger, the lower the gas bubble size. An increase of the effective shear rate, on the other hand, will increase the circulation of gas bubbles, giving them less chance to reach the free surface and, therefore, to burst. The consequence of this will be a reduction in the time needed to reach a given overrun.

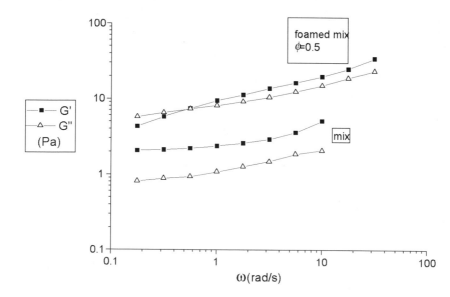

Figure 27.7 Mix and foamed mix mechanical spectra, ω = 10 rad/s at 4°C (1% strain).

This is illustrated in Figure 27.8, in the case of two impeller geometries for which, at a given effective shear rate, the time to reach 85% overrun is roughly the same. The relatively small differences in relative viscosity should be attributed to the bubble size and bubble size distribution in the foamed mix.

In another impeller geometry, still a helical ribbon impeller but smaller in size than the previous one and noted HR2, the effective shear rate in the whole vessel is calculated to be lower than those of the previous geometries at the same rotational speed. We will, however, compare the evolution of the relative viscosity of the foamed mix at a constant effective shear rate; for instance, $100 \ s^{-1}$, for three geometries, as illustrated in Figure 27.9. The rotational speeds of the impellers are indicated in Figure 27.9. Clearly, even if the effective shear rate is the same, the shear rate in the immediate vicinity of the gas sparger will be very different, and, the higher this shear rate, the smaller the bubble size will be. The pumping capacity of HR2 will be even larger than for the other two cases; consequently, the time to reach 85% overrun will be reduced (here, by a factor 2).

The significant difference in the relative viscosity obtained in the case of HR2 as compared with the values for other geometries cannot, however, be explained on the sole basis of bubble size considerations. Certainly, we should involve structural modifications of the mix, which is nothing but a macroemulsion. With the HR2 impeller, the true and local shear rate to which the macroemulsion is submitted should be calculated, and experiments should be designed so as to evaluate possible structural modifications of the unaerated mix with time.

Nevertheless, our preliminary results show a very complicated coupling between macromixing conditions and aeration conditions in the foaming of a mix. This will clearly influence the following freezing step. At this stage of our study, we will

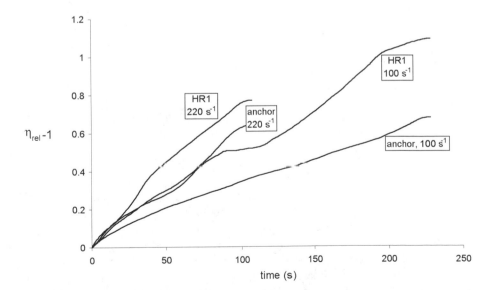

Figure 27.8 Foaming up to 85% overrun, with anchor and helical ribbon impellers.

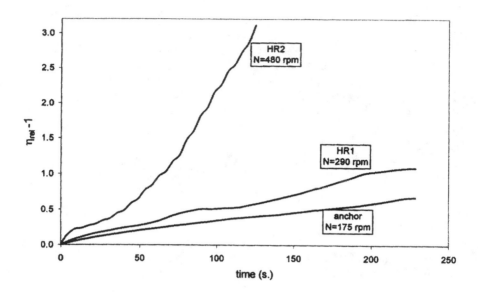

Figure 27.9 Mix aeration at 100 s⁻¹, 3°C, with different impellers.

concentrate on the freezing of a foamed mix obtained with a specific geometry (HR2) at a specific rotational speed (N = 480 rpm), or effective shear rate ($\dot{\gamma}_{eff}$ = 100 s⁻¹).

27.3.3 FREEZING

The protocol during the freezing step was as follows:

- While the temperature started to decrease, we rotated the vessel at an effective shear rate of 25 s⁻¹ to ensure a certain amount of renewal by macromixing in the vessel wall vicinity.
- From time to time, the macromixing action was interrupted, and a frequency sweep experiment (small amplitude oscillatory strain, in the linear viscoelasticity domain: 1% strain) was performed between 0.1 and 100 rad/s.

This procedure allows us to follow up *in situ* the evolution of the mechanical spectrum of the medium as the freezing proceeds. Because the moduli G' and G'' are essentially frequency independent in the experimental window considered, it suffices to obtain G' and G'' at a given frequency, 10 rad/s for instance, selected for a relatively short duration of the measurement (on the order of a few seconds).

Typical kinetics of the freezing step are shown in Figure 27.10. Up to –3°C, the moduli G' and G'' do not evolve significantly with time, although the temperature changes from 4 to –3°C. Crystallization, probably by nucleation and growth, starts

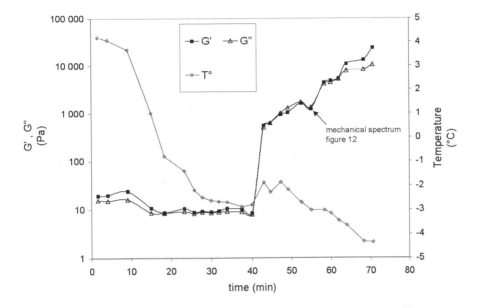

Figure 27.10 Structural development of ice cream and temperature profiles during freezing, ω = 10 rad/s (1% strain).

to occur around –3°C and is accompanied by a substantial increase in temperature due to its exothermicity.

The kinetics of crystallization can be divided into two steps, in which both moduli increase by two orders of magnitude in less than one minute, and a second and slower step with still a significant increase in both moduli by one order of magnitude over around 30 min. Note that, at –4°C the elastic modulus, G', seems to increase more than the loss modulus G''.

Even if the freezing and the freezing conditions prevailing in our rheo-reactor configuration do not necessarily represent a real freezing process, such a rheological and *in situ* follow-up is particularly interesting for evaluating how the freezing kinetics, and hence the crystal size and therefore the ice cream quality, can be influenced by operating conditions in the freezer.[8]

Another representation of the moduli can be used and is represented in Figure 27.11. These data are similar to those obtained by Goff et al.,[9] who studied the evolution of mechanical properties of ice cream while melting in a parallel plate rheometer.

In Figure 27.12, we report the mechanical spectra of the mix as well as the aerated (or foamed) mix of soft ice cream. This evolution is directly connected to the structural changes occurring during the crystallization process, provided the mechanical spectrum is the macroscopic mirror of the microstructure of ice cream, which determines its physical and sensory properties. This rheological follow-up realized *in situ* can be now studied as influenced by operating conditions during freezing.[10] This will be reported in a subsequent publication.

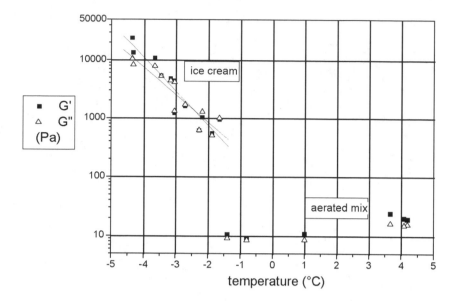

Figure 27.11 Viscoelasticity development during freezing of ice cream, $\omega = 10$ rad/s, $\gamma = 10^{-2}$.

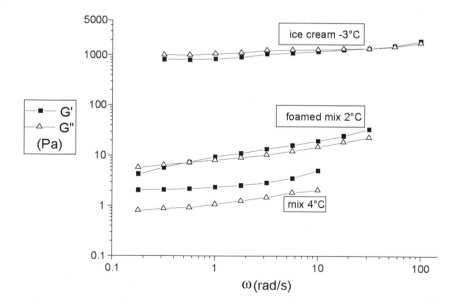

Figure 27.12 Mechanical spectra of mix, foamed mix, and ice cream (1% strain).

27.4 CONCLUSIONS

The rheo-reactor methodology appears to be a powerful tool for studying rheologically complex and evolving media like food products during their processing, as is the case for ice cream manufacturing. In this rheo-reactor, specially designed to manage the aeration and the freezing operations of this process in a consecutive sequence, we will be able to study in detail how the processing protocols and conditions could affect the structural development of ice cream and, consequently, its end-use properties.

27.5 ACKNOWLEDGEMENTS

The authors gratefully acknowledge partial financial support provided by CONACyT (Mexico) and SFERE (France).

REFERENCES

1. Ait-Kadi, A., P. Marchal, A. S. Chrissemant, M. Bousmina, and L. Choplin. 1997. "Mixer-Type Rheometry for Complex Fluids," in *Rheology and Fluid Mechanics of Nonlinear Materials*, Dallas TX, U.S.A., FED–vol. 243 / MD-Vol. 78, pp. 159–170.
2. Choplin, L. and P. Marchal. 1999. "Rheo-Reactor for *in situ* Rheological Follow-Up of Chemical or Physical Processes," *Annual Trans. Nordic Rheol. Soc.*, 7: 5–12.
3. Windhab, E. and S. Bolliger. 1995. "Combined Aerator/Freezer for Ice Cream Manufacture," *European Dairy Magazine*, 7(1): 28–34.
4. Marshall, R. T. and W. S. Arbuckle. 1996. *Ice Cream*. 5th ed. New York: Chapman & Hall.
5. Barnes, H. A., J. F.Hutton, and K. Walters. 1989. *An Introduction to Rheology*. Amsterdam: Elsevier.
6. Kraynik, A. M. 1988. "Foam Flows," *Ann. Rev. Fluid Mech*, 20: 325–357.
7. Carreau, P. J., D. De Kee, and R. D. Chabra. 1997. *Rheology of Polymeric Systems*. Munich: Carl Hanser Verlag.
8. Hartel, R. W. 1996. "Ice Crystallization During the Manufacture of Ice Cream," *Trends in Food Science and Technology*, 7(10): 315–321.
9. Goff, H. D., B. Frelson, M. E. Sahagian, T. D. Hauber, A. P. Stone, and D. W. Stanley. 1995. "Structural Development in Ice Cream. Dynamic Rheological Measurements," *Journal of Texture Studies*, 26: 517–536.
10. Russell, A. B., P. E. Cheney, and S. D. Wantling. 1999. "Influence of Freezing Conditions on Ice Crystallisation in Ice Cream," *Journal of Food Engineering*, 39(2): 179–191.

Part V

Food Structure

28 Levels of Structure and Mechanical Properties of Solid Foods

M. Peleg

CONTENTS

Abstract

Solid foods come in a great variety of sizes, shapes, and internal structures and microstructures. Their observed mechanical properties are determined simultaneously by events at very different levels, with a corresponding size from less than molecular dimensions to several centimeters, and time scales from less than microseconds to minutes or even hours. In a few cases, processes at the molecular levels have a direct manifestation in mechanical properties and, in these, mechanical tests can be used to monitor their kinetics. In most cases, however, changes in one structural level produce a cascade of changes in higher levels by what can be called a nonlinear process. Understanding the nonlinear interactions between the levels, therefore, becomes essential to controlling the texture of foods and improving the processes of their manufacture. This, however, may require the employment of complementary non-mechanical analytical methods such as biochemical assays, microscopy, X-ray diffraction, or nuclear magnetic resonance (NMR), to name a few. Certain properties are meaningful only at certain structural levels and morphology. Trying to explain them solely in terms of underlying changes in molecular states, or to control them through changes in formulation only, is not the most profitable approach. The internal organization of many foods (notably, meats, vegetable tissues, and many cereals) has been extensively studied, and the existence of structural levels is widely and well recognized. The role of the interactions between the levels, however, has rarely been the focus of the studies in the past but probably will be in the future.

28.1 INTRODUCTION

Everyone knows that the mechanical properties of a cheese block, a cheese slice, and a spoonful of shredded cheese are very different, despite that their composition and microstructure are basically the same. In this case, the differences are clearly due to size and geometry, whose effect can be anticipated in light of basic engineering principles, at least qualitatively. In other cases, the situation is more complicated. The main reason is that there are hierarchies of structures whose length scale can span over many orders of magnitude and whose interactions are not always known or fully understood. The composition–structure–properties relationship is obviously a core issue in material science and has received considerable attention in the food literature. But the interactions between different levels of structure have rarely been the focus or specific objective in food research; a study by Eads[1] is a notable exception. The objective of this particular presentation is not to survey the gamut of known effects of structural components on texture but to highlight the need to study and understand the intricate relationships between the different structural levels and how they may affect the interpretation of mechanical analyses of solid foods. We use selected examples from our own research simply because these are the most convenient; they are not more relevant or enlightening than similar studies published by other investigators. The reader can find numerous other examples where interactions are described and discussed, but usually in the context of the texture of specific foods rather than in the framework of a general concept.

28.2 STRUCTURAL LEVELS

With very few exceptions, mechanical tests are performed on specimens whose size is on the order of several millimeters to a few centimeters. Consequently, their result are the integrated record of events that take place at all smaller length scales.

The most fundamental level of matter known today are the quarks from which subatomic particles are made. But even the most ardent reductionist would not try to explain food texture in terms of quantum mechanics. The most likely basic level of interaction relevant to the mechanical properties of food is the molecule. Molecules, needless to say, determine food composition and hence its physical properties. They are also responsible for any chemical interactions that can affect the latter. Cross-linking and degradation (chemical or enzymatic) are two typical examples of how events at the molecular level can affect macroscopic properties.

Because most foods are composed of a variety of biopolymers, each having different combinations of monomers, molecular size distribution, and spatial configuration,[2] the theoretical number of possible ways in which they can interact is enormous. Nevertheless, natural foods, from meats and shellfish to vegetables and nuts, all have a characteristic texture. This, of course, is not surprising, given that the cells and tissues, and the natural biopolymers from which they are constructed, are all produced by generically controlled processes. But why man-made products such as gels, breads, and extruded cereals have a characteristic and reproducible texture is not so obvious. Possibly, this is because, under the conditions of their manufacture, the molecules cannot be ordered or aligned in any other way, due to steric or other physicochemical considerations. But an alternative, or complementary, explanation is that, once a certain degree of connectivity between the molecules is achieved (Figure 28.1), the order by which it has been reached is unimportant.[3] Connectivity is a property of an assembly, not of individual molecules, and it can explain, at least in principle, the ubiquitous phenomenon of phase transition in systems that bear not even the slightest similarity.[4,5] Whether and how the concept of connectivity can be used to explain the properties of specific food systems could be a very interesting topic for future research.

Perhaps the simplest case of molecular properties being directly responsible for the mechanical properties are food gels. But when different polymers (e.g., gluten and starch) form fibers, filaments, or films (as in baked products), the situation is by far more complicated, because the type, number, and strength of possible interactions increase dramatically. Also, the uniformity of the resulting structure at different length scales may vary considerably, and so do the actual physical dimensions of its components—the thickness and length of fibers, for example. These, however, can be strongly affected not only by the material's chemical composition but also by local mechanical and thermal histories. Because most foods are physically and, in some cases, biologically and biochemically active, changes at the molecular levels continue to take place when the food is stored, and they may be accelerated if it is allowed to absorb or lose moisture (see below).

Many foods are not composites only in the conventional sense; they can also have cellular or fibrous structures. In the first case, the cell geometry, cell size, and distribution, and the cell walls thickness and strength, determine the mechanical properties of the food. The principles of cellular solids mechanics and how they apply to natural and man-made materials are discussed in great detail by Gibson and Ashby.[6] But since the cell properties are determined primarily, although not exclusively, by the process of formation, structures with very different mechanical properties can be produced from the same raw material. For example, the texture of puffed extrudates strongly depends on the dough's initial moisture and shear history as well as on the temperature profile within the extruder. In baked goods, the spongy texture is produced and controlled by the gas release method and its ability to escape. Both the extrusion and baking processes also promote chemical reactions and, consequently, there is a nonlinear relationship between the different factors and the interactions at the different levels. The chemical reactions influence the physical properties of the system, which in turn change the environment in which the reactions take place.

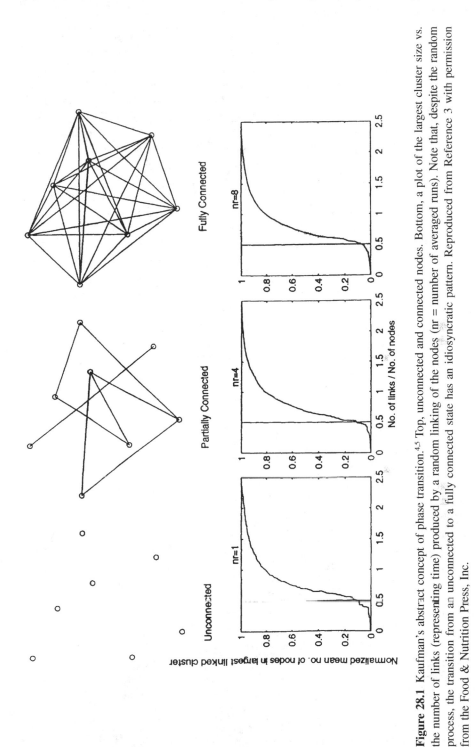

Figure 28.1 Kaufman's abstract concept of phase transition.[4,5] Top, unconnected and connected nodes. Bottom, a plot of the largest cluster size vs. the number of links (representing time) produced by a random linking of the nodes (nr = number of averaged runs). Note that, despite the random process, the transition from an unconnected to a fully connected state has an idiosyncratic pattern. Reproduced from Reference 3 with permission from the Food & Nutrition Press, Inc.

In many cases, the geometry of the tested specimen is primarily chosen for convenience and hence need not faithfully represent the food itself. For example, most solid foods are tested by compressing cylindrical specimens uniaxially. How the results are related to radial loading of the actual food is not always clear. Sometimes, the difference can be qualitative. The typical sigmoid stress-strain relationship of cellular foods, for example, may disappear if a spherical or cylindrical specimen of the latter is compressed radially. A more obvious example of a strong and qualitative scale effect are the properties of food particulates in bulk. Despite that the properties of the assembly are determined by those of its members, strength, geometry, and size distribution, to name a few, calculation of the bulk properties from those of the individual particles, or vice versa, if at all possible, may require a lengthy and complicated procedure.[7] Needless to say, such an undertaking would be totally impossible if all that is known is the particles' composition. Freeze dried and agglomerated spray dried instant coffee particles, for example, despite their similar composition, have very distinct morphology and fragmentation patterns as a result of the different ways in which they have been formed. The same applies to larger particle assemblies. The properties of bags of tortilla or potato chips, for example, are obviously very different from those of the individual chips. These, in turn, are determined by the thin morphology, and probably also by residual stresses, at least to a certain extent. Knowledge of the chips' material properties and microstructure, therefore, is clearly insufficient to predict the final product properties and its stability during handling.

28.3 DIRECT MANIFESTATION OF MICROSTRUCTURAL/MOLECULAR CHARACTERISTICS IN MECHANICAL PROPERTIES

There are food systems in which changes in the mechanical properties can be directly related to changes at the smallest microstructural level or even in the molecules themselves. In such cases, mechanical tests can be employed to monitor events in the most fundamental relevant levels of the material. An obvious example is melting, where softening of the material is a manifestation of the molecules' increased freedom to move, or textural collapse as a result of pectolitic enzyme activity. Another, but less clear, example is the phenomenon known as glass transition, wherein, as a result of heating and/or moisture sorption, biopolymers (so it is claimed) gain increased mobility. But as long as the order in which molecules or their parts are affected remains unknown, the mechanical changes cannot be traced to specific events at the molecular level. Moreover, different mechanical parameters need not change in unison, indicating that the transition itself has a "structure;" i.e., that there is a multiple effect on the different molecules, their parts, and their aggregates. Therefore, to reveal the origin of the mechanical changes, complementary analyses (NMR for example) are needed.

The tensile properties of squid mantle can serve as an example of a more intricate relation between mechanical and microstructural characteristics. The mantle is known to have a criss-crossing grid of fibers made mainly of collagen (see Figure 28.2). The purpose of this fiber array is to avoid buckling of the living animal. A similar method is used in water hoses. Because of the fibers' inclination angle (see figure), the mantle has very different tensile properties (stiffness, strength, and strain hardening pattern) in the longitudinal and lateral directions. But, after heating the mantle to 60°C, which causes melting of the collagen, not only is there a loss of overall strength, but also the differences between the properties along the two axes largely disappear.[8]

The biochemical microstructural changes that occur during rigor mortis or cooking are well known.[2] They can also be directly monitored by following the stress of a slightly stretched sample (Figure 28.3). Instead of the expected stress relaxation, a most common phenomenon in viscoelastic materials, the force actually increases in proportion to the material's tendency to shrink.[9] In fact, the phenomenon of strain hardening in flesh can also be traced to combined microstructural and molecular changes,[9] as can be seen in Figure 28.4. But, here again, although a model of the kind shown provides a conceptual framework to deal with the phenomenon, the details can be revealed only through different mechanical tests and complementary microscopic and biochemical analyses.

28.4 MACRO PROPERTIES

Some properties are only meaningful at the macroscopic level. Powder flowability is a particularly illustrative example; the particles themselves, under most circumstances, hardly deform during their transport. Their microstructure will play a decisive role, however, if in the process they undergo attrition, for example. Solid foods' strength, i.e., the stress at which they fail, is obviously determined by the foods' microstructure, structure and their integrity (see below), but also on the geometry of the imposed stresses. Thus, specimens compressed, bonded or strongly attached to the plates may exhibit very different properties than those of specimens compressed between lubricated plates.[10] Also, the apparent compressive properties inherently and strongly depend on the specimen aspect ratio[11,12] and hence, compressive tests are not always an effective tool to monitor interactions at the molecular level unless the latter result in drastic changes in the material's texture.

Failure events in most solids involve crack propagation from a weakened site, a point of stress concentration, structural defect, or the interface between structural elements, e.g., crystals. In theory and in practice, the larger the specimen, the larger the probability that such defects will be present and initiate failure. The dependence between the specimen volume and the distribution of failure strength measurements, at least in principle, can be used to estimate the density of such structural faults. The use of the statistical characteristic of mechanical measurements as a tool to monitor structural faults in foods is unknown to the author.

Moisture toughening is another example of a macro property with practical implications. Brittle cereal products are plasticized as a result of moisture sorption, and one would therefore expect that their stiffness and strength would monotonously decrease with moisture content. In at least some cereals, however, moderate levels

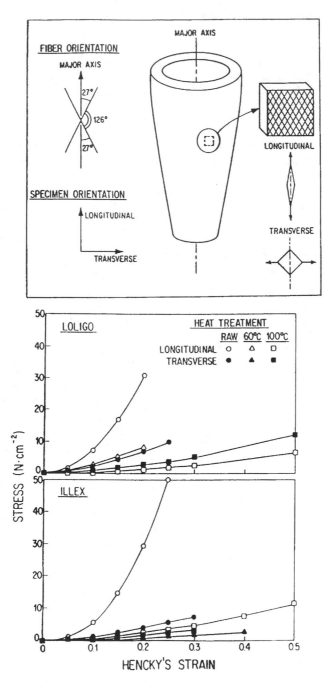

Figure 28.2 The structure and microstructure of squid mantle. Note the differences in the mechanical properties when measured in the longitudinal and lateral directions. The differences largely disappear when the collagen, which is the main ingredient of the fibers, is melted.

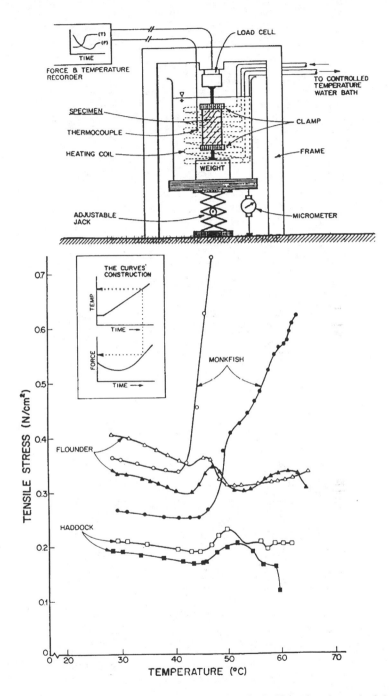

Figure 28.3 Force-time relationships of stretched fish flesh. Note that, instead of the usual force decrease accelerated by the rising temperature, the force can actually increase. Reproduced from Reference 9 with permission from the Food and Nutrition Press, Inc.

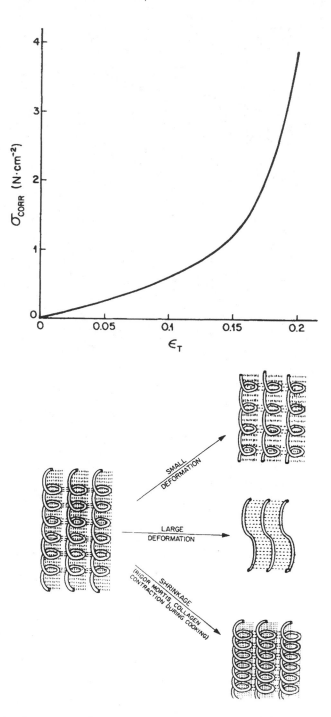

Figure 28.4 A microstructural/molecular model of strain hardening in fish flesh. Reproduced from Reference 9 with permission from the Food and Nutrition Press, Inc.

of moisture uptake result in an increase of stiffness and strength (Figure 28.5). The most plausible explanation of this phenomenon is that plasticization, which stems from the molecules' increased ability to reorient themselves, thus reducing the material brittleness, also inhibits the ability of cracks to propagate. This allows the material to absorb more mechanical energy and to develop higher stresses,[13] A similar phenomenon has been reported in gluten films[14] and probably has a similar explanation.

28.5 TYPES OF MECHANICAL PROPERTIES

Different mechanical tests reveal different properties, and these in turn are related to different structural properties. The most prominent parameter strength, the stress at failure, is determined by the presence of weak points in the structure and the ease of the failure to propagate. It should be complemented by determining the strain at failure, which is the measure of the specimen's ultimate deformability. The area under the stress-strain curve (work/volume units) is a measure of the material toughness, i.e., the amount of mechanical energy absorbed by the material prior to its failure. Stiffness, measured in terms of a modulus (stress units), reflects the structure's resistance to deformation. The rate dependence of these properties is indicative of the time scale at which different structural components respond. The change in these properties after cyclic loading can be used to probe not only the material's apparent elasticity, i.e., its tendency to recover its shape, but also the accumulation of structural damage. (Sometimes the latter is already evident in the shape of the stress-strain curve.[15]) But, with few exceptions, as already mentioned, the results of mechanical tests do not specifically reveal which structural levels are responsible for the material's response to stress. The same can be said about probes such as stress relaxation, creep, and dynamic tests. They all provide a spectrum of response times determined by reorientation and reconfiguration of structural components of different levels. Since mechanical tests are usually the easiest and least expensive to perform, tailoring the test conditions to probe the response of specific structural elements seems an attractive option. In most cases, however, solid food testing is performed under arbitrary conditions (some determined by the limitations of old testing machines) and, therefore, it is difficult to relate their results to the specimen's structural organization, let alone to interactions between different structural levels. There is no reason, however, why mechanical tests should not be tuned to probe the response of specific structural components, for example, through the selection of the deformation rate or oscillation frequency, and then confirming the results by additional tests, mechanical or otherwise. Examples are cyclic loading to different strains or testing before and after creep or stress relaxation. The number of available options is, of course, very large, and selection of the most useful combination of tests and their conditions will be a major challenge to researchers in the field.

28.6 ACKNOWLEDGEMENT

The contribution of the Massachusetts Agricultural Experiment Station in Amherst is hereby acknowledged.

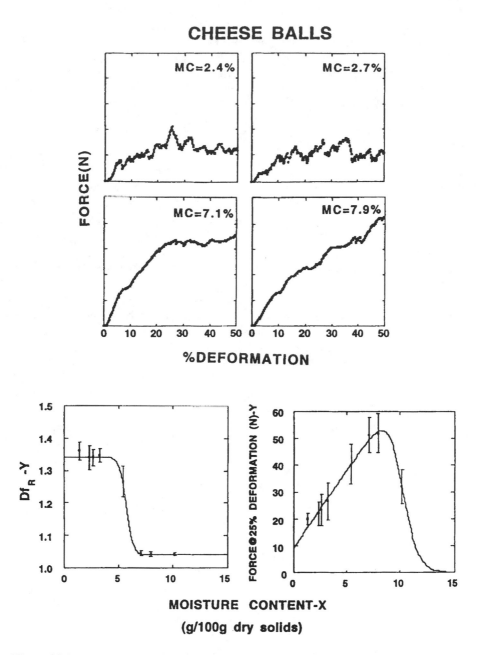

Figure 28.5 Moisture toughening in a puffed cereal snack. Note that, on moderate absorption of moisture, the loss of brittleness is accompanied by increased stiffness and toughness (top). The relationships between the jaggedness of the force-displacement curve (expressed in terms of an apparent fractal dimension), stiffness (expressed as the force at 25% displacement), and moisture contents are shown at the bottom. Reproduced from Reference 13 with permission from the Food & Nutrition Press, Inc.

REFERENCES

1. Eads, T. M. 1994. "Molecular Origins of Structure and Functionality in Foods," *Trends. Food Sci. Technol.,* 5: 147–159.
2. Belitz, H. D. and W. Grosch. 1999. *Food Chemistry.* Springer, Berlin.
3. Normand, M.D. and Peleg, M. 1998. "Kaufman's Abstract Model of Phase Transitions," *J. Texture Stud.* 29: 375–386.
4. Kauffman, S. A. 1993. *The Origin of Order-Self-Organization and Selection in Evolution.* Oxford: Oxford University Press.
5. Kauffman, S. A. 1995. *At Home in the Universe-The Search for the Self Organization and Complexity.* Oxford: Oxford University Press.
6. Gibson, L. J. and M. F. Ashby. 1988. *Cellular Solids.* Oxford: Pergamon.
7. Suwonsichon, T. and M. Peleg. 1997. "Estimation of the Mechanical Properties of Individual Brittle Particles from Their Bulk Compressibility," *J. Texture Stud.,* 28: 673–686.
8. Kuo, J-D., M. Peleg, and H. O. Hultin. 1990. "Tensile Characteristics of Squid Mantle," *J. Food Sci.,* 55: 369–371 & 433.
9. Segars, R. A., E. A. Johnson, J. G. Kapsalis, and M. Peleg. 1981. "Some Tensile Characteristics of Fresh Fish Flesh," *J. Texture Stud.,* 12: 375–387.
10. Bagley, E. B., W. J. Wolf, and D. D. Christianson. 1985. "Effect of Sample Dimensions, Lubrication and Deformation rate on Uniaxial Compression of Gelating Gels," *Rheol.Acta.,* 24: 265–271.
11. Lindley, P. B. 1979. "Compression Module for Blocks of Soft Elastic Materials Bonded to Rigid Plates," *J. Strain Anal.,* 14: 11–16.
12. Chu, C. F. and M. Peleg. 1985. "The Compressive Behavior of Solid Food Specimens with Small Height to Diameter Ratios," *J. Texture Stud.,* 16: 451–464.
13. Suwonsichon, T. and M. Peleg. 1998. "Instrumental and Sensory Detection of Simultaneous Brittleness Loss and Moisture Toughening in Three Puffed Cereals," *J. Texture Stud.,* 29: 255–274.
14. Gontard, N., S. Guilbert, and J-L. Cuq. 1993. "Water and Glycerol as Plactizer Affect Mechanical and What Vapor Barrier Properties of Edible Wheat Gluten Film," *J. Food Sci.,* 58: 206–211.
15. Peleg, M. 1987. "The Basics of Solid Foods Rheology," in *Food Texture,* H. R. Moskowitz, ed. New York: Marcel Dekker, pp. 3–33.

29 Structure–Functionality Relationships in Foods

A. H. Barrett

CONTENTS

Abstract

Very fortunately for consumers, food encompasses myriad forms, structures, and textures. Virtually any processed product is mutable through adjustment of production or formulation parameters, either of which will effect changes in product characteristics. Such market variety not only maintains consumer interest, it also keeps food scientists continually engaged in manipulating physical properties in the pursuit of new or optimized products.

In most instances, some structural characteristic of a food product bears a relationship to a functional property, particularly texture. Moreover, foods can be broadly grouped into both structural and textural categories in which characteristics particular to the category can be quantified. Relationships between structure and functionality are of value in that they provide predictive capability and thus facilitate the development of specific products; structure–functionality relationships can be additionally instructive by revealing fundamental underpinnings of important product characteristics.

29.1 ASPECTS OF STRUCTURE

29.1.1 DISTRIBUTIONAL

Food structure can be described, at the most basic level, by how much material exists in a unit volume of the product—by the product's bulk density. While variations in molecular structure give rise to differences in *intrinsic* density, *bulk* density—which is a function of overall pore volume—determines the overall, or gross, structure of foods.

However, a thorough analysis of porosity or cellularity necessitates characterization not only of total pore volume but also of the distribution of void; equal overall (percent) porosity can vary widely in how the pore volume is distributed. The material can be finely divided into a large number of spaces.

Take, for example, two cellular products with small cells, or relatively coarsely divided into a smaller number of large cells. Pore size distribution and cell number are critical structurally, because they affect the number, thickness, and length of cell wall supports (Figure 29.1) and, consequently, mechanical properties and texture. Resistance to deformation, for example, has been shown to increase with decreased average cell size as well as with increased density.[1,2]

Other kinds of food products also have distributional issues. Gels, for example, can be thought of as microcellular products in which cells contain water rather than

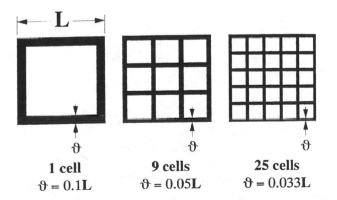

1 cell
$\vartheta = 0.1L$

9 cells
$\vartheta = 0.05L$

25 cells
$\vartheta = 0.033L$

Figure 29.1 Two-dimensional representation of increasing division of a porous structure with constant (64%) pore area. Within each illustration, pores are of equal size and uniformly spaced so that cell wall thickness is invariant; this demonstrates cell wall thinning with increasing number of pores.

air. And, as with cellular solids, the physical properties of gels are affected by the mass and distribution of the supporting phase. Macromolecular content strongly affects gel strength,[3–6] as does the fineness of the gel net: a number of studies[6–9] have demonstrated the effects of formation conditions (i.e., temperature, heating time, heating rate, pH) on gel structure, particularly on whether the gels formed as aggregated networks. The type and extent of macromolecular associations determine functional properties such as texture and water-binding capability. Barrett et al.,[6] for example, demonstrated tightening of whey protein gels of invariant solids concentration with increased heat treatment (Figure 29.2).

29.1.2 MOLECULAR

In solids, relationships observed between physical/organizational structure and mechanical properties in the dry, or glassy, state are generally not applicable once the medium absorbs moisture. Water is a powerful plasticizer and brings about a pronounced reduction of glass transition temperature[10,11] and a concomitant reduction in rheological parameters. Additionally, the plasticizing influence of other low-molecular-weight constituents is often manifest only in the presence of at least a small amount of water. Low concentrations of sucrose (2–10%) were found to reduce the average compressive stress and glass transition temperature of corn-based extrudates after equilibration of the samples at 43% RH or higher.[12] Such an effect is significant in light of the ubiquitous presence of this ingredient in bakery products. Analogously, softening due to the presence of glycerol or various sugars was observed in intermediate moisture extrudates prepared as jerky-type products.[13]

Glycerol, a common additive in shelf stable products due to its ability to reduce water activity, has additionally been shown to have a powerful effect on texture in several different systems, including shelf-stable bread[14] and intermediate-moisture meats.[15]

29.2 ASPECTS OF FUNCTIONALITY

29.2.1 TEXTURE

Texture is one of the most important functional properties of food and is a strong determinant of acceptance. Texture can be assessed either by instrumental means or sensory techniques. Instrumental measurement has the advantage of being non-arbitrary and inherently quantitative; sensory assessment has the advantage of identifying subtle differences in foods that elude mechanical testing. A complete description of product texture necessitates both techniques.

Instrumental determination of texture involves obtaining a stress-strain relationship (typically either by compressing or elongating a specimen of known dimensions), the general form of which is specific for individual material types or structures. Parameters can be derived from mathematical description of these relationships or from quantification of "failure" characteristics.

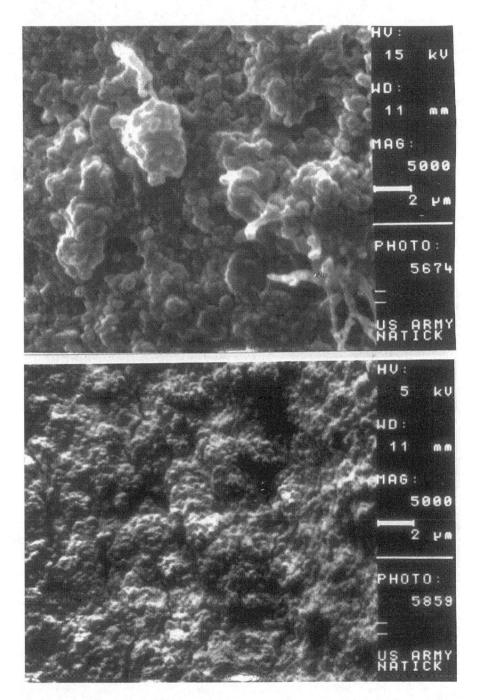

Figure 29.2 Whey protein gels formed using different heating times. Top, 20 min at 82°C, bottom, 60 min at 82°C. This shows increased network tightening with increased heating time. Reprinted from Reference 6 with permission of Elsevier Science.

Sensory description of texture involves the participation of panelists who provide answers to specific questions about the product. Panelists can be trained or untrained; their assessments can be quantitative or descriptive.

29.2.2 STABILITY

There are several facets of stability, depending on product type. In addition to the critical requirement for microbiological safety, there also exist different aspects of physical stability. For very moist foods, water associated properties are important. Gels, for example, can undergo syneresis and expel water. Rehydrated dried foods can have poor association between moisture and the matrix, rendering the product more akin to a sponge than a food with intimate incorporation of water.[16] In multilayer products, moisture can migrate between regions, depending on differences in local water activity.[6] Each of these processes will affect product quality after storage.

29.3 ASSESSMENT OF STRUCTURE

29.3.1 EXPANSION/PORE VOLUME

For porous foods, which encompass virtually all baked and most snack products, the most basic characteristics are (1) bulk density (D_B),

$$D_B = M/V_E \qquad (29.1)$$

where M = specimen mass
V_E = specimen volume

and (2) percent pore volume (V_P),

$$V_P = 1 - (D_B/D_S) \qquad (29.2)$$

where D_S = solid, or true density

All other aspects being equal, bulk density is a major determining factor for mechanical and textural properties. Gibson and Ashby[17] relate mechanical parameters such as modulus and crushing strength to a function of the ratio of solid to bulk density. Generally, materials increase in resistance to compression with increasing bulk density, since more material to deform results in more highly developed stresses. Similarly, many sensory attributes, such as firmness, are dependent on bulk density. In fact, true assessment of the effects of processing or formulation parameters on intrinsic (cell wall) strength necessitates normalization of bulk density differences.[14]

29.3.2 DISTRIBUTIONAL

Quantification of distributions can be accomplished by image processing/image analysis, using either an image from a microscope or, in the case of macroporous foods, an image directly captured from a video camera and macro lens. While different image analysis approaches have been described,[18–23] in general, this technique involves digitization of the image and designation to each pixel a value corresponding to its gray level, or darkness. A binary image is constructed by including only those pixels with gray levels between selected threshold values, and consists of silhouettes that correspond to specific features of the product. Such objects can then be counted, measured, and statistically or geometrically evaluated.

Essential to image processing is construction of a binary image that accurately represents the specimen. In particular, for cellular foods, good contrast between structural features and background is necessary. For example, measurement of cell size distributions can be obscured by the depth of the specimen, in which interior cells are only partially captured in the image. For that reason, it may useful to observe only one level in a sample at a time. Some samples can be microtomed. For others, darkening the cut surface with ink serves to highlight the cut plane; features deeper into the interior of the specimen will be relatively much lighter and can be eliminated during thresholding and construction of the binary image. Inverting the image (switching black and white pixels) allows cells to be presented as discrete black objects that can subsequently be evaluated.

For many cellular materials, such as expanded extrudates, the presence of small fractures in the structure may produce an artifact of inappropriately joined cells in the image. Application of an inadequate coating of ink to the cut surface will produce an equivalent effect. Objects that are in reality discrete, but in the image connected by even a single pixel, will be measured as one unit, producing an error in count and also in the calculation of cell size distribution. Automatic enhancement capability, such as dilation algorithms that iteratively add pixels to the periphery of objects, are now typical features of image analysis programs. Additionally, manual editing by correcting slight discontinuities in the image using a mouse can improve the quality and accuracy of the projected image.

Analysis of the enhanced (and inverted) binary image results in numerical data corresponding to each discrete object it contains. Size and geometrical data for specific objects can be extracted by selection of these features with the mouse. Image data are particularly useful with regard to distribution analysis, which is employed when a large number of cells or other objects need to be evaluated. Statistical parameters such as mean area size or perimeter, as well as various geometrical features, are thus automatically calculated.

Mathematical description of the distribution of analyzed objects will provide a more complete assessment of material structure. In particular, distribution modeling is a useful technique for processing large arrays of data such as that obtained from evaluations of cellular material. Many such distributions are wide and/or skewed. For example, the cell size distribution of extrudates has been determined to have standard deviations on the same magnitude as mean cell size values.[18] Furthermore, these distributions are significantly right-skewed,[1,18] i.e., they have a relatively greater number of small rather than large cells. Extrudate cell sizes were found to conform to both (1) the log normal distribution function,

$$F(Z) = (1/\sigma_z 2\pi)\exp\{-[(Z - \mathbf{Z})^2/2\sigma_z^2]\} \qquad (29.3)$$

where $Z = \ln(X)$
        X = area size
   $\mathbf{Z}$ and σ = respectively, the mean and standard deviation of Z

and (2) the Rosen–Rammler distribution function,

$$Gd = \exp-[(X/C)]^s \qquad (29.4)$$

where Gd = the fraction of cells larger than X
      C and S = constants[18]

29.4 ASSESSMENT OF FUNCTIONALITY

29.4.1 INSTRUMENTAL TEXTURE

Different food structures have different deformation characteristics, manifest in different shapes of stress-strain relationships. The variation in these force deformation functions is illustrated in Figure 29.3, which shows representative compressive relationships for porous-brittle, porous-plastic, and gel-like materials. As was first described by Gibson and Ashby,[17] compression functions of all cellular materials are typified by three-part curves containing a linear elastic region (at small strain levels), a region of generally non-rising stress indicating cell wall failure (at intermediate strain levels), and a densification region (at high strain levels). The type of cell wall collapse occurring at intermediate levels of deformation can consist of either plastic collapse (in the case of moist or spongy foods) or brittle failure (in the case of dry or fracturable foods). Foods undergoing plastic collapse have fairly smooth stress-strain relationships; foods undergoing brittle failure have extremely jagged appearing stress-strain relationships, in which stresses in the plateau region rapidly oscillate due to fracturing of the material (i.e., stress builds up and is iteratively released due to localized breakages in the structure). Gel-like materials have exponential force deformation functions because specimen cross-sectional area increases (barrels out) during compression, due to maintenance of specimen volume.[24]

All three types of relationships can be evaluated using appropriate mathematical functions or procedures. An expression for the three-part curve for porous-plastic foods,

$$Y = \frac{C1 \times X}{[1 + (C2 \times X)] \times (C3 - X)} \qquad (29.5)$$

was proposed by Swyngdau et al.[25] and subsequently used by Kou and Chinachoti et al.[26] and Barrett et al.[14] to describe different baked products.

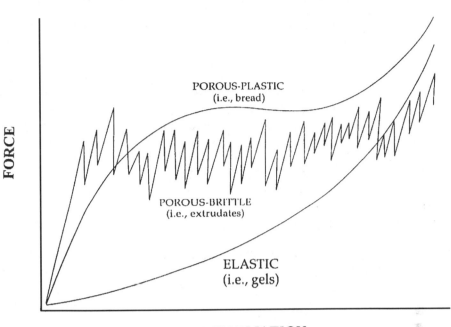

DEFORMATION

Figure 29.3 Different shapes of force deformation relationships corresponding to different types of food material.

For brittle fracturable products, techniques that quantify the ruggedness of the functions, by describing the frequency and/or magnitude of oscillations in force, were determined to be the most effective means of characterizing the materials.[1,27-31] These techniques include fractal analysis, accomplished through use of a blanket algorithm program.

$$X_{e+1}(I) = max\{X_e(I) + 1, max[X_e(I-1), X_e(I+1)]\} \qquad (29.6)$$

(in which X is magnitude at position, I, and iteration, e) that iteratively smooths the function. Fractal dimension is obtained by calculating 1 minus the slope of the relationship, log[(blanket area/2e] vs. log (2e). Extremely irregular curves will have a high negative slope, since blanket area changes rapidly with iteration as the curve is progressively filled in. Blanketing thus yields a relatively high fractal dimension, the limits of which are 2 for images; conversely, the blanket area surrounding smooth curves changes little from one iteration to the next. Thus, calculated fractal dimension is close to 1.

The fast Fourier transform can also be used to obtain power spectra of stress-strain functions, from which mean power magnitudes in specific frequency ranges can be used as indices of fracturability.[1,30]

Alternatively, the distribution of the intensity of fractures occurring during compression can be assessed using an exponential distribution,

$$Y = \text{frequency} = C \exp(-b\Delta X) \qquad (29.7)$$

in which X fracture intensity, equal to the reduction in developed stress (obtained by determining differences between subsequent stress levels in the data array), and C and b are fitted for each sample.[29] Parameters from all three fracturability techniques have been shown to be highly correlated, indicating that each method is useful in describing relative differences in fracture behavior.[29]

The force-deformation relationships for gels are typically curved rather than linear, due to barreling out of the material during compression. Since sample volume and density are maintained during deformation, a reduction in specimen height is accompanied by an increase in cross-sectional area. Modulus can be determined, however, since cross-sectional area, and thus stress, can be calculated at every point. Modulus is determined by

$$E = \frac{\text{corrected stress}}{\text{strain}} = \frac{\dfrac{F(t)[L_0 - \Delta L(t)]}{A_0 L_0}}{\ln\left[\dfrac{L_0}{L_0 - L(t)}\right]} \qquad (29.8)$$

where F = force
ΔL = deformation
L_0 = initial sample height
A_0 = initial sample cross-sectional area
t = time[24]

Other parameters of stress-strain functions can be used for comparative purposes to evaluate food texture. For example, single-point measurement of stress determined at specific strain levels is the simplest means of comparing the apparent firmness of different products. Many products also have a short linear elastic region from which a modulus can be determined. However, such procedures employ only a fraction of the stress-strain function and, while useful for a simple comparison of products, do not provide a complete description of deformation behavior.

29.4.2 SENSORY TEXTURE

Sensory texture attributes are determined, defined, and quantified by a sensory panel. Sensory panels can be trained or untrained (i.e., "consumer") and can use a variety of methods for evaluating attributes. Analysis of sensory results can, in turn, be accomplished using a variety of mathematical techniques.

One procedure for obtaining mathematically useful data from a panel involves magnitude estimation methodology.[32,33] According to this procedure, each panelist ratiometrically rates each attribute across a range of specimens. A numerical score is arbitrarily assigned to describe the intensity of the attribute in the first sample; then, the intensity of the attribute in the second sample is judged with respect to that in the first sample, and a proportional score is assigned. The same procedure is applied for all remaining samples and for all attributes. Magnitude estimate scores are then normalized across panelists (and across replicated sessions) to bring all judgments onto the same scale, and mean values are calculated for use as numerical parameters in sensory instrumental relationships.[33] Often, geometric means are employed to minimize the effect of outliers.[34] Furthermore, an array of mean magnitude estimates corresponding to a range of sensory attributes can serve as a descriptive numerical profile of the sample.

29.4.3 OTHER PHYSICAL PROPERTIES

As mentioned previously, moisture content controls the fluidity/plasticity of foods and strongly determines mechanical properties and sensory attributes. But, in some instances, textural perception is controlled by the physical binding of water as much as by total moisture content. Water binding is an important property of gelled foods, particularly of dried-and-rehydrated products (in which attachment of water to the solid matrix can be weaker than in native products). Water binding can be evaluated by a number of methods, such as expression or capillary suction,[35] or by measuring the relative diffusivity of water when the food is placed in contact with a dry material such as filter paper.[6,16]

29.5 TEXTURE/PHYSICAL PROPERTY/SENSORY PROPERTY RELATIONSHIPS

29.5.1 STRUCTURAL-MECHANICAL PROPERTY RELATIONSHIPS

Cellular Structures

The deformation properties of porous or cellular foods are strongly dependent on physical structure. In the simplest case of constitutionally similar materials, strength (i.e., modulus) is a function of density, with the specific form of the function influenced by parameters such as whether the structure is an open or closed cell.[17] Additionally, a strong dependency of strength and fracturability parameters on cell size has been reported. Barrett et al.[1] determined linear relationships between either average compressive resistance or fracturability parameters (obtained from either fractal or Fourier analysis techniques) and a function of mean cell size and bulk density (Figure 29.4). Specifically, both strength and fracturability increased with increasing density or with decreasing cell area size.

Strength quite obviously depends on density, since resistance to deformation, all else being equal, will increase with increasing mass being deformed. The negative effect of cell size on strength is most likely attributable to the relatively greater number of cell wall supports in a relatively finely divided structure, and also to the

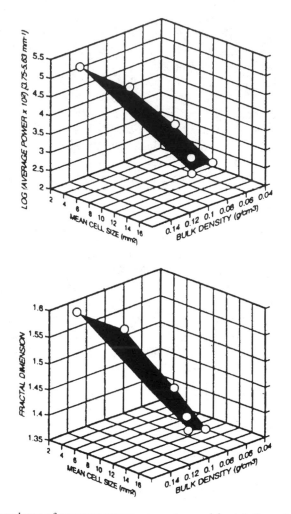

Figure 29.4 Dependence of power spectrum parameters and fractal dimension (derived from stress-strain relationships) on the mean cell size and density of corn-based extrudates. Reproduced from Reference 1 with permission from the Food & Nutrition Press, Inc.

fact that such supports are shorter than in a more coarsely divided structure. Analogously, measured fracturability may increase with density due to the relatively increased thickness of individual cell walls (which fail with greater energy); increases in fracturability corresponding to decreased cell size may be due to the presence of a greater number of cell walls (i.e., contributing to a greater frequency of fracturing).

In "nonglassy" cellular materials, i.e., those in which the structure has plasticity, texture is strongly affected by small compositional differences, as well as by physical structure. Moisture, in particular, is a critical determinant of deformation behavior, because it increases the free volume of macromolecules.[10] Specifically, increased moisture will reduce the glass transition temperature and soften the material. Moisture sorption will also change the deformation patterns of dry materials by causing the structure to fail by plastic deformation rather than by brittle fracture.[12,27,29,30] However, occasional increases in mechanical strength due to sorption of a low level of moisture, which does not completely eliminate fracturability, have been observed.[27,36] Moisture induced toughening may be attributable to differences in energy required for different failure mechanisms (i.e., brittle fracture may occur at lower levels of stress than that required for plastic deformation) or to macromolecular association that is permitted under conditions of enhanced mobility.

Other small-molecular-weight constituents have been demonstrated to have plasticizing capability, particularly in systems already containing moisture. In corn meal extrudates, sucrose was shown to lower compressive resistance and glass transition temperature (T_g) after samples were equilibrated at elevated relative humidity levels.[12] Analogous effects of plasticizers in extruded products were reported by Shogren et al.[37] and Olette et al.[38] Plasticizers, furthermore, have been demonstrated to help maintain the texture of products that typically firm during prolonged storage. For example, the elastic modulus and T_g of intermediate moisture extrudates (produced as jerky-type products) after accelerated storage were demonstrated to be lower in samples containing either glycerol or one of several sugars.[13] Similarly, glycerol was found to reduce the ultimate firmness of shelf stable bread.[14]

Gels

It is well established that the mechanical properties of gels are determined by process parameters that affect the nature and extent of macromolecular associations within the system. These interactions determine the microstructure and, ultimately, the physical properties of the gels. Molecular associations can include covalent and hydrogen bonds, and also hydrophobic interactions.[3,7,39,40]

The textural and functional properties of gels are influenced by many factors. A primary determinant of gel strength is macromolecular concentration,[3,4] and gelation parameters that affect molecular structure and interaction. These parameters include thermal treatment[4,7,8] and pH,[9] each of which influences molecular conformation and association with water. Also important is the compatibility of constituents in multicomponent gels.[41]

The mechanical strength of gels, all else being equal, will increase with increasing solids concentration.[6,42] Heat-set gels will also generally increase in hardness with increasing thermal treatment. Additionally, the rate of heating has been shown to affect macromolecular interactions and gel structure in some systems. Stading et al.[8] demonstrated that, all other factors being equal, fast heating produces fine-stranded whey protein gels, whereas slow heating produces relatively weaker, coarse-stranded structures.

Barrett et al.[6] also using whey protein gels, demonstrated that strength was a linear function of both protein concentration and heating time. Furthermore, the ability of the gels to absorb water (i.e., to swell) was directly related to gel modulus; i.e., more tightly interconnected structures, due either to greater solids content or to more extensive molecular interaction, had less ability to imbibe water.

29.5.2 SENSORY-INSTRUMENTAL RELATIONSHIPS

Many correlations linking specific sensory texture properties and mechanical or structural parameters exist. Both physical property testing and sensory analysis must be designed such that the quantitative information obtained describes pertinent properties of the food. In general, it is preferable to employ parameters from a function that describes the overall behavior of a stress-strain relationship than to use single-point parameters derived from the relationship. Textural perception, after all, relies on large deformation of the food during mastication.

For brittle cellular products, such as extrudates, sensory properties have been shown to be correlated with fracturability parameters; i.e., quantitative description of the rugosity of the stress-strain function was predictive of sensory perception.[1] Both sensory crunchiness, defined as the perceived extent (magnitude and frequency) of repeated incremental fractures during mastication, and sensory denseness, defined as the perceived amount of material per unit volume of material, were correlated with fractal dimension and power spectrum parameters derived from the stress-strain function (Figure 29.5).

Sensory firmness is a characteristic important to a variety of different food structures, and it has been shown to be correlated with parameters obtained from different types of stress-stain functions. For example, the perceived firmness of different bread samples was shown to be positively correlated with the C1 parameter obtained from fitting their stress-strain data to Equation (29.5)[14] (Figure 29.6). Similarly, the perceived firmness of shelf-stable meat sticks was shown to be positively correlated with their elastic modulus values after correction by Equation (29.7).[15]

Occasionally, functional rather than mechanical properties will determine sensory perception. In some high-moisture foods, for example, the relative strength of the association between water and the polymeric matrix is the basis for texture. Moisture can be released during mastication, thus increasing the mechanical strength and the perceived toughness of the specimen. For example, Barrett et al.[16] demonstrated that the sensory chewiness, rubberiness, and moistness of rehydrated freeze-dried eggs were positively related to how easily moisture could be removed from the samples through diffusion (Figure 29.7). Furthermore, these authors determined that the relative ease of moisture release, instead of a mechanical characteristic, influenced acceptance.

Each food system, then, should be characterized using approaches relevant to its particular deformation behavior and its particular structure. Quantification of mechanical, physical, and sensory properties is requisite for a full understanding of the effects of processing and formulation parameters on food quality.

29.6 CONCLUSIONS

Strategies for characterizing the structural and functional properties of foods are best tailored to the specific type of food under analysis. Structure can be quantified in terms of the number, arrangement, and distribution of support units; functionality

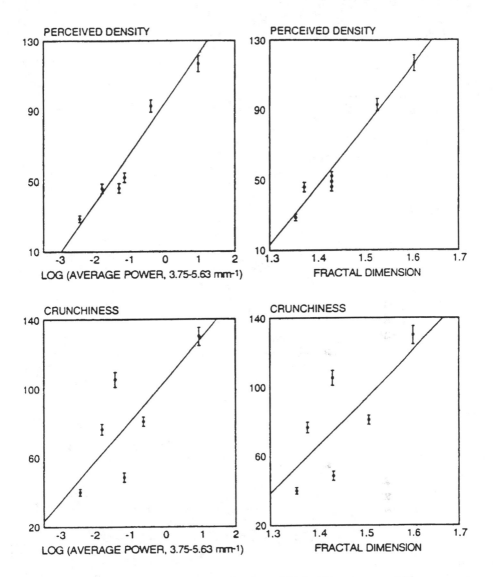

Figure 29.5 Dependence of (top) sensory density and (bottom) sensory crunchiness on power spectrum parameters and fractal dimension of stress-strain functions of corn based extrudates. Reproduced from Reference 1 with permission from Food & Nutrition Press, Inc.

can be expressed as a mechanical parameter or failure characteristic (either of which are particular to the structural category), or in some instances as a functional property such as water binding. Furthermore, relationships between structure and functionality can be determined for specific systems and employed as useful tools that streamline product development efforts.

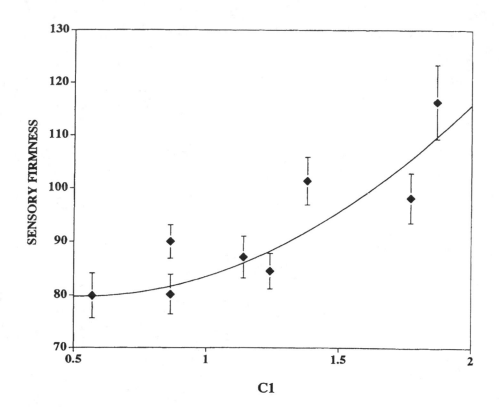

Figure 29.6 Sensory firmness of bread vs. C1 parameter. Reproduced from Reference 14 with permission from the American Association of Cereal Chemists.

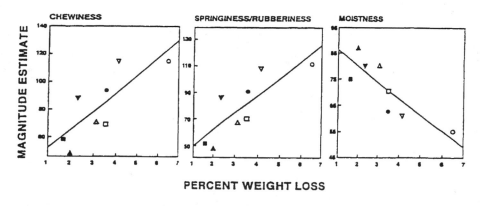

Figure 29.7 Sensory chewiness, springiness/rubberiness, and moistness of rehydrated freeze-dried eggs vs. measured moisture loss (as a percentage of original sample weight) during diffusion. Reproduced from Reference 16 with permission from Food & Nutrition Press, Inc.

REFERENCES

1. Barrett, A. H., A. V. Cardello, E. W. Ross, and I. A. Taub. 1994. "Cellularity, Mechanical Failure and Textural Perception of Corn Mean Extrudates," *J. Texture Studies*, 25: 77–95.

2. Barrett, A.H. and M. Peleg. 1992. "Cell Structure-Texture Relationships in Puffed Extrudates," *J. Food Science*.

3. Katsuta, K, D. J. Rector, and J. E. Kinsella. 1990. "Viscoelastic Properties of Whey Protein Gels: Mechanical Model and Effects of Protein Concentration on Creep," *J. Food Science*, 55: 516–521.

4. Tang, Q., P. A. Munro, and O. J. McCarthy. 1993. "Rheology of Whey Protein Concentrate Solutions as a Function of Concentration, Temperature, pH and Salt Concentration," *J. Dairy Research*, 60: 349–361.

5. Ross-Murphy, S. B. 1991. In *Polymer Gels: Fundamentals and Biomedical Applications*, D. De Rossi, K. Kajiwara, Y. Osada, and A. Yamauchi, eds. New York: Plenum Press, pp. 21–39.

6. Barrett, A. H., A. Prakash, D. Sakelakos, I. A. Taub, S. Cohen, and Y. Ohashi. 1998. "Moisture Migration in Idealized Bilayer Systems: Relationships Among Water-Associated Properties, Structure, and Texture," *Food Hydrocolloids*, 12: 401–408.

7. Katsuta, K. and J. E. Kinsella. 1990. "Effects of Temperature on Viscoelastic Properties an Activation Energies of Whey Protein Gels," *J. Food Science*, 55: 1296–1299.

8. Stading, M., M. Langton, and A. M. Hermansson. 1993. "Microstructure and Rheological Behavior of Particulate B-Lactoglobulin Gels," *Food Hydrocolloids*, 7: 195–212.

9. Langton, M. and A. M. Hermanson. 1992. "Fine-Stranded and Particulate Gels of B-Lactoglobulin and Whey Protein at Varying pH," *Food Hydrocolloids*, 5: 523–539.

10. Levine, H. and L. Slade. 1992. "Glass Transitions in Foods," in *Physical Chemistry of Foods*, H. G. Shwartzburg and R. W. Hartel, eds. New York: Marcel Dekker, Inc., pp. 83–221.

11. Roos, Y. and M. Karel. 1991. "Plasticizing Effect of Water on Thermal Behavior and Crystallization of Amorphous Food Models," *J. Food Science*, 56: 38–43.

12. Barrett, A. H., G. Kaletunc, S. Rosenberg, and K. Breslauer. 1995. "Effect of Sucrose on the Structure, Mechanical Strength and Thermal Properties of Corn Extrudates," *Carbohydrate Polymer*, 26: 261–269.

13. Barrett, A. H., M. Tsoubeli, P. Maguire, K. Conca, B. Baillargeon, Y. Wang, and I. A. Taub. 1998. "Minimizing Firming in Meat-Flour Extrudates," presented at the American Association of Cereal Chemists Meeting, September.

14. Barrett, A. H., A. V. Cardello, L. Mair, P. Maguire, L. L. Lesher, M. Richardson, J. Briggs, and I. A. Taub. 2000. "Textural Optimization of Shelf-Stable Bread: Effects of Glycerol Content and Formation Technique," *Cereal Chemistry*, in press.

15. Barrett, A. H., A. V. Cardello, P. Maguire, L. L. Lesher, I. A. Taub, A. Defao, J. Briggs, M. Richardson, and A. Senecal. 1998. "Optimization of Shelf-Stable Meatstick Texture by Adjustment of Water Activity and Fat Levels: Sensory-Instrumental Correlations," presented at the Institute of Food Technologist's Annual Meeting, June.

16. Barrett, A. H., A. V. Cardello, A. Prakash, L. Mair, I. A. Taub, and L. L. Lesher. 1997. "Optimization of Dehydrated Egg Quality by Microwave-Assisted Freeze Drying and Hydrocolloid Incorporation," *J. Food Processing and Preservation*, 21: 225–244.

17. Gibson, L. and M. F. Ashby. 1988. *Cellular Solids*. Pergamon Press.

18. Barrett, A. H. and M. Peleg. 1992. "Cell Size Distributions of Puffed Corn Extrudates," *J. Food Science*, 57: 146–148,154.

19. Moore, D., A. Sanei, E. Van Hecke, and J. M. Bouvier. 1990. "Effect of Ingredients on Physical/Structural Properties of Extrudates," *J. Food Science*, 55: 1383–1387, 1402.

20. Russ, J. C., W. D. Stewart, and J. C. Russ. 1988. "The Measurement of Macroscopic Images," *Food Technology*, February:94–102.

21. Gao, X. and J. Tan. 1996. "Analysis of Expanded Food Texture by Image Processing," *J. Food Process Engineering*, 19: 425–456.

22. Tan, J., X. Gao, and F. Hseih. 1994. "Extrudate Characterization by Image Processing," *J. Food Science*, 59: 1247–1250.

23. Smolarz, A., E. Van Hecke, and J. M. Bouvier. 1989. "Computerized Image Analysis and Texture of Extruded Biscuits," *J. Texture Studies*, 20: 223–234.

24. Nussinovitch, A., M. D. Normand, and M. Peleg. 1990. "Characterization of Gellan Gels by Uniaxial Compression, Stress-Relaxation, and Creep," *J. Texture Studies*, 21: 37–42.

25. Swyngdau, S., A. Nussinovitch, I. Roy, M. Peleg, and V. Huang. 1991. "Comparison of Four Models for the Compressibility of Breads and Plastic Foams," *J. Food Science*, 56: 756–759.

26. Kou, Y. and P. Chinachoti. 1991. "Structural Damage in Bread Staling as Detected by Recoverable Work and Stress-Strain Model Analysis," *Cereal Foods World*, 36: 888–892.

27. Barrett, A. H. and G. Kaletunc. 1998. "Quantitative Description of Fracturability Changes in Puffed Corn Extrudates Affected by Sorption of Low Levels of Moisture," *Cereal Chemistry*, 75: 695–698.

28. Barrett, A. H., M. D. Normand, M. Peleg, and E.W. Ross. 1992. "Application of Fractal Analysis to Food Structure," *J. Food Science*, 28: 553–563.

29. Barrett, A. H., S. Rosenberg, and E. W. Ross. 1994. "Fracture Intensity Distributions During Compression of Puffed Corn Meal Extrudates: Method for Quantifying Fracturability," *J. Food Science*, 59: 617–620.

30. Rhode, F., M. D. Normand, and M. Peleg. 1993. "Effect of Equilibrium Relative Humidity on the Mechanical Signatures of Brittle Food Materials," *Biotechnology Progress*, 9: 497–501.

31. Norton, C. R. T., J. R. Mitchell, and J. M. V. Blanshard. 1998. "Fractal Determination of Crisp or Crackly Textures," *J. Texture Studies*, 29: 239–253.

32. Stevens, S. S. 1953. "On the Brightness of Lights and the Loudness of Sounds," *Science*, 118: 576.

33. Moskowitz, H. 1977. "Magnitude Estimation: Notes on What, How, When, and Why to Use It," *J. Food Quality*, 3: 195–227.

34. Stevens, S. S. 1957. "A Comparison of Ratio Scales For the Loudness of White Light," Doctoral Dissertation, Harvard University.

35. Labuza, T. P. and P. P. Lewicki. 1978. "Measurement of Gel Water Binding Capacity by Capillary Suction Potential," *J. Food Science*, 43: 1264–1273.

36. Suwonsichon, T. and M. Peleg. 1998. "Instrumental and Sensory Detection of Simultaneous Brittleness Loss and Moisture Toughening in Three Puffed Cereals," *J. Texture Studies*, 29: 255–274.

37. Shogren, R. L., C. L. Swanson, and A.R. Thompson. 1992. "Extrudates of Corn Starch with Urea and Glycols: Structure/Mechanical Property Relations," *Starch/Starke*, 44: 335–338.

38. Ollette, L., R. Parker, and A. C. Smith. 1991. "Deformation and Fracture Behavior of Wheat Starch Plasticized with Glucose and Water," *J. Materials Science*, 26: 1351.

39. Sone, T., S. Dosako, and T. Kimura. 1983. "Microstructure of Protein Gels in Relation to Their Rheological Properties," in *Instrumental Analysis of Foods*, vol. 2, pp. 209–218, Orlando: Academic Press.

40. Mulvahill, D. M. and J. E. Kinsella. 1988. "Gelation of B-Lactoglobulin: Effects of Sodium Chloride and Calcium Chloride on the Rheological and Structural Properties of the Gels," *J. Food Science*, 53: 231–236.

41. Aguilera, J. M. and H. G. Kessler. 1989. "Properties of Mixed and Filled-Type Dairy Gels," *J. Food Science*, 54: 1213–1217, 1221.

42. Karleskind, D., I. Laye, F. I. Mei, and C. V. Morr. 1995. "Gelation Properties of Lipid-Reduced, and Calcium-Reduced Whey Protein Concentrates," *J. Food Science*, 60: 731–737, 741.

30 Structure and Food Engineering

J. M. Aguilera

CONTENTS

Abstract

Structure is a fundamental variable often disregarded by food engineers. In the future, application of materials science and microscopy will lead to further advances in the understanding of the role of structure on properties of foods and in processing. The relation of food microstructure to transport phenomena, rheology, mechanical properties, fabrication processes, stability, and surfaces is reviewed and discussed in this chapter.

30.1 MICROSTRUCTURE AND FOOD PROCESSING

Foods have a structure imparted either by nature or through processing. The relevance of structure in engineering is that practically all properties of foods are structure sensitive. Thus, the microstructural engineer understands food technology as a controlled effort to preserve, transform, destroy, or create structure (Figure 30.1).[1] Preserving structure is a major objective in postharvest technology of fruits and

Food Processing is a controlled effort to:

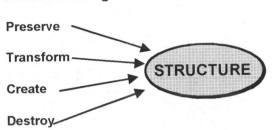

Preserve

Transform

Create

Destroy

STRUCTURE

Figure 30.1 Microstructural definition of food processing.

vegetables, as changes in structure lead to detrimental changes in texture, flavor, and even nutritional properties. The same objective is pursued by food scientists interested in preserving the quality of meats after slaughter, fish and crustaceans after capture, and cereals and legumes after harvest. Similarly, the shelf life of processed foods is largely determined by undesirable changes in the engineered structure imparted through processing.

Transforming structure is the bulk of the food processing industry. Raw materials changed into refined ingredients by primary processing of agricultural output (oils and fats, milk, cereal and grain flours, sugars and starches, among others) are then mixed and converted into traditional products representing the majority of processed foods consumed around the world, such as baked products, processed meats, dairy products, confectionery products, and many others.

Controlled destruction of structure in food processing is needed to release valuable components, facilitate handling of materials, reduce particle size and/or to prepare refined ingredients. Destroying structure during food processing is not trivial, since the microstructural characteristics often dictate the type of breakdown of the material. The food technologist also recognizes that extensive destruction of the food structure is achieved in the mouth prior to swallowing.

Creation of structure is a major task for the food industry in the next millennium. Product development and product improvement are largely based on creating structures in which nutrients are conveyed in textures and forms that appeal the consumer. While ice cream epitomizes creation of totally new and complex food structures, extrusion represents how a foreign technology can be adapted to make foams or fibrous structures from amorphous starchy and proteinaceous powders, respectively.

Unfortunately, the relevant structures are not evident to the naked eye in foods as they are in structural engineering. Linking food structure to properties is a difficult task, because foods possess complex structures, most of which have biological activity. However, the food engineer must admit that structure is a key variable that cannot be disregarded in problem solving.

This chapter attempts to summarize the role that structure will play in food engineering, be it in modeling or in the understanding of underlying phenomena, and how it can be introduced for our advantage. It also points out differences in approach that food engineering should have with respect to classical chemical engineering.

30.2 FOOD ENGINEERING AND CHEMICAL ENGINEERING

Major advances in food engineering in the twentieth century came from transfer of knowledge and technology from related fields such as chemical or mechanical engineering and physics. The impact of these technologies was largely at the macroscale through the adoption and adaptation of unit operations, mechanization, and automation of processes; use of protective packaging; etc., all of which led to the transformation of an artisan production into the high-volume and reliable food industry of today. This chapter postulates that, in the coming decades, materials science and biology will have a higher impact in food engineering and that the relevant scale of analysis will change to the submicron level. Modeling and optimization of current and future food processes will not only depend on better mathematical algorithms or faster and more powerful computers, but on a clearer comprehension of the underlying phenomena at the appropriate relevant scales.

Unfortunately, the concept of structure, which is fundamental in other engineering sciences, was largely disregarded by chemical engineers until the advent of bioprocessing. Table 30.1 lists some of the distinctive traits of classical chemical engineering and food engineering from a very broad perspective. The parental influence that classical chemical engineering has had on food engineering is disappearing, because further advances in food fabrication will be dictated by the nature of phenomena related to materials science and biology.

TABLE 30.1
Distinctive Aspects of Chemical and Food Engineering

Subject	Chemical Engineering	Food Engineering
Support sciences	PChem, chemistry, physics	PChem, biochemistry, (micro)biology, materials science
Focus	Macrolevel (unit operations)	Microlevel (properties)
Raw materials	Mostly inorganic or organic (synthesis)	Biological origin
Type of raw materials	Largely Newtonian liquids and gases	Deals mostly with solids and non-Newtonian liquids
Type of solids	Mostly homogeneous, sometimes porous, inert	Mostly structured, cellular, or phase-separated, perishable
Main products	High-volume, low-price, commodities	High number of units, specialties
Product characteristics	Purity	Desirable properties
Product stability	Largely inert or moderately stable	Highly unstable and perishable products
Packaging	Bulk packaging and handling	Unitized and protective packaging
Consumer evaluation	Price (gas), performance (plastics)	Eating properties, convenience, nutrition

As a matter of fact, the increasing role of structure in customized products from major chemical, pharmaceutical, and food industries, and the lack of exposure to the subject of graduates of chemical engineering departments, was recently addressed at a conference with participation of representatives from industry and academia.[2] It was recognized that several products from the above-mentioned industries (e.g., powders, creams, etc.) perform well in the market only because they have the right structure. In conclusion, they advocated product engineering based on structured materials, cross-fertilization from other disciplines, application of modern techniques to study structure, and a change of emphasis in the teaching of chemical engineering.

30.3 TRANSPORT PHENOMENA

Transport phenomena and quantitative modeling are central to food engineering. According to Bailey,[3] modeling does not make sense without defining beforehand what is intended to be resolved. Modeling should represent and interpret how things work, bring order to experience, clarify important interactions, and generate new hypotheses. How is it possible, then, that food engineers have paid so little attention to the physical model in the study of transport phenomena favoring black boxes over structural models? The relationship between structure and transport phenomena will be illustrated for the case of diffusion in diluted systems.

The diffusion coefficient, or diffusivity, is the main parameter in Fick's laws of molecular diffusion. Application of these laws to solid foods has been made routinely by food engineers using the homogeneous continuum approach with the objective of calculating an effective or apparent diffusion coefficient (D_{eff}), which supposedly characterizes (as the sole parameter) the mass transfer process. As revealed by data in the literature, values of D_{eff} vary by several orders of magnitude for the same material and process, and even during different stages of the same process.[4] What follows assumes that diffusivity is the main parameter in mass transfer related to foods, which may not be the case, as illustrated later.

One simple form of assessing the effect of structure in mass transfer is to compare the value of the diffusion coefficient of molecule A through a continuum B at high dilution (D_{AB}) with the effective diffusivity determined from experimental data (D_{eff}). As a first approximation, chemical engineers have done so in porous solids by correcting D_{AB} by the ratio of porosity (ε) to tortuosity (τ). For materials used in chemical engineering (adsorbents, catalysts, etc.), tortuosity varies between 2 and 6 and porosity between 0.3 and 0.8; thus, D_{eff} may be 6 to 15 times smaller than D_{AB}. In the case of leaching, experimental data suggest that D_{eff} is 0.1 to 0.9 times D_{AB}.[1]

Structural models should be preferred by engineers, because they are based on the architecture and materials properties of the intervening elements (walls, fillers, matrix, etc.). Suppose a soybean flake being solvent-extracted is modeled as a composite of impermeable walls arranged perpendicular and staggered to the flow and uniformly distributed through a continuous permeable matrix containing the oil. In this, D_{eff} depends on the volume fraction of walls (ϕ) and their aspect ratio (α),[5]

$$D_{eff} = D_{AB}\left[\cfrac{1}{1 + \cfrac{\alpha^2\phi^2}{1 - \phi}}\right] \tag{30.1}$$

Figure 30.2 shows that the ratio D_{eff}/D_{AB} can vary from 0 (when the volume fraction of impermeable walls and the aspect ratio are high) to near 1 (when walls are largely absent). Thus, in most cases where diffusion is the main mechanism of mass transfer in foods, the architecture and properties of the intervening elements may explain the magnitude of effective diffusivity. However, other effects such as those of shrinkage, change in prevailing controlling step, driving force, etc., may also contribute to a low instant diffusivity. We have studied some simple but realistic effects of microstructural architecture during extraction in a recent paper.[6]

In some cases, mass transfer occurs by mechanisms that differ from molecular diffusion, but structure continues to play a key role. One such case is washing during solvent extraction. Size reduction of cellular material necessary to achieve faster extraction rates (diffusion rates vary with the square of dimension) results in broken and damaged surface cells and a large increase in concentration of the extract on

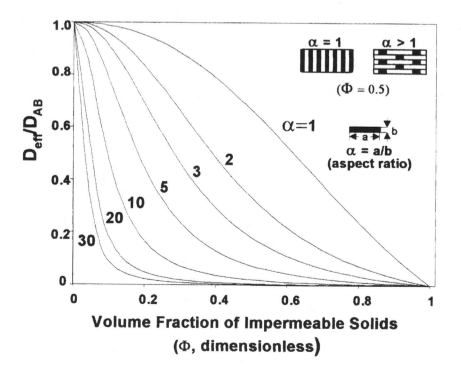

Figure 30.2 Effect of the architecture of a solid slab on the effective diffusion coefficient during extraction.

contact of the solid and the liquid. If the solvent is particularly selective for the extract, then structure can be obliterated to achieve faster and more complete extraction without impairing the downstream purification steps (e.g., oil extraction with hexane). However, when this is not the case (e.g., most aqueous extractions of biological materials), structure needs to be damaged only to the extent that facilitates the release of the desired product (e.g., sucrose extraction from sugar cane).

A notable case of mass transfer recently studied is that of deep fat frying of potato strips. After removal from the fryer, most of the oil in a fried product is wetting the surface, while some of it is retained in crevices and pores in the solid. Upon cooling, surface oil is suctioned into intercellular spaces by a negative pressure due to condensation of steam in the interior of the piece. Water, in turn, leaves the surface of the product as steam bubbles (another form of bulk transport). Confocal laser scanning microscopy has been instrumental in revealing, with minimal intrusion and large 3-D resolution, that cells in the crust of a fried potato strip remain largely intact and oil free, while oil forms an egg-box arrangement surrounding these cells.[7] This intercellular oil is easily exchanged if the fried product is fried again in another oil.[8]

In summary, transport phenomena in foods and biological material depend strongly on structure. Its study would be greatly advanced if structures were visualized and considered to derive physical models on which basic equations and mathematical methods are applied. Transport phenomena in plants, particularly mass transfer, should be understood at the tissue, cellular, and subcellular levels so that mechanisms prevailing during processing (e.g., dehydration, extraction, impregnation, etc.) are clearly identified and modeled.

30.4 RHEOLOGY

It may be hard to think that there is structure in flowing systems, but deviations from ideal Newtonian behavior can always be traced back to structural aspects of the dispersed elements in the solvent, whether macromolecules or particles, and their interactions. Rheology is extremely important in food engineering and, in practice, the engineer is confronted with multiple equations and dimensionless numbers that depend on viscosity. Most complex liquid foods are non-Newtonian fluids, among them two-phase liquid systems (emulsions), biopolymer solutions, particle suspensions, etc. Non-Newtonian behavior is characterized through a single parameter called the *apparent viscosity,* which takes different values depending on the local shear rate (γ). In other words, the instantaneous and local structure of the fluid determine its viscosity or resistance to flow, not to mention some time effects. Since interaction forces are potential forces, which are therefore elastic in nature, viscoelastic responses are also observed. Discussion of structural effects in suspensions is found in Macosko.[9]

All effects leading to nonlinear rheological behavior in nondilute systems (and sometimes in diluted systems) have a structural component as follows:

1. Hydrodynamic effects due to the modification of the velocity distribution of the flowing liquid by the presence of a particle. The extra energy

dissipated is reflected as an increase in viscosity. It is interesting to note that adding "nothing" in the form of gas bubbles actually increases the viscosity.

2. Brownian motion for particle size smaller than roughly 1 mm. In particular, rotary Brownian motion of nonspherical particles can lead to non-Newtonian effects such as shear thinning and viscoelasticity.

3. In emulsions, if hydrodynamic forces are strong enough, they can overcome interfacial tension and cause a droplet to change shape and even to break up, thus changing the structure of the system.

4. Particle shape and orientation. Hydrodynamic forces tend to align the major axis with the flow, while Brownian motion tends to randomize orientation. Again, a simple example of the effect of morphology is presented for the case of nondilute suspensions of solid particles of different shapes (Figure 30.3).[10]

5. Interparticle forces, which can be dispersion, electrostatic, and steric forces. Shear can induce gradual changes in aggregate structure, which give rise to gradual (i.e., time dependent) changes in viscosity or thixotropy.

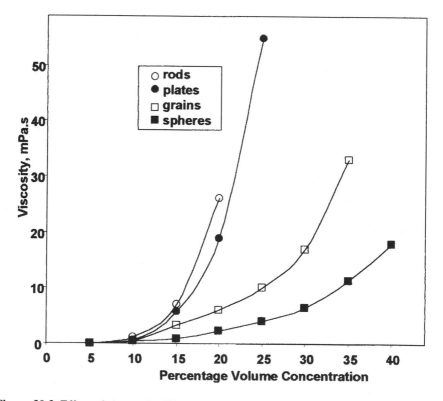

Figure 30.3 Effect of shape of solid particles on shear viscosity of suspensions.

6. Entanglements or segment to segment contacts in the case of semicon-
centrated and concentrated polymeric solutions. In this case, it is molecular structure that has to be related to rheology.

The simplest deviations from the Newtonian behavior are those presented by
so-called pseudoplastic and dilatant fluids, which have an apparent viscosity that
decreases or increases with increasing shear rate, respectively. As pointed out before,
there is a structural base for these nonlinear behaviors. For example, in pseudoplastic
fluids containing weak flocs, these could break up in high shear fields, and the
individual fragments may be capable of regrouping to give structures offering lower
resistance to shear. The specific surface of particles may be very high, and surface
forces play an important role in determining the structure of the fluid at any time.
Some suspensions may even have a finite yield stress.

Another example is polymeric suspensions in which the initial resistance to flow
is due to entanglements. As molecules become aligned along the shear direction,
the number of entanglements decreases, and macromolecules become elongated,
thus contributing to less resistance to flow. In all cases, pseudoplasticity is related
to a transition from a disordered structure into one ordered along the shear field. In
turn, dilation may occur in concentrated suspensions which, upon shearing in the
presence of insufficient liquid to fill the void spaces, create particle-to-particle solid
friction and the concomitant higher resistance to shear. Again, it is phase segregation
or a transition from a homogeneous structure into a more ordered structure that
causes dilatancy.

It should be clear by now that complex fluids possess a structure that varies with
flow conditions and over time and determines the rheological properties. Structural
models should become increasingly used in rheology, the idea being that rheological
behavior is the manifestation of the contribution of each element to flow and its
interactions. Hence, observation of structures at the molecular and supramolecular
level during flow should be correlated with rheological data to derive functional
models. Such an approach may have a high impact on product development and
ingredient substitution.

30.5 MECHANICAL PROPERTIES

Nowhere is the role of structure more apparent than in its effects on mechanical
properties of foods and, by extension, on texture. Foods having similar chemical
composition may exhibit totally different mechanical behavior, depending on the
structure that has been imparted through processing. Although mechanical models
are available for composites and cellular materials, their use in foods requires that
a structure be identified beforehand; this is where microsocopy plays a key role.
Since nonlinearity in structure-property relationships of gels arises largely by the
architecture at the microstructural level (e.g., composites or cellular gels), the effect
of structure and architecture on mechanical properties will be demonstrated for
composite gels.[11]

Protein gels filled with fat particles are composite systems that exhibit a compressive strength that depends on the size of the filler, the type of emulsion formed, and the amount of fat in the system. Fat homogenized in the presence of protein results in globules with adsorbed protein membranes, which then may interact with the protein network when the gel is formed.[12] Thus, it is possible to significantly increase the compressive strength of whey protein gels when they are filled with homogenized fat (Figure 30.4). The compressive strength also depends on the solid/liquid ratio of the fat phase at each temperature.[13]

A second case is that of mixed gels containing starch as filler and where gelatinization occurs *in situ* prior to protein gelation. For example, different sources of starch are added to surimi-based products to improve their gel strength. These filled gels show enhanced mechanical and rheological properties at low-volume fractions of starch, although starch itself has poor mechanical properties. This synergy is explained by the early removal of water from the system by swelling starch, which effectively concentrates the polymer before formation of the gel matrix. If the volume fraction of starch is such that a strong continuous network of the gelling polymer is formed, the compressive strength of the filled gel will be higher than

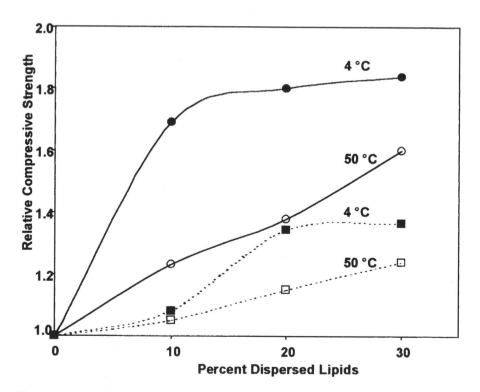

Figure 30.4 Effect of percent emulsified lipid filler and temperature on the compressive strength of whey protein gels. Dashed line = high liquid/solid ratio, solid line = low liquid/solid ratio.

that of the pure polymer (at equal total solids content). Hot-stage microscopy is instrumental in visualizing with minimum intrusion and, in real time, the initial hydration and swelling of starch granules (at ~65°C), followed by the continuous formation of a whey protein network (at 80°C) surrounding swollen granules. Since the mixed gel is a binary composite, models such as those proposed by Takayanagi have been used to predict the mechanical properties as a function of starch content and total solids.[14]

These two examples suggest a trend that studies on structuring of food products may follow in the future.[15] It begins with by a comprehension of the main mechanisms leading to structure formation studied by the most appropriate microscopy techniques and supported by other experimental data. Then, mathematical models should be derived based on available structural data and previous findings, which allow prediction of a property for any change in structure (e.g., those induced by a change in formulation).

30.6 STRUCTURAL CHANGES DURING PROCESSING

This section is not intended to review structural changes occurring during processing but to remind engineers that it almost inevitably induces microstructural changes that alter the properties of food materials. These microstructural transformations occur without a change in the gross composition of the food and may be unnoticed to the naked eye, but they can be detected by techniques normally used to evaluate physical properties (Table 30.2).

Kinetics of structural changes can be inferred from rheological or mechanical responses. Nondestructive methods such as dynamic oscillatory rheometry or mechanical spectroscopy are particularly suitable for this purpose.

TABLE 30.2
Some Structure-Related Changes Induced by Processing

Starting material	Formed structure
Native protein	Denatured proteins
Starch granules	Gelatinized or dextrinized starch
Polymeric solutions	Gels
Amorphous solids	(Semi)crystalline structures
Two-phase liquid system	Emulsion
Polymeric mixture	Phase-separated system
Polymorphic form A	Polymorphic form B
Dispersion of emulsifiers	Liquid crystals and gel phases
Elastic dough or cream	Foam
Fat globules w/natural membrane	Fat globules w/adsorbed membrane
Cells in turgor	Collapsed membranes and cells

However, the trend is toward direct observation of structural changes in real time by videomicroscopy in simulated experiments (miniaturization of processes). A hot stage mounted under the microscope lens permits controlled heating or cooling at a specified rate while structural changes are recorded (Figure 30.5). Further image processing and analysis give quantitative information of geometrical or morphological features as a function of time, the kind of data that engineers like to deal with. Conceivably, other testing devices can be miniaturized and placed under the microscope lens to directly assess changes in properties and structure with time.

30.7 STABILITY

The stability of a food system may also be controlled by structuring. As an example, many foods are stabilized at low moisture contents as metastable amorphous solids below their glass transition temperature (T_g). Chemical and structural stability are based on the extremely low mobility of the glassy state (ca. 10^{13} Pa·s). If, during storage or distribution, the temperature T exceeds T_g, then the rate of approach to equilibrium (the crystalline state) increases as $10^{[T-Tg]}$. Typical examples of materials stabilized in the amorphous state are freeze-dried bioproducts and encapsulated ingredients.[16]

Stabilization by structuring can also be achieved as water immobilization by gelation, compartmentalizing by edible polymer films, sorption of antioxidants in membranes of emulsified lipids, increased viscosity of continuous phase in emulsions, and controlled release of chemicals, among others. As fabrication techniques

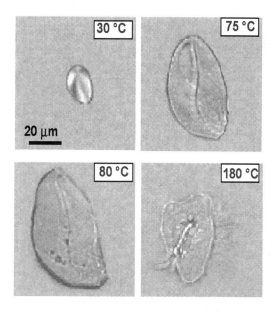

Figure 30.5 Gallery of images taken by video microscopy during frying of starch granules.

move down in scale, the food engineer may intervene and create structures that provide increased stability.

30.8   THE STRUCTURE OF SURFACES

Surface topography, or surface morphology, is an important physical property of solid foods, influencing not only their visual and sensorial aspects but also their behavior during processing and storage (Table 30.3). Surface topography is related to the specific surface area on which chemical and physical interactions may occur. Moreover, from a marketing viewpoint, the first impression a consumer gets from a food is that of its surface. Important as it may seem, surface characterization has not been addressed to a major extent by food engineers.

TABLE 30.3
Some Phenomena in which the Surface Topography of Foods Is Important

Phenomenon	Examples
Wetting	Uptake of oil in fried products
Friction	Sliding in conveyors
Optical properties	Color of freeze-dried coffee
Adhesion	Enrobbing and film coating
Sensorial properties	Licking of foods
Chemical deterioration	Oxidation of surface oils
Moisture adsorption	Caking of amorphous powders
Dissolution	Water migration into solids
Attrition or erosion	Breakdown of powders

The problem is that a surface may look smooth to the naked eye but be quite rough at higher magnifications and, at this level, roughness may have technological implications. For instance, the brain may perceive as "sandy" surfaces that look smooth but expose to the tongue particles in the order of 30 to 50 μm. The questions to be answered are

- At which scale does the smooth-to-rough transition occur?
- Beyond this point, how rough is the surface?

Material scientists have successfully characterized surfaces in an ample range of dimensions to understand topographically related phenomena in engineering materials by using recent advances in acquisition techniques of topographic data [e.g., scanning laser profilometry system (SLPS), interferometry, atomic probe, and confocal scanning microscopy, among others] that generate heights (z) as a function of position (x, y) at high levels of resolution. The process is completed by computing and data handling for analysis.[17]

Area-scale analysis by the patchwork method determines apparent areas of surfaces over a range of scales by repeated virtual tiling exercises with triangular patches of progressively smaller areas. After a sufficient number of iterations of the virtual tiling exercise, the result is expressed as a log-log plot of relative area (apparent area divided by the projected area) vs. the area (scale) of the triangular patch. This scale-sensitive fractal analysis allows the calculation of two kinds of characterization parameters (Figure 30.6). One parameter is the smooth-rough cross-over scale (SRC). The SRC is the scale above which the surface appears smooth and is easily characterized by Euclidean geometry, and below which the surface appears rough and can be characterized by fractal geometry. Figure 30.6 shows that at scales above the SRC the relative areas are close to 1, and below the SRC are markedly greater than 1. The SRC has a clear physical interpretation in that interactions with the surface at scales above the SRC see the surface as smooth, while interactions at scales below the SRC see a rough surface. The other parameter is complexity which for area-scale analysis is the area-scale fractal complexity, ASFC. The complexity parameter is equal to −1000 times the slope of the respective log-log area plots. It is related to the fractal dimension, which is 2 minus the slope of the area-scale plot. A large ASFC is an indication of higher complexity, intricacy, or roughness of the surface at scales somewhere below the SRC. If the data exhibits multifractal behavior, there may be different slopes that describe the plot over different scale ranges.[18]

Images derived from topographical data sets acquired by SLPS for the surface of the crumb of bread and chocolate are shown in Figure 30.7.[19] The smooth surface of chocolate and the porous surface of bread are evident to the naked eye, and this

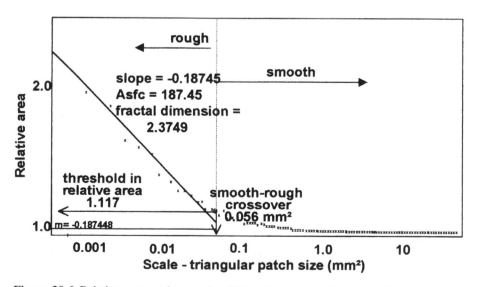

Figure 30.6 Relative area-patch area plot (Richardson's plot) of a series of tiling exercises from the surface of a raw potato slice. Smooth-rough crossover (SRC) and area-scale fractal complexity (ASFC) determined by SURFRAX.

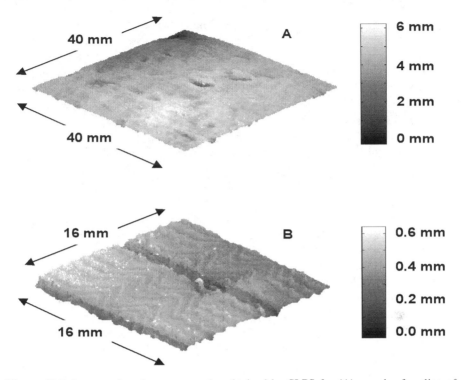

Figure 30.7 Images of surface topography obtained by SLPS for (A) crumb of a slice of white bread and (B) a chocolate bar. Scanned areas were 16 cm² and 2.56 cm², respectively.

difference is corroborated by the different values of the fractal parameters ASFC and SRC calculated by SURFRAX. Thus, the chocolate surface that appears smoother than that of a slice of bread has a smaller SRC (0.0385 mm² vs. 0.4752 mm²), indicative that roughness in chocolate appears at a lower scale than in the slice of bread. The ASFC values for chocolate and the slice of bread (45.84 and 271.90, respectively) indicate that the chocolate surface is smoother than that of bread at scales below their respective SRC.

30.9 TOWARD MICROPROCESSING

The desirable structures of many food products, among them bread, cheeses, yogurt and other gels, confectionery, etc., were created thousands of years ago based on a handful of available ingredients and by processes that are quite simple by present technological standards. In the age of the consumer, it is valid to think that foods will be engineered by designing the right architecture to satisfy specific commercial needs and trends.

Nature is an obvious source of ideas for structuring. Nature compensates for the limited types of molecules available by utilizing the same macromolecular design and varying only the hierarchical structure. The principle of hierarchy is based on increasing levels of organization which are progressively assembled from the molecular to the macroscale until the desired properties and functions are achieved.[20] An example is tendon, a uniaxial hierarchical structure that serves to transmit muscular force while being capable of absorbing large amounts of energy without fracturing.

Observing the hierarchical architectures of biological materials is rewarding and inspirational in several ways.

1. It provides means of understanding how natural food materials are assembled at different size scales and thus suggests structural models for analysis and alternatives to disassemble them (e.g., fish tissue fibers).
2. It provides ideas for mimicking in fabricated foods and to understand interactions at each level.
3. It reveals the extraordinary combinations of performance properties and may suggest routes for new structuring techniques.[21]

However, nature designs structures within a constraint context. Natural structures need to grow, so molding is not a preferred fabrication method. Neither is nature good at making flat surfaces or air-filled biomaterials, and it uses alloys and ceramics to a limited extent.[22]

Microprocessing and engineering intervention at the microstructural level should be greatly advanced by

- Application of *appropriate* microscopy techniques that really provide information in the quality and form in which it is needed (borrowed from biology and materials science).
- Description of microstructural elements and architectures at different scales that result in realistic physical models. Quantization of microstructural features by image processing and analysis (e.g., use of fractals).
- Methods that predict physical properties based on structural models and element or group-contribution approaches (e.g., the equivalent to UNIQUAC for gases and liquids). Development of the appropriate software.
- The development of scaling laws to predict structure-properties relationships in foods by extrapolation from the molecular and macromolecular to the macroscale (e.g., borrowed from polymer physics).

30.10 CONCLUSIONS

In the past, the application of physics and classical principles of chemical engineering, seasoned with a dose of empiricism, were sufficient to explain phenomena in food processing at the macrolevel. However, this approach has led to a marginal contribution the fundamental principles of engineering sciences, sometimes justified by the "complexity" of food materials.

Structure has been a most important variable disregarded by food engineers. Further advances in the understanding of properties of foods, as well as in processing and fabrication, need to reduce the scale of analysis to that of basic building elements and the cell. The myriad of microscopy, visualization, and image processing techniques available, including noninvasive and real-time methods, will greatly assist in this effort. Foods must also be understood in the way they are formed, transformed, and broken down in the mouth. This trend will be greatly assisted by concepts, techniques, and tools borrowed from biology and materials sciences. Last, but not least, food engineers should be trained at universities and industries along these lines, which tacitly imply multidisciplinary and teamwork.

30.11 ACKNOWLEDGMENTS

The research line at the Biomaterials Laboratory has been supported by grants from Fondecyt (Projects 2980058 and 196285), the Nestlé Research Center (Vers-chez-les-Blancs, Switzerland), the J.S. Guggenheim Foundation (New York), and the A. von Humboldt Foundation (Germany). Nestlé-Chile provided financial support for the graduate studies of Mr. F. Pedreschi.

References

1. Aguilera, J. M. and D. W. Stanley. 1999. *Microstructural Principles of Food Processing and Engineering* (2nd ed.). Gaithersburg, MD: Aspen Publishers, Inc.
2. Villadsen, J. 1997. "Putting structure into Chemical Engineering," *Chem. Eng. Sci.,* 52: 2857–2864.
3. Bailey, J. E. 1998. "Mathematical Modeling and Analysis in Biochemical Engineering: Past Accomplishments and Future Opportunities," *Biotechnol. Progr.,* 14: 8–20.
4. Gekkas, V. 1992. *Transport Phenomena in Foods and Biological Materials.* Boca Raton, FL: CRC Press.
5. Cussler, E. L. 1997. *Diffusion: Mass Transfer in Fluid Systems* (2nd ed.). Cambridge: Cambridge Univ. Press.
6. Crosley, J. J. and J. M. Aguilera. 2000. "Modelling the Effect of Microstructure on Extraction of Foods," *J. Food Process. Preserv.* (submitted).
7. Pedreschi, F., J. M. Aguilera, and J. J. Arbildua. 1999. "CLSM Study of Oil Location in Fried Potato Slices," *Microscopy and Analysis,* 37: 21–22.
8. Aguilera, J. M. and H. Gloria-Hernández. 2000. "Oil Absorption During Frying of Frozen Parfried Potatoes," *J. Food Sci.,* 63: 476–479.
9. Macosko, C. W. 1993. *Rheology: Principles, Measurements and Applications.* New York, NY: VCH Publishers.
10. Barnes, H. A., J. F. Hutton, and K. Walters. 1989. *An Introduction to Rheology.* New York: Elsevier Science Publisher B.V.
11. Aguilera, J. M. 1992. "Generation of Engineered Structures in Gels," in *Physical Chemistry of Foods*, H. G. Schwartzberg and R. W. Hartel, eds. New York: Marcel Dekker, pp. 387–421.
12. Aguilera, J. M. and H. G. Kessler. 1989. "Properties of Mixed and Filled-type Gels," *J. Food Sci.,* 54: 1213–1217, 1221.
13. Mor, Y., C. F. Shoemaker, and M. Rosenberg. 1999. "Compressive Properties of Whey Protein Composite Gels Containing Fractionated Fat," *J. Food Sci.,* 64: 1078–1083.

14. Aguilera, J. M. and P. Baffico. 1997. "Structure-Mechanical Properties of Heat-Induced Whey Protein/Starch Gels," *J. Food Sci.,* 62: 1048–1053, 1066.

15. Stanley, D. W., J. M. Aguilera, K. W. Baker, and R. L. Jackman. 1998. "Structure-Property Relationships of Foods as Affected by Processing and Storage," in *Phase/State Transitions in Foods,* M. A. Rao and R. W. Hartel, eds. New York: Marcel Dekker, pp. 1–56.

16. Franks, F. 1994. "Accelerated Stability Testing of Bioproducts: Attractions and Pitfalls," *TIBTECH,* 12: 114–117.

17. Brown, C. A. 1994. "A Method for Concurrent Engineering Design of Chaotic Surface Topographies," *J. Mater. Process. Technol.,* 44: 337–344.

18. Brown, C. A., P. D. Charles, W. A. Johnsen, and S. Chesters. 1993. "Fractal Analysis of Topographic Data by the Patchwork Method," *Wear,* 161: 61–61.

19. Pedreschi, F., J. M. Aguilera, and C. A. Brown. 1999. "Characterization of Food Surfaces Using Scale-Sensitive Fractal Analysis," *J. Food Process Engineering,* 23: 127–143.

20. Baer, E., J. J. Cassidy, and A. Hiltner. 1991. "Hierarchical Structures of Collagen Composite Systems," in *Viscoelasticity of Materials,* A. Glasser and H. Hatakeyama, eds. ACS Symposium Series 489. Washington DC: American Chemical Society, pp. 2–23.

21. NRC. 1994. *Hierarchical Structures in Biology as a Guide for New Technology Materials.* Washington, DC: National Research Council.

22. Vogel, S. 1992. "Copying Nature: A Biologist's Cautionary Comments," *Biomimetics,* 1: 63–79.

31 Viewing Food Microstructure

J. C. G. Blonk

CONTENTS

Abstract

The microstructure of food products determines, to a large extent, the properties of the food product. Processing of food changes the microstructure. A broad range of techniques to observe the microstructure are available to the food researcher. Based on studies performed at the author's laboratory, an overview is given on how microscopical techniques can be used to visualize the microstructure of foods.

31.1 INTRODUCTION

The study of the microstructure of food and food products is of great importance to understand their properties and to develop new product concepts. Nature offers food in large varieties, but the structure as found in nature is often unsuitable for direct consumption by man. Therefore, food needs to be processed. There are many reasons for processing: removal of unwanted ingredients, protection of goodies, improving digestibility, convenience, shelf life, boosting taste, suppression of off-flavors, control of appearance, and introduction of variance. Processing can also contribute to minimize spoilage of food by making as many as possible of the raw material components acceptable for consumption by man or animals. Food processing is also used to obtain constant quality of food by reducing natural variance introduced by varieties of species and seasonal influences.

Two important driving forces in food processing nowadays are health and convenience. In fat-containing products, both the amount and the composition of fats are important with respect to health. Saturated fats are important structuring components in products like butter and spreads. Reduction of fat accompanied by replacement of saturated by (poly)unsaturated fat has a large impact on structuring these products. Milk has a short shelf life when stored as a raw product. Acidification of milk and heat treatment yield products like cheese, yogurt, and quark in many varieties of which shelf life is increased and in which taste and appearance can be adapted to local preferences. Staple foods like wheat and rice can be processed in many ways to offer attractive food products.

In foods, the most important building blocks are lipids, proteins, carbohydrates, water, and air. It is the art of the food processing industry to use these building blocks to obtain the desired food products. Knowledge at the molecular level has to be combined with knowledge at the structural level to control the properties of the food products. Microscopic studies offer this information supplemented by compositional analysis and rheological experiments.[1-5] The microscopy of foods is performed at the interface area between material science and biological science. Many sample preparation techniques in food microscopy are derived from those used in preparation of biological tissue samples. The wide range of microstructure elements necessitates the use of a wide range of microscopic techniques, ranging from high-resolution transmission electron microscopy and scanning probe microscopy to a variety of light microscopy techniques. Although less powerful in resolution, light microscopy offers many ways to study the microstructure under environmental conditions.

Food products are becoming more complex. This is directly reflected in the complexity of the microstructure. Visualization of the structure and interpretation of what is observed in relation to product properties is often based on qualitative criteria. The fast developments in image analysis capabilities allow microstructures to be analyzed in more quantitative ways by correlating relevant physical measurements in the microstructure with product properties like consistency, but also consumer preferences. This way of performing microstructural measurements, especially when performed in 3-D data sets, is rather new, but the techniques are expected to grow rapidly.

In this chapter, a number of examples will be given of visualization of the microstructure of a variety of products, related to processing and composition. There will be an emphasis on the way in which microscopy is used to study the microstructure of the products. In some examples, quantitative measurements have been performed.

31.2 FAT-BASED PRODUCTS

Fats (lipids) are main sources of energy in food. The source of the fat can be from animal milk or meat, or from plant seeds. Tri-acylglycerols form the bulk of the lipids in foods, but mono-acylglycerols, fatty acids, and lecitines play an indispensable role in food structuring in the way they act as emulsifiers. Unsaturated fats in cis-configuration are generally accepted as being the healthy types of the tri-acylglycerides. Main sources for these molecules are oil seeds and fish.

The crystallization behavior of fat in various polymorphs offers the food scientist tools to control the microstructural elements in fat-based foods. Various microscopical techniques can be used to visualize the crystal structure of fats: polarized light and confocal and (scanning) electron microscopy (see Figure 31.1). Scanning electron microscopy performed at low temperature (–130°C) allows observation of the network after the liquid oil has been removed by a balanced series of washings of the product with suitable solvents.[6,7]

31.2.1 FAT SPREADS

Butter and margarine are the main source for fat in the human diet in many countries, both containing 80–85% fat. Butter is obtained by a churning process. During this process, the fat globules in milk are transformed from a dispersed phase into a fat continuous phase in which water areas are included, and butter fat crystals form a network, giving butter its strength. Margarine is processed by rapid cooling of a water-in-oil emulsion in which part of the fat is solidified, giving the product its consistency. The crystal network in both products are different, giving the two products a different rheological behavior when shear is applied. The network in margarine is a continuous one while, in butter, a discontinuous network is observed (see Figure 31.2). This property influences, to a large extent, the application of margarine and butter in, for instance, laminating fat in puff pastries. In this case, the rheological behavior of the fat should be matched with that of the dough with which it forms alternating layers. The fat crystal network cannot be visualized in a polarized light microscope, since the resolution of light microscopy is insufficient. However, observation in cryo-SEM after careful removal of the oil unveils the crystal network.

Based on the concept of margarine, fat spreads have been developed in which the amount of fat has been reduced. Replacement of fat by water cannot be done without affecting the product integrity. The aqueous phase of the product has to contribute to the product stability. Various solutions are possible: gelation of the water phase, filling the water phase with oil droplets (O/W/O emulsion), or distribution of the water in two distinct phases in the spread (W/O/W emulsion). For the sake of convenience, liquid margarines have been developed in which water droplets are suspended in the oil phase. The water droplets are stabilized by emulsifiers that can be visualized both in cryo-SEM after removal of the oil by the de-oiling technique and in the confocal microscope, yielding a 3-D image of the dispersion of the droplets in oil (see Figure 31.3).

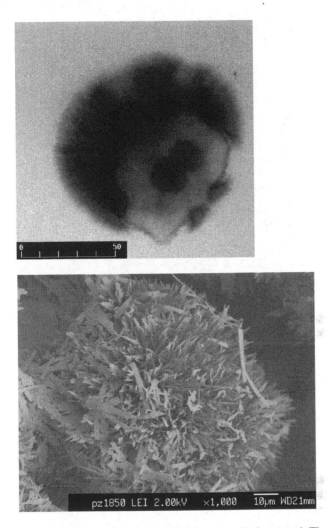

Figure 31.1 Fat crystal agglomerates. Top: confocal image of a fat crystal. The oil phase has been stained with Nile Red. Changes in density during growth induced by a changing crystallization process is clearly visible. Bottom: image obtained in the cryo-scanning electron microscope after de-oiling of the crystal-oil suspension. The shape and size of the single crystals are well exposed.

The water droplet size distribution of fat spreads influences the product properties, but also the microbiological stability of the product. Small droplets are preferred. The size distribution can be controlled by the production process and can be obtained by microscopical observation.

However, the visualization of the water droplets in a way that an unbiased distribution can be measured is rather complicated. Thin smears may destroy or flatten large droplets, and confocal optical sections present a chord distribution

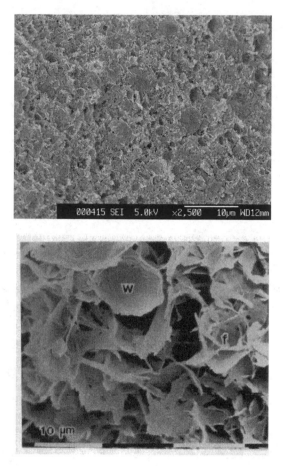

Figure 31.2 Fat crystal network observed in the cryo-scanning electron microscope. Top: butter, no continuous fat crystal network containing fat globules. Bottom: margarine, continuous fat crystal network (f) containing fat crystal shells in which water droplets (w) are dispersed.

instead of a diameter distribution. Electron microscopy is rather labor intensive with respect to the minimum number of droplets that must be analyzed to obtain statistically sound results. Diffusion measurements by NMR of a bulk sample offers a mean value of the droplets and a measure of the distribution width. A 3-D confocal data set is probably the best way to measure the real distribution.

31.2.2 DRESSINGS/SAUCES

Mayonnaise is probably the best known representative of dressings. It is an 80% oil-in-water emulsion stabilized with protein as emulsifier. In this composition, the dispersed phase is space filling. The amount of interaction between the emulsifying

Figure 31.3 Liquid margarine. Top: distribution of emulsifiers and proteins at the water/oil interface. Water droplets are situated within the shells. Bottom: cryo-SEM image after de-oiling. The shells of the water droplets are clearly visible.

proteins at the oil droplet/water interface in combination with the oil droplet size distribution controls the rheological behavior of the product. Low-caloric dressings can be obtained by reducing the fat content and increasing the viscosity of the aqueous phase by addition of filler components like starches and hydrocolloids.

Depending on the product application and composition, numerous mixtures have been developed. Especially in the area of ready-to-use hot and cold sauces, optimized mixtures are on the market. These products do not have a well defined product microstructure in which structural elements interact with each other. Specific viscosity of the continuous phase in relation to size and shape determines the main macroscopic properties (see Figure 31.4).

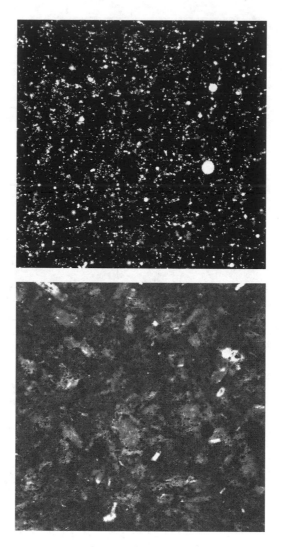

Figure 31.4 Confocal images of a low-fat dressing. Top: distribution of the fat droplets. Bottom: distribution of the carbohydrates (same field of view).

31.2.3 ZERO-FAT SPREADS

Based on the technology that mono-glycerides in combination with a co-surfactant can bind up to 95% water, food products have been made having the properties of fat spreads with respect to spreadability but containing almost no fat.[8] Figure 31.5 shows both a confocal image and a cryo-SEM image of such a system.

The properties of these products are determined by the thickness of the plates, size distribution of the pores between the plates, permeability of the structure, and

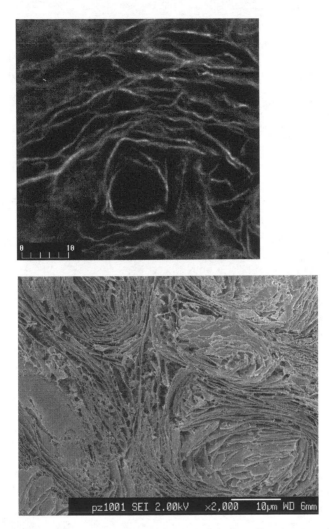

Figure 31.5 Zero-fat spread. Top: confocal section of the lamellar structure of the product. Bottom: lamellar structure as observed after freeze drying in the cryo-SEM.

concentration and properties of the monoglycerides and co-surfactants. Microscopy is a powerful tool but, in combination with 3-D image analysis of the confocal data set and NMR diffusion measurements, a more quantitative description of these systems can be obtained.

31.3 PROTEIN-BASED FOODS

Proteins can be used in various ways to structure food products. They can be present as particles (casein micelles), a network of macro-molecular strands (cheese), or sheets (gluten, foams). The nature of protein can be manipulated in a number of ways, of which denaturation induced by acidification, addition of salt, and heating are the most important. The large number of process parameters combined with the great variety of proteins in food material makes the number of combinations inexhaustible. Basic knowledge of the processes at (sub)micrometer scale can help the foods engineer to a large extent.

31.3.1 DAIRY-BASED FOODS

Dairy-based foods have their origin in the objective of our ancestors to store milk for a longer period than a few days. Cheese is one of the most important examples of how the main constituents in milk can be protected against spoilage. The number of cheese varieties is uncountable, but the basis of all is aggregation of casein proteins, dispersion of fat, and reduction of the water content. Acidification caused by microorganisms or enzymatic-induced aggregation are the main mechanisms. The influence of pretreatment of milk, processing conditions, and addition of components on the formation of the microstructure of dairy products can be studied successfully with microscopical techniques. The confocal microscope has many capabilities, since *in vivo* studies during the process can be performed.

An example is given in Figure 31.6, which shows the formation of the protein network after acidification during a subsequent heating process.[9] This study is performed in a specially designed vessel that can be placed onto an inverted confocal microscope. A drawing of the vessel is shown in Figure 31.7. Since the resistance on the stirrer caused by the protein network formation can be monitored, both microstructural and rheological information can be obtained from the same sample in real time. The increase of the viscosity in time is attributed to the contribution of denaturating whey proteins as a result of the heating process. It is suggested that these whey proteins precipitate onto the casein network and in this way fortify the network. In Figure 31.8, elongated particles are visible that show a brighter fluorescence compared to the casein network. It is suggested, and further evidence has been obtained from immuno-TEM studies, that these are the denaturated whey proteins.

Qualitative assessment of images of protein network, as shown in Figure 31.6, is not always satisfying. Quantitative analysis of these types of networks can be obtained, for instance, by the approach of Bremer,[10] based on the calculation of density correlation function of confocal slices. From this function, a number of structure-describing parameters can be derived, such as the fractal dimensionality of the network and the mean maximum size of the clusters that form the network. Results of these measurements are given in Figure 31.6. The mean maximum cluster size could be correlated with the strength of the clusters when compared with size measurements obtained by light scattering after dilution of the sample.

31.3.2 WHEAT

Wheat is an important ingredient in human food in many countries in the world. Bread, pasta, snacks, and sweets contain wheat. The composition of wheat varieties

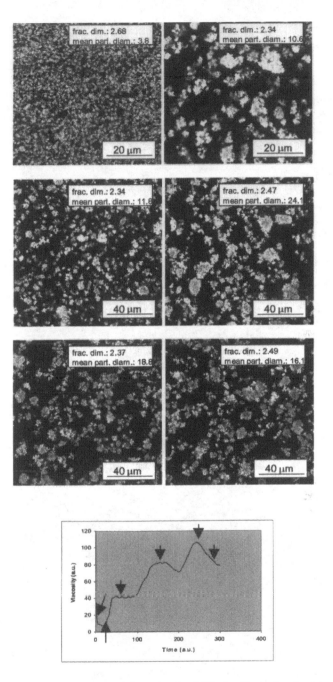

Figure 31.6 Time series of confocal images of acidified milk during heat treatment. The corresponding values for the fractal dimensionality and the mean cluster size are given in the image. The graph shows the changes in viscosity during the heat treatment. The arrows correspond with the six images (top left to bottom right).

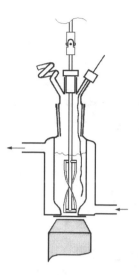

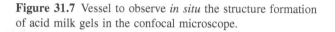

Figure 31.7 Vessel to observe *in situ* the structure formation of acid milk gels in the confocal microscope.

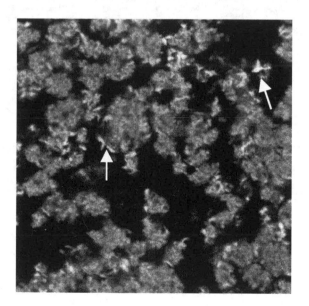

Figure 31.8 Confocal image of the structure of heat treated acidified milk. The bright fluorescent features (indicated by arrows) are suggested to be heat denatured whey proteins.

determines, to a large extent, their application. Convenience in use and preparation of fresh wheat products is one of the driving forces in studies of wheat dough. To find relations between dough microstructure, composition, additives, convenience, and consumer preference is a challenging task. The availability and localization of water in dough plays an important role. It is known that water migrates in dough even when it is in a frozen state, which results in inferior quality of the baked product. Both magnetic resonance imaging and cryo-scanning electron microscopy give insight into the water management. Figure 31.9 shows ice crystals that have been formed inside the gas bubbles during frozen storage of wheat bread.

In laminated systems, alternating layers of dough and fat are present. The layers are thinned and the rheological characteristics have to match to prevent breakup and fusion of the layers. The interface between the layers have been studied both by confocal microscopy and cryo-SEM. Figure 31.10 shows perfect matching thin-layered systems.

31.4 CARBOHYDRATE-BASED FOODS

Carbohydrates in food represent an important energy source. Starch is present in many food products as a filling component, acting in food structuring as an inert particle. However, heating of starch in the presence of water transforms these inert particles in sources for food structuring; in many cases, water is immobilized by the starch, which changes the rheological properties of starch.

Polymeric carbohydrates such as gums originating from plants, algae, fungi, and microorganisms can bind amounts of water that are many times more than their own mass. The viscosity of the water phase containing these gelling agents is related to the organization of the macromolecules that form specific networks. TEM can be used to get insight into the organization of the macromolecules and the way in which these biopolymers form a network (see Figure 31.11).

Products that contain large amounts of water that is loosely bound or weakly gelled need special preparation prior to observation of the microstructure. Electron microscopy is indispensable to obtain submicrometer resolution. The best sample preparation procedure is freezing of the sample and fracturing of the frozen sample, followed by preparation of a replica of the fractured surface. The replica can then be observed in the TEM. On freezing, ice crystals are formed at an extremely fast

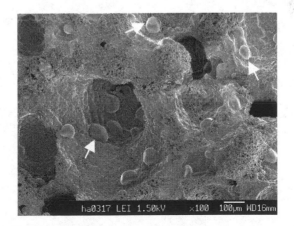

Figure 31.9 Ice crystals (see arrows) in gas cells in frozen dough after storage (cryo-SEM).

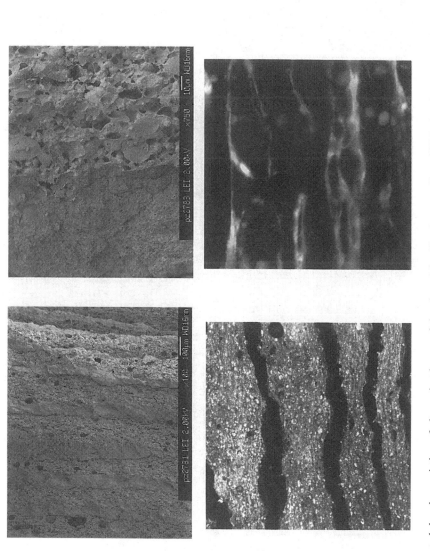

Figure 31.10 Laminated dough consisting of alternating layers of dough and fat. Top images: Cryo-SEM, showing the interaction between dough and fat. Bottom images: Confocal, showing the thickness variations in the dough and the localization of the gluten around the starch granules.

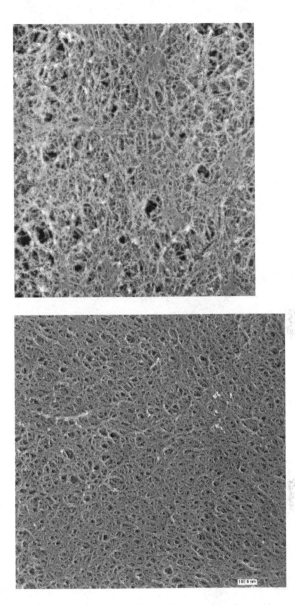

Figure 31.11 Replica after deep etching of biopolymer networks as observed in TEM. Top: extra cellular polysaccharide from Schyzophylan (1% w/v in water). Bottom: apple pectin network (0.1% w/v). Bar equals 100 nm.

rate. Very fast cryo-fixation techniques are required to obtain minimal sample disruption caused by ice crystal formation. Freezing speeds have been increased using new technologies, e.g., fast plunging, jet freezing, and metal mirror slamming. The combination of fast freezing and exposure to very high pressure of the sample using the high-pressure freezing technique has increased the depth in the sample, which can be cryo-fixed without visible ice-crystal formation.

Mixtures of various types of carbohydrates or with proteins often yield phase-separated systems, sometimes called water-in-water emulsions. These types of emulsions play an important role in tailoring water phases in water-in-oil emulsions and can replace, for instance, the gelatin. The formation of the phase-separated systems can be monitored easily in the confocal microscope, especially when the various components are labelled with specific fluorescent stains (see Figure 31.12).

31.5 IMAGE ANALYSIS/CONSUMER UNDERSTANDING

Although high-quality images contain a huge amount of information about the microstructure, food researchers like to extract quantitative data from the images that can be correlated with relevant physical characteristics of the products. This will offer tools to modify the production process or composition to obtain the required structure. Combined observation of the microstructure, extraction of quantitative data, and modeling of the microstructure based on characteristics of the structuring elements should point to the optimal product. 2-D image analysis is a well developed technique to extract quantitative data. The power of today's computer

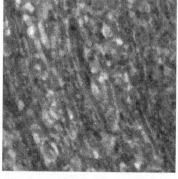
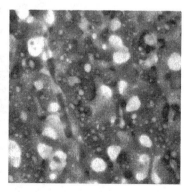
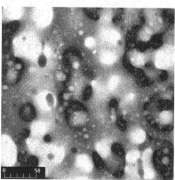

Figure 31.12 Change in the microstructure of a mixture of fluorescently labeled alginate and not-labeled caseinate after finishing mixing at room temperature. Confocal images obtained (top left) 1, (top right) 2, and (left) 3 min after cessation of mixing. Regions of various concentrations of alginate can be observed.

systems allows 3-D image analysis to be performed on large 3-D image sets. These can be acquired from tilt series of SEM images or high-resolution replicas in the TEM or from atomic force data measurements. However, these techniques show a 3-D surface and are therefore often referred to as 2.5-D images. Confocal microscopy, on the other hand, provides a true 3-D image of the specimen. After image deconvolution, the image can be further processed, and quantitative data can be extracted from this confocal 3-D data set.[11]

Understanding the microstructure of complex food products in relation to composition and processing conditions needs a strong cooperation and interaction between different microscopical techniques, 3-D image analysis, spectroscopic techniques, and rheological measurements. Since the consumer will finally judge whether food manufacturers have been successful in preparing a high-quality product, it is the art of the food researcher to discover the relationship between consumer acceptance, microstructure of the product, and the optimal ways of processing the food.

31.6 ACKNOWLEDGEMENTS

Many food researchers have contributed to this survey of microstructural research. Among them the author would like to mention Henrie van Aalst, Ruud den Adel, Johan Hazekamp, Ies Heertje, Eli Royers, Frank Kleinherenbrink, Eddy Esselink, Geert van Kempen, Cees Vooys, John van Duynhoven, Peter Zandbelt, Peter Nootenboom, and Aat Don.

REFERENCES

1. Heertje I., and M. Pâques. 1995. "Advances in Electron Microscopy," in *New Physico-Chemical Techniques for the Characterization of Complex Food Systems,* E. Dickinson, ed. London: Blackie Academic & Professional, Chapman and Hall, pp. 1–52.
2. Brooker, B.E. 1995. "Imaging Food Systems by Confocal Laser Scanning Microscopy," in *New Physico-Chemical Techniques for the Characterization of Complex Food Systems,* E. Dickinson, ed. London: Blackie Academic & Professional, Chapman and Hall, pp. 53–68.
3. Blonk, J. C. G., H. van Aalst. 1993. "Confocal Scanning Light Microscopy in Food Research," *Food Res. Int.,* 26(4): 297–311.
4. Heertje, I. 1998. "Fat Crystals, Emulsifiers and Liquid Crystals. From Structure to Functionality," *Pol. J. Food Nutr. Sci.,* 7/48 (2): 7(S)–18(S).
5. Blonk, J. C. G. 1998. "New Imaging Techniques and Future Developments," *Pol. J. Food Nutr. Sci.,* 7/48 (2): 19(S)–30(S).
6. Heertje, I., M. Leunis, W. J. M. van Zeijl, E. Berendse. 1987. "Product Morphology of Fatty Products," *Food Microstruct.,* 6: 1–8.
7. Heertje, I. 1993. "Microstructural Studies in Fat Research," *Food Struct.,* 12: 77–94.
8. Heertje, I., E. C. Roijers, H. C. M. Hendrickx. 1998. "Liquid Crystalline Phases in the Structuring of Food Products," *Lebensm.-Wiss. U.-Technol.,* 31: 387–396.
9. Kleinherenbrink, F. A. M. 1999. "Whey Protein Denaturation and Gelation in Acidified Milk Products," presented at the 2nd International Symposium on Industrial Proteins, March 4–5, 1999.

10. Bremer L. 1992. *Fractal Aggregation in Relation to Formation and Properties of Particle Gels.* Thesis. Wageningen Agricultural University.
11. Van Kempen, G.M.P. 1999. *Image Restoration in Fluorescence Microscopy.* Thesis Technical University Delft.

Part VI

Thermal Processing and Packaging

32 Active Packaging: Science and Application

M. L. Rooney

CONTENTS

Abstract

Packaging of foods usually requires some compromise in matching the properties of the package to the requirements of the food. Active packaging provides novel approaches to optimizing package properties to address specific needs of the foods. The removal of oxygen from packages is now possible commercially using both plastics-based and sachet technologies. Other technologies are in the development stage to address microbiological, physiological, and sensory problems. The rapid growth of this technology provides diverse opportunities for contributions from food scientists and engineers.

32.1 INTRODUCTION

Successful packaging is achieved when the properties of the package match the requirements of the food, from packaging until use. Packaged goods often have multiple requirements and, when considered with the limitations imposed by the packaging machinery, offer substantial challenges to existing packaging technology.

Traditionally, this goal is achieved only in part, with compromises often being made at substantial cost due to technical limitations of the material or the packing machinery. Examples are found in the sophisticated laminates used to provide high barriers to both oxygen and water while still allowing transparency and abrasion resistance.

Since foods lose quality or become unsafe by several different mechanisms, it is often necessary to prioritize the requirements of the food, resulting in a shelf life limited by the packaging solutions available. Research over the past two decades has made making it increasingly clear that active packaging can be designed to affect one or more of these technical limitations to meet the needs of the food in an optimal manner. Active packaging can be defined as performing some desired role in food quality or safety other than to provide an inert barrier to external conditions.[1,2]

Specific reasons for developing active packaging are summarized as follows:

- removal of an unwanted component
- addition of a desired ingredient
- package surface antimicrobial activity
- changed physical properties of the package

These general concepts may be considered best by looking at foods from the viewpoint of their requirements based on the conventional scientific and engineering disciplines. The most important of these are summarized in Table I and are discussed in detail below.

32.2 MICROBIOLOGICAL EFFECTS

The increased emphasis on minimally processed foods has demanded a higher level of sophistication in packaging processes that clear one or more microbiological "hurdles." This is particularly important in modified atmosphere packaging (MAP) in which the residual oxygen level can determine whether mold is able to grow within the desired shelf life.

32.2.1 OXYGEN SCAVENGING

The degree of difficulty in the replacement of air with the desired atmosphere strongly depends on the physical state of the food. Abe and Kondoh[3] reported that the economic limit for removal of oxygen from food packs is around 0.5%. Values of 2% or higher can be experienced, depending on food porosity. It is potentially valuable to provide a residual oxygen scavenger so that gas flushing can be applied for the minimal time commensurate with optimal line speed.

The most successful current approach to oxygen removal is inclusion of iron-based oxygen-scavenging sachets inside the pack.[4] These sachets are capable of reducing the headspace oxygen concentration to less than 0.1%. Despite these advances in refinement of iron-based systems, the most dramatic advances have occurred in the development of plastics that react with oxygen. Over the past year, worldwide patent applications for oxygen-scavenging packaging have appeared at the rate of 5 to 10 per week, with at least half of these based on plastics materials.

TABLE 32.1
Food Requirements Addressed by Active Packaging

Food requirement	Package activity	
	Release	Absorb
Microbial	Preservative	Oxygen, water
Chemical	Antioxidant	Oxygen
Physiological	Inhibitors	Gas balance
Presentation	Heat	Heat, water
Sensory	Inhibitors	Flavors, odors

32.2.2 ANTIMICROBIAL RELEASE

The release of antimicrobial agents and the generation of antimicrobial package surfaces are alternative approaches to modification of the headspace gas composition. At present, release of antimicrobials could well exceed the overall migration limit of 60 mg/kg of food in Europe. However, there is a substantial body of research that demonstrates the potential benefits of this approach.

Some of the research has been reviewed.[5] Agents used include quaternary ammonium compounds, imazalil, various food acidulants, and benzoic anhydride. Silver ions have been bound into porous zeolite particles, which in turn have been incorporated into coatings on glass or into polyethylene films. These compounds have been incorporated into typical food-contact plastics used as sealants.

Release of permitted food acids has been investigated, and it has been shown that their polarity makes them unsuitable for use in the common non-polar sealant plastics. Anhydrides are relatively stable to heating involved in plastics processing. Weng et al.[6] chemically bound benzoic acid to a Surlyn film by means of weak anhydride bonds that were hydrolyzed by water derived from the food. It was found that up to 1.6 mg/g of film could be released, suggesting an effective initial impact on surface contamination. This result needs to be evaluated in terms of food requirements and permitted limits.

The antifungal agent imazalil has also been shown to be suitable for release from a low-density polyethylene film in quantities that have been found to be effective in fruit and vegetables or for cheese.

Recent work in CSIRO and the University of Western Sydney, using a new release system for food acid, has shown that a *Pseudomonas* strain can be suppressed below spoilage levels on inoculated chilled beef for over two weeks.

32.2.3 ANTIMICROBIAL SURFACES

The alternative of generating a surface that is capable of killing microorganisms is of interest for packaging of beverages and where cut surfaces are subsequently subjected to tight-fitting, flexible packaging. Potential applications are cheese, meats, sauces, and particularly beverages, so there has been ongoing interest in generation of plastic surfaces with antimicrobial properties.

Early research involved binding of Benomyl, a fungicide, to the surface of Surlyn™ (DuPont) an ionomer film.[7] This process would need substantial modification if it were to be commercialized. The test results with strains of *Aspergillus flavus* and *Penicillium notatum* were positive but indicated some release of Benomyl from the polymer matrix.

Lysozyme is an enzyme that is particularly effective against gram-positive bacteria, and it has been bound to a variety of solid substrates.[5] When physically immobilized at up to 15% in cellulose triacetate films, the enzyme reduced the loading of *Micrococcus lysodeiticus* in water by 7 logs in 20 h. The work is important in that it shows that immobilization of a GRAS food enzyme in a plastic is possible without chemical bonding, with very little migration.

This pioneering work will need to be developed further to determine the breadth of its effects, its compatibility with industry standards, and methods of film manufacture.

The approach of rapid conversion of chemical groups on the surface of extruded plastic packaging to generate antimicrobial activity is attractive from an industrial viewpoint. Surface nitrogen-containing groups on the surface of a variety of polymers have been converted to amines by subjecting preformed films to strong from a laser at 193 nm. Exposure of films treated in this way to *Staphylococcus aureus*, *Pseudomonas fluorescens,* and *Enterococcus faecalis* suspensions in phosphate buffer resulted in a decrease in microbial count by over 4, 2, and 1 logs, respectively.[8] If the microbial kill results can be corroborated, the approach appears attractive for commercial use.

32.3 CHEMICAL EFFECTS

Food quality deterioration results from a wide range of chemical reactions, most of which are not amenable to treatment by means of active packaging. The variation in the sensitivity of foods to oxygen has been discussed by Koros,[9] who estimated that 1–200 mg/kg of food would need to be taken up to limit the shelf life to one year.

32.3.1 Oxygen Scavenging

The most common oxygen absorbers used in the food industry consist of porous sachets containing iron powder as described earlier. The subject of sachets and their use in packaging has been reviewed.[4] The more attractive alternative to the use of sachets is the incorporation of oxygen-scavenging capability into the plastics components of packaging. This has now grown from a curiosity to a small but rapidly developing commercial field within the past 20 years.[1]

The manner in which some of the plastics oxygen scavengers announced to the industry can be used is summarized in Table 32.2. Oxygen-scavenging polymers offer the prospect of being very versatile, as their layer thickness or blend composition can be varied to match the amount of oxygen to be removed.

The use of a scavenging plastic to provide a chemical oxygen barrier offers an opportunity to cheapen barrier packaging for relatively short shelf life products such as wholesale or export units.

TABLE 32.2
Oxygen-Scavenging Plastics for the Food Industry

Package	Activation	Company	Trade name	Polymer
Flexibles, bottles, rigids	UV light	CSIRO	ZERO2	Unrestricted
Flexibles	Metal/UV	CRYOVAC	OS1000	Polydiene
Bottles	Metal/UV	BP-AMOCO	OS3000	Polydienes-co-PET
Flexibles, rigids	Water	BP-AMOCO	TRI-SO2RB	Blends
Bottles	Metal	Crown, Cork & Seal	Ox-Bar	Promatic polyamides
Flexibles	Water	Toyo Seikan	OXY-GUARD	Blends
Bottles	?	Continental PET	Na	Aromatic polyamides

The need to achieve oxygen removal from foods and other sensitive products at a rate faster than that of the degradation reactions or the growth rate of microorganisms (or insects) means that only fast oxidation reactions of the polymer can be considered.

Oxygen-scavenging plastics require the use of an outer passive barrier layer to oxygen ingress, or one blended with a low-permeability polymer to minimize the load on the chemical system Polymers that provide suitable physical barriers are ethylene vinyl alcohol copolymer (EVOH) or a poly(vinylidene chloride)-coated layer. Substantial research and development has been devoted to incorporating iron powder into plastics. Some commercial products have been announced, but so far there has been little independent research to allow comparisons with sachet technologies.

Amoco Chemicals (now BP-AMOCO) developed a master batch system containing iron and other components for use in polyolefin or PET structures.The products are termed AMOSORB™ and are water activated, preferably by retorting. The first product incorporating this material is a closure liner termed TRI-SO₂B™. A patent application by Amoco Chemicals describes the use of iron in the presence of organic acids in polymer films, but the composition of the product AMOSORB™ is not revealed, being described only as consisting of substances GRAS in the U.S.A.[10]

Oxygen-scavenging plastics based on reactive polymers are still in their infancy, although the patent literature indicates a widespread commercial interest in their potential by companies such as Chevron Petroleum, Sealed Air Corporation, and Amoco-BP. Since such plastics need to be handled in air, the control of their reactivity by use of an activation step, such as by use of light, appears to offer a convenient way of overcoming this limitation.

It is in the field of plastic beverage bottles that the chemical barrier to oxygen permeation will be of most value. Beer flavor is particularly sensitive to oxidation, and PET alone does not provide a barrier for more than approximately one month at ambient temperatures. It is widely accepted that 1 ppm oxygen in a bottle leads to a shelf life of three months or less. Accordingly, a range of multilayer bottles with middle barrier layers have been produced, but these still allow only about five months shelf life.

The first oxygen-scavenging polymer useful as a blend with PET was named Ox-Bar™ (Crown Cork and Seal Ltd., U.K.). This process is based on the cobalt-catalyzed oxidation of MXD-6 polyamide.[11] When used with 200 ppm of cobalt as the stearate salt, this polyblend allows blowing of bottles with an oxygen permeability of essentially zero, for one year.

Continental PET Technologies Inc., of the U.S.A., has introduced commercially a five-layer beer bottle consisting of a central and two outer layers of PET sandwiching two layers of MXD-6 nylon. This package is an effective barrier to oxygen permeation, and the nylon layers appear to consume oxygen, giving the shelf life demanded by the consumer market.

The oxidation of polydienes can also be accelerated by cobalt salts. It has been shown that poly(1,2-butadiene) containing cobalt octoate can scavenge oxygen from package headspaces by oxidation of double bonds.[12] The process is activated by exposure of the polymer to UV light, electron beam, or corona discharge energy sources that cause more rapid consumption of the antioxidants. This type of film, manufactured as a coextrusion by CRYOVAC under the name OS1000™, is designed as both a headspace oxygen scavenger and an oxygen permeation barrier for short shelf life processed meats and bakery products.

AMOCO Corporation (now BP-AMOCO) has investigated use of polydiene oxidation for commercial use in PET plastics such as three-layer blow-moulded bottles for beer or other beverages.[13] The copolymer chosen for use is a block copolymer of PET with a 2 to 12% polybutadiene or other polyolefin with tertiary or allylic hydrogens and containing from 50 ppm to 300 ppm of cobalt octoate.

Oxygen-scavenging polymers with pendant benzylic esters made from ethylene-acrylic acid copolymers have been developed by Chevron Chemical Company.[14] The transition-metal-catalyzed oxidation results in the release of benzoic acid, which has food additive approval. In subsequent patent applications, the inventors have claimed autoxidation of styrene-butadiene copolymer and, in a separate application, polyterpenes dissolved in polymers, also with a transition metal catalyst. The terpenes include food components such as a-pinene and limonene.[15,16]

Plastics proposed as middle layer in a PET-based multilayer will not be rapid headspace scavengers. The plastics described above will have different scavenging rates and capacities for oxygen due to the partial barrier provided by the PET. Plastics in which the reactive layer is separated from the product by, at most, a polyolefin heat seal can be expected to scavenge headspace and dissolved oxygen rapidly within a few days.

32.4 PHYSIOLOGICAL EFFECTS

Ethylene causes ripening and senescence of respiring horticultural produce and is therefore removed whenever possible. The attempts made so far to incorporate absorbents and reagents into plastic films used for carton liners without any process has been free from drawbacks.[17]

An alternative approach utilizes the reaction of ethylene within tetrazines in commodity films such as plasticized PVC.[18] This approach has been shown to remove headspace ethylene concentrations to a few parts per billion, at which it is unable to cause petal loss in sensitive flowers like carnations. These films are unique in that they change from pink initially to colorless on reaction with ethylene, which occurs rapidly at concentrations that are physiologically important.

The ratio of the permeability of plastics to oxygen and carbon dioxide can be a critical factor in the packaging of horticultural produce. Unmodified plastics do not normally satisfy the requirements for permeation rate and permeability ratio, and some form of active packaging is needed to achieve optimum product condition.[19] A variety of semipermeable patches and techniques for generating controlled porosity have been developed without providing the atmospheric control required.

Besides the absolute value of the permeability and the various permeability ratios, there is a need for the permeabilities to oxygen and carbon dioxide to respond to temperature change in the same manner as does the produce's respiration rate. The approach taken by Landec Corporation (Menlo Park, CA, U.S.A.) offers promise if their Intellimer™ polymers can be manufactured more economically and with a wide range of gas permeability ratios, in addition to the temperature abuse properties that they already demonstrate. These copolymers have side chains that crystallize, and the melting point can be varied with a precision of ±2°C.[20] At the side-chain melting point, the gas permeability increases sharply, and the response to further temperature increases can also increase.

32.5 SENSORY EFFECTS

Early developments in odor-absorbent plastics occurred in Japan, where there was seen to be a need to remove amine smells from fish stored in domestic refrigerators.[21] An important example of removal of a low-volatility food component by utilizing diffusion is removal of naringin, the bitter principle of grapefruit juice. Recent research has shown that immobilization of the naringinase in cellulose triacetate film can cause hydrolysis of 60–80% of the naringin within 15 days at 7°C.[22] The naringin is postulated as diffusing into the polymer so as to reach the enzyme. This effect shows the potential for bitterness being reduced to acceptable levels during commercial distribution rather than by use of additional processing equipment to bring the enzyme into contact with the juice.

Aldehydes such as hexanal and heptanal are formed when fats in foods such as cereals, dairy products, and fish are oxidized. Odour and Taste Control Technology™ developed by DuPont is based on incorporation of molecular sieves with a pore diameter approximately 5 nm.[23] If this approach is proven to remove aldehydes effectively, it may be attractive to plastics processors and regulators alike, due to the simplicity of the materials.

There is also an approach based on reactive polymers, and this research demonstrates removal of aldehydes from package headspaces by use of polyimines containing amine groups pendant from the backbone.[24] This technology was further developed for application in recycled high-density polyethylene, which contains low-molecular-weight aldehydes coming from milk residues in the original milk jugs.[25] It will be necessary for industry and regulators to ensure that such processes are not used to conceal the marketing of substandard or even dangerous foods and other products if, for instance, microbial odors were to be scavenged.

32.6 CONCLUSIONS

Some of the most important forms of active packaging have been highlighted, indicating that this field is no longer merely a scientific curiosity. The growth has been largely in the areas of food quality as affected by microbiological and oxidative effects. Commercial uptake has been most dramatic in the case of oxygen-scavenging for processed foods, but the future looks bright for overcoming distribution of horticultural produce. Besides cost considerations, regulatory and environmental hurdles will apply and will provide challenges to scientists and engineers in the first decades of the millennium.

REFERENCES

1. Rooney, M. L., ed. 1995. *Active Food Packaging*, Glasgow and London: Blackie A & P, an imprint of Chapman & Hall.
2. Floros, J. D., L. L. Dock, and J. H. Han. 1997. "Active Packaging Technologies and Applications," *Food, Cosmetics and Drug Packaging.* January, 10–17.
3. Abe, Y. and Y. Kondoh. 1989. "Oxygen Absorbers in Controlled/Modified Atmosphere/Vacuum Packaging of Goods," A. L. Brody, ed. Trumbull, CT: Food & Nutrition Press, pp. 151.
4. Smith, J. P., J. Hoshino, Y. Abe. 1995. "Interactive Packaging Involving Sachet Technology," in *Active Food Packaging*, M. L. Rooney, ed. Glasgow and London: Blackie A & P, an imprint of Chapman & Hall, pp. 143–172.
5. Appendini, P., and J. H. Hotchkiss. 1997. "Immobilization of Lysozyme on Food Contact Polymers as Potential Antimicrobial Films," *Packag. Technol. Sci.,* 10: 271–279.
6. Weng, Y-M., M. J. Chen, and W. Chen. 1997. "Benzoyl Chloride Modified Ionomer Film as Antimicrobial Food Packaging Material," *Internat. J. Food. Sci. Technol.,* 32: 229–234.
7. Halek, G. W. and A. Garg. 1989. "Fungal Inhibition by a Fungicide Coupled to an Ionomeric Film," *J. Food Safety,* 9: 215–222.
8. Paik, J. S., M. Dhanasekharan, and M. J. Kelly. 1998. "Antimicrobial Activity of UV-Irradiated Nylon Film for Packaging Applications," *Packag. Technol. Sci.,* 11:179–187.
9. Koros, W. J. 1990. "Barrier Polymers and Structures: Overview," in *Barrier Polymers and Structure,* W. J. Koros, ed. Washington DC: American Chemical Society, pp. 1–21.
10. Tsai, B. 1996. "Amosorb: Oxygen Scavenging Concentrates for Package Structures," in *Proc. Future-Pack '96,* Skillman, NJ: Schotland Business Research, Inc., pp. 11.
11. Folland, R. 1990. "Ox-Bar. A Total Oxygen Barrier System for PET Packaging," in *Proceedings Pack Alimentaire '90.* Princeton, NJ: Innovative Expositions, Inc., Session B-2.

12. Speer, D.V., W. P. Roberts, and C. R. Morgan. 1993 "Methods and Compositions for Oxygen Scavenging," U.S. Patent 5211875.
13. Cahill, P. J. and S. Y. Chen. 1997. "Oxygen Scavenging Condensation Polymers for Bottles and Packaging Articles," PCT Application PCT/US97/1671.
14. Ching, T.Y., K. Katsumoto, S. Current, and L. Theard. 1994. "Ethylenic Oxygen Scavenging Compositions and Process for Making Same by Esterification or Trans-esterification in a Reactive Extruder," PCT Application PCT/US94/07854.
15. Katsumoto, K., T. Y. Ching, J. L. Goodrich, and D. Speer, 1998. "Photoinitiators and Oxygen Scavenging Compositions," PCT Application PCT/US98/07734.
16. Katsumoto, K. and T. Y. Ching. 1997. "Multi-component Oxygen Scavenging Composition," PCT Application PCT/US97/13015.
17. Zagory, D. 1995. "Ethylene-Removing Packaging," in *Active Food Packaging*, M. L. Rooney, ed. Glasgow and London: Blackie A & P, an imprint of Chapman & Hall, pp 38-54.
18. Holland, R. V. 1992. "Absorbent Materials and Uses Thereof," Australian Patent Application No. PJ6333.
19. Yam, K. L. and D. S. Lee. 1995. "Design of Modified Atmosphere Packaging for Fresh Produce," in *Active Food Packaging*, M. L. Rooney, ed. Glasgow and London: Blackie A & P, an imprint of Chapman & Hall, pp 55–72.
20. Stewart, R. F., J. M. Mohr, E. A. Budd, X. P. Lok, and J. Anul. 1994. "Temperature Compensating Films for Modified Atmosphere Packaging of Fresh Produce," in *Polymeric Delivery Systems-Properties and Applications,* ACS Symposium Series No.520, M. A. El-Nokaly, D. M. Pratt and B. A. Charpentier, eds. Washington D C: American Chemical Society, 232–243.
21. Labuza, T. P. and W. M. Breene. 1989. "Applications of 'Active Packaging' for Improvement of Shelf-Life and Nutritional Quality of Fresh and Extended Shelf-Life Foods," *J. Food Proc. Preservat.,* 13: 1–69.
22. Soares, N. F. F. and J. H. Hotchkiss. 1998. "Naringinase Immobilization in Packaging Films for Reducing Naringin Concentration in Grapefruit Juice," *J.Food Sci.,* 63: 61–65.
23. Anon. 1996b. "Odor Eater," *Packaging News*, August, 3.
24. Visioli, D. L. 1991. "Novel Packaging Compositions that Extend the Shelf Life of Oil-Containing Foods," U.S. Patent Application. Serial No. 07/724,421.
25. Visioli, D. L. and V. Brodie. 1994. "Method for Reducing Odors in Recycled Plastics and Compositions Relating Thereto," U.S. Patent 5,350,788.

33 Mass Transfer in Food/Plastic Packaging Systems

R. Gavara
R. Catalá

CONTENTS

Abstract

The packaging of foods is one of the major factors involved in food preservation. Packages are designed to meet the special requirements of each specific product. Many variables take part in the design (size, shape, etc.), but none is so important as the materials. In the last third of the twentieth century, plastics have been used in the packaging of many food products. They are utilized as an alternative to traditional paper, glass, and metal containers, and they have been present in the development of new products along with food technology industries. The continuous increase in plastic packaging applications is due to some attractive properties such as low cost, easy manufacture, low weight, versatility in size and shape, and a wide range of mechanical properties. However, plastic materials do allow mass transport of low molecular weight due to their peculiar morphology. Substances such as permanent gases, water vapor, food aroma components, odors, plastic residues, and additives are exchanged within the environment/package/food system. Such exchange processes are commonly called permeation, migration, and sorption. Due to the effect that mass transport has in food quality, many studies have been focused on the characterization and understanding of the process involved. In this chapter, we will introduce mass transport mechanisms, factors affecting its extent and kinetics, how they can be controlled and used in the design of a product package, and the beneficial effects that can be obtained from permeation, migration, and sorption in the preservation of food products.

33.1 ABBREVIATIONS

PE	polyethylene
LDPE	low-density polyethylene
LLDPE	linear low-density polyethylene
HDPE	high-density polyethylene
ULDPE	ultra-low-density polyethylene
ION	ionomer
EVA	ethylene-vinyl acetate copolymer
PP	polypropylene
OPP or BOPP	oriented polypropylene
PC	polycarbonate
PVC	polyvinyl chloride
EVOH	ethylene-vinyl alcohol copolymers
PVdC	copolymers of vinylidene chloride
PS	polystyrene
PA	polyamide
PAAr	aromatic polyamide
PAN	polyacrylonitrile
PET	polyester or polyethylene therephthalate
PVOH	polyvinyl alcohol
VTR	vapor transmission rate

33.2 INTRODUCTION

Food packaging is the science, art, and technology of protecting food products from the overt and inherent adverse effects of the environment. The package is the container in which the food product is contained.[1] It can be made of various types of materials ranging from glass, metal, wood, paper, or plastic. *Plastics* is the generic name of a numerous group of synthetic polymeric materials that present various common characteristics: molecules of different length constituted by the repetition of a chemical segment, high molecular weight, wide melting temperature range, similar transformation procedure, etc. Plastics are very young compared with traditional glass and metal; nevertheless, their presence in packaging as well as in other industrial sectors is very significant. Their extensive and increasing use is due to the conjunction of very attractive properties. Plastics are light and economical, and their mechanical, thermal, chemical, and optical properties vary widely and can be adjusted taking into account specific requirements. The enormous number of different materials grouped under the name "plastic" that can be mixed or combined among them or with other substances is responsible for this versatility. From a scientific point of view, plastic is a material that presents viscoelastic properties. Thus, many polymeric materials used in packaging are not plastics; e.g., polystyrene is rigid and fragile. In this contribution, however, the word "plastic" will be equivalent to polymeric material.

Due to their high and disperse molecular length, polymers cannot crystallize completely; they are amorphous or semicrystalline (crystal and amorphous regions are mixed in polymer matrices). In the amorphous regions, the polymeric molecules are coiled and physically intercrossed, leaving many empty spaces. These "holes" allow the movement of the polymer atoms (angle rotation and torsion), making the position of these empty spaces momentary. This feature constitutes the main cause of mass transport properties in food plastic packaging. Substances of low enough molecular weight can be sorbed into the holes and move through them. The results of this molecular transfer are the known permeation, migration, and sorption processes. The difference between permeation, migration, and sorption, three manifestations of a unique physicochemical phenomenon, is the origin and destination of the transferred molecule.[2] Permeation is the exchange of molecules between the external and internal atmospheres through the packaging materials. Sorption is the retention of food components by the package. Migration is the release by the package of polymeric residues and additives into the food product.

Indubitably, permeation to water vapor and oxygen are critical parameters for the preservation of many food products. The water content of foodstuffs affects both physicochemical stability (texture changes or color degradation) and microbiological stability. The presence of oxygen results in lipid rancidity, enzymatic browning, and the oxidation of vitamin C. Permeation and sorption of organic vapors are also relevant as regards food quality, since aroma components determine the aromatic quality and intensity of foods and are the main cause of consumer selection or rejection of a product. Migration of additives and residues may change organoleptic food quality and even promote toxicity.

In this chapter, we will introduce mass transport mechanisms, factors affecting its extent and kinetics, how they can be controlled and used in the design of a product package, and the beneficial effects that can be obtained from permeation, migration, and sorption in the preservation of food products.

33.3 MASS TRANSPORT MECHANISMS

Mass transport is the consequence of the natural tendency by which the chemical system advances towards equilibrium. When a food is packed in a container, the system comprises four phases: the external environment (*EE*), the package itself (*P*), the package headspace (*HS*, internal atmosphere), and the food product (*F*). Each phase is constituted by some components of a relatively high mobility, e.g., oxygen, carbon dioxide, water, or volatile organic compounds (food aroma, external odors, polymer residues), while others can be considered static (polymer and food matrices). The mobile components are transferred from the original phase into the others until all phases are equilibrated, that is, the chemical potential of substance i (μ_i^α) is equal in all phases ($\alpha = F$, HS, P, or EE).[3]

$$\mu_i^F = \mu_i^{HS} = \mu_i^P = \mu_i^{EE} \tag{33.1}$$

In practice, equilibrium is expressed in concentration (c) or pressure (p) values and the evaluation method depends on the phases involved. When a gaseous phase is involved in the process (*HS* or *EE*), equilibrium can be measured through the sorption isotherm. Depending on the profile of the isotherm, it can be described by Langmuir, Flory–Huggins, G.A.B., or B.E.T. equations. However, the isotherms for gases and vapors at low partial pressure range can be described by the simple Henry's law.[3]

$$c_i^{For P} = S_i \cdot p_i^{HS\ or\ EE} \tag{33.2}$$

where S is the solubility coefficient, which is considered dependent only on temperature for a given solute/substrate system. When equilibrium is established between two condensed phases, the partition coefficient (K) describes the distribution of the solute between phases [e.g., package (*P*) and food (*F*)]:

$$K_i = c_i^P / c_i^F \tag{33.3}$$

K, S, or μ measure the extent of the mass transport processes, since they determine the eventual equilibrium that would be achieved with time. The transport kinetics will determine the duration of the process. Gaseous and stirred liquid phases are assumed to be homogeneous while the substances move slowly within solid and viscous liquids. In these media, the kinetics is assumed to follow Fick's laws.[4]

$$J_i = \frac{qi}{A^{For\ P} \cdot t} = -D^{For\ P} \frac{\partial c_i^{For\ P}}{\partial x}; \quad \frac{\partial c_i^{For\ P}}{\partial t} = D \frac{\partial^2 c_i^{For\ P}}{\partial x^2} \tag{33.4}$$

J, the flux of substance (q is the amount of substance expressed in mass for condensable fluids or in volume for permanent gases) per unit of time (t) and surface (A), is proportional and opposite to the concentration gradient in the condensed medium. D is the diffusion coefficient, which is commonly assumed to be dependent only on temperature for a given substance/medium system. Both equilibrium and kinetics are involved in every mass transport process. Therefore, mass transport will be dependent on those variables affecting sorption or diffusion of substances in plastics.

The properties of the packaging material and the characteristics of the transferred substance affect both the extent and kinetics of mass transport.[5] Obviously, the chemical composition and structure of the polymer are related to mass transport. Chemical composition affects the level of compatibility with the penetrant and therefore the sorption. A nonpolar lipophilic polymer sorbs nonpolar organic substances preferentially, while a polar hydrophilic polymer will be avid for water and polar compounds. The chemical composition also affects the structure. Voluminous substituents and the irregularity of the chain (copolymers, tacticity, branching) will limit the crystallinity of the material, thus reducing density and increasing free volume. Crystalline regions are impermeable and act as knots between polymer chains restricting their movement. Hence, the greater the crystallinity, the lesser the value of sorption and the slower the diffusion. Large substituents increase the free volume and the sorption capacity, but they constrain segment rotation, and therefore the diffusion is slower.

The chemical composition of the penetrant is related to the compatibility with the polymer as mentioned above, but the molecule size and shape are also mass transport factors. There are fewer holes available for a large molecule than for a small one. The more rigid the molecular structure, the more difficult the movement; therefore, its diffusion is restricted.

33.4 PERMEATION

Due to the implications of permeation (to oxygen and water vapor) in the preservation of perishable foodstuffs, this manifestation of mass transport has been thoroughly studied.[6] Mass transport through plastics can occur through the package walls or through permanent holes present in package walls. Permeation through nonporous materials occurs by molecular transport through eventual itineraries. Transport through pores occurs through permanent itineraries (holes). For this reason, permeation through porous and nonporous materials are described separately.

33.4.1 PERMEATION THROUGH NONPOROUS MATERIALS

In a permeation process through nonporous package walls, the permeant molecules are first sorbed into the interface in contact with the *EE* (or *HS*), then they diffuse through the package material, and finally they are desorbed at the interface in contact with the *HS* (or *EE*). It is considered that both interfaces are always in equilibrium with the contacting gaseous phases. Thus, a different composition in *HS* and *EE* will imply a non-zero concentration gradient ($\partial c/\partial x$), and a flux of substance will be established through the package walls. When this gradient is constant, the package composition is homogeneous and, assuming that Henry's and Fick's first laws are applicable, the flux through the package thickness (ℓ) can be expressed as[4]

$$J_i = \frac{q_i}{A \cdot t} = -D_i^P \frac{\partial c_i^P}{\partial x} = D_i^P \frac{S_i(P_i^{EE} - P_i^{HS})}{\ell} = D \cdot S \frac{\Delta p}{\ell} \qquad (33.5)$$

Rearranging terms, Equation (33.5) can be used to define the permeability coefficient (*P*).

$$P = D \cdot S = \frac{q \cdot \ell}{A \cdot t \cdot \Delta p} \qquad (33.6)$$

The permeability coefficient will depend on the variables affecting D and S and, consequently, it is assumed to depend exclusively on temperature for a given permeant/plastic system. As regards barrier properties, P is an intensive magnitude useful in the design of packages. The values of P to oxygen and water vapor are common parameters in the technical specifications of polymers for food packaging applications. Contrary to glass, metal, and cellulose-based materials, the permeability values highly vary with the type of polymer, as shown in Figure 33.1 for some common materials.[6] As can be seen, the oxygen permeability values for plastic materials range nearly seven orders of magnitude. For this reason, the knowledge of P values is essential for adequate design of a food package.

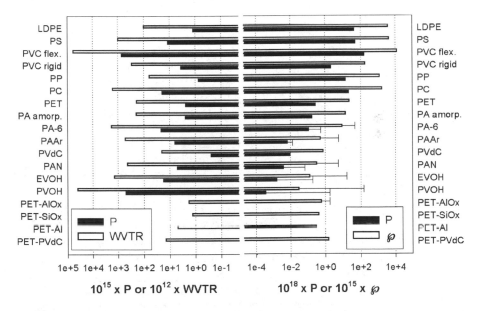

Figure 33.1 Values of oxygen permeability [P, m³ · m/(m² · s · Pa)] and permeance [$\wp$, m³ /(m² · s · Pa)] at 23°C, and water permeability [P, kg · m/(m² · s · Pa)] and water vapor transmission rate [WVTR, kg/(m² · s)] at 38°C and 85% RH for diverse plastic films and structures (12 µm thick). Errors show the effect of humidity (85% RH).

The permeability coefficient is a useful parameter for homogeneous materials. However, when the package wall is a multilayer structure, the value of permeance ($\wp$) is more useful.

$$\wp = \frac{P}{\ell} = \frac{q}{A \cdot t \cdot \Delta p} \tag{33.7}$$

Permeance is a parameter of a specific structure including thickness and therefore cannot be extrapolated to another structure made of the same basic materials. Besides permeability and permeance, permeation can be expressed in terms of transmission rates, e.g., water vapor transmission rate (*WVTR*) measured at a given temperature; and RH is a standard specification. *WVTR* is related to *P* through:

$$WVTR = \frac{P \cdot \Delta P}{\ell} = \wp \cdot \Delta p \tag{33.8}$$

Permeation values to water are expressed in both *P* and *WVTR* magnitudes in Figure 33.1 for some common food packaging materials.

In the final design of a package for a specific product, it is necessary to know the final flow of permeant considering the package as a whole. Bearing in mind that the container is made of a single structure, the final flow (*F*) would be expressed as

$$F = \frac{q}{t} = \frac{P \cdot \Delta p \cdot A}{\ell} = \wp \cdot \Delta p \cdot A = WVTR \cdot A \tag{33.9}$$

Equation (33.9) expresses the factors that can be changed to adjust the barrier of a package. Increasing or decreasing wall thickness can vary the total flow through the package. However, the value of thickness cannot be changed much without greatly affecting structural properties. Also, the package area affects the results, but the package design for a given volume hardly changes the surface in a Factor 10 (e.g., from a cubic container to a thin bag). However, the presence of this parameter in Equation (33.9) also indicates that a suitable package structure for a product cannot be adequate when the product volume changes. The environment effect is expressed as Δp. This parameter can be changed by varying the headspace composition. The last parameter to be analyzed is *P*. *P* can be varied by modifying the materials constituting the package. As seen in Figure 33.1, *P* values (or $\wp$) vary in a factor 10^7 between plastic materials, and therefore this is the most contributing factor in the final flow of vapors and gases through food packages.

The width of permeability ranges in plastic packages easily explains how these materials are applied to the packaging of very different foodstuffs, from high barrier structures for preserves to highly permeable packages for fresh produce. In some cases, the need of a barrier cannot be obtained exclusively with a single material without compromising its mechanical properties (too thick walls). The use of multilayer structures containing PVdC or EVOH is widely applied to the design of high-barrier containers. Low oxygen permeability can also be achieved by using plastic films (oriented polyester, polypropyelene, or polyamide) coated with aluminum or metal oxides such as silicon oxide (SiOx) or aluminum oxide (AlOx). The coating provides the required barrier, maintaining the flexibility of the plastic substrate.[6]

The packaging solution for a specific product can be predicted from estimations based on the equations describing the permeation processes considering the requirements of the food.

Selection of Packaging Materials for Oxygen-Sensitive Products

Many food products are sensitive to oxygen and therefore require a packaging structure capable of limiting the presence of this gas in the headspace. If the oxygen partial pressure is the key parameter in preservation, it is essential to know the maximum oxygen level that would be acceptable. Other packaging requirements such as size and type of package, packaging technology, and expected shelf life are also crucial. With this information, the maximum permeation of the package can be estimated. This value will be enough to select a few acceptable packaging materials for this specific product. The following example provides a better picture.

A food company wants to commercialize fried peanuts. They want to use their vertical form-fill-seal equipment to produce a flexible bag for 300 g of product. Such a bag will have a 600 cm² area and will inevitably contain a 100 cm³ headspace. The product requires very low oxygen content, since lipid oxidation is the main deteriorative mechanism. Therefore, the bag will be filled with modified atmosphere (nitrogen) using the flow procedure. With this technology, the initial atmosphere composition is around 3% of oxygen. The company knows that the product will be deteriorated if the oxygen partial pressure reaches 0.12 atm and would like to have a three-month shelf life. Using Equation (33.7) and substituting the above values, the maximum permeance of the packaging structure would be

$$\wp = \frac{P}{\ell} = \frac{q}{A \cdot t \cdot \Delta p} =$$

$$\frac{(9 \cdot 10^{-6} \text{m}^3)}{(600 \cdot 10^{-4} \text{ m}^2) \cdot (18000 \text{ Pa}) \cdot (3 \cdot 2592000 \text{ s})} = 1.07 \cdot 10^{-15} \frac{\text{m}^3}{\text{m}^2 \cdot \text{s} \cdot \text{Pa}}$$

This value can be achieved using various commercially available materials. A structure containing a layer of EVOH (high-barrier polymer) thicker than 1.5 μm will be sufficient. There are many commercial flexible films with 5–10 μm EVOH sandwiched between polyolefins. Using a medium-barrier polymer such as a polyamide (nylon-6), the structure should contain at least a 100-μm-thick polyamide layer so as to obtain the required permeance, but the package would lose its flexibility. Low-barrier materials such as LDPE are totally inadequate, as the layer should be thicker than 40,000 μm (4 cm). As can be seen, a simple calculation provides sufficient information to determine the range of materials suitable for our product.

Selection of Packaging Materials for Breathing Products

Fresh produce is being commercialized with the use of modified atmosphere packaging technology. To avoid anoxia, a package for fresh produce should have a permeation rate close to the respiration rate of the product. Therefore, the respiration rate is an essential parameter for making a good package design. The product, initially packaged in a container with the most beneficial atmospheric composition, breathes and exchanges gases with the package headspace, package permeability being the only mechanism to control the evolution of the gas mixture. The oxygen consumed by the product should be similar to the incoming flow through the package, while the carbon dioxide exhaled should be released to the external atmosphere through a similar permeation process. An ideal package design should maintain the initial atmosphere throughout product shelf life. In practice, the headspace composition keeps changing until an eventual equilibrium is reached. Most synthetic plastics present a ratio of oxygen permeability to carbon dioxide permeability $[r(O_2/CO_2)]$, ranging from 1/3 to 1/5. Taking into account that, during breathing, a molecule of oxygen consumed generates a molecule of carbon dioxide, this range is apparently inadequate for maintaining the proper headspace. From Equation (33.6), a film suitable to maintain the concentration of O_2 and CO_2 should have the following barrier properties:

$$F(O_2) = -F(CO_2) \Rightarrow r(O_2/CO_2) = P(O_2)/P(CO_2) = \Delta p(O_2)/-\Delta p(CO_2)$$

New developments have been made to expand this range. Nowadays, $r(O_2/CO_2)$ = 3/1 can be reached with the use of elastomers and natural polymers,[7,8] and $r(O_2/CO_2) = 1/10$ has been measured in films with different additives.[9]

In addition to the possibility of adjusting the transport of oxygen and carbon dioxide, the other problematic aspect of fresh produce packaging is the high respiration rate of some foods, which prevents any plastic film from providing sufficient permeation. This is the case with products such as berries (strawberries, raspberries, etc.), mushrooms, cauliflower, broccoli, lettuce, etc. Packaging solutions for these products are addressed by the use of plastic films with enhanced permeabilities, which include the use of perforation technologies.

33.4.2 PERMEATION THROUGH POROUS MATERIALS

The presence of pores in a food container is generally considered to be a packaging defect, since they can compromise its hermeticity. In practice, it is not unusual to find pores in the thermosealing area of a plastic package. Pores highly increase the exchange of gases between the external and internal atmospheres, and it has been proven that packaged foods are greatly at risk of microbial recontamination when the container pores are larger than 10 μm.[10,11]

With the development of MAP technologies for fresh produce, porous plastic films have become a widely used solution for food products with a high physiological activity. The perforations range from laser-made micropores to mechanically perforated macropores. Diffusion through pores is different from the type of diffusion explained above. The permeant molecules move by capillarity thanks to their thermal energy, diffusion being dependent on both pore and permeant sizes. The packaging material in this case is the passive substrate in which permanent holes are made. Permeation through pores is controlled by the collisions of gas molecules with the pore walls or other molecules. In the case of very small pores, collisions with the walls are more probable than collisions with gas molecules; the kinetics of the process is explained by Knudsen's law.[12]

$$J = -0.0224 \frac{\pi \cdot d^2}{4} D_k \frac{\Delta p}{R \cdot T \cdot \ell}; \quad D_k = \frac{d}{3} v$$

$$J = -6.8 \cdot 10^{-7} \frac{d^3 \cdot v \cdot \Delta p}{\ell \cdot T}; \quad v = \sqrt{\frac{8 \cdot R \cdot T}{\pi \cdot M}} \tag{33.10}$$

where D_k = Knudsen's diffusion coefficient, a function of the mean molecular veloc-
       ity, v (a term related to the thermal energy of the permeant)
       M = the molecular mass of the permeant
       R = the gas constant
       T = temperature

The flow of gas through a micropore ($d < 1 \times 10^{-7}$ m) depends on the dimensions of the pore (diameter, d, and length, ℓ) and the thermal energy of the gas molecules. The constant value is calculated using the international system units. The parameters of the packaging materials do not affect the permeation process. In practice, there is no commercial technology to make pores of such a diameter.

When the diameter of the pore is in the range 1×10^{-7} m $< d < 1 \times 10^{-5}$ m, the kinetics is governed by both diffusivity of gases and Knudsen's flow. Considering that two substances are permeating through the pore, the flow can be expressed as[12]

$$J_A = 2 \cdot 10^{-6} d^2 \frac{J_A}{J_A + J_B} \frac{D_{A,air} \cdot p_T}{\ell \cdot t} \ln \frac{\dfrac{J_A}{J_A + J_B}\left(1 + \dfrac{D_{A,air}}{D_{A,k}}\right) - \dfrac{P_A^0}{p_T}}{\dfrac{J_A}{J_A + J_B}\left(1 + \dfrac{D_{A,air}}{D_{A,k}}\right) - \dfrac{p_A^i}{p_T}} \tag{33.11}$$

where D_{air} and D_k = the diffusivity in the gas media and the Knudsen's diffusion
                  coefficient of each permeant (A and B)
       p_T = the total pressure

Values of diffusivity for various gases and vapors can be found in tables.[13] To solve Equation (33.11), it is necessary to utilize an iterative method. At present, there are commercial PP films with laser perforations. Pore diameter is within the micron range and the number of perforations can be as high as 1 million pores/m².

In macropores, collisions with the pore walls can be disregarded, and the permeation process is solely governed by the gas diffusivity in the environment.

$$
J_A = 2 \cdot 10^{-6} d^2 \frac{J_A}{J_A + J_B} \frac{D_{A,air} \cdot p_T}{\ell \cdot t} \ln \frac{\dfrac{J_A}{J_A + J_B} - \dfrac{P_A^0}{p_T}}{\dfrac{J_A}{J_A + J_B} - \dfrac{p_A^i}{p_T}}
\tag{33.12}
$$

Films with macropores (mechanical perforations of up to 4 mm dia.) are widely used for the commercialization of many fruits and vegetables. With the application of pores, the permeability range of plastic materials can be expanded in various orders of magnitude.

33.4.3 PERMEATION OF FOOD AROMAS

The aroma of a food product is a complex mixture of many volatile components of sufficiently low molecular weight as to undergo a mass transport process. In general, the loss of aroma does not alter the product in terms of its nutritional or toxicological properties. For this reason, this area of mass transport needs more research. Food industries are concerned with this issue, since it affects the perception of food quality by the consumer. In fact, sensorial properties are the main attribute of a food product that consumers use when selecting an item from among its competitors.

The study of food aroma/package interactions is very complex for different reasons: the aroma is a mixture of many components at very low concentration ranges and with different sensorial threshold levels. Moreover, the perception of a compound can be enhanced or occulted by other components.

From an analytical point of view, aroma mass transport is generally determined by using simplified systems, often by isolation of a single aroma compound. The values obtained are solely valid for that particular compound/package system, since the kinetics and the extent of the process differ between permeants due to differences in size, structure, polarity, vapor pressure, etc. Also, the presence of co-permeants affects mass transport.[18] Table 33.1 presents the permeation values of different aroma compounds through diverse plastics. These values show the variability of the process between permeants and polymers. The permeation of a permeant may vary several orders of magnitude depending on the selection of the package material.

The result of aroma mass transport in food quality is a gradual loss of volatile compounds, which will continue until the aroma is totally lost as permeated compounds are rapidly dispersed into the environment. Therefore, the loss of quality is a just a matter of time. It should also be noted that the compounds constituting the aroma permeate the package structure at different rates, which may promote an aroma unbalance. Table 33.1 shows the high ratio between the permeation of hexanol and limonene in polyethylene. In this material, the loss of the nonpolar aroma compounds is much faster than that of the polar ones. For those products with a

TABLE 33.1
Permeation of Aroma Components through Plastic Packaging Materials

Material	Permeant	Value	Transport process	Ref.
LDPE	ethanol	3×10^{-11} kg·m/(m²·s)	VTR through film	8
LDPE	ethyl acetate	6×10^{-10} kg·m/(m²·s)	VTR through film	8
LDPE	heptane	2×10^{-9} kg·m/(m²·s)	VTR through film	8
HDPE	Butanol	0.6%	Weight loss through bottles	8
HDPE	limonene	6.7%	Weight loss through bottles	8
HDPE	ethyl acetate	4.0%	Weight loss through bottles	8
HDPE	limonene	2×10^{-9} kg·m/(m²·s)	VTR through film	8
PP	limonene	1×10^{-10} kg·m/(m²·s)	VTR through film	8
PET	ethyl acetate	3×10^{-13} kg·m/(m²·s)	VTR through film	8
EVOH	limonene	5×10^{-12} kg·m/(m²·s)	VTR through film	8
PVdC	limonene	2×10^{-13} kg·m/(m²·s)	VTR through film	8
EVA	limonene	3×10^{-7} kg·m/(m²·s·Pa)	Permeability through film	14
LDPE	limonene	1×10^{-13} kg·m/(m²·s·Pa)	Permeability through film	14
HDPE	limonene	4×10^{-14} kg·m/(m²·s·Pa)	Permeability through film	14
PP	limonene	3×10^{-19} kg·m/(m²·s·Pa)	Permeability through film	15
PA	limonene	$<1 \times 10^{-20}$ kg·m/(m²·s·Pa)	Permeability through film	15
LLDPE	hexanal	7×10^{-16} kg·m/(m²·s·Pa)	Permeability through film	16
LLDPE	hexanol	5×10^{-15} kg·m/(m²·s·Pa)	Permeability through film	16
HDPE	hexanal	1×10^{-15} kg·m/(m²·s·Pa)	Permeability through film	16
HDPE	hexanol	3×10^{-15} kg·m/(m²·s·Pa)	Permeability through film	16
EVOH	hexanal	4×10^{-18} kg·m/(m²·s·Pa)	Permeability through film	16
EVOH	hexanol	2×10^{-17} kg·m/(m²·s·Pa)	Permeability through film	16
ULDPE	hexanal	7×10^{-16} kg·m/(m²·s·Pa)	Permeability through film	17
ULDPE	hexanol	3×10^{-15} kg·m/(m²·s·Pa)	Permeability through film	17
ULDPE	decane	5×10^{-15} kg·m/(m²·s·Pa)	Permeability through film	17
ULDPE	ethyl caproate	5×10^{-15} kg·m/(m²·s·Pa)	Permeability through film	17
ULDPE	limonene	6×10^{-15} kg·m/(m²·s·Pa)	Permeability through film	17
ULDPE	phenylethanol	3×10^{-15} kg·m/(m²·s·Pa)	Permeability through film	17
ION	hexanal	1×10^{-16} kg·m/(m²·s·Pa)	Permeability through film	17
ION	hexanol	2×10^{-15} kg·m/(m²·s·Pa)	Permeability through film	17
ION	decane	1×10^{-15} kg·m/(m²·s·Pa)	Permeability through film	17
ION	ethyl caproate	8×10^{-16} kg·m/(m²·s·Pa)	Permeability through film	17
ION	limonene	3×10^{-15} kg·m/(m²·s·Pa)	Permeability through film	17
ION	phenylethanol	3×10^{-16} kg·m/(m²·s·Pa)	Permeability through film	17
PET	hexanal	3×10^{-18} kg·m/(m²·s·Pa)	Permeability through film	17
PET	hexanol	1×10^{-16} kg·m/(m²·s·Pa)	Permeability through film	17
PET	decane	8×10^{-17} kg·m/(m²·s·Pa)	Permeability through film	17
PET	ethyl caproate	5×10^{-18} kg·m/(m²·s·Pa)	Permeability through film	17
PET	limonene	1×10^{-17} kg·m/(m²·s·Pa)	Permeability through film	17
PET	phenylethanol	2×10^{-17} kg·m/(m²·s·Pa)	Permeability through film	17

high-value aroma, permeation is reduced by the use of high-barrier plastic materials, metallized films, or aluminum foil.

33.5 MIGRATION

Migration is the release of substances initially present in the package into the food product. Plastics contain numerous low-molecular-weight substances that can be transferred. Some migrants are residues from polymer synthesis such as monomers, oligomers, catalizers, detergents, solvents, etc. Others are additives that are mixed with the polymers to enhance their properties, such as photostabilizers, antioxidants, lubricants, plasticizers, antifog agents, ink pigments and solvents, etc. These substances—already in the package—diffuse through the structure until they reach the interfaces. There, they are partially transferred to the headspace and/or dissolved in the food phase. When this occurs, the food product may be altered to produce off-flavors, color changes, etc. These processes result in a loss of quality of the product, which shortens its shelf life. Moreover, some of the transferable substances may be toxic by ingestion, and therefore they can induce toxicity to the content. Examples of such compounds are vinyl chloride, styrene, benzene, bisphenol A and BADGE, acrylonitrile, etc. For this reason, migration has been regulated by national and supranational laws.[19] The regulations establish maximum migration limits for plastics in contact with foods and limit plastic additives to those included in positive lists. In some cases, the regulations define the tests that the materials should pass to be acceptable for packaging applications, including compositional limits for some potential migrants.

The theoretical prediction or description of migration is necessary to understand the mechanisms of mass transport but, in practice, they are not very useful to determine the suitability of a package structure for food contact. It is more common to determine the migration values experimentally following regulatory procedures. In these, global and specific migration terms are defined. Global migration is the total mass released from the package into the food irrespective of its composition. Specific migration is the mass of a compound with special (toxic or organoleptic) significance. Analytical procedures are also listed in the regulations. Food products are substituted by simulants that are assumed to behave similarly to the specific food. Temperatures and time of exposure are also fixed depending on the product shelf life and the thermal profile that packaged food will suffer.[19]

As regards mass transport, migration processes present the same mechanisms as permeation, with the exception of the first sorption step since the migrants are already sorbed in the package and they are governed by the same laws. The diffusion of migrants in both package and food follows Fick's laws [Equation (33.4)]. The equilibrium can be described by Equations (33.2) and (33.3) for volatile and non-volatile migrants, respectively. Contrary to permeation of permanent gases, migrants desorbed from the package are sorbed into the food and diffuse throughout the food matrix. The theoretical description of a migration process considering all the steps mentioned above is complex and, in many cases, some steps are disregarded. One of the most frequent assumptions is to consider that the food is a well stirred liquid;

that is, migrant diffusion within the food phase is instantaneous and migrant concentration homogeneous. Although this consideration may cause discrepancies with respect to real situations, especially with solid foods, it results in an overestimation of the process (worst case) that is desirable to guarantee food quality. The migration is dependent on the diffusion kinetics within the package and the partition coefficient at the food/package interface.[3]

$$M(t) = \frac{A \cdot \ell \cdot K + V_s}{A \cdot \ell \cdot V_s \cdot c_i^P} \left\{ 1 - \sum_{n=1}^{\infty} \left[\frac{2\alpha(1 + \alpha)}{1 + \alpha + \alpha^2 q_n^2} \exp\left(\frac{D^P q_n^2 t}{\ell^2} \right) \right] \right\}; \qquad (33.13)$$

$$\tan(q_n) = -\alpha q_n$$

In Equation (33.13), the migrated mass at a time $t[M(t)]$ is a function of the initial migrant concentration in the package (c_i^P), D, and K.[4] This expression has been scarcely used when describing migration processes due to the number of variables and the necessity of using tables, as there is no explicit solution for q_n calculation. For this reason, migrants are considered to be easily solved in the food ($K \ll 1$). Taking into account both assumptions, the migration process solely depends on the migrant diffusion through the packaging material.

$$M(t) = 2c_i^P \sqrt{\frac{D^P t}{\pi}} \qquad (33.14)$$

Some studies have been carried out to characterize and describe migration. The data obtained are controversial, since the assumptions considered are not always realistic. The use of a $K \ll 1$ drives to controversial conclusions: non-Fickian processes, effect of film thickness, large discrepancies on D values, etc.[20] Much work is further needed to fully understand these complex processes.

It can be said that virgin plastic materials for food packaging applications are basically safe according to international legislation. However, the addition of recycled materials and/or additives to aid processing, as well as printing pigments and solvents and even by-products from degradation during transformation, can drive the final package to surpass maximum migration limits. Present trends are focused on the study and development of functional barriers. A functional barrier is a layer of an approved food-contact material that is applied as the innermost layer of the structure to delay or avoid the migration of compounds into food. Besides significant analytical efforts, most of the work is being carried out through theoretical predictions.

In general, migration processes are observed as a negative characteristic of food packages, since they represent the transference of undesired substances (e.g., vinyl chloride). However, some of the new trends in food packaging science (active and smart packaging) are based on migration.[21] These new concepts consider the addition to polymers of substances whose migration into the food product would cause a beneficial effect in food preservation. Examples of such substances are antioxidants, aromas, or antifungal agents. Instead of direct addition to foodstuffs, the introduction of these compounds within the package walls can provide some benefits: the initial concentration of the additive is highly reduced, the substance is released into food at the adequate dose, and the amount of substance in the food increases during storage as required. A film containing ethanol has been proved to reduce the microbiological spoilage of grapes.[22]

Present regulations do not take into account the above new packaging concepts. FDA regulations define packaging as a passive barrier to reduce food deterioration. Therefore, active migrants would have to be considered as food additives and be included in food compositional information. The EU legislation is broader in the definition and, legally, active packaging should comply with the current global and specific migration limits. Since active compounds are nonhazardous, no specific limits should be expected, although they should be included in the positive list of additives. However, these packages are likely to fail the global migration limits, as the active principle is released, this increasing the global migration result. An EU project is presently studying the necessary amendments to EU regulations with a view to including active packaging.

33.6 SORPTION

The third manifestation of mass transport is the retention of food components in the package structure. Components potentially sorbed by the package are main constituents such as water, oil, or fat, and minor constituents such as sugars, alcohols, aroma components, colorants, etc. Sorption processes are hardly implicated in food spoilage, since sorbed components are not related to microbiological or toxicological effects. For this reason, sorption is the least reported mass transport process in food packaging literature. However, sorption of certain compounds may result in consumer rejection. This is the case with a product that partially loses an active organoleptic compound (a key aroma compound). The sorption of other compounds (vitamins or sugars) cannot be easily detected by taste.

Main constituents are usually sorbed in significant amounts, since their concentrations in the food phase are high. From the food point of view, the loss of a 1% of water in an aqueous product is not critical. However, it can be critical for the package. If the food component and the packaging materials are compatible, sorption may result in polymer swelling and even dissolution (as occurs in PVOH films in the presence of water). In these cases, the packaging properties are affected, and the mass transport properties largely increase because of structure relaxation. The solubility of alcohols in EVOH increases by a 100 factor in the presence of water.[16] In general, this does not occur, since aqueous foods are packaged in hydrophobic materials. On the other hand, lipophilic materials are commonly utilized in the preservation of fatty foods. It has been shown that the sorption of oil in ULDPE results in an increment of the solubilities of ethyl caproate, limonene, and decane, while the retention of alcohols is significantly reduced.[17] Moreover, sorption may increase the mass transport of polymer additives and residues that are soluble in the food main constituent (values of global migration from polyolefins to fatty food simulants are commonly higher to heptane than oils, and those higher than ethanol).

Minor constituents are partitioned between the food and the package. The sorption level depends on the solubility of the sorbate in both food and package and is evaluated through the partition coefficient as expressed in Equation (33.3). Since the concentration of these components is low, the sorption in absolute terms is supposed to be low, although it may be significant in relative terms. An LDPE film (area = 2 dm², thickness = 50 µm thick) retained 40% of the limonene present in 250 ml of orange juice).[23] In addition to the effects on some properties of the foodstuff (sensorial, nutritional, etc.), sorption of minor constituents may alter package properties. This is the case with delamination of polymeric structures and complex packages caused by the sorption of acid compounds and aroma constituents.[24–26]

The main body of literature related to sorption processes deals with aroma scalping.[27] As regards consumer perception of quality, aromas are crucial food components. They are complex mixtures of organic compounds (mainly terpenes, esters, aldehydes, ketones, and alcohols) present in food in concentrations within the ppm range or below. Their sorption within the packaging materials may decrease the organoleptic properties of the product and results in an aroma imbalance, since their components are retained at different levels. Perhaps, this phenomenon is more significant in aqueous products such as juices. Juice is commonly packaged in hydrophobic materials to avoid the negative effect of water and inevitably, organic compounds such as aromas present a strong preferential sorption by the package. Table 33.2 shows the retention values of diverse aroma constituents in plastic materials. Values are representative of the wide range of flavor scalping that may be present as a function of the package structure. As an example, the content of d-limonene in orange juice stored in packaging structures using polyethylene as thermosealable layer in direct contact with food is 40% reduced due to scalping.[23]

Sorption processes can also be used beneficially when the sorbate is an undesired food component. At present, there are some developments of films with an ability to sorb lactose from milk or cholesterol from beaten egg. The films contain some additives, such as cyclodextrins, that greatly retain specific components. Other developments include the addition of enzymes to decrease the concentration of a specific component such as naringin.[28]

33.7 CONCLUSIONS

Without a doubt, mass transport processes in food packaging with plastic materials affect the preservation of food products. Such processes, known as permeation, migration, and sorption, have been considered negative, since they are present in all packages irrespective of the chemical composition of the structure and at a much greater level than in traditional materials such as tin plate or glass. The measurement and control of mass transport is essential to assure food quality and product shelf life. However, through the characterization and understanding of these processes,

TABLE 33.2
Retention of Aroma Components in Plastic Packaging Materials

Material	Sorbate	Value	Transport process	Ref.
ULDPE	hexanal	2×10^{-3} kg/(m³·Pa)	Solubility in film	17
ULDPE	hexanol	2×10^{-2} kg/(m³·Pa)	Solubility in film	17
ULDPE	decane	4×10^{-2} kg/(m³·Pa)	Solubility in film	17
ULDPE	ethyl caproate	3×10^{-2} kg/(m³·Pa)	Solubility in film	17
ULDPE	limonene	7×10^{-2} kg/(m³·Pa)	Solubility in film	17
ULDPE	phenylethanol	2×10^{-1} kg/(m³·Pa)	Solubility in film	17
ION	hexanal	2×10^{-3} kg/(m³·Pa)	Solubility in film	17
ION	hexanol	3×10^{-1} kg/(m³·Pa)	Solubility in film	17
ION	decane	7×10^{-2} kg/(m³·Pa)	Solubility in film	17
ION	ethyl caproate	6×10^{-2} kg/(m³·Pa)	Solubility in film	17
ION	limonene	4×10^{-1} kg/(m³·Pa)	Solubility in film	17
ION	phenylethanol	2×10^{-1} kg/(m³·Pa)	Solubility in film	17
PET	hexanal	1×10^{-3} kg/(m³·Pa)	Solubility in film	17
PET	hexanol	5×10^{-2} kg/(m³·Pa)	Solubility in film	17
PET	decane	2×10^{-2} kg/(m³·Pa)	Solubility in film	17
PET	ethyl caproate	3×10^{-3} kg/(m³·Pa)	Solubility in film	17
PET	limonene	1×10^{-2} kg/(m³·Pa)	Solubility in film	17
PET	phenylethanol	1×10^{-2} kg/(m³·Pa)	Solubility in film	17
LLDPE	hexanal	2×10^{-2} kg/(m³·Pa)	Solubility in film	16
HDPE	hexanal	8×10^{-2} kg/(m³·Pa)	Solubility in film	16
EVOH	hexanal	6×10^{-3} kg/(m³·Pa)	Solubility in film	16
LLDPE	hexanol	3×10^{-3} kg/(m³·Pa)	Solubility in film	16
HDPE	hexanol	2×10^{-2} kg/(m³·Pa)	Solubility in film	16
EVOH	hexanol	1×10^{-3} kg/(m³·Pa)	Solubility in film	16
EVA	limonene	2 kg/(m³·Pa)	Solubility in film	14
LDPE	limonene	2 kg/(m³·Pa)	Solubility in film	14
LLDPE	limonene	8×10^{-1} kg/(m³·Pa)	Solubility in film	14
HDPE	limonene	5×10^{-1} kg/(m³·Pa)	Solubility in film	14
PET	limonene	0.17	Partition between drink and bottle	29
PET	myrcene	0.26	Partition between drink and bottle	29
LDPE	hexanal	4.3	Partition between solution and film	30
LDPE	hexanol	0.7	Partition between solution and film	30
LDPE	hexyl acetate	19	Partition between solution and film	30

plastic packages can be designed so that mass transfer actively collaborates in the preservation of food products. Present and future trends in packaging science and technology are focused on the development of active packaging with selective permeability to certain gases, migration of substances with a beneficial activity, and retention of food components that may promote undesirable effects on human health.

REFERENCES

1. Brody, A. L. and K. S. Marsh. 1997. *The Wiley Encyclopedia of Packaging Technology*, New York: John Wiley and Sons, Inc., pp. xi.

2. Gavara R., R. Catalá, P. M. Hernández-Muñoz, and R. J. Hernández. 1996. "Evaluation of Permeability Through Permeation Experiments: Isostatic and Quasiisostatic Methods Compared," *Packag. Technol. Sci.,* 9(4): 215–224.

3. Hernández, R. J. and R. Gavara. 1999. *Plastic Packaging. Methods for Studying Mass Transfer Interactions*, Surrey, U.K.: Pira International, pp. 33 and 44.

4. Crank, J. 1975. *The Mathematics of Diffusion*, London, UK: Clarendon Press.

5. Salame, M. 1989. "The Use of Barrier Polymers in Food and Beverage Packaging, "in *High Barrier Plastic Films for Packaging*, K. M. Finlayson ed. Lancaster, PA: Technomic Pub. Co., Inc., pp. 132–145.

6. Catalá, R. and R. Gavara 1996. "Review: Alternative High Barrier Polymers for Food Packaging," *Food Sci. Technol. Int.*, 2: 281–291.

7. McHugh, T. H. and J. M. Krochta. 1994. "Permeability Properties of Edible Films," in *Edible coatings and Films to Improve Food Quality*. Lancaster, PA: Technomic Pub. Co., Inc., pp. 139–188.

8. Anonymous. 1995. *Permeability and Other Film Properties of Plastics and Elastomers*. New York: Plastics Design Library, pp. 485–487.

9. Watanabe, K. and R. Tanaka. 1997. "Thermoplastic Resin Compositions with High Permeability Ratio of Carbon Dioxide Gas/Oxygen Gas as Packaging Materials for Perishable Foods," *Jpn. Kokai Tokkyo Koho,* JP 11001624 A2.

10. Lee, D. S. and H. D. Paik. 1997. "Use of a Pinhole to Develop an Active Packaging System for Kimchi, a Korean Fermented Vegetable," *Packaging Technol. Sci.*, 10 (1): 33–43.

11. Hernández, R. J. 1995. "Mass Transfer Phenomena in Polymeric Materials, Permeability," presented at Technical seminar: High-barrier plastic materials for packaging design. Current and future trends, Valencia (Spain), June 22–23, 1995.

12. Hernández, R. J. "Food Packaging Materials, Barrier Properties and Selection," in *Handbook of Food Engineering Practice*, K. J. Valentas, E. Rotstein, and R. P. Singh, eds., Boca Raton, FL: CRC Press, pp. 291–360.

13. Perry, R. H. and D. W. Green, eds. 1984. *Perry's Chemical Engineer's Handbook*. New York: McGraw-Hill, pp. 3-256, 3-257.

14. Kobayashi, M., T. Kano, K. Hanada, and S. I. Osanai. 1995. "Permeability and Diffusivity of D-Limonene Vapor in Polymeric Sealant Films," *J. Food Sci.*, 60(1): 205–209.

15. Apostolopoulus, D. and N. Winters. 1991."Measurement of Permeability for Packaging Films to D-Limonene Vapour at Low Levels," *Packaging Technol. Sci.*, 4: 131–138.

16. Johansson, F. and A. Leufven. 1994. "Food Packaging Polymer Films as Aroma Vapor Barrier at Different Relative Humidities," *J. Food Sci.*, 59(6): 1328–1331.
17. Hernández-Muñoz, M. P. 2000. *Estudio de interacciones envase-alimento y su incidencia en el aroma de alimentos grasos envasados con materiales plásticos*, Ph.D. Dissertation, Univ. Valencia (Spain).
18. Delassus, P. T., J. C. Tou, M. A. Babinec, D. C. Rulf, B. K. Karp, and B. A. Howell. 1988. "The Transport of Apple Aromas in Polymer Films," *ACS Symp. Ser.*, 365: 11–27.
19. Ashby R., I. Cooper, S. Harvey, and P. Tice, eds. 1997. *Food Packaging Migration and Legislation*. Surrey, U.K.: Pira International.
20. Garde, J. A. 2000. *Efecto de los Tratamientos Térmicos Sobre la Migración de Antioxidantes en Poliolefinas*. Ph.D. Dissertation, Univ. Valencia (Spain).
21. Rooney, M. L., ed. 1995. *Active Food Packaging*. Glasgow, UK: Blackie Academic & Professional.
22. Hotchkiss, J. H. 2000. "Current and Future Trends in Active Packaging," presented at the II Food Packaging International Congress RISEA-2000, March 14–16, 2000.
23. Imai, T., B. R. Harte, and J. R. Giacin. 1990. "Partition Distribution of Aroma Volatiles from Orange Juice into Selected Polymeric Sealant Films," *J. Food Sci.*, 55(1): 158–161.
24. Pieper, G. and K. Petersen. 1995. "Free Fatty Acids from Orange Juice Absorption into Laminated Cartons and their Effects on Adhesion," *J. Food Sci.*, 60(5): 1088–1091.
25. Nielsen, T. J. and G. E. Olafsson. 1995. "Sorption of Beta-Carotene from Solutions of a Food Colorant Powder into Low Density Polyethylene and Its Effect on the Adhesion Between Layers in Laminated Packaging Material," *Food Chem.*, 54(3): 255–260.
26. Schroeder, M. A., B. R. Harte, J. R. Giacin, and R. J. Hernández. 1990. "Effect of Flavor Sorption on Adhesive and Cohesive Bond Strengths in Laminations," *J. Plastic Film Sheeting*, 6(3): 232–246.
27. Nielsen, T. and M. Jaegerstad. 1994. "Flavour Scalping by Food Packaging," *Td. Food Sci. Technol.*, 5(11): 353–356.
28. Soares, N. F. F. and J. H. Hotchkiss. 1998. "Naringinase Immobilization in Packaging Films for Reducing Naringin Concentration in Grapefruit Juice," *J. Food Sci.*, 63(1): 61–65.
29. Nielsen, T. J. 1994. "Limonene and Myrcene Sorption into Refillable Polyethylene Terephtalate Bottles, and Washing Effects on Removal of Sorbed Compounds," *J. Food Sci.*, 59(1): 227–230.
30. Nielsen, T. J., I. M. Jaegerstad, and R. E. Oeste. 1992. "Study of Factors Affecting the Absorption of Aroma Compounds into Low Density Polyethylene," *J. Sci. Food Agric.*, 60(3): 377–381.

34 Mass Transport within Edible and Biodegradable Protein-Based Materials: Application to the Design of Active Biopackaging

S. Guilbert
A. Redl
N. Gontard

CONTENTS

Abstract

Advantages, types, formation, and properties of agro-polymer based materials are presented in this chapter, with emphasis for protein based materials. The structuration and rheological properties of the network depend on the material-forming technology (casting or thermoforming), the process variables (plasticizer, temperature, shear rate, etc.), and the usage conditions of the film (relative humidity, temperature).

Some unique mass transport (gas selectivity, solute release) properties and their ability to be modified and controlled opens a new field of industrial application for these macromolecules in the field of edible or biodegradable "active packaging."

34.1 INTRODUCTION

Edible and biodegradable packaging materials produced from agricultural origin macromolecules offer numerous advantages (renewable and biodegradable materials) over conventional plastics.[1,2] Applications of bioplastics can be classified in three categories.

1. Plastics to be composted or recycled (fields where reuse or fine recovery are difficult)
2. Plastics used in a natural environment (fields where recovery is not economically or practically feasible)
3. Special plastics (fields with specific features were bioplastics possess preferential properties)

Different techniques[1,2] using fully renewable agricultural raw material to make what can be called *agro-polymers* are summarized in Figure 34.1.

Our studies demonstrated that proteins have interesting material-forming properties. Homogeneous, transparent, strong, water resistant, highly permeable to water vapor, and highly gas (CO_2/O_2) selective protein-based materials have been obtained

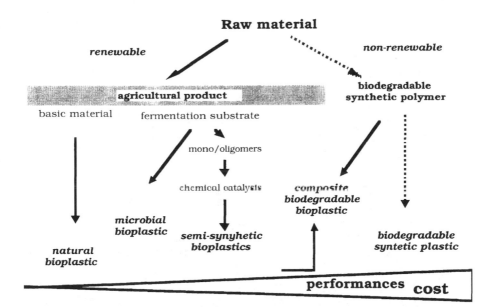

Figure 34.1 Different approaches in making bioplastics.

either by "casting" techniques or by "thermoplastic processing." Combination of proteins with lipids, natural fibers, paper, or conventional plastic materials was found to improve material water vapor barrier or mechanical properties. These properties can be exploited to form edible and/or biodegradable packaging that could provide a supplementary and sometimes essential means for controlling physiological, micro-biological, and physicochemical changes in food products.[3] Many protein materials can be used: collagen, zein, wheat gluten, ovalbumin, soybean, casein, etc.[4,5] We will expand on the use of protein films as active layers (i.e., when the film itself contributes to the food preservation) with some examples of potential uses of these biomaterials (e.g., modified atmosphere packaging or controlled release of food preservatives). Figure 34.2 gives a schematic representation of food preservation with edible/biodegradable films as active layers when the first mode of deterioration results from respiration, dehydration, or moisture uptake, or from surface microbial development or oxidation. The protective feature of the film is dependent on gas and water vapor barrier properties, on modification of surface conditions, and on its own antimicrobial properties.[6,7]

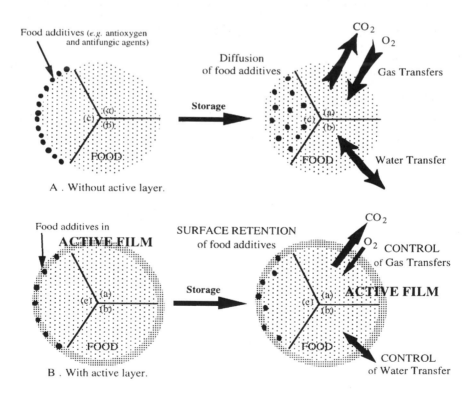

Figure 34.2 Schematic representation of food preservation with or without edible films and coatings as active layers, when the first mode of deterioration results from respiration (a), from dehydration or moisture uptake (b), or from surface microbial development or oxidation (c).

34.2 PROTEIN-BASED MATERIALS, ELABORATION, AND RHEOLOGY

Protein undergoes glass transition phenomenon,[8] and the controlled presence of water or other plasticizers lowers the glass transition temperature below the breakdown temperature (molecular degradation). Proteins are thus allowed to be shaped by thermoplastic processes such as extrusion (Figure 34.3) as standard synthetic polymers with similar transformation costs.

The *glass transition temperature* of various proteins has been studied as a function of water content and various other plasticizers.[9–11] The general behavior of the glass transition temperature broadly follows the Couchmann–Karasz relation. Comparing the plasticizer efficiency on a weight basis, a similar plasticization was obtained with plasticizers of different molecular structure (hydroxyl or amino groups). The plasticizing efficiency (i.e., decrease of T_g) at equal molar content is generally proportional to the molecular weight and inversely proportional to the percent of hydrophilic groups of the plasticizer. Migration rate of the plasticizers in the polymer is related to their physicochemical characteristics. In this study, it was assumed that polar substances interacted with readily accessible polar amino acids, whereas amphiphilic ones interacted with nonpolar zones that are, in most proteins, buried and accessible with difficulty.[11] The *rheological behavior* of a gluten material prepared in a two-blade counter-rotating batch has been studied by Redl et al.[12,13] (Figure 34.4). During batch mixing, the evolution of torque is characterized by a lag phase, followed by an exponential increase up to a maximum and thus a continuous decrease. Viscous heat dissipation is very important and leads to an increase of temperature of about 60°C. Maximum torque is obtained for a stable level of specific energy (500–600 kJ/kg) and temperature (50–60°C) independent of processing conditions.

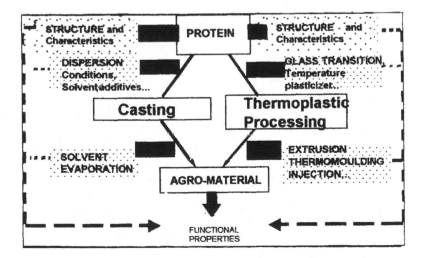

Figure 34.3 General mechanisms of formation of protein-based agromaterial.

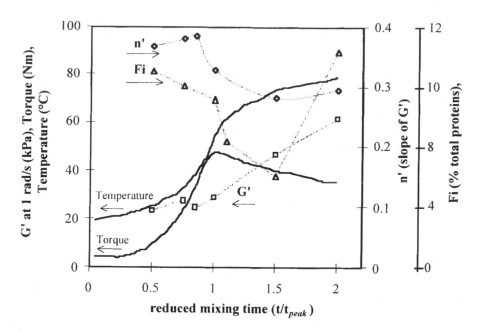

Figure 34.4 Evolution of torque, temperature, storage modulus G' at 1 rad/s (▫), the slope of the storage modulus n' (◊), and the high-molecular-weight fraction Fi (△) with reduced mixing time (t/t_{peak}) (80°C, 65% gluten, 45% glycerol, 30 rpm).

In the first stage of material structuration (Figure 34.4), plasticization occurs. The reason for the drastic change at maximum torque and 50–60°C may be an exposure of hydrophobic sites, reinforcing a gel-like behavior. Thereafter, the network structure might be stabilized with covalent cross links, probably via disulphide bonds through free-radical reaction mechanisms, leading to the observed increase of molecular size. This structuration mechanism is probably a temperature-controlled phenomenon, dependent on mechanical energy input, but quite independent of mixing conditions and glycerol content.

The influence of temperature and plasticizer content can be roughly determined with time/temperature and time/plasticizer shifts. A general expression was proposed by Redl et al.[12] for describing the viscous behavior of a gluten/glycerol mix, which allows the modeling of the flow behavior in a twin-screw extruder.

Extrusion of wheat proteins plasticized with glycerol appeared to be feasible in steady-state conditions. However, an important problem encountered within the extrusion process was the high viscosity of the "proteic melt" and its reactivity at high temperatures. Therefore, the possible processing conditions were restricted at a given glycerol content (35% w/w), and this window would be even smaller or vanish at a lower plasticizer content. Best results were obtained at low temperature (60°C) and low specific energy input, which led to smooth surfaced materials with a molecular weight distribution close to native gluten. Evolution of molecular size distribution along the screw shows a slight depolymerization in high shear rate zones, but polymerization is globally dominant and positively related to specific mechanical energy input.

The *mechanical behavior* of the protein material is network-like. This behavior has been characterized and modeled for wheat gluten (mainly composed of glutenins and gliadins) by using Cole–Cole distributions.[13]

The molecular weight between network strands or entanglement couplings as derived from the rheological properties, corresponds to 1–3 times the glutenin subunit and suggests that the building block of the network is the glutenin subunit. Changes in molecular size of proteins during extrusion have been measured by chromatography and appear to be correlated to the molecular size between network strands, as derived from the rheological properties of the obtained materials. Increasing network structure appeared to be induced by the severity of the thermomechanical treatment, as indicated by specific mechanical energy and the maximum temperature reached. Evolution of molecular size distribution along the screw shows a slight depolymerization in high shear rate zones, but polymerization is globally dominant and positively related to specific mechanical energy input. Gliadins might react as cross-linking agents.

34.3 CONTROL OF SOLUTE TRANSPORT

Aside the mechanical properties of "agro-plastics," the control of mass transfer properties may be important, especially for applications that are designated to be in contact with food. For such applications, the migration of plasticizers or additives into the food bulk is generally undesired, unless a controlled release of a chosen additive is wanted ("active packaging"). The improvement of food microbial stability can be obtained by using active layers such as surface retention agents to limit food additives (particularly antioxygen and antifungic agents) diffusion in the food core.[6,7,14–16] Maintaining a locally high effective concentration of preservative may allow to a considerable extent reduction of its total amount for the same effect (Figure 34.5). It is then important to be able to predict and control surface preservative migration. The mathematical theory of diffusion in isotropic substances is based on the hypothesis that the rate of transfer of a diffusing substance through unit area of a section is proportional to the concentration gradient normal to the section, i.e., Fick's first law. One method to quantify the diffusion coefficient is to study the desorption of a solute from a thin layer of sorbic into a stirred liquid medium.

In the case of Fickian diffusion, the initial desorption is linear versus the square root of time and a diffusion coefficient can be derived.[17] As for the rheological properties, morphological or physicochemical changes of the matrix are reflected in the migration properties of the diffusant. The study of the influence of variables such as temperature and composition of the matrix can therefore be of precious interest for structural insight. Fickian *diffusion* has been identified as the predominant release mechanism of sorbic acid desorption from thin protein layers. The diffusion coefficient of sorbic acid in a gluten based film was found to be 7.6×10^{-12} m^2/s (Table 34.1). Addition of lipidic components led to a 20–50% reduction in sorbic

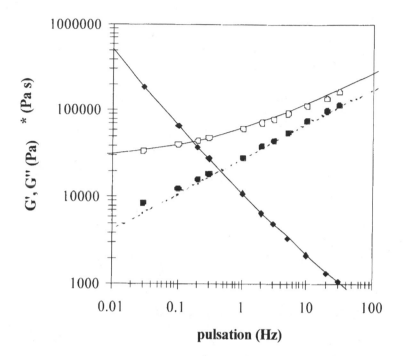

Figure 34.5 Evolution of storage G' (□) and loss G'' (■) modulus and of complex viscosity η^* (◆) with pulsation. Lines represent the fit of Cole–Cole functions. ($N = 100$ rpm, $Q = 4.8$ kg/h; $Tr = 80°C$); adapted from Reference 13 with permission from Cereal Chemistry.

TABLE 34.1
Diffusion Coefficient of Sorbic Acid from Gluten and Lipid–Based Films

	T = 20°C $D \times 10^{12}$ (m²/s)	T = 10°C $D \times 10^{12}$ (m²/s)	T = 4°C $D \times 10^{12}$ (m²/s)
Gluten	7.5	4.1	3.1
Gluten-beeswax	5.6	3.0	2.2
Gluten-acetylated monoglycerides	3.2	2.2	1.6
Acetylated monoglycerides	0.27	—	—
Beeswax	0.00024	—	—

acid diffusivity but remained far from the diffusivity in the corresponding pure lipidic films ($2.4 \ 10^{-16}$ m²/s and $2.7 \ 10^{-13}$ m²/s for beeswax and acetylated monoglycerides, respectively). The effect of temperature could be described by an Arrhenius-type law with activation energies ranging from 30.0 to 39.8 kJ/mole.

An explicit numerical method for the *modeling* of the *migration* of sorbic acid from a surface layer into food is proposed, allowing the tracing of the evolution of the concentration profiles of sorbic acid within the surface layer and the food. The model was validated with experimental data obtained with wheat gluten and beeswax films placed on agar-agar gels as model foods. Modeling showed that, in the case of a gluten film, the surface concentration drops below 10% of the initial value after 1 h whereas, in the case of a beeswax film, surface concentration remains above 75% after one week. Simulation lead to the conclusion that, to achieve a significant surface retention, sorbic acid diffusivity in the edible surface layer has to be less than 10^{-15} m^2/s.

Microbiological analysis has confirmed the efficiency of preservative retention within surface coatings. The antimicrobial effect of various gluten/sorbic acid based films was evaluated against *Penicillium notatum*. The study was conducted at two temperatures on an acid model food. Simple gluten-based films had no fungicidal effect; however, the addition of lipidic compounds (e.g., datem or beeswax) to these films delayed fungi growth. Sorbic acid retention by all films was highly dependent on temperature. At 4°C, gluten-based films containing sorbic acid delayed the *Penicillium notatum* growth for four days, while no effect was observed at 30°C. The gluten/lipid-based films showed a strong sorbic acid retention and a marked fungicidal effect, either at 30 or 4°C, delaying the *Penicillium notatum* growth for more than 21 days.

34.4 CONTROL OF GAS TRANSPORT

The development of biopackaging films with selective gas permeability (oxygen, carbon dioxide, ethylene) that allow control of respiration exchange seems very promising for achieving a "modified atmosphere" effect in fresh "living" products such as fruit, vegetables, and cheese microflora. An improvement of the storage potential of these products, as schematized in Figure 34.6, could then be expected.

Films formed with wheat gluten have particularly good oxygen and carbon dioxide barrier properties under low-moisture conditions. Oxygen permeability of wheat gluten film was found to be 800 times lower than low-density polyethylene and twice lower than polyamide 6, a well known high oxygen barrier polymer.

Increased a_w promotes both gas diffusivity (due to the increased mobility of hydrophilic macromolecule chains) and gas solubility (due to the water swelling of the matrix), leading to a sharp increase in gas permeability. With carbon dioxide, the sharp increase of permeability is more important than with oxygen permeability.

The selectivity coefficient between carbon dioxide and oxygen (defined as the ratio of the respective permeabilities of both gases) is sensitive to moisture variations. The selective coefficient of edible gluten films[18,19] varies from 4.0 at $a_w = 0.30$ to 25 at $a_w = 0.95$, whereas the selectivity coefficient for synthetic polymers remains relatively constant, at 4 to 6. This could be explained by the differences in water solubility of these gases (i.e., carbon dioxide is very soluble) but also to specific interactions between carbon dioxide and the water plasticized proteic matrix, as schematized in Figure 34.7.[18]

The exceptional selectivity of wheat gluten films can be qualified as sorption selectivity (while, for main conventional selective films, the mechanism of selectivity of diffusion is different).

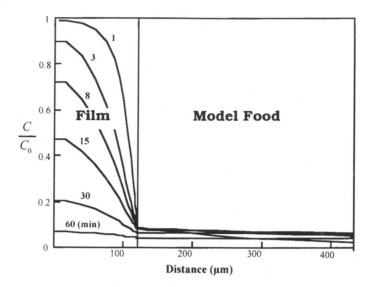

Figure 34.6 Theoretical evolution of concentration profile of sorbic acid in wheat gluten film placed on the model food as a function of time. Calculated values are obtained using $D_{Film} = 7.5 \times 10^{-12}$ m²/s, $D_{model\,food} = 9 \times 10^{-10}$ m²/s. (C sorbic acid concentration at time t, C_0 initial concentration of sorbic acid in film).

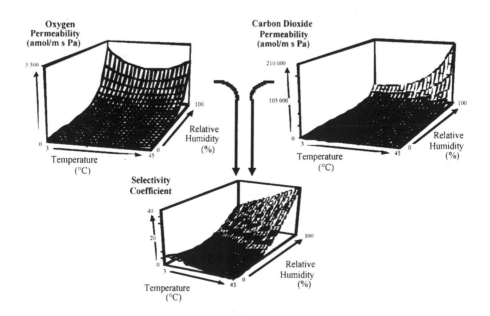

Figure 34.7 Evolution of O_2 and CO_2 permeability in gluten films as a function of temperature and relative humidity. Resulting effect on gas selectivity.

Some selective gluten-based films have been shown to lead to the creation of original atmospheres when used to wrap fresh vegetables. The evolution of atmosphere composition around fresh mushrooms placed in a glass jar covered with various films was studied. With a wheat gluten film, atmosphere composition rises to an equilibrium of 2–3% CO_2 and 2–3% O_2, i.e., a modified atmosphere within the semi-permeable package is produced after application. This high *in situ* selectivity leads to a modified atmosphere that is favorable to the mushrooms' overall quality.

34.5 CONCLUSION

The studies presented here have demonstrated a number of characteristics of protein-based materials that make them suitable for the formation of different types of edible or biodegradable packaging. Their ability to be modified and controlled, as well as their compatibility with paper, natural fibers, or even conventional (modified) plastics to form composite material, open new applications in the field of active edible/biodegradable wrappings or coatings.

REFERENCES

1. Gontard, N. and S. Guilbert. 1994. "Bio-Packaging: Technology and Properties of Edible and/or Biodegradable Material of Agricultural Origin," in *Food Packaging and Preservation*, M. Mathlouthi, ed. Blackie Academic & Professional, pp. 159–181.
2. Guilbert, S. 1999. "Biomaterials for Food Packaging: Applications and Future Prospects" in *Trends in Food Engineering*, G. Barbosa-Cánovas, ed. Aspen.
3. Guilbert, S. 1986. "Technology and Application of Edible Protective Films," in *Food Packaging and Preservation*, M. Mathlouthi, ed. Elsevier Applied Science Publishers, pp. 371–394.
4. Guilbert, S. and B. Cuq. 1998. "Les Films et Enrobages Comestibles," in *L'emballage des Denrées Alimentaires de Grande Consommation*, Technique et Documentation, Lavoisier, pp. 471–530.
5. Guilbert, S., B. Cuq., and N. Gontard. 1997. "Recent Innovations in Edible and/or Biodegradable Packagings," *Food Additives and Contaminants*, 14(6–7): 741–751.
6. Cuq, B., N. Gontard, and S. Guilbert. 1995. "Edible Films and Coatings as Active Layers," in *Active Food Packagings*, M. L. Rooney, ed. Blackie Academic & Professional, pp. 111–142.
7. Guilbert, S., N. Gontard, and L. G. M. Gorris. 1996. "Prolongation of the Shelf-life of Perishable Food Products Using Biodegradable Films and Coatings," *Lebens. Wiss. Technol.*, 29: 10–17.
8. Gontard, N. and S. Ring. 1996. "Edible Wheat Gluten Film: Influence of Water Content on Glass Transition Temperature," *J. Agric. Food Chem.*, 44(11): 3474–3478.
9. Pouplin M., A. Redl, and N. Gontard. 1999. "Glass Transition of Wheat Gluten Plasticized with Water, Glycerol or Sorbitol," *J. Agric. Food Chem.*, 47: 538–543.
10. Di Gioia, L. and S. Guilbert. 1999. "Corn Protein-Based Thermoplastic Resins: Effect of Some Polar and Amphiphilic Plasticizers," *Journal of Agricultural and Food Chemistry*, 47(3): 1254–1261.

11. Di Gioia, L., B. Cuq, and S. Guilbert. 1998. "Effect of Hydrophilic Plasticizers on Thermomechanical Properties of Corn Gluten Meal," *Cereal Chemistry*, 75(4): 514–519.

12. Redl, A., M. H. Morel, J. Bonicel, S. Guilbert, and B. Vergnes. 1999. "Rheological Properties of Gluten Plasticized with Glycerol: Dependence on Temperature, Glycerol Content and Mixing Conditions," *Rheol. Acta*, 38: 311–320.

13. Redl, A., M. H. Morel, J. Bonicel, B. Vergnes, and S. Guilbert. 1999. "Extrusion of Wheat Gluten Plasticized with Glycerol: Influence of Process Conditions on Flow Behaviour, Rheological Properties and Molecular Size Distribution," *Cer. Chem.*, 76(3): 361–370.

14. Torres, J.A. and M. Karel. 1985. "Microbial Stabilization of Intermediate Moisture Food Surfaces. III. Effects of Surface Preservative Concentration and Surface pH Control on Microbial Stability of an Intermediate Moisture Cheese Analog," *Journal of Food Processing and Preservation*, 9: 107.

15. Torres, J.A., M. Motoki, and M. Karel. 1985. "Microbial Stabilization of Intermediate Moisture Food Surfaces. I. Control of Surface Preservative Concentration," *Journal of Food Processing and Preservation*, 9: 79.

16. Redl A., N. Gontard, and S. Guilbert. 1996. "Determination of Sorbic acid Diffusivity in Edible Wheat Gluten and Lipid Based Films," *Journal of Food Science*, 61(1): 116–120.

17. Crank 1975. *The Mathematics of Diffusion,* 2nd ed., Clarendon Press.

18. Gontard, N., R. Thibault, B. Cuq, and S. Guilbert. 1996. "Influence of Relative Humidity and Film Composition on Oxygen and Carbon Dioxide Permeabilities of Edible Films," *Journal of Agricultural and Food Chemistry*, 44(4): 1064–1069.

19. Mújica Paz, H. and N. Gontard. 1997. "Oxygen and Carbon Dioxide Permeability of Wheat Gluten Film: Effect of Relative Humidity and Temperature," *J. Agric. Food Chem.*, 45(10): 4101–4105.

35 Cooling Uniformity in Overpressure Retorts Using Water Spray and Full Immersion

M. A. Tung
G. F. Morello
A. T. Paulson
I. J. Britt

CONTENTS

Abstract

Following the application of overpressure heating processes, pressurized cooling phases were studied in a steam/air retort with water spray cooling and a water immersion retort cooled by cold water exchange to assess the uniformity of cooling. The influence of operating procedures on cooling temperature profiles and heat transfer distribution during cooling was evaluated. Analyses of uniformity of the cooling rate index derived from cooling curves, as well as additional lethality accumulated at the geometric center of food-simulating transducers during the cooling cycle, were complementary in demonstrating differences in heat transfer conditions. Still cooling processes exhibited temperature and heat transfer variability between locations monitored in the upper and lower regions of test cars. Rotational processes exhibited temperature and heat transfer variability between internal and peripheral locations.

35.1 INTRODUCTION

Effective procedures for evaluating retort performance are required as thermal processors strive to produce more desirable, shelf-stable products. Heat transfer must be uniform throughout heating and cooling cycles to minimize variability in process lethalities and in the quality of heat labile product attributes. The design of a safe thermal process is based on the least lethal conditions experienced within a load. If cooling conditions can be applied in a reproducible manner throughout the retort load, it may be possible to apply lethality conferred during cooling toward the total process lethality. Overall processing time could thus be reduced while increasing throughput and improving the uniformity of product quality.

Variations in thermal treatment due to cooling procedures occur in most batch retorts, but especially in overpressure processes where air is introduced to maintain retort pressure when adding cooling water. A stable overpressure is critical during the early stages of cooling to counteract internal container pressure. Loss of hermetic integrity at the package closure or damage to container construction may otherwise occur.

Procedures to evaluate temperature distribution and stability within a retort system are well established for assessing the uniformity of heating in steam environments. However, several researchers[1–6] have pointed out the error of assuming that temperature uniformity of an overpressure heating medium will ensure uniform heat transfer. The same concerns apply with respect to the cooling medium.

Tung et al.[6] described procedures to evaluate temperature distribution during the constant temperature cook period of a process. At each time interval that data were recorded, temperature stability was studied by examining mean temperatures and standard deviations associated with each thermocouple location within the retort through the constant temperature period.

The concept of limiting surface heat transfer was developed in relation to heating packaged foods in overpressure heating media[7,8] but can apply equally well to heat transfer during cooling. During the heating cycle, calculation of heating rate indices (f_h) for food-simulating transducers reflect surface heat transfer conditions within the retort load. Similarly, cooling rate indices (f_c) may be calculated to assess heat transfer uniformity during cooling. An alternative approach to evaluate heat transfer conditions is to consider accumulated lethalities, calculated from temperature histories of food-simulating transducers within a retort load.

The objectives of this study were to consider procedures for assessing the cooling uniformity of retort systems and to demonstrate these methods by evaluating various cooling operations provided by common steam/air and water immersion overpressure retort systems.

35.2 MATERIALS AND METHODS

35.2.1 RETORT SYSTEMS

A single-car forced convection steam/air retort (Model SN#RA.106, J. Lagarde Autoclaves, Montélimar, France) and a commercial four-car full water immersion overpressure retort (Model RCS/1300, Stock America, Inc., Milwaukee, WI) were studied. Both retorts had 1300-mm dia shells, microprocessor-based control systems, and capabilities for operating in batch-type still and rotary modes. The steam/air retort used water spray cooling and the water immersion retort was cooled by gradual replacement of hot processing water with cold water.

Product cars for both retorts were equipped with metal supporting racks (Stock America, Inc., Milwaukee, WI) designed for semi-rigid plastic trays. Construction of the racks was the same for the two retorts except for the number of containers accommodated in each layer and the number of racks per car, as shown in Figures 35.1 and 35.2. The water immersion test car was placed in the position nearest the retort door for all experiments.

35.2.2 ENVIRONMENTAL THERMOCOUPLES

Retort temperatures were measured with 24-ga. copper/constantan thermocouples (Type T-TT-24 SLE, Omega Engineering, Inc., Stamford, CT) previously calibrated in a circulating oil bath against an ASTM traceable mercury-in-glass thermometer. Thermocouple sensing junctions were isolated in stainless steel springs (13 mm ID × 40 mm long) to prevent contact with packages and supporting racks to ensure that observed temperatures represented the cooling medium conditions.

35.2.3 FOOD-SIMULATING TRANSDUCERS

Heat transfer transducers were constructed from Teflon® (polytetrafluoro-ethylene) slabs measuring approximately $102 \times 152 \times 29$ mm ($4 \times 6 \times 1.125$ in). These transducers, more stable and uniform than foods in packages,[4,5,7] were designed to simulate the thermal properties of conduction-heating foods in thin-profile plastic trays. A hole was drilled lengthwise along the midplane to the centerpoint of each brick. Silicone rubber sealant was injected into the hole, and a calibrated 24-ga. copper/constantan thermocouple was inserted with the sensing junction positioned at the geometric center of the brick. The thermocouple was secured with a brass packing gland threaded into the end of the brick (Type C-5.2, Ecklund-Harrison Technologies, Inc., Fort Myers, FL). Two sets of heat transfer transducers were

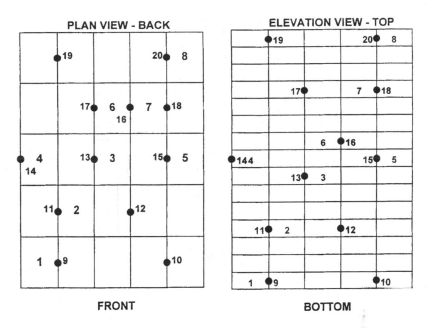

Figure 35.1 Placement of food-simulating transducer bricks (1–8) and environmental thermocouple junctions (dots 9–20) in the test car of the steam/air retort (not to scale).

constructed: set 1 for the steam/air study, and set 2 for the water immersion study. Transducers were calibrated using steam processes at 124°C to determine the heating characteristics of each transducer under high surface heat transfer conditions associated with condensing steam.

35.2.4 PROCESSING CONDITIONS

Heating cycle duration from steam-on was approximately 58 min for the brick centerpoint temperatures to be within 2°C below the retort temperature before starting the cooling cycle. Nominal retort conditions were 124°C (255°F) and 2.9 bar (42 psig) for all processes. To simulate commercial conditions, a complete ballast load was constructed with retortable plastic trays, nominally rated at 255 g (9 oz), and filled with a 5% bentonite suspension or a 5% starch suspension for the steam/air and water immersion systems, respectively.

Experiments were conducted for four cooling conditions:

- Still processes in which the ballast containers were positioned with their flat surfaces horizontal and the lidstock facing *up* for the steam/air retort experiments
- Still processes in which the ballast containers were positioned with their flat surfaces horizontal and the lidstock facing *down* for the water immersion retort experiments

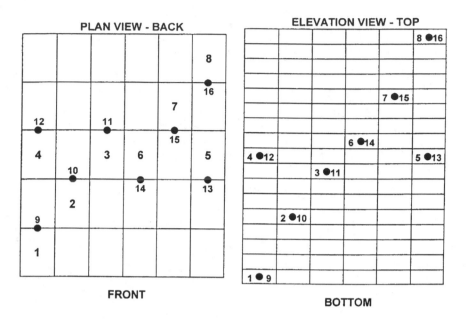

Figure 35.2 Placement of food-simulating transducer bricks (1–8) and environmental thermocouple junctions (dots 9–16) in the test car of the water immersion retort (not to scale).

- Still processes with containers rotated 90° to orient their flat surfaces in a vertical position
- Four- and 10-rpm rotary processes

The steam/air retort cooling cycle began with a four-minute pre-cool as cooling water was sprayed slowly into the circulating steam/air mixture through two nozzles positioned in an annular circulation channel inside the retort shell. This procedure allowed steam to collapse in a controlled manner while air was added concurrently to maintain the retort pressure. At the end of the pre-cool, the circulation fan was stopped, and water, accumulated in the bottom of the retort, was pumped through an external counterflow plate heat exchanger and cooled with line water to a programmed temperature. Water then flowed back into the processing vessel via a series of spray nozzles mounted in a manifold located longitudinally at the top surface of the retort. The microprocessor controller was programmed in five consecutive segments of 4, 10, 12, 10, and 20 min duration, to linearly ramp the retort cooling water temperature from 124 to 95°C (segment 1), from 95 to 20°C (segment 2), and maintain 20°C for the remainder of the cooling period (segments 3, 4, and 5). The controller was also programmed to maintain the retort pressure at 2.9 bar (segment 1), then linearly ramp down from 2.9 to 1.5 bar (segment 2), from 1.5 to 0.6 bar (segment 3), maintain 0.6 bar for segment 4, and ramp from 0.6 to 0.0 bar for the remainder of the cooling cycle. The retort car rotated continuously throughout the rotary processes.

The water immersion retort cooling cycle began with cooling water being added at the top of the processing vessel, while hot water was pumped from the bottom of the vessel into a storage tank. The microprocessor program consisted of five consecutive segments of 4, 5, 6, 27, and 6 min duration. The retort pressure was ramped from 2.9 to 2.25 bar (segment 1), 2.25 to 1.8 bar (segment 2), 1.8 to 1.4 bar (segment 3), 1.4 to 1.2 bar (segment 4), and decreased to 0.0 bar during the final drain segment. The test car remained stationary throughout the still processes; however, rotation was not continuous for the rotary processes. For the 4-rpm processes, rotation was maintained only during segments 1 and 2 (9 min) and was ramped down from 4 to 0 rpm during segment 3 (6 min). For the 10-rpm processes, rotation was initially maintained during segment 1, ramped down to 4 rpm during segment 2, and to 0 rpm during segment 3.

35.2.5 DATA ACQUISITION

The steam/air and water immersion retorts were equipped with 32- and 16-channel thermocouple slip-ring assemblies, respectively (Ecklund-Harrison Technologies, Inc., Fort Myers, FL). Temperature data were collected using a Doric Digitrend 235 data logger (Beckman Industrial Corporation, San Diego, CA) interfaced to a microcomputer for data storage. Data were collected at 1-min intervals for all experiments. Retort pressure was monitored using an electronic pressure transducer attached to the retort shell. Output from the transducer, previously calibrated with a deadweight tester, was recorded continuously on a strip chart for each process.

35.2.6 DATA ANALYSIS

Retort environment temperatures and centerpoint temperature histories for the transducers were analyzed using commercial (Pro Calc Associates, Surrey, BC) and in-house thermal processing software. Mean temperatures and standard deviations for environmental thermocouple locations at each time interval were calculated for each experiment. Temperature data from repeated runs were compared; readings of the thermocouples distributed through the load for the three replicate runs were combined to obtain grand mean temperatures and standard deviations at each time interval for each cooling condition.

Cooling rate indices and cooling lag factors[9] were derived from the centerpoint temperature histories. Cooling rate indices were calculated over the 15- to 40-min time range by plotting the difference between individual brick centerpoint temperatures and the mean cooling temperature at 40 min of cooling. Cooling lag factors were also calculated using the mean cooling temperature at 40 min.

Using the method derived by Schultz and Olson,[10] each centerpoint temperature data set was adjusted to an initial cooling temperature of 122°C to eliminate minor differences in initial temperature that would influence General Method lethality calculations. Accumulated lethalities (F_C) for the cooling phase of the process were calculated using the Improved General Method[9] with a reference temperature of 121.1°C and a z-value of 10°C. Statistical analyses utilized Tukey's multiple comparison test.[11]

35.3 RESULTS AND DISCUSSION

35.3.1 TEMPERATURE UNIFORMITY

Assessment of the temperature uniformity within the retort load during cooling is of great interest. It is important to know the rate of temperature decrease and the ability to replicate these conditions in successive retort runs. High variability in the retort cooling temperatures at any given time, and poor control over the rates of temperature decrease would be expected to result in nonuniform process lethality as well as variable product quality. Thus, retort environment temperatures were analyzed to determine whether cooling conditions were consistent from run to run. Mean cooling temperature profiles for the steam/air and water immersion retort experiments are summarized in Tables 35.1 and 35.2.

Figure 35.3 is a sample plot of retort cooling environment temperature for a water spray cooling cycle in the steam/air system in still cook mode. The curve illustrates the rate of decrease in mean retort temperature and the variability in temperature throughout the retort car at each minute during the cooling cycle. The

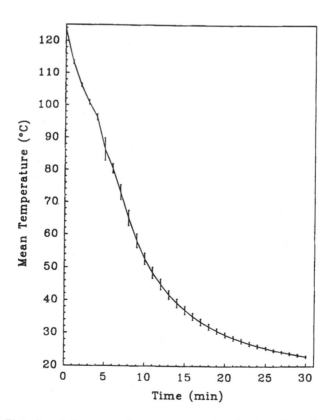

Figure 35.3 Plot of pooled mean cooling temperature data for the still cooling cycle of the steam/air retort (transducer bricks horizontal; vertical lines ±1 SD).

TABLE 35.1
Summary of Mean Cooling Temperatures (and Coefficients of Variation, %) during Water Spray Cooling in the Steam/Air Retort (*n* = 3)

Cooling time (min)	Mean cooling temperature (°C)			
	Still processes		Rotary processes	
	Horizontal, lid up	Vertical orientation	4 rpm	10 rpm
0	123.7 (0.1)*	123.7 (0.1)	123.7 (0.1)	123.6 (0.2)
4	96.3 (1.0)	95.8 (1.5)	96.6 (1.9)	96.7 (1.0)
10	52.7 (3.3)	52.5 (6.2)	53.6 (4.6)	55.4 (6.3)
15	37.0 (3.5)	36.7 (6.8)	36.8 (6.3)	39.9 (6.9)
20	29.6 (2.5)	29.3 (7.2)	29.3 (5.7)	31.5 (6.2)
25	25.5 (2.0)	25.0 (7.0)	24.8 (5.8)	26.5 (5.1)
30	22.9 (1.6)	22.3 (6.7)	22.1 (4.2)	23.4 (4.1)

*Each mean value is the pooled average of the 12 readings of thermocouples distributed through the load for 3 replicate runs.

TABLE 35.2
Summary of Mean Cooling Temperatures (and Coefficients of Variation, %) during Water Immersion Cooling in the Water Immersion Retort (*n* = 3)

Cooling time (min)	Mean cooling temperature (°C)			
	Still processes		Rotary processes	
	Horizontal, lid down	Vertical orientation	4 rpm*	10 rpm
0	†124.0 (0.1)	124.0 (0.1)	123.8 (0.1)	123.8 (0.2)
4	79.3 (10.1)	83.2 (5.7)	90.2 (5.6)	88.9 (13.6)
10	43.4 (12.8)	46.6 (5.9)	51.3 (9.1)	47.1 (5.9)
15	30.4 (8.7)	32.4 (5.1)	34.4 (5.8)	33.0 (6.2)
20	24.8 (5.2)	25.6 (3.8)	26.8 (5.2)	26.2 (5.1)
25	22.2 (2.8)	22.5 (2.5)	23.2 (3.5)	23.0 (3.5)
30	20.9 (1.2)	21.0 (2.2)	21.4 (1.9)	21.2 (2.6)

*Four-rpm processes have later cool starts than the rest

†Each mean value is the pooled average of the eight readings of thermocouples distributed through the load for three replicate runs.

slightly curved shape up to the 4-min point indicates that a more aggressive temperature profile could be programmed for the pre-cool segment. Temperature variability was very small during the pre-cool, then increased when the main water spray commenced, and progressively decreased through the remainder of the cool as the product load temperature approached the cooling water spray temperature.

For the still cook processes with the flat packages in a horizontal orientation, the upper thermocouple locations closest to the spray nozzles cooled more quickly than the lowest locations monitored. Lower parts of the retort car likely received cooling water that had picked up heat from packages in the upper layers. As well, greater nonuniformity was observed in the lower locations compared to the upper locations when the retort car was rotated 90° to orient the packages with their flat surfaces in the vertical direction. Rotational processes exhibited warmer temperatures at the center of the test car compared with the peripheral regions, attributable to a "windmilling" effect that inhibited cooling water penetration into the retort car load.

For the water immersion retort, some temperature stratification existed between the cooler upper half and warmer lower half of the test car during still processes with horizontally oriented flat packages. Figure 35.4 shows a smooth curve with large variabilities in temperature lasting more than 15-min into the cooling cycle. As expected, examination of specific thermocouple locations indicated that cooling occurred more quickly in the upper region of the retort car than in the lower region.

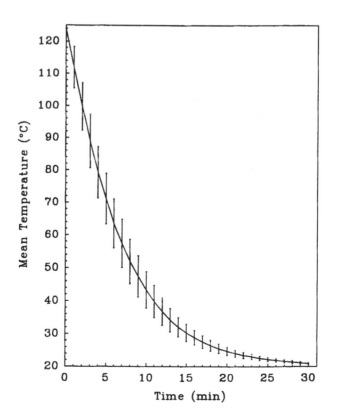

Figure 35.4 Plot of pooled mean cooling temperature data for the still cooking cycle of the water immersion retort (transducer bricks horizontal; vertical lines ±1 SD).

Temperature uniformity improved with the test car rotated 90°. The vertical flow channels created by a 90° orientation would improve the ability of the water to mix; however, thermocouples located at upper parts of the load were consistently cooler than the others. Rotational processes initially exhibited warmer temperatures at the center of the test car compared to the peripheral regions; however, when rotation stopped after 15 min, temperature distribution changed to again become slightly cooler in the upper regions compared to the lower regions of the retort test car.

Throughout these studies, temperature uniformity was poorest early in the cool and improved as the cooling cycle proceeded. In general, temperature decreased more quickly in the water immersion retort than in the steam/air retort. This difference is attributable to the more gradual programmed water spray cool compared to the cold water immersion procedure. However, temperatures were much less uniform in the water immersion processes.

35.3.2 Uniformity of f_c and j_c

Heat transfer uniformity during cooling can be indicated by the uniformity of cooling rate indices (f_c) and cooling lag factors (j_c) derived from cooling curves of the food-simulating transducers. The Teflon transducers have a repeatable thermal diffusivity and maintain nearly constant dimensions; thus, internal temperature history variability directly reflects changes in surface heat transfer conditions. Cooling rate indices of cooling curves are analogous to heating rate indices (f_h) of heating curves, so smaller f_c values indicate more effective heat transfer conditions during cooling. Cooling lag factors describe the shape of a cooling curve prior to the formation of the straight line portion of the curve in which the temperature difference between the brick transducer centerpoint and the cooling medium changes semilogarithmically with time.

In general, changes in the slope of the straight line portion of a cooling curve will affect the j_c of the curve as the pseudo-initial cooling temperature (T_{pic}) changes; however, for bricks with identical f_c values, variabilities observed among their j_c values will reflect different conditions during the early stages of cooling. In a single test with brick transducers located throughout the retort car with similar f_c values but differing j_c values, cooling conditions are likely to be nonuniform in the retort. Cooling rate indices and lag factors have important impacts on the application of formula methods for process calculations, where assumptions are made about the contribution of the cooling period to the overall thermal process lethality.[9] The use of actual f_c and j_c values would be of critical importance if formula methods are used to estimate the lethality delivered during cooling.

Inherent differences in heating (cooling) characteristics within and between the two sets of transducer bricks were examined during heating in saturated steam to promote very high and uniform surface heat transfer conditions. Heating rate indices were derived from the centerpoint temperature histories to compare the uniformity of their heating (and potential cooling) characteristics under optimal heat transfer conditions. Mean heating rate indices (and coefficients of variation) determined for brick sets used in the steam/air and water immersion systems were 28.35 min (1.1%)

and 27.01 min (2.4%), respectively. A t-test to compare mean values and Bartlett's test for homogeneity of variance[11] confirmed there were significant differences between mean f_h values and sample variances for the two sets of transducer bricks. Values of f_c derived in water spray or water immersion tests would be expected to be larger than f_h values in steam because of inherent limitations of surface heat transfer by sensible means compared to condensation heat transfer, respectively. Nevertheless, large deviations from these baseline means and variabilities would indicate reduced surface heat transfer and nonuniform heat transfer conditions.

The still, horizontal process for the steam/air system had a significantly higher overall mean f_c value, indicating the test load experienced reduced surface heat transfer conditions compared to the other three process conditions (Table 35.3). All transducer positions, except position 1 at the bottom front corner, and to some extent position 4 at the middle left edge, exhibited very high f_c values.

TABLE 35.3

Individual and Overall Mean Cooling Rate Indices (f_c, min) and Lag Factors (j_c) with Coefficients of Variation (CV) for the Brick-Shaped Transducers during Water Spray Cooling in the Steam/Air Retort (n = 3)

	Still Processes				Rotary Processes			
	Horizontal, lid up		Vertical orientation		4 rpm		10 rpm	
Position	f_c	j_c	f_c	j_c	f_c	j_c	f_c	j_c
1	35.9ᵃ	2.07ᶜ	46.1ᵈ	1.75ᵃ	35.0ᵃ	2.07ᵃ	35.6ᵃ	2.03ᵇᶜ
2	55.3ᶜ	1.56ᵃ	33.7ᵃᵇ	2.17ᶜ	35.6ᵃ	2.11ᵃ	37.5ᶜᵈ	2.05ᶜᵈᵉ
3	56.4ᶜ	1.60ᵃ	34.7ᵇ	2.14ᵇᶜ	37.1ᵃᵇ	2.08ᵃ	40.1ᶠ	1.99ᵃ
4	46.8ᵇ	1.70ᵇ	33.4ᵃ	2.16ᶜ	35.8ᵃ	2.10ᵃ	37.1ᵇᶜ	2.04ᶜᵈ
5	55.4ᶜ	1.61ᵃ	34.7ᵇ	2.16ᶜ	38.8ᵇ	2.02ᵃ	38.8ᵉ	2.00ᵃᵇ
6	56.7ᶜ	1.58ᵃ	34.1ᵃᵇ	2.16ᶜ	36.6ᵃ	2.08ᵃ	38.2ᵈᵉ	2.04ᶜᵈ
7	55.8ᶜ	1.59ᵃ	34.0ᵃᵇ	2.15ᶜ	35.7ᵃ	2.11ᵃ	36.7ᵇ	2.06ᵈᵉ
8	54.3ᶜ	1.56ᵃ	36.1ᶜ	2.07ᵇ	35.5ᵃ	2.08ᵃ	35.8ᵃ	2.08ᵉ
Mean	52.1ʸ	1.66ʳ	35.8ˣ	2.09ˢ	36.3ˣ	2.08ˢ	37.5ˣ	2.04ˢ
CV (%)	13.5	10	11.3	6.6	3.6	2.2	4.0	1.5

Values in columns or means with same letters do not differ significantly ($p > 0.05$).

The ability of the cooling water spray to penetrate the horizontally oriented test load and contact the flat surfaces of the transducer bricks and ballast packages is quite restricted. Cooling may initially be by an air/water mist. The surface heat transfer capability of air cooling is much poorer than water cooling and could explain the high f_c values observed. The bricks at position 1, and to some extent position 4, may have been splashed by cooling water hitting the bottom and sides of the retort shell and thereby experienced more favorable heat transfer conditions.

The still, vertically oriented test load showed quite uniform cooling, except for positions 1 and 8, located at the top front and lower back corners. These positions may be less favorably oriented in relation to the spray pattern of the water spray nozzles. The best f_c uniformity occurred during rotation of the test load, although internal positions in the 10-rpm process generally showed slightly higher f_c values than external positions.

The mean j_c value of the still, horizontal process was significantly lower than the other processes, as a larger f_c value resulted in a smaller T_{pic} and hence a smaller j_c. Position-to-position differences observed for both still processes were similar to the f_c results observed. The best uniformity was observed for the rotary processes, with essentially no differences among positions at 4 rpm, and only slight differences at 10 rpm.

For the water immersion retort, all four cooling processes showed some variability among positions, with the still, vertically orientated test load exhibiting the most uniform f_c values (Table 35.4). During the 4- and 10-rpm processes, rotation ceased after 15 min, with the load remaining stationary with the flat surfaces horizontal and the lidstock facing up. The patterns of f_c values and degree of variation for the two conditions were very similar, with the still, horizontal oriented test load displaying similar variation. For these three processes, f_c values were larger for bricks located in the lower half of the test load than those in the upper half, which followed the pattern of temperature stratification observed. Comparison of the f_c mean values indicated no differences among the four cooling processes,

---

TABLE 35.4
Individual and Overall Mean Cooling Rate Indices (f_c, min) and Lag Factors (j_c) with Coefficients of Variation (*CV*) for the Brick-Shaped Transducers during Water Cooling in the Water Immersion Retort ($n = 3$)

	Still Processes				Rotary Processes			
	Horizontal, lid down		Vertical orientation		4 rpm		10 rpm	
Position	f_c	j_c	f_c	j_c	f_c	j_c	f_c	j_c
1	33.9[abc]	1.96[a]	33.0[b]	2.03[ab]	34.4[d]	2.13[c]	34.7[ab]	2.03[ab]
2	33.2[abc]	1.99[a]	34.0[cd]	2.06[b]	37.2	2.01[b]	38.2[c]	1.95[ab]
3	32.3[ab]	1.97[a]	34.9	2.03[ab]	37.6	2.02[b]	38.2[c]	1.98[ab]
4	31.5[a]	1.99[a]	33.8[c]	2.03[a]	37.2	1.89[u]	37.4[be]	1.85[a]
5	32.3[ab]	2.07[a]	32.1[a]	2.17[e]	32.0[a]	2.23[d]	32.4[a]	2.11[b]
6	36.4[bcd]	1.97[a]	33.2[b]	2.13[cd]	32.7[bc]	2.26[d]	33.3[a]	2.16[b]
7	39.7[d]	1.92[a]	33.3[b]	2.14[de]	32.4[ab]	2.27[d]	33.0[a]	2.13[b]
8	37.1[cd]	1.86[a]	34.4[d]	2.11[c]	33.3[c]	2.11[c]	34.1[a]	1.98[ab]
Mean	34.6[x]	1.97[r]	33.6[x]	2.09[st]	34.6[x]	2.11[t]	35.1[x]	2.02[rs]
CV (%)	8.8	6.0	2.5	2.6	6.6	6.2	7.0	6.1

Values in columns or means with same letters do not differ significantly ($p > 0.05$).

which is not surprising, since loads were all static over the time range of the f_c calculation.

Some small but significant differences in mean j_c values were noted among the four processes. The still process with vertically oriented packages displayed the least variability but indicated some small significant differences among positions; no difference was noted among positions for the still, horizontally oriented process. The pattern of significant differences among positions, observed for f_c values during both rotational processes, held for j_c values.

35.3.3 UNIFORMITY OF ACCUMULATED F_C

Heat transfer uniformity during cooling was also indicated by the uniformity of accumulated lethality (F_C) derived from centerpoint temperature histories of the food-simulating transducers. As discussed, internal temperature history variability directly reflects changes in surface heat transfer conditions. This method of comparing temperature responses to changing cooling conditions focuses on the early stages of the cooling period when container temperatures were sufficiently high to result in further beneficial spore destruction, but also in detrimental nutritional and product quality degradation. Lethal rate effects accumulated during the cooling cycle must be uniform if any degree of sterilization calculated for a thermal process is to be attributed to the cooling cycle to minimize scheduled cook times.

Some variability in accumulated F_C values was observed between transducer positions for each process condition in both steam/air and water immersion systems (Tables 35.5 and 35.6). In general, F_C variability between transducer brick locations could be traced to observed differences in retort environment temperature. For both retort systems, variability among brick locations was smallest when the test load was vertically oriented in the still mode. Greater variability within the retort car occurred in the water immersion system, which was expected, since more pronounced temperatures variabilities were observed for those processes. The pre-cool contribution to high F_C values was more pronounced in the steam/air cooling system. Note the large contribution to overall lethality developed during the cooling cycle.

These data demonstrate the usefulness of analyzing F_C values to identify differences between the cooling process conditions. Lethal rates accumulated at the brick centerpoints were less than 0.1 at 12 min and 10 min of cooling for the steam/air and water immersion retort experiments, respectively. This period is only one-quarter of the 40-min cooling time and does not necessarily reflect changing conditions during the remainder of the cool. However, it is noteworthy that product quality is still affected by temperatures below which lethal rates can be recorded.

For the water immersion retort, comparison of mean accumulated F_C values of the four cooling conditions indicated similar significant differences among processes. It should be pointed out that the start of cooling for the 4-rpm runs was delayed approximately 30 s as compared with the other processes. Synchronization of the cool starts between experimental repetitions, and more frequent temperature recordings of 10 or 15 s time intervals should improve the precision of results when applying this methodology.

TABLE 35.5
Individual and Overall Mean Accumulated Lethalities (F_C, min) and Coefficients of Variation (CV) for the Brick-Shaped Transducers during Water Spray Cooling in the Steam/Air Retort ($n = 3$)

Position	Still processes		Rotary processes	
	Horizontal, lid up	Vertical orientation	4 rpm	10 rpm
1	8.9[ab]	9.3[b]	8.3[a]	7.9[a]
2	9.6[cd]	9.3[b]	9.5[bc]	9.5[bc]
3	10.0[d]	9.4[b]	9.6[c]	9.6[c]
4	8.8[ab]	8.1[a]	8.8[ab]	9.0[bc]
5	10.0[d]	9.4[b]	9.3[bc]	9.1[bc]
6	9.7[cd]	9.3[b]	9.5[bc]	9.2[bc]
7	9.2[bc]	9.1[b]	9.1[bc]	9.3[bc]
8	8.3[a]	9.0[b]	8.3[a]	8.7[ab]
Mean	9.3[x]	9.1[x]	9.1[x]	9.0[x]
CV (%)	6.5	5.1	6.1	6.5

Values in columns or means with same letters do not differ significantly ($p > 0.05$).

TABLE 35.6
Individual and Overall Mean Accumulated Lethalities (F_C, min) and Coefficients of Variation (CV) for the Brick-Shaped Transducers during Water Immersion Cooling in the Water Immersion Retort ($n = 3$)

Position	Still processes		Rotary processes	
	Horizontal, lid down	Vertical orientation	4 rpm	10 rpm
1	7.1[c]	7.0[a]	8.4[b]	7.4[bc]
2	7.2[c]	7.6[bcd]	9.0[cd]	9.1[d]
3	6.2[ab]	8.1[d]	9.8	9.9
4	5.8[a]	7.1[ab]	7.6[a]	7.5[c]
5	6.9[bc]	7.3[abc]	7.9[ab]	7.0[b]
6	8.4[de]	7.4[abc]	9.3[de]	9.2[d]
7	9.2	7.7[cd]	8.5[bc]	7.3[bc]
8	7.6[cd]	8.2[d]	7.7[a]	6.6[a]
Mean	7.3[x]	7.6[xy]	8.5[z]	8.0[yz]
CV (%)	14.7	6.1	9	14.7

Values in columns or means with same letters do not differ significantly ($p > 0.05$).

Note that 4-rpm processes have later cool starts than the rest.

35.3.4 PRESSURE

Strip chart recordings of retort pressure profiles during the cooling cycle of each process showed no major fluctuations or irregularities. Careful attention to pressure profiles is necessary when developing cooling procedures for overpressure processes to be certain that sharp pressure drops do not occur in the transition from heating to cooling. Continuous recordings of pressure are preferred over intermittent data logging to detect pressure spikes that may not be detected with discrete measurements.

35.4 CONCLUSIONS

This study illustrated both temperature and heat transfer distribution principles and practices that may be useful to evaluate the uniformity of cooling cycles in batch-type retorts. The procedures were effective in identifying variability in the cooling behavior of different operating modes during water spray cooling in a steam/air retort and cold water exchange cooling in a full immersion water retort, both used for overpressure processes. Monitoring retort environment temperatures alone will not indicate the impact of varying heat removal rates on packages distributed throughout a retort load. The impacts of temperature and heat transfer differences during cooling were expressed in the form of variable cooling rate indices and "product" centerpoint lethalities, with respect to food-simulating brick-shaped heat transfer transducers. These techniques may be used to identify major variabilities in cooling uniformity, assess the influence of seasonal changes in cooling water temperature, and evaluate the success that operational changes may have in improving cooling conditions.

35.5 ACKNOWLEDGMENTS

Support for these studies was provided by the Strategic Grants Program of the Natural Sciences and Engineering Research Council of Canada. Editorial assistance in the preparation of this manuscript by Sylvia Yada is appreciated.

REFERENCES

1. Pflug, I. J. 1964. "Evaluation of Heating Media for Producing Shelf Stable Food in Flexible Packages—Phase I," Final report, Contract No. DA19-AMC-145(N), U.S. Army Natick Laboratories, Natick, MA.
2. Yamano, Y. 1976. *Studies on Thermal Processing of Flexible Food Packages by Steam-and-Air Retort*, Ph.D. Thesis, Kyoto University, Kyoto, JPN.
3. Adams, J. P., W. R. Peterson, and W. S. Otwell. 1983. "Processing of Seafood in Institutional-Sized Retort Pouches," *Food Technol.*, 37(4):123–127, 142.
4. Tung, M. A., H. S. Ramaswamy, T. Smith, and R. Stark. 1984a. "Surface Heat Transfer Coefficients for Steam/Air Mixtures in Two Pilot Scale Retorts," *J. Food Sci.*, 49(3): 939–943.

5. Tung, M. A., G. F. Morello, and H. S. Ramaswamy. 1989. "Food Properties, Heat Transfer Conditions and Sterilization Considerations in Retort Processes," in *Food Properties and Computer- Aided Engineering of Food Processing Systems*, NATO ASI Series E: Applied Sciences vol. 168, R. P. Singh and A. G. Medina, eds. Dordrecht, NLD: Kluwer Academic Publishers, pp. 49–71.

6. Tung, M. A., I. J. Britt, and H. S. Ramaswamy. 1990. "Food Sterilization in Steam/Air Retorts," *Food Technol.*, 44(12): 105–109.

7. Tung, M. A., H. S. Ramaswamy, A. M. Papke, and R. Stark. 1984b. "Thermophysical Studies for Improved Food Processes," Final report, DSS File No. 35SZ.01804-9-0001, Agriculture Canada, Ottawa, ON.

8. Ramaswamy, H. S. and M. A. Tung. 1986. "Modelling Heat Transfer in Steam/Air Processing of Thin Profile Packages," *Can. Inst. Food Sci. Technol. J.*, 19(5): 215–222.

9. Stumbo, C. R. 1973. *Thermobacteriology in Food Processing*, 2nd ed. Orlando, FL: Academic Press, Inc.

10. Schultz, O.T. and F. C. W. Olson. 1940. "Thermal Processing of Canned Foods in Tin Containers. III. Recent Improvements in the General Method of Thermal Process Calculations—a Special Coordinate Paper and Methods of Converting Initial and Retort Temperatures," *Food Res.*, 5(4): 399–407.

11. Wilkinson, L. 1988. *SYSTAT: The System for Statistics*. Evanston, IL: SYSTAT, Inc.

36 Contribution to Kinetics for the Development of Food Sterilization Processes at High Temperatures: IATA-CSIC Results

M. Rodrigo
A. Martínez
C. Rodrigo

CONTENTS

Abstract

There is great interest in developing and applying HTST and UHT processes for food preservation. These processes provide microbiologically safe, very high quality foods, provided that enzyme inactivation is achieved. To establish these processes, it is necessary to determine kinetic parameters with sufficient precision to quantify the inactivation of microorganisms, enzymes, and quality factors in this temperature range. Given the almost total absence of these parameters, the IATA has been using new heating systems specially designed for high-temperature studies and applying the most appropriate kinetic models in each case so as to make a significant contribution in this field. Some results for the inactivation of *Cl. sporogenes*, peroxidase

(POD), texture, and weight loss at high temperatures in asparagus are described below.

36.1 INTRODUCTION

The optimization of food heat treatment processes requires, among other things, a knowledge of the distribution of temperatures in the food product throughout the process (heating and cooling) and of the parameters that define the inactivation and heat degradation kinetics of target microorganisms and of heat-sensitive quality factors that are of interest in each food product. The scarcity and lack of precision of the kinetic parameters available is well known. Most of the data published refer to model solutions and not to specific substrates or foods where the microorganisms, enzymes, and quality factors are really found, developed, inactivated, or destroyed.

When it is a question of optimizing high-temperature short-time (HTST) treatments, the absence of sufficiently precise and specific heat resistance kinetic data is much more marked.

To establish and validate the HTST processes currently applied commercially, *in situ* methods are used; i.e., determining the composition before and after heat treatment and applying container inoculation methods. Estimates are also employed, using kinetic parameters obtained in standard solutions or else by extrapolating the TDT curves obtained at lower temperatures. In all of these cases, the studies are laborious and, in some cases, the assumptions made are erroneous. In many cases, estimates lead, at best, to overtreatment, a solution that goes against the current trend of applying minimal processes to reduce costs and to ensure that the quality retained may be as similar as possible to the natural food.

Clearly, the work of determining kinetic parameters to obtain a databank that can be used for establishing optimum HTST treatments is essential but very arduous. Nevertheless, various food research centers throughout the world have recently become involved in these studies. To perform this task, in each case, it is vital to have new systems for rapid heating and cooling and to apply the most appropriate analytical models. Such is the case of the computer-controlled high-temperature thermoresistometer, which offers very short inertia, heating, and cooling times (0.4 s for a sample (10 to 50 μl) at ambient temperature to reach 130–150°C).

In recent years, with the aid of these resources, the IATA (CSIC) has made a considerable contribution to the development of this databank by studying inactivation or destruction kinetics of a series of microorganisms, enzymes, vitamins, and quality factors in HTST conditions.

The results obtained are, among other things, making it possible to accomplish the following:

1. To confirm the appropriateness of the rapid heating and cooling systems and the kinetic models used to develop parameters to interpret the inactivation or destruction, as appropriate, of microorganisms, enzymes, vitamins, and texture in HTST conditions in different foods.

2. To obtain kinetic parameters (D_T, Z, E_a, and K_T) that may enable us to reduce and optimize heat treatments in specific foods. Foods of higher quality can be produced more cheaply without compromising safety. In this field, work has been done on a series of substrates: crushed tomato, mushroom, low-acid artichokes, asparagus, vegetable salad with tuna, and fresh and pickled cucumber.

3. To develop microbiological temperature time integrators (TTI) to validate sterilization processes of foods with particles carried out continuously or in the can. Processes have been developed and validated for canned mushrooms, vegetable salad with tuna, and canned carrots.

4. To develop models for estimating D_T values of specific microorganisms in relation to the pH of the substrate during heating and the actual treatment temperature.

By way of example, the following are some of the results obtained in the IATA for one of the foods studied, asparagus.

36.2 ASPARAGUS

Until just over 20 years ago, agricultural production and industrial processing of asparagus in Spain focused exclusively on white asparagus. The increasing world interest in the production and industrial processing of canned green asparagus is basically due to a series of advantages in relation to white asparagus; it can be harvested mechanically, there is little postharvest development of fiber, and it does not need peeling.

In 1973, despite the lack of interest in agricultural and industrial sectors in Spain at the time, the IATA embarked on a series of studies for manufacturers within the framework of the Association of Vegetable Canning Industries (AICV). The work focused on suitability of varieties for the cultivation and mechanical harvesting of green asparagus, development of specifications of raw material for industrial processing, and canning process specifications. In Spain, cultivation and industrial processing of green asparagus began in the early 1980s, and current production may exceed that of white asparagus.

By far the most important problem in the industrial processing of green asparagus has to do with the losses of firmness and weight that it undergoes during sterilization.

To help alleviate this problem, the IATA has recently carried out various studies with a view to optimizing the sterilization process. The aspects studied include the following:

1. Heat resistance of *Cl. sporogenes*, PA 3679, in asparagus substrate; natural and acidified at different pH levels, with citric acid and with glucono delta lactone

2. Thermal degradation of green asparagus texture in two temperature ranges: 70–100°C and 100–130°C

3. Weight loss kinetic in green asparagus during the sterilization process
4. Kinetics of inactivation by HTST treatments and regeneration of peroxidase activity in asparagus

36.3 RESULTS AND CONCLUSIONS OF INTEREST

Heat Resistance of *Cl. sporogenes*, PA 3679, in Asparagus Substrate; Natural and Acidified at Different pH Levels, with Citric Acid and with Glucono Delta Lactone[1]

Five pH values were established in the substrate, from natural (5.9) to 4.5, with each of the two acidulants (citric and GDL). The treatment temperatures applied were: 110, 115, 118, and 121°C. The heat resistance study was performed with capillary tubes with an internal diameter of 0.7 mm, heated in a thermostatically controlled bath and cooled rapidly with ice.

The most notable results are the following:

- The heat resistance of the spores decreased as temperature increased and as the substrate was acidified (Figure 36.1).
- The reduction in D_T values with acidification was more noticeable at low temperatures. When the data for the two acidulants were compared, a greater reduction was observed in the case of citric at the lowest temperatures.
- A tendency for the Z values to increase with acidity was observed, more pronounced in the case of citric than in that of GDL, but, in any case, the differences were not significant.
- With the pH levels customary in these products (5.1–5.9), whether acidified or not, the Z values for PA 3679 were equal or very close to 10. This is the value considered as a reference for *Cl. botulinum*.

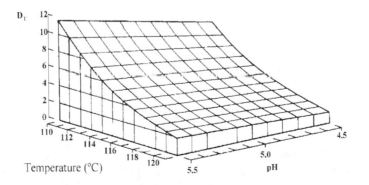

Figure 36.1 Relationship among pH, temperature and D values for PA 3679 heated in asparagus purée.

Asparagus is normally canned at temperatures on the order of 115°C. The results obtained show that, at these temperatures and with pH values of about 5.0–5.1, there was an appreciable decrease in the heat resistance of the spores of PA 3679 with respect to that of unacidified asparagus. This is the first step toward reducing the heat treatment for these products. It is clear that, before recommending a new minimum process, studies must be made of biological validation (by experimental inoculation of cans or by means of time temperature integrators) and the effect of acidification on the sensory and nutritional characteristics of these products.

Thermal Degradation of Green Asparagus Texture in Two Temperature Ranges: 70–100°C and 100–130°C

In the two temperature ranges, a study was made in steady-state conditions; in the second range (higher temperatures), an unsteady-state method was also applied. In the first case, the heating was carried out by introducing the asparagus spears directly into a thermostatically controlled bath, and, in the second case, each spear was inserted into a stainless steel tube that fit its diameter and that contained a small quantity of liquid so as to avoid occlusion of air, and the tube was pressure sealed and heated in the same thermostatically controlled bath. The determinations of texture were performed with two kinds of cell, wire and Kramer, fitted to an Instron texturometer.

The study of the texture of green asparagus has been presented in three articles. In the first,[2] a method for measuring asparagus was developed, and the thermal degradation of texture was evaluated in the range of 70–100°C. The results obtained are described below.

- A new cell was developed that measures resistance to being cut with a wire. The cell was used in conjunction with a universal texturometer, and it improves on the determination made by the Wilder fibrometer (Figure 36.2). With this cell, experimental working conditions were determined for measuring the texture of whole asparagus subjected to various heat treatments. Greater discrimination between samples is achieved with this cell than with the Kramer cell (Table 36.1). When the kinetic of texture degradation was studied with the new cell and with the Kramer cell in the temperature range 70–100°C, in both cases a biphasic (two-component) behavior was observed, with each phase corresponding to a first-order kinetic.
- The kinetic parameters for texture measured with the wire cell (cut 5 cm from the tip of the asparagus) and with the Kramer cell are summarized in Table 36.2.
- The rate constants increase with temperature. The rate constants for the degradation of component (a) were generally 10 times greater than those of component (b). The differences between the values of E_a obtained with the two cells can be explained as different ways of measuring texture. The information provided by the wire cell is more specific, and capable

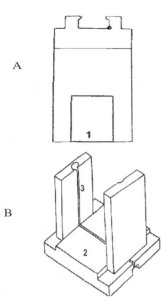

Figure 36.2 Diagram of the wire cell. (A) steel strip 1 mm thick with a steel wire (1) soldered onto the ends; (B) the base (2) and side guides (3) holding the cell.

TABLE 36.1
Comparative Study of the Wire Cell and the Kramer Shear–Press Cell

Sample	Wire (2.5 cm)	Wire (5.0 cm)	Wire (7.5 cm)	Kramer Cell
3 min*	6.2(1.7)a	5.0(1.0)a	8.4(2.5)a	1653(79)a
5 min*	4.6(1.2)b	6.5(2.0)a	8.4(3.2)a	1364(148)ab
9 min*	2.8(2.4)c	4.4(2.6)b	7.4(3.2)a	1182(187)b
A**	0.8(0.4)d	1.2(0.5)c	2.2(1.1)b	541(45)c
B**	0.7(0.3)d	0.7(0.2)c	1.4(0.6)b	454(27)c
C**	0.7(0.5)d	0.9(0.4)c	1.8(1.3)b	436(38)c
D**	0.9(0.2)d	0.7(0.2)c	1.2(0.3)b	238(23)c
E**	0.8(0.3)d	0.9(0.5)c	1.8(0.8)b	238(5)c

*Heating time at 100°C.

**Commercial samples.

Standard deviation between parentheses.

Means within columns followed the same letter are not significantly different at the 5% level.

of quantifying the individual fibrousness of each asparagus spear, and the value depends on the area of the cross section of the spear at the point where the cut is made. When the measurement made with the wire cell is compared with that made by the Wilder fibrometer, the advantages of the former are evident.

TABLE 36.2
Thermal Degradation of Asparagus Firmness at
A Reference Temperature of 85°C

	Wire cell (5 cm)	Kramer cell
E_{aa} (Kcal/mol)	9.6 ± 1.4	23 ± 3
K_{85a} (min^{-1})	1.05 ± 0.07	0.25 ± 0.03
r_a	0.9783	0.9873
E_{ab} (Kcal/mol)	20.4 ± 1.9	18 ± 4
K_{85b} (min^{-1})	0.057 ± 0.005	0.025 ± 0.005
r_b	0.9913	0.9531

In a second article,[3] texture degradation was studied in the temperature range of 100–130°C, using the two methods (wire cell and Kramer cell). With the wire cell, the cuts were made at distances of 2.5, 5, and 7.5 cm from the tip, and, with the Kramer cell, the measurements were made with 80 g of pieces obtained from section 1 (the cuts made with the wire cell at distances of 2.5, 5 and 7.5 cm from the tip) or from section 2 (cuts made 10 cm from the tip).

The following observations were made:

- In all the measuring conditions, the texture values decreased as time increased (Table 36.3). The ANOVA of the texture values indicated that the two methods (cutting with the wire cell at 2.5 and 5 cm from the tip and with the Kramer cell in sections 1 or 2) discriminate equally between samples subjected to the various thermal treatments. It must be emphasized that with the Kramer cell information was obtained about the texture of the asparagus in areas further from the tip.
- The thermal degradation of the texture followed a first-order pattern. The reaction rate constants, K_T, increased with temperature (Figure 36.3). The values of the kinetic parameters, E_a and $K_{115°C}$, corresponding to each cross section cut using the two methods are summarized in Table 36.4.
- In the earlier work,[2] the texture of the asparagus, which was previously blanched and then heated to temperatures below 100°C, showed a biphasic behavior ($n = 2$), when the texture was determined with the two cells, giving values of $E_{aa} = 9.6$ and $E_{ab} = 20.4$ Kcal/mol for the wire cell (cut at 5 cm) and $E_{aa} = 23.4$ and $E_{ab} = 18.3$ Kcal/mol for the Kramer cell (section 1). It was observed that the values of E_a in the second stage of inactivation (E_{ab}) were similar to those obtained in the second work for the two cells at temperatures above 100°C ($E_a = 20$ Kcal/mol for the wire cell, cut at 5 cm, and 18.3 Kcal/mol for section 1 with the Kramer cell). According to Huang and Bourne,[4] at high temperatures, the changes in fibrousness are marked mainly by the behavior of component (b) because it is more resistant to heat, and therefore the E_a values that represent the inactivation of this component are similar for the two temperature ranges.

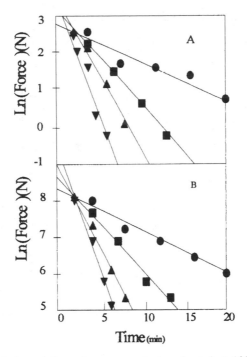

Figure 36.3 Thermal degradation of asparagus texture heated at 100°C (●), 110°C (■), 120°C (▲), and 130°C (▼), measured with A the wire cell (5-cm cutting) and B the Kramer cell (section 1).

In the third article,[5] the kinetic parameters were measured and an unsteady-state method was developed to estimate thermal degradation of asparagus texture in the temperature range of 100–130°C. The method used a mathematical model of heat transmission to estimate temperature distribution and a nonlinear regression of the texture measurements to estimate the kinetic parameters.

The specific heat, conductivity and convective coefficient were determined experimentally, along with the temperature profiles of the liquid in contact with the asparagus at various heating and cooling temperatures. The specific heat was calculated by DSC. The conductivity was obtained from the diffusivity, which in turn was derived from the slope of the heat penetration curve obtained in disks of asparagus heated in a high-temperature thermoresistometer,[6] which will be discussed later. The convective coefficient between the asparagus and the heating fluid was obtained by minimizing the residuals of the sum of squares (RSQ) between the experimental temperature values and those estimated with the model.

The values of the physical properties were considered constants for the temperature range studied and were introduced into the mathematical model to estimate the distribution of temperatures in the asparagus. Among the results, it is noteworthy that

1. The convective coefficient may take any value greater than 500 since, above this level, the RSQ does not vary.

TABLE 36.3
Mean Values of the Cutting Force Given by the Wire Cell Expressed in Newtons

Temperature (°C)	Time (min)	Cutting force (N)					
		$DT^* = 2.5^{**}$		$DT^* = 5^{**}$		$DT^* = 7.5^{**}$	
100	4	9.3	a	12	a	14	a
	8	6	b	5.6	b	5.9	bc
	12	4.4	c	4.8	bc	5.9	c
	16	3.3	d	4	c	6.5	c
	20	2.1	e	2	d	5	cd
110	4	9.9	a	10	a	9	a
	7	4.9	b	4.7	b	5.7	b
	10	1.9	c	1.8	c	3.4	c
	13	1.2	d	0.8	d	1.3	d
120	2	11	a	13	a	17	a
	4	6.9	b	8	b	8	b
	6	2.2	c	2.9	c	5.6	c
	8	1.2	d	1.0	d	1.8	d
130	2	11	a	145	a	17	a
	3	7	b	7	b	7.3	b
	4	4.4	c	4.9	c	7.1	b
	5	1.5	d	1.2	d	2.2	c
	6	1.0	d	0.7	d	1.2	c

*DT = distance from tip, in centimeters.

**Means within columns following the same letter are not significantly different at the 5% level according to Duncan's multiple range test.

TABLE 36.4
Kinetic Parameters at the Reference Temperature of 115°C of Degradation of Asparagus Texture for All Cutting of the Wire Cell, and All Sections of the Kramer Cell

Parameter	Site (cm)					
	2.5	5	7.5	S1	S2	S3
E_a (kcal/mol)	19 ± 3	20 ± 2	23 ± 3	18.3 ± 15	20.9 ± 1.7	24 ± 1
K_{115} (min^{-1})	0.27 ± 0.03	0.32 ± 0.03	0.24 ± 0.03	0.33 ± 0.02	0.25 ± 0.016	0.18 ± 0.007
r	0.9817	0.9877	0.9828	0.9933	0.9936	0.9984

2. The fit between the experimental temperature evolution and that estimated by the model is excellent (Figure 36.4).
3. In view of the evolution of texture loss with heating time at each temperature, a first-order kinetic is assumed.
4. The model can be used to estimate the temperature distribution at each point of the asparagus and at each heating or cooling time.

With these temperatures, the equation that determines the texture was integrated numerically and a nonlinear regression used to calculate the parameters $E_a = 76.91 \pm 0.13$ kJ/mol and $K_{115°C} = 0.00528 \pm 0.00005$ s^{-1}, minimizing the differences between experimental and estimated texture. Subsequently, and this time assuming isothermal heating, new kinetic parameters were obtained in the same way, giving $E_a = 56.4 \pm 1.3$ kJ/mol and $K_{115°C} = 0.00365 \pm 0.00005$ s^{-1}.

The texture values calculated are shown together with the estimated values, obtained in one case by means of the unsteady-state method and in the other by means of the parameters provided by the steady-state method (Figure 36.5).

The asparagus textures estimated by means of the unsteady-state method fit the experimental textures better. This is because the heating and cooling profiles inside the asparagus were included in the unsteady-state method. This result is more noticeable in short heating times.

Weight Loss Kinetic in Green Asparagus during the Sterilization Process

The heating of whole asparagus spears was performed by inserting them individually in stainless-steel tubes, which were sealed and heated in a thermostatically controlled oil bath. The values of asparagus weight were plotted against time at each heating temperature and fitted to a nonlinear regression. A reaction order of 0.99 was

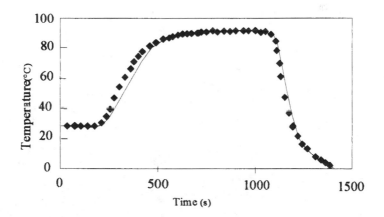

Figure 36.4 Experimental (♦) and theoretical (–) temperature profiles for asparagus during thermal processing at 9°C.

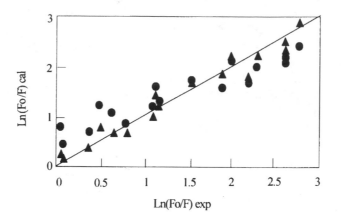

Figure 36.5 Comparison of $Ln(F_0/F)$exp with $Ln(F_0/F)$ cal using parameters obtained from the steady-state procedure (●) and unsteady-state procedures (▲).

obtained, and consequently the reactions were taken to be first-order. The reaction rate constants at each temperature were estimated from the curves, and the values obtained were used to plot ln K_T against $1/T$ (Figure 36.6). The estimated values obtained were $E_a = 8.9 \pm 1.3$ kcal/mol and $K_T = 0.0071 \pm 0.0004$ min^{-1}. The values of both parameters were smaller than those obtained for texture degradation.

Kinetics of Inactivation by HTST Treatments and Regeneration of Peroxidase Activity in Asparagus

It is known that HTST treatments can improve food quality, provided that enzyme inactivation is also achieved. The inactivation of any enzyme must be studied in the actual vegetable substrate in which it is found. POD is considered the most heat-resistant enzyme in vegetables.

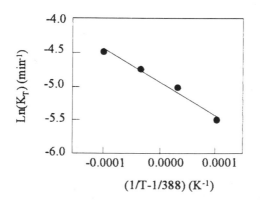

Figure 36.6 Arrhenius plot for asparagus weight loss.

The studies of inactivation at high temperatures and of regeneration have been summarized in two articles. In the first,[7] indirect heating was used, with the sample being placed in capillary tubes in a thermostatically controlled bath. In the second,[8] direct heating with steam was used. In this second case, the samples were heated in a high-temperature thermoresistometer.

In both cases, the quantity of vegetable sample to be analyzed was very small (10 to 50 μl). To determine POD activity, it was therefore necessary to use a simple method of the required sensitivity.

In the first study,[7] a spectrophotometric method was developed, based on the chromogen guaiacol. The method is capable of detecting at least two levels of decimal reduction of initial POD enzyme activity in unpurified asparagus (Figure 36.7).

It is known that POD can regenerate itself after thermal inactivation and that HTST treatments are less effective than conventional treatments for preventing regeneration. The ease of regeneration depends on the vegetable and even on the isoenzyme in question in a particular vegetable. Regeneration also depends on certain factors that are eliminated during purification of the extract. Consequently, the inactivation was performed with unpurified asparagus extract.

In the first study,[7] the substrates were inserted into capillary tubes and heated between 90 and 125°C for different periods of time. To study regeneration, samples

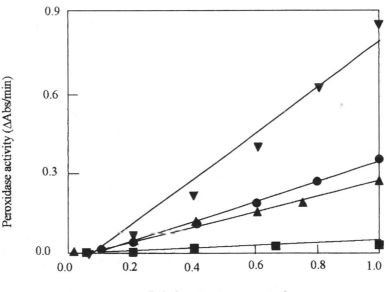

Figure 36.7 Change of peroxidase activity with enzyme concentration for different chromogens: guaiacol (▼), o-dianisidine (●), 2,2'-azino-bis-(3-ethylbenzothiazol-6-sulfonic acid) (ABTS) (▲), and 3-methyl-2-benzothiazoline hydrazone hydrochroride monohydrate and 3-(dimethylamino) benzoic acid (MBTH-DMAB) (■).

of asparagus extract treated at 90°C for 20 s and 110°C for 20 s were cooled and stored at 25°C for 6 days. Noteworthy results include the following:

1. The regeneration of POD activity did not follow a first-order or even second-order pattern. A rapid increase was observed during the first 100 h, followed by stabilization.
2. The regeneration was greater when the intensity of the inactivation is less strong.
3. After six days of storage, 85% of the maximum regenerable POD activity was attained. This period is therefore considered appropriate for the subsequent study of inactivation kinetics to prevent regeneration.
4. The POD inactivation curves were plotted by determining the residual activity regenerated after heating at 90, 100, 110, 120, and 125°C for different periods of time, cooling, and storing the samples for six days at 25°C.
5. The estimation of kinetic parameters was made by means of three different least squares methods: two-step linear regression, one-step linear regression, and nonlinear regression.
6. The kinetic parameters for preventing POD regeneration were obtained (Table 36.5).
7. The values corresponding to the two-step linear regression and the nonlinear regression were very similar, with typical deviations generally smaller in the nonlinear regression.
8. The regenerated POD followed a first-order inactivation kinetic, with E_a = 13.62 kcal/mol and $K_{100°C}$ = 2.07 min^{-1}.

TABLE 36.5
Kinetic Thermal Destruction Parameters, for Regenerated Asparagus Pod, Calculated by Using Three Least Square Methods at the Reference Temperature of 100°C.

Regression least square method	E_a (kcal/mol)	K_R (min^{-1})	z (°C)	D_R (min)
Two-step linear	14.51 ± 2.01	1.52 ± 0.16	45.49 ± 5.75	1.25 ± 0.13
One-step linear	11.31 ± 0.95	4.79 ± 0.20	58.31 ± 4.89	0.49 ± 0.02
Nonlinear	13.62 ± 0.96	2.07 ± 0.19	48.56 ± 3.46	1.13 ± 0.10

In the second article,[8] mentioned earlier, the heat treatment for the inactivation of asparagus POD was performed in a thermoresistometer (direct heating),[6] at temperatures of 110, 113, 115, 117, and 120°C.

When the ln of POD activity was plotted against heating time, straight lines with a single slope were obtained. At these temperatures, only the stable fraction of the enzyme remained. When the temperature was increased, the rate constant increased.

The Arrhenius representation also showed a linear behavior, with a value of E_a = 20 ± 3 kcal/mol for the stable fraction of the asparagus POD.

Heating with the thermoresistometer inactivated POD irreversibly, both in horse-radish and in asparagus. After heating, no regeneration was observed after storing for 7 days at 25°C. It must be pointed out that, when the heating was performed with capillary tubes (indirectly) in similar temperature-time conditions, regeneration occurred in the same storage conditions.

The results show, among other things, that the thermoresistometer used to study the heat resistance of microorganisms in liquid samples[6] is also suitable for studying enzymes in liquid and semi-solid samples[8] and that the heating conditions (direct or indirect) affect regeneration of POD.

REFERENCES

1. Silla, M. H., H. Nuñez, A. Casado, and M. Rodrigo. 1992. "The Effect of pH on the Thermal Resistance of *Cl. sporogenes* (PA 3679) in Asparagus Purée Acidified with Citric Acid and Glucono-δ-lactone," *Int. J. Of Food Microbiology*, 16: 275–281.

2. Rodrigo, C., M. Rodrigo, S. Fiszman, and T. Sánchez. 1997. "Thermal Degradation of Green Asparagus Texture," *J. Food Protection*, 60/3: 315–320.

3. Rodrigo, C., M. Rodrigo and S. Fiszman. 1997. "The Impact of High-Temperature, Short-Time Thermal Treatment on Texture and Weight Loss of Green Asparagus," *Z. Lebensm. Unters. Forsch.*, 205: 53–58.

4. Huang, Y. T. and M. C. Bourne. 1983. "Kinetics of Thermal Softening of Vegetables," *J. Texture Studies*, 14: 1–9.

5. Rodrigo, C., A. Mateu, A. Alvarruiz, F. Chinesta, and M. Rodrigo. 1998. "Kinetic Parameters for Thermal Degradation of Green Asparagus Texture by Unsteady-State Method," *J. F. Sci.*, 63/1: 126–129.

6. Rodrigo, M., A. Martínez, T. Sánchez, M. J. Peris, and J. Safón. 1993. "Kinetics of *C. sporogenes PA 3679* Spore Destruction Using Computer-Controlled Thermoresistometer," *J. F. Sci.*, 58/3, 649–652.

7. Rodrigo, C., M. Rodrigo, A. Alvarruiz, and A. Frígola. 1996. "Thermal Inactivation at High Temperatures and Regeneration of Green Asparagus Peroxidase," *J. Food Protection*, 59/10: 1065–1071.

8. Rodrigo C., A. Alvarruiz, A. Martínez, A. Frígola, and M. Rodrigo. 1997. "High-Temperature Short Time Inactivation of Peroxidase by Direct Heating with Five-Channel Computer-Controlled Thermoresistometer," *J. Food Protection*, 60/8: 967–972.

Part VII

Minimal Processing

37 Shelf-Life Prediction of Minimally Processed Chilled Foods

B. M. McKenna

CONTENTS

Abstract

Consumer demands for more convenient, less processed, but simultaneously safer foods has accelerated the development of minimally processed formulated foods with short shelf lives (generally no more than 7 to 10 days). The variation in type, formulation, and composition of such foods is as vast as is their processing. The one common factor is their use of the chilled distribution chain. A problem for the processor, both in terms of safety and distribution costs, is the prediction of the safe shelf life of the product.

This chapter outlines the current state of development of prediction systems and reviews their limitations. Although significant, these limitations have not inhibited the process and are mostly related to food and microbiological data deficiencies. Most serious among these is the lack of kinetic data for microbial growth and decay over a sufficiently wide range of substrate characteristics, along with an almost complete absence of data in the temperature range where growth has stopped and killing/decay has not yet started.

1-56676-963-9/02/$0.00+$1.50
© 2002 by CRC Press LLC

37.1 INTRODUCTION

37.1.1 BACKGROUND

In recent years, two consumer driven demands have arisen in the food industry. The first is for the provision of fresher, more natural foods requiring minimal preparation; the second is food safety. In particular, food safety considerations have rapidly become the primary consumer-driven concerns of the food industry. This is reflected in the emphasis now put on food safety by research funding agencies.[1]

To satisfy the first demand for fresh, natural foods needing minimal preparation, but which nevertheless have a substantial shelf life, minimal processing is becoming popular. Minimal processing is a vague concept and a misleading term, as it usually involves substantial processing but results in foods that have a fresh-like quality and contain only natural ingredients. A recent overview identifies two purposes for minimal processing: (1) keeping the produce fresh, yet supplying it in a convenient form without loss of nutritional quality; and (2) keeping the shelf life sufficient to make distribution to the consumer feasible.[2]

The processing aspects of minimal processing can be invisible to the consumer and can be applied at various stages of the food processing and distribution chain (processing, storage, and packaging).[3] However significant and severe the processing, minimally processed products are rarely sterile and can undergo rapid deterioration.

To counteract this, minimal processing is often associated with hurdle technology concepts where other systems (added salts, packaging systems, refrigeration systems) are used to inhibit spoilage. While not a new concept, hurdle technology has been given new impetus by the requirements of minimal processing. Through it, existing and novel preservation techniques are combined to give a series of preservative factors, called *hurdles,* that cannot be overcome by microorganisms.[4] Common hurdles are temperature (heating, cooling), water activity, pH, redox potential, and preservatives (chemical agents, bacteriocins). Novel preservative factors include gas packaging, ultra-high pressure treatment, edible coatings, and use of bacteriocins.[5] Hurdles are generally synergistic in their operation and use combinations of effects (e.g., the combination of pH and water activity). The concept of hurdle technology is widely used in the meat industry for the production of formulated products such as sausages and can be used now also in the production of fruits and vegetables and bakery, dairy, and fish products.[5]

A particular form of minimal processing involving hurdle technology is *sous vide,* or vacuum cooking in heat-stable vacuum pouches under controlled conditions of temperature and time.[6] This process is especially suitable for ready-made meals and is claimed to give better sensory and nutritional quality than conventional processing. It is characterized by long heating times and relatively low heating temperatures to avoid thermal damage. However, the use of a low heating temperature gives a small pasteurizing effect, so low-temperature storage and distribution is essential. A major disadvantage is the limited shelf life that is available even at 0°C. To improve the microbiological safety of sous-vide cooked products, the concept of hurdle technology is used. *Sous-vide* cooking is normally a semicontinuous process and is used primarily in catering. Strict temperature control is essential.[6]

With such products, accurate estimation of their shelf life is essential but, unfortunately, none of the previously available estimation systems is applicable. In fact, only frozen and sterilized foods have any mathematical estimation systems available.

37.1.2 Shelf-Life

Food products are usually nonequilibrium systems; i.e., they are in a state of thermodynamic instability. While severe processing conditions can bring the food to a pseudo stable state, this is not possible with minimal processing and several forms of spoilage/deterioration have to be considered in predicting shelf life.

- *Microbial spoilage.* Foods must be safe from a microbiological point of view. Pathogenic microorganisms (or sometimes their metabolites) should be removed or eliminated. However, it is spoilage organisms rather than pathogenic organisms that limit the shelf life of foods, and these should be inhibited. The most frequently used process in this respect is heating in the form of pasteurization (the minimal heat treatment to ensure absence of pathogens) or sterilization (no microorganisms left, complete microbial stability—unfortunately, rarely an option with minimal processing).
- *Chemical spoilage.* Some chemical reactions may limit the shelf life of foods (e.g., nonenzymatic browning reactions, causing discoloration, off-flavors, loss of nutritional quality, perhaps formation of toxicologically suspect compounds, and fat oxidation causing off-flavors and loss of nutritional quality). Obviously, it is necessary to minimize such unwanted reactions as much as possible.
- *Biochemical spoilage.* Enzymes present in raw materials from both plant and animal origin can cause deterioration, e.g., protein breakdown by proteases, fat breakdown by lipases, enzymatic browning by polyphenol oxidase. Destruction or inhibition of such enzymes is essential for significant shelf life.
- *Physical deterioration.* Foods should be physically stable; that is, they should not show phase separation, should not dry out, should have a certain consistency, etc. Migration of components within formulated products may determine a shelf-life limitation (e.g., moisture migration in a pasta product may make it tough and inedible).

Combine all these deterioration factors in multi-ingredient formulated foods, and it quickly becomes obvious that shelf-life prediction is a difficult problem for the processor. Trial and error shelf-life estimations have become the norm and for safety considerations, the processor has had to err on the side of caution and publish product expiration dates that are normally much earlier than the real one. Research has been limited in the area but, in recent years, the European Union and the funding agencies of several member states have invested significant funds in the subject.[7]

37.2   SHELF-LIFE PREDICTION METHODS

37.2.1   Methods

The European Union sponsored programme listed above involved 43 researchers from 17 countries covering both the EU member states and some Central and Eastern European states (Hungary, Poland, Bulgaria, Slovenia, Latvia). The research activities of the group covered three approaches to shelf-life modeling.

- Predictive modeling of shelf-life
- Predictive modeling of microbial growth and decay
- Shelf-life estimation on specific individual products

As the first of these, predictive modeling of shelf life, offers the best long-term benefits for the food processor, this paper will concentrate on this method and on its inherent problems. The steps outlined below are those used by several research groups in seeking a solution.

- Select the product recipe and from a database, estimate the probable microbial population and numbers of each species.
- Select the size and shape of the formulated product along with the container (if any) and package (if any) and the characteristics of both.
- Select the process conditions that will be used and the specific equipment types and characteristics (ovens, retorts, etc.).
- Select the shelf-life limiting factor. In most cases, this will be when the average number of bacteria exceeds a given level.
- Use finite difference modeling to predict the heat transfer to/from the product during heating and cooling, and estimate the temperature at selected nodes in the product at selected but short time intervals.
- Use predictive microbiology to estimate the bacteria numbers of each species at each time interval.
- Repeat the modeling exercise for the storage, distribution, retailing, and home storage process (maximum home storage and conditions must be specified).
- When the predicted number exceeds the chosen limit, the end of safe shelf life has been reached.
- If suitable kinetic data is available, other limiting factors (enzyme activity, desired physical characteristics, etc.) can be easily incorporated into the software.

37.2.2   Limitations

While computer systems are quite capable of handling the complexities of the above system, a significant number of limitations exist and are considered below.

Product Recipe and Initial Bacterial Loads

While there are some proprietary recipe formulation programmes, these are limited in nature, and a company may have to prepare a database of its own ingredients and characteristics. This could be a major complication for the smaller company. A secondary important issue is the estimation of initial bacterial contamination levels. Only estimates can be given and, for safety reasons, these will err on the high side. Consequently, the predicted numbers will always be higher than the real ones (except in the rare case when sterility is achieved). In addition, there is currently no mechanism for estimating the effects of mixing ingredients on the initial bacterial population and loads.

Shape, Size, and Characteristics of Containers and Equipment

While container shapes and physical characteristics are available from the manufacturers, they are not in database form and will have to be entered manually in the system. However, for maximizing the shelf life of a product, an iterative process may be necessary with the procedure tested for different container shapes and materials (e.g., a thinner product in a metal foil container may permit greater heat penetration and bacterial killing than the same quantity of product in a deeper board container).

More serious is the lack of equipment characterization data. Often, this is not known by the equipment manufacturers themselves. For example, while the temperature variation within an oven cavity may be specified by the manufacturer, they will seldom have measured the air flow rates and directions at various parts of the cavity, nor can they provide data on how either temperature or flow characteristics will vary with the degree of product loading in the oven (obviously, air flow will be impeded by the degree of filling of the oven).

Process Conditions

This is a simple matter for existing products where the set points of the process equipment have been determined by experience. However, for new products or modified products, shelf-life optimization it remains a problem. Here again, several iterations of the programme will be necessary to determine the optimum conditions.

Shelf-Life Limiting Factor

As previously stated, this is likely to be the number of bacteria present. Consequently, when shelf life is determined, it is normally, the "safe shelf life" that is predicted. However, other factors often render the product inedible long before this safe limit is reached! Therefore, it is suggested that the concept of *total quality modeling,* or TQM should be used, with total quality as the controlling parameter in shelf-life prediction.[8,9]

Modeling of Temperature within the Food at All Stages of Processing and Distribution

Lack of data remains the major inhibiting factor at this stage in the programme. The lack of physical property data for foods has long been a problem in process calculations. However, this is slowly being rectified through another European Union sponsored project. Despite this, several modeling groups have been able to predict temperatures to within a few degrees. Sensitivity analysis of the input variables have shown that the convective surface heat transfer coefficient during heating and cooling remains the variable of highest sensitivity. Values are never available in databases but are often calculated from dimensionless number relationships. However, these normally require air velocity data and, as stated above, such data is seldom available, nor is its variation with the degree of filling of the equipment.

A more serious deficiency is that the heat transfer coefficient relationships are normally for transfer of heat to or from inert surfaces. In the heating and cooling of foods, both evaporation from and condensation on the surface will change the effective heat transfer coefficient and lead to incorrect estimation of the internal temperatures. It is essential that values used allow for this mass transfer complication.

A third problem arises in predicting conditions that are outside the control of the processor, namely the retailing, home storage, and final heating/cooking conditions (if any). Here, the processor can do no more than specify the desired retail, home storage, and final heating conditions and allow for the worst conditions of transfer between them (e.g., significant time in a hot car).

Predictive Microbiology

During minimal processing, distribution, and use, foods will experience both microbial growth and decay at different stages. Consequently, both growth and decay kinetics are required. Once again, problems arise through the lack of suitable data. Predictive modeling has long been used in microbiology, and database packages are available (see, for example, Reference 10). However, since safety has been the primary consideration in development of such models, pathogenic organisms are well represented, but many common food spoilage organisms are missing.

A second problem is the limitation of growth and killing data to a small range of substrate conditions. For complete use in a shelf-life prediction system, kinetic growth and decay data are required on a wide range of pH values in the product, a range of moisture contents, a range of salt contents, and, ideally, a vast range of ingredient contents.

A third problem is the variation in microbiological kinetic data. This is often the result of substrate variation but is also related to the inherent inexact nature of microbiological measurements. Consequently, it is difficult to put exact confidence limits of the kinetic data.

Yet a fourth problem arises from poor data at two stages in the growth and decay system. Variation in the lag phase has long been recognized and complicates the prediction process. Less obvious is the lack of data at the shoulder period, when temperatures reach a level where growth is stopping but killing has not yet started (for mesophilic bacteria, in the temperature range 45 to 60°C).

Other Shelf-Life Limiting Factors

Kinetic data on texture variation with time and temperature is very limited. Similarly, diffusion will affect this (as it could also affect microbial kinetics), and, while some diffusion kinetics are available for water and NaCl, it is deficient for many other components.

General

Computer systems are powerful enough to handle the computing complexities of the prediction systems. However, the systems developed to date require at least work-station power, whereas the food industry is largely PC equipped.[11,12] Acceptable running times are just now being achieved with Pentium III machines.

37.3 CONCLUSION

It might be concluded from the list of limitations in the above section that shelf-life prediction remains an impossibility. This is not the case. Current computer systems are working well. The limitations certainly restrict their use to a small range of products. The main message is that the computer systems are working well but are being hindered by inadequate food and microbiological data. However, even in the short few years since the first developments, there have been significant improvements, and these are continuing.

REFERENCES

1. Anon. 1998. "European Union 5th Framework Programme for Research," *Official Journal of the European Union.*
2. Ahvenainen, R. 1996. "New Approaches in Improving the Shelf Life of Minimally Processed Fruit and Vegetables," *Trends Food Sci. Technol.,* 7: 179–187.
3. Ohlsson, Th. 1994. "Minimal Processing-Preservation Methods of the Future: An Overview," *Trends Food Sci. Technol.,* 5: 341–344.
4. Leistner, L. L. 1992. "Food Preservation by Combined Methods," *Food Res. Intern.,* 25:151–158.
5. Leistner, L. L. and L. M. Gorris. 1995. "Food Preservation by Hurdle Technology," *Trends Food Sci. Technol.,* 6: 41–45.
6. Schellekens, W. 1996. "New Research Issues in sous-vide Cooking," *Trends Food Sci. Technol.,* 7: 256–262
7. European Union (EU). 1994. "Shelf-Life Prediction for Improved Safety and Quality of Foods," Copernicus project CIPA-CT94–0120.
8. Nicolai, B. M. and J. Baerdemaeker. 1998. "Food Quality Modeling (COST Action 915)," Office of Official Publications of the European Communities.
9. McKenna, B. M. 1999. "Shelf-Life Prediction for Improved Safety and Quality of Foods," in *Proceedings of the Final Symposium of COPERNICUS Project CIPA-CT94–0120, Wageningen.*
10. Food Research Association. 1999. *Food MicroModel Database,* Leatherhead, U.K.
11. Keane, G. 1997. Private communication.

12. Nicolai, B. M., P. Van den Broeck, M. Schellekens, G. De Roeck, T. Martens, and J. De Baerdemaeker. 1995. "Finite Element Analysis of Heat Conduction in Lasagna During Thermal Processing," *International Journal of Food Science and Technology*, 30(3): 347–364.

38 Update on Hurdle Technology

L. Leistner

CONTENTS

Abstract

In this chapter, recent developments are discussed related to the hurdle concept (i.e., sustainable agriculture, additional hurdles, lines of defense), to basic aspects (i.e., homeostasis, metabolic exhaustion, stress reactions, multi-target preservation), and to the application of hurdle technology (i.e., minimally processed foods, chilled

1-56676-963-9/02/$0.00+$1.50
© 2002 by CRC Press LLC

foods, fermented foods, healthful foods, nonthermal preserved foods, Indian hurdle foods, Chinese fusion foods, food packaging in Japan, and the design of hurdle-technology foods). This update does not aim for completeness but only presents the highlights that have emerged since 1994.

38.1 INTRODUCTION

Many preservation methods are traditionally used for making foods stable, safe, and tasty; e.g., heating, chilling, freezing, drying, curing, salting, sugar addition, acidification, oxygen removal, and fermentation. However, these methods are based on relatively few parameters, i.e., high temperature (F value), low temperature (t value), water activity (a_w), acidification (pH), redox potential (Eh), competitive flora, and preservatives, which might be called *hurdles*. In some of the mentioned preservation methods, these parameters are of major importance; in others, they are only secondary hurdles.[1,2] The critical values of these parameters for death, survival, or growth of microorganisms have been determined in recent decades and are now the basis of food preservation. However, it must be kept in mind that the critical value of a particular parameter changes if other preservative hurdles are present. For instance, the heat resistance of bacteria increases at low a_w and decreases at reduced pH or in the presence of some preservatives, whereas a low Eh increases the inhibition of microorganisms caused by a reduced a_w. The simultaneous effect of different hurdles in foods could be additive or even synergistic.

The microbial stability and safety of most traditional and novel foods are based on a combination of several preservative factors (hurdles) that the microorganisms present in the food are unable to overcome. This is illustrated by the so-called *hurdle effect,* first introduced by Leistner.[3] The hurdle effect is of fundamental importance for the preservation of foods, since the hurdles in a stable product control microbial spoilage, food poisoning, and desired fermentation processes.[1,3] From an understanding of the hurdle effect, hurdle technology[4] was derived, which allows improvements in the safety and quality of foods using deliberate and intelligent combinations of hurdles.

The application of hurdle technology for a gentle but efficient preservation of foods started in the 1980s, in Germany, with meat products[4] and is now advancing worldwide. Various expressions are used for the same concept in different languages: *Hürden-Technologie* in German, *hurdle technology* in English, *technologie des barrières* in French, *barjernaja technologija* in Russian, *technologia degli ostacoli* in Italian, *tecnología de obstáculos* (or *métodos combinados*) in Spanish, *shogai gijutsu* in Japanese, *zanglangishu* in Chinese, etc. However, the term *hurdle technology* is most often used worldwide.

In June 1994, the International Symposium on the Properties of Water (ISO-POW)-Practicum II was held at the Universidad de las Américas-Puebla, Mexico. At this meeting the state-of-the art of hurdle technology up to that time was presented,[5] and these data will not be repeated here. But, in the present update, additional data will be summarized that have emerged since the ISOPOW-Practicum II.

38.2 HURDLE CONCEPT

Hurdle technology proved successful, since an intelligent combination of hurdles secures microbial stability and safety as well as the sensory and nutritional quality of foods,[6-8] it provides convenient and fresh-like foods to the consumer, and it is cost efficient, since it demands little energy during production and storage. In industrialized countries, hurdle technology is currently of particular interest for minimally processed and ready-to-eat foods, whereas, in developing countries, foods storable without refrigeration, due to stabilization by hurdle technology, are of most concern.

38.2.1 SUSTAINABLE AGRICULTURE

Food preservation based on hurdle technology meets consumer demands, is gentle on resources (energy, capital investment), and is applicable in industrialized as well as in developing countries. These are requirements for food processing in compliance with issues of sustainability and food security in the future. Therefore, food preservation by hurdle technology has been chosen to be part of an *Encyclopedia of Life Support Systems (EOLSS)*, which is designed to be a guide to sustainable global life in the next millennium. The EOLSS will be published in 2000/2001 under the guidance of UNESCO.[9] Furthermore, with similar intentions, a contribution on food preservation by hurdle technology was invited to be presented at an international conference on Sustainable Animal Production, Health and Environment: Future Challenges, which was held in November 1999 at the C.C.S. Haryana Agricultural University at Hisar, India.[10]

38.2.2 ADDITIONAL HURDLES

Better understanding of the occurrence and interactions of different preservative factors (hurdles) in foods is the basis for improvements in food preservation. If the hurdles for a food are known, the microbial stability and safety of this food might be optimized by changing the intensity or quality of these hurdles. Using an intelligent mix of hurdles, it is possible to improve not only the microbial stability and safety but also the sensory and nutritional quality as well as the economic aspects of a food.

In a previous publication, about 50 hurdles of potential use for foods (of animal or plant origin, which improves the stability and/or quality of these products) were listed.[5] In the meantime, the number of suitable hurdles for application in advanced hurdle-technology foods has increased to more than 60,[11] but this list is by no means complete.

Today, in particular, physical and nonthermal preservation methods of foods (high hydrostatic pressure, high-intensity pulsed electric fields, oscillating magnetic fields, light pulses, osmo-dehydro-freezing, food irradiation) are receiving attention, since, in combination with conventional hurdles, they have potential benefits for the microbial stabilization of fresh-like food products with little induced degradation of sensory and nutritional properties.[12]

Another group of hurdles of special interest at present in industrialized as well as in developing countries are natural preservatives (spices, spice extracts, lysozyme, chitosan, pectine hydrolysate, protamine, paprika glycoprotein, hop extracts, etc.). The reason for a replacement of chemicals by such natural preservatives is that, in many countries, the preference of the consumer for green preservatives. But in some countries that are short of foreign currency, natural preservatives such as spices or spice extracts are preferred, because they are readily available and less expensive.

38.2.3 LINES OF DEFENSE

A medical aspect related to food safety relates to the hurdles in the human body, which determine whether foods containing pathogens will lead to food poisoning. This aspect of hurdle technology has recently attracted attention. The first line of defense is the saliva in the mouth of the consumer, which contains bacteriostatic substances (e.g., lysozyme). The low pH in the stomach is a very strong hurdle for the passage of pathogenic bacteria, and its effectiveness is influenced by the residence time of the food in the stomach. The next lines of defense are the high pH in the small intestine and the competitive flora (probiotics) in the large intestine. The condition of the intestinal mucosa and the state of the immune defense within the body influence whether pathogens might cause bacteremia. It is feasible that the scope of the hurdle concept is enlarged to encompass the health of the animal from which the food comes (in some food animals, similar hurdles are effective as in humans), the hurdles for pathogenic microorganisms in a well designed food product, and, finally, the hurdles that are active in the body of the consumer; these three links of a chain determine whether food poisoning occurs or is prevented.

38.3 BASIC ASPECTS

Food preservation implies exposing microorganisms to a hostile environment to inhibit their growth, shorten their survival, or cause their death. The feasible responses of microorganisms to such a hostile environment determine whether they grow or die. A better understanding of the physiological basis for growth, survival, and death of microorganisms in food products could open new dimensions for food preservation.[7] Furthermore, such an understanding would be the scientific basis for the effective application of hurdle technology in the preservation of foods. Recent advances have been made by considering the homeostasis, metabolic exhaustion, and stress reactions of microorganisms as well as by introducing the innovative concept of multi-target preservation for a gentle, yet effective, preservation of hurdle technology foods.[13,14]

38.3.1 HOMEOSTASIS

Homeostasis is the strong tendency to uniformity and stability in the internal status of organisms. For instance, the maintenance of a defined pH in narrow limits is a prerequisite and feature of all living cells, and this applies to higher organisms as well as to microorganisms.[15] Much is already known about homeostasis in higher organisms at the molecular, subcellular, cellular, and systemic levels in the fields of pharmacology and medicine.[15] This knowledge should be transferred to microorganisms important for the poisoning and spoilage of foods. In food preservation, the homeostasis of microorganisms is a key phenomenon that deserves more attention.[16] If the homeostasis of microorganisms, i.e., their internal equilibrium, is disturbed by preservative factors (hurdles) in foods, they will not multiply but will remain in the lag-phase or even die before their homeostasis is repaired (reestablished). Thus, disturbing the homeostasis of microorganisms in a food temporarily or permanently[7] achieves food preservation.

During their evolution, a wide range of more or less rapidly acting mechanisms (e.g., osmoregulation to counterbalance a hostile water activity in the food) have developed in microorganisms that act to keep the important physiological systems operating, in balance, and unperturbed even if the environment around them is greatly perturbed.[17] In most foods, the microorganisms operate homeostatically so as to react to stresses imposed by the preservation procedures applied. The most useful procedures employed to preserve foods are effective in overcoming the various homeostatic mechanisms the microorganisms have evolved so as to survive under extreme environmental stress.[17] The repair of a disturbed homeostasis demands much energy, and thus the restriction of energy supply inhibits repair mechanisms in microbial cells, and leads to a synergistic effect of preservative factors (hurdles). Energy restrictions for microorganisms are, for example, caused by anaerobic conditions such as vacuum or modified-atmosphere packaging of foods. Therefore, low a_w (and/or low pH) and low redox potential act synergistically.[17] Such interference with the homeostasis of microorganisms or entire microbial populations provides an attractive and logical focus for improvements in food preservation techniques.[17]

38.3.2 METABOLIC EXHAUSTION

Another phenomenon of practical importance is the metabolic exhaustion of the microorganisms, which could lead to an autosterilization of foods. This was first observed in experiments with mildly heated (95°C core temperature) liver sausage that was adjusted to different water activities by the addition of salt and fat. The products were inoculated with *Clostridium sporogenes* and stored at 37°C. Clostridial spores surveying the heat treatment vanished in the products during storage, if the products were stable due to the reduced a_w.[18] Later, this behavior of *Clostridium* and *Bacillus* spores were regularly observed during storage of shelf stable meat products (SSP) if the products were stored at ambient temperature.[19] The most likely explanation is that bacterial spores that survive the heat treatment are able to germinate in these foods under less favorable conditions than those under which vegetative bacteria are able to multiply.[2] Therefore, the spore counts in stable hurdle-technology foods actually decrease during storage of the products, especially if stored without refrigeration. The same phenomenon was also observed by Latin

American researchers[20,21] in studies of high-moisture fruit products (HMFP) because the counts of a variety of bacteria, yeasts, and molds that survived the mild heat treatment decreased quickly in the products during refrigerated storage; the hurdles applied (pH, a_w, sorbate, sulfite) did not allow growth.

A general explanation for this surprising behavior might be that vegetative microorganisms that cannot grow will die, and they die more quickly if the stability of the food is close to the threshold for their growth, storage temperature is elevated, antimicrobials are present, and the microorganisms are sublethally injured.[7] Apparently, microorganisms in stable hurdle technology foods strain every possible repair mechanism for their homeostasis to overcome the hostile environment; by doing this, they completely use up their energy and die if they become metabolically exhausted. This leads to an autosterilization of such foods.[13] Due to autosterilization, hurdle technology foods that are microbiological stable become safer during storage, especially at ambient temperatures. So, for example, salmonellae that have survived a long ripening process in fermented sausages will vanish more quickly if the products are stored at ambient temperature, and they will survive longer and possibly cause food poisoning if the products are stored under refrigeration.[7] Thus, refrigeration is not always beneficial for the microbial safety and stability of foods. However, this is true only if the hurdles present in a food also inhibit the growth of microorganisms without refrigeration; if this is not the case, then refrigeration is beneficial. Certainly, the survival of microorganisms in stable hurdle technology foods is much shorter without refrigeration.[14]

38.3.3 STRESS REACTIONS

Some bacteria become more resistant or even more virulent under stress, since they generate stress shock proteins. The synthesis of protective stress shock proteins is induced by heat, pH, a_w, ethanol, oxidative compounds, etc., and also by starvation. Stress reactions might have a nonspecific effect since, due to a particular stress, the microorganisms become more tolerant to other forms of stress; i.e., they acquire a *cross-tolerance*. Stress reactions of microorganisms have recently been the topic of several international symposia, e.g., "Stress et Viabilité des Microorganismes dans les Aliments," held June 1999 at Quimper, France; more pertinent information on this topic is found in the proceedings of these events. Herein, only the relationship of stress reactions to hurdle technology will be briefly discussed.

The various responses of microorganisms under stress might hamper food preservation and could be problematic for the application of hurdle technology. On the other hand, the activation of genes for the synthesis of stress shock proteins, which help microorganisms to cope with stress situations, should be more difficult if different stresses are received at the same time. Simultaneous exposure to different stresses will require energy consuming synthesis of several or at least much more protective stress shock proteins, which in turn may cause the microorganisms to become metabolically exhausted. Therefore, multi-target preservation of foods could be the answer to avoid synthesis of stress shock proteins, which otherwise might jeopardize the microbial stability and safety of hurdle technology foods.[13,14]

38.3.4 MULTI-TARGET PRESERVATION

Leistner[7,13] has introduced the concept of multi-target preservation of foods. Multi-target preservation should become the ambitious goal for a gentle but effective food preservation in the future. It has been suspected for some time that different hurdles inherent in a food might not only have an additive effect on the microbial stability, but they could act synergistically.[3] A synergistic effect could be achieved if the hurdles in a food hit, at the same time, different targets (e.g., cell membrane, DNA, enzyme systems, pH, a_w, Eh) within the microbial cells and thus disturb the homeostasis of the microorganisms present in several respects. If so, the repair of homeostasis and the activation of stress shock proteins become more difficult.[7,14] Therefore, simultaneously employing different hurdles in the preservation of a particular food should lead to an optimal microbial stability. In practical terms, this could mean that it should be more effective to employ different preservative factors (hurdles) of small intensity than one preservative hurdle of larger intensity, because different preservative factors might have a synergistic effect.[22]

It is anticipated that the targets within microorganisms of different preservative factors for foods will be elucidated, and that hurdles could then be grouped in classes according to their targets. A mild and effective preservation of foods, i.e., a synergistic effect of hurdles, is likely if the preservation measures are based on intelligent selection and combination of hurdles taken from different target classes.[7,11] This approach is probably not only valid for traditional food preservation procedures but as well for modern processes such as food irradiation, ultra-high pressure, and pulsed technology.[12] Food microbiologists could learn from pharmacologists, because the mechanisms of action of biocides have been studied extensively in the medical field. At least 12 classes of biocides are already known that have different targets, and sometimes more than 1 target, within the microbial cell. Often, the cell membrane is the primary target, becoming leaky and disrupting the organism, but biocides also impair the synthesis of enzymes, proteins, and DNA.[23] Multidrug attack has proven successful in the medical field to fight bacterial infections (e.g., tuberculosis) as well as viral infections (e.g., AIDS), and thus a multi-target attack on microorganisms should also be a promising approach in food microbiology.[13]

38.4 APPLICATIONS OF HURDLE TECHNOLOGY

Foods based on combined preservation methods (hurdle technology) are prevalent in industrialized as well as in developing countries. In the past, and often still today, hurdle technology has been applied empirically without knowing the governing principles in the preservation of a particular food. However, with a better understanding of the principles and improved monitoring devices for hurdles (e.g., rapid measurement of a_w in foods), deliberate application of hurdle technology has advanced worldwide.

A major contribution to widespread application of hurdle technology was a 3-year research project on Food Preservation by Combined Processes, supported by the European Commission, to which scientists from 11 European countries contributed. The final report[24] of this project was requested by 2200 scientists and technologists, and, since there was continued interest worldwide, this report was put on the Internet. Another publication that spread the gospel was a picture book called *Food Design by Hurdle Technology and HACCP*,[19] of which the author of this contribution handed out ~4000 copies during invited lectures in numerous countries. The state of the art until the International Symposium of ISOPOW- Practicum II held in 1994 in Mexico has been outlined in a scientific publication on Principles and Applications of Hurdle Technology,[7] and these data will not be repeated here. In the present contribution, more recent developments related to the applications of hurdle technology are discussed by briefly presenting some examples and trends.

38.4.1 MINIMALLY PROCESSED FOODS

Minimally processed, convenient, ready-to-eat, high-moisture but ambient stable foods are the trend in industrialized as well as in developing countries. The consumer prefers minimally processed foods, since these foods have appealing fresh-like characteristics and thus a superior sensory quality. However, at the same time, these foods must be microbiologically safe and stable. These somewhat conflicting goals are achievable by the application of advanced hurdle technology. Therefore, this concept is now used in the production of a variety of minimally processed foods, and, depending on the type of food, different hurdles are employed and combined.[25] For instance, various approaches are successfully used for minimally processed meats, fruits, or vegetables.

Minimally processed and ambient stable meat products, with superior sensory properties, were introduced in Germany in the 1980s. Today, four types of these shelf stable products (SSPs) (i.e., F-SSP, a_w-SSP, pH-SSP, and Combi-SSP) are known.[19] These products are mildly heated in sealed containers, and the surviving bacterial spores are inhibited by carefully selected hurdles (e.g., a_w, pH, Eh, nitrite), which are applied in combination. SSP meats have been available for nearly two decades in large quantities and varieties on the market and have not caused spoilage or food poisoning problems, assuming that established guidelines for their production were observed.[26] In such products, the number of viable bacterial spores decreases during storage due to metabolic exhaustion of the microorganisms, especially if the products are stored without refrigeration.[13] Since SSPs are tasty and ambient stable, they are also attractive as military rations.[27]

The application of advanced hurdle technology in Latin America has created a new line of minimally processed, fresh-like fruits that for several months are microbiologically safe and stable at ambient temperature.[28,20,21] The hurdles that proved suitable for this group of food products are a mild heat treatment (blanching), slight reduction of a_w and pH, and the moderate addition of preservatives (sorbate and sulfite). The blanching of the fruits (partial decontamination with steam) is important for microbial stability because, even though vegetative microorganisms might survive this mild heat treatment, their number is reduced, and thus only fewer and lower hurdles are essential. The number of surviving bacteria, yeasts, and molds decreases rapidly during ambient storage of the products, probably due to metabolic exhaustion, since they are not able to multiply in stable hurdle technology fruits. However, also, the added sulfite and sorbate deplete during storage of the fruits, and this is beneficial for the consumer but diminishes microbial stability. Therefore, a recontamination of the fruits during storage should be avoided by suitable measures.[5] It is anticipated that this innovative process will gain ground worldwide. The advances in Latin America in fruit preservation are very impressive and have already been confirmed by researchers from India.[29,30]

Minimally processed vegetables (e.g., raw, sliced vegetables or cooked *sous vide* dishes) are heated only mildly or not at all and subsequently must be stored under refrigeration.[25] Additional hurdles (such as modified-atmosphere packaging or vacuum packaging, possibly the addition of bacteriocins or bacteriostatic spices, or treatment with ultra-high pressure) are often applied.[31] For the raw, sliced vegetables *Listeria monocytogenes* is of major concern,[32] whereas, for the *sous vide* dishes, this is non-proteolytic *Clostridium botulinum*.[33] Both of these product groups are safe and stable only under strictly controlled processing and storage conditions, and these requirements for minimally processed vegetables are also applicable to minimally processed meats and fruits. However, minimally processed vegetables are more risky with respect to safety compared to these other product groups, because they are not ambient stable hurdle technology foods and thus have to be stored under strict refrigeration. This has the consequence that pathogenic bacteria survive longer in minimally processed vegetables than in minimally processed meats or fruits, because metabolic exhaustion hardly takes place during storage of these vegetables. It would be an ambitious goal to design minimally processed vegetables that are ambient stable due to the application of proper hurdles, because then storage and distribution would be facilitated, and the safety of the products could be improved if metabolic exhaustion of the pathogenic bacteria present would take place.[25]

38.4.2 Chilled Foods

The advantages of hurdle technology are most obvious in high-moisture foods that become shelf stable at ambient temperature due to an intelligent application of combined methods. However, the use of hurdle technology is also appropriate for chilled foods, because, in the case of temperature abuse, which can easily happen during food distribution, the stability and safety of chilled foods break down, especially if low-temperature storage is the only hurdle. Therefore, it is advisable to incorporate into chilled foods (e.g., ready-to-eat meats, *sous vide* dishes, salads, fresh-cut vegetables) some additional hurdles (e.g., a_w, pH, competitive bacteria) that will act as a backup in case of temperature abuse. This type of safety precaution for chilled foods is called *invisible technology*,[19,8] implying that additional hurdles act as safeguards in chilled foods, ensuring that they remain microbiologically stable and safe during storage in retail outlets as well as in the home.

38.4.3 Fermented Foods

In fermented foods (e.g., fermented sausages, raw hams, ripened cheeses, pickled vegetables), a sequence of hurdles (i.e., the hurdles act one after the other and not at the same time) leads to a stable and safe product, because spoilage and pathogenic bacteria are inhibited, and the desired competitive flora is selected. The sequence of hurdles acting in fermented sausages (salami) has been studied well[7,34,35] and therefore will be used as an example.

Important hurdles in the early stages of the ripening process of salami are nitrite and salt, which inhibit many of the bacteria present in the initial product. However, other bacteria are able to multiply, use up the oxygen, and therefore cause the redox potential of the product to decrease. This in turn enhances the Eh hurdle, which inhibits aerobic organisms and favors the selection of lactic acid bacteria. These are the competitive flora, which flourish by metabolizing the added sugars, causing the pH to decrease (i.e., an increase of the pH hurdle). In long-ripened salami, nitrite is depleted and the lactic acid bacteria vanish, while the Eh and pH increase again. Only the water activity hurdle (a_w), due to the drying process, is strengthened with time, and it is then largely responsible for the stability of long-ripened raw sausage. Since this sequence of hurdles was revealed, the production of fermented sausages has become less empiric and more advanced, and this knowledge has been used to achieve required inhibition of *Clostridium botulinum, Listeria monocytogenes, Salmonella* spp., and *Staphylococcus aureus* in salami during fermentation and ripening.

The sequence of hurdles that secure the microbial stability and safety of raw hams is also well known.[36] Sequences of hurdles for the proper fermentation processes of other foods (e.g., ripened cheeses, pickled vegetables) are also important, and it would be challenging to elucidate them.

38.4.4 HEALTHFUL FOODS

In healthful foods, the nutritional properties are the first priority; however, their microbial stability and safety must not be neglected. Novel healthful food products derived from meat, poultry, or fish contain less fat and/or salt and therefore are more prone to spoil or to cause food poisoning. The reduction of salt and fat, as well as the substitutes for these traditional ingredients, for muscle foods diminish the microbial stability, since several hurdles (a_w, pH, preservatives, and possibly Eh and microstructure) will change.[37] The compensation could come from an intelligent application of hurdle technology. Other trends are functional foods that combine nutritional and medical benefits, but they also need appropriate hurdles. In the design of healthful and functional foods, microbial stability and safety are sometimes not sufficiently observed. However, if "healthy foods" spoil easily, or even cause food poisoning, the consumer will hardly become convinced that they are indeed healthy.

38.4.5 NONTHERMAL PRESERVED FOODS

Novel, emerging technologies for food preservation (i.e., nonthermal, physical preservation methods such as high hydrostatic pressure, high-intensity pulsed electric fields, oscillating magnetic fields, light pulses, etc.) are often most effective in combination with traditional food preservative hurdles. Barbosa-Cánovas et al.[12] pointed out that an improved understanding of the mechanisms underlying the effectiveness of the nonthermal processes and the combinations with the traditional hurdles is therefore urgently required, so that the new preservation possibilities can move forward with a sound scientific basis because, most likely, combined technologies is the future of food preservation. Hurdle technology is also successfully used with food irradiation, because a decreased treatment requirement for irradiation or both hurdles can improve the sensory properties and microbiologically quality of the food thus treated.[38,39]

38.4.6 INDIAN HURDLE FOODS

In December 1997, a national seminar called Preservation of Food by Hurdle Technology was organized by the Defense Food Research Laboratory at Mysore, India. About 100 Indian scientists attended, and 23 papers on food preservation based on hurdle technology were presented. These papers gave an introduction to hurdle technology; were related to hurdle technology for the preservation of fruits, vegetables, meat, fish, poultry, cereals, and dairy products; and discussed packaging, storage, and quality aspects of hurdle technology foods. In 1999, the proceedings of this landmark meeting were published.[40] They also contain recommendations for the future course of action necessary to further refine hurdle technology to make it more competitive and acceptable for processing food products of mass consumption and for minimizing post harvest losses and seasonal gluts in the market. Today, India is one of the most active countries in the field of hurdle technology.

38.4.7 CHINESE FUSION FOODS

Nowadays in China, traditional meat products (rou gan, la chang, yan la, etc.) and Western meat products are on the market side by side.[41] The traditional products, which have been known for centuries, still prevail, because they are simple to prepare, have a typical flavor, are ready-to-eat, and are ambient stable. Western-type meat products are also attractive because of their high yield, nice appearance, and pleasant taste. But they still cause difficulties with respect to storage and microbial shelf life, because refrigeration is scarce and costly in China.

However, a third group of meat products has emerged in China, which might be called *fusion products*. They are derived from a German minimally processed, ambient stable meat product (F-SSP, see section on "Minimally Processed Foods") but have been adapted to suit Chinese meat processing. German F-SSP is a hurdle technology product, since it is mildly heated ($F = 0.4$), and the a_w is adjusted (<0.97/0.96). In China, no counterpressure autoclaves are used, and it is also not feasible to adjust the a_w of these sausages. Therefore, the fusion product is heated to $F \sim 10$ but, due to the addition of starch, soybean, phosphate, and carrageenan, the Chinese variety is acceptable from the sensory and nutritional point of view and is ambient stable. These sausages are produced in huge quantities and are called *tourist sausage* (since the Chinese eat them while travelling) or *ham sausage* (since these products resemble cooked ham). Chinese fusion sausage is a good example of how hurdle technology products from one region can be adapted to suit the needs and conditions of another region.

38.4.8 FOOD PACKAGING IN JAPAN

Packaging is an important hurdle for foods, since it supports microbial stability and safety as well as the sensory quality of food.[11] Industrialized countries have the tendency to overpackage their foods. This is especially true for Japan, where "active packaging" (using scavengers, absorbers, antimicrobials, or antioxidative packaging material, etc.) has been developed to perfection. These smart packaging systems are sophisticated but wasteful, too. Japanese experts are aiming now for less packaging of food.[42] Future packaging shall provide only necessary information and some convenience to the consumer; however, the required shelf life of the products should not come from packaging but should be based on (1) super-clean packaging procedures with just-in-time delivery or (2) the development of advanced hurdle technology foods that are stable and safe in spite of minimal packaging.[43]

38.4.9 DESIGN OF HURDLE TECHNOLOGY FOODS

Hurdle technology, as a concept, has proved useful in optimization of traditional foods as well as in the development of novel products. However, it is beneficial to combine hurdle technology in food design with hazard analysis critical control point (HACCP) or good manufacturing practice (GMP), and possibly with the modeling of growth and survival of microorganisms (i.e., predictive microbiology). For the proper design of hurdle technology foods a ten-step procedure has been suggested[44] that comprises hurdle technology, predictive microbiology, and HACCP. This user guide to food design has proved suitable for solving development and processing tasks in the food industry. However, recently, some doubts have emerged since, strictly speaking, with HACCP only the biological, chemical, and physical hazards are controlled, but not microbial stability (spoilage) and sensory food quality. Since, for hurdle technology foods, microbial safety and stability, as well as sensory quality (i.e., the total quality of the food) are essential, the HACCP concept might be too narrow for this purpose if it relates to hazards only. Therefore, the HACCP concept should be broadened to cover safety (food poisoning) and stability (spoilage) of foods as well as their sensory quality. If this is not acceptable, the production process should be controlled by good manufacturing practice (GMP), and rules or guidelines for the production of each food must be quantitatively defined. For foods of developed countries, GMP guidelines are more acceptable, because the application of HACCP poses practical difficulties where small producers prevail.[26]

38.5 CONCLUSIONS

Over the years, numerous data related to hurdle technology have been accumulated worldwide, and it might become difficult to know and evaluate them all. Therefore, it was timely that two comprehensive review papers on the principles and applications of hurdle technology were published in 1999. The update on hurdle technology given here summarizes only some of the recent highlights.

The first review article[11] presents a general outline of hurdle technology, and the second review article[41] presents information on advanced applications of hurdle technology in some developing countries (Latin America, China, India, and Africa). Furthermore, a comprehensive but well focused book is in preparation,[45] and it offers a fresh look on the present and future perspectives of hurdle technology.

REFERENCES

1. Leistner, L., W. Rödel, and K. Krispien. 1981."Microbiology of Meat and Meat Products in High- and Intermediate-Moisture Ranges," in *Water Activity: Influences on Food Quality*, L. B. Rockland, and G. F. Stewart, eds. New York, NY: Academic Press, pp. 855–916.

2. Leistner, L. 1992. "Food Preservation by Combined Methods," *Food Res. Int.*, 25: 151–158.

3. Leistner, L. 1978. "Hurdle Effect and Energy Saving," in *Food Quality and Nutrition*, W. K. Downey, ed. London, UK: Applied Science Publishers, Ltd., pp. 553–557.

4. Leistner, L. 1985. "Hurdle Technology Applied to Meat Products of the Shelf Stable Product and Intermediate Moisture Food Types," in *Properties of Water in Foods*, D. Simatos and J. L. Multon, eds. Dordrecht, NL: Martinus Nijhoff Publishers. pp. 309–329.

5. Leistner, L. 1995. "Use of Hurdle Technology in Food Processing: Recent Advances," in *Food Preservation by Moisture Control, Fundamentals and Applications*, G. V. Barbosa-Cánovas and J. Welti-Chanes, eds. Lancaster, PA: Technomic Publishing Co., Inc., pp. 377–369.

6. Leistner, L., and L. G. M. Gorris. 1995. "Food Preservation by Hurdle Technology," *Trends Food Sci. Technol.*, 6: 41–46.

7. Leistner, L. 1995. "Principles and Applications of Hurdle Technology," in *New Methods of Food Preservation*, G. W. Gould, ed. London, UK: Blackie Academic & Professional, pp. 1–21.

8. Leistner, L. 1996. "Food Protection by Hurdle Technology," *Bull. Jpn. Soc. Res. Food Prot.*, 2(2): 2–27.

9. Leistner, L. 2000/2001. "Hurdle Technology," in *Encyclopedia Of Life Support Systems, Article No. 5.10.4.12*. Oxford, UK: EOLSS Publishers Co. Ltd., in press.

10. Leistner, L. 1999. "Hurdle Technology and HACCP for Shelf Stable Animal Products," presented at the International Conference on Sustainable Animal Production, Health and Environment: Future Challenges, held November 24–27, 1999 at Hisar, India.

11. Leistner, L. 1999. "Combined Methods for Food Preservation," in *Handbook of Food Preservation*, M. Shafiur Rahman, ed. New York, NY: Marcel Dekker, Inc., pp. 457–485.

12. Barbosa-Cánovas, G. V., U. R. Pothakamury, E. Palou, and B. G. Swanson. 1998. *Nonthermal Preservation of Foods*. New York, NY: Marcel Dekker, Inc., pp. 235–268.

13. Leistner, L. 1995. "Emerging Concepts for Food Safety," *Proceedings 41st Internat. Congress of Meat Sci. & Technol.*, held at San Antonio, TX, U.S.A., pp. 321–322.

14. Leistner, L. 2000. "Basic Aspects of Food Preservation by Hurdle Technology," *Int. J. Food Microbiol.*, in press.

15. Häussinger, D. ed. 1988. *pH Homeostasis-Mechanisms and Control*. London, UK: Academic Press.

16. Gould, G. W. 1988. "Interference with Homeostasis-Food," in *Homeostatic Mechanisms in Microorganisms*, R. Whittenbury, G. W. Gould, J. G. Banks, and R. G. Board, eds. Bath, UK: Bath University Press, pp. 220–228.

17. Gould, G. W. 1995. "Homeostatic Mechanisms During Food Preservation by Combined Methods," in *Food Preservation by Moisture Control, Fundamentals and Applications*, G. V. Barbosa-Cánovas and J. Welti-Chanes, eds. Lancaster, PA: Technomic Publishing Co., Inc., pp. 397–410.

18. Leistner, L. and S. Karan-Djurdjic. 1970. "Beeinflussung der Stabilität von Fleischkonserven durch Steuerung der Wasseraktivität," *Fleischwirtschaft*, 50: 1547–1549.

19. Leistner, L. 1994. *Food Design by Hurdle Technology and HACCP*. Printed and distributed by the Adalbert-Raps-Foundation, Kulmbach, Germany.

20. Alzamora, S. M., P. Cerrutti, S. Guerrero, and A. López-Malo. 1995. "Minimally Processed Fruits by Combined Methods," in *Food Preservation by Moisture Control, Fundamentals and Applications*, G. V. Barbosa-Cánovas and J. Welti-Chanes, eds. Lancaster, PA: Technomic Publishing Co., Inc., pp. 463–492.

21. Tapia de Daza, M. S., S. M. Alzamora, and J. Welti-Chanes. 1996. "Combination of Preservation Factors Applied to Minimal Processing of Foods," *Crit. Rev. Food Sci. & Nutr.*, 36: 629–659.

22. Leistner, L. 1994. "Further Developments in the Utilization of Hurdle Technology for Food Preservation," *J. Food Eng.*, 22: 421–432.

23. Denyer, S. P. and W. B. Hugo, eds. 1991. *Mechanisms of Action of Chemical Biocides: Their Study and Exploitation*. London, UK: Blackwell Scientific Publications.

24. Leistner, L. and L. G. M. Gorris. 1994. *Food Preservation by Combined Processes*. Brussels, Belgium: Final Report of FLAIR Concerted Action No. 7, Subgroup B. European Commission, Directorate-General XII, EUR 15776 EN.

25. Leistner, L. 2000. "Hurdle Technology in the Design of Minimally Processed Foods," in *Minimally Processing Fruits and Vegetables*, S. M. Alzamora, M. S. Tapia, and A. López-Malo, eds. Gaithersburg, MD: Aspen Publishers, Inc., pp. 13–27.

26. Leistner, L. 2000. "Minimally Processed, Ready-To-Eat and Ambient-Stable Meat Products," in *Shelf Life Evaluation of Foods, 2nd Edition*, D. Man and A. Jones, eds. Gaithersburg, MA: Aspen Publishers, Inc., in press.

27. Leistner, L. and H. Hechelmann. 1993. "Food Preservation by Hurdle Technology," *Proceedings of Food Preservation 2000 Conference,* held at the Army Natick, Research, Development and Engineering Center, Natick, MA, U.S.A., vol. II, pp. 511–520.

28. López-Malo, A., E. Palou, J. Welti, P. Corte, and A. Argaiz. 1994. "Shelf-stable High Moisture Papaya Minimally Processed by Combined Methods," *Food Res. Int.*, 27: 545–553.

29. Rastogi, N. K., J. S. Sandhi, P. Viswanath, and S. Saroja. 1995. "Application of Hurdle/Combined Method Technology in Minimally Processed Long-Term Non-Refrigerated Preservation of Banana and Coconut," *Abstract for ICFoST '95, held at Mysore, India,* p. 109.

30. Jayaraman, K. S., H. S. Vibhakara, and M. N. Ramanuja. 1999. "Preservation of Fruits and Vegetables by Hurdle Technology," in *Preservation of Food by Hurdle Technology*, Defence Food Research Laboratory, Mysore, India, pp. 48–56.

31. Gorris, L. G. M. 1999. "Hurdle Technology, a Concept for Safe, Minimal Processing of Foods," in *Encyclopedia of Food Microbiology*, R. Robinson, C. Batt, and P. Patel, eds. London, UK: Academic Press, Ltd., in press.

32. Bennik, M. H. J., E. J. Smid, F. M. Rombouts, and L. G. M. Gorris. 1995. "Growth of Psychrotrophic Foodborne Pathogens in a Solid Surface Model System Under the Influence of Carbon Dioxide and Oxygen," *Food Microbiol.,* 12: 509–519.

33. Gorris, L. G. M. and M. W. Peck. 1998. "Microbiological Safety Considerations When Using Hurdle Technology with Refrigerated Processed Foods of Extended Durability," in *Sous Vide and Cook-Chill Processing for the Food Industry,* S. Ghazala, ed. Gaithersburg, MA: Aspen Publishers, Inc., pp. 206–233.

34. Leistner, L. 1987. "Shelf-Stable Products and Intermediate Moisture Foods Based on Meat," in *Water Activity: Theory and Applications to Food,* L. B. Rockland and L. R. Beuchat, eds. New York, NY: Marcel Dekker, Inc., pp. 295–327.

35. Leistner, L. 1995. "Stable and Safe Fermented Sausages World-Wide," in *Fermented Meats,* G. Campbell-Platt and P.E. Cook, eds. London, UK: Blackie Academic & Professional, pp. 160–175.

36. Leistner, L. 1986. "Allgemeines über Roschinken," *Fleischwirtschaft,* 66: 496–510.

37. Leistner, L. 1997. "Microbial Stability and Safety of Healthy Meat, Poultry and Fish Products," in *Production and Processing of Healthy Meat, Poultry and Fish Products,* A. M. Pearson and T. R. Dutson, eds. London, UK: Blackie Academic & Professional, pp. 347–360.

38. IAEA. 1981. *Combination Processes in Food Irradiation.* International Atomic Energy Agency, Vienna, Austria.

39. Thomas, P. 1999. "Combined Treatments of Irradiation and Other Hurdles in Food Preservation," in *Preservation of Food by Hurdle Technology,* Defence Food Research Laboratory, Mysore, India, pp. 17–19.

40. D.F.R.L. 1999. *Preservation of Food by Hurdle Technology.* Siddarthanagar, Mysore, India: Defence Food Research Laboratory.

41. Leistner, L. 1999. "Use of Combined Preservative Factors in Foods of Developing Countries," in *The Microbiological Safety and Quality of Food,* B. M. Lund, A. C. Baird-Parker, and G. W. Gould, eds. Gaithersburg, MA: Aspen Publishers, Inc., in press.

42. Ono, K. 1994. "Packaging Design and Innovation," presented at a Third World Country Training Course in the field of food packaging, conducted at Singapore, February 20 - March 5, 1994.

43. Ono, K. 1996. Snow Brand Tokyo, Japan. Personal communication.

44. Leistner, L. 1994. "User Guide to Food Design," in *Food Preservation by Combined Processes,* L. Leistner and L. G. M. Gorris, eds. Brussels, Belgium: Final Report of FLAIR Concerted Action No. 7, Subgroup B. European Commission, Directorate-General XII, EUR 15776 EN, pp. 25–28.

45. Leistner, L. and G. W. Gould. 2001. *Hurdle Technology: Combination Treatments for Food Stability, Safety and Quality.* Gaithersburg, MA: Aspen Publishers, Inc., in preparation.

39 Microbial Behavior Modeling as a Tool in the Design and Control of Minimally Processed Foods

S. M. Alzamora
A. López-Malo

CONTENTS

Abstract

Consumers are concerned about the safety and stability of their food supply. However, their major concerns are identified as substances incorporated as aids during food harvesting, processing, or distribution, such as additives and antimicrobials or to residual pesticides and heavy metals. This fact is contrary to industrial and government relative priority, microbiological hazard being a major food safety concern. Synthetic antimicrobial and additive-free, minimally processed, natural, and low-salt-content products are currently associated with safer foods. This is challenging research and development, academic, and industrial activities to search for new or improved methodologies to satisfy consumer demands while assuring microbiological safety, even during home kitchen manipulation, where the most likely source of food abuse occurs. Within this frame of risks, hazards, and consumer trends, predictive microbiology emerges as a powerful tool to quickly explore the microbiological impact of varying conditions within food formulation, processing, and/or distribution and retail conditions. Selected examples of microbial mathematical models applied to the design and optimization of minimally processed fruits are addressed herein.

39.1 INTRODUCTION

Food quality involves a sum of characteristics that determines consumer satisfaction and compliance to legal standards.[1] Microbial, chemical, biochemical, and physical reactions may result in food quality loss. Safety is an important and essential component of quality, being the first requisite of any food. Any harmful microbial or chemical contaminant cannot be allowed at food consumption. Consequently, preservation techniques must prevent the growth of toxigenic microorganisms and avoid the growth/presence of infective bacteria as well as the harmless but undesirable activity of spoilage microorganisms.

Knowledge of quality-loss kinetics is important in estimating product shelf life and optimizing preservation procedures. A considerable amount of research has been conducted to determine physicochemical and sensory deterioration reaction rates in foods. Recently, food microbiologists have relied on non-kinetic empirical data obtained by challenge testing with specific microorganisms to predict the safety of foods. The traditional approach, rather than kinetic, was used to fix intrinsic and extrinsic factors governing microbial growth and then establish maximum and minimum limits in which organisms will grow. In the last 15 years, a number of models to mathematically describe the growth or inactivation/decline of microorganisms under specific environmental conditions have been developed.[2–4] These models are based on data generated in laboratory media at various combinations of environmental stress factors such as a_w, pH, antimicrobials, and incubation temperature, among others. The effect of each factor is assumed to be independent whether the selected organism is growing in broth or in a food.[3] Thus, the models can be used to predict the behavior of a specific microorganism in foods, provided their pH, a_w, antimicrobial concentration, and storage temperature are within the range included in the model.

Predictive modeling as defined by the National Advisory Committee on Microbiological Criteria for Foods[5] is the use of mathematical expressions to describe the likely behavior of biological agents. Mathematical modeling (also called *predictive microbiology*) of microbial growth or decline is receiving a great deal of attention because of its enormous potential within the food industry The driving forces for the impressive progress in microbial modeling have been mainly three.

1. The advent of reasonably priced microcomputers, which has facilitated multi-factorial data analysis.
2. The great improvement in techniques for establishing mathematical models in the area of predictive microbiology.

3. The assessment of safety of foods that have been reformulated because of changes in lifestyles and consumer demands for natural and minimally processed products. To meet these demands, shelf-stable and not shelf-stable innovative foods both rely on a combination of preservation factors and are being manufactured in increasing numbers. Rather than inactivating the microorganisms, these combined techniques inhibit their growth through the manipulation of various growth-controlling factors. Thus, microbiological safety and stability margins are being reduced, and improved control in manufacture and distribution are required to assure the recommended shelf lives. Temperature abuse, emergent pathogens, induced tolerance of microorganisms to stress factors, and food matrix-stress factor interactions have caused concern about the safety of some of these nonsterile products.[6] In this type of products, end-product testing is useless, and mathematical modeling can help developing qualitative and quantitative information to describe microbial behavior, allowing a better control and prediction of the shelf life.

Therefore, potential advantages of using microbial behavior models are significant. Table 39.1 shows some current and suggested applications reported by Whiting,[7] Skinner and Larkin,[8] and Ratkowsky and Ross.[9] Quite different predictive models have been developed, some of which are based on modeling the effect of fixed or variable parameters on microbial growth, while others are concerned with microbial death or fluctuating populations.[3,7–14] Within each category, models can be classified as being at the primary, secondary, or tertiary level.[3] Primary-level models describe the changes of the microbial response (i.e., colony-forming units per milliliter, toxin formation, metabolic products, colony diameter, absorbance, and impedance) with time. Secondary-level models describe the responses by parameters of primary models to changes in environmental factors (i.e., a_w, pH, temperature, antimicrobials). Application software and expert systems developed from primary-

TABLE 39.1
Major Current Applications of Mathematical Models

- Calculation of the shelf life of foods or prediction of the time span in which significant microbial growth may occur
- Estimation of microbial stability of newly developed food products or new formulations
- Prediction of risk of growth or survival of a pathogen after normal or abuse storage; establishing "pull dates" by estimating the growth of pathogens and spoilage flora
- Education of nontechnical people and food microbiology students
- Increase of laboratory efficiency by targeting critical areas for research, thereby reducing expensive time-consuming experimentation
- Better development of HACCP programs through the setting of CCPs and evaluation of microbial consequences of out-of-compliance events
- Estimation of the likelihood that consuming a particular food will make someone sick (risk assessment)

and secondary-level models are designated as tertiary-level models, which constitute a user-friendly form in software for personal computers.[15–17] Examples of the three types of models reported by Whiting,[3] McMeekin et al.,[4] Skinner and Larkin,[8] Ratkowsky and Ross,[9] Peleg,[13] Zwietering et al.,[14] McClure et al.,[17] Peleg,[18] Zwietering et al.,[19] Cole et al.,[20] are listed in Table 39.2. Scientific literature on predictive models is expanding rapidly, with models varying greatly in theory and complexity. They evolve from simple probabilistic models to linear and logistic regression based models and sigmoidal functions for fitting microbial growth and death curves. Some of them are empirical or phenomenological, while others have a true physicochemical or kinetic basis, with thermophysical or biological assumptions.[8]

TABLE 39.2
Selected Models for Mathematical Modeling Microbial Behavior

Primary-Level Models for Growth

First order kinetics

Sigmoid growth functions
   Gompertz, logistic, Richards, Stannard and Schnute models

Probabilistic model

Primary-Level Models from Growth to Lethality

Peleg model

Primary-Level Models for Thermal and Nonthermal Inactivation

Thermal death time model
Cole et al. model
Buchanan linear model
Fermi equation

Secondary-Level Models

Arrhenius based models for temperature effect
   Simple Arrhenius, Hinshelwood, Schoolfield, Johnson and Lewin, Adair et al. models
Square root or Bêlehrádek type models for temperature effect
   Ratkowsky et al. four parameter square root model for the full biokinetic temperature range model
Response surface models for the effect of various factors

Tertiary-Level Models

USDA Pathogen Modeling Program, USA
Food MicroModel of Leatherhead Food Research Association, UK

39.2 DESIGN OF MINIMALLY PROCESSED FOODS AND PREDICTIVE MICROBIOLOGY

The objective of hurdle technology is to select and combine preservation factors or hurdles in such a way that microbial stability and safety can be assured while retaining sensory acceptance and nutritional characteristics. Information provided by the models can be used advantageously in the design of combined technologies for obtaining minimally processed foods in many ways:

1. Although these minimally processed foods are preserved by more than one hurdle, there is a lack of quantitative date available to allow predicting the adequate and necessary levels of each hurdle. Thus, if we can predict with accuracy the growth kinetics (lag phase duration, growth rate) or the interface between microbial growth and no growth for an identified target microorganism under several combinations of factors, the selection of such factors can be made on a scientific basis, and the selected hurdles can be kept at their minimum levels.
2. Modeling has application in products where the perceived "naturalness" prevents the use of large quantities of chemical preservatives and/or solutes and/or acids. Sensory selection of hurdles and their levels may be done between several "safe" equivalent combinations of interactive effects determined by the models.
3. While much relevant information is available in the scientific literature concerning factors/interaction of factors that influence microbial activities in foods, it is seldom usable in formulating combined techniques. Many times, information involves only data from traditional challenge testing in particular food conditions, microbial presence or absence tests, or growth or no growth determination. These isolated results do not allow us to compare quantitatively what happens in a food system when the levels of the independent preservation variables are changed. Neither can the sensitivity of the key microorganisms to the different factors be inferred. Experience gained in developing predictive models has resulted in improved design, efficiency of data collection, and precision of results. Therefore, the methodology of predictive microbiology provides a systematic approach for planning, collection, and appropriate processing of good-quality data describing the effects of different factors in combination on microbial development. The secondary models, such as surface plots, are a very useful means of visualizing interacting effects of inhibiting factors.

In conclusion, combined factors technologies developed originally through observations of cardinal parameters for growth as well as trial and error can be improved and/or scientifically designed by using mathematical modeling.

39.3 APPLICATION OF MODELING IN MINIMALLY PROCESSED FRUIT DESIGN

New fruit processing methods are being investigated and developed to meet demands for maximum quality retention, fresh-like characteristics, and added safety. Combined preservation technologies have been proposed in the last decade for obtaining fresh-like high-moisture fruit products (HMFPs). Hurdles used for the stabilization of HMFP have generally included reduction of a_w and pH, mild heat treatment, and addition of preservatives. The preservation process combining these factors is very simple and consists of fruit blanching followed by an a_w depression step by osmotic dehydration with simultaneous incorporation of additives, achieving final values of a_w 0.94–0.98, pH 3.0–4.1 (adjusted with citric or phosphoric acid), 400–1,000 ppm potassium sorbate or sodium benzoate, and generally 150 ppm sodium bisulfite.

Optimization of these combined technologies is on course, and, from the microbiological point of view, three aspects are receiving consideration.

1. Selection of minimum levels of hurdles that assure safety and keepability of the HMFP but maintain their quality
2. Use of mixtures of preservatives
3. Use of antimicrobials of natural origin as replacement (total or partial) of sorbates and other synthetic additives

Predictive modeling of microbial growth or survival is applicable to bacteria, yeast, and molds. Even though most emphasis has been made on pathogenic bacteria as an essential safety tool, spoilage microorganisms are also important in products where safety is ensured by the preservation factor combination. Next, some examples of these mathematical developments applied to the design and optimization of minimally processed fruits are addressed.

39.3.1 PROBABILISTIC MODELING

The goal of analysis using any statistical model-building technique is to find the best suited and most parsimonious, yet biologically reasonable, model to describe the relationship between a dependent variable and a set of independent variables. Modeling the probability of a growth/no-growth interface is a valuable approach to microbial modeling that has not yet been fully explored and exploited. As conditions become less favorable for growth, lag time increases, and this behavior is captured by the various approaches to model growth.[4] However, at the same time, no-growth results begin to appear and, in most microbial studies, are ignored. Ignoring these no-growth events is an oversimplification and misinterpretation of what is occurring. Probabilistic models based on logistic regression analysis provide a useful way to describe the growth/no-growth boundary.[9,21,22]

Statistically, a logistic regression model relates the probability of occurrence of an event, Y, conditional on a vector, x, of explanatory variables.[9] The quantity $p(x)$ $= E(Y \mid x)$ represents the conditional mean of Y given x when the logistic distribution is used (in this case, the probability that growth will occur). The specific model of the logistic regression is as follows:

$$p(x) = \frac{\exp[\sum \beta_i x_i]}{1 + \exp[\sum \beta_i \bar{x}_i]} \tag{39.1}$$

being the logit transformation of $p(x)$ defined as

$$\text{logit}(p) = g(x) = \ln\left[\frac{p(x)}{1 - p(x)}\right] = \sum \beta_i x_i \qquad (39.2)$$

The logit, $g(x)$, is linear in its parameters, may be continuous, and may vary depending on the range of x.[23] In the examples presented next, a_w, pH, antimicrobial type and concentration, as well as incubation temperature and incubation time, can be included as the independent variables. The dependent variable is the microbial response ("growth" or "no-growth"). To fit the logistic model, lineal polynomial models were selected.

$$g(x) = \beta_0 + \beta_1 V_1 + \beta_2 V_2 + \beta_3 V_3 + \dots + \beta_j \text{interaction}_1 + \beta_{j+1} \text{interaction}_2 + \dots$$

where V_i are the independent variables as well as their interactions, and the coefficients (β_i) are the parameters to be estimated by fitting the model to experimental data. Predictions of the interface (growth/no-growth) were made at selected probability levels by substituting the value of logit (p) in the model and finding the value of one independent variable maintaining fixed the others. Also, the probability of growth was calculated using the logistic model under the evaluated conditions in each case.

Example 1: Probabilistic Modeling of *Saccharomyces cerevisiae* Inhibition by Controlling a_w, pH, and Potassium Sorbate Concentration

Probabilistic modeling using logistic regression was used to predict the boundary between growth and no growth of *Saccharomyces cerevisiae* after 350 h of incubation, in the presence of the following growth-controlling factors: water activity (0.97, 0.95, and 0.93), pH (6.0, 5.0, 4.0, and 3.0), and potassium sorbate concentration (0, 50, 100, 200, 500, and 1000 ppm).[22] The data set was obtained from 72 growth/survival curves of the yeast in Sabouraud Broth adjusted to the different combination of a_w, pH, and potassium sorbate concentration values. Some of these combinations gave observable growth, while, in others, growth was not detected,[22,24,25] as observed in Table 39.3. Growth data was represented as "1," and no growth under each evaluated combination was recorded as "0".

In this particular case, a_w, pH, and potassium sorbate (KS) were the independent variables, and the dependent variable was the response of *S. cerevisiae*. Predictions of the interface (growth/no-growth) were made at a probability level of 0.05 by substituting the value of logit (p) in the model and finding the value of one independent variable maintaining fixed the other two. Also, the probability of growth was calculated using the logistic model under the evaluated conditions, as has been previously reported.[9,21,22,26]

The proposed model (Table 39.4) predicts the probability of growth under a set of conditions and calculates critical values of a_w, pH, and potassium sorbate concentration needed to inhibit yeast growth for different probabilities. Main effects

TABLE 39.3
Effect of pH, a_w, and Potassium Sorbate (KS) Concentration on Growth of *Saccharomyces cerevisiae* after 350 h of Incubation at 27°C

KS (ppm)	pH											
	6			5			4			3		
a_w	0.97	0.95	0.93	0.97	0.95	0.93	0.97	0.95	0.93	0.97	0.95	0.93
0	1	1	1	1	1	1	1	1	1	1	1	1
50	1	1	1	1	1	1	1	1	1	–	0	0
100	1	1	1	1	1	1	1	0	0	0	0	0
200	1	1	1	1	1	1	–	0	0	0	0	0
500	1	1	1	1	1	1	0	0	0	0	0	0
1000	1	1	0	0	0	0	0	0	0	0	0	0

1, growth observed; 0, no growth observed; –, factor combination not evaluated.

TABLE 39.4
Logistic Model Coefficients

	Estimated Coefficient		
		Z. bailii (Example 2)	
Variable	*S. cerevisiae* (Example 1)	Case A	Case B
Constant	–70.8877	–315.6960	–7.9341
a_w	62.9415	317.6854	
pH	3.7438		
T			0.1446
KS	–0.0413	–0.1664	
pH KS	0.0053		
a_w KS		0.1556	
T pH		3.7311	
T NaB		–0.0007	
a_w T pH		–3.5474	
T KS NaB		3.30×10^{-7}	
a_w T pH KS		-5.60×10^{-5}	
$t \times T \times a_w$			0.0328
$t \times T \times A$			–0.0108

were significant as well as the pH-potassium sorbate concentration interaction; the remaining terms were not significant ($p > 0.10$). The log likelihood ratio and Chi-square test were in both cases significant, indicating that the model was useful to predict the outcome variable (growth or no-growth) and tested the model goodness of fit. Figures 39.1 and 39.2 present the predicted probability of growth for *S. cerevisiae* in the variable range tested after 350 h incubation. The interaction between pH and potassium sorbate concentration is demonstrated in Figure 39.1. Similar

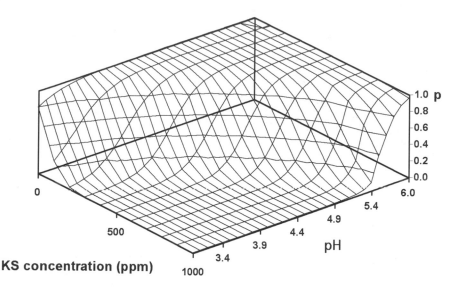

Figure 39.1 *Saccharomyces cerevisiae* probability of growth (*p*), potassium sorbate (KS), and pH effects at water activity (a_w) 0.97.

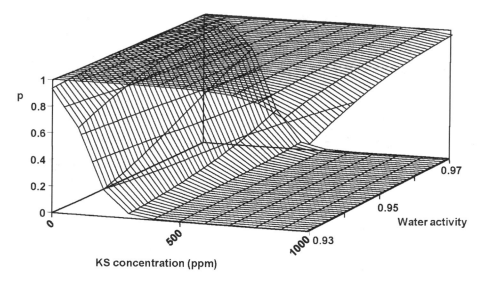

Figure 39.2 Saccharomyces *cerevisiae* probability of growth (*p*), potassium sorbate (KS), and water activity effects at pH 6.0 (top) and pH 4.0 (bottom).

patterns were obtained at a_w 0.95 and 0.93. Figure 39.2 presents a_w and potassium sorbate concentration effects at selected pH values. The top surface represents data for pH 6.0 and the bottom surface for pH 4.0. Interesting conclusions can be obtained as follows:

1. At pH 6.0 (Figures 39.1 and 39.2), potassium sorbate addition has no effect on yeast inhibition with probabilities of growth > 0.50. Reduction of pH to 5.0, 4.0, or 3.0 (Figure 39.1) gradually increases the number of combinations of potassium sorbate concentration with probabilities to inhibit yeast growth lower than 0.05.

2. Selecting a probability of growth of 0.05 and using the logistic model, critical pH values can be calculated. As a_w decreases and potassium sorbate concentration increases, the critical pH value increases (Figure 39.2). For a selected pH value (lower than 6.0), decreasing a_w and increasing antimicrobial concentration result in lower probabilities of yeast growth.

Example 2: Probabilistic Modeling of *Zygosaccharomyces bailii* Inhibition by Hurdle Technology in Model Systems and Mango Purée

The boundary between growth and no growth of *Zygosaccharomyces bailii* in the presence of selected combinations of controlling factors was evaluated using logistic regression. Two cases are presented next.

Case A.　Yeast response after 30 days of incubation in model systems for evaluating a_w (0.99, 0.98, or 0.97), pH (4.0, 3.5, or 3.0), incubation temperature (15 or 25°C), concentration (0, 250, 500, up to 1500 ppm) of potassium sorbate (KS) and/or sodium benzoate (NaB).[27] The model was built on observing a total of 2,646 tubes corresponding to triplicate observations of the total possible combinations of factors. Incubation temperature, a_w, pH, SK, and NaB concentration were selected as independent variables, and the outcome or dependent variable was the response of *Z. bailii*. Predictions of the interface (growth/no-growth) were made at selected probability levels, and the probability of growth was calculated. The proposed model (Table 39.4) predicts the probability of growth under a set of conditions and can be used to calculate critical values of a_w, pH, incubation temperature, and antimicrobials concentration needed to inhibit yeast growth for different probabilities. The log likelihood ratio and Chi-square test were in both cases significant, indicating that the model is useful to predict *Z. bailii* growth or no-growth. Figure 39.3 presents the predicted probability of yeast growth under a selected combination of factors. Reducing incubation temperature dramatically affected probability of yeast growth. KS was more effective than NaB to inhibit *Z. bailii*. Combinations of KS and NaB as antimicrobials did not show a synergic effect on yeast inhibition. The reduction of pH increased the number of combinations of a_w and antimicrobials concentration with probabilities to inhibit yeast growth lower than 0.05. For low a_w values and increasing SK or NaB concentration, critical pH values to inhibit yeast growth increased.

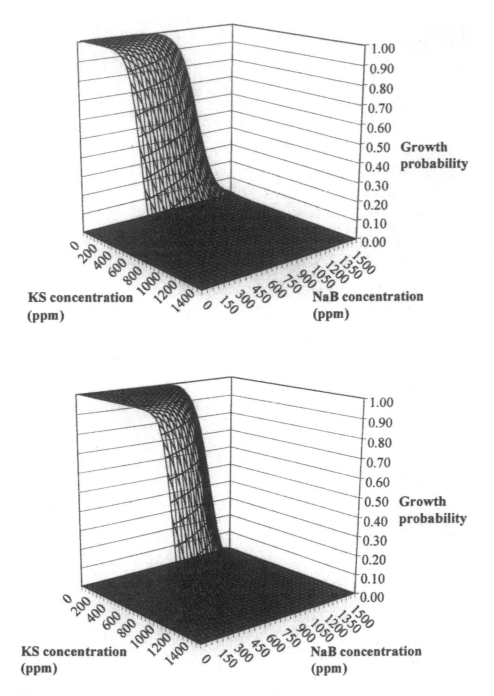

Figure 39.3 *Zygosaccharomyces bailii* probabilities of growth after 30 days of incubation at 15 (top) and 25°C (bottom) in model systems formulated with a_w 0.99, pH 4.0, potassium sorbate (KS), and sodium benzoate (NaB).

Case B. Z. *bailii* growth response in a pH 3.5 mango purée formulated with
1000 ppm of potassium sorbate or sodium benzoate at selected a_w (0.99,
0.98, or 0.97) was evaluated, and the growth/no-growth interface was modeled using logistic regression.[28] If an independent variable is discrete, then
it is inappropriate to include it in the model as if it were interval scaled. In
this situation, the method of choice is to use a collection of design or dummy
variables.[23] For antimicrobial type (A), which is a discrete variable, the
codification we used was as follows: "1" for potassium sorbate and "–1"
for sodium benzoate. Fitting the logistic model to the growth/no-growth
data by logistic regression and eliminating nonsignificant ($p > 0.10$) terms
resulted in the reduced model presented in Table 39.4. The log likelihood
ratio and Chi-square tests, both being significant, tested the goodness of fit
of the model, which indicates that the model is useful to predict the outcome
variable (growth or no growth). Model goodness of fit was also evaluated
comparing predicted values and experimental observations. A predicted
probability of growth (cut value) ≥ 0.50 was considered as a growth prediction. Using this criterion, the overall correct observation was 92.2%,
with 14 misclassified predictions from a total of 180 observations. In only
6 of these 14 disagreements was growth predicted when no growth was
observed, and in 8 cases the model predicted no growth for growth observations.

The probability of Z. *bailii* growth in a_w 0.99 and 0.98 mango purée formulated
with sodium benzoate or potassium sorbate is presented in Figures 39.4 and 39.5.
For both antimicrobials, an increase in storage temperature or storage time increased
yeast growth probability. However, significant differences in the probability of
growth were observed between antimicrobials, with potassium sorbate being more
effective (Figure 39.5), in terms of delaying growth, than sodium benzoate (Figure
39.4). Similar patterns were observed for mango purées formulated with a_w 0.98 or
0.99, Z. *bailii* probability of growth was minimally affected by water activity, but
lowering a_w diminished probability of yeast growth when combined with antimicrobials and reduced storage temperature. The interaction between storage time, temperature, and antimicrobial type is demonstrated in Figures 39.4 and 39.5, where
the probability of growth increases rapidly for higher storage temperatures and longer
storage times, and even more dramatically when sodium benzoate is used as the
antimicrobial. A delicate balance among the applied preservation factors (a_w, pH,
reduced temperature, antimicrobials) permits reduction in the probability of growth
of Z. *bailii* and guarantees a reasonable product shelf life, based on Z. *bailii* growth
delay.

The presented examples demonstrate how the growth/no-growth interface can
be defined and modeled using logistic regression. This quantitative approach is very
useful to optimize combined technologies, not only by maintaining hurdles at their
minimum levels to inhibit/delay growth, but selecting the combination of factors
that minimize sensory changes. The experimental data transformed to mathematical
expressions detailing the effects of environmental factors on yeast growth can be

0.99 - BNa

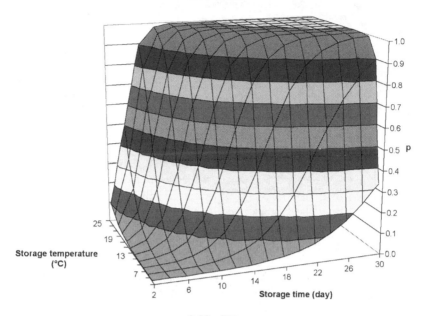

0.98 - BNa

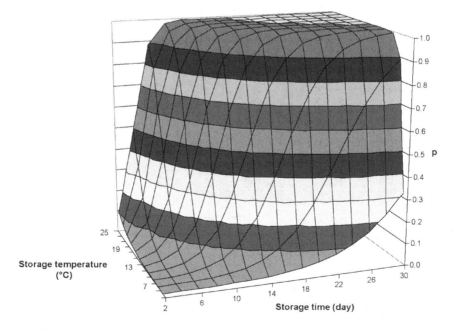

Figure 39.4 *Zygosaccharomyces bailii* probabilities (*p*) of growth in mango purée formulated at a_w 0.98 or 0.99, pH 3.5, with 1000 ppm of sodium benzoate, during storage at selected temperatures.

0.99 - KS

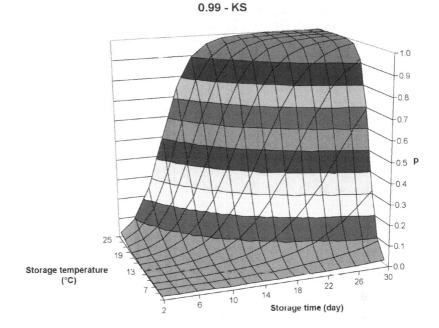

0.98 - KS

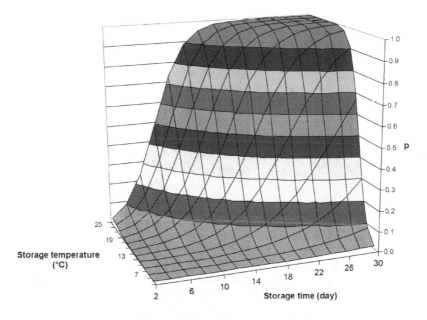

Figure 39.5 *Zygosaccharomyces bailii* probabilities (*p*) of growth in mango purée formulated at a_w 0.98 or 0.99, pH 3.5, with 1000 ppm of potassium sorbate, during storage at selected temperatures.

easily used make predictions about stability and factor combination adequacy even when the specific combination of factors have not been tested. This is an important advantage that makes predictive models very powerful tools for ensuring stability of food.[2]

Example 3: Assessment of Direct and Interactive Effect of New and/or Different Combinations of Food Preservatives and pH

Paracelsus, a German chemist from the 16th Century, wrote, *"All substances are poisons, there is none which is not a poison, the right dose differentiates poisons and remedies."* Modern food safety, toxicological evaluation, and dose response microbial curves are based on this assumption.[29] Data acquisition for elucidating the interactions of multiple variables associated with food systems has been underway for several decades, particularly in relation to the determining of how the activity of antimicrobials is affected by other parameters. There is a need for better understanding of the effects of combinations of traditionally used preservatives on microbial growth. This lack of research is even greater when the combination includes a natural antimicrobial. When a combination of antimicrobial agents is used, three situations can occur: synergism, antagonism, or additive effects.[30] In many cases, the effects of antimicrobial combinations may produce growth increasing the lag time; this effect can be modeled using response surface techniques. The aim of this example is to present the combined effects of a_w, pH, incubation temperature, sodium benzoate, and vanillin concentration on *Aspergillus flavus* growth in potato-dextrose agar. PDA was prepared with commercial sucrose to reach a_w 0.98, 0.96, or 0.94, and acidified with citric acid to attain pH 5.0, 4.0, or 3.0. The individual and combined concentrations of vanillin and sodium benzoate varied from 200 to 1600 ppm in 200-ppm increments. The inoculated plates were incubated for 1 month at 15 or 25°C and were periodically observed to determine if the mold grew. Inhibition was defined as no observable mold growth after one month of incubation.[31]

Of the 648 total evaluated combinations of factors, *A. flavus* grew in only 181, from which 78 corresponded to a 15°C and the rest to 25°C. From these growth conditions, and with the recorded lag times, the model was constructed. Significant terms and estimated coefficients are presented in Table 39.5. Goodness of fit was tested by a significant lack of fit ($p > 0.10$), predicted values error ($<15\%$) and a regression coefficient > 0.90.[31]

Figure 39.6 presents contour plots for fixed lag times obtained with the predictive model. *A. flavus* lag time increased as sodium benzoate concentration increased and pH decreased, maintaining the other variables fixed. The effects of pH, incubation temperature, and sodium benzoate concentration were the most important in delaying mold growth. Reducing pH increased the mold lag time, being more noticeable at 25°C than at 15°C. Incubation temperature greatly affected mold response; when incubated at 15°C, 400–200 ppm sodium benzoate with 600 ppm vanillin was enough to notably delay mold growth. A combination including 1000-ppm vanillin significantly decreased sodium benzoate concentration needed to cause the same effect on

TABLE 39.5
Estimated Coefficients for the Polynomial Equation that
Describes *Aspergillus flavus* **Lag Time**

Term	Range	Estimated coefficient
Intercept		−109673.400
Individual terms		
Incubation temperature (*T*)	15–25	34.406
Water activity (*a*$_w$)	0.94–0.98	217583.720
pH	3.0–5.0	2340.495
Sodium benzoate concentration (NaB)	0–1600	22.595
Vanillin concentration (*V*)	0–1600	3.412
Interactions and quadratic terms		
T × *a*$_w$		−88.309
a$_w$ × *a*$_w$		−104937.400
a$_w$ × pH		−3052.011
pH × pH		71.421
a$_w$ × NaB		−25.066
pH × NaB		0.445
NaB × NaB		2.038E−03
T × *V*		−0.016
a$_w$ × *V*		−3.198
NaB × *V*		1.303E−03
V × *V*		2.566E−04

mold growth delay. Water activity also affected mold lag time; when lowering a_w, the antimicrobial concentration required to inhibit at least during 30 days *Aspergillus flavus* growth decreased.[31]

39.4 FINAL REMARKS

Food safety and stability have been controlled by the food processing industry using a number of methods. Vegetative microbial cells and spores are destroyed or restricted by the individual or combined use of the following traditional preservation methods: heat sterilization, pasteurization, refrigeration, freezing, water activity control, cooking, acidification, fermentation, vacuum packaging, and/or adding anti-microbial agents. The extensive use of these procedures has maintained the low-risk status of the food supply. However, the new generation of minimally processed, fresher, antimicrobial and additive–free food products have less resistance to spoilage and allow greater risks of foodborne illness if some abuse occurs. The potential of food poisoning related to consumption of this kind of foods is great, while they have not been implicated in foodborne illness outbreaks.[32] Traditional preservation procedures generally base their intensity on microbial control, so after destroying,

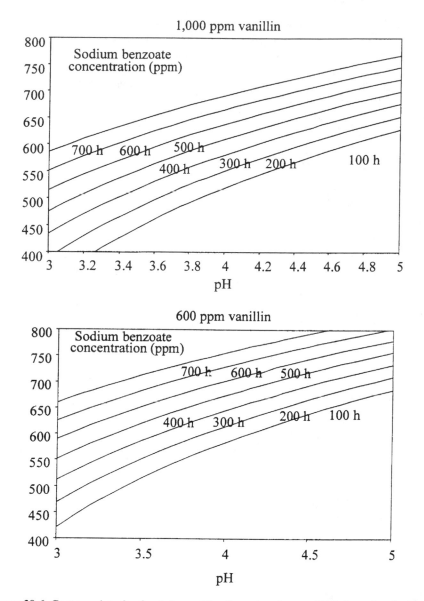

Figure 39.6 Contour plots for fixed *Aspergillus flavus* lag time on PDA formulated with a_w 0.98, 1000 or 600 ppm of vanillin, selected pHs, and sodium benzoate concentration, and incubated at 25°C.

eliminating, or substantially reducing initial microbial counts, measures to prevent recontamination should be taken, which often conduces to a safe and stable product.[33] If the microbial hazard or spoilage cannot be totally eliminated from the food, microbial growth and toxin production must be inhibited. Microbial growth can be inhibited by combining intrinsic food characteristics with extrinsic product storage and packaging conditions. If confident judgments about food quality, stability, and safety are to be made, it is essential to rely on accurate quantitative data about combined factors affecting growth or survival of selected microorganisms.[2] Data about any microbial response under selected combination of factors is a useful source of information. The wider spectrum of microbial type growth or survival models enables resources to be concentrated more efficiently and avoid problems concerning time-consuming microbial experimentation. The limiting factor for probability-based models has been their adaptation for use by nontechnical personnel; one of the key questions is what is a realistic probability of failure that one should be willing to accept.

39.5 ACKNOWLEDGEMENTS

The authors acknowledge the financial support from Universidad de Buenos Aires and CONICET of República Argentina and from CONACyT and Universidad de las Américas, Puebla, México.

REFERENCES

1. Zwietering, M. H., T. Wijtzes, J. C. de Witt, and K. van't Riet. 1992. "A Decision Support System for Prediction of the Microbial Spoilage in Foods," *J. Food Prot.*, 55(12): 973–979.
2. Gould, G. W. 1989. "Predictive Mathematical Modelling of Microbial Growth and Survival in Foods," *Food Sci. Technol. Today*, 3(2): 89–92.
3. Whiting, R. C. 1995. "Microbial Modeling in Foods," *Crit. Rev. Food Sci. Nutr.* 35(6): 467–494.
4. McMeekin, T. A., J. N. Olley, T. Ross, and D. A. Ratkowsky. 1993. *Predictive Microbiology: Theory and Application*. England: Research Studies Press, Ltd.
5. National Advisory Committee on Microbiological Criteria for Foods. 1998. "Principles of Risk Assessment for Illness Caused by Foodborne Biological Agents," *J. Food Prot.* 61:1071–1074.
6. Alzamora, S. M. "Application of Combined Factors Technology in Minimally Processed Foods," in *Sous Vide and Cook Chill Processing for the Food Industry*, S. Ghazala, ed. Maryland: Aspen Publishers, Inc.
7. Whiting, R. C. 1997. "Microbial Database Building: What Have We Learned?" *Food Technol.* 51(4): 82–86.
8. Skinner, G. E. and J. W. Larkin. 1998. "Conservative Prediction of Time to *Clostridium botulinum* Toxin Formation for Use With Time-Temperature Indicators to Ensure the Safety of Food," *J. Food Prot.*, 61(9): 1154–1160.
9. Ratkowsky, D. A. and T. Ross. 1995. "Modeling the Bacterial Growth/No Growth Interface," *Lett. Appl. Microbiol.*, 20: 29–33.
10. Baranyi, J. and T.A. Roberts. 1994. "A Dynamic Approach to Predicting Bacterial Growth in Food," *Int. J. Food Microbiol.*, 23: 277–294.

11. Buchanan, R. L., R. C. Whiting, and S. A. Palumbo. 1993. "Proceedings of Workshop on the Application of Predictive Microbiology and Computer Modeling Techniques to the Food Industry," *J. Ind. Microbiol.*, 12: 137–360.

12. Gibson, A. M. and A. D. Hocking. 1997. "Advances in the Predictive Modelling of Fungal Growth in Food," *Trends Food Sci. Technol.*, 8: 353–358.

13. Peleg, M. 1995. "A Model of Temperature Effects on Microbial Populations From Growth to Lethality," *J. Sci. Food Agric.*, 68: 83–89.

14. Zwietering, M. H., I. Jorgenburger, F. M. Rombouts, and K. van't Riet. 1996. "Modeling of the Bacterial Growth Curve," *Appl. Environ. Microbiol.*, 56(6): 1875–1881.

15. Adair, C. and P. A. Briggs. 1993. "The Concept and Application of Expert Systems in the Field of Microbiological Safety," *J. Ind. Microbiol.*, 12: 263–267.

16. Buchanan, R. L. 1993. "Developing and Distributing User-Friendly Application Software," *J. Ind. Microbiol.*, 12: 251–255.

17. McClure P. J., C. W. Blackburn, M. B. Cole, P. S. Curtis, J. E. Jones, J. D. Legan, I. D. Ogden, M. W. Peck, T. A. Roberts, J. P. Sutherland., and S. J. Walker. 1994. "Modelling the Growth, Survival and Death of Microorganisms in Foods: the UK Food Micromodel Approach," *Int. J. Food Microbiol.*, 23: 265–275.

18. Peleg, M. 1996. "Evaluation of the Fermi Equation as a Model of Dose-Response Curves," *Appl. Microbiol. Biotechnol.*, 46: 303–306.

19. Zwietering, M. H., J. C. de Witt, H. A. G. M. Cuppers, and K. van't Riet. 1994. "Modeling of Bacterial Growth with Shifts in Temperature," *Appl. Environ. Microbiol.*, 60(1): 204–213.

20. Cole, M. B., K. W. Davies, G. Munro, C. D. Holyoak, and D. C. Kilsby. 1993. "A Vitalistic Model to Describe the Thermal Inactivation of *Listeria monocytogenes*," *J. Ind. Microbiol.*, 12: 232–239.

21. Presser, K. A., T. Ross, and D. A. Ratkowsky. 1998. "Modelling the Growth Limits (Growth/No Growth Interface) of *Escherichia coli* as a Function of Temperature, pH, Lactic Acid Concentration, and Water Activity," *Appl. Environ. Microbiol.* 64: 1773–1779.

22. López-Malo, A., S. Guerrero, and S. M. Alzamora. 2000. "Probabilistic Modeling of *Saccharomyces cerevisiae* Inhibition Under the Effects of Water Activity, pH and Potassium Sorbate," *J. Food Prot.*, 63: 91–95.

23. Hosmer, D. W. and S. Lemeshow. 1989. *Applied Logistic Regression*. New York: John Wiley and Sons.

24. Cerrutti, P., S. M. Alzamora, and J. Chirife. 1990. "A Multi-Parameter Approach to Control the Growth of *Saccharomyces cerevisiae* in Laboratory Media," *J. Food Sci.*, 55: 837–840.

25. Cerrutti, P. 1988. Efectos Combinados de a_w, pH, Aditivos y Tratamiento Térmico en el Crecimiento y Supervivencia de *Saccharomyces cerevisiae*. Ph.D. Dissertation. Universidad de Buenos Aires, Argentina.

26. Bolton, L. F. and J. F. Frank. 1999. "Defining the Growth/No-Growth Interface for *Listeria monocytogenes* in Mexican-Style Cheese Based on Salt, pH and Moisture Content," *J. Food Protection*, 62:601–609.

27. López-Malo, A. and E. Palou. 2000. Growth/No Growth Interface of *Zygosaccharomyces bailii* as a Function of Temperature, Water Activity, pH, Potassium Sorbate and Sodium Benzoate Concentration. Presented at *Predictive Modeling in Foods*, Leuven, Belgium, 12–15 September.

28. López-Malo, A. and E. Palou. 2000. "Modeling the Growth/No Growth Interface of *Zygosaccharomyces bailii* in Mango Purée," *J. Food Sci.*, 65:516–520.

29. Miller, S. A. 1992. "History of Food Safety Assessment: From Ancient Egypt to Ancient Washington," in *Food Safety Assessment*, J. W. Finley, S. F. Robinson, and D. J. Armstrong, eds. Washington, D.C.: ACS series. American Chemical Society.

30. Parish, M. E. and P. M. Davidson. 1993. "Methods of Evaluation," in *Antimicrobials in Foods*, P. M. Davidson and A. L. Branen, eds. New York: Marcel Dekker, pp. 597–615

31. Carrillo-Inungaray, M. L., E. Palou, and A. López-Malo. 2000. "Fungistatic Effect of Vanillin and Sodium Benzoate in Combination." Presented at *Predictive Modeling in Food*s, Leuven, Belgium, 12–15 September.

32. Buchanan, R. L. 1992. "Predictive Microbiology: Mathematical Modeling of Microbial Growth in Foods," in *Food Safety Assessment*, J. W. Finley, S. F. Robinson, and D. J. Armstrong, eds. Washington, D.C.: ACS Series. American Chemical Society, pp. 250–260.

33. Baird-Parker, A. C. 1995. "Development of Industrial Procedures to Ensure the Microbiological Safety of Food," *Food Control,* 6(1): 29–36.

40 The Role of An Osmotic Step: Combined Processes to Improve Quality and Control Functional Properties in Fruit and Vegetables

D. Torreggiani
G. Bertolo

CONTENTS

Abstract

Fruits and vegetables play a very important role in the daily diet. As there is a clear interaction between consumption of fruits and vegetables and good health, particular attention has not only been paid to their composition, vitamin content, microelements, antioxidant substances, etc., but also to the continual development of technological processes needed to preserve these products. When the aim is to improve quality of fruit and vegetable products, combined processes are the methods of choice. These processes use a sequence of technological steps to achieve controlled changes of the original properties of the raw material. While some treatments, such

1-56676-963-9/02/$0.00+$1.50

as freezing, have primarily a stabilizing effect, other steps, namely, partial dehydration and osmotic dehydration, allow the properties of the material to be modified. A partial dehydration step is useful to set the product in the required moisture range, while a finer adjustment of water activity and structural, nutritional, sensory, and other functional properties is better achieved by what is generally called an osmotic step: *dewatering impregnation soaking* (DIS) in concentrated solutions. The numerous positive effects of a DIS pretreatment on texture, pigment, vitamin, and aroma retention, and formulation of processed fruits and vegetables, along with a global and critical view of the whole process, are underlined in this chapter.

40.1 INTRODUCTION

For many centuries, fruit and vegetables have been an important part of the human diet. In many parts of the world, wherever these foods are widely available and in great variety, they play a fundamental role in the daily diet. As there is a clear interaction between the consumption of fruits and vegetables and good health, more and more research has been made into their composition over the last few decades. Particular attention has been paid to the vitamin content, microelements, antioxidant substances, etc.

The discovery of the above-mentioned substances, having such a beneficial effect on the human organism, has given, even where the diet is based mainly on animal products, renewed interest into the so-called "Mediterranean diet," which consists of a high consumption of fresh fruits and vegetables. The consumption of fresh fruit and vegetables is of great importance, but so is the continual development of the technological processes needed to preserve these products. Over the centuries, we have come from the primitive techniques of solar drying and preserving in salt, up to the most refined processing that guarantees longer periods of storage, higher quality, and safer hygienic conditions.

To have products to be used like fresh fruits and vegetables or ingredients in the preparation of other foods, it is inevitable that fruits and vegetables undergo some form of treatment. The type of technological treatment chosen must be able to maintain not only the sensory characteristics and physical structure, but also the original composition as much as possible.

There is a vast literature that refers to the deterioration in sensory quality; to vitamin, microelement, and aroma losses; to oxidation, etc., due to a severe treatment or one not suitable for the nutritional characteristics of a certain product. Every type of treatment should have as its goal the preservation of the sensory characteristics that are an important aspect of the processed fruit and vegetables, and of the nutritional elements associated with it. It is common knowledge that, today, not all the technological processes of preserving are adequate to reach this objective.

As a means of preserving biological materials, temperature reduction is generally superior to that of the primary competing methods such as heating, drying, or use of chemical preservatives. In fact, when the aim is to extend the storage life of living materials, temperature reduction is the method of choice. Over the past few years, many new frozen fruit and vegetable based products have appeared on the market.

They are widely requested as basic materials or as additional components in many food formulations, giving them an additional value appreciated by the consumer and providing higher margins for the producer. But the apparent simplicity of the fruit and vegetable freezing process hides some puzzling complexities. Chemical and physical actions of freezing are detrimental to fruits and vegetables, since their texture is mainly ensured by structural integrity of the cell wall and the middle lamella as well as by the turgescence, which is the ability to retain water inside the cells.[1,2] Rupture of cell walls due to the growth of ice crystals and/or enzymatic actions during freezing prevents a return to the initial state. The product will exhibit a loss of cellular structure, which can manifest itself in increased drip loss while thawing, loss of shape, and less definite structure.[3–5]

There is a vast amount of literature describing the numerous factors that could influence the quality of frozen fruit and vegetables. Among these factors, so well summarized by Fennema in his "Lesson of Roses,"[6] and Reid,[7] I would like to highlight the following:

- The importance of the raw materials. Plant species and cultivars differ greatly in their tolerance to freezing conditions; thus, selection of proper species or cultivar has a great influence on the end product quality, along with the stage of development at harvest.
- The importance of improving the process itself through the improvement of heat transfer and the study of new ice crystal nucleation processes.
- The importance of identifying the properties of frozen systems that have an influence on frozen storage stability, and thus the relevance of freeze concentration and molecular mobility to an understanding of frozen storage stability.

40.2 ROLE OF AN OSMOTIC STEP

Considering all these factors influencing frozen food quality, the question that arises is, "How can an osmotic step, such a simple process, help in such a complex matter as the quality improvement of frozen fruit?"

This process involves placing the solid food (whole or in pieces) into solutions of high sugar or salt concentration. Le Maguer,[8] Raoult-Wack,[9] and Torreggiani[10] have recently reviewed the basic principles, modeling and control, and specific applications of osmotic dehydration on fruit and vegetables. Additionally, the most recent research advances in this field can be obtained from the European-founded network on osmotic treatments (FAIR, 1998).

As is known, soaking gives rise to at least two major countercurrent flows: an important outflow of water from the food into the concentrated solution, and a simultaneous transfer of solute from the concentrated solution into the food. Some of the osmotic syrup may not actively migrate into the cell but simply penetrate into the intercellular space. This impregnation effect may be important, and the term dewatering-impregnation-soaking in concentrated solutions (DIS) instead of osmotic dehydration has been recently proposed.[11] The main unique feature of DIS, compared with other dehydration processes, is the penetration of solutes into the food material. So it is possible, to a certain extent, to change nutritional, functional, and sensory properties of the food system, making it more suitable to the freezing process by

- Adjusting the physicochemical composition of food by reducing water content, or adding water activity lowering agents
- Incorporating ingredients or additives with antioxidant or other preservative properties into the food
- Adding solutes of nutritional or sensory interest
- Providing a larger range of food consistency

The characteristics of frozen fruits and vegetables that could be improved through the application of an osmotic step are numerous and will be here underlined. So far, the technique has been developed primarily for fruits and, to a lesser extent, vegetables. Therefore, the following examples relate mainly to fruits.

40.2.1 TEXTURE IMPROVEMENT

The use of pre-freeze treatments can help to reduce (or avoid) the detrimental phenomenon of loss of shape and texture reduction of fruits at thawing. Over the last few years, there has been a renewed interest in dehydrofreezing, a combined process proposed by Lazar in the 1960s,[12] in which freezing is preceded by partial dehydration.

Partial water removal from the food prior to the freezing process leads to the concentration of cytoplasmatic components within the cells, to the depression of the freezing point, and to an increase of supercooling. Thus, there are relatively fewer large ice crystals and a lower ratio of ice crystals to unfrozen phase, with a consequent reduction of structural and sensory modifications. When the refrigeration load is reduced, there are savings in packaging and distribution costs.[13] Partial dehydration is generally achieved by convective air dehydration; however, whatever the air drying technique, the color of certain fruits, such as kiwifruit, is susceptible to heat modification. For example, kiwifruit shows a definite yellowing of the typical green color when air dehydrated to 50% weight reduction, even at 45°C.[14] For these fruits, air drying must be replaced by DIS, which is effective at room temperature and operates away from oxygen.

Data on the usefulness of partial water removal prior to freezing were previously reviewed[10,15] and have been confirmed by recent research. Peaches processed under vacuum in natural juice or sucrose solutions showed better texture following frozen storage than fresh thawed fruits, and had better color and flavour characteristics.[16] As for kiwifruit, three dehydration processes were compared in the production of dehydrofrozen products: DIS, convective air dehydration, and a combination of DIS after partial water removal with convective air dehydration.[17] The highest force value for dehydrofrozen kiwifruit was found for fruits where 25% of water was removed by air dehydration followed by DIS to 40% moisture content.

Partial reduction of water through air dehydration, DIS, or a combination thereof has proven to be effective in reducing structural collapse after thawing-rehydration of strawberry slices.[18] Data reported in Figure 40.1 indicate that a reduction in moisture content of at least 60%, corresponding to a 50% weight reduction, is needed to improve the texture characteristics of thawed-rehydrated fruits, irrespective of the dehydration method used.

The application of DIS alone or in combination with air dehydration prolongs the processing time but could, through the incorporation of sugars, improve color, flavour, and vitamin retention during frozen storage, as will be described further on.

The light photomicrographs of freeze-thawed strawberry tissues, reported in Figure 40.2, are a further clear indication of the reduction of freezing damage due to the decrease in moisture content.[19]

In fact, predehydrated strawberry slices have a greater retention of the tissue organization after thawing than untreated ones, which show a definite continuity loss and thinning of cell wall. These findings may account for the significant lower decrease of texture and exudate loss observed in the dehydrofrozen strawberry slices.

DIS using selective solutes can also allow cryoprotection of the cell during freeze-thawing. Cryoprotective agents, especially high-molecular-weight compounds, are also thought to lower intermembrane stresses,[20] inhibit protein denaturation in membranes,[21] and, in the case of sugars, increase hydrophobic interactions in the membrane, increasing stability. The presence of sugars in apples has been shown to provide protection during freezing, allowing firmer tissue than for fruit treated in the same manner without sugars.[22] SEM revealed that cells protected by sugars exhibited less damage to the middle lamella and less severe shrinking of the cell content.

To reduce ice crystal damage in frozen apple cylinders, the use of vacuum infusion (VI) with cryoprotectants (sugars from concentrated grape must)/cryostabilizers (HM pectin) has been proposed.[23] The influence of the treatment on product

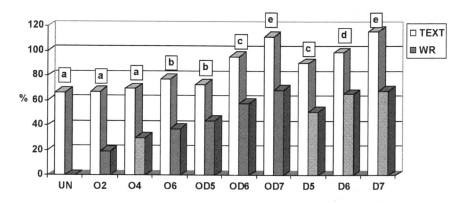

Figure 40.1 Texture modification (percent of raw fruit texture, TEXT) and weight reduction (WR) of thawed-rehydrated strawberry slices, untreated (UN), osmodehydrated (O2, O4, O6), osmo-air dehydrated (OD5, OD6, OD7), and air-dehydrated (D5, D6, D7) before freezing. Different letters (a, b, c, and d) indicate significant difference (P ≤ 0.05) among different treatments.

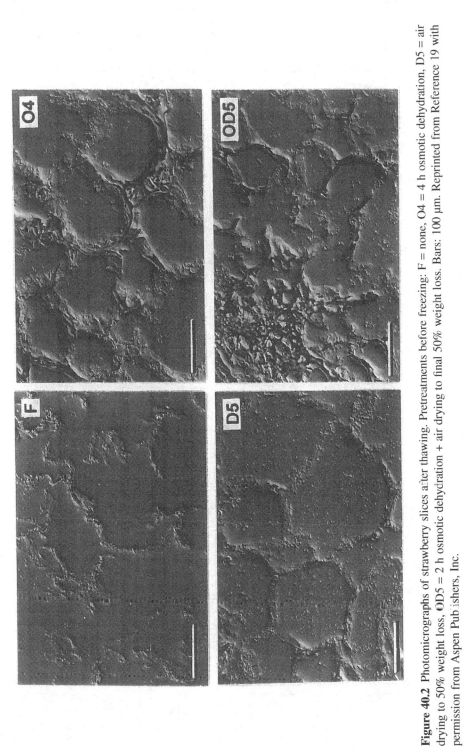

Figure 40.2 Photomicrographs of strawberry slices after thawing. Pretreatments before freezing: F = none, O4 = 4 h osmotic dehydration, D5 = air drying to 50% weight loss, OD5 = 2 h osmotic dehydration + air drying to final 50% weight loss. Bars: 100 μm. Reprinted from Reference 19 with permission from Aspen Publishers, Inc.

microstructure was analyzed by cryo-SEM observations and by measuring sample mechanical properties. Vacuum infusion promoted fast compositional changes of porous fruits, especially when concentrated cryoprotectant solutions were used. In these cases, a notable reduction of freezable water was obtained, which could improve resistance to freezing damage. The use of pectins could reinforce the structural matrix by means of intercellular bridges formed from polysaccharide gels.

Furthermore, Torreggiani et al.[24,25] have ascertained that partial dehydration before freezing could enhance the resistance of texture of frozen strawberry slices and apricot cubes to a thermal treatment, which is unavoidable before the addition of these fruits as ingredients, for example, to yogurt. In fact, for yogurt production, fruit pieces have to be frozen at the harvest season to be available all year round, and then, at the production stage, be heat treated to avoid yogurt contamination. Both freezing and heat treatments cause texture damage. To reach a texture improvement after the proposed heat treatment, a moisture reduction before freezing of at least 50% is needed both for strawberries and apricots, irrespective of the dehydration method used (Figure 40.3).[18]

This percentage of moisture reduction is required to reduce the freezing damage of the fruits at thawing,[18] thus indicating the freezing step as the crucial point of the production process of thermally stabilized strawberry and apricot ingredients. Limiting the freezing damage can improve fruit texture, even after heat treatment.

40.2.2 PIGMENT, VITAMIN, AND AROMA RETENTION

Texture improvements of frozen fruit and vegetables, as previously described, can be achieved by water removal through convective air dehydration, except for fruits whose color is susceptible to heat modification and that require the application of

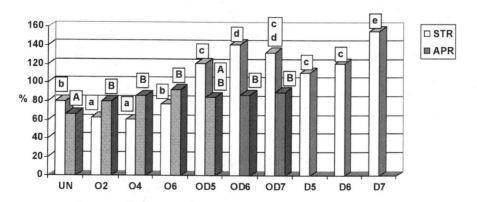

Figure 40.3 Texture modification (percent of raw fruit texture) of heat-treated strawberry slices (STR) and apricot cubes (APR) not pretreated (UN), osmodehydrated (O2, O4, O6), osmo-air dehydrated (OD 5, OD6, OD7), and air-dehydrated (D5, D6, D7) before freezing. Different letters (a, b, c, d, e, A, B) indicate significant difference (P ≤ 0.05) among different treatments. Upper case = apricot cubes, lower case = strawberry slices.

DIS. So why propose DIS as dehydration step even for nonsensitive fruits? The reason is its main unique feature: the penetration of solutes, which is combined with a dehydration effect and could modify the unfrozen phase composition.

Considering a fruit or vegetable during freezing, as the temperature drops, more ice forms, and the remaining unfrozen phase increases in concentration. As the temperature is lowered, viscosity increases. As concentration increases, viscosity increases. As first described by Levine and Slade,[26] these factors work together to produce a matrix from which no more ice may be separated. There is extensive literature discussing the implications of this observation.[27–32] The temperature at which no more ice can separate, often called *maximal freeze concentration,* is the glass transition temperature labeled T_g' by Levine and Slade. Glass transition occurs when the small, supersaturated unfrozen phase of a partially frozen biological material is cooled sufficiently to cause conversion to a glass. Glass is a metastable, noncrystalline solid with exceptionally high viscosity. When the temperature is at or below T_g', diffusion limited changes occur at very slow rates; i.e., stability, if based on diffusion-limited events, is excellent. Yet it must be kept in mind that many chemical changes are not diffusion limited. Also, the rates of presumed diffusion controlled reactions are proportional to the difference between the T_g', called also *mobility temperature,*[7] and the temperature of study. Manipulation of mobility temperatures through composition could therefore influence reaction rates. It has been hypothesized that modifying, through DIS, the fruit "formulation" to increase its glass transition temperature could also increase the storage stability.

Color is one of the most important attributes of foods both for its aesthetic value and quality judgment, and it may easily change if not properly protected.[33] The incorporation by DIS of different sugars into kiwifruit slices modified their low-temperature phase transitions and significantly influenced chlorophyll and color stability during frozen storage at different temperatures (–10, –20, and –30°C).[34] There was a correlation between the storage temperature and chlorophyll retention: the lower the temperature, the higher the retention (Figure 40.4a).[34] At the same storage temperature, kiwifruit pretreated in maltose, and thus having the highest T_g' values, showed the highest chlorophyll retention.

Kiwi is known also for its high nutritional value, related mainly to its high content of vitamin C. Therefore, the influence of a DIS pretreatment on vitamin C stability during frozen storage has been studied. Vitamin C content was stable at –20 and –30°C, while there was a significant decrease at –10°C (Figure 40.4b).[35] At –10°C, the vitamin C content showed the highest retention in the kiwifruit pretreated in maltose.

While the kinetic interpretation, based on the glass temperature, holds for chlorophyll and vitamin C stabilization in kiwifruit, for the anthocyanin pigments in strawberry, a simple relationship did not exist between the pigment loss and the amplitude of the difference between the storage temperature and the glass transition temperature of the maximally freeze-concentrated phase. These results were obtained by analyzing both strawberry halves subjected to DIS[36,37] and strawberry juices added with different carbohydrates.[38] The latter were used as a model of anthocyanin pigments degradation during frozen storage.

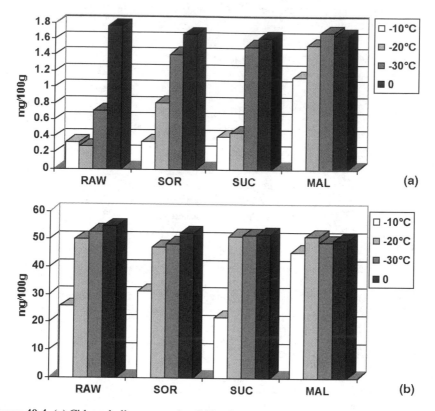

Figure 40.4 (a) Chlorophyll *a* content (mg/100 g fr. wt.) and (b) ascorbic acid content (mg/100 g fr. wt.) of kiwifruit slices, not pretreated (raw) and pretreated in sorbitol (SOR), sucrose (SUC), and maltose (MAL), after nine months of frozen storage, 0 = content before freezing.

In strawberry halves, there was a correlation between the storage temperature and anthocyanin pigment retention: the lower the temperature, the higher the retention, as observed for chlorophyll pigments.[37] The osmodehydrated strawberry halves showed pigment losses significantly lower than those observed in the fruits frozen without a concentration pretreatment (Figure 40.5).[37] No differences in pigment retention were observed among the fruits osmodehydrated in the different sugars.

As for strawberry juices added with different carbohydrates and stored at −10°C, juices having similar sugar composition at different concentration, and thus having similar T_g' values, showed significantly different pigment losses (Figure 40.6).[38]

Perhaps another parameter that could be taken into account is the freeze concentration level. At −10°C, the ice quantity was nearly twice as high as in the raw juice than in the sugar added juices. Parker and Ring[39] showed a large reaction rate enhancement for diffusion-controlled reactions in low concentrated solutions between 0 and −10°C. The freezing effect of increasing the initial sugar concentration was to reduce the effect of the freeze concentration on the reactants. Furthermore,

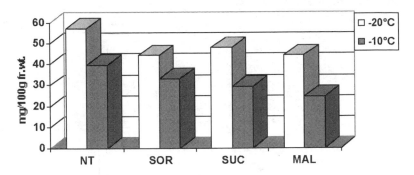

Figure 40.5 Anthocyanin contents of strawberry halves not pretreated (NT) and pretreated in sorbitol (SOR), sucrose (SUC), and maltose (MAL) after six months of storage at different temperatures.

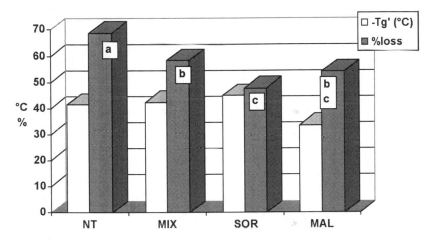

Figure 40.6 Anthocyanin content percent loss (% loss) and T_g' values of strawberry juices not added (NT) and added with glucose-fructose-sucrose mixture (MIX), sorbitol (SOR), and maltose (MAL) after four months at −10°C. Different letters (a, b, c) indicate significant difference (P ≤ 0.05) among different treatments.

the sorbitol added juice, which had the lowest T_g' values, showed the highest pigment retention. Other important factors such as the pH of the unfrozen phase could influence anthocyanin pigment stability. As reported by Bell and Labuza,[40] the pH of reduced-moisture solutions is not the same as that of dilute solutions and is dependent on the chemical nature of the solute. It could be hypothesized that sorbitol could alter the nucleophilic power of the water or could play a specific protective role, due to its chemical nature, in the enzymatic breakdown of the anthocyanin pigments.

The modification of sugar composition due to an osmotic step could also improve the stability of vitamin C and color during air drying and frozen storage of osmo-dehydrofrozen apricot cubes.[41] As shown in Figure 40.7, the higher the sugar enrich-

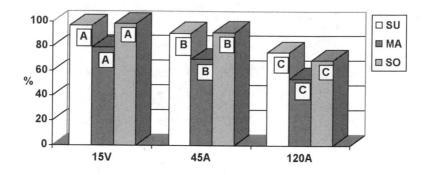

Figure 40.7 Percentage degradation after drying of ascorbic acid of apricot cubes osmode-hydrated for 15 min under vacuum (15V), 45 and 120 min at atmospheric pressure (45A, 120A) in sucrose (SU), maltose (MA), and sorbitol (SO). Different letters (A, B, C) indicate significant difference ($P \leq 0.05$) among different osmosis times for the same osmotic syrup.

ment, the higher the protective effect on vitamin C both during air drying at 65°C and during frozen storage at –20°C, with maltose being the most effective carbohydrate.

These findings were confirmed by the results of research carried out on the influence of the syrup composition on color stability of osmodehydrofrozen apricot cubes at $a_w = 0.80$.[42] Apricot cubes pretreated in a sucrose solution at 13% (wt/wt), isotonic with the fresh fruits were used as control and compared with fruits pretreated in concentrated (60% wt/wt) sucrose and maltose solutions. Both the concentrated osmotic solutions and the isotonic solution were added with 1% ascorbic acid (vitamin C) and 0.5% NaCl as antioxidant. The browning effect, expressed as browning index, was calculated from the Hunter[43] coordinates L^*, a^*, and b^* as follows:

$$\frac{L_f^* \times b_f^*}{70 \times a_f^*} - \frac{L^* \times b^*}{70 \times a^*}; \quad x_f = \text{fresh fruit values}$$

The value was significantly lower in the cubes pretreated in the concentrated solutions of both sucrose and maltose (Figure 40.8).[42]

Apricot cubes pretreated in the sucrose solution isotonic with the fresh fruit before drying had the same vitamin C content as the cubes pretreated in concentrated solutions, but lower sugar concentration. They showed the same browning effect as the non-pretreated apricot cubes, confirming the protective effect of sugar concentration. The sugar uptake, due to the osmotic step, reduced not only the browning phenomenon but also the vitamin C degradation during air drying. The lower ascorbic acid degradation observed in apricot cubes treated in sucrose and maltose could also be related to the fact that these fruits have T_g' values higher than those of the untreated ones, hence lower $T - T_g'$ values. As a consequence, they could have lower structural collapse during drying. In fact, several collapse phenomena, such

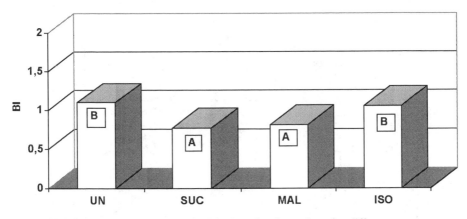

Figure 40.8 Browning index (BI) or air dehydrated apricot cubes after different pretreatments, not pretreated (UN), osmosis in sucrose (SUC), maltose (MAL), and isotonic solution (ISO). Different letters (A, B) indicate significant difference (P $\leq$ 0.05) among different treatments.

as recrystallization of freeze-dried sugars or structural collapse, have been shown to be governed by T_g'.[29,44,45] However, the physical state of dehydrated food material has been suggested as one of the rate-defining factors of diffusion-controlled deteriorative changes, for example non-enzymatic browning in low- and intermediate-moisture foods.[46] Structural collapse, which would increase material density, may also affect phenolase activity and thus ascorbic acid degradation rate. To provide evidence of this hypothesis, further research is in progress to determine the influence of different sugars on the level of structural collapse during drying.

As for vegetable dehydration, by incorporating sorbitol into red pepper cubes, a lower color degradation could be obtained when they are subjected to air drying to produce reduced moisture red pepper ingredients.[47] So, as already observed in strawberries, even in vegetables and during air drying at 65°C, sorbitol showed a significant protection of red color and thus of the anthocyanin pigments. Red pepper cubes, osmodehydrated in a new type of syrup (HLS = hydrolyzed lactose syrup from cheese whey ultrafiltration permeate), added with sorbitol, showed the lowest color differences,

$$\Delta E = \sqrt{\Delta L^{*2} + \Delta a^{*2} + \Delta b^{*2}}$$

when compared with fresh ones (Figure 40.9).[47] The results also suggested that for red pepper at low water activities the combination of osmo- with air-dehydration could even be detrimental to color characteristics if a nonprotective sugar other than sorbitol is utilized as osmotic medium.

The influence was also studied of an osmotic step on the volatile retention during air drying at 60°C of strawberry slices.[48] The concentration of strawberry slices obtained through DIS in concentrated sucrose solutions improved the volatile retention during the drying stage. By increasing the drying level, there was a greater

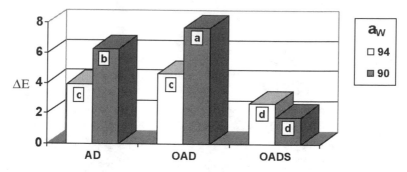

Figure 40.9 Color differences of processed red pepper cubes, referred to fresh samples, air-dehydrated (AD), osmo-air-dehydrated without sorbitol (OAD), osmo-air-dehydrated with sorbitol (OADS). Different letters (a, b, c, d) indicate significant difference (P ≤ 0.05) among different treatments.

volatile decrease, so previously osmodehydrated strawberry slices could be dried at higher drying levels as compared with the not-pretreated fruits.

40.2.3 FORMULATION

Soluble solids uptake due to DIS, besides improving color, aroma, and vitamin stability, could also play a very important role in the preparation of new types of ingredients at reduced water activity.[49] Owing to the soluble solid intake, the overall effect of DIS is a decrease in water activity, with only a limited increase in consistency. Consistency is actually associated with the plasticizing and swelling effect of water on the pectic and cellulosic matrix of the fruit tissues. Hence, it depends primarily on the insoluble matter and water content rather than on the soluble solids and water activity. In this way, low water activities may be achieved while maintaining an acceptable consistency.

A general representation of the extent to which physical changes can be induced and functional properties controlled in practical processing can be obtained by developing, according to Maltini et al.,[50] a "functional compatibility map." These maps illustrate that functional properties gained by using single or combined steps are related to the water activity, which is the main parameter making the ingredients compatible with the food. In the maps, the relationship is reported between the phase composition (i.e., the relative amount of insoluble solids, soluble solids, and water), the texture index, and the water activity of the fruits after processing. The two sets of data, referring to partial dehydration of raw fruit and partial dehydration of osmotically treated fruit, are presented in the same diagram and give a pair of curves for phase composition and for texture. The difference between the upper and lower curve for phase composition and texture at equal water activity is the result of the solid gain after osmotic treatment. The higher the solid uptake, the higher the difference in texture (Figure 40.10).[50] Compared to simple air dehydration, the combination of DIS and air dehydration can produce a softer product at low water

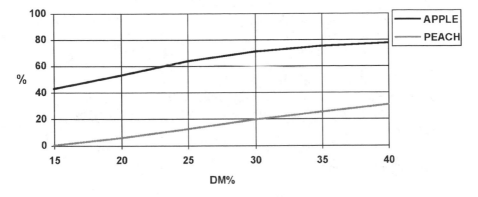

Figure 40.10 Texture decrease (percent of not-osmodehydrated fruit texture) of apple and peach cubes, air dehydrated to different dry matter contents. Solid uptake: apple = 18%, peach = 4.8%.

activity, which is more pleasant to eat by hand or to incorporate into pastry, ice cream, cheese, yogurt,[51] etc.

The choice of the osmotic syrup plays a very important role, and the specific effect of the solution has to be taken into account. The choice depends mainly on taste, cost, and a_w-lowering capacity along with the possible kinetic hindering of diffusion controlled reaction during frozen storage. Fruit juice concentrates have similar osmotic properties to high-fructose syrups,[52] and resulting products are of total fruit origin. If a concentrated fruit juice is used as osmotic solution, an even softer product could be obtained because of the higher content of monosaccharides in the fruit juice, compared with the amount contained in syrup from starch hydrolysis, and because of the higher relative water content at a determined water activity.[49] If a fructose syrup contains sorbitol, softer osmodehydrated apricot, clingstone peach cubes, and sweet cherry halves can be obtained as compared with the same fruit osmodehydrated in fructose alone.[53,54] The presence of sorbitol in HLS (hydrolyzed lactose syrup) leads to a lower texture also in osmodehydrated red pepper cubes.[47] Moreover, as previously reported in this chapter, sorbitol has a specific protective effect on color during the air-drying step.

40.3 CONCLUSIONS

After having underlined all the numerous positive effects of a DIS pretreatment on processed fruits and vegetables, a global and critical view of the whole setup should be made using an effective tool such as a "SWOT" analysis as described below.

Strengths (S)

- Enhancement of nutritional and sensory quality of fruit and vegetable products
- Improvement of suitability to further processing

- Small investments, as relative simple equipment are required
- Low-cost processing
- Reduced energy input over traditional drying processes (40 to 50% energy cost reduction)
- Addition of solutes of nutritional or sensory interest
- Providing a larger range of product consistencies at the same water activity
- Incorporating natural ingredients or additives with preservative properties
- Minimal processing
- Minimization of thermal stress
- Compatibility with environmental aspects of food processing

Weaknesses (W)

- Difficulty in defining a general predictive processing model, owing to the great variability of plant materials (species, cultivar, maturity stage, etc.)
- Lack of adequate responses to problems related to the management of the osmotic solutions (reconcentration, reuse, microbial contamination, re-utilization, and discharge of the spent solution, etc.)
- Difficulty developing continuous processing equipment

Opportunities (O)

- Tremendous market potential for high-quality fruit and vegetable end products and ingredients
- Increased variety of products offered on the market
- Tool to develop new products
- Introduction of value added fruit and vegetable products on the market
- Preparation of fruit and vegetable ingredients with functional properties tailored for specific food systems
- Utilization of fruits and vegetables in excess of, or not suitable for, the fresh market (size, appearance, etc.)

Threats (T)

- Change in government legislation on sugar and salt solution disposal
- Change in dietetic trends in western populations (salt and sugar intake)
- Delay in the progress of research projects aimed at a full development of the technique at the industrial level

REFERENCES

1. Ilker R. and A. S. Szczesniak. 1990. "Structural and Chemical Bases for Texture of Plant Foodstuffs," *J. Texture Stud.*, 21: 1–36.

2. Jackman R. L. and D. W. Stanley. 1995. "Perspectives in the Textural Evaluation of Plant Foods," *Trends Food Sci. Technol.*, 6: 187–194.

3. Sterling, C. 1968. "Effect of Low Temperature on Structure and Firmness of Apple Tissue," *J. Food Sci.*, 33: 577–580.

4. Szczesniak A. S. and B. J. Smith. 1969. "Observations on Strawberry Texture. A Three-Pronged Approach," *J. Texture Stud.*, 1: 65–89.

5. Mohr, W. P. 1974. "Freeze-Thaw (and Blanch) Damage to Vegetable Ultrastructure," *J. Texture Stud.*, 5: 13–27.

6. Fennema, O. 1995. "Chilled and Frozen Biological Material: The Lesson of Roses," in *Proc. 19th Int. Cong. of Refrigeration*, vol. 1, pp. 13–20, The Hague, The Netherlands, 20–25 August 1995.

7. Reid, D. S. 1999. "Factors Which Influence the Freezing Process: an Examination of New Insights," in *Proc. 20th Int. Cong. of Refrigeration,* Sidney, Australia, 19–24 September 1999, in press.

8. Le Maguer, M. 1988. "Osmotic Dehydration: Review and Future Directions," in *Proc. Int. Symposium "Progress in Food Preservation Processes"*, CERIA, Bruxelles, Belgium, vol. 1, pp. 283–309.

9. Raoult-Wack, A. L. 1994. "Recent Advances in the Osmotic Dehydration of Foods," *Trends Food Sci. Technol.,* 5 (8): 255–260.

10. Torreggiani, D. 1995. "Technological Aspects of Osmotic Dehydration in Foods," in *Food Preservation by Moisture Control: Fundamentals and Applications,"* ISOPOW PRACTICUM II, G. V.Barbosa-Cànovas & J.Welti-Chanes, eds. Lancaster, PA (USA): Technomic Publishing Co. Inc., pp. 281–304.

11. Raoult-Wack, A. L. and S. Guilbert. 1990. "La Déshydratation Osmotique ou Procédé de Déshydratation-Imprégnation par Immersion dans des Solutions Concentrées," in *Les Cahiers de L'Ensbana "L'eau dans les Procédés de Transformation et de Conservation des Aliments,"* 7: 171–192.

12. Lazar, M. E. 1968. "Dehydrofreezing of Fruits and Vegetables," in *The Freezing Preservation of Foods*, D. K.Tressler, W. B. Van Arsdel, and M. J. Copley, eds., Westport, Conn. (USA): AVI Publishing Company, pp. 347–376.

13. Huxsoll, C. C. 1982. "Reducing the Refrigeration Load by Partial Concentration of Food Prior to Freezing," *Food Technol.*, 5: 98–102.

14. Forni E., D. Torreggiani, G. Crivelli, A. Maestrelli, G. Bertolo, and F. Santelli. 1990. "Influence of Osmosis Time on the Quality of Dehydrofrozen Kiwifruit," *Acta Horticulturae* n. 282, A.R. Ferguson, ed., ISHS, Wageningen, The Netherlands, pp. 425–34.

15. Torreggiani, D. 1993. "Osmotic Dehydration in Fruit and Vegetable Processing," *Food Res. Int.,* 26: 59–68

16. Hout, S. A., C. C. Huxoll, and D.W. Sanshuck. 1995. "Partial Dehydration of Peaches in Vacuum Osmotic Evaporation to Improve Texture," *Abstracts IFT Annual Meeting*, Anaheim, CA (U.S.A.), 3–7 June 1995.

17. Robbers, M., P. R. Singh, and L. M. Cunha. 1997. "Osmotic-Convective Dehydrofreezing Process for Drying Kiwifruit," *J. Food Sci.*, 5: 1039–1042 (1047).

18. Maestrelli A., G. Giallonardo, E. Forni, and D. Torreggiani. 1997. "Dehydrofreezing of Sliced Strawberries: a Combined Technique for Improving Texture," in ICEF 7 *"Engineering & Food"*, R. Jowitt, ed. Sheffield (UK): Sheffield Academic Press, Part 2, pp. F37–40.

19. Sormani, A., D. Maffi, G. Bertolo, and D. Torreggiani. 1999. "Textural and Structural Changes of Dehydrofreeze-Thawed Strawberry Slices: Effects of Different Dehydration Pre-treatments," *Food Sci. Technol. Int.*, 5(6): 484.

20. Wolfe, J. and G. Bryant. 1992. "Physical Principles of Membrane Damage due to Dehydration and Freezing," in *Mechanics of Swelling*, NATO ASI Series H64, T. K. Karalis ed., Berlin: Springer-Verlag, pp. 205–224.

21. Burke, M. J., L. V. Gusta, H. A. Quamme, C. J. Weiser, and P. H. Li. 1976. "Freezing and Injury in Plants," *Ann. Rev. Plant Physiol.*, 27: 507–528.

22. Tregunno, N. B. and H. D. Goff. 1996. "Osmodehydrofreezing of Apples: Structural and Textural Effects," *Food Res. Int.*, 29: 471–479.

23. Martínez-Monzó, J., N. Martínez-Navarrete, A. Chiralt, and P. Fito. 1998. "Mechanical and Structural Changes in Apple (Var. Granny Smith) Due to Vacuum Impregnation with Cryoprotectants," *J. Food Sci.*, 3: 499–503.

24. Torreggiani, D., B. Rondo Brovetto, A. Maestrelli, and G. Bertolo. 1999. "High Quality Strawberry Ingredients by Partial Dehydration Before Freezing," in *Proc. 20th International Congress of Refrigeration*, Sidney, Australia, 19–24 September 1999, in press.

25. Torreggiani, D., G. Giallonardo, B. Rondo Brovetto, and G. Bertolo. 1999. "Impiego della Deidrocongelazione per il Miglioramento Qualitativo di Semilavorati di Albicocca in Pezzi per Yogurt," in *Ricerche e innovazioni nell'industria alimentare*, 4th CISETA (Congresso Italiano di Scienza e Tecnologia degli Alimenti), S. Porretta, ed., Chiriotti Editori, Pinerolo, Italy, in press.

26. Levine, H. and L. Slade. 1988. "Principles of 'Cryostabilization' Technology from Structure/Property Relationships of Carbohydrate/Water Systems—A Review," *Cryo-Letters*, 9: 21–63.

27. Le Meste, M., A. Voilley, and B. Colas. 1991. "Influence of Water on the Mobility of Small Molecules Dispersed in a Polymeric System," in *Water Relationship in Food*, H. Levine and L. Slade, eds. New York: Plenum Press, pp. 123–138.

28. Simatos, D. and G. Blond. 1991. "DSC Studies and Stability of Frozen Products," in *Water Relationship in Food*, H. Levine and L. Slade, eds. New York: Plenum Press, pp. 139–155.

29. Slade, L. and H. Levine. 1991. "Beyond Water Activity: Recent Advances Based on an Alternative Approach to the Assessment of Food Quality and Safety," *Crit. Rew. Food Sci. Nutr.*, 30: 115–360.

30. Karel, M., M. P. Buera, and Y. Roos. 1993. "Effects of Glass Transitions on Processing and Storage," in *Glassy State in Foods*, J. M. V. Blanshard. and P. J. Lillford, eds., Loughborough, U.K.: Nottingham University Press, pp. 13–85.

31. Roos, Y. and M. Karel. 1993. "Effects of Glass Transitions on Dynamic Phenomena in Sugar Containing Food Systems," in *Glassy State in Foods*, J. M. V. Blanshard, and P. J. Lillford, eds. Loughborough, U.K.: Nottingham University Press, pp. 207–222.

32. Champion, D., G. Blond, and D. Simatos. 1997. "Reaction Rates at Subzero-Temperature in Frozen Sucrose Solutions: a Diffusion-Controlled Reaction," *Cryo-Letters*, 18: 251–260.

33. Chichester, C. O. and R. McFeeters. 1970. "Pigment Degeneration During Processing and Storage," in *The Biochemistry of Fruit and their Products*, Vol.II, Academic Press London, pp. 707–718.

34. Torreggiani, D., E. Forni, and L. Pelliccioni. 1994. "Modificazione della Temperatura di Transizione Vetrosa Mediante Disidratazione Osmotica e Stabilità al Congelamento del Colore di Kiwi," in *Ricerche e innovazioni nell'industria alimentare*, 1st CISETA (Congresso Italiano Scienza Tecnologia Alimenti), S. Porretta, ed. Pinerolo, Italy: Chiriotti Editori, pp. 621–630.

35. Torreggiani, D. and G. Bertolo. 1999. "High Quality Fruit and Vegetable Products Using Combined Processes," in *Proc. Industrial Seminar "Osmotic Dehydration and Vacuum Impregnation: Application of New Technologies to Traditional Food Industries,"* Concerted Action FAIR-CT96-1118 "Improvement of Overall Food Quality by Application of Osmotic Treatments in Conventional and New Process," Valencia (Spain) 15–17 March 1999, in press.

36. Torreggiani, D., E. Forni, A. Maestrelli, G. Bertolo, and A. Genna. 1995. "Modification of Glass Transition Temperature by Osmotic Dehydration and Color Stability of Strawberry During Frozen Storage," in *Proc. 19th International Congress of Refrigeration,* vol. 1, pp. 315–321, The Hague, The Netherlands, 20–25 August 1995.

37. Forni, E., A. Genna, and D. Torreggiani. 1998. "Modificazione della Temperatura di Transizione Vetrosa Mediante Disidratazione Osmotica e Stabilità al Congelamento del Colore delle Fragole," in *Ricerche e Innovazioni nell'Industria Alimentare,* 3rd CISETA (Congresso Italiano di Scienza e Tecnologia degli Alimenti), S. Porretta, ed., Pinerolo, Italy: Chiriotti Editori, pp. 123–130.

38. Torreggiani, D., E. Forni, I. Guercilena, A. Maestrelli, G. Bertolo, G. P. Archer, C J. Kennedy, S. Bone, G. Blond, E. Contreras-López, and D. Champion. 1999. "Modification of Glass Transition Temperature Through Carbohydrates Additions: Effect Upon color and Anthocyanin Pigment Stability in Frozen Strawberry Juices," *Food Res. Int.,* 32: 441–446.

39. Parker, R. and S. G. Ring. 1995. "A Theoretical Analysis of Diffusion-Controlled Reactions in Frozen Solutions," *Cryo-Letters,* 16: 197–208.

40. Bell, L. N. and T. P. Labuza. 1994. "Influence of the Low-Moisture State on pH and Its Implication for Reaction Kinetics," *J. Food Eng.,* 22: 291–312.

41. Forni, E., A. Sormani, S. Scalise, and D. Torreggiani. 1997. "The Influence of Sugar Composition on the color Stability of Osmodehydrofrozen Intermediate Moisture Apricots," *Food Res. Int.,* 30: 87–94.

42. Camacho, G., G. Bertolo, and D. Torreggiani. 1998. "Stabilità del Colore di Albicocche Disidratate: Influenza del Pre-Trattamento Osmotico in Sciroppi Zuccherini Diversi," in *Ricerche e Innovazioni nell'Industria Alimentare,* 3rd CISETA (Congresso Italiano di Scienza e Tecnologia degli Alimenti), S. Porretta, ed., Pinerolo, Italy: Chiriotti Editori, pp. 553–561.

43. Hunter, R. S. 1975. "Scales for Measurement of color Difference," in *The Measurements of Appearance,* New York: Wiley, pp. 133–140.

44. Ross, H. Y. and M. Karel. 1992. "Crystallization of Lactose," *J. Food Sci.,* 57: 775–777.

45. Levi, G. and M. Karel. 1995. "Volumetric Shrinkage in Freeze-Dried Carbohydrates Above Their Glass Transition Temperature," *Food Res. Int.,* 28: 145–151.

46. Karmas, R., M. P. Buera, and M. Karel. 1992. "Effect of Glass Transition on Rates of Nonenzymatic Browning in Food Systems," *J. Agric. Food Chem.,* 40: 873–879.

47. Torreggiani, D., E. Forni, M. L. Erba, and F. Longoni. 1995. "Functional Properties of Pepper Osmodehydrated in Hydrolysed Cheese Whey Permeate With or Without Sorbitol," *Food Res. Int.,* 28 (2): 161–166.

48. Di Cesare, L. F., D. Torreggiani, and G. Bertolo. 1999. "Preliminary Study of Volatile Composition of Strawberry Slices Air Dried With or Without an Osmotic Pre-Treatment," in *Proc. V Plenary Meeting of Concerted Action FAIR-CT96-1118 "Improvement of Overall Food Quality by Application of Osmotic Treatments in Conventional and New Process,"* pp. 39–44, Valencia (Spain) 15–17 March, 1999.

49. Torreggiani, D., E. Maltini, G. Bertolo, and F. Mingardo. 1988. "Frozen Intermediate Moisture Fruits: Studies on Techniques and Products Properties," in *Proc. Int. Symposium "Progress in Food Preservation Processes,"* Bruxelles, Belgium: CERIA, vol. 1, pp. 71–78.
50. Maltini, E., D. Torreggiani, B. Rondo Brovetto, and G. Bertolo. 1993. "Functional Properties of Reduced Moisture Fruits as Ingredients in Food Systems," *Food Res. Int.*, 26: 413–419.
51. Giangiacomo, R., D. Torreggiani, M. L. Erba, and G. Messina. 1994. "Use of Osmo-dehydrofrozen Fruit Cubes in Yogurt," *Itl. J. Food Sci.*, 3: 345–350.
52. Maltini, E., D. Torreggiani, E. Forni, and R. Lattuada. 1990. "Osmotic Properties of Fruit Juice Concentrates," in *Engineering and Food: Physical Properties and Process Control*, W. L.E. Spiess & H. Schubert, eds. London: Elsevier Science Publishing Company, 1, pp. 567–573.
53. Erba, M. L., E. Forni, A. Colonnello, and R. Giangiacomo. 1994. "Influence of Sugar Composition and Air Dehydration Levels on the Chemical-Physical Characteristics of Osmodehydrofrozen Fruit," *Food Chem.*, 50: 69–73.
54. Torreggiani, D., E. Forni, and F. Longoni. 1997. "Chemical-Physical Characteristics of Osmodehydrofrozen Sweet Cherry Halves: Influence of the Osmodehydration Methods and Sugar Syrup Composition," in *Proc. Food Eng. 1st Int. Cong." Food Ingredients: New Technologies. Fruits & Vegetables,"* Allione Ricerca Agroalimentare S.p.A., pp. 101–109, Cuneo (Italy), 15–17 September 1997.

41 Approaches for Safety Assessment of Minimally Processed Fruits and Vegetables

M. S. Tapia
J. Welti-Chanes

CONTENTS

Abstract

Fruit and vegetables, and minimally processed fruits and vegetables (MPFV), cannot be excluded from the application of any of the modern tools for ensuring safety of foods. Approaches to safety assessment of produce and MPFV, either from the industry side or from the regulatory perspective, are being revised. An attempt to place in a coordinated frame GMP, HACCP, risk assessment, and predictive microbiology for this type of products is made herein.

41.1 INTRODUCTION

Traditionally, to ensure safe food, industry and regulators have relied on inspection, spot checks of manufacturing conditions (which may fail to detect contaminated batches), and random end-product testing. This approach is expensive, time consuming, inefficient, and tends to be reactive rather than preventive. The reliance on microbial testing and certification to maintain microbiological quality is not practical and often is of limited value.[1]

This situation was improved by the introduction of good manufacturing practices (GMPs), which are preventive measures that have evolved from general principles of hygiene, based on practical experience over a long period of time. This rather subjective, qualitative, and inspectional approach, relying on personal opinions and expertise, is not adequate for an objective risk assessment of foods. This is particularly true when there is a markedly increased desire and need for quantitative data on the microbial risks associated with different classes of foods. These data would help regulatory authorities in reliable decision making for control programs of foodborne microbial hazards.[2,3]

GMP and process control, based on the Hazard Analysis and Critical Control Point (HACCP) system, are important tools used by the food industry for production of microbiologically safe food.[2,3] Also, statistics play a very important role in helping to make such decisions. There are a variety of new, and not so new, methods that have arisen over the last few decades to help decision makers organize scientific information about risk. Risk analysis is one such method.[1]

The food industry applies GMP and HACCP to control the food production process. Identification of critical control points (CCPs) may lead to a quantifiable reduction that a hazard will occur. More directly related to the health of consumers is risk analysis. Since regulatory agencies now face the task of inspection of control systems as the introduction of the HACCP becomes frequent and integrated into legislation, they are primarily responsible for setting realistic microbiological specifications for HACCP by application of risk assessment, which is one step of risk analysis.[4]

Fruits and vegetables and MPFV cannot be excluded from the application of any of the modern tools for ensuring the safety of foods. These approaches to safety assessment of produce and MPFV, either from the industry side or from regulatory directives, which together are interrelated and have the final goal of assuring consumers' ultimate health and well being, will be discussed in this chapter.

41.2   GOOD MANAGEMENT PRACTICES

Fresh fruits and vegetables are important components of the human diet. The potential for microbiological contamination of fruits and vegetables is high because of the wide variety of conditions to which produce is exposed during growth, harvest, processing, and distribution. Over the last several years, the detection of outbreaks of foodborne illness associated with fresh fruits and vegetables has increased, and produce has been identified as an area of food safety concern. In 1997, the president of the United States, as part of the Food Safety Initiative (FSI) to improve the safety of the nation's food supply, directed the secretary of Health and Human Services, in partnership with the secretary of Agriculture, and in close cooperation with the agricultural community of that country, to issue guidance on good agricultural practices (GAPs) and good manufacturing practices (GMPs) for fruits and vegetables. In response to this directive, the FDA and USDA issued the document titled "Guidance for Industry-Guide to Minimize Microbial Food Safety Hazards for Fresh Fruits and Vegetables."[5] This guidance document addresses microbial food safety hazards and good agricultural and management practices common to the growing, harvesting, washing, sorting, packing, and transporting most fruits and vegetables sold to consumers in an unprocessed or minimally processed (raw) form. Good management practices refer to general practices to reduce microbial food safety hazards. The term may include both "good agricultural practices" and "good manufacturing practices" used in sorting, packing, storage, and transportation operations.[5] Table 41.1 presents the basic principles of this document.

TABLE 41.1
Basic Principles and Practices Associated with Minimizing Microbial Food Safety Hazards from the Field through Distribution of Fresh Fruits and Vegetables

Principle 1	Prevention of microbial contamination of fresh produce is favored over reliance on corrective actions once contamination has occurred.
Principle 2	To minimize microbial food safety hazards in fresh produce, growers, packers, and shippers should use good agricultural and management practices in those areas over which they have control.
Principle 3	Fresh produce can become microbiologically contaminated at any point along the farm-to-table food chain. The major source of microbial contamination with fresh produce is associated with human or animal feces.
Principle 4	Whenever water comes in contact with produce, its source and quality dictates the potential for contamination. Minimize the potential of microbial contamination from water used with fresh fruits and vegetables.
Principle 5	Practices using animal manure or municipal biosolid wastes should be managed closely to minimize the potential for microbial contamination of fresh produce.
Principle 6	Worker hygiene and sanitation practices during production, harvesting, sorting, packing, and transport play a critical role in minimizing the potential for microbial contamination of fresh produce.
Principle 7	Follow all applicable local, state, and federal laws and regulations, or corresponding or similar laws, regulations or standards for operators outside the U.S., for agricultural practices.
Principle 8	Accountability at all levels of the agricultural environment (farm, packing facility, distribution center, and transport operation) is important to a successful food safety program. There must be qualified personnel and effective monitoring to ensure that all elements of the program function correctly and to help track produce back through the distribution channels to the producer.

The guide concludes that analyzing the risk of microbial contamination should include a review of five major areas of concern:

1. water quality
2. manure/municipal biosolids

3. worker hygiene
4. field, facility, and transport sanitation
5. traceback

Growers, packers, and shippers should consider the variety of physical characteristics of produce and practices that affect the potential sources of microbial contamination associated with their operation and decide which combination of good agricultural and management practices are most cost effective for them. Once good agricultural and manufacturing practices are in place, it is important that the operator ensure that the process is working correctly.

Operators should follow up with supervisors to be sure that regular monitoring takes place, equipment is working, and good agricultural and management practices are being followed. Without accountability to ensure that the process is working, the best attempts to minimize microbial food safety hazards in fresh fruits and vegetables are subject to failure. This opens the door to HACCP in fruit and vegetable production operations. HACCP is a very important aspect of GMP. The industry is urged to take a proactive role to minimize those microbial hazards over which they have control. Operators are encouraged to utilize the guide to evaluate their own operations and assess site-specific hazards so they can develop and implement reasonable and cost-effective agricultural and management practices to minimize microbial food safety hazards.

Identifying and supporting research priorities designed to help fill gaps in food safety knowledge is another focus of the food safety initiative. Over the long term, research and risk assessment on fresh produce will be incorporated in the multi-year food safety initiative research planning process.[5]

41.3 HACCP

The HACCP is a systematic approach to the identification, assessment, and control of hazards in a food operation by identifying problems before they occur and establishing measures for their control in those stages critical for safety.[6] HACCP has been extensively discussed elsewhere, and its seven steps as set out by the Codex Alimentarius Commission[7] in 1991 have evolved and become part of the common terminology of food industry and government.

1. hazard analysis
2. determination of critical control points (CCPs) in the process
3. specification of criteria
4. implementation of CCP monitoring systems
5. corrective actions
6. verification
7. documentation

The first step, or hazard analysis, is to identify potential hazards associated with food production at all stages to the point of consumption. It consists of hazard identification, assessment of the likelihood of occurrence of hazards, and identification of preventive measures for their control. In relation to the HACCP concept, a hazard had been identified as "any aspect of the food production chain that is unacceptable because it is a potential cause of food safety problems." More recently, WHO[8] defined the HACCP-hazard as "a biological, chemical, or physical agent with the potential to cause an adverse health effect when present at an unacceptable level." This definition complies with the actual HACCP approach of becoming a quantitative system.

Microbial hazards (occurrence of microorganisms that have the potential to cause illness or injury) are almost always present, and acceptable levels should be defined. Recognition of this increased regulatory responsibility has been a major factor in the endorsement of the HACCP system for food control. HACCP has been endorsed by the NAS of United States, the Codex Alimentarius Commission, and the National Advisory Committee on Microbiological Criteria for Foods. The Food and Drug Administration has adopted HACCP and intends to eventually apply it to much of the U.S. food supply. Many of its principles are already in place in the FDA-regulated low-acid canned food industry. The FDA established HACCP for the seafood industry in a 1995 final rule that took effect in December 1997. Also, the FDA has incorporated HACCP into its Food Code, a document that gives guidance to and serves as model legislation for state and territorial agencies that license and inspect food service establishments, retail food stores, and food vending operations in the United States. A number of U.S. food companies already use the system in their manufacturing processes, and it is in use in other countries as well. In addition, the U.S. Department of Agriculture has established HACCP for meat and poultry processing plants. Most of these establishments were required to start using HACCP by January 1999; very small plants were given until Jan. 25, 2000.[9]

To help determine the degree to which such regulations would be feasible, the FDA is considering whether to propose regulations, under existing legislative authority, that would require food manufacturers to provide food safety controls based upon the seven principles of HACCP. The FDA has concluded that its determination on whether to adopt HACCP regulations will be facilitated by the operation of an HACCP pilot program. The FDA is conducting a pilot program with a number of volunteers from the food manufacturing industry who are using HACCP to control food hazards. The pilot program is intended to provide information that food science professionals can use in determining whether HACCP should be expanded beyond seafood as a food safety regulatory program. The information is being gathered from firms that produce several different types of food products and from firms that control a variety of potential food hazards. The programs have involved cheese, frozen dough, breakfast cereals, salad dressing, fresh and pasteurized juices, bread, flour, and other products. The pilot program is also intended to provide the FDA with additional experience in working with the audit-type inspection necessary for verifying an HACCP program.[10]

HACCP is a food safety assurance program that shares common characteristics with recognized quality assurance (QA) programs such as total quality management (TQM) and ISO 9000. These common characteristics include

1. identifying the safety attributes of the product the customer requires
2. establishing points of control
3. monitoring these control points
4. verifying that the program is being implemented as designed

The firms with QA and/or sanitation control programs in place have incorporated HACCP into these programs. They have reported that HACCP was not a major departure from what was done before and required refinements more than substantial changes. Confusion can develop over the integration of HACCP with preexisting QA systems because, on a practical basis, it may be difficult to separate controls of safety hazards from controls that deal with quality and productivity factors. A firm's system of controls to ensure safety is, in many respects, identical to systems that ensure quality and efficient production. As a result, most firms have found that it is a challenge to separate safety hazards that were to be controlled at CCPs from quality factors that were to be controlled at other types of control points.[10] Application of the principles of HACCP has become mandatory for food companies in the European Community.[11]

Vasconcellos[12] conducted an in-depth revision of regulatory and safety aspects of minimally processed fruits and vegetables. In addition to participating in working groups of the National Advisory Committee on Microbiological Criteria for Foods, the industry has been conducting other significant efforts. The National Food Processors Association (NFPA) recognizes the concerns about minimally processed foods whose shelf lives are extended by refrigeration in combination with other preservative procedures or techniques, and it has suggested certain factors to consider in establishing good manufacturing practices.[13] The committee recommended the use of 4.4°C in place of 7.2°C as the upper limit for refrigerated products. While it is recognized that 4.4°C may be unrealistic for practical applications, the committee endorsed the concept as a desirable goal. The "Manufacturing Guidelines for the Production, Distribution and Handling of Refrigerated Foods"[14] represents a significant industry effort. These guidelines represent a responsible approach to food safety when product and process development, predistribution, retailing, and food services rely on HACCP principles clearly emphasized. In addition to HACCP, the industry is initiating other steps to ensure the safe and economical production, distribution, and storage of minimally processed foods, including consumer awareness programs. The International Fresh-Cut Produce Association suggested HACCP guidelines for fresh produce, which are shown in Table 41.2,[14,15] as well as the corresponding HACCP principle.

In April 1998, the FDA proposed requiring HACCP controls for fruit and vegetable juices in two regulations designed to improve the safety of both fresh and processed fruit and vegetable juices. Once the HACCP regulation is finalized, implementation would allow one year for large manufacturers, two years for small businesses, and three years for very small businesses. The manufacturing changes to be defined by the regulations would increase the protection of consumers from illness-causing microbes and other contaminants in juices. The proposal came after the number of consumer illnesses associated with juice products rose during the past

TABLE 41.2
Suggested Guidelines[14] for an HACCP Program for Fresh Produce and Its Correspondence with the Seven HACCP Principles

Suggested Guidelines for an HACCP Program for Fresh Produce	HACCP Principles
Designate a person responsible for the HACCP plan and members of an HACCP "team" for the food facility and target products.	
List the target food products, describe each product, list raw materials and ingredients, and prepare a preliminary flow diagram.	
Document the hazard analysis associated with the target products, their ingredients, and the hazards of the entire product manufacturing chain.	Principle 1: Conduction of a hazard analysis
Develop individual flow diagrams for each product that document the location and type of CCPs for identified hazards.	Principle 2: Determination of critical control points (CCPs) in the process
Document descriptions of each CCP, including the type of hazard, procedures, or processes to control the hazard and definition of the critical limits or tolerances that apply to each CCP.	Principle 2 and Principle 3: Specification of criteria (establishment of target levels—critical limits and tolerances for preventive measures associated with each identified CCP)
Document monitoring procedures for the CCPs and critical limits, monitoring frequency and the person(s) responsible for specific monitoring activities.	Principle 4: Implementation of CCP monitoring systems
Document deviation procedures for each CCP that specify the actions to be taken if monitoring determines that a CCP is out of control. Actions must include the safe disposition of the affected food product and correction procedures for the conditions that caused the situation.	Principle 5: Corrective action (establish corrective actions to be taken when monitoring indicates that a particular CCP is not under control)
Develop and document record-keeping systems for the HACCP program using Principle 6. Designate trained and responsible company personnel for management and sign-off of records.	Principle 6. Verification (establish procedures for verification that HACCP system is working correctly)
Develop and document verification procedures based on Principle 7. Designate responsible company personnel to conduct verification of compliance with the HACCP program on a scheduled basis. Designate responsible persons to conduct verification who are not generally involved in the HACCP functions (such as corporate or division quality assurance personnel).	Principle 7. Documentation (establish documentation concerning all procedures)

TABLE 41.2 (CONTINUED)
Suggested Guidelines[14] for an HACCP Program for Fresh Produce and Its
Correspondence with the Seven HACCP Principles

Suggested Guidelines for an HACCP Program for Fresh Produce	HACCP Principles
Document procedures for the revision and updating of the HACCP program anytime there is a change of ingredients, products, manufacturing conditions, evidence of new potential or actual hazard risks, or any other reason that may influence the safety of the product(s). Otherwise, specify scheduled revision and updating.	
Consult with the appropriate regulatory agency or agencies regarding company intention to develop an HACCP program and involve the agency or agencies in the development and approval of the HACCP program.	

several years, including *Salmonella* infection from orange juice and a 1996 *E. coli* O157:H7 outbreak associated with apple juice products.[15]

The first proposed regulation would require processors of packaged fruit and vegetable juices, both domestic and foreign, to implement hazard control programs at their plants to prevent microbiological, chemical, and physical contamination of their products. Part of this proposal would require manufacturers of unpasteurized juices to adjust their processes to achieve a 100,000-fold reduction in the numbers of harmful microbes in their finished products compared with levels that may be present in untreated juice. The second proposed regulation would require warning labels on all packaged juice products that have not been pasteurized or otherwise treated to eliminate harmful microbes to ensure that consumers are adequately informed of the risks from consumption, especially by children, the elderly, and persons with weakened immune systems. Warning labels would not be required on juice products processed under HACCP programs or that are treated to reduce harmful microbes by 100,000-fold (i.e., processed to achieve a 5-log reduction in the most resistant pathogen of public health concern).[15]

The new HACCP regulation would apply to juice manufacturers who sell and distribute packaged products to consumers in the United States. As part of their HACCP plans, fresh juice processors would be required to have processes in place to reduce the number of harmful microbes to the same level achievable by pasteurization. So, the FDA did not propose a specific intervention technology (e.g., pasteurization) to be used to meet the performance standard; instead, the agency proposed a flexible 5-log reduction standard that theoretically could be met for some fruits (e.g., citrus) through cumulative steps. Juice processors would be free to employ microbial reduction methods other than pasteurization, including washing, scrubbing, antimicrobial solutions, alternative technologies, or a combination of techniques. In the April 1998 proposal, the FDA stated that steps such as culling, washing, brushing, and sanitizing, followed by extraction processes that minimize contact with the peel, could be used cumulatively to attain the 5-log reduction, as long as processors could validate bacterial reduction under their HACCP systems.[15,16]

Prior to the 1998 rulings, the FDA had sought the advice of the National Advisory Committee on Microbiological Criteria for Foods (NACMCF), which recommended the proposed 5-log performance standard to ensure the safety of juice. At the time of the proposal, the FDA believed that such cumulative steps could be appropriate for juices produced from citrus fruit, because pathogens were not reasonably likely to be found in the interior of such fruit. The agency also stated that the acidic nature of citrus fruits could further inactivate any pathogens that may be present in the fruit. However, comments on the proposed rule have questioned the assumption that pathogens are not likely to be found in the interior of citrus fruit. In consequence, the FDA has conducted a search of the relevant published literature and has undertaken research, results of which suggest that pathogens indeed could be internalized into citrus fruit and survive under certain conditions once inside the fruit. A pathogen that has become internalized within a fruit, including citrus fruits, or vegetable must be able to survive in the product until it reaches the consumer to become a public health hazard.[16] Survival of pathogens, both plant and human, has been demonstrated in both produce and juice. In laboratory studies, human pathogens have been found in or on tomatoes, cantaloupe, watermelon, honeydew melon, and apples. There appears to be no published information on human pathogens in citrus fruits; however, the presence of other bacteria internalized in citrus has been noted. Numerous studies have shown that human pathogens can survive in both apple and orange juice, despite their natural acidity.[16]

HACCP is not yet mandatory for the fresh-cut produce industry, although the Food and Drug Administration (FDA) published the guidance document for producers discussed above.[5] In 1992, Stier[17] published an HACCP model developed for shredded lettuce packaged in gas-permeable bags, based on an actual food operation used by California lettuce packers. By the time of the publication, the model was still under consideration. In the model, there are CCPs for chemical hazards (pesticides, chemicals, fertilizers), and physical hazards (metals and foreign objects), but the majority are of microbial nature. Lettuce is a product extremely sensitive to abuse conditions, so controls to ensure safety will also help maintain quality. CCPs are identified at the coring operation, trimming and sorting, and choppers, to control potentially harmful organisms and physical hazards. The keys to the former are proper equipment design and cleaning, which includes rinsing during operation. There are other CCPs related to maintaining chlorine levels in water systems, which are monitored continuously and adjusted as needed. Basket loading and centrifuge operations are timed and considered as CCPs, and, if the established times are exceeded, the product should be discarded. The system also includes a general sanitation CCP, which might be considered as a good manufacturing practice and not a CCP; since it includes education of cleaning crews, compliance with cleaning protocols, maintenance staff, etc., the operation cannot be started until established cleaning and sanitizing protocols have been completely reviewed by management. There are also two CCPs related to packaging and coding. Final CCPs deal with checking of package integrity and with maintenance and monitoring of refrigerated van temperature and, at the retail level, with possible use of time/temperature indicators on the package. As can be seen, many of these CCPs have a strong qualitative component.

The model described by Stier is a generic model for many fresh fruits and vegetables. Most large produce companies are using some form of HACCP, but most have been developed by their own staff. Most companies have only adopted two critical control points: chlorine concentration in wash and flume waters, and metal detection after bagging. Other important control points (temperature control, film type, etc.) are covered under SOPs.[18]

41.4 RISK ASSESSMENT (RA)

Risk assessment, by the 1995s definition of Codex Alimentarius,[19] is the estimation of the severity and likelihood of harm or damage from exposure to hazardous agents or situations. In 1996, another definition by Codex[20] states that risk assessment is a scientifically based process consisting of the following steps: hazard identification, hazard characterization, exposure assessment, and risk characterization.

Risk assessment has been utilized for managing several types of risks including radiation control, chemical diseases, contamination of the environment and foods, water quality, and cancer prevention. Application to microbiological food safety is a recent focus, especially for specific food safety issues like the hazard of *Listeria monocytogens* in milk, *E. coli* O157:H7 in ground beef, and *Salmonella* in eggs. This latter is the first quantitative farm-to-table microbial risk assessment and is expected to serve as a prototype for future risk assessments.[1]

Risk analysis can be either qualitative or quantitative. Qualitative risk analysis is based on data that (even being an inadequate basis for numerical risk estimation, when they are conditioned by prior expert knowledge and indication of attendant uncertainties) permits risk ranking or separation into descriptive categories of risk.[21] On the other hand, quantitative risk analysis (QRA) should be a mean of making consistent, objective, and reliable assessment of risks, relying on numerical expression of the risk by quantifying the probability of occurrence of an adverse health effect. QRA is based on quantitative data and models and consists of six activities:

1. hazard identification
2. exposure assessment
3. dose-response assessment
4. risk characterization
5. risk management
6. risk communication

Steps (1) through (4) are termed *risk assessment*.[11,22] Table 41.3 presents definitions of risk assessment with examples related to fruit and vegetables.

TABLE 41.3
Steps of Risk Assessment (RA) and Some Examples Related to Fruits and Vegetables

Steps of RA and definitions	Examples in fruits and vegetables
1. *Hazard identification:* Data from consumer complaints, results of epidemiological studies, microbiological data, predictive models, etc.	Qualitative acknowledgement in this step, for instance: • *Clostridium botulinum* may cause botulism, if present, and the toxin is formed in modified atmosphere packaged cabbage. • *Shigella sonnei* may be present in shredded lettuce and *Salmonella chester* may contaminate the surface of melons and cause shigellosis and salmonellosis, respectively.
2. *Exposure assessment:* Determines the extent of human exposure before or after application of regulatory or voluntary controls. Data can be obtained from product surveillance, storage testing, challenge tests, and mathematical models that predict the likely number of microorganisms present in a food at the time of consumption.	What is the distribution of *Salmonella* spp. in fruits like melons, and how is it affected by the quality of the irrigation water, the growing zone, and the season, etc.?
3. *Dose-response assessment:* Defines the relationship between the magnitude of exposure and the probability of occurrence of the spectrum of possible health effects. The quantitative estimation of risk at the time of consumption is made, based on the information obtained from dose-response relationships determined in human volunteer and animal model studies as well as from epidemiological analysis of foodborne diseases.	What is the likelihood of becoming ill if ten *Salmonella o Shigella* cells are consumed, and how severe will the illness be? (This is one of the weakest components of risk analysis. Because of ethical considerations, it is unlikely that adequate human dose-response data will become available for highly infectious agents, and animal models must be carefully reviewed for applicability to humans because of the inherent variability to host/microorganism interaction).
4. *Risk characterization:* Intended to integrate the steps described above into a quantitative estimate (probability) of adverse effects likely to occur in a given population, and may identify additional economic and social impacts of human risk.	No example available.

This risk analysis approach is helpful way to face controversial and complex situations involving public health. Risk assessment helps to organize scientific information and to characterize the nature and likelihood of harm to the public. It helps to identify and define uncertainties, identify data gaps, and identify what assumptions must be made. Thus, risk assessment provides the decision maker with a level of confidence in that decision.[1]

HACCP is often used qualitatively and subjectively. A quantitative approach to HACCP should provide a better way to set proper criteria for critical process steps (critical control points, CCPs), to execute control measures, and to optimize process. The quantitative approach can be created by the implementation of quantitative risk analysis (QRA) in existing HACCP systems.[3,11,23,24] QRA is based on quantitative data and models. Hazard identification is the first activity in both QRA and HACCP. Within HACCP plans, risk assessment involves two basic components: the identification and the assessment of hazards. The former may consist of a literature review of likely pathogens, epidemiological data, surveys of the microbial composition of raw materials, etc. Once identification has been performed, it is necessary to determine which hazards can be present in raw materials and at the point of consumption. Bearing in mind that all steps from production through consumption will affect the food microflora, an assessment should be made of the impact of intrinsic, extrinsic, and other preservation factors used with regard to the growth and survival of identified hazards and, more importantly, which associated risks are acceptable. Although a hazard may be present, the risk of illness related to that hazard may not be great.[25] Therefore, the new task of the food industry is to maintain the level of risk at a minimum that is practically and technologically feasible.[26]

A hazard is always potential rather than actual and may be associated with any agent in food, and even a property of a food may have an adverse effect on human health; risk is a statistical concept directly related to a hazard. If the risk is zero (it will never be zero), then no hazards will arise (hazards are always possible, and small risks must be accepted). Thus, for safety purposes, the use of GMP will reduce the risks but, by applying the HACCP system, the risks should be reduced in a quantifiable manner. HACCP in terms of RA should focus on those operations (practices, procedures, etc.) that can be managed so that a desired level of safety (with an acceptable risk) can be attained. HACCP has a direct influence on the safety of the product, and the benefits should be quantifiable. Growth of microorganisms and recontamination of the product should no longer be considered as hazards but as risk-increasing events. Thus, processing conditions should result in an acceptably safe product for which the risk of occurrence of adverse effects can be calculated in advance.[4]

Terms and concepts originating from quantitative risk analysis are now being introduced, although there has been little experience applying these terms in practice, and available information is insufficient at present for quantitative risk analysis to be applied in controlling food safety. This approach, however, is not yet well developed in relation to the microbiological safety of food and has had little impact on foodborne hazards. The reason for this is that, even if the goal is to be quantitative in the assessment of risk, this is often hindered by the lack of available data on microbiological food safety risks.[4,22] This is particularly true for fruit and vegetable products.

As stated by Beuchat,[27] a quantitative microbiological risk assessment of human infections and intoxications that can be linked to the consumption of contaminated raw fruits, vegetables, and plant materials should be undertaken. The development of a highly efficient, international epidemiological surveillance system for better understanding the role of raw fruits and vegetables as vehicles for disease is critical.

Information generated by such a system would be valuable in establishing more meaningful practices and guidelines for preventing contamination and for decontamination.

The importance of risk assessment has been recognized by the Food and Drug Administration's Center for Food Safety and Applied Nutrition (CFSAN) when identifying and supporting research as focus of the Food Safety Initiative (FSI), for its mission of conducting science-based programs to effectively prevent or control the occurrence of microbial pathogens and their toxic metabolites in food, and to respond efficiently to foodborne disease outbreaks. Scientific knowledge will provide a systematic approach (i.e., risk assessment) to evaluate commodities and their associated pathogens and toxins, from initial production through processing and consumption.[28] Risk assessment further provides a means for elucidating major areas of uncertainty in scientific knowledge (or outright gaps), thus identifying priority research that can lead to the development of cost-effective means for improving public health. Integration of research and risk assessment are fundamental to the development of prudent guidelines and regulations.

Identified in the Food Safety Initiative (FSI) and affirmed under the Produce and Imported Foods Safety Initiative (PIFSI), there are three broad areas in risk assessment development where significant knowledge gaps or the lack of appropriate scientific data, methods, or models require a concerted interagency research effort. They are presented in Table 41.4, along with some priority research areas identified by CFSAN, related to produce and fruit and vegetable products.[28]

Risk assessment concerns overall product safety and is applied to analysis of the food product as presented to the consumer (analysis at end point), while HACCP enhances overall product safety by assuring day-to-day process control and may be applied at any point in the processing/handling chain.[24] Confusion often arises between risk assessment and HACCP on the point of hazard identification being the first activity in both QRA and HACCP: recognition of microorganisms of concern. As mentioned above, Notermans et al.[3,29] and Buchanan[24] have suggested the use of the four steps of risk assessment for specifying the microbiological criteria of HACCP systems (in hazard analysis and in the setting of critical limits); that is, to use QRA as a part of HACCP. Foegeding[22] broadens this approach by considering both risk assessment and HACCP as part of the risk analysis, with HACCP representing one management strategy. On the other hand, the Codex Alimentarius Commission has focused its attention on risk analysis in the elaboration of standards and guidelines for the international trade in food for all classes of foodborne hazards.[30]

There are some weaknesses in the risk assessment process as well. Risk assessment has its roots in toxicology and carcinogenicity studies, and its application to other disciplines poses significant challenges. For food safety, one challenge relates to the fact that, unlike chemical, environmental, or toxicological contaminants, bacteria can multiply and produce toxins as conditions change while food moves through the farm-to-table continuum. In addition to the difficulties in applying risk assessment to pathogens, risk assessment is also subject to two types of uncertainties: those related to data, and those associated with any assumptions that are required when directly applicable data are not available.[1]

TABLE 41.4
**Areas in Risk Assessment Development and Some Research Activities
Incorporated in the Multi-year Food Safety Initiative Research Planning
Process of CFSAN for Produce and Fruit and Vegetable Products in the
Context of FSI**

Risk Assessment
- Development and validation of microbial exposure models, based on probabilistic methodology, for use in risk assessment
- Development validation of dose-response models for use in risk assessment
- Identify improved modeling techniques for quantitatively describing microbial risks

Produce, General
- Analyzing produce samples for pathogenic microorganisms more effectively/cost-efficiently
- Guidance for industry on "good agricultural practices" and "good manufacturing practices" to help assure the safety of fresh/fresh-cut produce
- Identify/evaluate the strategies/technologies to prevent, reduce, or effectively eliminate pathogenic microorganisms on produce; incidence/prevalence of foodborne pathogens on fruits/vegetables associated with outbreaks of illness (e.g., berries, cantaloupes, lettuce)
- Infiltration into plant tissues

Sprouts
- Rapid techniques to reliably determine the level of viable *Cryptosporidium parvum* oocysts on produce
- Role of hand sorting in cross-contamination
- Guidance on safe sprout production
- Evaluate the plausibility/safety of proposed approaches for the prevention, reduction, or effective elimination of pathogenic bacteria on seeds or sprouts; assess sprout contamination with foodborne pathogens

Juices
- Simple, nonthermal treatments for reduction/elimination of pathogens in fruit and vegetable juices amenable for use by small processors
- Study nonpathogenic microorganisms (e.g., *Lactobacillus fermentans*, *Enterobacter aerogenes*) with similar behavior to pathogens (e.g., *Salmonella hartford* or *E. coli* 0157:H7) for validating the effectiveness of antimicrobial treatments/processes on the reduction of pathogens during commercial juice production

Surrogates
- Fill gaps in the safety data with the most promising surrogate organisms. Reduction achieved through antimicrobial treatments (e.g., pulsed light, ultraviolet light, pulsed electric fields, ozone, etc.)
- Relevant treatment parameters (e.g., temperature, pH, radiation dose, field strength, concentrations, etc.,) needed to achieve specified reductions (e.g., 1 log, 3 logs)
- Develop modeling techniques for assessing human exposure to foodborne pathogens and contaminants *(Listeria monocytogenes, Vibrio parahaemolyticus,* and methylmercury)

41.5   HAZARD IDENTIFICATION

The importance of hazard identification as the first common step of HACCP and QRA is recognized elsewhere. It is considered a step of both risk assessment and

HACCP, where there exists a lot of comprehensive knowledge. General information exits about foodborne microorganisms that may cause disease and the severity of the disease, even though there are still obscure areas in emerging pathogens as well as on viral pathogens.[22]

Notermans et al.[3] present an approach to identify potential hazardous bacteria in a given situation, based on a list of all bacteria that are known to cause foodborne disease. This list is used for determining whether or not the microorganisms are likely to be present in the raw materials used, deleting those organisms that have never been found. Of the remaining organisms, it is to be established whether they are destroyed by processing (if so, they will be removed) and if re-contamination is likely to occur with a pathogen (if so, it must be included in the list). The question that ensues is whether the listed organisms have ever caused a foodborne disease involving either an identical or related product. If this is not the case, the organism is deleted. All remaining organisms are discriminated into infectious or toxinogenic. All infectious organisms are regarded as potentially hazardous, while only those toxinogenically capable of growing are considered potentially hazardous. A more precise evaluation of the hazards is to be made during the identification of the CCPs, the setting of control criteria at each one of them, and during the step of verification.

Van Gerwen et al.[11] implemented a computer-programmed hazard identification procedure based on the general approach of Notermans et al.,[3] described above, differing in its stepwise identification of important hazards and its interactive character. It is the first step of a procedure for quantitative risk assessments to be completely developed as a computer-aided system. Van Gerwen[31] has recently applied the hazard identification procedure to vegetables and fruits. The following organisms were identified as hazards in vegetables: *B. cereus, E. coli, L. monocytogenes, Salmonella* spp., *Shigella* spp., *Staphylococcus* spp. and *Y. enterocolitica*. Particularly for fruit, the authors agreed on the limitations of the hazard identification procedure, since apparently there are very little data in the food database for fruits or fruits and nuts.

A major concern with minimally processed fruits and vegetables is the survival and growth of several pathogenic organisms that can be traced to raw produce, plant workers, and the processing environment. Abdul-Raouf et al.[32] studied the effect of storage temperature, use of modified atmosphere, and storage time on the survival and/or death of *E. coli* O157:H7 in vegetable salads prepared with cucumber, shredded carrots, and lettuce. *E. coli* 0157:H7 decreased rapidly and remained only a short time in salads stored at 5°C, while salads stored at 12 and 21°C showed an increase in the population of the pathogen. The atmospheric composition did not affect the survival of *E. coli* O157:H7 in the conditions of their experiment. The rapid growth of psychrophiles and mesophilic microorganisms did not have an important effect on the growth rate of the pathogen.

In 1993, Zhao et al,[33] showed that unpasteurized apple cider was an appropriated substrate for growing *E. coli* O157:H7. The microorganism was able to survive 10 to 31 days at 8°C. The survival and growth of *E. coli* O157:H7 in cantaloupe and watermelon has been studied, and *E. coli* O157:H7 was observed in both types of fruit, reaching levels of 6.8 $\log_{10}$ CFU/g in melon and 8.51 $\log_{10}$ CFU/g in watermelon stored at 25°C, but the number of viable cells did not suffer changes at 5°C until 34 h. The microorganism was able to survive in the peel of both fruits when they were stored at 25 or 5°C.

Diaz and Hotchkiss[34] compared the microbiological spoilage with survival of *E. coli* O157:H7 in modified atmosphere stored shredded lettuce and observed that growth rates of the aerobic plate count and *E. coli* O157:H7 were higher in air at 22°C. Temperature and level of CO_2 had no significant effect on the growth of *E. coli* O157:H7. The extended shelf life provided by the modified atmosphere allowed *E. coli* O157:H7 to grow to higher numbers compared with air-held shredded lettuce. Similar results have been reported by other authors.[35,36] Challenge tests in freshly peeled Hamlim[37] orange inoculated with selected pathogenic bacteria (*Salmonella* spp., *E. coli* O157:H7, *L. monocytogenes,* and *Staphylococcus aureus*) to study the survival and growth of these microorganisms showed that *E. coli* O157:H7 counts remained constant throughout storage at refrigeration temperatures (4 and 8°C). Growth was observed with all tested pathogens only when stored at abuse temperature (24°C). Itoh et al.[38] emphasized the importance of using seeds free from *E. coli* O157:H7 in the production of radish sprouts because, according to their results, they found *E. coli* O157:H7 not only in the outer surface but also in the inner tissues and stomata of cotyledons of radish sprouts grown from seed experimentally contaminated with the pathogen. The tolerance of *E. coli* O157:H7 to acid pH has been widely studied. The pH can play an important role on the survival of *E. coli* O157:H7 in vegetables. Weagant et al.,[39] working with salads with mayonnaise dressing and acidified with acetic, citric, and lactic acid, and stored at 5, 21, or 30°C for 72 h, found that the storage temperature played an important role in the effectiveness of the acids; at low temperature (5°C), all acids had the same effectiveness, while at 5°C, the population of *E. coli* O157:H7 was significantly reduced during the first four hours, and the mortality of the microorganisms was higher when the pH of the media decreased. The tolerance of some strains of *E. coli* O157:H7 has been studied in yogurt[40] apple cider,[41] and mayonnaise.[39,42,43]

Recently, it has been shown that *E. coli* O157:H7 was able to survive more than 35 days in salads with mayonnaise-based dressing, when stored at 5°C.[40] The survival and growth characteristics of acid-adapted, acid-shocked, and control cells of *E. coli* O157:H7 inoculated into tryptic soy broth acidified with acetic acid and lactic acid in apple cider and orange juice indicates that the pathogen shows an extraordinary tolerance to the low pH of apple cider and orange juice held at 5 or 25°C for up to 42 days.[44]

L. monocytogenes has been isolated from intact vegetables such as asparagus,[45] tomatoes,[46] cabbage, cucumbers, lettuce and lettuce juice, potatoes, radishes, mushrooms,[47,48] bean sprouts and leafy vegetables,[48] raw celery,[49] and other unprocessed vegetables.[50,51] *L. monocytogenes* can survive and grow at refrigeration temperatures on many raw or processed vegetables such as ready-to-eat fresh salad vegetables including cabbage, celery, raisin, onion, and carrot salad; lettuce, cucumber, radish, fennel watercress, and leek salad,[52] asparagus, broccoli and cauliflowers,[53] lettuce, lettuce juice, and minimally processed lettuce,[53-56] butterhead lettuce salad; broad-leaved endive and curly-leaved endive,[58,58] minimally processed fresh endive,[58–60] freshly peeled Hamlin orange,[37] and vacuum-packaged pre-peeled potatoes.[61]

Odumer et al.[62] assessed the microbiological quality of ready-to-use (RTU) vegetables for health care food service, including chopped lettuce, salad mix, carrots sticks, cauliflower florets, and sliced green peppers before and after processing and after seven days storage in hospital coolers. The authors also determined the effects of storage temperature and time on the microbial population of RTU vegetables. Microbial profiles were obtained 24 h after processing and on days 4, 7, and 11 after storage at 4° and 10°C to simulate temperature abuse. *L. monocytogenes* was isolated from 6 of the 8 vegetable types tested, representing 2.8% of samples stored at 4°C and 13 of 129 samples at 10°C. Recommendations regarding processing, distribution, and storage of such products is presented by the authors. Temperature abuse resulted in a significantly higher incidence of *L. monocytogenes* in RTU vegetables. Temperature is a critical control point for the maintenance of the quality and safety of these products. Under optimal storage conditions, these results do not suggest an increased risk of *L. monocytogenes* isolation from RTU vegetables, but the need for continued surveillance for this pathogen and strict monitoring of critical control points during processing and handling is stressed.

Aeromonas hydrophila demonstrates characteristics of concern in vegetables. It is a psychrotrophic and facultative anaerobe.[56,63] *A. hydrophila* is considered to be a ubiquitous microorganism, and it has been isolated from many sources. Callister and Agger[64] evaluated the incidence of *Aeromonas* spp. in vegetables such as parsley, spinach, celery, alfalfa sprouts, lettuce, broccoli, cauliflower, escarole, and endive. Levels of *Aeromonas* spp. ranged from 1.00×10^2 to 2.30×10^4 CFU/g at the moment of purchase. Levels of *Aeromonas* spp. reached more than 10^5 CFU/g after 14 days of storage at 5°C. *A. hydrophila* has been found in fresh asparagus, broccoli, and cauliflower[65] and in commercial vegetable salads.[66] *Aeromonas* spp. were also recovered from commercial mixed vegetable salads.[67] Incidence of *Aeromonas* spp. in vegetable products was studied in Venezuela by Díaz et al.[68] The products analyzed were lettuce, watercress, cabbage, escarole, and parsley. *Aeromonas* spp. was detected in all products. The highest incidence of *Aeromonas* spp. was detected in watercress followed by parsley, escarole, lettuce, and cabbage. At the moment of purchase, levels of *Aeromonas* spp. ranged from $<1.00 \times 10^2$ CFU/g and <3.00 MPN/g to $2.50 \times 1\,0^6$ CFU/g and 4.60×10^6 MPN/g. After 7 days of storage at 3°C, the levels of *Aeromonas* spp. increased 10 to 100 times, with 6.75×10^7 CFU/g and 4.60×10^7 MPN/g as the lowest levels detected. The main species identified were *A. hydrophila* (20.5%), *A. caviae* (43.6%), and *A. sobria* (2.6%).

Nottermans et al.[29] proposed an approach for setting criteria at critical control points (CCPs) for bacterial hazards associated with foodborne disease and addressed such methods as tools to calculate or predict the numbers of organisms expected to be present in final food products, leading to decision made on the basis of levels of acceptability. Thus, human exposure to potentially hazardous pathogens can be evaluated to provide a quantitative risk assessment. In relation to quantitative risk analysis, CCPs can be considered as operations (e.g., steps, processes) where risks can be reduced through control of procedures, practices, etc. CCPs will be meaningful if they can be managed in a way that a risk is reduced and the reduction can be quantified. In this context, the authors present an approach to establish the identity of quantitative CCPs. A list of operations and variables that exert a quantitative control on potential hazards should be made so that parameters important for a particular product and its production process are defined and will lead to decisions of whether they can be utilized to reduce or stabilize a potential hazard. In case it is utilized, the next question will be whether the effect is nullified by a subsequent process or procedure. If this is not the case, it is necessary to decide whether the operation controls the hazard in a quantifiable manner. Control does not necessarily mean a decrease in the number of hazardous organisms. In some cases, stabilization of numbers, as in prevention of growth, may be sufficient.[29]

Potential hazards in minimal processing of fruits and vegetables can be controlled by application of some operations or variables that exert quantitative effects. Some examples are shown in Table 41.5.

41.6 PREDICTIVE MICROBIOLOGY

The use of mathematical models to describe the growth, survival, and inactivation responses of foodborne microorganisms under specific environmental conditions have given rise to one of the most rapidly advancing subspecialties in food microbiology: predictive microbiology. Microorganisms are cultured under a variety of intrinsic factors such as a_w, pH, etc., and extrinsic factors such as temperature and gaseous atmosphere; their response is measured, and the resulting data are fitted to a mathematical equation. A database can be generated, and the growth response can be predicted, under conditions not specifically tested in the laboratory.[81] Predictive microbiology can be used to evaluate the effects of composition of the product, processing, storage conditions, etc., on the final contamination of the product at the time of consumption.

Current and newly obtained knowledge, facts, and expert opinion, along with mathematical models, can be used to build computerized systems related to food safety and quality issues. As stated by Foegeding,[22] useful information for application in all steps of quantitative risk assessment can be generated by predictive microbiology. Research can be focused to fill gaps in the knowledge, especially in an area like risk assessment. Predictive microbiology should help to improve the estimates and allow quantification of food safety risks. It is recognized, for instance, that one of the difficulties associated with microbial risk assessment is determining the number of microorganisms in a food at a given time and estimating the exposure of an individual to the microorganism. Under specified conditions, predictive microbiology can be used for such purpose, providing estimates of potentially hazardous microorganisms in a product at various stages during distribution. A decision can then be made on whether the levels are acceptable, and what action should be taken in order to minimize the risk to the consumer.[81]

In 1998, CFSAN awarded risk assessment grants for research projects including establishment of a correlative dose-response model for human cryptosporidiosis and development of a risk assessment dose-response model for foodborne *Listeria mono-*

TABLE 41.5
Examples of Effects Observed by Application of Some Operations or Variables that Control Potential Hazards Quantitatively in Minimal Processing of Fruits and Vegetables

- Dipping lettuce in water containing 300 ppm chlorine reduced total microbial counts by about 3 $\log_{10}$ CFU/g on lettuce; no effect on carrots or red cabbage.[69]
- Dipping brussels sprouts in to a 200 ppm chlorine solution for 10 s decreased the number of viable *Listeria monocytogenes* cells ($\log_{10}$ 6.1 CFU/g) by about 2 $\log_{10}$.[70]
- Dipping shredded lettuce and cabbage in chlorine (200 ppm for 10 min) allowed a maximum $\log_{10}$ reduction of 1.3–1.7 $\log_{10}$ CFU/g for lettuce and 0.9–1.2 $\log_{10}$ CFU/g for cabbage from initial populations of *L. monocytogenes* of $\log_{10}$ 5.4 to 5.7 CFU/g; reductions were greater when treatment was at 22°C rather than at 4°C.[70]
- Dipping tomatoes for two minutes in a solution containing 60 or 110 ppm chlorine, respectively, allowed significant reductions of populations (ca. 05 to 1.0 $\log_{10}$ CFU/g of *Salmonella montevideo* on the surface and in the stem core tissue of mature-green fruits).[71]
- Treatment of alfalfa seeds inoculated with *Salmonella stanley* (10^2 to 10^3 CFU/g) in 100 ppm chlorine solution for 10 min caused a significant reduction in population; treatment in 290 ppm chlorine solution resulted in a significant reduction compared with treatment with 100 ppm chlorine.[72]
- Treatment of seeds containing 10^1 to 10^2 CFU of *S. stanley* per gram for 5 min in a solution containing 2040 ppm chlorine reduced the population to <1 CFU/g.[*72,73]
- Dipping alfalfa sprouts inoculated with a five-serovar (*S. agona, S. enteritidis, S. hartford, S. poona, S. montevideo*) cocktail of *Salmonella* in chlorine solutions (200, 500, or 2000 ppm) for 2 min reduced the pathogen by about 3.4 $\log_{10}$ CFU/g after treatment with 500 ppm chlorine and to an undetectable level (<1 CFU/g) after treatment with 2000 ppm chlorine.[74]
- Treatment of cantaloupe cubes inoculated with a five-serovar (*S. agona, S. enteritidis, S. hartford, S. poona, S. montevideo*) cocktail of *Salmonella* in chlorine solution (2000 ppm) resulted in less than 1 $\log_{10}$ reduction in viable cells.[74†]
- Exposure of lettuce for 10 min to 5 ppm ClO_2 caused a maximum reduction of 1.1 and 0.8 logs in numbers of *L. monocytogenes* at 4 and 22°C, respectively, as compared with a tap water control (0 ppm ClO_2).[75]
- Dipping tomatoes in 15% trisodium phosphate (TSP) solution for 15 s allowed complete inactivation of *Salmonella montevideo* (5.18 $\log_{10}$ CFU/cm²) on surface and in core tissue of inoculated mature-green fruits.[76‡]
- Application of lemon juice to the surface of papaya (pH 5.69) and jicama (pH 5.97) cubes inoculated with *Salmonella typhi* reduced populations compared to the control, but growth resumed after several hours.[77]
- Washing parsley leaves inoculated with 10^7 CFU/g of *Yersinia enterocolitica* in a solution of 2% acetic acid or 40% vinegar for 15 min allowed reduction in counts to <1 CFU/g.[78]
- Dipping alfalfa sprouts with 200 ppm chlorine or 2% H_2O_2 allowed reductions in populations of *Salmonella* by slightly more than 1 $\log_{10}$ CFU/g.[79]
- Dipping cantaloupe cubes with 200 ppm chlorine or 2% H_2O_2 allowed reductions in populations of *Salmonella* by slightly less than 1 $\log_{10}$ CFU/g.[81§]
- Treatment of *Cryptosporidium parvum* oocysts with 1 ppm ozone for 5 min allowed more than 90% inactivation of *C. parvum*.[80]

*The FDA has since recommended that alfalfa seeds destined for sprout production be treated with 2000 ppm chlorine for thirty minutes.[73]

†The very high level of organic matter in the juice released from cut cantaloupe tissue apparently neutralizes the chlorine before its lethality can be manifested.[74]

‡The use of TSP as a sanitizer for removal of *Salmonella* from the surface of mature-green tomatoes has good potential.

§Results from studies on fruits and vegetables indicate that H_2O_2 has potential for use as a sanitizer.

cytogenes, as well as dose-response to *Vibrio* species. Among the Government Performance and Results Act (GPRA) research related goals for food safety for the year 2000 are

1. Work with industry and academic to develop new techniques for eliminating pathogens on sprouts and in citrus juice and apple cider.
2. Develop modeling techniques to assess human exposure and dose response to certain foodborne pathogens, develop and make available an improved method for the detection of hepatitis A virus, *Cyclospora cayetanensis,* and *Escherichia coli* O157:H7 on additional fruits and vegetables, and provide knowledge and technologies needed to develop guidance and methods for the control and elimination of pathogens on particular fruits and vegetables, such as *Escherichia coli* O157:H7 and *Salmonella.*
3. Develop more rapid and accurate analytical methods for foodborne chemical contaminants (including bacterial toxins).

Predictive microbiology can be a useful tool in all of these research topics.

41.7 CONCLUSIONS

Produce and MPFV are not absent of hazards, and the level of risk should be assessed and maintained at a minimum that is practically and technologically feasible. A major concern with minimally processed fruits and vegetables is the survival and growth of several pathogenic organisms that can be traced to raw produce, plant workers, and the processing environment. The food safety challenge that represents minimal processing demands maximum and continuous investigation efforts by scientists, sustained research, and the cooperation and education of all parties involved. It is also of paramount importance that food processors become familiar with the different tools that allow handling the elements of processing, packaging, and distribution that should be optimized to deliver safe MPFV products. Such tools include GMP, HACCP, risk assessment, and predictive microbiology. Fruit and vegetable growers and handlers should be informed of the risks associated with pathogenic microorganisms related to raw fruits and vegetables to control microbiological hazards that may be influenced by current and changing practices in agronomy, processing, marketing, and preparation. Approaches for safety assessment of produce and MPFV, either from the industry side or from regulatory instances, converge in the same final goal: the well being of consumers.

REFERENCES

1. Woteki, C. 1998. "Nutrition, Food Safety, and Risk Assessment—A Policy-Maker's Viewpoint," in *FDA/CFSAN National Food Safety Initiative.* Risk Assessment.(www.FoodSafety.gov).

2. Notermans, S. and J. L. Jouve. 1995. "Quantitative Risk Analysis and HACCP: Some Remarks," *Food Microbiol.,* 12: 425–429.

3. Notermans, S., G. Gallhoff, M. H. Zwietering, and G. C. Mead. 1994. "The HACCP Concept: Identification of Potentially Hazardous Micro-organisms," *Food Microbiol.,* 11: 203–214.

4. Notermans, S., G. C. Mead, and J. L. Jouve. 1996. "Food Products and Consumer Protection: A Conceptual Approach and a Glossary of Terms," *Int. J. Food Microbiol.,* 30: 175–185.

5. Food and Drug Administration U.S. Department of Agriculture. FDA and USDA. 1998. "Guidance for Industry. Guide to Minimize Microbial Food Safety Hazards for Fresh Fruits and Vegetables Food Safety Initiative Staff," HFS-32 U.S. Food and Drug Administration. Center for Food Safety and Applied Nutrition. 200 C Street. S.W. Washington, DC 20204, October 26, 1998 (http://www.fda.gov.).

6. Notermans, S., G. Gallhoff, M. H. Zwietering, and G. C. Mead. 1995. "The HACCP Concept: Specification of Criteria Using Quantitative Risk Assessment," *Food Microbiol.,* 12: 81–90.

7. Codex Alimentarius Commission, Committee on Food Hygiene. 1991. "Draft Principles and Application of Hazard Analysis Critical Control Point (HACCP) System," Aliform 93/13 VI. FAO/WHO.

8. World Health Organization (WHO). 1995. "A Proposal for Amendment of the Codex Guidelines for the Application of Hazard Analysis Critical Control Point System," Geneva.

9. Food and Drug Administration. 1999. U.S. Department of Agriculture. "FDA Backgrounder. HACCP: A State-of-the-Art Approach to Food Safety," August 1999. http://www.fda.gov/opacom/backgrounders/haccp.html.

10. Food and Drug Administration U. S. Department of Agriculture. Center for Food Safety and Applied Nutrition. 1997. "Second Interim Report of Observations and Comments. Hazard Analysis and Critical Control Point (HACCP) Pilot Program for Selected Food Manufacturers," October 31, 1997.

11. Van Gerwen, S. J. C., J. C. de Bit, S. Notermans, and M. H. Zwietering. 1997. "An Identification Procedure for Foodborne Microbial Hazards," *Int. J. Food Microbiol.,* 38: 1–15.

12. Vasconcellos, J. A. 2000. "Regulatory and Safety Aspects of Refrigerated Minimally Processed Fruits and Vegetables: A Review," in *Minimally Processed Fruits and Vegetables*, S. M. Alzamora, M. S. Tapia, and A. López-Malo, eds. Gaithersburg, MD: Aspen Publishers, Inc., pp. 319–343.

13. National Food Processors Microbiology and Food Safety Committee. 1989. "Guidelines for the Development, Production and Handling of Refrigerated Foods," Washington, DC.

14. IFPA. 1996. *Food Safety Guidelines for the Fresh-Cut Produce Industry*, 3rd. ed. Alexandria, VA: International Fresh-Cut Produce Association (IFPA).

15. Food and Drug Administration U.S. 1998. Department of Agriculture Department of Health and Human Services. HHS NEWS P98, April 21, 1998. http://www.fda.gov/bbs/topics/NEWS/NEW00635.html.

16. Food and Drug Administration. Center for Food Safety and Applied Nutrition. 1999. "HACCP for Land Foods. Potential for Infiltration, Survival and Growth of Human Pathogens within Fruits and Vegetables," November, 1999. http://vm.cfsan.fda.gov/~lrd/haccpsub.html.

17. Stier, R. 1992. "Practical Application of HACCP," in *HACCP Principles and Applications,* M. D. Pearson and D. A. Corlett, eds. New York: AVI Book, pp. 127–167.
18. Brackett, R. E. 1999. Personal communication.
19. Codex Alimentarius. 1995. "Guidelines on the Application of the Principles of Risk Assessment and Risk Management to Food Hygiene Including Strategies for their Application," CX/FH 95/8.
20. Codex Alimentarius. 1996. "Hazard Analysis and Critical Control Point (HACCP) System and Guidelines for its Application," ALINORM 97/13A. Apendix II.
21. Codex Alimentarius. 1998. "Draft Principles and Guidelines for the Conduct of Microbiological Risk Assessment," ALINORM 99/13A.
22. Foegeding, P. M. 1997. "Driving Predictive Modelling on a Risk Assessment Path for Enhanced Food Safety," *Int. J. Food Microbiol.,* 36: 87–95.
23. Corlett, D. A. and R. F. Stier. 1991. "Risk Assessment within the HACCP System," *Food Control.,* 2: 71–72.
24. Buchanan, R. L. 1995. "The Role of Microbiological Criteria and Risk Assessment in HACCP," *Food Microbiol.,* 12: 421–42.
25. Panissello, P. J. and P. C. Quantick. 1998. "Application of Food Micromodel Predictive Software in the Development of Hazard Analysis Critical Control Point (HACCP) Systems," *Food Microbiol.,* 15: 425–439.
26. World Health Organization. 1995. "Applications of Risk Analysis to Food Standard Issues," Report of the Joint FAO/WHO Expert Consultation, 13–17 March 1995. Geneva, Switzerland.
27. Beuchat, L. R. B. 1999. "Problems Associated with Pathogenic Microorganisms on Raw Fruits and Vegetables," II Congreso Venezolano de Ciencia y Tecnología de Alimentos. Dr. Asher Ludin. Caracas, Venezuela April 24–28, (proc. abstract)
28. Food and Drug Administration. Center for Food Safety and Applied Nutrition National Food Safety Initiative. 1999. "Produce and Imported Foods Safety Initiative. 1999–2001 Update." August 1999 U.S. Three Year Research Plan.
29. Nottermans. S., G. Gallhoff, M. H. Zwietering, and G. C. Mead. 1995. "Identification of Critical Control Points in the HACCP System with a Quantitative Effect on the Safety of Food Products," *Food Microbiol.,* 12: 93.
30. Hathaway, S. C. and R. L. Cook. 1997. "A Regulatory Perspective on the Potential Uses of Microbial Risk Assessment in International Trade," *Int. J. Food Microbiol.,* 36: 127–133.
31. Van Gerwen. 1999. Personal communication.
32. Abdul-Raouf, U. M., L. R. Beuchat, and M. S. Ammar. 1993. "Survival and Growth of *Escherichia coli* O157:H7 on Salad Vegetables," *Appl. Environ. Microbiol.,* 59: 1999–2006.
33. Zhao, T., M. P. Doyle, and R. E. Besser. 1993. "Fate of Enterohemorrhagic *Escherichia coli* O157:H7 in Apple Cider with and without Preservatives," *Appl. Environ. Microbiol.,* 59: 2526–2530.
34. Diaz, C. and J. H. Hotchkiss. 1996. "Comparative Growth Shredded Iceberg Lettuce Stored under Modified Atmospheres," *J. Sci. Food Agric.,* 70: 433–438.
35. Barriga, M. I., G. Trachy, C. Willemot, and R. E. Simard. 1991. "Microbial Changes in Shredded Iceberg Lettuce Stored under Controlled Atmospheres," *J. Food Sci.,* 56: 1586–1588, 1599.
36. Hao, Y. Y. and R. E. Brackett. 1993. "Influence of Modified Atmosphere on Growth of Vegetable Spoilage Bacteria in Media," *J. Food Prot.,* 56: 223–228.

37. Pao, S., G. E. Brown, and K. Schneider. 1998. "Challenge Studies with Selected Pathogenic Bacteria on Freshly Peeled Hamlim Orange," *J. Food Sci.*, 63: 359–362.

38. Itoh, Y., Y. Sigita-Konishi, F. Kasuka, and M. Iwaki. 1998. "Enterohemorrhagic *Escherichia coli* O157:H7 Present in Radish Sprouts," *Appl. Environ. Microbiol.*, 64: 1532–1535.

39. Weagant, S. D., J. L. Bryant, and D. H. Bark. 1994. "Survival of *Escherichia coli* O157:H7 in Mayonnaise and Mayonnaise-based Sauces at Room and Refrigerated Temperatures," *J. Food Prot.*, 57: 629–631.

40. Massa, S., C. Altieri, V. Quaranta, and R. DePace. 1997. "Survival of *Escherichia coli* O157:H7 in Yogurt During Preparation and Storage at 4°C," *Lett. Appl. Microbiol.*, 24: 347–350.

41. Zhao, T., M. P. Doyle, and R. E. Besser. 1993. "Fate of Enterohemorrhagic *Escherichia coli* O157:H7 in Apple Cider with and without Preservatives," *Appl. Environ. Microbiol.*, 59: 2526–2530.

42. Zhao, T. and M. P. Doyle. 1993. "Rate of Enterohemorrhagic *Escherichia coli* O157:H7 in Commercial Mayonnaise," *J. Food Prot.*, 57: 780–783.

43. Hathcox, A. L., L. R. Beuchat, and M. P. Doyle. 1995. "Death of Enterohemorrhagic *Escherichia coli* O157:H7 in Real Mayonnaise and Reduced-caloric Mayonnaise Dressings as Influenced by Initial Population and Storage Time," *Appl. Environ. Microbiol.*, 61: 4172–4177.

44. Ryu, J. H. and L. R. Beuchat. 1998. "Influence of Acid Tolerance Responses on Survival, Growth, and Thermal Cross-protection of *Escherichia coli* O157:H7 in Acidified Media and Fruit Juices," *Int. J. Food Microbiol.*, 45: 185–193.

45. Langlois, B. E., S. Bastin, K. Akers, and I. O'Leary. 1987. "Microbiological Quality of Foods Produced by an Enhanced Cook-Chill System in a Hospital," *J. Food Prot.*, 60: 655–661.

46. Heisick, J. E., D. E. Wagner, M. L. Nierman, and J. T. Peeler. 1989. "*Listeria* spp. Found on Fresh Market Produce," *Appl. Environ. Microbiol.*, 55: 1925–1927.

47. Vahidy, R. 1992. "Isolation of *Listeria monocytogenes* from Fresh Fruits and Vegetables," *HortScience*, 27: 628.

48. Arumgaswamy, R. K. G., G. Rusul Rahamat Ali, and S. Nadzriah Bte Abd. Hamid. 1994. "Prevalence of *Listeria monocytogenes* in Foods in Malasya," *Int. J. Food Microbiol.*, 23: 117–121.

49. Odumuru, J. A., S. J. Mitchell, D. M. Alves, and J. A. Lynch. 1997. "Assessment of the Microbiological Quality of Ready-to-Use Vegetables for Health-Care Food Services," *J. Food Prot.*, 954–960.

50. Wong, H. C., W. L. Chao, and S. J. Lee. 1990. "Incidence and Characterization of *Listeria monocytogenes* in Foods Available in Taiwan," *Appl. Environ. Microbiol.*, 56: 3101–3104.

51. Tortorello, M. L., K. F. Reineke, and D. S. Stewart. 1997. "Comparison of Antibody-Direct Epifluorecent Filter Technique with the Most Probable Number Procedure for Rapid Enumeration of *Listeria* in Fresh Vegetables," *J AOAC Int.*, 80: 1208–1214.

52. Sizmur, K. and C. W. Walker. 1988. "*Listeria* in Prepacked Salads," *Lancet.*, i: 1167.

53. Brión, D. 1994. "Incidencia de *Listeria monocytogenes* y *Aeromonas hydrophila* en productos vegetales mínimamente procesados," Caracas, Venezuela: Universidad Central de Venezuela; thesis.

54. Steinbruegge, E. G., R. B. Maxcy, and M. B. Liewen. 1988. "Fate of *Listeria monocytogenes* on Ready to Serve Lettuce," *J. Food Prot.*, 51: 596–599.

55. Beuchat, L. R. and R. E. Brackett. 1990. "Survival and Growth of *Listeria monocytogenes* on Lettuce as Influenced by Shredding, Chlorine Treatment, Modified Atmosphere Packaging and Temperature," *J. Food Sci.,* 55: 755–758, 870.

56. Francis, G. A. and D. O. Beirne. 1997. "Effect of Gas Atmosphere, Antimicrobial Dip and Temperature on the Fate of *Listeria innocua* and *Listeria monocytoges* on Minimally Processed Lettuce," *Int. J. Food Sci. Technol.,* 32: 141–151.

57. Carlin, F. and C. Nguyen. 1994. "Fate of *Listeria monocytogenes* on Four Types of Minimally Processed Green Salads," *Lett. Appl. Microbiol.,* 18: 222–226.

58. Carlin, F., C. Nguyen, and A. Abreu da Silva. 1996. "Factors Affecting the Growth of *Listeria monocytogenes* on Minimally Processed Fresh Endive," *J. Appl. Bacteriol.,* 78: 636–648

59. Carlin, F., C. Nguyen, and C. E. Morris. 1996. "Influence of Background Microflora on *Listeria monocytogenes* on Minimally Processed Fresh Broad-Leaved Endive (*Cichorium endivia* var. *latifolia*)," *J. Food Prot.,* 59: 698–703.

60. Vankerschaver, K., F. Willocx, C. Smout, and M. Hendrickx. 1996. "The Influence of Temperature and Gas Mixture on the Growth of the Intrinsic Microorganisms on Cut Endive: Predictive Versus Actual Growth," *Food Microbiol.,* 13: 427–440.

61. Juncja, V. K., S. T. Martin, and G. M. Sapers. 1998. "Control of *Listeria monocytogenes* in Vacuum-Packaged Pre-Peeled Potatoes," *J. Food Sci.,* 63: 911–914.

62. Odumer, J. A., S. J. Mitchell, D. M. Alves, and J. A. Lynch. 1997. "Assessment of the Microbiological Quality of Ready-to-Use Vegetables for Health-Care Food Services," *J. Food Prot.,* 6: 954–960.

63. Donnelly, C. W. 1994. "*Listeria monocytogenes,*" in Y. D. Hui, J. R. Gorham, K. D. Murrel, and D. O. Cliver, eds. *Foodborne Disease Hanbook,* vol. 1., New York: Marcel Dekker, Inc., pp. 215–252.

64. Callister, S. and W. Agger. 1987. "Enumeration and Characterization of *Aeromonas hydrophila* and *Aeromonas caviae* Isolated from Grocery Store Produce," *Appl. Environ. Microbiol.,* 53: 249–253.

65. Berrang, M. E., R. E. Brackett, and L. R. Beuchat. 1989. "Growth of *Aeromonas hydrophila* on Fresh Vegetables Stored Under a Controlled Atmosphere," *Appl. Environ. Microbiol.,* 55: 2167–2171.

66. Marchetti, R., M. A. Casadei, and M. E. Guerzoni. 1992. "Microbial Population Dynamics in Ready-to-Use Vegetable Salads," *Ital. J. Food Sci.,* 2: 97–108.

67. García-Gimeno, R. M., M. D. Sánchez-Pozo, M. A. Amaro-López, and G. Zurera-Cosano. 1996. "Behavior of *Aeromonas hydrophila* in Vegetable Salads Stored under Modified Atmosphere at 4 and 15°C," *Food Microbiol.,* 13: 369–374.

68. Díaz, R., A. Martínez, M. Tapia, and A. Tablante. 1989. "Enumeración y Caracterización de *Aeromonas* sp. en Productos de Origen Animal y Vegetal," presented at II Congreso Latinoamericano de Microbiología de Alimentos. November, Caracas. Venezuela.

69. Garg, N., J. J. Churey, and D. F. Splittstoesser. 1990. "Effect of Processing Conditions on the Microflora of Fresh-Cut Vegetables," *J. Food Prot.,* 53: 701–703.

70. Brackett, R. E. 1987. "Antimicrobial Effect of Chlorine on *Listeria monocytogenes,*" *J. Food Prot.,* 50: 999–1003.

71. Zhuang, R-Y., L. R. Beuchat, and F. J. Angulo. 1995. "Fate of *Salmonella montevideo* on and in Raw Tomatoes as Affected by Temperature and Treatment with Chlorine," *Appl. Environ. Microbiol.,* 61: 2127–2131.

72. Jaquette, C. B., L. R. Beuchat, and B. E. Mahon. 1996. "Efficacy of Chlorine and Heat Treatment in Killing *Salmonella stanley* Inoculated onto Alfalfa Seeds and Growth and Survival of the Pathogen During Sprouting and Storage," *Appl. Environ. Microbiol.*, 62: 2212–2215.

73. Department of Health and Human Services. 1996. "Memorandum: Microbiological Safety of Alfalfa Sprouts," From J. Madden to S. Altekrause, March 1, 1996: 2 pp.

74. Beuchat, L. R. and J-H. Ryu. 1997. "Produce Handling and Processing Practices," *Emerg. Infect. Dis.*, 3: 459–465.

75. Zhang, S. and J. M. Farber. 1996. "The Effects of Various Disinfectants Against *Listeria monocytogenes* on Fresh-Cut Vegetables," *Food Microbiol.*, 13: 311–32.

76. Zhuang, R-Y. and L. R. Beuchat. 1996. "Effectiveness of Trisodium Phosphate for Killing *Salmonella montevideo* on Tomatoes," *Lett. Appl. Microbiol.*, 22: 97–100.

77. Escartin, E. F., A. Castillo Ayala, and J. S. Lozano. 1989. "Survival and Growth of *Salmonella* and *Shigella* on Sliced Fresh Fruit," *J. Food Prot.*, 52: 471–472.

78. Karapinar, M. and S. A. Gonul. 1992. "Removal of *Yersinia enterocolitica* from Fresh Parsley by Washing with Acetic Acid or Vinegar," *Int. J. Food Microbiol.*, 16: 261–264.

79. Beuchat, L. R. and J-H. Ryu. 1997. "Produce Handling and Processing Practices," *Emerg. Infect. Dis.*, 3: 459–465.

80. Peeters, J. E., E. A. Mazas, W. J. Masschelein, I. V. M. de Maturana, and E. Debacker. 1989. "Effect of Disinfection of Drinking Water with Ozone or Chlorine Dioxide on Survival of *Cryptosporidium parvum* Oocysts," *Appl. Environ. Microbiol.*, 55:1519–1522.

81. Walls, I. and V. Scott. 1997. "Use of Predictive Microbiology in Microbial Food Safety Risk Assessment," *Intl. J. Food Microbiol.*, 36: 97–102.

Part VIII

Emerging Technologies

42 Membrane Permeabilization and Inactivation Mechanisms of Biological Systems by Emerging Technologies

D. Knorr
V. Heinz
A. Angersbach
D.-U. Lee

CONTENTS

Abstract

Membrane permeabilization is a key feature of emerging technologies such as high hydrostatic pressure and high-intensity electric field pulses. To quantify membrane permeabilization, a measure of cell disintegration based on relative changes of sample conductivity has been developed. The kinetics of membrane permeabilization of animal, plant, and yeast cells have been evaluated in detail by following the development and breakdown of transmembrane potentials and by identifying the permeabilization characteristics during and after pulsed application.

Mechanisms of microbial inactivation, including the impact of high pressure, process temperature, and the magnitude of injured vegetative cells, have been examined. Based on transmission X-ray data, it has been postulated that pressure/temperature inactivation of *Bacillus subtilis* spores is due to inactivation of the enzymatic germination system.

Data on the impact of high electric field pulses on vegetative microorganisms under pressure suggest that alterations in the membrane structure imposed by compressive forces do not immediately increase susceptibility to pulsed electric fields. However, time-dependent reactions in response to sublethal pressure stress can produce a destabilization.

42.1 INTRODUCTION

One of the key features of emerging technologies such as high hydrostatic pressure or high-intensity electric field pulses is their membrane related effect impacting on the reversible or irreversible permeabilization of biological cells. The consequences of membrane permeabilization include affecting heat- and mass-transfer-related processes of treated food systems, influencing viability, biosynthetic activity such as stress and wound responses, as well as changes of physicochemical properties like texture.

Extensive reviews on voltage modulated[1–5] or pressure-related[6] membrane permeabilization have been provided. However, limited quantitative information exists regarding the impact of membrane permeabilization on food models and real food systems.

Microbial inactivation as related to high hydrostatic pressure and to high-intensity electric field pulse treatment has been studied extensively,[3,7–10] and survival/inactivation models have been provided for vegetative microorganisms in the case of high pressure[11,12] as well as for pulsed electric field treatment[13,14] following characteristic sigmoid shapes when plotted in linear coordinates.

As for electric field pulse,[13]

$$S(V, n) = \frac{100}{\left\{ 1 + \exp\left[\dfrac{V - V_c(n)}{a(n)} \right] \right\}} \tag{42.1}$$

where $S(V, n)$ = the percentage of surviving organisms
V = field intensity
n = number of pulses
$V_c(n)$ = critical field intensity corresponding to 50% survival
$a(n)$ = constant representing the curve's steepness

The three-state model[15] developed for high-pressure treatment suggested the existence of a hypothetical intermediate state that comprises the sublethally damaged portion of microbial populations exposed to high hydrostatic pressure. The main characteristic of this model is the non-autonomous nature of the transition step from the unaffected to the intermediate state, which is modeled by the Weibull distribution. The actual rate of the subsequent inactivation reaction is assumed to depend on the accumulation level of the intermediate state. This concept makes it possible to consider the diversity of active or passive resistance mechanisms within a population against external stress (homeostasis) as well as the "single-hit" approach for the probability of cell death. The mathematical representation of this hypothesis yields a three parameter model which is given in Equation (42.2). The model has been validated for *Listeria, EHEC,* and *Bacillus subtilis* in simple buffers and in real food systems.

$$\log(N) = \log(N_0) \cdot \left[\exp\left(-\frac{t}{b}\right)^c - 1 \right] + \exp-(k \cdot t) \cdot \int_0^t W_d(s) \cdot \exp(k \cdot s)ds \quad (42.2)$$

where N = survivor count after time t
N_0 = initial microbial count
W_d = density function of Weibull distribution with characteristic parameters
b (scale parameter) and c (shape parameter)
k = rate constant of inactivation step

Interestingly, the phenomenon of recovery of microorganisms after sublethal pressure treatment has already been reported by Chlopin and Tamman in 1903,[16] who stated, "Under high pressure, the majority of microorganisms evaluated transfers into a state of fainting, from which they completely recover only after a certain period of time."

Overall, most of the data on microbial inactivation by high pressure or pulsed electric fields accumulated so far suggest the potential for sufficient inactivation of vegetative systems (pathogens and spoilage organisms). However, despite the pioneering work on temperature and pressure induced germination of bacterial spores by Sale et al.,[17] and despite the data provided by Isaacs[18] that indicate that a combination of moderate pressures (150 MPa) and temperatures (70 to 95°C) led to sterilization effects in model systems, limited information is available in the public domain on the impact of high pressure/heat processes on the sterilization of real food systems or on the mechanisms of pressure/heat inactivation of bacterial spores.[14]

Grahl and Märkel[19] reported no inactivation of bacterial spores in model milk systems by high-intensity electric field pulses and, according to Barsotti and Cheftel,[3] they appear to resist electric pulse conditions that inactivate vegetative cells. A process combining moderate heat treatment (80°C, 10 min), lysozyme application, and electric pulses (80 × 2 µs pulses, 60 kV/cm, 60°C) suggests thermal induction of germination.[20]

Within the framework of this contribution, we attempt to contribute parts of our ongoing work on the better understanding of membrane permeabilization by emerging technologies, especially by high-intensity electric field pulses and its impact on food systems and on subsequent processing operations. Further, we aim to provide data on the inactivation mechanisms of pressure /heat treated bacterial spores and to report on a pH induced germination effect during pulsed electric field treatment.[21] Finally, preliminary data on the inactivation of vegetative miccroorganisms by electric field pulse treatment under high hydrostatic pressure conditions[22] are provided.

42.2 PERMEABILIZATION OF BIOLOGICAL MEMBRANES

Data on high-pressure permeabilization of cultured plant cells[6,23] suggest that up to approximately 100 MPa cell viabilities can be maintained, and up to 150 to 175 MPa reversible permeabilization of the tonoplast occurs. Beyond these pressure levels, irreversible permeabilization of the tonoplast and permeabilization of the plasma membrane (175 to 250 MPa) have been identified via release of intracellular liquids, suggesting decompartmentalization of the cells.

Systematic studies on the extent of permeabilization require appropriate tools for evaluation. Consequently, to effectively quantify the impact of emerging technologies on membrane disintegration, a quantifiable measure was needed.

Based on a study of the electrophysical behavior of intact cells with insulated biological membranes, a model of a cell system with different ratios of intact, ruptured cells and extracellular compartments was developed.[5] Based on the relative changes of sample conductivities, which were obtained within a characteristic low- and high-frequency range of the characteristic β-dispersion, a measure of cell permeabilization, the cell disintegration index (p_o), was established:

$$p_0 = \frac{\dfrac{\sigma_h^i}{\sigma_h^t} \cdot \sigma_l^t - \sigma_l^i}{\sigma_h^i - \sigma_l^i} \tag{42.3}$$

Determination of p_0 in a processed cell system is carried out by measuring the conductivity of initial intact (σ_l^i and σ_h^i) and of treated (σ_l^t and σ_h^t) samples at low and high frequencies within the band of β-dispersion. (For most plant and animal cells in suspension culture or tissue, the characteristic low-frequency range is on the order of 10^3 Hz, and the high-frequency range is on the order of 10^7 Hz). For intact cells, $p_0 = 0$; for total cell disintegration, $p_0 = 1$.

Conductance spectra of cell systems of intact plant and animal tissues are given in Figure 42.1.[24] The impact of high-intensity electric field pulses (HELPs) on plant membranes was evaluated in detail. Cell membranes are inherently electrical in nature, as exemplified in Figure 42.2.

This method was used for the identification of the efficiency of HELP impact parameters on biological cells[6] and for the recording of rapid permeabilization changes during and after pulse application. Data on cell permeabilization kinetics after HELP treatment of potato cell cultures are given in Figure 42.3.[5]

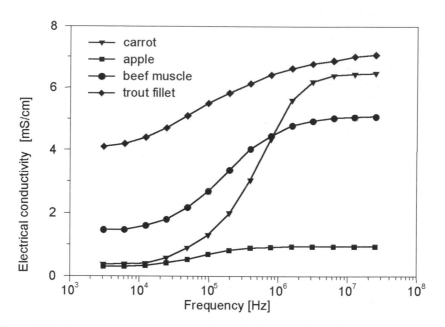

Figure 42.1 Frequency-dependent electrical conductivity data for cell systems in plant and animal tissues.

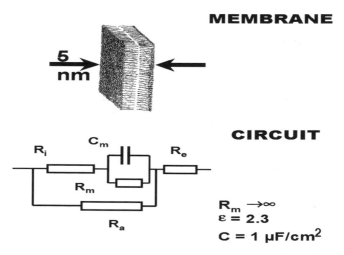

Figure 42.2 Simplified presentation of a cell membrane and related electrical circuit C_m. Membrane capacitance = R_m, membrane resistance = Ra.

The buildup of the membrane potential and pore formation in cells of potato tissue in the first microsecond after the initiation of pulsing with supercritical field strength is demonstrated in Figure 42.3.[5] Initially, the transmembrane potential ϕ increases exponentially with the time constant $\tau = 0.7$ μs, the charging time constant

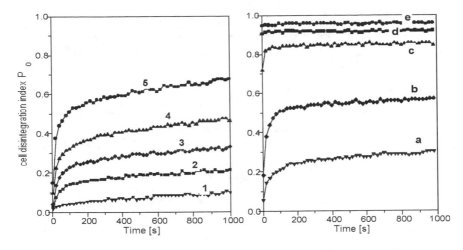

Figure 42.3 The kinetic of cell permeabilization after application of a single pulse on potato cell cultures with various energy input (left) and after 1–20 pulses with constant energy input per pulse of 0.6 kJ/kg (right). Energy input (kJ/kg): 1 = 0.07; 2 = 0.17; 3 – 0.6; 4 = 1.7; 5 = 5.6. Number of pulses: a = 1; b = 2; c = 5; d = 10; e = 20. The first measurement for p_0 calculation was taken 2 s after the pulse application.

for intact membrane in potato cells. At near 0.4 μs, φ reached 1.7 V and later decreased to a residual value of 0.1 V, consequent to pore formation and a drastic increase in membrane conductance. The progressive permeabilization of potato tissue by repeated pulsing is demonstrated in Figure 42.4.[5]

Monitoring of the current during the first 8 μs after starting of repeated super-critical pulse applications resulted in reversibility after the first pulses and irreversible cell membrane rupture after several pulses (Figure 42.4). At repeated pulses with intervals of one second, the oscillographic tracking of the charging current of the cell membranes for the first and the second pulse were reproduced with high accuracy. This means that structural properties of the membrane after the first pulse were identical to those of an intact membrane. A reversible breakdown of the cell membranes was observed. This finding is essential for the appropriate use of pulsed electric fields. Only if the membrane structure is restored will the induction of critical membrane voltage and membrane breakdown occur again. Also, for the third pulse, the current course showed only a slight deviation from the exponential decaying process. This means that membrane structure was restored to a great extent within one second after the second pulse. The current course of the pulse indicated that an irreversible membrane permeabilization occurred.

To clearly identify the onset of membrane permeabilization characteristic conductivity, changes for various real food systems during the initial microseconds after application of electric fields with sub- and supercritical strength were obtained. As indicated by the breakdown of the membrane potential (breakdown point), cell permeabilization took place within less than one microsecond after the beginning of the pulse (Figure 42.5).[5]

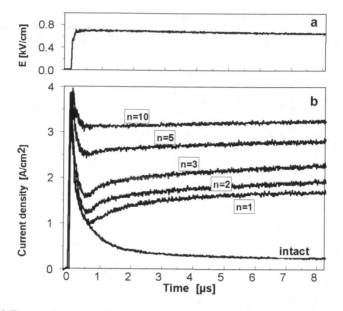

Figure 42.4 Progressive permeabilization of potato tissue by repeated (repetition rate 1 s⁻¹) pulsing with constant amplitude of field strength (E_o = 0.7 kV/cm). (a) Time dependency of field strength. (b) Current density in the tissue in the first 8 μs after pulse start. n = pulse number. Current density trace, for example, without membrane permeabilization (intact) is reconstructed from data obtained by application in the same sample of a pulse with subcritical field strength level, E = 0.08 kV/cm.

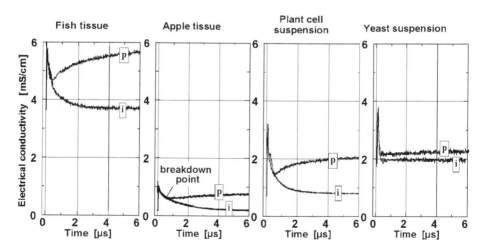

Figure 42.5 Conductivity changes in real food systems within a few microseconds after initiation of electric fields at subcritical (curves i, E = 0.07 kV/cm) and supercritical (curves p, E = 1.0 kV/cm for fish, apple, plant cell suspension; E = 1.7 kV/cm for yeast suspensions) field strength.

The characteristic development of the transmembrane voltage of the cell membranes as achieved by the application of supercritical field strength and presenting breakdown points (critical transmembrane voltage) is shown in Figure 42.6. The exponential increase of the electrical potential at the intact membrane interfaces develops with the charging time constant τ, which depends on cellular size, membrane capacitance, and conductivity of cell and extracellular electrolytes. After the transmembrane voltage reached a critical value, it later decreased as a consequence of pore formation and a drastic increase in membrane conductance.

The critical transmembrane potential (breakdown voltage) of potato, apple, fish tissue cells, and plant cell suspension over a field amplitude range from $E = 0.2$ kV/cm to $E = 2.0$ kV/cm are summarized in Figure 42.7.[5]

This is essential for process development, since optimal repetitive pulse sequences for irreversible permeabilization can be developed based on these data. Furthermore, the possible induction of indigenous enzyme activities, evident in substantial increase of the cell disintegration index within a few hours after HELP treatment, resulting in irreversible permeabilization, offers the potential for unique process and product development concepts. Figure 42.8 provides a detailed insight into the buildup and decrease of the membrane potential, the pore formation, and the time-dependent resealing after the application of a single pulse.

42.3 MECHANISMS OF MICROBIAL INACTIVATION

The interactions between pressure, temperature and physiological state of vegetative microorganisms have been explored with *Escherichia coli* inoculated in liquid whole egg. In Figure 42.9,[28] some of the results are exemplified. An increase of processing

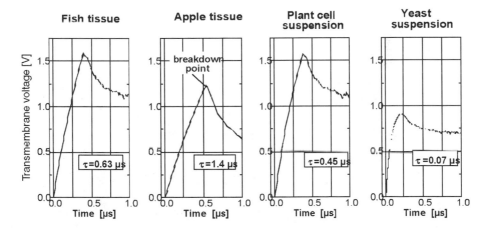

Figure 42.6 Characteristic development of the transmembrane voltage of animal, plant, and yeast cell membranes caused by the application of supercritical field strength ($E_o = 1.0$ kV/cm for fish, apple, and plant cell suspension; $E_o = 1.7$ kV/cm for yeast). τ is the charging time constant of the cell membrane.

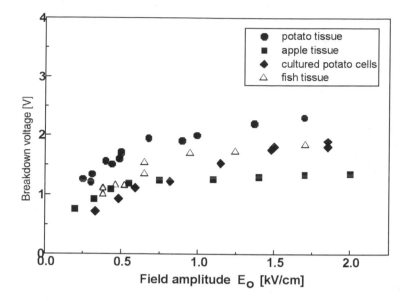

Figure 42.7 Critical transmembrane potentials (breakdown voltage) of potato, apple, fish, and plant cells in cultured suspension (*Solanum tuberosum*) for field amplitudes from E_o = 0.2 to 2.0 kV/cm. Critical voltage is defined as the maximal electrical potential across the cell membrane.

temperatures from 5 to 25°C did not result in markedly enhanced microbial inactivation at the given pressure (i.e., 350 MPa). Furthermore, at lower pressures (250 and 300 MPa), pressure treatments at 5°C showed increased *E. coli* inactivation over that at 25°C (data not shown). A more detailed evaluation of the survived portion can offer another viewpoint for understanding the effects of pressure on microorganisms. An increase of temperature from 5 to 25°C renders the degree of injury in the surviving cells increased by an order of magnitude (from approximately 90 to 99%). For example, at 25°C and the processing time of 200 s, about 2 log reduction of viable counts can be reached (Figure 42.9B). At this point, log degree of injury indicates about 1.0; that is, 90% of surviving cells are in a sublethally injured state. This is of significance for process development and for the design of combination processes, because further 1 log cycle or even more microbial inactivation could be achieved by addition of other stressing factors.[25,26] In some instances of low pH food like fruit juice, only subsequent storage after pressure treatment enables such further microbial inactivation.[27]

Attempts have been initiated to identify mechanisms involved in the inactivation of bacterial spores as demonstrated at pressure/temperature combinations (150 MPa/70°C) using transmission X-ray microscopy.[29] As evident from data in Figure 42.10, pressure/temperature inactivated spores showed no differences as compared with untreated controls, while germinated spores clearly show a disintegration of the cores. We postulated[14,21] that vital parts of the enzymatic germination system are

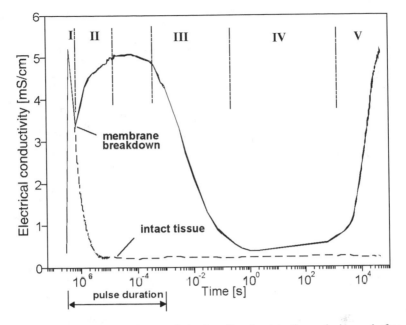

Figure 42.8 Permeabilization characteristics in cells of potato tissue during and after pulse application. Peak field strength and duration of the exponential pulse was 0.88 kV/cm and 1 ms, respectively. The conductivity changes of intact tissue (discontinued line) is obtained by application of pulse at subcritical field strength of 0.1 kV/cm. I = membrane charging process; II = development of conductance membrane due the pore formation; III = resealing of the pore; IV = irreversible pore formation; V = stress induced indigenous (possibly enzymatic) membrane permeabilization.

affected by the combined action of pressure and elevated temperature, which prevents the spore system from outgrowing. Applying pressure temperature cycles so as to use high-pressure technologies for rapid increase and even distribution of temperatures in small pressure vessels (Figure 42.11) resulted in similar findings (data not shown).

Due to the occurrence of heat formation during the compression of most foods, it is possible to reach the required temperature level for spore inactivation within a comparably short time. Different from heat transfer across the boundary layer, warming by compression heating produces an evenly distributed temperature field because of the instantaneous and homogeneous pressure propagation within high-pressure vessels. This is a unique procedure to produce a rapid increase of the core temperature of bulky products with high viscosity and low thermal conductivity (Figure 42.11).

42.4 COMBINATION OF EMERGING TECHNOLOGIES

Next to combinations of emerging technologies with heat or chemical antimicrobial agents,[3,9,10] Heinz and Knorr[15] demonstrated the combined impact of high pressure

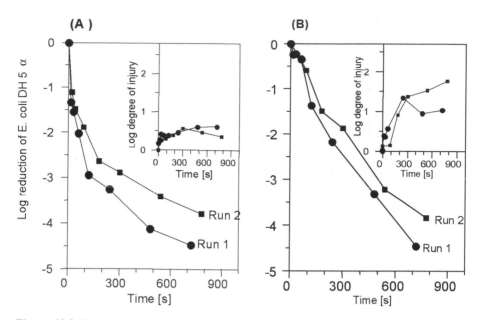

Figure 42.9 Pressure dependent inactivation curves of *E. coli* at 350 MPa with processing temperatures of 5°C (A) and 25°C (B).

and high-electric-field pulses on microbial systems. In a specially designed apparatus square, wave pulses at a peak field strength of ca. 25 kV/cm were applied to microbial suspensions pressurized to 200 MPa. With *Bacillus subtilis*, no synergistic lethal effect was observed unless the pressure treatment time prior to the pulsed electric field treatment was extended to 10 min (Figure 42.12). This suggests that alterations in the membrane structure imposed by compressive forces do not immediately increase the susceptibility to pulsed electric fields, but that time-dependent reactions in response to sublethal pressure stress can produce a destabilization.

42.5 CONCLUSIONS

Although significant advances have been achieved in the field of emerging advanced food technologies, there is still a need for substantial research efforts to better understand structure-function relationships and to quantify product-process interactions on a cellular and molecular level.

42.6 ACKNOWLEDGMENTS

Support by the European Commission (FAIR96-1175, FAI97-3044) and the German Ministry for Education and Research (BMBF-0339598/5) is acknowledged.

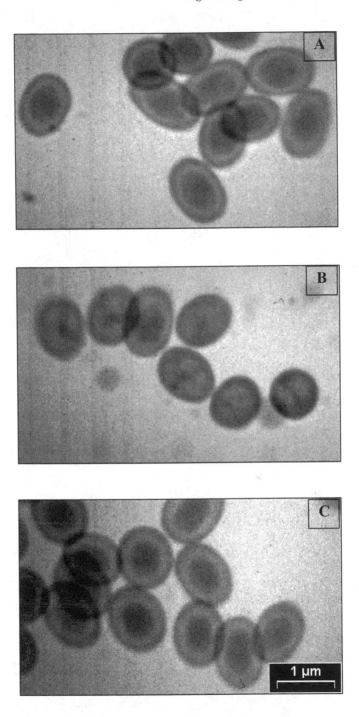

Figure 42.10 X-ray microscopic images of *Bacillus subtilis* spores; (A) control, (B) after treatment, 30°C, 150 MPa, 5 min, and (C) after treatment, 70°C, 150 MPa, 5 min.

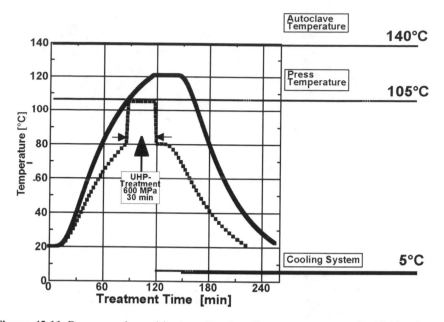

Figure 42.11 Pressure enhanced heat sterilization. Center temperature of a highly viscous cylindrical sample (diameter 5 cm) with low thermal conductivity. Whereas an autoclave temperature of 140°C is required to reduce 12 log-cycles *B. stearothermophilus* spores at ambient pressure, the same lethal effect can be achieved by a combined pressure temperature treatment at an autoclave temperature of 105°C within a reduced total treatment time.

REFERENCES

1. Tsong, T.Y. 1983. "Voltage Modulation of Membrane Permeability and Energy Utilization in Cells," *Bioscience Reports,* 3: 487–505.
2. Ho, S. Y. and G. S. Mittal. 1996. "Electroporation of Cell Membranes: A Review," *Critical Rev. in Biotechnol.,* 16(4): 349–362.
3. Barsotti, L. and J. C. Cheftel. 1999. "Food Processing by Pulsed Electric Fields. II. Biological Aspects," *Food Rev. Int.,* 15(2): 18–213.
4. Knorr, D. and A. Angersbach. 1998 "Impact of High-Intensity Electrical Field Pulses on Plant Membrane Permeabilization," *Trends Food Sci. & Technol.* 9:185–191.
5. Angersbach, A., V. Heinz, and D. Knorr. 1999. "Electrophysiological Model of Intact and Processed Plant Tissues: Cell Disintegration Criteria," *Biotechn. Prog.,* 15: 753–762.
6. Dörnenburg, H. and D. Knorr. 1999."Monitoring the Impact of High-Pressure Processing on the Biosynthesis of Plant Metabolites Using Plant Cell Cultures, "*Trends in Food Sci. & Technol.* 9(10): 355–361.
7. Cheftel, J. C. 1995. "Review: High-Pressure, Microbial Inactivation and Food Preservation," *Food Sci. and Technol. Internat.,* 1: 75–90.
8. Knorr, D. 1995. "Hydrostatic Pressure Treatment," in *New Methods of Food Preservation,* G. W. Gould, ed. London: Blackie Academic & Professional.

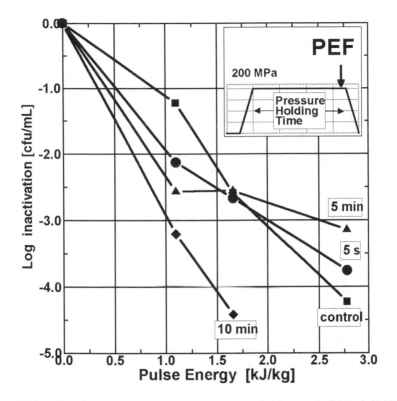

Figure 42.12 Effect of pressure holding time. At constant field strength (24.7± 0.61 kV/cm), 100 pulses were applied to suspensions of *Bacillus subtilis* at pH 7. Except for the control ■, all samples were exposed to high-pressure treatments (200 MPa) with different holding times prior to the PEF treatment: ● = 5 s; ▲ = 5 min, and ◆ = 10 min. The pulses were applied immediately before the system was decompressed (see pressure profile, top right).

9. Palou, E, A. López-Malo, G. V. Barbosa-Cánovas, and B. G. Swanson. 1998. "High Pressure Treatment in Food Preservation," in *Handbook of Food Preservation*, Rahman, M. S., ed. New York: Marcel Dekker.

10. Wouters, P. and J. P. P. M Smelt. 1997. "Inactivation of Microorganisms with Pulsed Electric Fields: Potential for Food Preservation," *Food Biotechnol.*, 11(3): 193–229.

11. Heinz, V and D. Knorr. 1996. "High Pressure Inactivation Kinetics of *Bacillus subtilis* Cells by a Three-State Model Considering Distribution Resistance Mechanisms," *Food Biotechnol.*, 10: 149–161.

12. Smelt, J. P. P. M. 1998. "Recent Advances in the Microbiology of High Pressure Processing," *Trends Food Sci. Technol.*, 9: 152–158

13. Peleg, M. 1995. "A Model of Microbial Survival After Exposure to Pulsed Electric Fields," *J. Sci Food Agric.*, 67: 93–99.

14. Heinz, V., S. T. Phillips, M. Zenker, and D. Knorr. 1999. "Inactivation of B*acillus subtilis* by High Intensity Pulsed Electric Fields Under Close to Isothermal Conditions," *Food Biotechnol.*, 13(2): 155–168.

15. Heinz, V. and D Knorr. 1999. "Effect of Medium Composition and Compression on the Inactivation of *Bacillus subtilis* by Pulsed Electric Fields," *Innovative Food Science and Emerging Technologies*, submitted.

16. Chlopin, G. W. and G. Tamman. 1903. "Ueber den Einfluss hoher Drucke auf Mikroorganismen," *Z Hygiene u. Infektionskrankheiten*, 45: 193–204.

17. Sale, A. J. H., G. W. Gould, and W. A. Hamilton. 1970. "Inactivation of Bacterial Spores by Hydrostatic Pressure," *J.Gen.Microbiol.*, 60: 323–334.

18. Isaacs. 1998. Personal Communication.

19. Grahl, T. and H. Märkl. 1996. "Killing of Microorganisms by Pulsed Electric Fields," *Appl. Microbiol. Biotechnol.*, 45: 148–157.

20. Barbosa-Cánovas, G. V., V. R. Pothakamury, E. Palou, and B. Swanson. 1997. *Nonthermal Preservation of Foods*, New York: Marcel Dekker Inc.

21. Knorr, D. 1999. "Novel Approaches in Food-Processing Technology: New Technologies for Preserving Foods and Modifying Function," *Current Opinion Biotech.*, 10: 485–491.

22. Knorr, D. and V. Heinz. 1999. "Recent Advances in High Pressure Processing of Foods," *Innovative Food Science and Emerging Technologies,* submitted.

23. Dörnenburg, H. and D. Knorr. 1993. "Monitoring the Impact of High-Pressure Processing on the Biosynthesis of Plant Metabolites using Plant Cell Cultures," *Trends Food Sci & Technol.*, 9: 355:261.

24. Angersbach, A., V. Heinz, and D. Knorr. 1997 "Elektrische Leitfähigkeit als Maß des Zellaufschlußgrades von zellulären Materialien durch Verarbeitungsprozesse," *Lebensmittel- und Verpackungstechnik*, 42: 195–200.

25. Hauben, K. J. A., E. Y. Wuytack, C. F. Soontjens, and C. W. Michiels. 1996. "High-Pressure Transient Sensitization of *Escherichia coli* to Lysozyme and Nisin by Disruption of Outer-Membrane Permeability," *Journal of Food Protection*, 59(4): 350–355.

26. Kalchayanand, N., M. B. Hanlin, and B. Ray. 1992. "Sublethal Injury makes Gram-Negative and Resistant Gram-Positive Bacteria Sensitive to the Bacteriocins, Pediocin AcH and Nisin," *Letters in Applied Microbiology*, 15: 239–243.

27. Linton, M., J. M. J. McClements, and M. F. Patterson. 1999. "Survival of *Escherichia coli* O157:H7 During Storage in Pressure-Treated Orange Juice," *Journal of Food Protection,* 62(9): 1038–1040.

28. Lee, D.-U. 1999. Unpublished data, Department of Food Biotechnology and Process Engineering, Berlin University of Technology, Germany.

29. Mönch, S., V. Heinz, P. Guttmann, and D. Knorr. 1999. "X-Ray Microscopy in Food Sciences," presented at the VIth International Conference on X-Ray Microscopy–XRM 99, August, 1999.

43 Innovative Fruit Preservation Methods Using High Pressure

CONTENTS

Abstract

Fruit processing and preservation technologies must keep fresh-like characteristics while providing an acceptable and convenient shelf life as well as assuring safety and nutritional value. Processing technologies include a wide range of methodologies to inactivate microorganisms, improve quality and stability, and preserve and minimize changes of fruit fresh-like characteristics. Nonthermal processes applied to food preservation without the collateral effects of heat treatments are being studied and tested. High pressure (HP) as a food preservation technique inactivates microorganisms and can be applied without raising temperature; thus, sensory and nutritional characteristics can be maintained. Selected practical examples of fruit processing using high pressure will be presented and discussed.

1-56676-963-9/02/$0.00+$1.50
© 2002 by CRC Press LLC

43.1 INTRODUCTION

Some of the current industrial applications of HP include fruit jams, dressings, jellies and yogurts, citrus fruit juices, sugar-impregnated tropical fruits, and *guacamole* (Mexican dressing made from avocado paste). In general, HP-processed fruit products have a vivid and natural color and flavor.[1,2] Many commercial HP-treated products are acid foods; hence, they have an intrinsic safety factor. Also, some of the products are stored and sold refrigerated and, consequently, enzymatic and oxidative reactions are retarded.[2–5]

Applications of HP technology have focused on the application of pressure as an alternative technology to heat treatments, especially for heat sensitive foods, and as nonthermal processes to assure safety and quality attributes in minimally processed foods. This contribution explores the potential of HP treatments for fruit preservation, emphasizing its potential as an innovative method for fruits, using as examples banana and avocado purees, along with *guacamole*.

43.2 HIGH-PRESSURE PROCESSING

The food industry utilizes raising pressure, a technique called *isostatic pressing*, which can be operated with a controlled temperature, above or below ambient temperature. Several reviews on HP[3–6] describe the available methods for generating high pressure in more detail, the hydrostatic pressure generation system being the most common in the food science and technology literature. In hydrostatic pressing systems, the press chamber or vessel is loaded and closed, filled with the pressure-transmitting medium, degassed, and, using a high-power pump, the high pressure is then generated. HP pressure treatment is by definition a batch or semicontinuous process. The technical advantage of the batch-type pressure vessel is the simplicity of fabrication when compared to semicontinuous flow pressure vessels operating at pressures as high as 500 to 900 MPa. A semicontinuous system with a processing capacity of 600 L/h of liquid food at a maximum operating pressure of 400 MPa is used to commercially produce grapefruit juice in Japan.[2,3] For pumpable foods, a sequence of pressing vessels or repeated pressing cycles using an isolator is now commercially available;[7] the food product is pumped into the vessel, pressurized, and discharged into an aseptic line. The process is repeated for as long as needed to meet production requirements. Multiple units can be sequenced so that, while one unit is being filled, others are in various stages of operation.[8]

Microbial inactivation, modification of chemical and enzymatic reactions, as well as changes in biopolymer structure and functionality as a result of HP are demonstrated.[4,5,8–10] During HP treatments, the mechanical force accompanying pressure generation can deform or noticeable modify solids integrity, especially in porous products. HP processors need to understand the influence of pressure on quality deteriorative enzymes, microorganisms, and sensory attributes[10–17] to generate innovative HP processes as well as acid food products.

43.3 HIGH PRESSURE AND FRUIT-BORNE MICROORGANISMS

HP treatment often focuses on fruit products such as juices, jams, and jellies, and on diced forms blended or immersed with other ingredients, as in syrups or yogurt. A characteristic shared by most fruits is their high acidity; a pH below 4.0 is not uncommon, being an important factor determining the microbial types that can spoil fruit products. While most species of bacteria are inhibited by low pH, lactic acid bacteria, yeast, and molds find these pH values tolerable for growth. Although the acid concentration and low pH of fruit juices may inhibit pathogenic bacteria, recent outbreaks of food poisoning from *Salmonella* and *Escherichia coli* O157:H7 were traced to nonpasteurized orange and apple juices, respectively.[18] Nonthermal treatments emerge as options to public demand for effective pasteurization of fruit juices to assure safety. In general, spoilage and pathogenic microbial vegetative forms can be inactivated if pressure is applied with sufficient intensity.[4,10,19,20] A number of interacting factors, including type and number of microorganisms, magnitude and extent of HP treatment, temperature, and food composition define inactivation rate and extent.[4,13,21] Microbial response to a particular pressure treatment can be divided into inactivation and survival. Microbial cell damage will determine survival or death and the extent of damage depends on pressure level, time of exposure, temperature, and food composition. The possibility of cell recovery after pressure treatment is a very important consideration for process efficacy and death kinetics assessment.[21] If the repair mechanism remains intact, the microorganisms may be capable of regeneration and growth.[20,21] The ultimate fate of the injured cells will depend on the conditions after pressure treatment; microbial cell injury may be advantageous when HP is combined with other preservation methods.

For vegetative cells, it has been reported[2,4] that the increase in pressure treatment holding time can be noticed only when pressures above 200–300 MPa are used. Microbial pressure sensitivity varies with growth phase, cells in the exponential phase being in general more sensitive to pressure treatments than cells in the lag or stationary phases of growth.[21,22] Palou et al.[23] demonstrated pressure sensitivity differences between *Zygosaccharomyces bailii* cells from the exponential and stationary growth phases. Viability of *Z. bailii* cells from stationary phase were reduced 50% with a pressure around 300 MPa, while, for cells from the exponential phase, a pressure < 172 MPa was needed to cause the same effect on yeast viability. Cells in the stationary phase are under less optimal conditions than during exponential phase due to nutrient depletion or to formation of toxic metabolites that contribute to expose cells to a stress condition.[24] Cells from the stationary phase will be more resistant to adverse conditions.

Baroprotective effects of reduced a_w have been reported.[13–15,23] Increased microbial resistance at high pressures, when suspended in low-a_w media, may be attributed to partial cell dehydration, which may render smaller cells and thicker membranes.[25,26] For *Saccharomyces cerevisiae* inoculated in concentrated fruit juices, the number of survivors depends on soluble solids concentration. The inactivation effect at pressures lower than 200 MPa decreased as juice concentration increased.[27] Temperature during pressurization can have an important effect on the inactivation of microbial cells. Several authors[27–30] have observed that microbial pressure resistance is maximal at 15–30°C and decreases significantly at higher or lower temperatures. Freezing temperatures (–20°C) in combination with 100–400 MPa for 20 min enhance microbial inactivation when compared with HP treatments at 20°C.[31,32] A decreasing pressure resistance at low temperatures (<5°C) may be due to changes in membrane structure and fluidity.[2] Moderate heating (40–60°C) can also enhance the pressure microbial inactivation, resulting in some cases in a low minimal inactivation pressure. Ogawa et al[27,33] reported an enhanced inactivation of natural and inoculated microorganisms in mandarin juice when treated at 40°C in combination with pressures in the range of 400–450 MPa. Spores from yeast and molds are easily inactivated at pressures of 300–400 MPa,[2] while ascospores of heat resistant molds such as *Byssochlamys nivea* are extremely pressure resistant.[34,35] For the inactivation of ascospores, pressures above 600 MPa and temperatures above 60°C are needed. HP treatments in combination with moderate temperatures may be an alternative for fruit juice processing. Palou et al.[35] reported that *B. nivea* ascospores suspended in a_w 0.98 and 0.94 apple and cranberry juices survived a 25 min treatment at 60°C and 689 MPa. To inactivate *B. nivea* and *B. fulva* ascospores, a 20-min treatment at 20°C and 900 MPa was needed.[36] Advantageous HP treatments to inactivate *B. nivea* ascospores were three to five oscillatory compression-decompression cycles at 689 MPa in combination with 60°C.[35] Oscillatory pressure treatments combined with heat are potentially a sound approach to inactivate heat resistant molds in diluted fruit juices. However, for concentrated fruit juices, higher pressures, more cycles of compression-decompression, and/or higher temperatures need to be evaluated.

43.4 HIGH-PRESSURE AND FRUIT-QUALITY-RELATED ENZYMES

Food quality characteristics such as flavor and vitamins are unaffected or only minimally altered by high-pressure processing, while microorganisms as well as enzymes related to food safety and quality can be inactivated.[37] Enzymatic reactions are one of the key problem areas to address in high-pressure processing of fruits. HP treatment can fulfill the requirements of traditional hot-water blanching, diminishing mineral leaching and wastewater.[38,39] The pH, substrate concentration, subunit structure of enzymes, and temperature during pressurization influence pressure inactivation of enzymes.[37,40] Pressure effects on enzyme activity are expected to occur at the substrate-enzyme interaction and/or active site conformational changes.[37,40,41] Enzyme inactivation occurs at very high pressures. Enzyme activation that could be present at relatively low pressures is attributed to reversible configuration and/or conformation changes of the enzyme and/or substrate molecules.[42] Studies on catalase, phosphatase, lipase, pectinesterase, lipoxygenase, peroxidase, polyphenoloxidase, and lactoperoxidase revealed that peroxidase is the most barostable enzyme with 90% of residual activity after 30 min treatment at 60°C and 600 MPa,[11] suggesting that peroxidase could be selected as an enzyme indicator for HP treatments.

Polyphenoloxidase (PPO) is the enzyme responsible for many undesirable color changes during fruit product storage, and it can be considered as an HP target for fruit product processing. PPOs respond differently during and after HP treatments depending on the food source.[16,37,40,43] Cano et al.[44] studied the combination of HP and temperature on peroxidase and PPO activities of fruit derived products. Strawberry peroxidase inactivation was obtained by combining 43°C and 230 MPa. Combinations of high pressure and 35°C effectively reduced peroxidase in orange juice. Fruit residual PPO activity after HP treatments suggest that undesirable enzymatic reactions such as browning require the combination of HP with one or more additional factors to inhibit activity. Low pH, blanching, and refrigeration temperatures[16,17,43] can be used in combination with HP treatments. López-Malo et al.[16] reported that avocado PPO activity can be reduced to less than 20% when treated at 689 MPa at pH 3.9, 4.1, or 4.3. This reduction in PPO activity is not enough to avoid browning during storage. Brown color development was delayed more than 30 days in HP treated avocado puree if stored at 5°C.[16] Residual PPO activity in the range of 86–63% in guava purée treated at 400–600 MPa for 15 min and stored at 4°C caused a continuous decreasing in lightness and greenness of guava purée, but maintained an acceptable quality for at least 20 days.[12]

43.5 FRUIT PRESERVATION METHODS USING HIGH PRESSURE

Technical challenges of commercial application of high-pressure technology are listed in Table 43.1.[45] Despite current limitations, to further and widely apply HP in the food industry research must clarify potential applications in fruit processing. The results of our research, which give emphasis to the use of high pressure as an additional hurdle in the manufacture of high-moisture fruit products, are described next.

TABLE 43.1
High-Pressure Food Treatment Technical Challenges

Equipment	*HP treatment*
Sanitation	High pressure, short time
Cleaning	Low pressure, long time
Disinfection	Combination with heat
Processing	*Product*
Bulk-aseptic	Material handling
In-container	Package design
	Storage
	Refrigeration

43.6 HIGH-PRESSURE PROCESSED AVOCADO PUREE

Although avocados are mainly consumed as fresh fruit, freezing, drying, and freeze-drying have been attempted to preserve this delicate fruit. Several important changes may occur during avocado heating. Pasteurization treatments result in the development of off-flavors. HP treatments can preserve delicate avocado sensory attributes and assure a reasonable safe and stable shelf life. López-Malo et al.[16] evaluated the effects of HP treatments at 345, 517, or 689 MPa, for 10, 20, or 30 min on avocado puree.

Avocado adjusted to selected initial pHs (3.9, 4.1, or 4.3) was HP treated. PPO activity, color, and microbial counts were periodically evaluated after HP and during further storage at 5, 15, or 25°C. Standard plate count, as well as yeast and mold counts of HP-treated samples, were <10 cfu/g during 100 days of storage at 5–25°C. Residual PPO activity significantly (p < 0.05) decreased with increasing pressure and decreasing initial pH. However, several color changes were observed during storage, mainly brown color development and loss of characteristic green color was noticed in avocado puree. Browning was mainly related with changes in the red-green contribution to color (*a*). Avocado puree with a residual PPO activity < 45% and stored at 5°C maintained an acceptable color for at least 60 days, and the product's acceptable shelf life was assured for 35 days when stored at 15°C.

Table 43.2 presents a technical data summary for HP processing of avocado puree. Application of these HP conditions and product characteristics assures a reasonable refrigerated shelf life based on color changes as reported by López-Malo et al.[16]

TABLE 43.2
Technical Data for Treatment of Avocado Puree

Product	*HP treatment*
Avocado puree	High pressure, long time
1.5% sodium chloride	689 MPa
Phosphoric acid to attain pH < 4.1	30 min
Equipment	*Storage*
HP press EPSI	Refrigeration (5°C)
Operated at 21°C	
Processing	*Shelf life*
In container	Color changes after 60
100–g plastic bags	days

43.7 HIGH-PRESSURE PROCESSED GUACAMOLE

Palou et al.[17] evaluated the effects of continuous or oscillatory HP treatments on PPO and lipoxygenase (LOX) activities in a Mexican sauce called *guacamole*. *Guacamole* is prepared with avocado pulp, diced onion, salt, and citric acid (or lime juice) to attain a pH around 4.8–4.3, while coriander and *serrano* hot pepper are optional ingredients. Palou et al.[17] prepared *guacamole* adjusting avocado puree pH to 4.3 with citric acid and adding dried onion (1% w/w) and sodium chloride (1.5% w/w). Two HP treatments were evaluated: continuous (689 MPa with holding times of 5, 10, 15, or 20 min) and oscillatory (two, three, or four cycles at 689 MPa with holding times of 5 or 10 min each). It has been suggested that the efficiency of HP treatments on enzyme inactivation can be improved by applying oscillatory compression-decompression cycles. Oscillatory application of HP treatments resulted in higher inactivation of several enzymes.[37] The enzyme activity retention after a multi-cycle process was lower than that of a single-cycle process with the same total duration.[46] Palou et al.[17] obtained significantly less (p < 0.05) residual PPO and LOX activities by increasing process time and number of HP cycles. LOX was inactivated with a 15-min treatment or oscillatory HP.

PPO activity was reduced at most to 15% after four HP cycles of 5 min each, suggesting that to avoid quality deteriorative enzymatic reactions during storage, a combination of HP with additional factors such as a chilling be required. Residual PPO activity on guacamole resulted in browning during storage. Color changes were related, as for avocado puree,[16] with changes in the green contribution to the color, which promote a decreasing hue. Microbial counts (standard plate, yeast, and molds) of HP treated guacamole were < 10 cfu/g. Sensory acceptability and color of HP processed guacamole were not significantly different (p > 0.05) from that of guacamole controls (without HP treatment). Table 43.3 presents the technical data for HP processing of guacamole. An acceptable shelf life of 20 days at <15°C can be obtained using 4 HP-cycles at 689 MPa.

TABLE 43.3
Technical Data for Pressure Treatment of Guacamole

Product	*HP treatment*
Guacamole	High pressure, short time cycles
Avocado	4 cycles
Sodium chloride (1.5% w/w)	689 MPa
Citric acid to attain pH 4.3	5 minutes
Dried onion (1% w/w)	
Equipment	*Storage*
HP press–EPSI	Refrigeration (5°C)
Operated at 21°C	
Processing	*Shelf life*
In container	Color changes after 20 days
100-g plastic bags	

43.8 HIGH-PRESSURE PROCESSED BANANA PUREE

As stated above, enzymatic reactions are the key problem to address in HP processing of fruits. Several studies indicate that blanching is a prerequisite. The effects of blanching pretreatments and HP on PPO activity, color, and natural flora evolution of banana puree adjusted to pH 3.4 and water activity 0.97 were evaluated during storage at 25°C by Palou et al.[47] Microbiological analyses determined periodically during 14 days revealed standard plate, yeast, and mold counts of HP treated purees lower than the level of detection (10 cfu/g). PPO activity was reduced during vapor blanching (1, 3, 5, or 7 min) and further reduced after HP treatments for 10 min at 517 or 689 MPa. Puree browning during storage was diminished extending the acceptable shelf life when combining a 5–7-min vapor blanching with 10 min at 689 MPa. Longer browning induction times and slower browning rates were obtained as a result of reducing PPO activity. Although an important reduction of banana purée PPO activity was observed when a 7-min steam blanch was combined with a 689 MPa treatment for 10 min, the residual PPO activity (<5%) was sufficient to initiate browning, noticeable after 7 days of storage at 25°C. Table 43.4 contains the technical data for HP processing of banana puree.

TABLE 43.4
Technical Data for Pressure Treatment of Banana Puree

Product	*HP treatment*
Banana puree	High pressure, short time
5–7 minutes vapor blanching	689 MPa
Sucrose to attain a_w 0.97	10 minutes
Phosphoric acid to attain pH 3.4	
Equipment	*Storage*
HP press–EPSI	Room temperature (25°C)
Operated at 21°C	
Processing	*Shelf life*
In container	Color changes
100–g plastic bags	after 7 days

43.9 FINAL REMARKS

The unique physical and sensory properties of the high-pressure process suggest it as a promising preservation method. Identification of commercially feasible applications is probably the most difficult challenge. Future work must find synergistic combinations of high pressure with other preservation factors to improve quality and stability. Oscillatory pressure treatments increased process effectiveness. However, the characteristics of the oscillatory pressure treatments need to be further studied to determine commercial feasibility. For a better and more efficient use of high pressure in fruit processing, further research is needed on the mechanism of action of this emerging preservation technology on microorganisms, enzymes, and deteriorative reactions. A better understanding regarding these areas will help identify key factors and their combined effect on product safety, stability, and quality. Innovative fruit preservation methods using high pressure will bring exotic tropical fruit products to international markets soon.

43.10 ACKNOWLEDGEMENTS

The authors acknowledge financial support of the Universidad de las Américas-Puebla and the Project XI.15 of CYTED-Program.

REFERENCES

1. Hayashi, R. 1995. "Advances in High Pressure Processing Technology in Japan," in *Food Processing: Recent Developments*, A. G. Gaonkar, ed. London: Elsevier: pp. 185–195.
2. Cheftel, J. C. 1995. "Review: High-Pressure, Microbial Inactivation and Food Preservation," *Food Sci. Technol. Int.*, 1: 75–90.
3. Barbosa-Cánovas G. V., U. R. Pothakamury, E. Palou, and B. G. Swanson. 1998. *Nonthermal Preservation of Foods*. New York: Marcel Dekker.
4. Palou, E., A. López-Malo, G. V. Barbosa-Cánovas, and B. G. Swanson. 1999. "High Pressure Treatment in Food Preservation," in *Handbook of Food Preservation*, M. S. Rahman, ed. New York: Marcel Dekker, pp. 533–576.
5. Knorr, D. 1995. "High Pressure Effects on Plant Derived Foods," in *High Pressure Processing of Foods*, D. A. Ledward, D. E. Johnston, R. G. Earnshaw, and A. P. M. Hasting, eds. Nottingham: Nottingham University Press, pp. 123–136.
6. Ledward, D. A. 1995. "High Pressure Processing—The Potential," in *High Pressure Processing of Foods*, D. A. Ledward, D. E. Johnston, R. G. Earnshaw, and A. P. M. Hasting, eds. Nottingham: Nottingham University Press, pp. 1–6
7. Anonymous. 1998. Flow International Co., Technical data.
8. Hoover, D. G, C. Metrick, A. M. Papineau, D. F. Farkas, and D. Knorr D. 1989. "Biological Effects of High Hydrostatic Pressure on Food Microorganisms," *Food Technol.*, 43: 99–107.
9. Farr, D. 1990. "High Pressure Technology in the Food Industry," *Trends Food Sci. Technol.*, 1: 14–16.
10. Knorr, D. 1995. "Hydrostatic Pressure Treatment of Food: Microbiology," in *New Methods of Food Preservation*, G. W. Gould, ed. New York: Blackie Academic and Professional, pp. 159–175.
11. Seyderhelm, I., S. Bouguslawski, G. Michaelis, and D. Knorr. 1996. "Pressure Induced Inactivation of Selected Enzymes," *J. Food Sci.*, 61: 308–310.
12. Yen, G. C. and H. T. Lin. 1996. "Comparison of High Pressure Treatment and Thermal Pasteurization Effects on The Quality and Shelf-Life of Guava Purée," *Int. J. Food Sci. Technol.*, 31: 205–213.
13. Palou, E., A. López-Malo, G. V. Barbosa-Cánovas, J. Welti-Chanes, and B. G. Swanson. 1997. "High Hydrostatic Pressure as a Hurdle for *Zygosaccharomyces bailii* Inactivation," *J. Food Sci.*, 62: 855–857.
14. Palou, E., A. López-Malo, G. V. Barbosa-Cánovas, J. Welti-Chanes, and B. G. Swanson. 1997. "Effect of Water Activity on High Hydrostatic Pressure Inhibition of *Zygosaccharomyces bailii*," *Lett. Appl. Microbiol.*, 24: 417–420.
15. Palou, E., A. López-Malo, G. V. Barbosa-Cánovas, J. Welti-Chanes, and B. G. Swanson. 1997. "Kinetic Analysis of *Zygosaccharomyces bailii* Inactivation by High Hydrostatic Pressure," *Lebensm. –Wiss. u-Technol.*, 30: 703–708.
16. López-Malo, A., E. Palou, G. V. Barbosa-Cánovas, J. Welti-Chanes, and B. G. Swanson. 1998. "Polyphenoloxidase Activity and Color Changes During Storage of High Hydrostatic Pressure Treated Avocado Purée," *Food Res. Int.*, 31: 549–556.
17. Palou, E., C. Hernández- Salgado, A. López-Malo, G. V. Barbosa-Cánovas, B. G. Swanson, and J. Welti-Chanes. 2000 "High Hydrostatic Pressure Treated Guacamole," *Innovative Food Science and Emerging Technologies*, 1: 69–75.
18. Parish, M. E. 1997. "Public Health and Non Pasteurized Fruit Juices," *Crit. Rev. Microbiol.*, 23: 109–119.

19. Gould, G. W. 1996. "Industry Perspectives on the Use of Natural Antimicrobials and Inhibitors for Food Applications," *J. Food Prot.*, Supplement: 82–86.

20. Isaacs, N. S., P. Chilton, and B. Mackey. 1995. "Studies on the Inactivation by High Pressure of Micro-Organisms," in *High Pressure Processing of Foods*, D. A. Ledward, D. E. Johnston, R. G. Earnshaw, and A. P. M. Hasting, eds. Nottingham: Nottingham University Press, pp. 65–79.

21. Patterson, M. F., M. Quinn, R. Simpson, and A. Gilmour. 1995. "Sensitivity of Vegetative Pathogens to High Hydrostatic Pressure Treatment in Phosphate-Buffered Saline and Foods," *J. Food Prot.*, 58: 524–529.

22. Earnshaw, R. G. 1995. "Kinetics of High Pressure Inactivation of Microorganisms," in *High Pressure Processing of Foods*, D. A. Ledward, D. E. Johnston, R. G. Earnshaw, and A. P. M. Hasting, eds. Nottingham: Nottingham University Press, pp. 37–46.

23. Palou, E., A. López-Malo, G. V. Barbosa-Cánovas, J. Welti-Chanes, P. M. Davidson, and B. G. Swanson. 1998. "High Hydrostatic Pressure Come-Up Time and Yeast Viability," *J. Food Prot.*, 61: 1657–1660.

24. Smelt, J. P., P. C. Wouters, and A. G. F. Rijke. 1998. "Inactivation of Microorganisms by High Pressure," in *The Properties of Water in Foods. ISOPOW 6*, D. S. Reid, ed. London: Blackie Academic and Professional, pp. 398–417.

25. Oxen, P. and D. Knorr. 1993. "Baroprotective Effects of High Solute Concentrations Against Inactivation of *Rhodotorula rubra*," *Lebensm. –Wiss. u- Technol.*, 26: 220–223.

26. Knorr, D. 1993. "Effects of High-Hydrostatic Pressure Process on Food Safety and Quality," *Food Technol.*, 47: 156–61.

27. Ogawa, H., K. Fukuhisa, and H. Fukumoto. 1992. "Effect of Hydrostatic Pressure on Sterilization and Preservation of Citrus Juice," in *High Pressure and Biotechnology*, C. Balny, R. Hayashi, K. Heremans, and P. Masson, eds. France: INSERM & John Libbey, pp. 269–278.

28. Smelt, J. P. and G. Rijke. 1992. "High Pressure Treatment as a Tool for Pasteurization of Foods," in *High Pressure and Biotechnology*, C. Balny, R. Hayashi, K. Heremans, and P. Masson, eds. France: INSERM & John Libbey, pp. 361–363.

29. Carlez, A., J. P. Rosec, N. Richard, and J. C. Cheftel. 1993. "High Pressure Inactivation of *Citrobacter freundii*, *Pseudomonas fluorescens* and *Listeria innocua* in Inoculated Minced Beef Muscle," *Lebensm. –Wiss. u- Technol.*, 26: 357–363.

30. Carlez, A., J. P. Rosec, N. Richard, and J. C. Cheftel. 1994. "Bacterial Growth During Chilled Storage of Pressure-Treated Minced Meat," *Lebensm. –Wiss. u- Technol.*, 27:48–54.

31. Takahashi, Y., H. Ohta, H. Yonei, and Y. Ifuku. 1993. "Microbicidal Effect of Hydrostatic Pressure on Satsuma Mandarin Juice," *Int. J. Food Sci. Technol.*, 28: 95–102.

32. Hashizume, C., K. Kimura, and R. Hayashi. 1995. "Kinetic Analysis of Yeast Inactivation by High Pressure Treatment at Low Temperatures," *Biosci. Biotech. Biochem.*, 59: 1455–1458.

33. Ogawa, H., K. Fukuhisa, Y. Kubo, and H. Fukumoto. 1990. "Inactivation Effect of Pressure Does Not Depend on the pH of the Juice," *Agric. Biol. Chem.*, 54: 1219–1225.

34. Butz, P., S. Funtenberger, T. Haberdtzl, and B. Tauscher. 1996. "High Pressure Inactivation of *Byssochlamys nivea* Ascospores and Other Heat Resistant Moulds," *Lebensm. –Wiss. u- Technol.*, 29: 404–410.

35. Palou, E., A. López-Malo, G. V. Barbosa-Cánovas, J. Welti-Chanes, P. M. Davidson, and B. G. Swanson. 1998. "Effect of Oscillatory High Hydrostatic Pressure Treatments on *Byssochlamys nivea* Ascospores Suspended in Fruit Juice Concentrates," *Lett. Appl. Microbiol.*, 27: 375–378.

36. Maggi, A., S. Gola, E. Spotti, P. Rovere, and P. Mutti. 1994. "Tratamenti ad Alta Pressione di Ascospore di Muffe Termoresistenti e di Patulina in Nettare di Albicocca e in Acqua," *Ind. Conserve.*, 69: 26–29.

37. Hendrickx, M., L. Ludikhuyze, I. Van den Broeck, and C. Weemaes. 1998. "Effects of High Pressure on Enzymes Related to Food Quality," *Trends Food Sci. Technol.*, 9: 197–203.

38. Eshtiaghi, M. N. and D. Knorr. 1993. "Potato Cubes Response to Water Blanching and High Hydrostatic Pressure," *J. Food Sci.*, 58: 1371–1374.

39. Knorr, D. 1999. "Process Assessment of High-Pressure Processing of Foods: an Overview," in *Processing Foods. Quality Optimization and Process Assessment*, F. A. R. Oliveira, and J. C. Oliveira, eds. New York: CRC Press, pp. 249–267.

40. Weemaes, C., L. Ludikhuyze, I. Van den Broeck, and M. Hendrickx. 1998. "High Pressure Inactivation of Polyphenoloxidases," *J. Food Sci.*, 63: 873–877.

41. Cano, P. M., A. Hernández, and B. De Ancos. 1999. "Combined High-Pressure/Temperature Treatments for Quality Improvement of Fruit-Derived Products," in *Processing Foods. Quality Optimization and Process Assessment*, F. A. R. Oliveira, and J. C. Oliveira, eds. New York: CRC Press, pp. 301–312.

42. Anese, M., M. C. Nicoli, G. Dall'Aglio, and C. R. Lerici. 1995. "Effect of High Pressure Treatments on Peroxidase and Polyphenoloxidase Activities," *J. Food Biochem.*, 18: 285–293.

43. López-Malo, A., E. Palou, G. V. Barbosa-Cánovas, B. G. Swanson, and J. Welti-Chanes. 2000. "Minimally Processed Foods and High Hydrostatic Pressure," in *Trends in Food Engineering*, J. Lozano, C. Añon, E. Parada-Arias, and G. V. Barbosa-Cánovas, eds. Lancaster: Technomic Publishing, pp. 267–286.

44. Cano, P. M., A. Hernández, and B. De Ancos. 1997. "High Pressure and Temperature Effects on Enzyme Inactivation in Strawberry and Orange Products," *J. Food Sci.*, 62: 85–88.

45. Mertens, B. 1995. "Hydrostatic Pressure Treatment of Food: Equipment and Processing," in *New Methods of Food Preservation*, G. W. Gould, ed. New York: Blackie Academic and Professional, pp. 135–158.

46. Ludikhuyze, L., I. Van den Broeck, C. A. Weemaes, and M. E. Hendrickx. 1997. "Kinetic Parameters for Pressure-Temperature Inactivation of *Bacillus subtilis* α-Amylase Under Dynamic Conditions," *Biotechnol. Prog.*, 13: 617–623.

47. Palou, E., A. López-Malo, G. V. Barbosa-Cánovas, J. Welti-Chanes, and B. G. Swanson. 1999. "Polyphenoloxidase Activity and Color of Blanched and High Hydrostatic Pressure Treated Banana Purée," *J. Food Sci.*, 64: 42–45.

44 Production Issues Related to UHP Food

E. Y. Ting
R. G. Marshall

CONTENTS

Abstract

The ability of high hydrostatic pressure to inactivate food pathogens and spoilage microorganisms is now well demonstrated in scientific literature. The quality advantages of ultrahigh pressure (UHP) produced foods are numerous and have been reviewed elsewhere. This paper summarizes the common economic and related issues of interest to food producers.

For over 100 years, scientists have studied the use of pressure as a potential extra control variable in manipulating and understanding biological systems such as foods, proteins, bacteria, and viruses. In comparison with today's equipment, the initial UHP hardware utilized in the 1890s by Hite are prehistoric. Today, with advances in computational stress analysis and new materials, large-capacity pressure systems can be manufactured to create reliable UHP conditions needed for food production. The last remaining hurdle for this technology is that of cost justification. This chapter reviews issues related to product suitability, UHP hardware selection, food packaging, value-added, and production costs.

44.1 PRODUCT SUITABILITY

Many products are suitable for UHP treatment. UHP is a low-energy, selective, inactivation approach. Unlike high-temperature treatment or irradiation, UHP does not induce covalent bond breakage and the formation of new compounds. As a result, taste and nutrition are unaffected. The microbiological behavior of microorganisms under high pressure has been extensively studied. The following discussion will focus on mechanical effects of high-pressure exposure.

The physical structure of most high-moisture products remains unchanged after exposure to UHP, since no shear forces generated by hydrostatic pressure and fluids are isotropic in nature. For gas-containing products, the color and texture may be changed due to gas displacement and liquid infiltration. For example, cut cantaloupe fruit will experience a darkening of color and a slight change in texture due to the infiltration of liquid into the gas pores. Other cut fruits, if of higher liquid content (e.g., oranges), do not show this effect. An example of pressure treated cut fruit is shown in Figure 44.1. As an extreme illustration, a marshmallow after UHP will be physically smaller and somewhat distorted when compared to an untreated one. Physical shrinkage is due to mechanical collapse of air pockets, and shape distortion is related to anisotropic behavior. For foods not containing air voids, UHP results in minimal or no permanent change in characteristics.

The ability of pressure to infiltrate liquids into a product can be exploited as an advantage in a number of situations. Pressure can be used to enhance the effect of marinades in both cooked and uncooked products. A uniform flavoring or other tenderizing agents can be rapidly distributed in the product by brief pressure exposure. Along with infiltration, extended shelf life can be obtained.

For some raw protein based products, and depending on the pressure-time exposure, research indicates that some degree of protein denaturization can take place. This can result in a change of physical functions and/or color change for raw products. For semicooked or cooked products, this effect is not observed, since their proteins have already been denatured. Depending on pressure and product, some meat products may also experience some modification in texture. Both tenderizing and toughening of meats have been reported after UHP exposure.

Figure 44.1 Fresh cut fruit prepared by UHP treatment. Very high-quality, extended shelf life products can be produced without heating or chemical additives.

Tenderization appears to be related to the destruction of connective muscle fibers, whereas toughening appears to be more related to mechanical compaction effects. It is also recognized that, for certain proteins and starches, gels can be formed by UHP with unique characteristics. These are formulation and product specific. Some of the most significant nonmicrobiological benefits have been reported with dairy products such as yogurts, in which significantly improved textures have been achieved by UHP. For many of the prepared foods products evaluated, such as ready-to-eat deli meats, no significant negative texture effects have been observed.

The best method to explore the potential of UHP is to conduct quick feasibility tests. A number of universities have established significant research programs around high pressure. Industrial labs and equipment for UHP product development are also becoming commonplace.

44.2 UHP EQUIPMENT SELECTION

A basic illustration of UHP food treatment is shown in Figure 44.2. The food product to be pressurized is placed into a pressure vessel capable of sustaining the required pressure. An external pump is used to push water into the pressure vessel until the desired pressure is achieved. Due to the compressibility of water, the amount of water added to the pressure vessel during pressurization is approximately 15% at 600 MPa. Between every cycle, the pressure vessel must be unloaded and loaded with product. Human operators or a machine can perform this task, depending on the degree of automation desired.

The selection of equipment for UHP treatment depends on the kind of food product to be processed. Foods with large solid particles can be treated only in a batch mode. Liquids, slurries, and other pumpable products have the additional option of an in-line production mode. For both modes, a system typically contains at least one high-pressure cylinder with end closures, and an end closure retention structure. The end closure retention structure must be able to resist the internal pressure-induced force acting on the end closure. Some lower pressure, intermittent use system might use machined threads in the main pressure vessel and end closures to retain the closure. However, the use of large-diameter threads is typically undesirable in heavy-duty UHP equipment due to their historical poor service reliability. Commercial production systems use an external yoke to contain the end closures. Prestressing of the yoke is frequently also performed to create a more fatigue-resistant structure.

In batch UHP production, in-package product is accumulated in a carrier and then inserted into the pressure vessel. The size of the pressure vessel determines the amount of product that can be treated per cycle. The largest pressure vessel built for food application is approximately 400 liters. A common vessel size for in-package food treatment is shown in Figures 44.3 and 44.4. Depending on product density and package size and shape, this 215-liter capacity machine typically produces from 300 to 400 lb of products per cycle.

If the product is pumpable, it may be more advantageous to pump it into and out of the processing vessel through special high-pressure transfer values and iso-

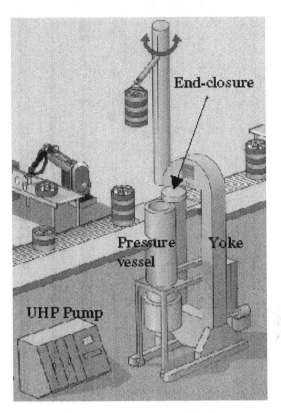

Figure 44.2 Basic pressure vessel system for food. Product handling approach depends on degree of automation desired.

lators (Figure 44.5). In this mode, the system alternates between filling, pressurizing, holding, depressurizing, and emptying cycles without the need to move the yoke or a large end closure. The system operates much like an automobile engine, with each isolator cylinder cycling out of sequence. In addition to faster turnaround time, this in-line approach uses 100% of the available volume of each cylinder for production. To package the product after treatment, additional subsystems, such as an aseptic filling station, are required. Because of the increased operational efficiency provided by pumpable products, moderate vessel sizes can be used in more commercial production. A typical flow rate for fresh juice processing is 10 gpm (37 l/min) using three 25-liter isolators. For much higher output, transfer valves can be used with larger vessels. A system of three 215-liter isolators can achieve an output of approximately 85 gpm (320 l/min). The in-line approach is ideally suited for continuous operation of a single product type. Frequent production product changes are not well suited for the in-line process, due to decreased productivity because of system cleaning and system priming.

The design of a pressure vessel is independent from its use as either an in-package or in-line production unit. However, in-package production equipment tends

Figure 44.3 A 215-liter capacity food pressurization vessel under construction. Large pressure vessels are typically constructed using wire-winding technology to overcome heat-treating related strength losses encountered with thick metal cross sections.

toward larger vessel sizes and thus is frequently made using wire-winding technology. By using wires rather than monolithic metal, large-capacity vessels can be constructed with consistent mechanical properties. For smaller vessels, monolithic metal alloy technology is a more cost-effective approach. Extensive advances in fatigue engineering and manufacturing techniques have been used to significantly increase the fatigue life of monolithic materials. These techniques frequently involve high-purity alloys, autofrettage, and other proprietary processing methods. These methods, however, are still limited to moderate vessel sections due to heat transfer limitations of thick sections. As a result of reduced cooling, high-performance mechanical properties cannot be achieved in thick sections. The properties of wire-wound vessels are not sensitive to the thickness of the wire layers, since heat treating is not required after construction. Multiple-layer vessels designs using thick sections can extend the size of non-wire-wound vessels, but, even for large multiple layer vessels, manufacturing difficulties quickly make them impractical for large sizes.

44.3 FOOD PRODUCT PACKAGING

Using an in-package approach, the packaging structure or material must allow the compression of the product during UHP exposure. Most foods are compressible to roughly the same extent as water. Figure 44.6 shows the change in volume for water as determined by Bridgman. At 600 MPa, water is compressed to 15% lower volume.

Figure 44.4 A commercial UHP food operation using simple product carriers. Preloaded carriers are used to minimize cycle time. Photo courtesy of Avomex.

Packaging used must allow for this deformation without compromising its sealing characteristics. For example, a rigidly closed glass bottle will certainly fail to serve as a package for UHP. However, a glass cup with an elastic membrane will work well. A commercial package using this approach is shown in Figure 44.7. Since the size and shape of the product will have major effects on the stacking effectiveness of the product carrier, product size and shape must be optimized for the most cost-effective process. It is, of course, uneconomical to treat empty space. Figure 44.8 illustrates the theoretical stacking efficiency for a 215-liter system as a function of using sample cylinders of a different diameter. Except when using a single, large sample cylinder, by using cylinders of a uniform size, a maximum packing density of about 75% can be achieved.

In production, the use of flexible pouches like the one shown in Figure 44.9 can achieve high packing ratios. The use of semi-rigid trays is also possible (Figure 44.10). Vacuumed-packed products are ideally suited for UHP. A gas-packed package can also work, depending on the product. The use of dissolved CO_2 gas can have significant synergistic inactivation effects with UHP. However, it is generally not desirable to introduce a large amount of gases into the product package, as it will add additional time to the pressurization process.

The production of pumpable products using the isolator concept allows the separation of packing requirements from pressure exposure. Any product-compatible

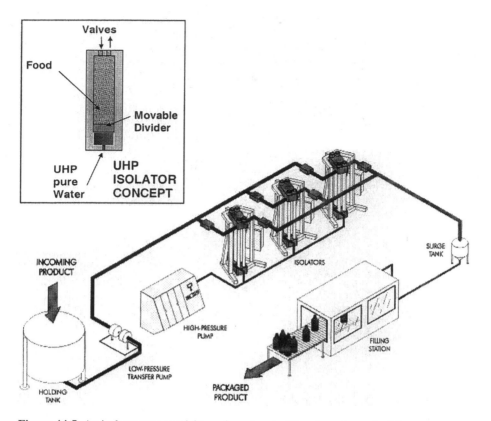

Figure 44.5 An isolator uses special transfer valves to fill and empty product from a pressure vessel (see insert). An internal divider prevents product mixing with UHP water. A number of coordinated isolators can be combined to produce a continuous or semicontinuous output.

packaging can be used. For example, glass bottles or gable cartons can be used if filling is performed after UHP exposure. This further allows innovative package shapes and printing graphics.

Since only high-quality products are typically produced using UHP, and extended quality is desired, the use of high-quality barrier film materials packaging materials should be considered. The use of vacuum packaging where possible also enhances loading effectiveness and avoids product degradation due to oxygen reactions.

44.4 VALUE

Commercial feasibility of any technology rests on business profitability. The production cost of a process must, of course, be lower than the value added to the product. The value added by UHP can be measured in terms of higher product quality, increased safety, and longer shelf life. These issues further can translate into less transportation cost, lower storage costs, lower labor costs, consumer conve-

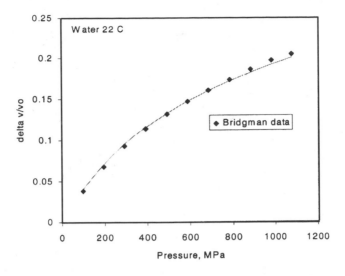

Figure 44.6 Pressure and temperature determine the compressibility of pure water. Most low-fat foods have similar behavior to water under UHP compression.

Figure 44.7 An elastic surface makes this rigid glass container suitable for UHP treatment. Plastic lid is installed after pressure process.

nience, consumer safety, or lower insurance costs. For example, the quality contrast between a premium fresh juice and a heat-pasteurized juice is so dramatic that the premium product can be valued at 5 to 10 times that of a "commodity" juice. This fresh juice strategy, though, is not feasible without the use of an active intervention method such as UHP, due to food safety concerns. Fundamentally, the strategy for

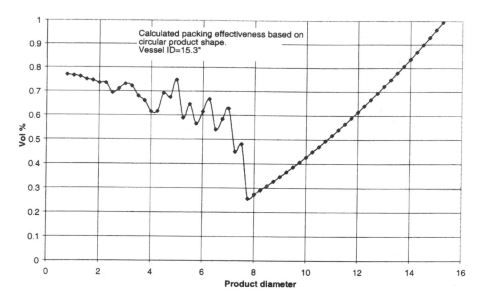

Figure 44.8 Calculated loading efficiency of a 215-liter capacity vessel as a function of inserted product diameter. Stacking efficiency is highest when using a single sample the same diameter as the pressure vessel, and roughly 75% when using a number of uniform very small sized (<1-in.) diameter samples.

Figure 44.9 A flexible self-standing display pouch makes an attractive UHP package for sauces and other viscous products.

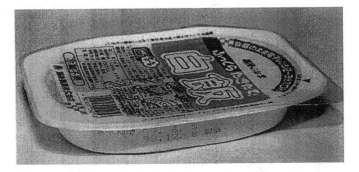

Figure 44.10 A semi-rigid tray can be used for in-package treatment. However, its loading efficiency is typically not as high as with flexible packages.

superior products is based on a higher perceived added value rather than on absolute cost.

Consumers are changing their habits. Consumers are spending less on basic foods. Premium product sales, restaurant spending, and ready-to-eat food spending are all increasing worldwide. Health- and nutrition-driven product sales have also increased significantly. New technologies such as UHP can allow producers to create new markets not possible with old technologies. These benefits are only now being exploited. Consumers are willing to pay more for greater perceived value.

The greatest value of UHP to producers is related to safety. Based on CDC statistics, foodborne pathogens are responsible for 76 million illnesses and 5,000 deaths yearly in the United States. In 1998, 21 deaths were attributed to *Listeria monocytogenes* in one outbreak. The vast majority of food illness cases are not reported. In effect, the public has paid for this significant cost. As microbiological standards become more widely mandated, and sensitive assessment techniques become available in food production, the financial cost of bad products will be transferred back to the producer. Massive product recalls will provide a quantifiable value to using an active intervention method such as UHP. The value of food safety is frequently difficult to quantify prior to an incident. However, as has been observed in recent pathogen contamination events, a producer's reputation or brand name may not ever recover from a single food safety related incident. Having an always-safe product is now part of the cost of doing business.

44.5 PRODUCTION COST

Two main factors make up production cost: capital cost and operations cost. Some approximate values are estimated for a twin 215-liter in-package (batch) production system in Table 44.1. These estimates show that production cost can be as low as approximately between 3 and 4 cents per pound for operations. Actual cost will depend on many factors ranging from operating pressure, cycle time, and product geometry to labor skills and energy costs. As with all capital equipment, the greater

TABLE 44.1
Rough Calculation Showing Estimated Cost per Pound of UHP Technology for Typical Applications

Yearly	
Production volume in liters (two machines)	430
Cycles per hour	6
Loading effectiveness	0.70
Hours per year	6,000
Pounds per hour	3,972
Total production, pounds	23,829,664
Equipment/facilities cost	$5,000,000
Total depreciation cost (10 years)	$500,000
Labor cost per year	$180,000
Energy-water cost per year	50,000
Preventive maintenance parts, cost per year	150,000
Total operating cost	$380,000
Per pound	
Total dep. cost per pound	0.021
Total operating cost per pound	0.016
Total cost per pound	$0.037

the utilization, the more cost-effective it is. As the technology matures and producers gain experience, lower equipment and operations costs can be anticipated.

The production cost for a pumpable product system using isolators can be expected to be 25–50% less due to increased volumetric efficiency and filling/emptying speeds. Additional costs for pumpable operation not encountered with in-package batch operation are related to cleaning equipment (CIP system) and packaging equipment. The magnitude of these added costs depends on the type of product to be produced.

44.6 THE FUTURE OF UHP FOOD

The use of UHP to produce high-quality high-acid and refrigerated low-acid products is now no longer new. The next major step in technology development will be toward the production of shelf-stable low-acid foods. Using synergistic approaches (hurdle concepts) such as biopreservatives, additives, moderate temperatures, germinate-then-kill cycles, and other strategies, current research is targeting the total inactivation of bacteria and spores without resorting to the high temperatures currently used in retort canning. Future applications will also expand into nontraditional applications involving pharmaceutical, medical, and biotechnology products. It is anticipated that this will lead to the production of many high-quality products not currently possible using today's methods.

44.7 CONCLUSION

UHP food production is now entering production on an ever-increasing commercial basis. As a new tool, opportunities abound for innovative applications and new food product development. The future of UHP food is now transitioning from the food scientist to the product and market development teams.

45 Enhanced Thermal Effects under Microwave Heating Conditions

H. S. Ramaswamy
T. Koutchma
S. Tajchakavit

CONTENTS

Abstract

This chapter brings into perspective issues related to the thermal and nonthermal effects of microwave heating. To distinguish thermal and nonthermal effects, most studies rely on experimental evaluation of existing differences in microbial or enzyme destruction under somewhat identical heating conditions, using some form of conventional heating and a microwave oven. Studies supporting additive microwave contribution to lethal effect over that caused by heat have been mainly criticized for lack of reliable temperature measurements or uncontrolled heating. Surprisingly, studies supporting only thermal effects don't suffer from such criticisms, although they are carried out under similarly questionable conditions.

Results of several studies carried out in our laboratory under continuous flow microwave and conventional heating systems are provided to demonstrate meaningful comparison of microbial destruction (*E. coli, L. plantarum,* and *S. cerevisae*) and enzyme inactivation (pectin methyl esterase) under the two heating modes. To quantify the thermal and/or nonthermal effects of microwave on microbial destruction and enzyme inactivation, traditional kinetic parameters in the form of *D* and *z* values were evaluated based on come-up and come-down time corrected temperature data obtained under experimental conditions. Modeling studies indicated that the time-temperature distribution in helical coils under continuous flow microwave heating systems closely matched a plug-flow distribution. *D* values for microorganisms and enzymes computed assuming such a distribution, effective residence times and microbial survivor data were almost an order of magnitude lower for microwaves than those obtained under comparable water bath heating conditions. Such consistent results obtained for microorganisms and enzymes in different systems clearly demonstrate that lethal contributions under microwave heating conditions cannot be solely explained on the basis of time-temperature distribution, thus implying additional contributory effects.

45.1 INTRODUCTION

Microwave heating involves conversion of electromagnetic energy into heat by selective absorption and dissipation. Microwave heating is attractive for heating of foods due to its volumetric origin, fast temperature rise, controllable heat deposition, and easy clean up. It is currently being used for a variety of domestic and industrial applications involving food. Destruction of microorganisms by microwave heating has had considerable interest since the 1940s, when the first work by Fleming[1] was reported, and the argument between nonthermal and thermal effects was born. Still, doubts exist, and research proceeds. Research studies have focused on different experimental procedures, approaches, experimental designs, techniques, and biological systems to distinguish thermal or nonthermal effects of microwave heating and to resolve the controversial question. Several theories have been advanced to explain how electromagnetic fields might kill microorganisms without heat as summarized in a review by Knorr.[2] On the other hand, some researchers[3] refute any molecular effects of electric fields compared with thermal energy using classical axioms of physics and chemistry. Palaniappan and Sastry[4] concluded that the effects of microwave and dielectric heating are clearly fields where there is a knowledge gap and that further studies are needed.

The objective of this chapter is to bring together the available information on microwave heating with a focus on distinguishing thermal and nonthermal effects. To make meaningful comparison, a quantitative approach is used to integrate the time-temperature effects under comparable microwave and conventional heating conditions. The equivalent or thermal time approach used in classical thermal processing application can be adopted to microwave heating situation. These aspects will be discussed based on the research carried out at McGill University involving continuous flow microwave heating applications.

45.2 MICROWAVE THERMAL, NONTHERMAL, AND ENHANCED EFFECTS

Microwaves used for domestic and industrial heating applications are part of the electromagnetic spectrum with a specific frequency of 915 or 2450 MHz. The waves have the capacity to penetrate the food and create heat by friction of dipole molecules of water, which will try to orient and align with the field. This is the macroscopic thermal effect of increasing temperature within the material. Traditionally, nonthermal effects under the application of electromagnetic radiation refers to lethal effects without involving a significant rise in temperature, as in the case of ionizing radiation. One of the effects of such quantum energy is breakage of chemical bonds. Roughly one electron volt (1 eV) of energy is required to break a covalent bond from a molecule to produce one ion pair;[5] this is referred to as a *direct nonthermal effect*. Electromagnetic radiation above 2500×10^6 MHz, which possesses such a capability, is mostly referred to as ionizing radiation (for example, X-rays, gamma rays, etc.). As the wavelength increases and frequency decreases, not enough energy is available to break chemical bonds. Ultraviolet, visible, and possibly infrared rays have energy to break weak hydrogen bonds, but microwaves do not have sufficient energy to break any chemical bonds and therefore belong to the group of nonionizing forms of radiation.

Since the beginning of the use of microwaves in chemistry and biology, the argument between nonthermal and thermal effects of microwaves has existed. First of all, it should be clearly defined what characterizes thermal, athermal or nonthermal, enhanced, or specific microwave effects. The review of Stuerga and Gaillard[3] provides some definitions used in biological studies based on irradiation or power flux density in W/m² or specific absorption rate in W/kg, and it does not involve any assumptions relating to mechanisms of interaction. Three domains of power density in comparison with the thermal capacity of biological metabolism were defined. The threshold selected for standard definition of biological metabolism corresponds to 4 W/kg of living matter; hence, thermal effects were defined for irradiation power density greater than 4 W/kg, when the organism can not dissipate the energy supplied by the irradiation; athermal effects correspond to density between 0.4 and 4 W/kg. In such conditions, the thermoregulation system is able to compensate for the effect of irradiation.

What must characterize nonthermal microwave effects? According to Risman,[6] any nonthermal effect must not be explicable by macroscopic temperatures, time-temperature histories, or gradients. This means that any effects that can be explained by applying verified theories to experimental data and macroscopic temperatures are not nonthermal. The cases in which microwave heating gives a particular time-temperature profile and gradients, which can not be achieved by other means, are only microwave specific. The effect of higher temperatures achieved by microwave heating was explained by Gedye.[7,8] According to Gedye, it is possible that microwave reactions could produce different products from reactions achieved by using conventional reflux techniques. Since microwave heating significantly increases the reaction temperature, it is possible that the microwave reaction temperature could exceed the temperature required for a new reaction that was not possible at the lower reflux temperature. According to Gedye, these results are important, because they confirm that microwave heating does not alter the reaction but simply provides a much faster and more efficient (higher temperature) method of carrying out organic reactions. These effects are defined as *enhanced microwave effects.*

45.2.1 MICROWAVE THERMAL EFFECTS ON MICROBIAL DESTRUCTION

In connection with studies supporting thermal effects of microwave heating, the results of Goldblith and Wang[9,10] showed no difference in inactivation of *E. coli* and *B. subtilis* exposed to the same time-temperature conditions of microwave and conventional heating. However, microwave treatments were made in transient-state and temperature gradient as well as cooling conditions that were not discussed. Lechowich et al.[11] studied the exposure of *S. faecalis* and *S. cerevisiae* to microwave at 2450 MHz and under conventional heating, and concluded that the inactivation by microwaves could be explained solely in terms of heat generated during the exposures. Kinetics of destruction were not evaluated in the study. Vela and Wu[12] used a different approach by exposing various bacteria, actinomycetes, fungi, and bacteriophages to microwaves in the presence and absence of water. They found that microorganisms were inactivated only in the presence of water and killed by thermal effect. There was no temperature increase when lyophilized cells were irradiated. The destruction of *E. coli*, *S. aureus*, *P. fluorescens*, and spores of *B. cereus* by microwave irradiation at three power levels were studied by Fujikawa et al.[13,14] Assuming uniform temperature distributions, these authors didn't find any remarkable difference and concluded that mostly thermal effects can interpret the destruction profiles by microwave exposure. No comparison between microwave and conventional heating temperature profiles was included in the study. Welt and Tong[15-17] introduced an apparatus to evaluate possible athermal effects of microwaves on biological and chemical systems and explained the limits and requirements of comparative studies between kinetics under microwave conventional heating. They compared inactivation of *C. sporogenes* and of thiamin under equivalent time-temperature treatments by perfectly stirred batch treatment by conventional and microwave heating and did not observe athermal effect. Diaz-Cinco and Martinelli[18] performed experiments to see if the lethality of microorganisms is due to microwave radiation or heat generated by microwave radiation with four types of microorganisms treated, as indicated by the authors, to the same exposures of time and temperature. They concluded that the killing effect was due to the heat generated by microwave energy, based on the percentage of the survivors. Heddlesson et al.[19] showed that the relationship between time of heating, temperatures, and microbial destruction achieved by microwave heating followed conventional reaction kinetics, and that this illustrated its thermal origin.

45.2.2 MICROWAVE ATHERMAL AND SPECIFIC EFFECTS

In connection with additional enhancement of microbial and enzymes destruction by microwave energy, the following studies indicate and support the phenomena of nonthermal and specific effects of microwaves. First, the study of Culkin and Fung[20] should be mentioned. They exposed single portions of soups inoculated with *E. coli* and *S. typhimurium* to 915-MHz microwaves and found that, for any exposure time, the closer the sampled organisms were to the top, the lower was their level of survival. They suggested that the heat generated during microwave exposure alone is inadequate to fully account for the nature of the lethal effects of microwaves. Dreyfuss and Chipley[21] characterized some of the effects of sublethal microwave heating on cells of *S. aureus*. They determined higher enzymatic activities in microwave-treated cells that cannot be explained solely by thermal effects. However, the results cannot be compared due to the lack of thermal control. Mudgett[22] calculated the lethality of *E. coli* strain in the continuous system and compared it with experimentally measured values. Experimental microbial lethality was somewhat greater than that predicted by numerical integration. The author believed that it could have resulted from sensitivity of the kinetic model to small differences in temperature, or possibly from the second-order kinetic effects from the selective absorption of microwave energy by the test organism based on high intracellular conductivity.

A comparative study of Khalil and Villota[23] showed consistently higher lethality when exposed to microwave irradiation and lower D values ($D_{212} = 157$ min for microwave and $D_{212} = 171$ min for conventional heating in distilled water) for *B. stearothermophilus* spores. Another study conducted by Khalil and Vilotta[24] on injury and recovery of *S. aureus* using microwave and conventional heating at a sublethal temperature (50°C) was carried out in which kerosene was chosen as a cooling medium to keep the sample temperature constant. They concluded that microwave-heated cells suffered greater injury as well as greater membrane damages. Kermasha et al.[25,26] analyzed inactivation of wheat germ lipase and soybean lipoxygenase at various temperatures using conventional and microwave batch heating, and found higher enzyme destruction rates under microwave heating conditions. Odani et al.[27] used microwave irradiation to kill *E. coli*, *S. aureus* and *B. cereus* in frozen shrimp, refrigerated pilaf, and saline. Microwave irradiation was shown to result in the release of proteins from *E. coli*, as detected by gel electrophoresis of cell-free supernatants using sensitive silver staining. They suggested that the mechanisms for killing bacteria depend not only on temperature but also on other effects of microwave irradiation. Comparing microwave and conventional methods of inactivation of *B. subtilis*, Wu and Gao[28] showed that the D_{100} of microwaves was 0.65 min, but, for conventional heating, it was 5.5 min and demonstrated athermal effects of such energy on microorganisms. In a review of contribution to the study of microwave action, Joalland[29] considered the action of microwaves upon bacteria by taking into account the different cellular components, such as genetic material, enzymatic activity, mitochondrias, membranes, and cytoplasma, and concluded that, besides thermal affects, there could exist some nonthermal effects. River et al.[30] reported different z values for *Enterobacter cloacae* and *Streptococcus faecalis*, in batch conventional heating (4.9 and 5.8°C) and microwave heating (3.8 and 5.2°C). The results were explained, rather, due to different heating kinetics and nonuniform local temperature distributions during microwave heating than existence of specific athermal effects. The results of a study by Aktas and Ozilgen[31] of the injury of *E. coli* and degradation of riboflavin during pasteurization with microwaves in a tubular flow reactor also indicated the effect of the flow behavior and other experimental conditions on the death mechanism in the microwave field. They suggested that microbial death might be caused through damage to a different subcellular part under each experimental condition.

There are a number of studies made recently using different approaches for estimation of microwave heating effects. Koutchma[32] studied bactericidal effects of microwaves and hypothermia on *E. coli* cells in batch-mode conditions. No differences were found in survival of those bacteria between the two heating modes with identical time-temperature profiles. It was shown that microwaves caused greater damages to the cell genome and resulted in different survival and interaction coefficients under combined applications with low concentrations of hydrogen peroxide. Kozempel et al.[33,34] developed a pilot-scale nonthermal flow process using microwave energy to inactivate *Pediococcus sp.* NRRLB-2354. A cooling tube within the process line to maintain temperatures below 40°C removed the heat generated by the application of microwave energy to the system. A significant reduction in microbial count was reported. Shin and Pyun[35] published the results of a comparative study of the inactivation of *L. plantarum* using microwave and conventional heating at 50°C for 30 min and indicated significant nonthermal effects under continuous and pulsed microwave heating. Many earlier interpretations and a majority of reports on the nonthermal effects of microwaves have been criticized for their lack of proper temperature control, and that undetected temperature rise could greatly enhance a lethal effect; however, similar arguments are never proposed to those who refute the nonthermal effects.

45.3   KINETICS OF MICROBIAL DESTRUCTION

In the absence of nonthermal effects, the destruction of microorganisms and inactivation of enzymes may be generally modeled as an *n*th order chemical reaction.

$$\frac{dC}{dt} = -kC^n \tag{45.1}$$

where dC/dt = the time rate of change concentration C
         k = the reaction rate constant
         n = the order of reaction

For many foods, a first-order kinetics model adequately describes the destruction. The thermal resistance of microorganisms is also traditionally characterized in the food processing by means of the D and z values.[36]

$$D_{T_{ref}} = \frac{2.303}{k}$$

$$z = \frac{T_2 - T_1}{\log \dfrac{D_1}{D_2}} \tag{45.2}$$

When the thermal resistance of a microorganism is known, it is possible to calculate the equivalent time necessary for thermal treatment by integration of the time-temperature history using Equation (45.3).

$$F = \int_0^t 10^{\frac{T(t) - T_R}{z}} dt \tag{45.3}$$

This approach has been traditionally used in thermal process calculations. A similar concept can be applied to determine kinetics parameters during microwave heating; however, nonisothermal heating conditions are involved in this case. Resulting D values can be computed using Equation (45.4).

$$D = \frac{t_{eff}}{\log\left(\dfrac{C_0}{C}\right)} \tag{45.4}$$

where t_{eff} = an effective time [same as F in Equation (45.3)] with T_R as exit temperature, obtained using either model predicted or experimentally determined time-temperature profiles

C_o and C = the initial and final concentration of microbial cells

The use of this approach is rare in studies concerning microwave effects. There have been only a few studies describing kinetics during microwave heating, and most results refer to evaluation of lethality effects without considering the thermal history of the product.[37] But, as in thermal destruction, microwave destruction kinetics of food constituents such as quality attributes, enzymes, and microorganisms are required for establishing microwave processing.

45.3.1 COME-UP TIME AND COME-DOWN PROFILE CORRECTIONS

Continuous-flow microwave heating has been used to avoid some of the indicated limitations of nonuniform microwave heating under batch processing. It allows us to maintain the time and temperature achieved in steady state, to record mean temperatures during heating, to minimize the temperature gradient by mixing of the liquid, and to cool liquid immediately at the exit. The microwave treatment involves only nonisothermal come-up time (CUT) inside the oven during microwave heating and a come-down period outside during cooling. The procedure for gathering kinetic parameters during continuous-flow heating was detailed in Reference 38. Briefly, the D values at the exit temperatures can be first calculated from the regression of log residual numbers of survivors vs. uncorrected heating time (residence time), and then z value is obtained as the negative reciprocal slope of log D vs. temperature. Using the calculated z value, the heating times can be corrected using Equation (45.3), D values and subsequently the z can be corrected. This step can be repeated as many times as necessary until the convergence of z value. To accommodate the CDT, which is carried out outside the microwave oven, it is necessary to determine the cooling (thermal) contribution to microbial destruction and to subtract it from total destruction. To do this, the effective cooling time (t_c) can be computed using Equation (45.3) with the z value obtained from thermal destruction studies. The extent of logarithmic thermal destruction (LTD) during cooling can be calculated using the following relationship:

$$\text{LTD} = \frac{t_c}{D} \tag{45.5}$$

where D = the D value at the exit temperature obtained from thermal destruction studies

This calculated value is then subtracted from the combined destruction of microbial population due to microwave heating plus cooling. Microbial destruction data of test samples can thus be corrected for both the come-up and come-down periods contribution to lethality. This approach has been used in all of our studies for comparative evaluation of microbial lethality under conventional and microwave heating. The following sections highlight the research carried out in our food processing group at McGill University.

45.4 MICROWAVE DESTRUCTION KINETICS OF MICROORGANISMS

45.4.1 BATCH HEATING (CONVENTIONAL)

For comparative purposes, thermal destruction kinetics of microorganisms in apple juice were first established. Aliquots of juice inoculated with spoilage microflora (*S. cerevisiae* and *L. plantarum*) were exposed to various time-temperature treatments in a well stirred water bath (50–80°C), and the survivors were evaluated. As previously described,[39] the heating times were corrected, taking into account the effective portion of come-up and come-down periods. Figure 45.1 shows typical survivor plots of *S. cerevisiae* before and after making time-temperature corrections. As

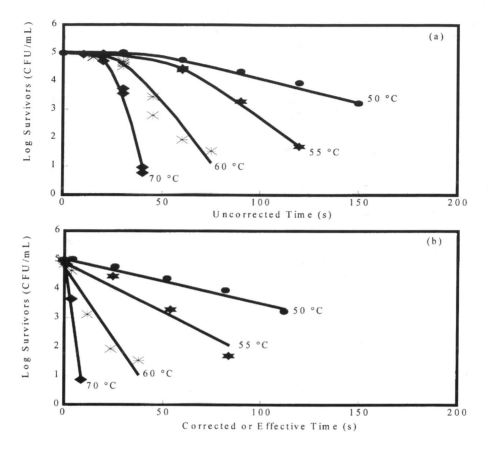

Figure 45.1 Thermal resistance curves of *S. cerevisiae* in apple juice: (a) uncorrected and (b) corrected.

expected, the destruction behavior indicated characteristic first-order rate kinetics. The results of D values calculations are summarized in Table 45.1.

45.4.2 CONTINUOUS-FLOW MICROWAVE HEATING

Microwave Treatment System

The destruction kinetics of *S. cerevisiae*, *L. plantarum*, and pectin methyl esterase (PME) using microwave heating in batch and continuous-flow and conventional heating conditions have been detailed elsewhere by Tajchakavit and Ramaswamy,[38–42] LeBail[43] and Koutchma.[44] Both microwave and conventional heating systems were used for subjecting test samples to continuous-flow heat treatments.

Briefly, the suspension was pumped through glass coils placed in one or two domestic microwave ovens (700 W, 2450 MHz) and cooled at the exit using a water-

TABLE 45.1
Kinetic Parameters (D and z values) of Spoilage Microorganism in Apple Juice during Conventional Heating

| | D values, s | | | |
| | S. cerevisiae | | L. plantarum | |
Temperature (°C)	Uncorrected	Corrected	Uncorrected	Corrected
50	60	57.8	–	–
55	22.2	25.1	50.8	52.5
60	14.3	10.1	24.9	21.9
70	5.03	1.89	13.5	8.44
80	–	–	7.6	1.2
z value	19.4	13.4	31.9	15.9

cooled condenser. Copper-constantan thermocouples were inserted into the tubes at inlet, between the ovens and at exit of the cavity and holding tube, and temperature data were gathered using an HP data logger. The system was sanitized by circulating water at 70–80°C through the system before and after treatment. Prior to heating, the cell suspension was precirculated in the system to establish steady-state flow conditions, and then the microwave ovens were turned on. Test samples were withdrawn during steady-state heating periods with exit temperatures in the range from 50 to 70°C and were collected into precooled sterile tubes immersed in an ice-water bath at the exit. Each exit temperature was achieved by preadjusting and changing the flow rate. For conventional heating, both coils were located inside a single steam cabinet or immersed into a water bath. The flow rates were manually adjusted to give the same exit temperatures.

Apple juice inoculated with *Saccharomyces cerevisiae* and *Lactobacillus plantarum* was subjected to microwave heating under continuous flow conditions at 700 W to selected exit temperatures 52.5–65°C.[41] Typical time-temperature profiles of test samples were gathered and corrected for both come-up and come-down time as previously described. The effectiveness of CUT and CDT ranged from 8 to 14%, depending on the sample size and temperature employed. The D and z values were corrected as detailed earlier Reference 39, and the results are summarized in Table 45.2.

The D values under microwave heating were significantly lower and were much more sensitive to changes in temperature as compared with thermal destruction. Semilogarithmic regression of D values vs. temperature yielded a D_{60} of 0.41 s for *S. cerevisiae* and 3.7 s for *L. plantarum* with z values of 6.1 and 4.5°C, respectively, in the microwave heating mode, and D_{60} of 10 and 26 s, respectively, with z values of 13 and 16°C in the conventional heating mode. This gives a relative microwave/thermal destruction ratio of 19 for *S. cerevisiae* at 60°C and 43 for *L. plantarum* at 65°C. Thus, the time required to destroy a given microbial population in the

TABLE 45.2
Kinetic Parameters of Spoilage Microorganism in Apple Juice during Continuous-Flow Microwave Heating

| Temperature (°C) | D values, s | | | |
| | S. cerevisiae | | L. plantarum | |
	Uncorrected	Corrected	Uncorrected	Corrected
52.5	34.0	4.75	–	–
55	15.6	2.08	–	–
57.5	8.25	1.07	154	14.1
60	4.79	0.378	44.3	3.83
62.5	–	–	9.58	0.794
65	–	–	4.11	0.327
z value	8.83	6.97	4.64	4.48

thermal mode is much longer than required in the microwave mode. These results suggest that there probably exist some enhanced thermal effects associated with microwaves that cannot be explained by conventional integration of thermal kinetics over the time-temperature regime.

45.5 MICROWAVE INACTIVATION KINETICS OF PME IN ORANGE JUICE

These studies are detailed in References 38 through 40 under continuous-flow and batch heating modes. Thermal resistance of pectin methylesterase (PME) implicated in the loss of cloudiness of citrus beverages has been recognized to be greater than that of common bacteria and yeast in citrus juices, and PME activity has been used to determine the adequacy of pasteurization.

Conventional heat treatment of orange juice (2 ml) was performed in water bath at 60, 70, 80, 85, and 90°C at pH 3.7. For batch-mode microwave heating, a 700-W microwave oven was used to heat test samples in a 100-mL cylindrical glass container to yield target temperatures 50, 55, 60, and 65°C. To obtain the final bulk temperature, the test sample was mixed immediately after heating, and temperature was measured in a well insulated container to prevent heat loss during temperature measurement. For continuous-flow microwave heating, the setup described earlier was used. Different lengths of heating tubes were employed to get different effective times at the exit temperatures at 55, 60, 65, and 70°C.

Time-temperature profile during microwave batch heating indicated a linear come-up similar to the continuous system. Heating times were corrected, and kinetic data analysis was performed as detailed earlier. The kinetics parameters (D and z values) obtained from the slopes of linear sections of the different curves are included in Table 45.3 for thermal as well as microwave heating conditions.

TABLE 45.3
D and *z* Values (Thermal and Microwave) of PME in Orange Juice

Temperature, °C	D value, s		
	Thermal	MW batch	MW continuous
50.3	–	40.5	
55.3	–	11.7	38.5
60.0	154	7.37	12.4
64.9	–	2.96	3.98
70	37.2		1.32
80	8.45		
85	6.53		
90	2.85		
z value	17.6	13.4	10.2

Figures 45.2 and 45.3 show the corrected PME inactivation curves at various temperatures under conventional and microwave heating conditions. The curves demonstrate that PME inactivation kinetics was of first order, and the rate at a higher temperature was more rapid. With the common temperature of 60°C as the basis, the effectiveness of the two systems could be compared by their *D* values: 154 s during thermal and 7 to 12 s during microwave heating experiments, the two differing by more than an order of magnitude. This difference again shows the possibility of some contributory *nonthermal* effects of microwaves for enzyme inactivation.

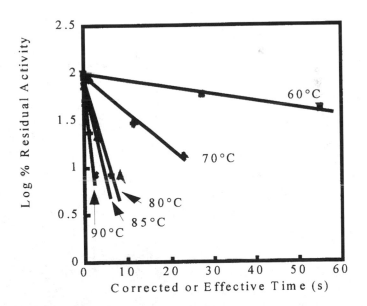

Figure 45.2 Inactivation of PME in orange juice by conventional heating.

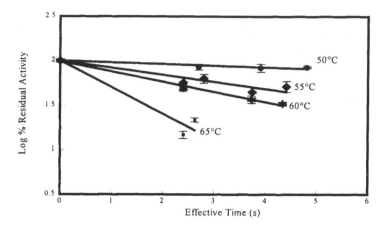

Figure 45.3 Inactivation of PME in orange by microwave heating.

45.6 NONTHERMAL OR ENHANCED THERMAL EFFECTS?

To differentiate the two, the following concept was used: the effects were considered nonthermal if they were observed independent of sample temperature; if they existed but also depended on temperature, they were considered enhanced thermal effects. To evaluate the nonthermal or enhanced thermal effects of microwaves contributed to PME inactivation in orange juice and destruction of *S. cerevisiae* in apple juice, three different techniques as described in Reference 42 were used to maintain temperatures below 40°C under continuous and batch-mode heating (nonthermal effects) as well as under progressively increasing temperature conditions (enhanced thermal effects). In the first setup, test sample temperatures were maintained below 40°C while being subjected to full-power microwave heating (700 W) conditions by surrounding the helical coil with a jacket through which cold kerosene as a microwave-transparent liquid was circulated. Inlet (~15°C) and outlet (<35°C) temperatures of the test sample and kerosene (~10°C) were monitored continuously using copper-constantan thermocouples positioned within the tubing just outside the microwave cavity.

During the 90-min treatment time, 1200 ml of test sample was circulated through the oven, but only a 110-ml portion was continuously exposed to microwaves. Some enzyme inactivation occurred during such exposure, but the extent was relatively small (68%) as compared with that at higher temperatures. In terms of added microwave energy, the 90-min heating would give 3780 kJ of heat to the 110 mL of orange juice, which would be sufficient to completely boil off the juice, let alone inactivate the enzyme. Hence, the nonthermal effect at the sublethal temperature was not considered to be significant.

In the second setup, microwave heating was carried out in a batch mode with a larger size of sample, with continuous flow maintained for mixing and sample removal. To maintain a low temperature (<40°C) for the juice, a stainless cooling coil grounded to the cavity wall was fully submerged in the juice inside the test beaker for rapid removal of heat generated in the sample. Ice-chilled water (0 to 2°C) was circulated through the cooling coil at a constant flow rate. Temperature uniformity in the beaker measured using fiber-optic probes indicated good stability (±1°C). Results indicated that, over a 3-h heating period, about 22% inactivation of PME occurred, and it was only 0.3% up to 1.5 h. The absorbed energy during the treatment time was about 10 MJ! With respect to destruction of *S. cerevisiae* in apple juice, the effect was even smaller, showing less than one log cycle reduction in microbial survivors. Again, it was concluded that there were no temperature-independent nonthermal effects associated with microwave heating when samples were held at temperatures below 40°C. Similar findings were reported in the studies (see References 9, 11, and 16).

To characterize the difference observed during the kinetic studies with respect to enzyme inactivation and microbial destruction between microwave and thermal heating modes, further studies were performed using the above batch system but without the cooling coil. Temperatures within the test beaker were observed to be relatively uniform (±1°C). Test samples were varied by size and temperature of the samples linearly increased from 20 to 40, 50, 60, 65, and 70°C under non-isothermal heating conditions. The temperature difference achieved in the heated sample was plotted against mass normalized time, which is microwave heating time expressed per unit mass of sample (Figure 45.4). Results indicated some deviation with the smaller size of samples at higher temperatures due to variations in energy absorption for different sample sizes. The inactivation of PME in orange juice and destruction of *S. cerevisiae* in apple juice are shown in Figure 45.5.

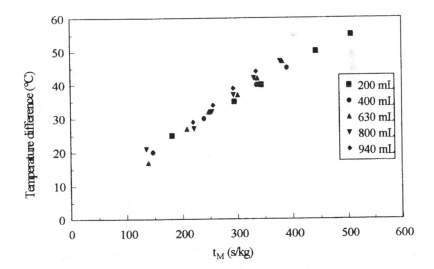

Figure 45.4 Temperature rise as function of mass normalized microwave heating time at various sample sizes.

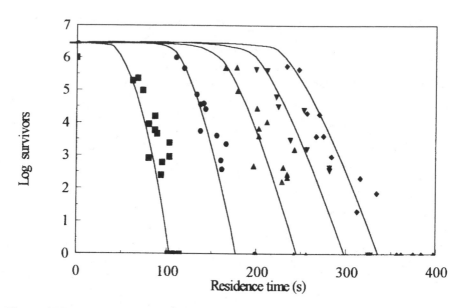

Figure 45.5 Microwave destruction curves of *S. cerevisiae* in apple juice as a function of residence time at various temperatures and sample sizes.

To assess the relative effects due to microwave heating, the equivalent thermal effects for similar heat treatments needed to be calculated and deducted from the total contribution. The thermal contribution was calculated from effective portion of the heating time based on thermal batch z value. To characterize the relative magnitude of enhanced effects, a microwave enhancement ratio (MER) was defined as the ratio of total inactivation/destruction under microwave heating conditions (thermal plus microwave effects) to that calculated to be due thermal effects. A value of the microwave enhancement ratio (MER) greater than 1.0 indicates the existence of enhanced thermal effects due to microwaves, while MER value of 1.0 or below shows its nonexistence. The enhanced microwave effects (MER) of PME inactivation and destruction of *S. cerevisiae* are illustrated as a function of sample size and temperature in Figure 45.6.

The enhanced effects were clearly more pronounced with smaller size of samples and at higher temperatures. The destruction of *S. cerevisiae* appeared to be slightly less enhanced with the microwave heating as compared with enzyme inactivation. The MER for PME was as high as 20 at higher temperature and smaller sample size, while, for bacterial destruction, the ratio was about 10.

D values computed based on single time-residual activity and survival data using the effective portion of heating times and the average values are shown in Figure 45.7 for PME (with dotted lines showing D values from thermal batch kinetics).

Similar, but slightly less pronounced, results were observed fore *S. cerevisiae and L. plantarum,* which clearly indicate that D value was a function of sample size. It increased with an increase in sample size, whereas the thermal effects were

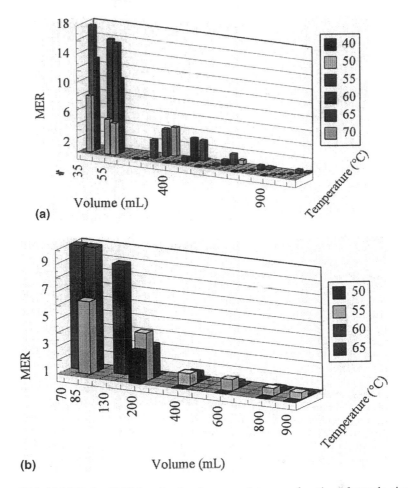

Figure 45.6 (a) MER for PME inactivation in orange juice as a function of sample size and temperature and (b) MER for destruction of *S. cerevisiae* in apple juice as a function of sample size and temperature.

independent of sample size. Microwave heating time for any given set-point temperature or a given level inactivation/destruction will have to be longer for a sample of larger size.

Figure 45.8 compares calculated *D* values as a function of temperature for microwave heating of 1-kg test samples at various power levels with those obtained under conventional heating. The figure indicates a region in which microwave heating would offer an advantage over conventional heating. It is important to recognize the dependence of microwave heating time on power level for a given sample size (or sample size at a given power level), without which it would be difficult to perceive the presence of enhanced microwave effect. Discrepancies observed in the literature could have arisen due to nonrecognition of these issues.

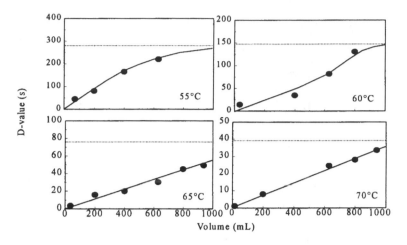

Figure 45.7 *D* values of PME in orange juice as influenced by sample volume during microwave heating conditions at various temperatures.

Figure 45.8 indicates that microwave heating would be beneficial for enzyme inactivation and microbial destruction at power levels of about 1.0 kW and higher for a sample size of 1.0 kg in the temperature range of 55 to 70°C. In a 700-W oven, the sample size will have to be lower than 1.0 kg for the enhanced effect to be felt in the temperature range of 50 to 60°C. Within the microwave oven, the absorbed microwave power increases with sample size. Clearly, a sample size yielding microwave absorption efficiency below 70% makes it less attractive from an economic standpoint. Using this value as the limiting condition, the zone for microwave advantage is indicated by the shaded region.

45.7 MODELING STUDIES AND RESULTS WITH *E. coli*

Prediction of time-temperature profiles under continuous-flow conditions are desirable so as to get a better understanding of the effects of flow distribution on process characteristics and on the resulting microbial lethality during microwave and conventional treatments of fluids. Simple mathematical models were developed based on perfectly mixed flow (PMF), piston flow with heat diffusion (PFHD), and laminar flow (LF) approaches to predict liquid temperature history and lethality developing under microwave and steam heating. Transient and steady-state mean temperatures of water were experimentally measured at the exit and were compared with the predictions from the mathematical models for both systems. The PFHD and the LF models better described temperature profiles during the initial transient period, while the PMF model showed a better agreement with experimental data during steady-state conditions.

Modeled steady-state temperature profiles along the length of the coils (or temperature rise) (Figure 45.9) were compared with experimental data for both

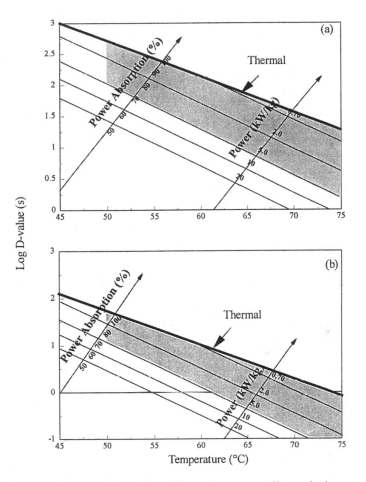

Figure 45.8 The optimal conditions for positive enhancement effects of microwave heating in a 700W microwave oven: (a) for PME inactivation in orange juice, and (b) destruction of *S. cerevisiae* in apple juice.

heating conditions. Predicted temperature rise (PMF model #1) showed that the uniform volumetric power resulted in a quasi-linear temperature profile for microwaves, while a marked curvature profile was observed for steam condition. The LF model overestimated the curvature of the axial temperature. The slight curvature experimentally observed for microwave was probably due to the fact that the loss factor of water decreases with increasing temperature. For the steam heating mode, the experimental observation was between the PMF model and the PFHD. These profiles again were much closer to those predicted with the PMF model. Since the flow rates employed were fully in the laminar region, the presence of secondary effects due to the use of helical coils (with Dean numbers from 200 to 300) appears to be the best reason to explain fluid mixing and closely approximating the ideal perfectly mixed flow behavior in steady state.

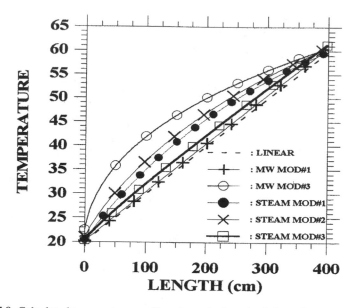

Figure 45.9 Calculated temperature profiles along the length of the coils.

Numerical models were used to predict the lethality based on thermal resistance of *E. coli* obtained from batch heating experiments. All three models failed to predict the lethality under microwave heating. Experimental microbial lethality was greater than that predicted by numerical integration, while the experimental lethality of steam heating was in acceptable agreement with calculated values using the PMF model.

Assuming perfectly mixed distribution in continuous-flow heating systems by steam or hot water and microwaves, the effective times, effectiveness, destruction, and *D* values of *E. coli* were computed for the different heat transfer modes. The comparison of conventional heating by steam, hot water, and microwaves showed differences in effective times and effectiveness (Table 45.4). Hot water continuous-flow heating was the most effective heat-transfer mode, with the effectiveness ranging from 20.5 to 25.5% in comparison with 7.5 and 5.7% for the least effective microwave heating (due to its predominantly linear temperature come-up profile).

Based on PMF flow distribution and thermal resistance in water bath, the survival ratio of *E. coli* was well predicted for hot water and steam heating (with holding) of the coils, while it was largely underpredicted for microwave heating. The lower calculated value of microbial survival ratio under microwave heating again indicates some additional effects of microwaves that cannot be accommodated by temperature profile and thermal resistance of bacteria.

Table 45.5 shows *D* values estimated from the experimental survivor data at different temperatures based on the computed effective times from time-temperature data (experimental or PMF model predicted) for microwave, hot water and steam heating conditions. A comparison of data between the *D* value in thermal batch

TABLE 45.4
Comparison of *E. coli* Destruction under Continuous-Flow Heating by Different Heat-Transfer Modes

Heating mode	Steam			Hot water			Microwave		
Temperature (°C)	55	60	65	55	60	65	55	60	65
Residence time (s)	27.7	34.6	42	28.5	30	40	28	32	37
Effective time (s)	2.6	3.6	6.4	5.84	6.9	10.2	2.1	2.1	2.1
Effectiveness (%)	9.4	10.4	15.2	20.5	23	25.5	7.5	6.5	5.7
Predicted survival ratio	0.90	0.22	3.0×10^{-7}	0.92	0.41	7.2×10^{-6}	0.97	0.77	0.09
Experimental survival ratio	0.78	0.27	2.0×10^{-5}	0.86	0.55	8.0×10^{-6}	0.61	0.44	1.4×10^{-4}

TABLE 45.5
Experimental *D* Values of *E. coli* K-12 during Microwave and Conventional Heating

	D values (s)		
Temperature, °C	55	60	65
Thermal batch method	173.0	18	1.99
Continuous-flow, hot water	44.70	26.80	2.00
Continuous-flow, steam	72.71	15.61	2.98
Continuous-flow, microwave	12.98	6.31	0.78
Continuous-flow, microwave + holding	19.89	8.33	1.98

heating, microwave, steam, and hot water continuous-flow heating shows that lowest *D* values were observed with the continuous-flow microwave heating section and ranged from 13 s at 55°C to 0.78 s at 65°C. The experimental *D* values in the microwave system with hold section (20, 8.3, and 2.0 s at 55, 60, and 65°C) were also lower than those obtained for steam heating (73, 16, and 3.0 at 55, 60, and 65°C) and hot water heating. Furthermore, values in continuous systems were considerably lower than those in batch heating systems.

45.8 CONCLUSIONS

The destruction of *Escherichia coli, Lactobacillus plantarum,* and *Saccharomyces cerevisae* and pectin methyl esterase (PME) enzyme was evaluated under continuous-flow microwave heating and compared with conventional heating. Contributions of lethality during CUT and CDT periods were accommodated for obtaining kinetic parameters. The destruction rate in microwave heating was found to be more than an order of magnitude higher than that in conventional batch heating systems. Modeling of time-temperature history in continuous-flow conditions showed the perfectly mixed flow model yielding better predictions than the laminar approach and piston flow models with heat diffusion. Model calculated lethality on the basis of thermal history and thermal resistance data were comparable between the three modes. However, while the predictions for steam and hot water heating were in good agreement with experimental data, those under microwave heating showed a large underestimation, indicating the presence of additive effects of microwave heating which cannot be explained by the time-temperature history. A comparison among thermal batch heating, microwave, and steam continuous-flow heating showed that the lowest D values were observed with the continuous-flow microwave heating. The time-temperature corrected D values obtained from these studies thus suggest significant of microbial and enzyme destruction to be associated with microwave heating. However, no nonthermal effect was observed at sublethal temperatures.

REFERENCES

1. Fleming, H. 1944. "Effect of High Frequency on Microorganisms," *Electrical Engineering,* 63 (18).
2. Knorr, D., M. Geulen, T. Grahl, and W. Stitzman. 1994. "Food Application of High Electric Fields Pulses," *Trends in Food Science & Technology,* 5(3): 71–75.
3. Stuerga, D. A. C. and P. Gaillard. 1996. "Microwave Athermal Effects in Chemistry: A Myth's Autopsy," *Journal of Microwave Power and EME,* 31(2): 87–113.
4. Sastry, S. and Palaniappan. 1991. "The Temperature Difference Between a Microorganism and a Liquid Medium During Microwave Heating," *Journal of Food Processing and Preservation,* 15: 225–230.
5. Loock, W. V. 1996. "Electromagnetic Energy for Pasteurization and Sterilization: Another Viewpoint," *Microwave World,* 17(1): 23 –27.
6. Risman, P. 1996. "Guest Editorial," *Journal of Microwave Power and Electromagnetic Energy* 31(2): 69–70.
7. Gedye, R.N., F. E. Smith, and K. C. Westaway. 1988. "The Rapid Synthesis of Organic Compounds in Microwave Ovens," *Can. J. Chem.,* 66: 17–26.
8. Gedye, R.N., F. E. Smith, and K. C. Westaway. 1991b. "Microwaves in Organic and Organometallic Synthesis," *Journal of Microwave Power and Electromagnetic Energy,* 22: 199–207.
9. Goldblith, S. A. and D. I. C. Wang. 1967. "Effect of Microwaves on *Escherichia coli* and *Bacillus subtilis,*" *Appl. Microbiol.,* 15: 1371–1375.
10. Goldblith, S. A., S. R. Tannenbaum. and D. I. C. Wang. 1968. "Thermal and 2450 MHz Microwave Energy Effect on the Destruction of Thiamine," *Food Technology,* 22: 1267–1268.
11. Lechowich R. V., L. R. Beuchat, K. I. Fox, and F. H. Webster. 1969. "Procedure for Evaluating the Effects of 2,450–MHz Microwaves upon *Streptococcus faecalis* and *Saccharomyces cerevisiae,*" *Appl. Microbiology,* 17: 106–110.
12. Vela, G. and J. Wu. 1979. "Mechanism of Lethal Action of 2450-MHz Radiation on Microorganisms," *Appl. Env. Microbiology,* 37(3): 550–553.

13. Fujikawa, H., H. Ushioda, and Y. Kudo. 1992. "Kinetics of *Escherichia coli* Destruction by Microwave Irradiation," *Appl. and Env. Microbiology*, 58 (3): 920–924.

14. Fujikawa, H. 1994. "Patterns of Bacterial Destruction in Solutions by Microwave Irradiation," *Journal of Applied Bacteriology*, 76: 389–394.

15. Tong, C. H. 1996. "Effect of Microwaves on Biological and Chemical Systems," *Microwave World*, 17 (4): 14–23.

16. Welt, B. and C. Tong. 1993. "Effect of Microwave Radiation on Thiamin Degradation Kinetics," *Journal of Microwave Power and Electromagnetic Energy*, 28 (4): 187–195.

17. Welt, B. A., C. H. Tong, J. L. Rossen, and D. B. Lund. 1994. "Effect of Microwave Radiation on Inactivation of *Clostridium sporogenes* (PA 3670) Spores," *Appl. Env. Microbiology*, 60: 482–488.

18. Diaz, M., and S. Martinelli. 1991. "The Use of Microwaves in Sterilization," *Dairy, Food and Environmental Sanitation*, 11 (12): 722–724.

19. Heddleson, R. L., S. Doores, R. Anantheswaran. 1994. "Parameters Affecting Destruction of *Salmonella spp.* by Microwave Heating," *Journal of Food Science*, 59 (2): 447–451.

20. Culkin K. A., Y. C. Daniel, and Fung. 1975. "Destruction of *E. coli* and *S. typhimurium* in Microwave-Cooked Soups," *Journal of Milk Food Technology*, 38 (1): 8–15.

21. Dreyfuss, M. and J. Chipley. 1980. "Comparison of Effects of Sublethal Microwave Radiation and Conventional Heating on the Metabolic Activity of *Staphylococcus aureus*," *Appl. and Env. Microbiology*, 39 (1): 13–16.

22. Mudgett, R. E. 1986. "Microwave Properties and Heating Characteristics of Foods," *Food Technology*, June: 84–93.

23. Khalil, H. and R. Vilotta. 1988. "Comparative Study on Injury and Recovery of *Staphylococuss aureus* Using Microwaves and Conventional Heating," *J. of Food Protection*, 51 (3): 181–186.

24. Khalil, H. and R. Vilotta. 1985. "A Comparative Study on the Thermal Inactivation of *Bacillus stearothermophilus* Spores in Microwave and Conventional Heating," *Food Engineering and Process Application*, 583–594.

25. Kermasha, S., B. Bisakowski, H. S. Ramaswamy, and F. R. Van de Voort. 1993a. "Comparison of Microwave, Conventional and Combination Treatments Inactivation on Wheat Germ Lipase Activity," *Int. J. Food Sci. Technol.*, 28: 617–623.

26. Kermasha, S., B. Bisakowski, H. S. Ramaswamy, and F. R. Van de Voort. 1993b. "Thermal and Microwave Inactivation of Soybean Lipoxygenase," *Lebensm.-Wiss.u. Technol.* 26: 215–219.

27. Odani, S., T. Abe, and T. Mitsuma. 1995. "Pasteurization of Food by Microwave Irradiation," *Journal-of-the-Food Hygienic of Japan*, 36(4): 477–481.

28. Wu H., and K. Gao. 1996. "Mechanisms of Microwave Sterilization," *Science and Technology of Food Industry*, 3: 31–34.

29. Joalland, G. 1996. "Contribution a L'etude de L'effect des Micro-ondes: Etude Bibliographique," *Viandes-et-Produits-Carnes*, 17 (2): 63–72.

30. Riva, M., M. Lucisano, M. Galli, and A. Armatori. 1991. "Comparative Microbial Lethality and Thermal Damage During Microwave and Conventional Heating in Mussels (*Mytilus edulis*)," *Ann. Microbiol*,. 41 (2): 147–160.

31. Aktas, N. and M. Ozligen. 1992. "Injury of *E. coli* and Degradation of Riboflavin during Pasteurization with Microwaves in a Tubular Flow Reactor," *Lebensmittel Wissenschaft unt Technologie*, 25 (5): 422–425.

32. Koutchma, T. 1997. "Modification of Bactericidal Effects of Microwave Heating and Hyperthermia by Hydrogen Peroxide," *Journal of Microwave Power and Electromagnetic Energy,* 32 (4): 205–214.

33. Kozempel, M., O. J. Scullen, R. Cook, and R. Whiting. 1997. "Preliminary Investigation Using a Batch Flow Process to Determine Bacteria Destruction by Microwave Energy at Low Temperature," *Lebenm.-Wiss. u. –Technol.,* 30, 691–696.

34. Kozempel, M., B. A. Annous, R. Cook, O. J. Scullen, and R. Whiting. 1998. "Inactivation of Microrganisms with Microwaves at Reduced Temperature," *Journal of Food Protection,* 61 (5): 582–585.

35. Shin, J. K. and Y. R. Pyun. 1997. "Inactivation of *Lactobacillus plantarum* by Pulsed-Microwave Irradiation," *Journal of Food Science,* 62(1): 163–166.

36. Stumbo. 1973. *Thermobacteriology in Food Processing.* 2nd ed., New York: Academic Press.

37. Riva, M., L. Franzetti, A. Mattioli, and A. Galli. 1993. "Microorganisms Lethality During Microwave Cooking of Ground Meat, 2. Effects of Power Attenuation," *Ann. Microbiol. Enzimol.,* 43(2): 297–302.

38. Tajchakavit, S. and H. S. Ramaswamy. 1997. "Continuous-Flow Microwave Inactivation Kinetics of Pectin Methylesterase in Orange Juice," *Journal of Food Processing and Preservation,* 21: 365–378.

39. Tajchakavit, S. and H. Ramaswamy. 1995. "Continuous-Flow Microwave Heating of Orange Juice: Evidence of Non-Thermal Effects," *Journal of Microwave Power and Electromagnetic Energy,* 30(3): 141–148.

40. Tajchakavit, S. and H. S. Ramaswamy. 1996. "Thermal vs. Microwave Inactivation Kinetics of Pectin Methylesterase in Orange Juice Under Batch Mode Heating Conditions," *Lebensm.-Wiss.u. Technol.* 2: 85 – 93.

41. Tajchakavit, S., H. S. Ramaswamy, and P. Fustier. 1998. "Enhanced Destruction of Spoilage Microorganisms in Apple Juice During Continuous Flow Microwave Heating," *Food Research International,* 31 (10): 713–722.

42. Tajchakavit, S. 1997. "Continuous-Flow Microwave Heating of Orange Juice: Evidence of Non-Thermal Effects," Ph.D. thesis. McGill University, Food Science Department.

43. LeBail, A., T. Koutchma, and H. Ramaswamy. 1999. "Modeling of Temperature Profiles Under Continuous Tube-Flow Microwave and Steam Heating Conditions," *Food Process Engineering,* in press.

44. Koutchma, T., A. LeBail, and H. S. Ramaswamy. 1998. "Modeling of Process Lethality in Continuous-Flow Microwave Heating-Cooling System," in Proceedings of the International Microwave Power Institute, Chicago (July 1998), pp. 74–77.

46 Applications of Low-Intensity Ultrasonics in the Dairy Industry

A. Mulet
J. A. Cárcel
J. Benedito
N. Sanjuan

CONTENTS

Abstract

Ultrasonic applications in the dairy industry are greatly increasing due to the fact that ultrasonic measurements are nondestructive, rapid, and easy to automate. Most of the existing applications relate ultrasonic velocity or attenuation to properties of products such as milk or cheese. Ultrasound has been used to detect milk adulteration, to assess the optimal cut time for cheese making, and to detect any buildup of fouling in milk pipes. It has been shown that ultrasonic imaging is an effective technique in assessing microbiological contamination in milk. In the cheese industry, the rheological and textural properties of cheese, as well as its composition, have been evaluated by using ultrasonic techniques, allowing estimation of cheese maturity. New applications are expected to appear in the near future, due to the valuable information that ultrasonics provide and the advantages over other existing destructive analytical techniques.

46.1 INTRODUCTION

Nowadays, food technology requires the development of new on-line sensors to monitor food properties during production and storage. Improvement in data acquisition and processing is allowing new nondestructive on-line sensors used for process and product control to be applied. Nuclear magnetic resonance (NMR), near infrared spectroscopy, ultrasound, electronic noses, and vibration rheometers are examples of techniques that are increasingly being applied in food technology. Ultrasound has been widely used in the past for metal testing and medicine, but only recently have its applications for nondestructive testing been extended to the food industry.[1]

Ultrasonics are elastic waves whose frequency is above the threshold of human hearing ($\approx$20 kHz). As long as they are mechanical waves, they need a physical medium to propagate through. Three types of waves can be distinguished: longitudinal waves, which move in the direction of the particle displacements; shear waves, which move perpendicular to the particle displacement; and Rayleigh waves travelling very close to the surface.[2] Acoustic waves are characterized by their frequency (f), velocity (v), and amplitude or intensity (A). This latter characteristic is used to divide ultrasonic industrial applications into two different general groups: high-intensity ultrasound (HIU) and low-intensity ultrasound (LIU). HIU applications are used to modify a process or a product, whereas LIUs are applied to monitor a process or a product.

The absorption of acoustic energy is especially important in food materials mainly due to the air content and to the highly structured materials. As long as higher frequencies are more attenuated than lower ones, HIU is applied at low frequencies (20–300 kHz) to obtain high power levels (10–1000 W/cm²). On the other hand, LIU uses higher frequencies (250 kHz to 1 MHz or more) to achieve good resolution, but lower power levels, typically less than 1 W/cm².[1]

HIU can be used in the food industry for cleaning[3] and for enhancing microbial destruction,[4,5] enzymatic inactivation,[6] chemical reactions,[7] and extractions.[8,9] Sastry et al.[10] studied the increase of heat convective coefficients by ultrasound. Mass transfer can also be accelerated by ultrasound.[11–15]

LIU applications address three main aspects of food control technology: assurance of product quality, process control, and nondestructive food inspections. Thus, industrial applications include concentration, texture, and viscosity measurements;[16,17] fish, meat, fruit, vegetables, and egg composition assessment;[18–21] process control such as thickness, flow, level and temperature measurements;[22–25] and nondestructive inspection of potatoes and kamabokos and food containers.[26–28]

Application of ultrasound in the dairy industry is notably increasing, due to the great amount of products and processes involved. This chapter reviews and discusses some of the most important applications related to milk, cheese, and other dairy products. Before these applications are addressed in detail, some basic concepts about ultrasound parameters, equipment, and techniques are considered.

46.2 ULTRASONIC PARAMETERS

In LIU applications, what is important is not the amount of energy introduced into the medium but the information that ultrasound provides when propagating through it. Ultrasonic waves are characterized by several parameters such as velocity, the attenuation coefficient, and the frequency spectrum composition. The influence of the medium on these parameters is the basis of many ultrasonic applications in food and process control.

By performing a fast Fourier transform (FFT) to the time-domain signal, the energy distribution for each frequency can be obtained.[29] The way the material affects each frequency of the spectrum is characteristic of the sample and can be used to estimate its properties.

Despite the interest of frequency spectrum analysis, the most frequently used parameters are velocity and attenuation.

46.2.1 Velocity

Ultrasonic velocity is characteristic of the propagation medium and therefore provides unique information about its structure and composition. Most ultrasonic instruments currently used in the food industry are based on measurements of the ultrasonic velocity, because it is the simplest and most reliable measurement.[30] Phase and group velocities are the main types of velocity. The former is the velocity of each single frequency wave, while the latter is the group velocity of several single frequency waves. If ultrasound travels through a nondisperse medium, phase and group velocity have the same value.

Acoustic impedance is defined as the product of ultrasonic velocity and the medium density. When an ultrasonic wave encounters a boundary between two different materials, it is partly reflected and partly transmitted. The fraction of ultrasound either reflected or transmitted depends on the difference between the acoustic impedance of both materials. The extent of reflection at the boundary has important practical implications for many ultrasound applications in the dairy industry, such as imaging, detection of foreign bodies or internal cracks in cheese, and detection of fouling in pipes, among others.

Ultrasonic velocity depends on the elastic modulus and the density of the medium. This modulus depends on the physical state and dimensions of the material being tested, as well as the type of ultrasonic wave used.

Solids

Longitudinal, shear, and surface waves can travel through a solid medium such as cheese or butter. The relationship between ultrasonic velocity of longitudinal waves (v_L) and elastic constants for a homogeneous, isotropic, and elastic medium is given by

$$v_L = \sqrt{\frac{K + \frac{4G}{3}}{\rho}} \tag{46.1}$$

where K = the bulk modulus
G = the shear modulus
ρ = density of the medium

When the sample diameter is shorter than the wavelength, such as in the case of a bar whose diameter is smaller than its length, the velocity can be calculated from Young's modulus (E) using

$$v_L = \sqrt{\frac{E}{\rho}} \tag{46.2}$$

Shear waves are greatly attenuated by foodstuffs, which explains why their use is restricted to solids of very small thickness. The ultrasonic velocity of this type of waves (v_S) can be calculated from the following equation:

$$v_S = \sqrt{\frac{G}{\rho}} \tag{46.3}$$

Gases and Liquids

Only longitudinal waves can propagate in gases and liquids. Assuming adiabatic conditions, the ultrasonic velocity in a gas or liquid is given by

$$v_L = \sqrt{\frac{1}{\rho\beta}} \tag{46.4}$$

where β represents the adiabatic compressibility (inverse of the temperature for perfect gases).

46.2.2 Attenuation

When an ultrasonic wave travels through a medium, part of its energy is lost and transmitted to the medium (absorption). Furthermore, in heterogeneous bodies, the ultrasonic beam is scattered in many different directions from that of the incident wave (scattering). Both phenomena (absorption and scattering) attenuate the ultrasonic wave. This attenuation is characteristic of the material and gives information about its physical properties[31] and irregularities within the samples.[26] If it is assumed that attenuation is only a function of friction owing to the shear viscosity of the liquid, the classical equation used to estimate attenuation (α) in liquids is[30]

$$\alpha = \frac{2\omega^2 \mu_T}{2\rho v} \tag{46.5}$$

where ω = ultrasonic radial frequency (=$2\pi f$)
   μ_t = shear viscosity
   ρ = density
   v = velocity

It can be assumed that other mechanisms are responsible for the energy losses, such as viscosity losses, heat conduction losses, and losses associated with molecular heat exchange.[32] The attenuation coefficient can be experimentally computed using Reference 1.

$$A = A_0 e^{-\alpha x} \tag{46.6}$$

where A_0 = initial amplitude of the wave
   x = distance traveled by the wave
   A = amplitude at this distance

46.3 EQUIPMENT

46.3.1 Transducers

In food technology applications, piezoelectric transducers are most commonly used. When an electric current is applied over two opposite faces of a piezoelectric ceramic, it vibrates, thereby producing longitudinal waves. To emit shear waves, specially cut ceramics can be used, or a wedge can be placed on the transducer's surface to introduce the waves into the material at an incidence angle. Factors that must be considered when choosing a transducer are, among others, the central frequency and bandwidth, its diameter, and its acoustic matching. The frequency for a particular application depends on the type of information required from the material being tested. The diameter determines the area of the material to be analyzed. The degree of acoustic matching between the transducer and the sample will determine the amount of energy introduced into the material.

46.3.2 PULSERS-RECEIVERS

Pulsers provide transducers with the electrical input of a specific frequency, amplitude, and duration to obtain the optimum ceramic vibration. The applied input can consist of a unique pulse (spike) or a burst (two or more pulses). Sometimes one pulse is not enough to produce ultrasonic waves with enough energy to cross a material, and in these cases a tone burst must be considered. Spike pulsers are less expensive, have a higher time resolution, and are of particular interest for detecting defects near the surface such as cracks in cheese. Burst pulsers are more complex and expensive and have a worse time resolution. In return, they have a higher penetration capability, thereby providing more energy to the transducer and, as a result, being more adequate for use in dairy liquids or solids with high air content. The receiver is the device that amplifies and conditions (filters) the signal from the transducer. The pulser and receiver are usually integrated into the same device. Commercial receivers range from a few decibels to 70 to 100 dB.

46.3.3 DIGITIZERS

The electrical signal from the transducer can be directly displayed on an analog oscilloscope or digitized for further examination. The sampling rate is the number of digitized points per second (measured in million samples per second, or Msamples/s) and is characteristic of each digitizer. The sampling rate must be at least twice the highest frequency being analyzed. Sampling rates range from a few million samples per second to 200 Msamples/s and more, in real time.

46.4 TECHNIQUES

Ultrasonic measurements can be performed using different setups, although the most commonly used are through transmission (TT) and pulse-echo techniques (PE).

46.4.1 THROUGH-TRANSMISSION METHODS

In this type of method, two transducers are placed on opposite faces of the material under study.[33] One of the transducers (Figure 46.1a) emits the ultrasonic signal, while the other receives the energy transmitted through the material. When the ultrasonic wave reaches a discontinuity in its path (for example, a crack inside a cheese or bubbles in milk or yogurt), part of the energy is reflected or scattered, thus reducing the signal amplitude. With this technique, the location of the discontinuity cannot be determined.[34] Therefore, through-transmission methods are mainly used to measure velocity and attenuation.[31] In the case of solids, to avoid energy losses in the sample-transducer interface, a couplant (gels, oils, and others) must be used to decrease the acoustic impedance difference. The couplant to be used must leave the samples unaffected.

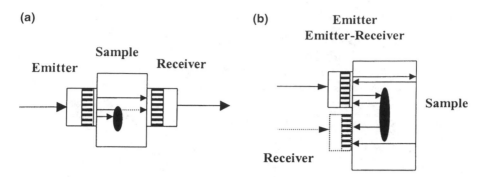

Figure 46.1 (a) Through-transmission and (b) pulse-echo techniques.

46.4.2 PULSE-ECHO METHODS

To carry out these measurements, the emitter and receiver are placed on the same face of the sample (Figure 46.1b). When the vibration of the emitter does not mask the received signal, the emitter and receiver can be the same transducer. For signals that come from reflections close to the surface, two transducers must be used. Sometimes, if only one transducer can be used, a delay line can be applied to separate the initial vibration of the emitter from the received signal. This technique is mainly used to detect defects, although it can also be used to determine velocity and attenuation in low attenuant materials.

46.4.3 ULTRASONIC IMAGING

Ultrasonic imaging uses the ultrasonic pulse-echo technique by displacing the transducer over the surface of an object or using an array of transducers. Ultrasonic imaging uses short-duration ultrasonic pulses so that reflection from objects within a sample can be distinguished from one another. The ultrasonic wave is reflected when it encounters a boundary between two materials of different acoustic imped-ance.[1] From the ultrasonic velocity and the time delay between transmission and receiving, the distance is computed. Thus, it is possible to determine the location of different layers or objects within materials. Several methods have been developed to generate and display ultrasonic images. The most common are A-scan, B-scan, and C-scan. A-scan provides a one-dimensional image along the depth of the sample, from which the location of objects can be deduced. More detailed information about the internal structure of samples is obtained by physically moving the transducer over the surface of the sample.[1] In a B-scan, one selects a particular time of flight along the sample depth and displays the amplitude of the signal as a grey scale. This procedure is carried out for each A-scan acquired over the sample surface, providing a two-dimensional image of the sample at a particular slice across the depth of the sample. In a C-scan, one focuses on a certain feature within an A-scan and measures the change in its time of flight or amplitude when the transducer is moved on the surface plane. Again, a grey scale is used to represent the time of flight or amplitude.

46.5 APPLICATIONS TO THE DAIRY INDUSTRY

The dairy industry is one of the most important food industries, since, from the same raw material (milk), a great number of solid, semisolid, and liquid products are manufactured. As also happens in other industries, a main concern is process assurance and control. Commonly, destructive inspections involve the waste of hundreds of tons of products and do not guarantee process control or final quality. Therefore, great effort has been made to develop on-line and nondestructive sensors to assure the quality control of dairy products. An on-line sensor for the dairy industry should provide rapid, precise measurements and be nondestructive, noninvasive, robust, of relatively low cost, and hygienic. It should also have a computer interface so that a process control unit can directly use the information. Ultrasound meets all of these attributes and is therefore highly suitable for applications involving dairy products.

46.5.1 MILK

Testing Milk Adulteration

Milk is designed by nature to be a nourishing food for the young, since it is capable of providing almost all the ingredients necessary for the growth of the body. Milk adulteration has usually consisted of adding water. To detect this fraud, Bathi et al.[35] studied the variation of ultrasonic velocity in two types of milk, cow and buffalo, adulterated with different percentages of water. Ultrasonic velocity of cow and buffalo milk was found to be different due to the differences in composition. In both cases, velocity decreased in line with the water addition and was dependent on temperature, as were the density and the viscosity of the samples. As long as the water and fat composition depends, for the same type of milk, on the season and the feeding of the cattle, the ultrasonic velocity can assess only the water content. However, it cannot be used to distinguish whether the water was added or was already present in the raw milk.

Quality Control of Packed UHT Milk

Aseptic packaging of UHT milk has many advantages, although it requires a significantly more complex technique, as the microbiological sterility of the product is the most important factor to be controlled. Ultrasound can penetrate all the commonly used plastic materials (LDPE, HDPE, PP, etc.) and aluminum foil. Ahvenainen et al.[36] evaluated the use of ultrasonic imaging to detect microbiological spoilage. Tetra-Brik cartons were contaminated by *Pseudomonas fluorescens*. Milk coagulation caused by proteases was perceived by ultrasonic imaging as white areas that did not disappear from the image after shaking. In a later work, Ahvenainen et al.[37] applied ultrasonic imaging in different milk products (soft ice cream, vanilla sauce, and chocolate sauce) and different contaminant bacteria (*B. cereus, E. coli, S. aureus,* and *P. aeruginosa*). Ultrasonic images were digitalized, and the numerical values of images, obtained from density histograms calculated in the measured area, correlated very well with the bacterial counts. These experiments allowed differentiating incubation periods in different products. Age gelification, typical phenomena produced by heat-stable lipases or proteases, which can even resist UHT-treatment, can also be detected by ultrasound signal.[28]

Wirtanen et al.[38] concluded that the frequencies needed to carry out imaging techniques in packed milk should not exceed 5 MHz. They found that a frequency of 3.75 MHz provided the best sensibility. The ultrasonic Doppler technique has also been used to detect spoilage in packed milk. An ultrasonic wave transmitted into a liquid induces a streaming motion. When moving particles reflect sound, the frequency of the sound is changed due to the Doppler effect. Hence, the velocity of the induced acoustic streaming in the liquid is proportional to the frequency change. Gestrelius[39] studied this technique in packed milk contaminated with *Staphylococcus epidermis* and *Bacillus subtilis* A. A clear decrease in streaming velocity could be seen 4–5 days after inoculation in the case of *Staphylococcus epidermis* and after 6 days in the case of *Bacillus subtilis* A. This means that the contamination is detected before coagulation traces can appear. Measurements are relatively fast ($\approx$10 s), which allows on-line use.

Evaluating Milk Coagulation

In cheese manufacturing, milk proteins are coagulated to form a continuous, solid curd in which milk fat globules, water, and water-soluble materials are entrapped.[41] The most common method of coagulating milk is by either using an enzyme extracted from calf stomachs (rennet) or microbial-produced enzymes. When the established curd firmness has been achieved, the gel is cut, which promotes the drainage of whey (syncresis) from the curd. Cutting the curd while too soft decreases cheese yield due to increased loss of fat and curd fines.

Cutting the gel while too firm delays syneresis and results in high-moisture cheese. Therefore, in cheese making it is important to determine the optimal cut time. The methods used to test curd firmness are destructive methods (penetrometers, suspended bodies, torsion viscometers, and rotational viscometers) that are not easy to automate. The study of ultrasonic propagation during coagulation may be an alternative to the traditional methods. Benguigui et al.[41] used a pulse-echo technique to determine variations in ultrasonic attenuation and velocity during the coagulation process. The two types of coagulation, enzymatic and acid, were distinguished, and two different stages were detected in the first case. Ay and Gunasekaran[42] studied the changes during enzymatic coagulation by ultrasound. They used a 1-MHz transducer and the pulse-echo technique. Ultrasonic velocity did not show any variation during coagulation, but ultrasonic attenuation decreased when coagulation progressed (Figure 46.2).

This change in attenuation is mainly linked to changes in viscosity, which increases the viscous attenuation of ultrasound [Equation (46.5)]. The change in the bulk modulus during coagulation is probably not big enough to produce a detectable velocity variation. In the first stage of coagulation, attenuation changed quickly, and, in the second, it changed slowly. From this data, the optimal cut time was determined. Nonsignificant differences were found for ultrasonic time determination both using

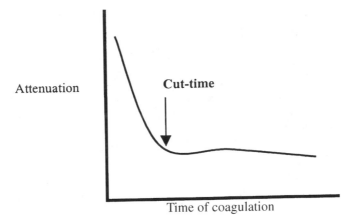

Figure 46.2 Ultrasonic attenuation versus coagulation time. Optimal cut-time is indicated.

this technique and also using conventional methods. This method is useful in evaluating the effectiveness of different enzymes under varying conditions.[43]

46.5.2 CHEESE

Ultrasonic techniques have been used for different applications on the evaluation of cheese characteristics. These applications include the use of ultrasound to evaluate rheological and textural properties of cheese, to estimate cheese composition, and to detect anomalies within cheese blocks.

Rheological and Textural Properties of Cheese

Lee et al.[44] studied the rheological properties of cheese and dough using ultrasonic spectroscopy. In this work, shear transducers were used and, from the time-domain signal, the frequency spectrum was obtained (fast Fourier transform). From the spectrum, velocity and attenuation were calculated for each frequency and, from these values, the storage and loss modulus were computed for each frequency. A good qualitative agreement was found between data obtained using the traditional rheometer and the ultrasonic measurements. The values of the modulus were several orders of magnitude higher for the ultrasonic measurements due to the several orders of higher magnitude of frequencies. On the other hand, ultrasonic velocity was higher for cheeses with a lower moisture content and consequently denser material. Phase velocity increased in line with increasing frequency, thus indicating the disperse nature of the cheese samples.

Several studies have been carried out to determine textural properties of cheese, mainly related to textural changes during maturation. Maiorov and Ostroumov[45] followed cheese maturation ultrasonically. Benedito et al.[46] related the ultrasonic velocity in pieces of Mahon cheese (20 × 20 × 8 cm) with textural parameters determined by uniaxial compression and puncture tests. The ultrasonic velocity increased during maturation, ranging from 1630 m/s for the fresh cheese to 1740 m/s for the highly ripened one. The explanation for this increase can be found in the relationship between the ultrasonic velocity and the elastic properties of the medium, as well as its density, as shown in Equation (46.1).

Because the differences in the modulus of food materials are greater than those in the density, the ultrasonic velocity variations are more heavily influenced by the elastic modulus than by the density,[1] this being the case in cheese maturing. As long as the density of Mahon cheese increases during maturation, the main influence on the velocity increase is due to the changes in the elastic properties of cheese. It should also taken into account that the bulk modulus of cheese greatly exceeds the shear,[47] and therefore the changes in velocity will be mainly linked to the differences in the bulk modulus of the samples. Benedito et al.[46] found that the decrease in water content during maturation caused an increase both in the deformability modulus and also probably in the bulk modulus, as the texture (uniaxial compression and puncture) measurements and the ultrasonic velocity are indirectly related.

The close relationship between ultrasonic velocity and textural parameters is shown in Figure 46.3 for the deformability modulus. As suggested by Equation (46.1), velocity was related ($r^2 > 0.83$) to the square root of all the textural properties determined by uniaxial compression and puncture, and also to the water content ($r^2 = 0.86$). Nevertheless, out of all the textural parameters, the best explained variance corresponds to the deformability modulus (92%) and the slope in puncture (90%). This may be due to the fact that these parameters involve small displacements for their measurement and also that ultrasonic waves involve very small particle displacements when travelling through a material.

As textural parameters and moisture are objective indicators of Mahon cheese maturity, ultrasonic velocity could be used for assessing maturity. Similar results

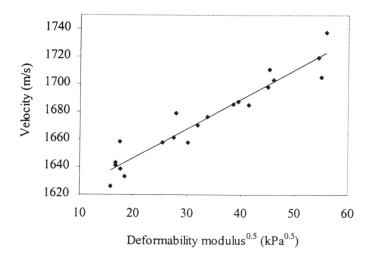

Figure 46.3 Relationship between the ultrasonic velocity and the deformability modulus in Mahon cheese.

have been found for avocados, in which a velocity decrease in line with ripening was linked to the loss of firmness.[48] Miles et al.[49] also reported the same type of relationship for adipose tissue, determining that the ultrasonic measurements could be used to assess the suitability of the tissue for bacon manufacturing.

The influence of moisture on velocity could be assessed by examining nonvarying moisture cheeses such as cheddar. Cheddar cheese blocks are matured vacuum wrapped, and therefore no water loss takes place. Benedito et al.[50] found that the deformability modulus and the slope in puncture increased with the storage time, while other textural properties such as hardness, maximum in puncture, compression work, and puncture work, tended to decrease. Creamer and Olson[51] also reported an increase in the slope of the first part of the TPA curve (related to the deformability modulus) and a decrease in hardness.

In cheddar cheese, there are no noticeable density variations during maturation as long as the moisture remains constant; thus, the main influence on velocity will be due to the change in the elastic properties of cheese. For this type of cheese, the ultrasonic velocity also increases with time, as with Mahon cheese (Figure 46.4).[50] Figure 46.4 shows that, for the two storage temperatures considered, ultrasonic velocity increases quickly during the first days of maturation, following the same increase pattern with time as the deformability modulus and the slope in puncture. After this period, the increase in velocity is slower. As in the case of Mahon cheese, the ultrasonic velocity was closely related to the square root of the deformability modulus ($r^2 = 0.88$ at 12°C) and the slope in puncture ($r^2 = 0.94$ at 12°C). The other textural parameters related to higher deformations (hardness, compression work, maximum in puncture, and puncture work) gave poor relationships (avg. $r^2 < 0.5$). These results show that the deformability modulus and the slope in puncture are

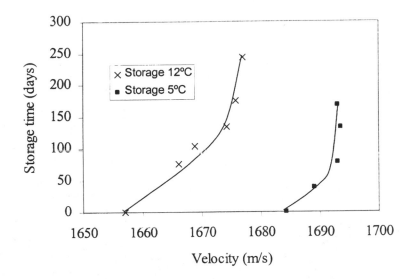

Figure 46.4 Increase of ultrasonic velocity in cheddar cheese during storage.

closely related textural parameters to the bulk modulus and consequently to velocity [Equation (46.1)].

Bulk modulus is temperature dependent; therefore, the temperature of the product affects velocity. This fact has been proven for food products such as oils and cod fillets.[49,52] This influence was also established for cheddar and Mahon cheese.[46,53] Figure 46.5 shows the variation of ultrasonic velocity with the sample temperature for Mahon and cheddar cheese. For both types of cheese, there is a decrease in velocity at higher temperatures, and three parts can be distinguished with different slopes. Miles et al.[49] also found changes in the slope of the velocity-temperature curve for adipose tissue. The moderate velocity decrease in line with the increase in temperature in the first part (a) is due to the same type of variation for the solid and liquid fat and also to the increase in the liquid content (fat melting), which has lower ultrasonic velocity. In the second part of the curve, the increase in the liquid content (high melting) and the phenomenon known as "oiling off" or "fat leakage" cause a more abrupt fall in the slope.[53] The slope of each part of the curve represents the accuracy of the ultrasonic measurements; therefore, a value of X ms^{-1}°C^{-1} for the slope indicates that a change of ±1°C would produce a variation of ±X m/s.

Therefore, for cheese texture (maturity) assessment, it is necessary to consider the influence of temperature on the ultrasonic measurements. For the particular case of Mahon cheese, Benedito et al.[46] developed Equation (46.7) by fitting a multiple regression model to the experimental data plotted in Figure 46.6. Equation (46.7) is the result of using the deformability modulus (DM) as the maturity indicator; other parameters could have been considered.

$$DM = (0.47v + 2.33T - 779.59)^2 \qquad (46.7)$$

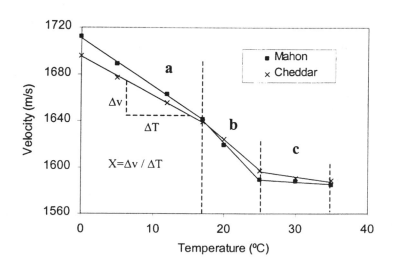

Figure 46.5 Variation of the ultrasonic velocity with temperature.

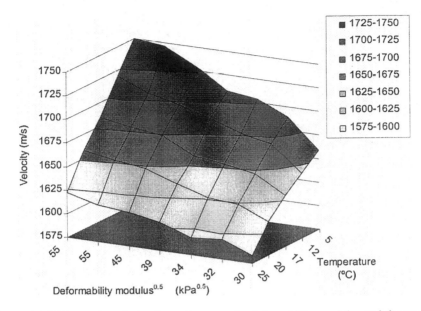

Figure 46.6 Ultrasonic velocity dependence on the deformability modulus and the temperature.

This equation allows assessment of cheese texture from the measurement of the ultrasonic velocity and the temperature at which it has been performed.

Cheese Composition

The ultrasonic velocity of a multicomponent material depends on the composition and velocity of each single constituent.[52] As long as the ultrasonic velocity of each constituent reacts in a different way to temperature changes, the study of the velocity-temperature curve can be used to assess the composition of some food products. This has been the case for oils and fish.[49,18]

Cheese is mainly composed of water, fat, and protein. Benedito et al.[50] related the moisture content to the slope of the first part (a) of the velocity-temperature curves (Figure 46.7). The absolute value of the slope can be seen to increase in line with the decrease in moisture content. The relationships between the moisture content and the slope in the second (b) and third (c) parts gave poor results, probably due to the fact that the fat melting was accompanied by structural changes and oil loss from the casein matrix produced by the "oiling off." Ultrasonic velocity in water increases with temperature; this is not the case for fat, where ultrasonic velocity decreases as temperature increases. For the same type of cheese, as long as the water content increases, the relative fat content decreases, and therefore the change of velocity according to temperature (slope of the curve) must be higher. Therefore, the measurement of the ultrasonic velocity at different temperatures (at least two) can be used to assess the moisture content of all the blocks in one batch. In the case

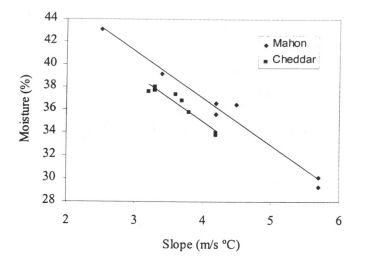

Figure 46.7 Ultrasonic assessment of cheese moisture through the slope of the velocity-temperature curves.

of cheddar cheese, these measurements at different temperatures could be carried out during the cooling process of the cheese blocks, which would greatly simplify the measuring process. To this end, more work should be carried out considering temperature profiles within the blocks.

Crack Detection

Resonant techniques (sonic frequencies) have been used to classify defects in cheese based on the differences found in the spectrum of cheeses with and without defect.[54] Ultrasonics were also used by Orlandini and Annibaldi[55] for the same purpose. Even more effective for this purpose can be the use of pulse-echo techniques commonly used in metallurgy.

Figure 46.8 shows the ultrasonic signal for a Mahon cheese with cracks, the dotted line being obtained by increasing the maximum signal values for 200 Mahon cheeses without cracks to 50%. As can be observed, the ultrasonic signal for a cheese with cracks crosses the dotted line as a consequence of the energy that is reflected on the crack surface and returns to the transducer. Takai et al.[27] also used the pulse-echo technique to evaluate the size and number of voids in kamaboko by counting the number of ultrasonic echo pulses on the oscillograms. Using this technique, cracked cheeses can be identified. Furthermore, it is also possible to determine the distance of the crack from the surface by assuming a range of velocities that include the maximum and minimum values found for the particular type of cheese. The calculated range for the cheese shown in Figure 46.8 was 1.84–1.98 cm (velocity range 1620–1740 m/s), which coincided with the distance measured with a digital gauge (1.9 cm).

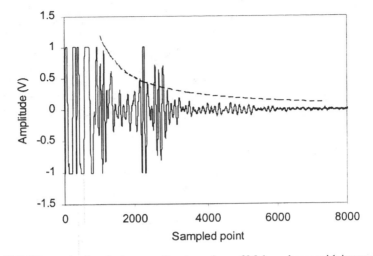

Figure 46.8 Ultrasonic signal corresponding to a piece of Mahon cheese with internal cracks.

46.6 OTHER APPLICATIONS

46.6.1 FOULING DETECTION

When dairy products are processed in continuous high-temperature processing plants, the internal walls of the plant can become fouled by burnt-on or chemically deposited material. The fouling layer will affect flow rate and also heat flow to or from the product. Withers[22] developed an ultrasonic sensor to detect and measure the thickness of these films in a dairy plant. Two techniques were considered for the sensor, pulse-echo and through transmission. The latter technique was found to be the most suitable. The sensor operated by transmitting a pulse of ultrasound across the pipe being tested. The received signal was analyzed in the time domain to determine film presence and thickness. Thickness measurement was possible over a range of 0.5–6.0 mm. Product temperature compensation over a temperature range of 20–140°C was implemented. Changes in product flow rate from 0 to 25 l/min and pressure from 0 to 3 bar had no effect on the ultrasonic measurements. The sensor offers the possibility of on-line, noninvasive detection and measurement of fouling films. It can be used to determine the optimum time for cleaning or to decide when a clean-in-place (CIP) operation is completed.

The main disadvantage of this method is that inspection is possible only point by point, requiring removal of the insulation. Another possibility is the use of several transducers located in strategic fixed positions on the plant and a signal multiplexer to combine all the measurements in only one pulser-receiver device.

Alternative ultrasonic methods to measure fouling are guided waves, acoustic emission and acoustic impact.[24] Guided waves can inspect long distances of pipe, including underneath insulation, but it is difficult to penetrate pipe joints. Acoustic emission relies on an applied stress to create acoustic emissions within the specimen.

It allows inspection of large pipe areas with only one test, although over-pressurization of the piping system is necessary. Acoustic impact uses a mechanical impacting device, such as a hammer tap or a coin tap, to generate sonic and ultrasonic waves in the specimen. Ultrasonic transducers are used to receive the waveform, and the velocity, attenuation, and frequency content are analyzed. Acoustic impact is the most adequate method for penetration beyond pipe joints and over long distances, although localization of fouling is not possible.

46.6.2 ULTRASONIC SPECTROSCOPY

Ultrasonic spectroscopy is becoming an increasingly popular analytical technique for characterizing food properties[1] and has been used to determine the size and concentration of casein micelles in aqueous solutions.[56]

This characterization is highly interesting, because whey proteins are widely used in food products to enhance the texture and stability of emulsions, foams, and gels. The functional attributes of proteins are ultimately determined by their molecular structure and interactions, as well as by changes in these molecular characteristics caused by alterations in environmental conditions such as temperature, pH, ionic strength, and mechanical force.[57]

Bryant and McClements[58] studied the influence of pH on the fast chemical reactions and aggregation of whey protein molecules in aqueous solutions. Ultrasonic attenuation spectra (1–100 MHz) of aqueous solutions of whey proteins were measured as a function of pH (2–12). Two types of whey solutions were used, one native and the other alkali-denatured. At 7 MHz, the solution containing the native proteins had two distinct peaks in the attenuation coefficient at pH 2.8 and 11.6, which were due to proton transfer equilibrium ($-CO_2H \leftrightarrow -CO_2^- + H^+$ and $-NH_2 + H^+ \leftrightarrow -NH_3^+$ respectively). Attenuation at other pH values was attributed to a hydration relaxation mechanism. Relaxation times for the equilibrium were of the order of 10^{-8} s. At 60 MHz, for solutions containing alkali-denatured protein, there was an additional attenuation peak at the isoelectric point of the proteins (pH 5). This was attributed to the aggregation of protein molecules close to their isoelectric point and the consequent scattering of the ultrasonic wave. Alkali-denatured proteins are more unfolded than native proteins and therefore have a higher tendency to aggregate. In the case of native protein, the concentration of protein aggregates was too small to be detected by the ultrasonic technique, which is only sensitive to particle concentrations greater than $\approx 0.5\%$. The particle size distribution of the aggregates was also determined using the ultrasonic scattering theory to analyze the attenuation spectra. These results show that this technique could be used to assess changes in protein properties under changing environmental conditions, which may lead to a better understanding of the relationship between the molecular and functional properties of proteins.

46.7 CONCLUSIONS

Ultrasonic velocity and attenuation are related, among other things, to physicochemical properties such as density, structure, texture, viscosity, and composition. Therefore, from ultrasonic measurements, the value or variation of these properties could be nondestructively assessed. The detection of different layers or bodies within a product can also be accomplished by ultrasonic imaging. Ultrasonic techniques are nondestructive and therefore allow control over the whole production. This advantage means that new applications will be developed and applied to the dairy industry.

46.8 ACKNOWLEDGMENT

The authors would like to acknowledge the financial support of ALI96-1180 from the Comisión Interministerial de Ciencia y Tecnología, Ministerio de Educación y Ciencia, Spain.

REFERENCES

1. McClements, D. J. 1997 "Ultrasonic Characterization of Foods and Drinks: Principles Methods and Applications," *Critical Reviews in Food Science and Nutrition*, 37 (1): 1–46.

2. Mulet, A., J. Benedito, J. Bon, and N. Sanjuan. 1999. "Low Intensity Ultrasonics in Food Technology," *Food Sci. Technol. Intl.*, 5: 285–397.

3. Mott, I. E. C., D. J. Sticler, W. T. Coackley, and T. R. Bott. 1998. "The Removal of Bacterial Biofilm from Water-Filled Tubes Using Axially Propagated Ultrasound," *J. Appl. Microbiol.*, 84: 509–514.

4. Sanz, B., P. Palacios, P. López, and J. A. Ordóñez. 1985. "Effects of Ultrasonic Waves on the Heat Resistance of *Bacillus stearothermophilus* Spores," in *Fundamental an Applied Aspects of Bacterial Spores*, G. J. Dring, D. J. Ellar, and G. W. Gould, eds., San Diego, U.S.A.: Academic Press, pp. 251–259.

5. Pagán, R. 1997. "Resistencia al Calor y los Ultrasonidos Bajo Presión de *Aeromonas hydrophila*, *Yersinia enterocolitica* y *Listeria monocytogenes*," Ph.D. Thesis. Universidad de Zaragoza, Spain.

6. Vercet, A., P. López, and J. Burgos. 1997. "Inactivation of Heat-Resistant Lipase and Protease from *Pseudomonas fluorescens* by Manothermosonication," *J. of Dairy Sci.*, 80 (1): 29–36.

7. Wang, D. Z., M. Sakakibara, N. Kondoh, and K. Suzuki. 1996. "Ultrasound-Enhanced Lactose Hydrolysis in Milk Fermentation with *Lactobacillus Bulgaricus*," *J. Chem. Technol. and Biotechnol.*, 65(1): 86–92.

8. Vinatoru, M., M. Toma, O. Radu, P. I. Filip, D. Lazurca, and T. J. Mason. 1997. "The Use of Ultrasound for the Extraction of Bioactive Principles from Plant Materials," *Ultrasonics Sonochemistry*. 4: 135–139.

9. Jun, C., Y. Kedie, C. Shulai, T. Adschiri, and K. Arai. 1997. "Effects of Ultrasound on Mass Transfer in Supercritical Extraction," The 4th International Symposium on Supercritical Fluids, May 11–14, Sendai, Japan.

10. Sastry, S. K., G. Q. Shen, and J. L Blaisdell. 1989. "Effect of Ultrasonic Vibration on Fluid-to-Particle Convective Heat Transfer Coefficients," *J. Food Sci.*, 54(1): 229–230.

11. Gallego-Juárez, J. A., G. Rodriguez-Corral, J. C. Gálvez-Moraleda, and T. S. Yang. 1999. "A New High-Intensity Ultrasonic Technology for Food Dehydratation," *Drying Technology*, 17 (3): 597–608.

12. Simal, S., J. Benedito, E. S. Sánchez, and C. Rosselló. 1998. "Use of Ultrasound to Increase Mass Transport Rates during Osmotic Dehydration," *J. Food Eng.*, 36: 323–336.

13. Sánchez, E. S., S. Simal, A. Femenia, J. Benedito, and C. Rosselló. 1999. "Influence of Ultrasound on Mass Transport During Cheese Brining," *Eur. Food Res. Technol.* 209: 215–219.

14. Cárcel, J. A., J. Benedito, P. Llull, G. Clemente, and A. Mulet. 1999. "Influence of Ultrasound in Meat Brining," in *6th Conference of Food Engineering (CoFE'99)*, G. Barbosa-Cánovas and S. P. Lombardo, eds. AICHE, pp. 163–170.

15. Mulet, A., J. A. Cárcel, J. Benedito, C. Rosselló, and S. Simal. 1999. "Ultrasonic Mass Transfer Enhancement in Food Processing," in *Proc. 6th Conference of Food Engineering (CoFE'99)*. G. Barbosa-Cánovas and S. P. Lombardo, eds. AICHE, pp. 74–85.

16. Nielsen, M. and H. J. Martens. 1997. "Low Frequency Ultrasonics for Texture Measurements in Cooked Carrots (*Daucus carota* L.)," *J. Food Sci.*, 62(6): 1167–1175.

17. Shore, D. and C.A. Miles. 1988. "Experimental Estimation of the Viscous Component of Ultrasound Attenuation in Suspensions of Bovine Skeletal Muscle Myofibrils," *Ultrasonics*, 26: 31–36.

18. Ghaedian, R., J. Coupland, E. Decker, and D. J. McClements. 1998. "Ultrasonic Determination of Fish Composition," *J. Food Eng.*, 35: 323–336.

19. Chanet, M. 1998. "Use of Ultrasound Reflection for Fresh Hams Classification," *Sensorial* (Montpellier-Narbone, France, February 23–27).

20. Mizrach, A., N. Galili, and G. Rosenhouse. 1989. "Determination of Fruit and Vegetable Properties by Ultrasonic Excitation," *American Society of Agricultural Engineers*, 32 (6): 2053–2058.

21. Abbott, J. A., D. R. Massie, B. L. Upchurch, and W. R. Hruschka. 1995. "Nondestructive Sonic Firmness Measurement of Apples," *Transactions of the American Society of Agricultural Engineers*, 38 (5): 1461–1466.

22. Withers, P. 1994. "Ultrasonic Sensor for the Detection of Fouling in UHT Processing Plants," *Food Control*, 5(2): 67–72.

23. Samari, S. 1994. "Ultrasonic Inspection Methods for Food Products," *Lebensm.-Wiss. u.-Technol.*, 27: 210–213.

24. Lohr, K. R. and J. L. Rose. 1999. "Acoustic Methods for Pipe Fouling Detection in the Food Industry," in *6th Conference of Food Engineering (CoFE '99)*. G. V. Barbosa-Cánovas and S. P. Lombardo, eds. AICHE, pp. 675–678.

25. Tavossi, H. M. and B. R. Tittmann. 1999. "Ultrasonic Sensor for Bulk-Temperature Determination of a Mixture In a Steam-Jacketed Kettle with Double Agitator," in *Proc. 6th Conference of Food Engineering (CoFE '99)*. G. V. Barbosa-Cánovas and S. P. Lombardo, eds. AICHE, pp. 716–720.

26. Cheng, Y. and C. G. Haugh. 1994. "Detecting Hollow Heart in Potatoes Using Ultrasound," *Transactions of the American Society of Agricultural Engineers*, 37 (1): 217–222.

27. Takai, R., T. Suzuki, T. Mihori, S. Chin, Y. Hocchi, and T. Kozima. 1994. "Non Destructive Evaluation of Voids in Kamaboko by an Ultrasonic Pulse-Echo Technique," *J. Japanese Society of Food Sci. and Technol.*, 41(12): 897–903.

28. Ahvenainen, R., G. Wirtanen, and T. Mattila-Sandholm. 1991. "Ultrasound Imaging—A Non-destructive Method for Monitoring Changes Caused by Microbial Enzymes in Aseptically-packed Milk and Soft Ice-cream Base Material," *Lebensmittel Wissenschaft und Technologie*, 24 (5): 397–403.

29. Oppenhein, A. V. and R. W. Schafer. 1989. *Discrete-Time Signal Processing*. New Jersey: Prentice Hall International, Inc.

30. Povey, M. J. W. and D. J. McClements. 1988. "Ultrasonics in Food Engineering. Part I: Introduction and Experimental Methods," *J. Food Eng.*, 8: 217–245.

31. Povey, M. J. W. 1998. "Rapid Determination of Food Material Properties," in *Ultrasound in Food Processing*. M. J. W. Povey and T. J. Mason, eds. New York, NY: Thomson Science, pp. 30–65.

32. Kinsler, L. E., A. R. Frey, A. B. Coppens, and J.V. Sanders. 1982. *Fundamentals of Acoustics*, 3rd ed. New York: John Wiley & Sons.

33. Kocis, S. and Z. Figura. 1996. *Ultrasonic Measurements and Technologies*. London: Chapman and Hall.

34. Kuttruff, H. 1991. *Ultrasonics. Fundamentals and Applications*. London: Elsevier Applied Science.

35. Bhatti, S. S., R. Bhatti, and S. Singh. 1986. "Ultrasonic Testing of Milk," *Acoustica*, 62: 96–99.

36. Ahvenainen, R., T. Mattila, and G. Wirtanen. 1989. "Ultrasound Penetration through Different Packaging Materials—A Non-destructive Method for Quality Control of Packed UHT Milk," *Lebensmittel Wissenschaft und Technologie*, 22 (5): 268–272.

37. Ahvenainen, R., G. Wirtanen, and M. Manninen. 1989. "Ultrasound Imaging—A Non-Destructive Method for Monitoring the Microbiological Quality of Aseptically-Packed Milk Products," *Lebensmittel Wissenschaft und Technologie*, 22 (6): 382–386.

38. Wirtanen, G., R Ahvenainen, and T. Mattila Sandholm. 1992. "Non-Destructive Detection of Spoilage of Aseptically-Packed Milk Products: Effect of Frequency and Imaging Parameters on the Sensitivity of Ultrasound Imaging," *Lebensmittel Wissenschaft und Technologie*, 25(2): 126–132.

39. Gestrelius, H. 1994. "Ultrasonic Doppler: a Possible Method for Non-Invasive Sterility Control," *Food Control*, 5 (2): 103–105.

40. Perdonet, G. 1990. "Tecnología Comparada de los Diferentes Tipos de Cuajada," in *El queso*, Eck, ed. Barcelona. Spain: Ediciones Omega S.A., pp. 199–224.

41. Benguigui, L., J. Emery, D. Durand, and J. P. Busnel. 1994. "Ultrasonic Study of Milk Clotting," *Lait*, 74 (3): 197–206.

42. Ay, C. and S. Gunasekaran. 1994. "Ultrasonic Attenuation Measurements for Estimating Milk Coagulation Time," *Transactions of de ASAE*, 37 (3): 857–862.

43. Gunasekaran, S. and C. Ay. 1996. "Milk Coagulation Cut-Time Determination Using Ultrasonics," *J. Food Process Engineering*, 19 (1): 63–72.

44. Lee, H. O., H. Luan, and D. G. Daut. 1992. "Use of an Ultrasonic Technique to Evaluate the Rheological Properties of Cheese and Dough," *J. Food Eng.*, 16: 127–150.

45. Maiorov, A. A. and L. A. Ostroumov. 1977. "Method of Monitoring the Degree of Cheese Maturation," USSR patent 586128.

46. Benedito, J., J. A. Cárcel, G. Clemente, and A. Mulet. 1999. "Cheese Maturity Assessment Using Ultrasound," *Journal of Dairy Science*. In press.

47. Povey, M. J. W. 1989. "Ultrasonics in Food Engineering. Part II: Applications," *J. Food Eng.*, 9:1–20.

48. Mizrach, A., N. Galili, S. Gan-mor, U. Flitsanov, and I. Prigozin. 1996. "Models of Ultrasonic Parameters to Asses Avocado Properties and Shelf Life," *J. Agricultural Engineering Research*, 65: 261–267.

49. Miles, C. A., G. A. J. Fursey, and R. C. D. Jones. 1985. "Ultrasonic Estimation of Solid/Liquid Ratios in Fats, Oils and Adipose Tissue," *J. Sci. Food and Agriculture*, 36: 215–228.

50. Benedito, J., J. A. Cárcel, N. Sanjuan, and A. Mulet. 1999. "Use of Ultrasound to Assess Cheddar Cheese Characteristics," *Ultrasonics*. In press.

51. Creamer, L. K. and N. F. Olson. 1982. "Rheological Evaluation of Maturing Cheddar Cheese," *J. Food Sci.*, 47(2): 631–636, 646.

52. Ghaedian, R., E. A. Decker, and D. J. McClements. 1997. "Use of Ultrasound to Determine Cod Fillet Composition," *J. Food Sci.*, 62: 500–504.

53. Mulet, A., J. Benedito, J. Bon and C. Rossello. 1999. "Ultrasonic Velocity in Cheddar Cheese as Affected by Temperature," *J. Food Sci.* In press.

54. Giangiacomo, R., G. Messina, and F. Abbiati. 1989. "Confronto Tra la Risonanza Acustica Strumentale e la Battitura per la Valutazione Commerciale di Formaggio Grana," *Latte*, 14:126–129.

55. Orlandini, I. and S. Annibaldi. 1983. "New Techniques in Evaluation of the Structure of Parmesan Cheese: Ultrasonic and X-Rays," *Sci. Latiero-Caseria*, 34: 20–30.

56. Griffin, W. G. and M. C. A. Griffin. 1990. "The Attenuation of Ultrasound in Aqueous Suspensions of Casin Micelles from Bovine Milk," *J. Acoust. Soc. Am.*, 87: 2541–2550.

57. Damoran, S. 1996. "Amino Acids, Peptides and Proteins," in *Food Chemistry*, O. R. Fennema, ed. New York: Marcel Decker, pp. 321–429.

58. Bryant, C. M. and D. J. McClements. 1999. "Ultrasonic Spectroscopy Study or Relaxation and Scattering in Whey Protein Solutions," *J. Sci. Food and Agriculture*. 79: 1754–1760.

47 Ohmic Heating and Moderate Electric Field (MEF) Processing

S. K. Sastry
A. Yousef
H-Y. Cho
R. Unal
S. Salengke
W-C. Wang
M. Lima
S. Kulshrestha
P. Wongsa-Ngasri
I. Sensoy

CONTENTS

Abstract

Ohmic heating was revived in the 1980s, because it showed promise in particulate sterilization. Although that dream has not yet been fully realized, a number of advances have been made regarding fundamental understanding of the process. This has included fundamental fluid mechanics and heat transfer phenomena, microbial death kinetics, and the monitoring of temperatures, microbiological, and chemical changes within solids.

1-56676-963-9/02/$0.00+$1.50
© 2002 by CRC Press LLC

Interest in ohmic heating has resulted in its use being extended to a wide array of processes that show great promise in future applications. These applications have been termed moderate electric field (MEF) processes, since they involve varying degrees of heat input and are characterized by moderate, arbitrary waveform fields, which distinguish them from pulsed electric field (PEF) processing. Applications include the acceleration of fermentations; detection of starch gelatinization in solutions and pastes; pretreatments for drying, extraction, and expression; and reduction in water use during blanching.

47.1 INTRODUCTION

47.1.1 OHMIC HEATING

The concept of ohmic (Joule) heating has been in existence since James Prescott Joule elucidated, in 1840, how heat was generated in an electrical conductor. This resulted in a number of patents in the latter part of the nineteenth century. The technology has since been revived periodically, having seen industrial application for milk pasteurization in the 1930s, before falling out of favor. In the 1980s, the technology was once again revived, and some industrial applications have resulted, including the pasteurization of liquid eggs and the processing of fruit products, among others.

The basic principle of ohmic heating is the well known dissipation of electrical energy into heat, resulting in internal energy generation, which is proportional to the square of the electric field strength and the electrical conductivity. The basic equations have been discussed in prior publications on several occasions (e.g., Sastry and Palaniappan,[1] Sastry and Li[2]) and will not be detailed here. Since electrical conductivity increases with temperature, ohmic heating becomes more effective at higher temperatures. Furthermore, for materials of uniform electrical conductivity, the energy generation is far more uniform than microwave heating. The basic principles have been addressed in a number of publications including Palaniappan and Sastry,[1] de Alwis and Fryer,[4] Halden et al.,[5] and Sastry.[6]

Applications of Ohmic Heating

The above advantages have resulted in great expectations for the technology for sterilization of particulate foods. However, a number of issues have had to be addressed for particulate sterilization to be realized. The foremost issue is the worst-case heating scenario and the location of cold zones. If the solid and liquid phases are of equal electrical conductivity, the heating is relatively uniform, and the technology shows itself to its best advantage. The problem arises when individual inclusion particles of electrical conductivity significantly differ from its surroundings. Under such conditions, one of the phases will lag the other.

The question of which phase lags which one is dependent on the electrical conductivity of the respective phases, and the extent of fluid motion (fluid-solid convective heat transfer coefficient). The static situation has been analyzed by Davies et al.[7] and Sastry and Salengke[8] have shown that the worst case does not necessarily conform to the static situation. When the solid is less conductive than the fluid, a situation involving mixed fluids is the worst; however, when fluid is less conductive than the solid, the static situation appears to be the worst case. A number of other studies have attempted to address the use of chemical markers[9] or temperature measurement within ohmic heaters.[10]

Microbial Death Kinetics

The question of whether ohmic heating results in a nonthermal contribution to microbial lethality has been addressed in a number of studies in the literature. Early literature on this topic has been inconclusive,[11] since most studies either did not specify sample temperature or failed to eliminate it as a variable. It is critically important that any studies comparing conventional and ohmic heating be conducted under temperature histories that are as identical as possible. Palaniappan et al.[12] attempted to compare ohmic and conventional heat treatments on the death kinetics of yeast cells (*Zygosaccharomyces baillii*) under identical histories, and they found no difference. However, a mild electrical pretreatment of *Escherichia coli* decreased the subsequent inactivation requirement in certain cases.

More recent studies suggest that a mild electroporation-type mechanism may be at play during ohmic heating. The presence of pore-forming mechanisms on cellular tissue has been confirmed by recent work.[13–15] Cho et al.,[16] conducting under near-identical temperature conditions, indicated that the kinetics of inactivation of *Bacillus subtilis* spores can be accelerated by an ohmic treatment. A two-stage ohmic treatment (ohmic treatment, followed by a holding time prior to a second heat treatment) was found to further accelerate death rates. A recent study (Lee and Yoon[17]) has indicated that leakage of intracellular constituents of *Saccharomyces cerevisiae* was found to be enhanced under ohmic heating as compared with conventional heating in boiling water.

The principal reason for the additional effect of ohmic treatment may be due to the low frequency (50–60 Hz) of ohmic heating, which allows cell walls to build up charges and form pores. This is in contrast to high-frequency methods such as radio frequency or microwave heating, where the electric field is essentially reversed before sufficient charge buildup (Figure 47.1). Some contrary evidence has also been noted; in particular, the work of Lee and Yoon[17] has indicated that greater leakage of *Saccharomyces cerevisiae* constituents occurs under high frequencies. However, the details of temperature control within this study are not available at this time; thus, it is not known if these researchers have adequately eliminated temperature effects.

Electrolytic Effects

When a current flows through an electrolyte, electrolytic reactions may occur at the electrode-solution interface. In a direct-current situation, different reactions occur at the cathode and anode; however, in an alternating-current situation, the cathode

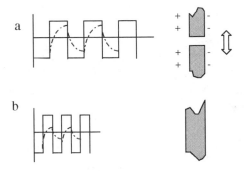

Figure 47.1 Illustration of square waves, showing the effect of frequency on cell wall pore formation. (a) Low-frequency fields allow membrane potential (dotted line) to build up to sufficient levels to cause pore formation. (b) High-frequency fields do not permit time for pore formation to occur.

and anode are periodically reversed, and both cathodic and anodic products may occur at either electrode. If the potential drop at the electrode-solution interface can be kept below the critical electrode potential for the system, Faradaic electrolytic processes can be prevented. This implies that either the frequency is increased sufficiently, or the electrode capacitance is increased for such minimization to occur. A detailed analysis of these phenomena has been presented by Amatore et al.[18] A key challenge is to address such issues economically. Electrodes will also undergo wear over time, and the use of wear-resistant, inert materials is desirable.

Fouling by Proteinaceous Materials

One advantage claimed for ohmic heating has been its ability to heat proteinaceous materials, which represent a challenge to conventional processing. However, ohmic heating of such materials also requires the overcoming of significant challenges, since protein deposits can adhere to electrode surfaces and create a high electrical resistance film at the interface. This in turn increases the voltage drop across such surfaces and may result in arcing. Reznik[19] discusses this issue in one of his patents. Studies by Wongsa-Ngasri[20] have shown that the incidence of arcing depends on the protein content, initial temperature, and flow rate, while carbohydrates (starch) were found to have a limited effect. Wongsa-Ngasri also presents arcing diagrams that show the range of conditions under which arcing could be expected to occur.

Seafood Processing

Research at Oregon State University and in Japan indicate that ohmic heating is a useful method for protease inactivation in restructured seafood products, permitting rapid heating to inactivation temperatures while minimizing the time spent in the temperature zone of maximum protease activity. Balaban and co-workers at the University of Florida have successfully thawed large blocks of shrimp using a specially designed ohmic heating process.

47.1.2 MEF Technology

The effects of ohmic heating are not purely thermal; various researchers have noted that a permeabilization effect occurs within both prokaryotic and eukaryotic cells that are exposed to an electrical field. The character of the permeabilization is different from thermal effects. For example, Imai et al.[13] reported that radish treated by ohmic heating shows superior texture to conventional thermally treated radish. Depending on the relative electrical and thermal effects of a particular treatment, it is possible to obtain a wide variety of effects in biological materials, and a large number of potential applications are waiting to be discovered in this area. Since the intent of these processes is in controlled permeabilization and other nonthermal effects, and because it is characterized by relatively moderate electrical fields, we call them moderate electric field (MEF) processes to distinguish them from the common high-voltage pulsed electric field (PEF) process.

We note that the permeabilizing effects of electricity (electropermeabilization or electroplasmolysis) have been known for some time. Zimmermann[21] produced a detailed body of work, which describes the mechanism of electropermeabilization of cells exposed to high voltage pulses. More recently, Knorr and co-workers at the University of Berlin[22,23] observed that both pulsed electric fields (PEFs) and ultra high pressure (UHP) processing have a permeabilizing effect on cellular materials. What makes MEF processes particularly interesting is that they are far less expensive than either PEF or UHP processes, while achieving many of the same effects. In many cases, the imposition of a simple alternating current from a readily available power supply is sufficient to cause permeabilization without the need for specialized switching circuits or heavy devices for pressure containment. Several examples of highly promising processes follow.

Pretreatments for Water Removal

Wang[14] showed that a short pretreatment of vegetable tissue with ohmic heating up to 80°C resulted in a significant acceleration of the drying process when compared to control (untreated), conventionally heat pretreated, and microwave pretreated samples. He showed that the sorption isotherms of such samples were significantly shifted at the higher levels of water activity, resulting in the more rapid drying rate. Wang[14] also found that apple juice expression could be made more efficient by such a permeabilization process, similar to an electoplasmolysis process previously developed in the former Soviet Union.[24] Furthermore, the mechanical energy required for juice expression was significantly reduced.

Lima and Sastry,[25] following up on Wang's work, showed that the waveform and frequency also had significant effects.

MEF-Assisted Fermentations

Studies on fermentation of L. acidophilus under the presence of a mild electric field[26] have indicated that the lag phase can be significantly reduced under the presence of an electric field. However, the ultimate productivity of the fermentation was found to be reduced by the presence of the field. It has been hypothesized that this may be due to the presence of mild electroporation, which improves nutrient transport at the early stages of fermentation, thereby accelerating it. At the later stages, the electroporation effect improves the transport of metabolites into the cell, thereby causing an inhibitory effect. Studies on other lactic acid bacteria (*Lactococcus lactis*) by Unal et al.[27] have shown similar results. These findings suggest that fermentation can be greatly accelerated by the application of MEF during the initial stages of fermentation; however, the field should preferably be absent when approaching the stationary phase. This has significant economic implications in improving process throughput.

MEF/Ohmic Heating for Detection of Starch Gelatinization

Wang[14] and Wang and Sastry[28] found that the electrical conductivity-temperature curve of starch solutions showed negative peaks at temperatures corresponding to similar peaks of a differential scanning calorimeter thermograph (Figure 47.2).[14] This led to the possibility of monitoring gelatinization from electrical conductivity measurements and suggests that on-line techniques could be developed for this purpose.

Water Use Reduction by MEF/Ohmic Blanching

Blanching is an operation that uses large quantities of water. In particular, the blanching of mushrooms on a canning line is a particularly lengthy and severe treatment aimed at ensuring that sufficient shrinkage occurs before canning. Sensoy et al.[29] found that mushrooms could be preshrunk at significantly lower temperatures, and with considerably lower water use, compared with conventional blanching. These results, too, suggest a process with economic potential.

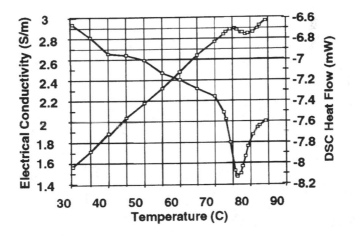

Figure 47.2 Superimposed DSC curve and electrical conductivity-temperature relationship for cornstarch suspension.

Extraction Enhancement

Russian work,[30,31] among others, has shown the effectiveness of MEF pretreatment in improvement of sugar beet extraction. Work in Korea[32] has indicated that soy milk extraction can be improved by ohmic/MEF treatment. Schreier et al.[33] reported the increase of diffusion by ohmic heating. Recent work by Kulshrestha and Sastry[15] shows that mild treatment of beet tissue (temperature rise of 1–2°C) results in leaching of dyes from cells.

47.2 CONCLUSIONS/OPPORTUNITIES

The effects of processes such as ohmic heating/MEF processing appear to have wide-ranging impact for a variety of processes. A number of key desirable and a few undesirable effects may occur. However, the harnessing of the positives, while minimizing and eliminating negative aspects, represents interesting engineering challenges for the future.

REFERENCES

1. Sastry, S. K. and S. Palaniappan. 1992. "Ohmic Heating of Liquid-Particle Mixtures," *Food Technol.*, 46 (12): 64–67.
2. Sastry, S. K. and Q. Li. 1996. "Modeling the Ohmic Heating of Foods," *Food Technol.*, 50(5): 246–248.
3. Palaniappan, S. and S. K. Sastry. 1991. "Electrical Conductivity of Selected Solid Foods During Ohmic Heating," *J. Food Proc. Engr.*, 14: 221–236.
4. De Alwis, A. A. P. and P. J. Fryer. 1990. "A Finite Element Analysis of Heat Generation and Transfer During Ohmic Heating of Food," *Chem. Engr. Sci.*, 45(6): 1547–1559.
5. Halden, K., A. A. P. de Alwis, and P. J. Fryer. 1990. "Changes in the Electrical Conductivity of Foods During Ohmic Heating," *Int. J. Food Sci. & Technol.*, 25: 9–25.
6. Sastry, S. K. 1992. "A Model for Heating of Liquid-Particle Mixtures in a Continuous Flow Ohmic Heater," *J. Food Proc. Engr.*, 15: 263–278.
7. Davies, L. J., M. R. Kemp, and P. J. Fryer. 1999. "The Geometry of Shadows: Effects of Inhomogeneities in Electrical Field Processing," *J. Food Engr.*, 40: 245–258.
8. Sastry, S. K. and S. Salengke. 1998. "Ohmic Heating of Solid-Liquid Mixtures: A Comparison of Mathematical Models Under Worst-Case Heating Conditions," *J. Food Proc. Engr.*, 21: 441–458.
9. Kim, H-J., Y-M. Choi, T. C. S. Yang, I. A. Taub, P. Tempest, P. Skudder, G. Tucker, and D. L. Parrott. 1996. "Validation of Ohmic Heating for Quality Enhancement of Food Products," *Food Technol.*, 50(5): 253–261.
10. Chang, K. and R. R. Ruan. 1999. "Fast MRI Technique for Study of Simultaneous Heat and Moisture Transfer in Foods Undergoing Ohmic Heating," Abstract No. 79B-7, 1999 IFT Annual Meeting, Chicago, IL, July 24–28.
11. Palaniappan, S., S. K. Sastry, and E. R. Richter. 1990. "Effects of Electricity on Microorganisms: A Review," *J. Food Proc. Pres.*, 14: 393–414.

12. Palaniappan, S., S. K. Sastry, and E. R. Richter. 1992. "Effects of Electroconductive Heat Treatment and Electrical Pretreatment on Thermal Death Kinetics of Selected Microorganisms," *Biotech. Bioeng.* 39: 225–232.

13. Imai, T., K. Uemura, N. Ishida, S. Yoshizaki, and A. Noguchi. 1995. "Ohmic Heating of Japanese White Radish *Rhaphanus sativus L.*," *Int. J. Food Sci. Technol.*, 30: 461–472.

14. Wang, W-C. 1995. *Ohmic Heating of Foods: Physical Properties and Applications.* Ph.D. dissertation, The Ohio State University, Columbus, OH.

15. Kulshrestha, S. A. and S. K. Sastry. 1999. "Low-Frequency Dielectric Changes in Vegetable Tissue from Ohmic Heating," Abstract No. 79 B-3, presented at the 1999 Annual IFT Meeting, Chicago, IL, July 24–28, 1999.

16. Cho, H-Y., A. E. Yousef, and S. K. Sastry. 1999. "Kinetics of Inactivation of *Bacillus subtilis* Spores by Continuous or Intermittent Ohmic and Conventional Heating," *Biotechnol. & Bioeng*, 62(3): 368–372.

17. Lee, C.H. and S. W. Yoon. 1999. "Effect of Ohmic Heating on the Structure and Permeability of the Cell Membrane of *Saccharomyces cerevisiae*," Abstract No. 79 B-6, presented at the 1999 Annual IFT Meeting, Chicago, IL, July 24–28, 1999.

18. Amatore, C., M. Berthou, and S. Hebert. 1998. "Fundamental Principles of Electrochemical Ohmic Heating of Solutions," *J. Electroanal. Chem.*, 457: 191–203.

19. Reznik, D. 1996. "Electroheating Apparatus and Methods," U.S. Patent No. 5,583,960.

20. Wongsa-Ngasri, P. 1999. "Effects of Inlet Temperature, Electric Field Strength and Composition on the Occurrence of Arcing During Ohmic Heating," M. S. Thesis, The Ohio State University.

21. Zimmermann, U. 1986. "Electrical Breakdown, Electropermeabilization and Electrofusion," *Rev. Physiol. Biochem. Pharmacol.*, 105: 176–256.

22. Knorr, D., M. Guelen, T. Grahl, and W. Sitzmann. 1994. "Food Application of High Electric Field Pulses," *Trends Food Sci. Technol.*, 5: 71–75.

23. Knorr, D. 1994. "Plant Cell and Tissue Cultures as Model Systems for Monitoring the Impact of Unit Operations on Plant Food," *Trends Food Sci. Technol.*, 5: 328–331.

24. Grishko, A. A., M. Kozin, V. G. Chebanu. 1991. "Electroplasmolyzer for Processing Raw Plant Material," U.S. Patent No. 5,031,521.

25. Lima, M. and S. K. Sastry. 1999. "The Effects of Ohmic Heating Frequency on Hot-air Drying Rate and Juice Yield," *J. Food Engr.*, 41: 115–119.

26. Cho, H-Y., A. E. Yousef, and S. K. Sastry. 1996. "Growth Kinetics of *Lactobacillus acidophilus* Under Ohmic Heating," *Biotech. & Bioengr.*, 49: 334–340.

27. Unal, R., A. E. Yousef, and S. K. Sastry. 1998. "Growth Kinetics of *Lactococcus lactis* subsp. *Lactis* Under Ohmic Heating," Abstract No. 59C-5, IFT Annual Meeting, Atlanta, GA, June 20–24.

28. Wang, W-C. and S. K. Sastry. 1997. "Starch Gelatinization in Ohmic Heating," *J. Food Engr.* 20(6): 499–516.

29. Sensoy, I., S. K. Sastry, and R. B. Beelman. 1999. "Ohmic Blanching of Mushrooms," Abstract No. 79B-1, 1999 IFT Annual Meeting, Chicago, IL, July 24–28.

30. Bazhal, I.G. and I. S. Guly. 1983a. "Effect of Electric Field Voltage on the Diffusion Process," *Pishchevaya Promyshlennost*, 1: 29–30.

31. Bazhal, I. G. and I. S. Guly. 1983b. "Extraction of Sugar from Sugarbeet in a Direct Current Electric Field," *Pishchevaya Tekhnologiya*, 5: 49–51.

32. Kim, J-S. and Y-R. Pyun. 1995. "Extraction of Soymilk Using Ohmic Heating," Abstract No. P125; 9th. World Congress of Food Science and Technology, Budapest, Hungary, July 30–August 4, 1995.
33. Schreier, P. J. R., D. G. Reid, and P. J. Fryer. 1993. "Enhanced Diffusion During the Electrical Heating of Foods," *Int. J Food Sci. Technol.,* 28: 249–260.

48 Design, Construction, and Evaluation of a Sanitary Pilot Plant System Pulsed Electric Field

Q. H. Zhang
C. B. Streaker, Jr.
H. W. Yeom

CONTENTS

Abstract

A sanitary fluid-handling system has been designed and constructed for the treatment of fluid and particulate (up to 1/8 in.) food products in an integrated pilot plant scale pulsed electric field (PEF) processing and aseptic packaging system. The system offers flexibility of processing parameters to explore the synergistic effects of heat and PEF to produce an extended-shelf-life product. The key parameters for this fluid-handling system are that it is capable of being sterilized; the product flow rate is constant, pulseless, and with minimal shear; and thermal and electrical treatment parameters are monitored. The system components were selected to meet 3-A standards and allow for clean-in-place (CIP) sanitation. An operational manual was developed for the safe and efficient operation of this pilot plant system.

For evaluation purposes, single-strength orange juice was treated with PEF in this pilot plant system to optimize PEF processing conditions for maximum inactivation microorganisms and pectin methyl esterase (PME). Electric field strengths of 20, 25, 30, and 35 kV/cm and total treatment times of 39, 49, and 59 μs were tested. The higher electric field strength and the longer total treatment time were more effective in inactivation. PEF treatment of orange juice at 35 kV/cm for 59 μs caused 7 log reduction in the total aerobic plate counts and the yeast and mold counts. A reduction of 90% PME activity was achieved by this PEF processing condition. This PEF treatment also prevented the growth of microorganisms and the recovery of PME activity at 4, 22, and 37°C for 16 weeks.

48.1 INTRODUCTION

Pulsed electric field processing of foods has been conducted with static treatment systems using a batch mode of processing[1] as well as with continuous systems.[2-4] The basis for any continuous food processing operation is a means of product transfer for processing and packaging. For liquid and flowable products, a fluid-handling system composed of pumps, tubing, and accessories is used. For any calculated food process, the pumping must provide a constant product flow rate to ensure uniformity of treatment conditions.[5]

Qiu et al.[3] described a pilot plant scale fluid-handling system for PEF processing of fresh orange juice followed by aseptic packaging. This new fluid-handling system is a second-generation, sanitary design that allows for processing and packaging of viscous and particulate foods. The system employs sets of tubular heat exchangers for product heating and cooling, as in a traditional aseptic process, as well as a cooling unit for PEF temperature regulation. Zhang et al.[6] note that cooling may be used to reduce the effects of product temperature increase caused by the application of high-voltage electric pulses. The overall function of the system is to allow for flexibility of treatment conditions necessary to produce assorted fresh-like extended-shelf-life products, including juices, milk, and salsa. After construction of the fluid-handling system, an operating guide was prepared to facilitate the safe and efficient operation of this system.

48.2 MATERIALS AND METHODS

48.2.1 PROCESS DESIGN

There are several key design features for the fluid-handling system. These include the ability to process fluid, viscous (0.5 Pa·s maximum), and particulate (1/8-in. maximum) food products with a combination of heat and PEF conditions; sterilize the system; monitor system temperatures, pressures, and flow rate; meet 3-A standards for design; and clean with CIP procedures and readily disassemble for inspection. A system block diagram is illustrated in Figure 48.1.

The system pump is capable of pumping 1/8-in. particles with minimal shear and providing a stable, pulseless flow rate against a system pressure of up to 150 psi. Processing calculations were based on the desired PEF processing flow rate of 100 l/h to provide adequate total treatment time. This system is able to heat the product from 15°C (59°F) to 121°C (250°F), hold the product for 30 s, and then cool it to 30°C (86°F) for PEF processing. The heat process is comparable to commercial UHT operations, thereby allowing comparison of UHT and PEF treated extended-shelf-life products for microbial, enzymatic, and sensory analysis. The system uses recirculating refrigerated water chillers to regulate product temperature during PEF treatment.

A tubular type of counterflow heat exchanger was selected to provide optimal heating and cooling of particulate products, without the expense of scraped surface heat exchangers.[5] The first objective of the design phase was to calculate the minimum heat exchanger length needed to provide product heating to 121°C (250°F) and subsequent cooling to 30°C (86°F). Using 1-in. o.d. (0.902-in. i.d.), the length was calculated to be 4.42 m, based on Equation (48.1).[7]

$$L = \frac{A}{\pi D} \tag{48.1}$$

where L = length of pipe (m)

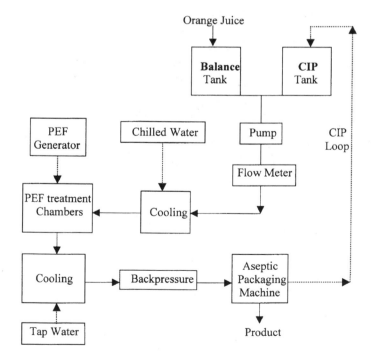

Figure 48.1 Pilot plant system PEF block diagram.

A = area of pipe (m²)

D = internal diameter of the pipe (m)

The objective for fabricating the heat exchangers was standardization. Thus, seven identical heat exchangers were used, each holding a set of 26-in. long tubes inside a 24-in. block. To reduce product fouling, 30% extra heating length was allowed, so a series of three vertically mounted heat exchangers provides 7.4 m of heating length. Heating water entered the top of each block and flowed downward across the product being pumped upward through the tubes. The cooling and PEF cooling systems each provided 4.9 m of cooling length from two heat exchangers arranged vertically for cooling and horizontally for PEF cooling.

The holding tube length was calculated for a 30-s residence time. The mean velocity of the product was 9 m/min based on a flow rate of 100 l/h in the 1-in. o.d. tubing, as calculated in Equation (48.2).[7]

$$u = \frac{m}{60\rho A} \tag{48.2}$$

where u = mean velocity (m/min)

m = mass flow rate (kg/h)

ρ = product density (kg/m³)

A = cross sectional area of tubing (m²)

Then, the product of u (9 m/min) times residence time (0.5 min) gives a length of 4.5 m. The design for the holding tube uses six sections of the standardized 26-in. long tubing, with five pairs of connecting elbows to provide a length of 4.77 m, giving an actual holding time of 31.8 s for a 100 l/h flow rate.

The Reynolds number (N_{Re}) was calculated[7] to determine the flow profile in the tubing, where

$$N_{Re} = \rho Du/\mu \tag{48.3}$$

For all products being processed in the system, a laminar flow was expected, as all calculated N_{Re} were below 2100. The pressure loss in the system was calculated to determine the compatibility of the available pump (150-psi maximum pressure) with the new system. Using the formula[7]

$$P = 32\ \mu v l/d^2 \tag{48.4}$$

where l = pipe length,

the maximum pressure loss for the entire system based on 0.5 Pa·s viscosity is 85 psi.

Another major design factor was the measurement of the critical processing parameters. System temperature measurement was made at these locations: product inlet to heating, mid-heating, holding tube inlet and outlet, cooling outlet, PEF cooling outlet, and the return line to the product tanks. The temperature at the outlet of the holding tube is the thermal processing temperature,[5] and the temperature at the return line to the product tanks is the minimum system sterilization temperature. System pressure measurements were made at the inlet to heating and prior to each of the backpressure valves. Two backpressure valves were used to provide a system overpressure of 15–50 psi for system sterilization and PEF processing.

48.2.2 COMPONENT SELECTION

The two groups of system components are (1) non-sanitary heating, cooling, and PEF cooling equipment (2) and sanitary product handling hardware.

Non-sanitary Fluid-Handling Systems

The principle component of the heating system is an 85-gal., 36-kW industrial electric water heater (model 6E744A, Rheem Mfg., Montgomery, AL). The tank is ASME rated for 400°F water, but the maximum thermostat setting was 82°C (180°F), and the temperature/pressure relief valve was set to release at 99°C (210°F)/150 psi. To achieve necessary product and sterilization temperatures, the hot water system must generate 129–132°C (265–270°F) water, so several changes were made.

First, a process controller (model CT 16164-942-992, Minco, Minneapolis, MN) was used in place of the factory thermostat. A type K thermocouple (Fisher Scientific, Pittsburgh, PA) was attached to the hot water exit of the tank and served as the temperature sensor for the process controller. The original temperature/pressure relief valve was then replaced with a pressure relief valve rated for 75 psi (model 790-75, Bell and Gossett, Morton Grove, IL). A severe-duty pump (model GA5-1π, Burks, Piqua, OH) equipped with a cooling jacket surrounding the pump housing was used for hot water circulation. This pump circulated hot water at a flow rate of 1500 l/h (15 times the product system flow rate) to provide adequate product heating.

The cooling system utilized cold tap water to cool the product after the holding tube, while PEF cooling water was supplied by two refrigerated recirculating chillers (Affinity, Ossipee, NH; and model CT-150, Neslab Instruments, Newington, NH).

Sanitary Fluid-Handling System

All components on the sanitary side of the fluid-handling system met 3-A standards of materials and workmanship. The first component of the sanitary product handling system was a pair of covered, 25-gal, flat-bottom, stainless steel tanks with welded 1-in. tri-clamp connections at the base and top-front side. The system pump was a 1-hp, sanitary progressive cavity pump (model CFB 2C SSV3SAA, Moyno® Industrial Products, Springfield, OH) with a 2-in. tri-clamp inlet port, and 1.5-in. tri-clamp outlet port and CIP bypass. The system flow rate was monitored with a magnetic flow meter (model AM202AG, Johnson-Yokogawa, Newman, GA).

Another major system constituent was a set of seven baffled-box tubular heat exchangers. The basis for each of these heat exchangers was a 24-in./15.5-in./2.5-in. aluminum block sliced into a 1.5-in. thick base and a 1-in. cover. The block was cut to accommodate four 26-in. long, 1-in. o.d. type 304 stainless steel tubes with tri-clamp connectors on each end. A 3/4-in. female NPT inlet and outlet for the heating or cooling water was cut through the base, and channels grooved in the block to allow the water to flow sequentially over the product tubes (Figure 48.2). The stainless steel tubes were placed in the base, and then all edges and bolt holes were coated with a high-temperature silicone gasket sealer (model 765-1480, Balkamp, Indianapolis, IN). The cover was then pressed onto the base, and the series of 35 bolts were securely tightened to clamp the heat exchanger together. The heat exchangers and all tubing were then insulated with fiberglass insulation.

An additional measure of processing flexibility was provided as the heating and holding, cooling, and PEF operations may be arranged in any sequence, depending on research needs and objectives. Several pictures were taken to show the results of the design and construction process. The product and CIP/SIP tanks are to be used when processing liquid products such as juice or milk (Figure 48.3). When process-

Figure 48.2 Heat exchanger base with product tubes inserted.

Figure 48.3 Product and CIP/SIP tanks for processing fluid foods.

ing salsa or yogurt, a vertical hopper is attached directly to the Moyno® pump inlet to improve product flow.

The high-voltage, ceramic PEF treatment chambers have a gap distance of 1 cm, and a 0.635-cm internal diameter.[8] The low-voltage, Delrin® isolation chambers have a gap distance of 1.2 cm and 0.635 cm internal diameter. Both types of chambers have stainless steel tubular electrodes, which thread into the body on one end, and at the opposite end have 1/2-inch tri-clamp connectors to attach to the fluid-handling system. These chambers are connected to the high voltage pulse generator and to ground. Figure 48.4 illustrates the PEF heat exchangers integrated with the PEF pulse generator and computer monitoring system chambers. After processing, a Benco® Asepack aseptic packaging machine (Piacenza, Italy) was used to pack the product (Figure 48.5).

System pressures were monitored using sanitary pressure gauges (Anderson Instrument Co., Fultonville, NY). The system pressures was controlled with a sanitary 1/2-inch diaphragm valve (no. 8842-A2306-BR-04, ITT, Lancaster, PA) and a non-sanitary screw-type valve. The sanitary valve was modified from an air-actuated control valve to a manually adjusted spring valve to provide 15–50 psi of system pressure. To monitor system temperatures, a series of seven sanitary RTD probes

Figure 48.4 Heat exchangers connected to the PEF unit (blue box) and data logger.

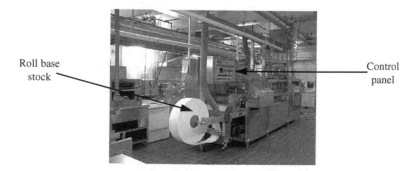

Figure 48.5 BENCO® aseptic packaging machine.

with dual sensing elements (model R1T285L4801, Inotek, Bensenville, IL) were placed in short outlet t-pieces at the necessary locations. These probes were connected to an RTD input module (777518-122) connected to a network module (777517-0) wired to a PC with the LabView® data logging software (776670-03, National Instruments, Austin, TX).

48.2.3 PROCEDURAL DEVELOPMENT

The methodology for the development of operating procedures is based on the following unit operations: start-up, sterilization, product processing, and CIP sanitation.

Preparation of Orange Juice for Optimization of PEF Conditions

Freshly squeezed single strength orange juice from Valencia oranges, quickly frozen in a 208-1 drum, was provided by Minute Maid (Houston, TX) and stored in a freezer at –25°C until processing. Unpasteurized orange juice was obtained by thawing the frozen orange juice at refrigeration temperature for ten days. To optimize PEF processing conditions for maximum microbial inactivation, the normal flora of orange juice were allowed to proliferate before PEF treatment by incubation of thawed orange juice at room temperature for two days. The number of microorganisms in orange juice was about 10^7 cfu/ml before PEF treatment.

PEF Treatment of Orange Juice for Optimization of PEF Conditions

The orange juice containing 10^7 cfu/mL of microorganisms was processed in the pilot plant PEF system using square wave and monopolar pulses. Electric field strengths of 20, 25, 30, and 35 kV/cm and total treatment times of 39, 49, and 59 μs were used. For the total treatment times of 39, 49, and 59 μs, frequency (pulses per second) was set at 400, 500, and 600, respectively. Pulse duration time was fixed at 1.4 μs. As a control, orange juice was pumped through the fluid-handling system without any PEF treatment. Control and PEF treated orange juices were aseptically packaged in 180-ml cups and stored at 4°C for two weeks to determine microbial and PME activity. To validate the results, orange juice was processed with the same PEF conditions three times. Five containers per PEF treatment condition were collected and analyzed.

Preparation of Orange Juice for Shelf-life Study

Unpasteurized single-strength orange juice was obtained by thawing frozen orange juice (Minute Maid, Houston, TX) at refrigeration temperature for ten days. To determine the shelf life of PEF treated fresh orange juice, the thawed orange juice was immediately processed in the pilot plant PEF system without incubation at room temperature. The number of microorganisms in orange juice was 10^3 cfu/ml before PEF treatment.

PEF Treatment of Orange Juice for Shelf-Life Study

The orange juice containing 10^3 cfu/ml of microorganisms was processed in the pilot plant PEF system at 35 kV/cm for 59 μs according to the results of PEF optimization. Control and PEF processed orange juice were stored at 4, 22, and 37°C for refrigerated, room temperature, and accelerated shelf-life studies, respectively. Two containers per storage condition were periodically taken after selected storage time and analyzed for the microbial and PME activity. For validation of data, the PEF processing was duplicated.

Microbial Assay

Microbial inactivation by PEF treatment was tested using total aerobic plate counts. Peptone water and potato dextrose agar (PDA) were purchased from Difco (Detroit, MI). Samples were diluted in 0.1% sterile peptone water up to 10^{-4} dilution and spiral plated by an Autoplater (Model 3000, Spiral Biotech Inc., Bethesda, MD) onto PCA for total aerobic plate counts. For each dilution, duplicate samples were plated. PCA plates were incubated at 30°C for 48 h.

PME Activity Assay

PME activity was measured using the method described in Reference 9. Citrus pectin and sodium chloride were purchased from Sigma (St. Louis, MO). 10-ml of orange juice was mixed with 40 ml of pectin-salt substrate (1%) and incubated at 30°C. The solution was adjusted to pH 7.0 with 2.0N NaOH, then the pH of solution was readjusted to pH 7.7 with 0.05N NaOH. After the pH reached 7.7, 0.10 ml of 0.05N NaOH was added. Time was measured until pH of the solution regained the pH 7.7. PME activity unit (PEU) and the relative PME activity (%) were calculated by the following formulas:

$$PEU = \frac{(0.05N \ NaOH)(0.10 \ ml \ NaOH)}{(10 \ ml \ sample)(time \ in \ minutes)} \tag{48.5}$$

$$Relative \ PME \ activity \ (\%) = \left(\frac{PEU \ of \ PEF \ treated \ orange \ juice}{PEU \ of \ control \ orange \ juice \ at \ day \ 0}\right)100 \tag{48.6}$$

Statistical Analysis

Analysis of variance (ANOVA) and Tukey's multiple comparisons method were used to determine statistical significance at the 5% level. All statistical analyses were conducted with Minitab 12.1 (Minitab, Inc., State College, PA).

48.3 RESULTS AND DISCUSSION

The result of procedural development is a manual for the safe and efficient operation of the fluid-handling system. The first section addresses the startup safety issues, including wearing protective eyewear, footwear, and gloves. After starting, the sterilization procedure provides a 30-min cycle at a minimum of 105°C (221°F) and 30 psi at a flow rate of 450 l/hr. The sterilizing solution is a salt-water solution prepared to the same conductivity as the product to be processed, so the sterile solution can be used as the starter for PEF processing. The next section of the procedures is the transfer to product processing, which must be done very carefully to avoid contamination of the system. In this stage, the product flow is not started until the desired flow rate, pressure, and thermal/PEF treatment parameters are achieved with the circulating sterile salt solution.

For product processing, the actual parameters vary, depending on the product and the level of treatment desired. Thus, a general procedure for processing is presented. When multiple conditions are used, the most severe treatment is applied first, followed by progressively lesser conditions. The processing steps can be challenging as they involve the collaboration of the operators of the fluid-handling system, pulse generator, and aseptic packaging machine. The final section of the manual describes CIP sanitation methods to ensure that the system is properly cleaned. After cleaning, the system is filled with a sanitizing solution of Quats to inhibit microbial growth in the system during storage. Ideally, the system should be drained completely after running; however, there are sections of the heat exchangers that are not readily drainable.

48.3.1 EFFECTS OF PEF PROCESSING PARAMETERS ON MICROORGANISMS

Effects of electric field strength on the total aerobic plate counts are shown in Figure 48.6. The total aerobic plate counts were significantly reduced at 25, 30, and 35 kV/cm ($p < 0.05$). There was 7 log reduction in the total aerobic counts by PEF treatment at 35 kV/cm for 59 μs. It is known that electric field strength and total treatment time are the major factors determining microbial inactivation in PEF processing.[10] This study showed that PEF treatment at higher electric field strength caused more inactivation of microorganisms in the orange juice. It also indicated that PEF treatment of orange juice might achieve commercial safety level in juice processing. At least 5 log reduction of microorganisms in juice processing is suggested by the Food and Drug Administration.[11]

48.3.2 EFFECTS OF PEF ON THE SHELF-LIFE OF ORANGE JUICE

To determine shelf-life of PEF-treated fresh orange juice, orange juice containing 10^3 cfu/ml of microorganisms was processed in a pilot plant PEF system at 35 kV/cm for 59 μs based on the results of PEF processing optimization. Total aerobic plate counts of PEF treated orange juice during storage time are shown in Figure 48.7. PEF treatment of orange juice at 35 kV/cm for 59 μs prevented the growth of microorganisms at 4, 22, and 37°C for 16 weeks, while control orange juice was spoiled after 2 weeks, even at 4°C.

The temperature of orange juice was increased to 60.1 ± 1.8°C during PEF treatment at 35 kV/cm for 59 μs. The retention time of orange juice at 60.1 ± 1.8°C

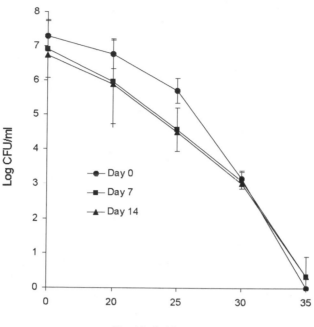

Figure 48.6 Effects of electric field strength on the total aerobic plate counts of orange juice with total treatment time of 59 μs during storage time at 4°C.

was less than 9 s. The thermal inactivation of microorganisms, especially yeast cells, was not the major cause.

48.3.3 Effects of PEF on PME Activity

Effects of electric field strength and total treatment time on the PME activity are shown in Figure 48.8. PME was significantly inactivated at the highest electric field strength, 35 kV/cm, with total treatment time of 49 and 59 μs ($p < 0.05$). Shorter treatment time, such as 39 μs, did not cause any inactivation of PME at 35 kV/cm. There was about 90% reduction in PME activity by PEF treatment at 35 kV/cm for 59 μs. A 90–100% reduction of PME activity is normal in commercial heat pasteurized orange juice, while total inactivation of PME is not always considered necessary.[12]

48.3.4 Inactivation of PME by PEF

PME is known to be more heat resistant than the spoilage microorganisms of orange juice.[13] During PEF treatment at 35 kV/cm for 59 μs, the temperature of orange juice was increased up to 60.1 ± 1.8°C. However, the retention time of orange juice at 60.1 ± 1.8°C was less than 9 s as it was immediately cooled to below 30.6 ±

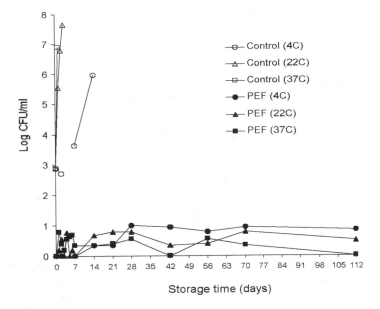

Figure 48.7 Total aerobic plate counts of the PEF treated orange juice at 35 kV/cm for 59 μs during storage time at 4, 22, and 37°C.

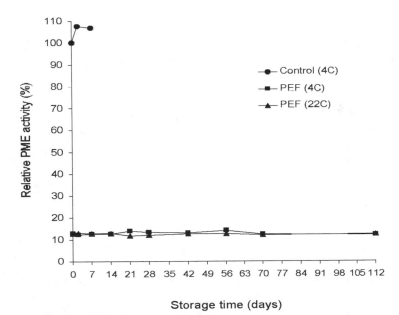

Figure 48.8 Effects of electric field strength and total treatment time on the relative pectin methyl esterase activity of orange juice.

0.6°C by tap water. According to Tajchakavit and Ramaswamy,[14] decimal reduction time of PME in orange juice for 90% reduction of PME activity was 153 s at 60°C. Thermal pasteurization of orange juice at 90°C for 1 min was recommended to inactivate pectinesterase.[15] Therefore, increase of temperature to 60.1 ± 1.8°C for 9 s during PEF treatment was not enough to cause 90% reduction of PME activity. Conformational change of enzymes has been suggested as the enzymatic inactivation mechanism of PEF by several researchers.[4,16,17]

48.4 CONCLUSIONS

The design and construction of a sanitary PEF pilot plant system was successfully evaluated with orange juice. PEF was effective for the control of microorganisms and pectin methyl esterase in orange juice. PEF treatment at 35 kV/cm for 59 µs reduced the population of microorganisms in orange juice by 7 log cycles and inactivated PME by 90%. PEF treated orange juice showed extended shelf life in terms of microbial and PME activity at 4, 22, and 37°C during storage time for more than 16 weeks. A continuous PEF treatment of orange juice on the pilot plant scale in combination with aseptic packaging proved to be useful for the preservation of orange juice.

48.5 ACKNOWLEDGMENTS

Equipment and operational funds for this project were provided by the US Army Natick RD&E Center and the NSF CAPPS Center. Borden Company provided aseptic packaging equipment. Minute Maid provided products. The authors are thankful to Dr. C. Wang for operating the pilot plant PEF system, Dr. W. Dantzer for operating the fluid-handling system, and Ms. Z. Ayhan and Ms. Y. Xu for operating the aseptic packaging machine. The authors also thank Dr. S. Palaniappan of Minute Maid for his technical advice.

REFERENCES

1. Dunn, J. E. and J. S. Pearlman. 1987. "Methods and Apparatus for Extending the Shelf Life of Fluid Food Products," U.S. Patent 4,695,472.
2. Qin, B., U. R. Pothakamury, H. Vega, O. Martin, G. V. Barbosa-Cánovas, and B. G. Swanson. 1995. "Food Pasteurization Using High Intensity Pulsed Electric Fields," Food Tech., 49(12): 55–60.
3. Qiu, X., S. Sharma, L. Tuhela, M. Jia, and Q. H. Zhang. 1998. "An Integrated PEF Pilot Plant for Continuous Nonthermal Pasteurization of Fresh Orange Juice," Trans ASAE., 41(4): 1069–1074.
4. Yeom, H. W., Q. W. Zhang, and C. P. Dunne. 1999. "Inactivation of Papain by Pulsed Electric Fields in a Continuous System," Food Chem., 67: 53–59.
5. David, J. R. D., R. H. Graves, and V. R. Carlson. 1996. Aseptic Processing and Packaging of Food, New York: CRC Press.

6. Zhang, Q. H., G. V. Barbosa-Canovas, and B. G. Swanson. 1995. "Engineering Aspects of Pulsed Electric Field Pasteurization," *J. Food Engineering*, 25: 261–281.

7. Singh, R. P. and D. R. Heldman. 1993. *Introduction to Food Engineering*, San Diego, CA: Academic Press, pp. 63–64, 133–180.

8. Yin, Y., Q. H. Zhang, and S. K. Sastry. 1997. "High Voltage Pulsed Electric Field Treatment Chambers for the Preservation of Liquid Food Products," U.S. Patent 5,690,978.

9. Kimball, D. A. 1991a. "Juice Cloud," Ch. 8 in *Citrus Processing-Quality Control and Technology*, D. A. Kimball, ed. New York: Van Nostrand Reinhold, p. 117–125.

10. Jeyamkondan, S., D. S. Jayas, and R. A. Holley. 1999. "Pulsed Electric Field Processing of Foods: a Review," *J. of Food Protection.*, 62(9): 1088–1096.

11. Sizer, C. E. and V. M. Balasubramaniam. 1999. "New Invention Processes for Minimally Processed Juices," *Food Technol.*, 53(10): 64–67.

12. Irwe, S. and I. Olsson. 1994. "Reduction of Pectinesterase Activity in Orange Juice by High Pressure Treatment," Ch. 3 in *Minimal Processing of Foods and Process Optimization*, R.P. Singh and F. A. Oliveira, eds. Ann Arbor: CRC press, pp. 35–42.

13. Chen, C. S. and M. C. Wu. 1998. "Kinetic Models for Thermal Inactivation of Multiple Pectinesterase in Citrus Juices," *J. Food Sci.*, 63(5): 747–750.

14. Tajchakavit, S. and H. S. Ramaswamy. 1995. "Continuous-Flow Microwave Heating of Orange Juice," *J. of Microwave Power and Electromagnetic Energy*, 30(3): 141–148.

15. Eagerman, B. A. and A. H. Rouse. 1976. "Heat Inactivation Temperature-Time Relationships for Pectinesterase Inactivation in Citrus Juices," *J. Food Sci.*, 41: 1396–1397.

16. Vega, H. M., J. R. Powers, G. V. Barbosa-Cánovas, and B. G. Swanson. 1995. "Plasmin Inactivation with Pulsed Electric Fields," *J. Food Sci.*, 60(5): 1132–1136.

17. Ho, S. Y., G. S. Mittal, J. D. Cross, and M. W. Griffiths. 1997. "Effects of High Field Electric Pulses on the Activity of Selected Enzymes," *J. of Food Eng.*, 31: 69–84.

Part IX

Process Control

49 Recent Advances in Simulation for the Process Industries

D. Bogle

CONTENTS

Abstract

This chapter discusses recent advances in simulation with particular reference to potential capabilities for the food industry and some case histories, and it focuses in particular on fermentation-based processes. Recent advances have been in two directions: systems level simulation and more detailed unit operations modeling. This contribution concentrates on the former. Because computer power can almost be taken for granted, it is possible to consider alternatives for systems encompassing large numbers of traditional unit operations and can be used more as a strategic decision making tool for design and operation under uncertainty. The computational requirements for this are much more demanding, but the capabilities and some of the methodologies, many based on optimization approaches, are available.

49.1 INTRODUCTION

It is of course a cliché to say that the power of computers has increased enormously over recent decades. While this has helped us to undertake more complex calculations and to do them more often, it cannot be said to have directly clarified physical phenomena that govern processes. This is just as true in the chemical industry as in the food industry.

For a long time yet, this will be up to the ingenuity of scientists and engineers. However, the technologies are now beginning to be in place for providing decision support to help us clarify such issues, to help guide process operations on the basis of a quantifiable degree of uncertainty in the operations, and in due course to build this systematically into the design process. We have traditionally coped with uncertainty through significant overdesign, but this can cause problems as well as benefits. If there are significant process variations, it is often difficult to control large units to be flexible enough to rapidly respond to necessary changes.

The purpose of this chapter is to introduce recent trends in simulation methodologies that can be of value to the food industry, particularly in dealing with uncertainty, and to highlight where the difficulties may arise in their implementation. Process simulation has historically covered a wide range of activities: detailed unit operation modeling, design of whole processes, and simulation of the scheduling of operations of plant. I have concentrated here on the use of simulation in the design and operation of the process itself, while considering the process as a whole.

The food industry is beset with problems of uncertainty. The raw materials come from diverse natural sources, which are never consistent. The processing operations have variation in their operation, which exacerbates any feed inconsistencies. The consumer is ever fickle in both quantity and quality requirements, depending on the trends of the moment. Perhaps the only certain issues are that the product should be delivered on time to the customer in the right quantity, of the right quality, and guaranteed to be safe for consumption.

Our historical way of tackling uncertainties has been by sensitivity analysis. The development of efficient and robust optimization techniques has permitted a systematic way of tackling problems with uncertainties in a comprehensive manner. One of the most significant recent developments in simulation technologies has been the swathe of approaches for dealing with the uncertainties inherent in flexible processing. These result in a range of mathematical formulations that can be solved reliably, giving quantitative approaches for managing the design and operations of process plant of diverse nature: continuous, batch, or a combination of the two.

This chapter is an overview of recent simulation trends that may be of value to food engineering in helping to design and operate plants more efficiently by coping with some of these problems of uncertainty. The issues are being tackled both by using more detailed models to account for some uncertainty; for example, with the use of computational fluid dynamics, which will be mentioned only briefly, and also business supply chain issues, where uncertainty in the market is handled but with very crude influence on the physical and chemical mechanisms of food production.

Much of the material deals with coping with uncertainty in the simulation and control of processes with some case study material coming from biochemical processes. Some food processes are biochemical, and so there are common issues. The issue of microbial safety is modeled in a way similar to biochemical production processes,[1,2] which also provides a common link. Likewise, some food processes are more akin to the chemical industry, with defined molecules but a wide range of potential chemical reactions producing another area of uncertainty. For example, the large-scale processing of starch to make food additives has much in common with the manufacture of specialty chemicals.

In the following section, some recent developments in the state of the art of simulation, particularly of nonchemical processes, are presented. In the "Process Simulation" section, we discuss how these could be used in decision support of operations of a plant with both batch and continuous units, in particular referring to a process for the production of an animal protein. Recent ideas for dealing with uncertainty in design and operation using simulation techniques are presented in the section titled "Decision Support." The main method is the multi-scenario approach and its use in a method for supervisory control is discussed in the section called "Simulation under Uncertainty." The issue of robustness of proposed solutions in the presence of hard constraints on quality and safety is discussed briefly. Finally, there is some discussion of future developments and conclusions on the state of this technology.

49.2 PROCESS SIMULATION

Mechanistic models of many operations within the food industry have been published. Zweitering and Hastings[1,2] have provided a valuable contribution demonstrating how the application of chemical engineering principles can be used to analyze and develop food processes. They developed their models in MATLAB, showing how valuable conclusions could be drawn using only simple models. McGrath et al.[3] have proposed the use of generic simulation technologies for use in process control of food processing plants. And there are plentiful examples in the literature of simulations of food processing operations where models and simulations have been used to generate control strategies and design improvements. More recently, the idea of incorporating the phenomena occurring within units and batches is growing in popularity with the emergence of reliable computational fluid dynamics techniques. An interesting project is the MesoDyn project,[4] the aim of which is to develop a new class of phenomenological mesoscopic models to simulate the phase separation dynamics in three-dimensional complex liquids.

Of course, there are still many deficiencies. The processing of food is a heterogeneous operation, and many of the models are very approximate. The transport coefficients on which many calculations are based are both inaccurate and by no means consistent for all the operations of even one unit. Also, the prediction of the properties of heterogeneous ill-defined mixtures is a major problem and constitutes a major contribution to uncertainty in the operations of the processes. But the basis exists for building on these models to use them to their full potential.

What about the state of simulation technologies itself? The processes are often dynamic, hence dynamic simulation is essential. Such systems are available, even for problems where batch and continuous units are part of the process,[5,6] although model libraries do not yet exist for the food industry. Optimization of dynamic processes can be handled by new simulation systems. Some recent work within the Mitsubishi Chemical Company has resulted in a CFD model for a crystallization process based on significant amounts of experimentation. It is interfaced to an equation-based process simulation system such that, when this level of complexity is required, it is available for simulation.[7] Systems for obtaining the batch schedule are also available,[8] and even links with the whole supply chain can be made so that the demands of the consumer on the food industry can be addressed using computational systems.

But the uncertainties referred to above introduce a major element of doubt into the results of the simulations, and it is on this aspect that I wish to concentrate. There have been major advances in developing methodologies that can find optimal solutions, given a range of possible uncertainties. These aim to provide designs and operating strategies that will work in spite of major variations in, for example, the feed conditions. These methodologies often aim purely to highlight the trades-offs that are inevitable or, in some cases, to provide a "robust" set of operating conditions; i.e., conditions that can guarantee that the quality and safety objectives of the plant will be met in spite of disturbances. Some methodologies have been developed and tested, in some cases industrially, but only in the petrochemical industry; perhaps there are opportunities for the food industry as well.

Before tackling uncertainty, the reported results on how complex deterministic simulations (i.e., without considering uncertainty) can be used in decision support for process operations using examples from the food and biochemical sectors will be discussed, in the following section.

49.3 DECISION SUPPORT

Simulation has a role to play in supporting the modeling of operating decisions. With a simulation, key decisions can be identified and the best operating strategy determined and tested, within the accuracy limits of the models that make up the simulation.

Georgiadis and co-workers[9,10] developed a mechanistic simulation of heat exchangers for milk treatment using transport coefficients from Toyoda and Fryer,[11] which were validated against available experimental data. They were able to confirm trends such as the demonstration of the effects of milk concentration, Reynolds number, and having denatured protein in the feed on fouling in the exchangers. They also optimized their model to obtain cost-optimal solutions. Perhaps most interesting is that they identified that the solutions were very dependent on the tube diameter and on the milk feed composition, but not with respect to the activation energies of the reactions. Such information provides guidelines as to the quality of information that might be required when incorporating uncertainty into the analysis.

Bogle et al.[12] gave three examples as to how this idea might be utilized in the control of fermentation processes. We reported how a fermentation model in conjunction with on-line measurements can be used to detect major changes in the internal enzyme levels and hence trigger events such as harvesting or feed changes. The respiratory coefficient, a variable that can be derived from easily measured variables, was used to detect a shift in model parameters, which mirrors the metabolic state and hence requires a modification of the values of manipulated variables. Also, the fermentation model with on-line measurements and models of other unit processes can be used to determine optimal operating conditions for other downstream units.

This idea was explored for a complete process in the production of an animal protein (Bovine Somatotropin or BST) as inclusion bodies within *E.coli* followed by recovery and purification of the product. Simulation of this process shows that obtaining the greatest inclusion body recovery and the highest product purity are conflicting demands.[13] This behavior results in an optimization problem: only one of the objectives can be optimized at one time, with the manipulation of the influential variables yielding low values for the other objectives. This indicates that no overall process optimum with acceptable values for each objective can be attained with these specified variables. For best process performance, an overall process optimum is needed, and it is of particular interest what levels of recovery, purity, and productivity can be achieved when the entire process is considered simultaneously. Hence, the objectives were optimized separately, considering optimization variables of the whole process, and examining the values of all objectives.

The process is a batch process consisting of a series of batch and semi-batch units. The problem is to optimize a dynamic process that is now achievable with tools on the market. The gOPT facility of the gPROMS simulation package was used for the optimization. The partial differential and integral equations of the models are converted to differential and algebraic equations (DAEs) by discretizing all nontemporal domains. Methods for the discretizations are based on finite difference approximations and orthogonal collocations on finite elements.[6] The system is optimized using the successive reduced quadratic programming technique.

The objective functions for the optimization of the BST production process are IB recovery, product purity, and process productivity. The variables shown in Table 49.1 have been considered simultaneously for the optimization problem.

The results of the optimization of the overall inclusion body recovery are given in Table 49.2. The results show that a maximum of 95% of all inclusion bodies can be recovered. The purity is low, at 74% when the IB recovery is optimized. A productivity of 1.0 kg BST/h can be achieved due to the high recovered mass of product. Each optimization variable has taken the value of one of its limits except the separator-centrifuge settling area. The upper limits of these variables are thus the expected values.

The values of the objectives and the optimization variables for maximizing product purity are also listed in Table 49.2. A product purity of 92% in the separator centrifuge sediment can be achieved with a concomitant low overall IB recovery of 13%, and a productivity of 0.16 kg BST/h. These results prove again the main feature

TABLE 49.1

Overview of the Used Optimization Variables, Their Initial Guesses and Their Boundaries

Process unit	Optimization variable	Initial guess	Lower boundary	Upper boundary
Harvester-centrifuge	Volumetric throughput (L/h)	1300	1200	10000
	Settling area (m²)	155000	10000	250000
Homogenizer	Number of passes (–)	3	1	3
Centrifuge-mixer	Dilution rate (–)	3.1	1.1	4
Separator-centrifuge	Volumetric throughput (L/h)	1400	1200	10000
	Settling area (m²)	105785	10000	250000

TABLE 49.2

Three Sets of Results for the Maximization of IB Recovery, Purity, and Productivity Respectively (to Two Significant Figures)

Overall IB recovery (%)	Purity (%)	Productivity (kg BST/h)	Overall process time (h)	Final process volume (L)	Recovered mass of BST (kg)
95	74	1.0	37	1300	38
13	92	0.16	31	370	5.1
94	56	1.2	33	820	38

Harvester settling area (m²)	Harvester throughput (L/h)	Number of passes (–)	Dilution rate	Separator settling area (m²)	Separator throughput (L/h)
250000	1200	3	4	106000	1200
250000	1200	3	1.1	92000	3200
250000	1200	2	2.5	250000	1800

of this type of process: the higher the purity, the lower the IB recovery and productivity. The values of the optimization variables show that the highest possible settling area and the lowest throughput minimize the loss of cells in the harvester-centrifuge.

The final row in Table 49.2 gives figures for maximizing productivity. In this optimization, an overall productivity of 1.2 kg BST and IB recovery of 94% is obtained, whereas the purity is low, at 56%. Again, the discrepancy between the IB recovery, productivity, and purity is demonstrated: high productivity and recovery are accompanied by low purity.

Since a desirable target for industrial realization is to maximize the IB recovery, product purity, and productivity, a solution that introduces an acceptable compromise has to be found. It has been shown that this solution cannot be reached by considering the most influential operational and design variables because of the contrasting behavior of IB recovery and purity. Thus, the optimization strategy has to be changed, taking structural decisions into account. Since the overall IB recovery has been high for this simulation, with 91%, any structural variation should mainly improve the purity with the smallest possible loss of IBs. Such a variation is suggested by batch-wise recycling the sediment of the last centrifuge back to the last mixer. After dilution, the broth is fed through the separator-centrifuge for the second time, using the optimal feed rate of the centrifuge for highest productivity. The separation efficiency of the second batch through the separator-centrifuge is different from the first batch, due to a different throughput and a different viscosity.

The results (Table 49.3) show that high levels of all three objectives are achievable; after the second batch, a high purity of 95% is reached with an acceptable IB recovery of 84% and a productivity of 0.95 kg BST/h.

TABLE 49.3
Results of a Separator-Centrifuge Batch Recycle to the Separator-Centrifuge Mixer and a Second Feed through the Centrifuge

Separator-centrifuge (sediment)	Overall process time (h)	Process volume (L)	Overall IB recovery (%)	Purity in the sediment (%)	Productivity (kg BST/h processing time)
First batch	34	1000	91	81	1.1
Second batch	35	370	84	95	0.95

By posing and solving these optimization problems, the trade-offs can be fully explored. In this way, the simulation can be used as a decision support tool for providing quantitative measures of the effects of taking certain operational decisions. Of course, this work has been done purely on the basis of a simulation model and not taking into account uncertainty of the parameters in the model. Below, we discuss how this uncertainty can be taken into account using optimization-based approaches and hence provide a more robust approach to operational decision support using simulation.

49.4 SIMULATION UNDER UNCERTAINTY

To deal with simulation under uncertainty, a number of approaches have been developed based on optimization formulations. One approach is to design to allow for the best match for a certain amount of overdesign that matches with the perceived uncertainty. Defining a flexibility index and using it in an optimization formulation does this. A second way is to consider a statistical sample of expected uncertainties in the parameters or operating variables and to optimize some aggregate, perhaps the mean, of these scenarios.

49.4.1 DESIGN FOR FLEXIBILITY

This approach has been to define potential ranges for the variables for a proposed new design or retrofit of a plant and to maximize the potential range of scenarios that the new design can handle. This is a robust, but also conservative, approach, and it has been applied to the design of a plant and its operating conditions where plants need to be designed to process a wide range of possible feed conditions.[14]

In this problem, an index is defined that characterizes the flexibility of the process. A possible index to maximize the objective function over the feasible region is

$$F = \max(\delta) \text{ subject to equality and inequality constraints}$$

$$T(\delta) = \{\theta_j | (\theta^N - \delta\Delta\theta^-) \leq \theta \leq (\theta^N + \delta\Lambda\theta^+)\} \tag{49.1}$$

where $T(\delta)$ is the largest possible region of uncertainty inside the feasible region where all constraints are satisfied. This can be seen in Figure 49.1[14] for a two-variable problem. θ^N is the nominal value of the uncertain parameter, and $\Delta\theta^+$ and $\Delta\theta^-$ are the expected deviations in the uncertain parameters. The calculations are done assuming that perfect control is achievable.

For a given design, if $F \geq 1$, then feasible operation can be guaranteed for all the possible solutions within the specified uncertain parameter ranges, and the feasibility test is passed (all constraints are satisfied). If $F < 1$, its value represents the maximum fraction of the expected parameter deviations that the design can handle. F may be optimized by computationally searching for the design, which maximizes the feasible "hyper-rectangle," i.e., the box shown in Figure 49.1.

However, this problem cannot be solved in this form because of the infinite number of possible solutions for the uncertainties within the range. Halemane and Grossmann[15] developed a formulation in which the problem is transformed into a max-min-max problem: a cascaded set of optimization problems.

The feasibility index is defined as

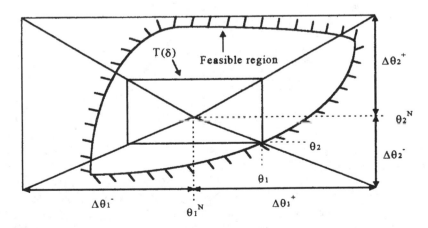

Figure 49.1 Maximum hyper-rectangle, T, inside the feasible region of uncertain parameter space.

$$F = max \ (\delta) \tag{49.2}$$

subject to

$$\begin{array}{ccc} max & min & max \ f(d, u, \theta) \leq 0 \\ \theta \in T(\delta) & u & j \in J \end{array}$$

and $T(\delta)$

where d are the disturbances, u the control variables, and θ the uncertain parameters. This formulation aims to maximize the objective function over the ranges of the uncertain parameters, while minimizing the control action, for all j independent variables. Again, if F is less than 1, then the proposed design is feasible for all $\theta \in T$. Otherwise, there exist some uncertain conditions within T for which no feasible control exists. This index defines the operating region that is possible with a proposed set of operating conditions for a plant. The limits of the uncertain variables, of course, must be defined.

This approach is applied in two stages.

1. A design stage in which a design is selected with the overall objective of ensuring feasibility for all possible realizations of the uncertain variables (by proper manipulation of the control variables during operations), and which also optimizes the objective function (usually an expected economic function) evaluated inside the feasible region associated with each design
2. An operating stage in which a design is assumed and the objective is that, for every realization of the uncertain variables in the associated feasible region of uncertainty space (of the design), a set of control variables can be selected that maintain feasible operation and optimize the expected value of the objective function

This leads to an iterative procedure. Simultaneous economic (or environmental) optimization and design feasibility under uncertainty can be achieved such that product and plant specifications can be met for any values of the uncertain parameters.

If only ranges of variables are defined and the largest feasible region determined, this is referred to as a *deterministic approach.* If probabilistic information about the uncertainties, in the form of probability distributions, is available, then *stochastic approaches* have been developed.

Given the large degree of uncertainty involved with food processes, this is likely to produce conservative designs. A useful development they proposed is to determine the critical parameter point(s) for which the feasible region of operation is smallest. Attention to this point will improve bottlenecks and potential operating problems. In this way, this approach can be used sensibly to guide design modifications and change control strategies to improve performance without compromising safety.

This is a robust approach in that, if the uncertainty predictions are correct or conservative, then the designs will work. However, given the likelihood of providing conservative estimates, it will be a very conservative approach and give no further insights to the problem, since it will produce just one answer. Also, it has not been tried in problems involving dynamic models or combinations of batch and continuous units, which are likely to confront the food industry. As in all the methods discussed, here the computational requirements are considerable.

49.4.2 MULTI-SCENARIO APPROACHES

A second approach has been to find the optimal solution to an operational or design problem by considering a statistical sample of expected uncertainties in the parameters or operating variables and optimizing some aggregate, perhaps the mean, of these scenarios.

We may define the expected value of the simulation expressed in terms of the expected value, E, of the overall yield of the process, C, which is a function of the control variables u, the state variables $\tilde{x}$, and the uncertain parameters, $\tilde{\theta}$ (the superscript ~ indicates uncertain variables).

$$\max_{u} E[C(u,\tilde{x},\tilde{\theta})]$$

(49.3)

subject to
$$h(u,\tilde{x},\tilde{\theta}) = 0$$
$$g(u,\tilde{x},\tilde{\theta}) \leq 0$$

A set of equality and inequality constraints characterize the balance equations and the limits arising from safety and plant capacity issues, etc. This formulation has been examined by many authors[15-18] and has been extended to consider the expected value when the parameters can be described by a probability distribution function f_θ for the uncertain parameters θ.

$$\max_{u} \int_{-\infty}^{=\infty} C(u,\tilde{x},\tilde{\theta}) f_\theta(\theta) d\theta$$

(49.4)

subject to
$$h(u,x,\theta) = 0$$
$$g(u,x,\theta) \leq 0$$

In the form given above, this cannot be solved except by considering a spectrum of operating scenarios for the uncertain parameters, $\tilde{\theta}$, and an average can be calculated for the range of scenarios (sometimes known as *periods*). This formulation is a multi-scenario (or multi-period) formulation of the problem, and the following formulation, which can be solved, results:

$$\max_{u} \sum_{i=1}^{n_{\theta}(\bar{\theta})} w_{\theta}^{i} C(u, x^{i}, \theta^{i}) \tag{49.5}$$

subject to
$$h(u, x^{i}, \theta^{i}) = 0$$
$$g(u, x^{i}, \theta^{i}) \leq 0$$

where w is a set of weighting factors. This general formulation can be used to find the combination of operating variables or set points, which optimizes the process with respect to the expected values of the parameters.

By assuming some probability distribution for the uncertainties, the problem becomes a conditional problem (in the jargon of the stochastic literature) such that we are optimizing the problem conditional on the expected values of the parameters.[19]

This approach has been used in a number of operational problems. One example is described below in which a process consisting of a number of stages is optimized subject to an expected range of disturbances. The process has batch and continuous units, and the product is an enzyme that occurs naturally in yeast.

49.5 SUPERVISORY CONTROL

Since the operation of a plant is subject to many large changes in the expected operating parameters, one use of simulation is in accounting for such changes at the earliest opportunity and taking appropriate action throughout the rest of the process. The uncertainty in the measurements must at all times be considered, but the simulation system should help manage this information. As more process information is gathered, this uncertainty would be reduced. The operating changes would be implemented through the process control system. Hence, a significant use for the uncertainty multi-scenario methodology outlined in the previous section is the use in supervisory control structures for a complex process plant.

Operation of a plant is dependent on the feed stream condition and quality. It is expected that this would be measured or known (perhaps subject to contractual obligations). The simulation can be used to devise optimal operating conditions for processing the feed of a certain quality by optimizing the simulation model. This is routinely done by the oil refining industry, for example, and appropriate simulations should be available that could then be undertaken by the food industry. The first unit of the process will subsequently process the feed, and the operation may well not achieve the targets expected or set by the model. An appropriate measurement or set of measurements could then be taken and used to re-estimate the model parameters, giving a more accurate representation of what happened and therefore the full state of the processed stream, allowing a reoptimization of the rest of the process. This has the effect of reducing the uncertainty in the model, although the uncertainty can never be completely removed.

This is best demonstrated by an example. Fermentation and subsequent purification in a number of unit operations often involve the biological production of enzymes. We have developed a simulation, using MATLAB, of a process for the production of an enzyme in yeast followed by subsequent cell breakage, cell debris removal, and two-stage fractional precipitation to obtain the product.[20,21] The supervisory control strategy referred to above is illustrated in Figure 49.2, showing the first two unit operations: fermentation and homogenization. The oval boxes refer to the components that must be undertaken as part of the strategy: O refers to an optimization, M to a measurement, and E to an estimation of one or more model parameters.

On the basis of the feed conditions, the strategy uses the model to optimize the conditions for the fermenter (and the downstream operations): problem O1. In particular, the optimization will determine the feed rate (D, often known as the dilution rate) so as to optimize the final production of product from the whole process. It will also predict the expected condition of the fermenter product. There is considerable batch-to-batch variation in fermentation. Certain key measurements, such as substrate or enzyme concentration, of the outlet stream should be taken, and these are used along with the model to estimate the parameters of the fermenter model. This can be used to calculate all other parameters within the model and to reoptimize the whole process for all subsequent downstream operations.

One feature of this process is that the first stage is the reaction stage of the process where the main chemical change has taken place. This is common of many processes in the process industries and where many of the major contributions to uncertainty arise. As a result, the handling of uncertainty in this way will help to reduce the overall uncertainty of the process.

One of the significant benefits of this approach is that it enables the operation to be always targeted at optimizing the whole process and not just at individual unit operations. Much literature has demonstrated the need to focus on the primary overall process goals.

The formulation of the supervisory control strategy optimizes the expected value of the process objective. The formulation becomes

$$\max_{u^*|m} E[C(u^*,\tilde{x}^*,\tilde{\theta}|m)] \tag{49.6}$$

subject to
$$h(u^*,\tilde{x}^*,\tilde{\theta}\ |m) = 0$$
$$g(u^*,\tilde{x}^*,\tilde{\theta}^*|m) \leq 0$$

The notation here indicates that a specific uncertainty profile for the uncertain variables x and parameters θ in the downstream sequence of operations is available, given that a set of measurement values m for a set of process stream variables has been found. The * indicates variable values found on the condition of the occurrence

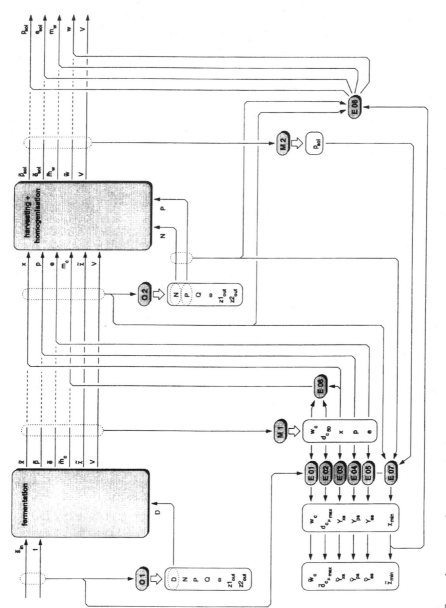

Figure 49.2 Supervisory control structure for fermentation and homogenization stages of enzyme production process from yeast.

of the measurements m. The problem is solved as a multi-scenario problem maximizing the objective for the whole process initially and then after each stage of the process has been completed. Again, the problem is transformed as shown in Equation (49.5).

For the yeast-based process mentioned above, the supervisory scheme was implemented on a simulation. An economic objective function based on the main costs and sales revenue was defined. A large number of simulations were run using a stochastic formulation. Eighty runs of the process were permitted, randomly generated according to the most recently generated probability distributions. The simulations were undertaken with MATLAB. The operation of 1 year of pilot plant operation required approximately 200 h of simulation time. With the increasingly efficient performance of computers, this time will be significantly reduced in coming years. Results show that, by using the new supervisory control strategy, the revenue for the process could be significantly improved. The spread of points in Figure 49.3 indicates how the uncertainty can affect performance of the optimized process, but in all cases the revenue was considerably higher.

Again, this type of supervisory control structure can be achieved only because large simulations can be achieved in short times. The re-estimation and reoptimization of the model must be done within the limits of the batch schedule. If a particular optimization or measurement is unsuccessful, then the existing operating strategy can still be implemented, enabling the process to achieve its schedule at the expense of perhaps improving performance.

49.6 ROBUSTNESS

The approach used above for optimizing operations aims to optimize the operations subject to uncertainty and to gradually improve the control of the plant as more up-to-date and comprehensive information becomes available. It is often necessary to include different types of constraints in the solution procedure.

The constraints a plant must achieve are of a varying nature. Some are soft constraints, where small violations can be permitted for periods, while others are hard constraints, which cannot be violated under any circumstances. Safety constraints on plant operations, for example, would come into this second category. Safety conditions of food products are more complex in that the plant would continue to operate, but the product would be unsalable. It would be useful to discriminate between these types of constraints, and an approach to this has been proposed by Samsatli et al.[22] by incorporating measures of robustness in the design of the operating conditions.

It is typical to measure robustness as the variance of one of many performance indicators of the ability of the process to meet its target. The Taguchi performance metrics have been used for this. For hard constraints, however, the variance of the output is not useful. Samsatli et al.[22] proposed a one-sided performance measure that corresponds to the variance measure in the permissible range but as a constraint for the inadmissible region. This results in a difficult numerical problem but has been dealt with by various smoothing techniques, in particular one that transforms Tagu-

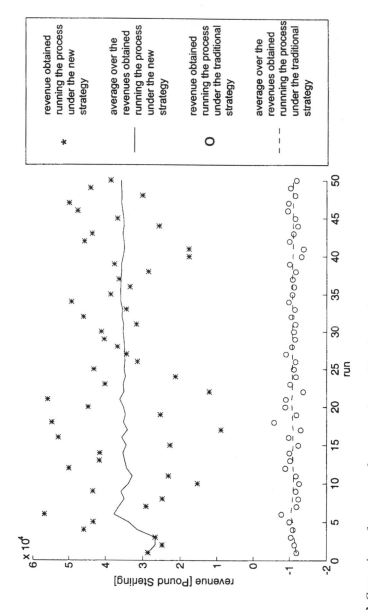

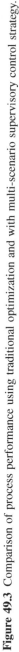

Figure 49.3 Comparison of process performance using traditional optimization and with multi-scenario supervisory control strategy.

chi's quality loss function. The resulting design problem is a multi-scenario dynamic optimization problem with scenarios generated by a Monte Carlo procedure.

Samsatli et al.[22] have applied this robustness approach to a fermentation problem where the product is expressed intracellularly, but a by-product is excreted. There is considerable uncertainty in the parameters and often in the composition of feed streams for fermentation. The operating policy generated by the multi-scenario method without incorporating robustness resulted in significant violations of the product quality constraint during the operation, although the final mean product quality was satisfactory. Their outcome when incorporating the robustness as an objective resulted in a more conservative strategy, with lower initial volumes and longer processing times, but reduced variance of the output quality conditions. As would be expected, the operation is taken more slowly to ensure that quality is maintained. However, through this mathematical procedure it is possible to quantify the extent required and the price to be paid.

49.7 SCHEDULING

Much of the food industry is heavily automated in the scheduling of batch recipes. This is one of the big contributions that systems engineering has made to improving efficiency and quality management. There have been some recent reviews of this (e.g., Reference 8), and they will not be referred to here. These problems result in large combinatorial optimization problems. Recent developments have concentrated on extending the developments to multi-product plants and to multi-site operations.

However, there has been little work attempting to blend design under uncertainty or adaptive scheduling resulting from mechanistic modeling approaches into the scheduling algorithms. One problem is the difficulty of the optimization problems that arise; they are large, combinatorial, nonlinear, and "convex," resulting in problems with many local minima, which cause problems for conventional optimization algorithms.

49.8 FUTURE DEVELOPMENTS

What can we expect from these new developments and what are the weaknesses that will limit their implementation? By systematically incorporating a way of allowing for uncertainty in the models and model parameters, these techniques will make it possible to incorporate quality measures explicitly in developing operating scenarios building on any design or operating information that has been incorporating into any sort of model available for a process. The solution methods are reliable enough to count on efficient and robust solutions. However, they are not readily available in commonly used simulation and design tools, but this can be expected in the future.

Any method for reducing uncertainty is desirable, provided it can be done at reasonable cost. We showed how incorporating measurements into these schemes can further enhance and optimize operations. We can anticipate further work on measurement and instrumentation of food processes and, provided that the measuring devices are reliable, they can be used to directly enhance production operations. For measurements that are uncertain, the benefits, given the level of uncertainty, can be determined by simulation to determine whether the new measurement can provide any worthwhile benefit.

There have been advances in modeling of food processes in recent years, but the heterogeneous nature will remain a problem for a long time to come. The incorporation of CFD techniques will enhance the accuracy of models, as mentioned above. However, the issue of properties will remain a big problem for the foreseeable future.[23] These issues all contribute to the uncertainty that the techniques mentioned above aim to manage. The supervisory control strategy requires fast, automated measurements, and improved measurement techniques will certainly help to make the most of this approach. The methodologies are, after all, just a way of managing the uncertainty; the unexpected can never be predicted, but at least some of it can be mitigated.

Finally, since many of the methods described here are based on solving optimization problems, it must be mentioned that most methods in common use cannot guarantee finding a global minimum, but much progress has been made in this direction. The new methods for seeking the global optimum are more computationally demanding, but, again, the rapid development of computing power is ensuring that they will be in reach of routine calculations in the near future.

49.9 CONCLUSIONS

I have tried to demonstrate how new developments in methodologies and simulation tools can be used in design and operation of processing plants. In particular, I have focused on dealing with uncertainties in parameters and disturbances. Many new techniques are based on optimization techniques, not necessarily to optimize the output of the process, but to minimize the effect of the uncertainty on the operation of the process. The approach using the flexibility index is likely to produce unnecessarily conservative solutions, but the multi-scenario approach provides both proposed solutions and more detailed information because of the large number of runs necessary for the stochastic solution. However, there is a computational price to pay. All of these techniques are expensive computationally but, with the incredible advances in computer power, this seems to be the least of our problems. The underlying explanation of the physical phenomena remains a big challenge, and these tools can help provide the basis for exploring physical systems and acting as the memory that records the explanations through mathematical modeling.

49.10 ACKNOWLEDGEMENTS

The author would like to acknowledge the following, whose discussions helped form the opinions in this paper: Mark Groep, Lester Kershenbaum, Malcolm Gregory, David Johnson, Carlos Lopez Urueta, Holger Graf, and Lazaros Papageorgiou.

Engineering and Food for the 21st Century

REFERENCES

1. Zwietering, M. H., and A. P. M. Hasting. 1997. "Modeling the Hygienic Processing of Foods—A Global Process Overview." *Food and Bioproducts Processing*, 75(C3): 159–167.

2. Zwietering, M. H. and A. P. M. Hasting. 1997. "Modeling the Hygienic Processing of Foods—Influence of Individual Process Stages," *Food and Bioproducts Processing*, 75(C3): 168–173.

3. McGrath, M. J., J. F. O'Connor, and S. Cummins. 1998. "Implementing a Process Control Strategy for the Food Processing Industry," *Journal of Food Engineering*, 35(3): 313–321.

4. Altevogt, P., O. A. Evers, J. G. E. M. Fraaije, N. M. Maurits, and B. A. C. van Vlimmeren. 1999. "The MesoDyn Project: Software for Mesoscale Chemical Engineering," *Journal of Molecular Structure*, 463(1–2): 139–143.

5. Engell, S. and K. Wollhaf. 1994. "Dynamic Simulation of Batch Plants," *Computers and Chemical Engineering*, 18 Suppl., S439–S444.

6. Oh and Pantelides. 1996. "A Modeling and Simulation Language for Combined Lumped and Distributed Parameter Systems," *Computers and Chemical Engineering*, 20(6–7): 611–633.

7. Urban, Z. and T. Ishekawa. 1999. "Dynamic Simulation Coupled with CFD for Detailed Modeling of a Reactor—Incorporate Complex, Distributed Simulation Calculations for Complex Reactors within CFD Simulations," *Chemputers Europe 5, Duesseldorf*, Oct. 21–23.

8. Shah, N. 1998. "Single and Multi-Site Planning and Scheduling: Current Status and Future Planning," Presented at FOCAPO-98, Snowbird, U.S.A., CACHE.

9. Georgiadis, M. C., G. E. Rotstein, and S. Macchietto. 1998. "Modeling and Simulation of Shell and Tube Heat Exchangers Under Milk Fouling," *AIChEJ*, 44(4): 959–971.

10. Georgiadis, M. C., G. E. Rotstein, and S. Macchietto. 1998. "Optimal Design and Operation of Heat Exchangers Under Milk Fouling." *AIChEJ*, 44(9): 2099–2111.

11. Toyoda, I. and P. J. Fryer. 1997. "A Computational Model for Reaction and Mass Transfer in Fouling from Whey Protein Solids," in *Fouling Mitigation of Industrial Heat Exchange Equipment*, Begell House, New York.

12. Bogle, I. D. L., A. R. Cockshott, M. Bulmer, N. Thornhill, M. Gregory, and M. Dehghani. 1996. "A Process Systems Engineering View of Biochemical Process Operations," *Computers and Chemical Engineering*, 20(6–7): 643–649.

13. Bogle, I. D. L. and H. Graf. 1997. "Simulation as a Tool for At-Line Prediction and Control of Biochemical Processes," in *Proc. ECCE1*, First European Congress on Chemical Engineering, Florence, May, pp. 1933–1936.

14. Swaney, R. E. and I. E. Grossmann. 1985. "An Index for Operational Flexibility in Chemical Process Design," *AIChEJ*, 31(4): 621–630.

15. Halemane, K. P. and I. E. Grossmann. 1983. "Optimal Process Design Under Uncertainty," *AIChEJ*, 29(3): 425–433.

16. Grossmann, I. E and R. W. H. Sargent. 1978. "Optimum Design of Chemical Plants with Uncertain Parameters," *AIChEJ*, 24(6): 1021–1028.

17. Grossmann, I. E. and D. A. Straub. 1991. "Recent Developments in the Evaluation and Optimization of Flexible Processes," in *Proc. COPE*, Computer Oriented Proc. Eng., Elsevier.

18. Pistikopoulos, E. N. 1995. "Uncertainty in Process Design and Operations," *Computers and Chemical Engineering*, 19 Suppl. S553–S563.

19. Mohidenen, M. J., J. D. Perkins, and E. N. Pistikopoulos. 1996. "Optimal Design of Dynamic Systems Under Uncertainty," *AIChEJ*, 42(8): 2251–2272.

20. Groep, M. 1997. *A Supervisory Control Strategy for Performance Optimization of Sequences of Interacting Biochemical Operations*. Ph.D. thesis University of London.

21. Groep, M., M. Gregory, L. S. Kershenbaum, and I. D. L. Bogle. 1999. "Performance Modeling and Simulation of Biochemical Process Sequences with Interacting Unit Operations," *Biotechnology and Bioengineering*, in press.

22. Samsatli, N. J., L. G. Papageorgiou, and N. Shah. 1998. "Robustness Metrics for Dynamic Optimization Models Under Parametric Uncertainty," *AIChEJ*, 44/9 1993–2006.

23. Saravacos, G. D. and A. E. Kostaropoulos. 1996. "Engineering Properties in Food Processing Simulation," *Computers and Chemical Engineering*, vol. 20, Suppl. pp. S461–S466.

50 Simulation-Based Design of Food Products and Processes

A. K. Datta

CONTENTS

Abstract

With major changes in its capabilities, computation is moving from research to design. Simulation-based engineering design or computer prototyping can partially replace

1-56676-963-9/02/$0.00+$1.50
© 2002 by CRC Press LLC

the older physical prototype-based engineering for quicker turnaround in the design cycle for food products and processes. Research in academia can also benefit from the availability of commercial software, as it may relieve researchers from the drudgery of code writing. The composition, temperature dependence of properties, and simultaneous occurrence of several phenomena (e.g., heat transfer, moisture transfer, bacterial and nutrient changes, cracking) make food processing an almost ideal candidate for such computer-aided engineering (CAE). Several application examples are discussed, along with the need for further development in food applications.

50.1 SIMULATION-BASED ENGINEERING AS A DESIGN TOOL

Computer models of food processes have been developed mostly for research purposes in the past,[1–3] starting with perhaps the earliest study by Teixeira et al.[4] Design typically requires more frequent computations with changes in parameters. Such intensive computations have only recently been practical, due to the availability of high-power desktop workstations, PCs, and advanced user-friendly software. These advances have made computer models a practical tool for product and process design. Consequently, computation for design purposes has been suggested for food applications.[5–10] Computer models can be thought of as a computer prototype, in contrast to a physical prototype. In computer prototyping, one builds a model that is as close to the physical model as possible—the exact shape and size, and the exact physical process. An accurate computer model works just like a physical prototype, but its engine is mathematical rather than physical. A computer model involves solving the set of partial differential equations that exactly describe the physics of the model. Use of computer prototyping has several advantages in a food product and process development environment.

1. Testing of "what-if" scenarios is quick and inexpensive, thus shortening the design cycle (quicker turnaround), which should result in reduced costs and increased profits.
2. It can provide insights into complex processes that are otherwise difficult to understand. It provides a clearer understanding of the interactions between the physical processes and their sensitivity to various operational parameters. This can enable the designer to be more creative.
3. It allows front-end engineering before prototyping, making the prototypes closer to the optimum and also reducing the number of prototypes.
4. It makes possible concurrent design and analysis, also shortening the design cycle. While an experiment is underway, results can be used simultaneously to further optimize the process on the computer, also reducing the amount of experimentation.

Computer-aided engineering, simulation-based engineering, and computer prototyping refer to the use of computers to build and test computer models of products and processes to reduce the extent of physical prototypes. It is useful to note that computer-aided design (CAD) has somewhat the same goals as CAE but is typically limited to only geometric manipulations without any physics of the process.

Computer prototyping can help today's competitive product and process design by reducing cost, reducing the time needed to market, and making more dramatic changes possible. Computer prototyping can reduce the disadvantages associated with physical prototypes, e.g.,

1. They are typically expensive to build and modify and can lead to lengthy design cycles.
2. Repeatability can be difficult, since the prototype goes through irreversible changes during testing and therefore may need to be discarded.
3. Measurement is often difficult or impossible and rather expensive.
4. Dramatic changes can be harder to conceive.

Industries such as automotive and chemical processing have been exploiting the advantages of CAE in a very significant way. The food industry also stands to benefit from the use of this tool. Although the total application is still somewhat small, use of computational software is also rapidly increasing in food process engineering. During the Seventh International Congress on Engineering and Food, 13 out of 30 presentations involving modeling of heat and mass transfer used some commercial computational software. Several of these presentations can be categorized as design use. The food industry, however, is generally behind other processing industries in such CAE applications.[11]

50.2 CHARACTERISTICS OF FOOD PROCESSING PROBLEMS

The following are some of the unique aspects of food processing problems:

1. In addition to the temperature changes during a heating or cooling process, there are biochemical (nutrient, color, flavor, etc.) or microbial changes that are important to know.
2. The moisture in food is constantly undergoing either loss (due to evaporation, especially when heated) or gain (from humid surroundings).
3. The properties of foods (e.g., density, thermal and electrical conductivity, specific heat, viscosity, permeability, and effective moisture diffusivity) are often a function of composition, temperature, and moisture content, and therefore keep changing during the process. The system is also quite nonhomogeneous. Such detailed input data are not available.
4. The hygroscopicity of food materials can cause food to shrink upon significant loss of moisture or swell when gaining moisture.
5. Often, irregular shapes are present.
6. Processes such as temperature, moisture, and mechanical changes are often coupled.

50.3 TYPICAL STEPS IN COMPUTER-AIDED ENGINEERING

The computation process typically consists of three steps—preprocessing, processing, and postprocessing. Preprocessing typically defines the geometry and computational grids. In a finite-element-based software, this step involves the generation of the geometry, grid, and elements, telling the computer about governing equations, boundary conditions, properties, and methods to be used. In the processing step, the computer solves the problem. In the postprocessing step, the solution is visualized by the user using shaded contours, movies, etc.

50.3.1 GOVERNING EQUATIONS AND BOUNDARY CONDITIONS

In this perhaps most important preprocessing step, a mathematical analog (equations) of the physical process is developed in terms of a set of equations called the *governing equations* and the *boundary conditions*. The mathematical description is often an intelligent simplification, specially in the initial stages. This step is often the most important and most difficult. The goal is to keep as many details of the process as possible, without creating unnecessary complexities. CAE software programs are able to solve a general set of governing equations and boundary conditions to cover a wide range of processes. In choosing software, availability of the needed governing equations is only one of many factors. Each software program has its strengths and weaknesses. Ease of use, availability for a particular computer platform, and cost are some of the main concerns. Typically, any software program based on numerical solutions is highly versatile.

50.3.2 MESH GENERATION

In this preprocessing step, after the geometry has been defined, the geometry needs to be broken down into smaller pieces for a numerical solution. The more parts the geometry is broken into, the more accurate is the final solution, but the computation time increases (sometimes dramatically) and can eventually make it unrealistic to compute. Thus, this step is a careful balance between providing enough elements (or nodes) such that all the essential physics are captured (but not too many). Developing automatic meshes in complex 3-D geometry that also adapts to the solution is a major research topic in itself.[12,13] Ease of mesh generation is constantly being improved.

50.3.3 MATERIAL PROPERTY DATA

A very important step in simulation-based engineering is to use material property data as accurately as possible. However, this is also where food processes are at a slight disadvantage as compared with processes that do not use natural materials. Properties of steel, for example, are more easily available and perhaps have a lot less variability than chicken soup. It is not necessary, however, in simulation-based design to have absolutely accurate data, as discussed below under sensitivity analysis.

Measure Data, If Possible

This is the ideal scenario, particularly considering how food product formulation will vary. Industry, for example, would like to measure the specific property of it formulation in-house, to keep confidentiality and to use it in product or process design. Sometimes this may not be possible due to the need for detailed expertise, expensive instrumentation, or the necessity for longer time. For example, dielectric property measurement for microwave food product development requires instrumentation costing around $70,000, and it still may not be able to measure the properties of a frozen food, necessary for developing products such as microwaveable frozen dinners.

Use Computerized Databases

In other manufacturing areas, huge materials property databases are currently available in CD-ROM format. For example, Cen-BASE/ Materials,[14] on the WWW, is a searchable database on over 35,000 plastics, metals, composites, and ceramics from over 300 manufacturers' product catalogs worldwide. Other property database examples include personal computer-based thermochemical and physical properties of compounds and phase diagrams,[15] and properties of engineering materials and polymers.[16] In food processing, such databases are only beginning to appear.[17,18] An important source that is sometimes overlooked is property data in databases outside the food context, but for biomaterials. An example is property data in the context of biomedical uses.[19,20]

Use Handbooks and Property Data Books

This is perhaps the most popular source of data for researchers. Although not as exhaustive as other materials in data handbooks, considerable data are available in printed format.[21–23] There are also summary papers that carry large amounts of data on a specific property, such as the dielectric property of foods.[24,25]

Use Prediction Formulas

Correlations between property and composition developed from measured data[26,24] are likely to be more useful for the purpose of estimating property in computer-aided engineering, provided the available correlation is for a material that is close. Accurate composition data becomes critical in using the correlations accurately. For example, dielectric properties can change drastically with slight changes in salt. Books dealing with food properties often provide prediction formulas or correlations.[21,22,27,28]

Perform Sensitivity Analysis Using Reasonable Estimates

Since the idea of simulation-based engineering is to try many "what-if" scenarios, a natural part of this exercise is to examine the effect of property variations on the process. In this way, we can bracket the properties and see the effect of a range of properties on the process. This process is also called *sensitivity analysis*. Such sensitivity analysis is not limited to just property variations but is extended to variations in the process itself. Different boundary conditions, different geometries, etc., can be tried relatively easily on the computer. If the process outcome turns out to be quite sensitive to property data variations in a particular case, it signals the need to obtain very accurate property data. Sensitivity analysis is one of the most important steps in computer-aided engineering.

50.3.4 SOLUTION TECHNIQUE

Once the equations and the properties are defined, the next step is to select a solution method (choice of time discretization, matrix solution method, etc.) that provides the most efficient solution. Today, there are a large number of software programs available in any given area, such as mechanics or heat transfer. For example, more than 75 commercial computational fluid dynamics (CFD) software programs are available. The advantage of using available software is that one can often choose a solution method without having to be skillful about the details of such methods. Commercial software tends to have good convergence property for a wide variety of situations. The software manufacturer also typically has recommendations for solution techniques for various types of problems. Some knowledge of numerical methods is helpful at this stage.

Once the solution methods are specified, the computer takes over the solution procedure. This is the processing step. The equations are discretized, and large linear systems of algebraic equations (matrices) are formed. Although the user can have some control over this matrix formation, it is mostly performed automatically by the software without much user intervention. Typically, this matrix formation represents the most time-consuming stage of the computation and, for many problems, can account for up to 80% of the total computer resources. Matrices are solved using well known procedures.

50.3.5 POSTPROCESSING

Postprocessing is the important step of visualizing the results and making further computations from the raw data generated by the solution. Most commercial CFD packages are able to show nice contour, history, vector, and other plots. These spatial or temporal profiles in the whole 3-D region can provide insight and understanding of the process that is not possible using experimentation. For example, the spatial distribution of moisture or oil during a drying or a deep-frying process can be obtained computationally; however, it requires elaborate and expensive experimentation using the MRI. Such visualization is one of the greatest advantages of computer-aided engineering.

Several specialized postprocessing software programs are now available that have significantly enhanced capabilities for visualization. For example, Ensight (Computational Engineering International, Inc., Research Triangle Park, NC) is designed to read the data output from many commercial software programs, and it provides a powerful display of the results, including animation of transient processes.

50.3.6 Trusting Computational Results

Computational results should never be trusted blindly, and the software should never be used as a "black box." However, it is also important to accept that there is no foolproof way to confirm the computational results. Several steps may be taken to minimize the chances of obtaining a wrong solution. Some of these include checking for mesh convergence, checking the input file for accuracy of problem definition, using common sense about the process physics, comparison with experimental data, and checking against the results of a simpler problem.

50.4 APPLICATION EXAMPLES

Computer-aided engineering can include computational fluid dynamics (CFD) and heat transfer to solve flow/heating/cooling problems, computational mechanics to solve rheological and stress related problems, computational electromagnetics to solve microwave and other heating problems, and so on. Examples of food applications in some of the computational areas are provided below.

50.4.1 Electromagnetics: Microwave Heating

Microwave heating of food is quite common, either by itself or combined with hot air or other heating mechanisms. Microwaves are electromagnetic waves similar to light waves. In the home microwave oven, microwaves are generated outside the oven (called a *cavity*) by a device called the *magnetron* and fed into the box using a tube called the *waveguide*. Once inside the oven, the microwaves are reflected from the metallic surface of the six walls of the box, setting up a pattern of standing waves with high and low amplitude, much like the standing waves in a string. When food is placed inside the box, water, the major component of food, interacts and absorbs energy from these waves. This energy appears as heat.

The microwave absorption in food depends in a complex way on the shape, size, and dielectric properties of the food, its placement inside the oven, and the oven design. For example, the standing waves generated inside the microwave oven amount to pockets of high and low energy. Depending on where the food is placed in the oven, some areas in the food will heat more than others. Since microwaves are electromagnetic waves, they can be focused by a curved surface, such as that of a cylinder or sphere, much like a lens focusing light. Exhaustive experimentation covering all of the parameters can often be frustrating—simulations can reduce the amount of experimentation considerably and provide insight into the process at far greater detail than is possible experimentally.

The physics of the microwave heating of a solid food, in the absence of moisture transport, is described by Maxwell's equations of electromagnetics combined with the energy equation. For many situations, due to the rapid heating, temperature distributions follow the patterns of absorption of electromagnetic energy, obtained by solving Maxwell's equations. The boundary conditions are set for the metallic walls only. The inside volume of the entire oven (air + food) is treated as the dielectric with respective properties. The dielectric properties can be obtained from a number of sources such as Datta et al.,[24] although it would be hard to find the property for an exact food formulation. In this case, the discussion of material property data, as mentioned above, is to be followed. Also, temperature variation of dielectric properties, important when heating a salty material or a frozen material is difficult to find. Several commercial software programs are available to numerically solve the Maxwell's equations with the boundary conditions. In one study,[29] the software program EMAS was used. An example of results found through electromagnetic calculation is shown in Figure 50.1. Note that this focussing effect cannot be obtained without solving the equations of electromagnetics for the appropriate geometry and properties, although this has been reported in the literature.[30] More results on volume, temperature, and other effects and applications to sterilization can be found elsewhere.[29]

50.4.2 MECHANICS: THERMAL CRACKING DURING FREEZING

Moisture- and temperature-induced shrinkage and swelling during processing often lead to cracking of the food material, such as during the cooling of eggs, rapid freezing, drying, baking, and microwave heating. The stresses and strains that lead to cracking depend in a complex way on processing parameters such as heating or cooling medium temperatures, size of the food, moisture content, and the mechanical and thermal properties of the material. Optimizing these parameters to obtain improved food quality (by reducing breakage) requires precise knowledge of their

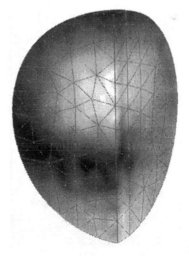

Figure 50.1 Focusing effect in a spherical food, causing areas near the center (shown in white) to heat relatively more than other areas.

influence on the process. A simulation-based approach can greatly reduce the needed experimentation.

Freezing of biomaterials is extremely important in preserving foods. Rapid freezing is frequently desired for multiple reasons, such as reduced extracellular migration of water. Cracks are often reported in rapid freezing, as when using liquid N_2, unless the size is too small, i.e., individual cells. Because foods are mostly water, it is postulated that the cracks are due to expansion during the phase change of water and to the resulting stresses. Accurate prediction and control of cracks during freezing is critical for successful ultra-rapid freezing.

Thermal stresses were studied in freezing of foods containing significant amounts of water.[31] The governing equations are the energy equation with an apparent specific heat formulation to consider the phase change process and the equations of viscoelasticity for the mechanics problem. A transient axisymmetric freezing process was used for simplicity. Viscoelastic properties were measured in an Instron machine, although detailed results show that, for very rapid freezing, elastic properties are sufficient, as the material does not have time to relax. Properties for partially frozen material were measured. An example of application of the model to designing a freezing process can be seen in Figure 50.2. The maximum principal tensile stress during phase change increases with a decrease in cooling medium temperature (faster cooling). This is consistent with experimentally observed fractures at a lower medium temperature.

50.4.3 MOISTURE AND HEAT TRANSFER: COMBINED MICROWAVE AND INFRARED HEATING

It is mostly the water component in food that makes microwave heating of food possible. Thus, all foods, when heated in a microwave, will experience a varying amount of evaporation of water due to increased energy levels (absorbed from

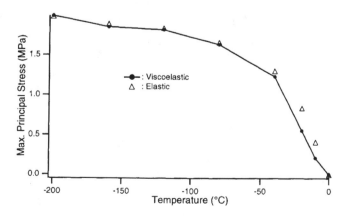

Figure 50.2 Increase in maximum principal stress as a function of boundary temperature (which affects the cooling rate).

microwaves). Depending on the internal resistance of the solid matrix to the transport of liquid water or vapor, and the outside resistance to moisture loss, a varying amount of moisture can be lost from any food being heated. In microwave heating, new mechanisms of internal transport and a drastically different external condition can make moisture transport quite different from that due to conventional surface heating. Internal pressure gradients arise from internal heating and vaporization and significantly enhance the transport. Even a moderate internal pressure gradient in a low-moisture material can cause more moisture to move to the surface than can be removed, leading to a soggy surface. Use of infrared (IR) heat is a possible way to remove surface moisture, keeping it crisp.

Simulation of this moisture transport process was not possible using commercial software due to, among other reasons, the strong internal evaporation present in microwave heating, leading to coupling of vapor and liquid phases. Governing equations include energy conservation and mass conservation for the three components: vapor, water, and air. A finite difference code was developed to solve the equations, as described in Ni et al.[32] The boundary conditions are also complex for this problem, due to the possibility of pumping out of the liquid at high pressure. Microwave and infrared heat are used as source terms, with different penetration depths, respectively. It is harder to find the penetration depth data for infrared heating. Results show that the ability of infrared heat to remove surface moisture depends on how far from the surface the infrared penetrates the material. Figure 50.3 shows how surface moisture can be kept very low compared to the microwave heat when all the infrared heat is absorbed very near the surface, as for foods with small penetration depth (<1 mm). Depending on the material, however, not all of the infrared heat can get absorbed at the surface. If the infrared heat penetrates significantly, it behaves closer to microwave heating and can increase the surface moisture instead. Thus, as shown in Figure 50.3 for foods with IR penetration depths comparable to microwaves (>4 mm), IR can actually increase surface wetness. When

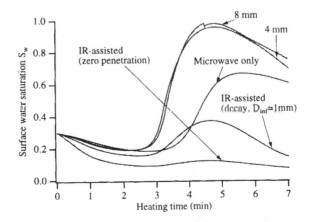

Figure 50.3 Effects on the surface moisture of a food heated by infrared and microwaves. The various infrared penetration depths shown are representative of different foods.

IR is assisted in a food with larger IR penetration, appropriately reducing MW power level is likely to lower surface moisture, creating a better end product. These results can be used in the design of combination ovens, as in microwave ovens that also have a radiative heating element.

50.4.4 HEAT AND MOISTURE TRANSFER: MISC. EXAMPLES

Freezing and Thawing of Food

Since only energy transfer (as opposed to fluid flow and mass transfer) is of interest, implementation of such problems is fairly straightforward in a CFD software program for arbitrary shape, size, and properties. Only enthalpy vs. temperature relations as material properties will be needed, as available in Mott.[33] Freezing and thawing time can be calculated and optimized for various food compositions efficiently and accurately, without the inaccuracies of the approximate semi-analytical formulas[34] currently used. Modification, for example, of using microwave heat for thawing can be easily added in a CFD software program, whereas the approximate semi-analytical formulas would completely fail in such situations.

Canning of Liquid and Solid Foods

In the canning industry, empirical correction factors are used to adjust processing time when the process undergoes deviations such as a drop in the steam temperature. But a CFD software program can automatically handle arbitrary process deviations such as time-varying boundary temperatures and thereby provide accurate process time. Different geometries, such as cylindrical or ham shaped, can be handled routinely with the grid generation capabilities of most CFD software. In the past, only conduction heating computation was feasible on a personal computer, but natural convection computations have now been successfully performed on a personal computer.[35] It would only get easier with more powerful computers. If internal heating, as in a microwave pasteurization, needs to be added, this can be accomplished quite readily with a CFD software program.

Continuous Sterilization of Liquids

When large particulates are not involved, process optimization can be routinely accomplished for continuous sterilization of liquid flowing through a pipe.[36] The CFD software can provide velocity, temperature, lethality, and nutrient changes, and how they are affected by changing pipe diameter, flow rate, viscosity, etc. As shown later in this contribution, adding ohmic heating to continuous flow is also relatively simple.

Extrusion

Polyflow (Polyflow S. A., Louvain-la-Neuve, Belgium) can be used to design the die shape and to model the extrusion and coextrusion processes. One can be interested in screw analysis and design by studying the flow between the rotating screw and the barrel to obtain better mixing or a more homogeneous temperature distribution. It is also important to balance the flow at the die exit to minimize the extrudate deformations in the free flow, and finally to obtain the right shape of the extrudate after the deformations. This software also provides the tools to address these issues, and the simulations provide information over the whole flow domain, which few affordable experiments could reveal. Other CFD software, such as FIDAP, can also compute similar extrusion processes.

Simple Moisture Transfer Processes

When drying and moisture transfer processes can be approximated as a simple diffusional process with an effective diffusivity, such processes can be very easily optimized using a CFD software program. For example, drying air temperature, velocity, product size, and shape can be optimized by changing input data and making several quick runs. Note that, although a simple analytical (series) solution to the diffusion equation is widely used in the drying literature, such a solution is highly limited to constant diffusivity, constant initial conditions, constant boundary temperature, and a regular geometry. In reality, diffusivity is a strong function of moisture content, and in multi-stage drying, the initial moisture distribution for a stage is not likely to be constant. The approximate analytical solutions can produce very erroneous results under such conditions, whereas a CFD software would handle those conditions routinely.

50.5 FUTURE NEEDS FOR CAE SOFTWARE IN THE FOOD INDUSTRY

50.5.1 Linkages between Software

Food processes quite often involve more than one discipline of engineering. For example, microwave food processing involves at least the two subject areas—electromagnetics and heat transfer. Sometimes, there is strong coupling between electromagnetics and heat transfer, since heating changes the dielectric properties, which can change the electromagnetics. The two calculations need to go on simultaneously, but, since the software is written by two different companies, exchanging solution variables between the two software can be a major problem or often impossible. Better linkages between software will be necessary.

50.5.2 Customized Software for Food Processing

Software such as CFD can be further customized for a specific application. For example, the software icepak™ is specific to the design of cooling systems for electronic components. It analyzes air flow and thermal distribution in enclosed spaces. The features of this particular software include a library of standard objects (PCBs, fans, resistances, etc.), automatic and optimized mesh generation, and specialized result viewing capability. As use of software increases in the food industry, one hopes that such customized software will become available for food processing applications. As an example, such a software program can have a module or button for calculating F_o values, needed in describing quality changes during a heating or cooling process.

50.5.3 INCLUSION OF APPROPRIATE PHYSICS

Physics appropriate for food processing applications need to be included. Examples of processes that are difficult to accomplish using today's commercial software include, for example, sterilization in rotating cans and sterilization (aseptic processing) of a flowing solid-liquid mixture with large particulates. The commercial CFD software of today still lacks the ability to handle processes in which internal evaporation is significant and can develop pressures, such as high-intensity drying, microwave heating, and deep frying. The research community is typically ahead of commercial software in developing specialized codes. Under normal courses of development, these specialized capabilities eventually get incorporated in the commercial software, depending on market demand.

50.6 RESOURCE NEEDS FOR COMPUTER–AIDED ENGINEERING

Use of computer-aided engineering in food and other process and product design will only increase in the future. In the immediate future, there are two ways to bring this expertise into a company: by utilizing outside consulting firms to perform this modeling or getting the in-house engineers trained by the software vendors and investing on the hardware and software. Sufficient knowledge of the use in the underlying physics is always critical.

In the long run, engineers and scientists need to be trained more in such computer-aided engineering tools. In colleges, CAE software has been included in courses for some time, but typically at a graduate level.[37] Use of CFD software in undergraduate teaching is beginning to appear.[38–40] and is expected to rise in the future. The graphics driven interface has almost revolutionized the ease of use, making it possible to be introduced to undergraduates. Thus, in the future, more of the entry-level engineers and scientists will be expected to be familiar with this tool.

Computational resources required for computer-aided engineering tools are highly dependent on the particular physical problem being solved, and few general guidelines can be given. As the computing power has gone through revolutions, most software is now available under the Windows NT operating system for Pentium-based machines. For the computing time to be reasonable, only the high-end Pentium-based machines are appropriate for these computationally intensive applications. As the prices of these high-end machines come tumbling down, hardware expenses are becoming a relatively small part of the total expense. The CFD software, on the other hand, typically needs to be leased, and the yearly lease can be quite substantial, particularly for industrial users. However, cost comparisons should be made with the time and expenses required for building physical prototypes and associated experimentations, which is also quite expensive and which will only increase in the future.

50.7 A CAUTIONARY STATEMENT

Computer-aided engineering to simulate products and processes on the computer reduces the drudgery of hand calculations and/or code development for real-life applications. It is, however, not a substitute for the knowledge of the technical subject matter. Such software are susceptible to GIGO (garbage in, garbage out). It is very easy to obtain colorful "results" that are totally meaningless (in such cases CFD would be "Color Full Dynamics"). Most of these programs do not have built-in intelligence to judge whether a computation is physically reasonable, so it will always compute something—which may or may not be what the user intended. Thus, it is dangerous to use such software as a complete black box without sufficient training in the software and, more importantly, without the proper background in the underlying engineering science.

REFERENCES

1. Puri, V. M. and R. C. Anantheswaran. 1990. "Finite Element Method in Food Processing: A Review," Paper 90-6523, ASAE, St. Joseph, MI.
2. Kumar, A. and K. R. Swartzel. 1993. "Selected Food Engineering Problems and Their Solutions Through FEM," in *Advances in Finite Element Analysis in Fluid Dynamics*, M. N. Dhaubhadel, M. S. Engelman, and W. G. Habashi, eds. American Society of Mechanical Engineers. FED-vol. 171.
3. Singh, R. P. and A. G. Medina. 1989. *Food Properties and Computer-Aided Engineering of Food Processing Systems*. Boston, MA: Kluwer Academic Publishers.
4. Teixeira, A. A., J. R. Dixon, J. W. Zahradnik, and G. E. Zinsmeister. 1969. "Computer Optimization of Nutrient Retention in the Thermal Processing of Conduction-Heated Foods," *Food Technology*, 23(6): 137–142.
5. Torok, D. F. 1991. "Computational Thermofluid Modeling in the Food Processing Industry," presented at the Conference on Food Engineering, Mar 10–12, 1991.
6. Quarini, J. 1995. "Applications of Computational Fluid Dynamics in Food and Beverage Production," *Food Science & Technology Today*, 9 (4): 234–237.
7. Khankari, K. and S. V. Patankar. 1996. "Computer Simulation of a Conveyer Dryer," *Cereal Foods World*, 681–685.
8. Scott, G. and P. Richardson. 1997. "The Application of Computational Fluid Dynamics in the Food Industry," *Trends in Food Science and Technology*, 8: 119–124.
9. Datta, A. K. 1998. "Computer-Aided Engineering in Food Process and Product Design," *Food Technology*, 52(10): 44–52.
10. Datta, A. K. 1998. "Simulation-Based Engineering as a Tool for Computer-Aided Engineering," Paper No. 96-6068 presented at the ASAE International Meeting, Phoenix, AZ.
11. Russel, M. 1995. "A Model of Efficiency," *Food Engineering*, 5: 63–70.

12. Carey, G. F. 1997. *Computational Grids: Generation, Adaptation, and Solution Strategies*. Bristol, PA: Taylor & Francis.

13. Owen, S. 1997. *Meshing Research Corner* [WWW], URL: http://www.ce.cmu.edu/sowen/mesh.html (accessed May 14, 1997).

14. *CenBASE/Materials on WWW* [WWW document], CenTOR Software Corporation, Garden Grove, CA. URL: http://www.centor.com/cbmat/index.html (accessed May 15, 1997).

15. TAPP 3.0. 1997. *ESM Software, Inc.* [WWW], URL: http://www.esm-software.com (accessed May 20, 1997).

16. MAPP 1.0. 1997. *ESM Software, Inc.* [WWW], URL: http://www.esm-software.com (accessed May 20, 1997).

17. Nesvadba, P. 1999. *Construction of a Database of Physical Properties of Foods.* [WWW document], URL: http://www.nel.uk/fooddb (accessed December 1999).

18. Singh, R. P. 1995. *Food Properties Database v2.0 for Windows.* Boca Raton, FL: CRC Press.

19. O'Brien, W. J. 1997. *Biomaterials Property Database* [WWW], University of Michigan, Ann Arbor, Michigan. URL: http://www.lib.umich.edu/libhome/Dentistry.lib/Dental tables/intro.html (accessed March 1997).

20. Chato, J. C. 1985. "Selected Thermophysical Properties of Biological Materials," in *Heat Transfer in Medicine and Biology: Analysis and Applications*, A. Shitzer and R. C. Eberhart, eds. New York, NY: Plenum Press.

21. Rao, M. A. and S. S. H. Rizvi. 1995. *Engineering Properties of Foods*, New York, NY: Marcel Dekker, Inc.

22. Okos, M. R. 1986. *Physical and Chemical Properties of Food.* St. Joseph, Michigan: American Society of Agricultural Engineers.

23. Perry, R. H., D. W. Green, and J. O. Maloney. 1984. *Perry's Chemical Engineers Handbook*, New York, NY.: McGraw-Hill.

24. Datta, A. K, E. Sun, and A. Solis. 1995. "Food Property Data and Their Composition-Based Prediction," in *Engineering Properties of Foods*, M. A. Rao and S. S. H. Rizvi, eds. New York, NY: Marcel Dekker.

25. Tinga, W. R. and S. O. Nelson. 1973. "Dielectric Properties of Materials for Microwave Processing—Tabulated," *J. Micro. Power*, 8(1): 24–65.

26. Choi, Y. and M. R. Okos. 1986. "Thermal Properties of Liquid Foods—Review," in *Physical and Chemical Properties of Foods*, M. R. Okos, ed. St. Joseph, MI: American Society of Agricultural Engineers.

27. Rahman, S. 1996. *Food Properties Handbook*, Boca Raton, FL: CRC Press, Inc.

28. Mohsenin, N. N. 1980. *Thermal Properties of Foods and Agricultural Materials*, New York, NY.: Gordon and Breach Science Publishers, Inc.

29. Zhang, H. and A. K. Datta. 2000. "Electromagnetics of Microwave Heating: Magnitude and Uniformity of Energy Absorption," submitted to *Handbook of Microwave Technology for Food Applications*, A. K. Datta and R. C. Anantheswaran, eds. New York, NY: Marcel Dekker, Inc.

30. Chen, D. S., R. K. Singh, K. Haghighi, and P. E. Nelson. 1993. "Finite Element Analysis of Temperature Distribution in Microwaved Particulate Foods," *J. Food Engineering*, 18: 351–368.

31. Shi, X., A. K. Datta, and Y. Mukherjee. 1998. "Thermal Stresses from Large Volumetric Expansion During Freezing of Biomaterials," Accepted in the *J. Biomech. Eng., Trans. of the ASME*.

32. Ni, H., A. K. Datta, and K. E. Torrance. 1998. "Moisture Transport in Intensive Microwave Heating of Biomaterials: A Multiphase Porous Media Model," Submitted to the *International Journal of Heat and Mass Transfer.*

33. Mott, L. F. 1964. "The Prediction of Product Freezing Time," *Aust. Refrig., Air Cond. and Heating,* 18: 16–18.

34. Cleland, A. C. 1990. *Food Refrigeration Processes: Analysis, Design, and Simulation.* London, UK: Elsevier Applied Science.

35. Yang, W. H., A. K. Datta, and M. A. Rao. 1997. "Rheological and Calorimetric Behavior of Starch Gelatinization in Simulation of Heat Transfer," in *Proceedings of the Seventh International Congress on Engineering and Food,* Brighton, UK.

36. Kumar, A. and Bhattacharya, M. 1991. "Numerical Analysis of Aseptic Processing of a Non-Newtonian Liquid Food in a Tubular Heat Exchanger," *Chemical Engineering Communications,* 103: 27–51.

37. Jaluria, Y. and K. E. Torrance. 1986. *Computational Heat Transfer.* Washington, D. C.: Hemisphere Pub. Corp.

38. Datta, A. K. 1999. *Computer-Aided Engineering: Applications to Biomedical and Food Processes* [WWW document]. URL: http://courseinfo.cit.cornell.edu/ courses/ aben453 [accessed December 1999].

39. *FEHT (Finite Element Analysis Package).* S. A. Klein, W. A. Beckman, and G. E. Myers. [WWW], URL: http://www.engr.wisc.edu/centers/sel/feht/feht.html (accessed May 21, 1997).

40. Ribando, R. J. and G. W. O'Leary. 1994. "Numerical Methods in Engineering Education: An Example Student Project in Convection Heat Transfer," *Computer Applications in Engineering Education,* 2(3): 165–174.

51 Modeling and Simulation of Reactive and Separation Food Bed Processes

F. López-Isunza

E. S. Pérez-Cisneros

S. I. Flores y De Hoyos

CONTENTS

Abstract

The computational advances gained in numerical methods, software, and hardware allow us to perform rigorous modeling and simulation of complex phenomena taking place in food solid beds. Reactive extraction food processing (e.g., pectin from protopectin), ion-exchange chromatography, reactive distillation, and immobilized enzyme reactors are good examples. A general approach proposed herein is modeling in terms of their multicomponent, multiphase, and heterogeneous nature, coupling fluid phase constitutive equations (transport) via interface transport, to those describing the solid (reaction) phase. In this way, a common mathematical structure for different problems is obtained. The linear differential equations from the fluid phases, linked to the nonlinear terms of the solid phase, can be analyzed and solved using, wherever possible, a similar implemented numerical scheme.

51.1 INTRODUCTION

For a long time, the design of processes in which a chemical reaction takes place prior to or simultaneously with separation has been considered difficult. This was mainly so because such processes demanded a better understanding of the phenomena complexity involved. However, to improve our knowledge about separation, reaction, and reaction-separation processes, it is necessary to consider two additional fundamental aspects.

The appropriate modeling of heat and mass transfer
The description of the hydrodynamics of two fluid phase flow in fixed-bed and
     fluid-bed operations

The influence of a system's pressure on gas-solid hydrodynamics is known to be significant and will affect regime transitions. It is a fact that the hydrodynamics have not yet been described well enough to allow prediction of mass transfer rates to a sufficient degree of accuracy. On the other hand, mass transfer is the basis for separations. Most mixtures encountered in practice are multicomponent in nature, containing three or more species in dilute or concentrated solutions. Due to unsteady-state mass transfer present in actual non-equilibrium situations, many workers have to use Fick's dimensionless group, $T = D_s \Theta / a^2$, as a design parameter. Θ is the extraction time, a is the characteristic dimension of the particles, and the in-solid diffusion coefficient is D_s. Spaninks[1] suggests the use of the following general expression for the diffusion partial differential equation (DPDE):

$$\frac{\partial X}{\partial \Theta} = \frac{r^{1-v}\left(D_s r^{v-1}\frac{\partial X}{\partial r}\right)}{\partial r} \tag{51.1}$$

where X is the mean solute concentration in solid; r varies from 0 to a; and the particle geometry is accounted for by v, which is 1 for infinite slabs, 2 for infinite cylinders, and 3 for spheres.

Hougen and Marshall[2] developed a solution for isothermal adsorption with linear equilibrium (e.g., $Y^* = mX$) and a special condition for extract solute concentration: (a) $Y = Y_{in}$ at $z = 0$ for all values of Θ, and (b) $X = 0$ at $\Theta = 0$ for all values of z. In dimensionless form the solution is

$$\frac{Y}{Y_{in}} = 1 - e^{-b\Theta} \int_{\frac{z}{H}}^{\frac{z}{H}} e^{-\frac{z}{H}} J_0\left[2i\sqrt{\frac{b\Theta_z}{H}}\right] d\left(\frac{z}{H}\right) \tag{51.2}$$

where J_0 = a Bessel function of the first kind and zero order

   z = the position in the bed in the flow direction

   $b = mG/\rho_s H$

   G = the mass velocity of adsorbate-free fluid

   ρ_s = the solids density

   H = the height of the transfer unit for mass transfer (HTU), which can be estimated by well known correlations depending on the importance of the resistances present

The assumption that the external mass-transfer resistance is the controlling factor rather than the in-solid diffusivity has also been applied in continuous countercurrent extractors where the concept of HTU is used in the design.[1,3] Spaninks recognizes the importance of the diffusive nature of the extraction phenomena in his lumped mass-transfer resistance model, which requires the first eigenvalue, q, for the continuous extraction series solution, D_s, and the stripping factor.[1] He uses this approach in the design of a diffusing battery and belt-type extractors. When using Spaninks' model for predicting the yields of diffusing batteries, one obtains values ranging from 50% smaller to 3% larger than those obtained experimentally. The reliability of the belt extractor model is suspect, since it is not clear if the eigenvalue can be validly applied, and the model does not take into account the non-uniformity of X from the second stage on.

For a single fixed bed with a uniform initial concentration, constant inlet concentration, a bed filling time Θ_f and $t = (\Theta - \Theta_f)/\Theta_f$, there is a dimensionless Y_{out} solution, $S(t) = (mX_0 - Y_{out})/(mX_0 - Y_{in})$. $S(t)$ will also be used to represent the functional value in terms of the dimensionless operative parameters, which would yield a given value of $(mX_0 - Y_{out})/(mX_0 - Y_{in})$ if we were dealing with a single fixed bed operating under the previously cited conditions.

Thomas[4] developed a PDE solution for fixed bed extractions for spherical solids and ε as void fraction. The solution was claimed to be applicable for the $T_f (= D_s \Theta_f/a^2)$ range commonly encountered in cross-flow extractors. However, it appears that Thomas' solution does not converge. Therefore, it cannot be used for numerically determined $S(t)$.

Solutions for finite values of Bi (= $k_Y ma/D_s$, where k_Y is the mass transfer coefficient from the fluid side) and without limitations with respect to T_f were developed by Rosen.[5,6] His solutions are quite complex but can be greatly simplified when $T_f > 50$. The solutions which apply in this case are

$$S(t) = 0.5\{1 + erf[\phi]\} \tag{51.3}$$

$$\phi = 0.5[\alpha't - 1]\sqrt{\frac{15T_f}{\alpha'}} \qquad (51.4)$$

When Bi is finite, $\sqrt{\dfrac{15T_f}{\alpha'}}$ should be replaced by $\sqrt{\dfrac{15T_f}{[\alpha'(1+5Bi)]}}$.

Both Thomas' and Rosen's solutions were developed based on the assumption that axial dispersion, D_A, is negligible, and $\alpha' = m\varepsilon/(1-\varepsilon)$. Rasmuson and Neretnieks[7] obtained a PDE solution for adsorption in fixed beds, which includes the effect of longitudinal dispersion. This solution is based on extract concentration constraints similar to those used in the heat transfer equations of Deisler and Wilhelm,[8]

$$\frac{\partial Y}{\partial \Theta} + V\frac{\partial Y}{\partial Z} - D_A\frac{\partial^2 Y}{\partial Z^2} = -\frac{1-\varepsilon}{\varepsilon}\frac{\partial X}{\partial \Theta} \qquad (51.5)$$

and the diffusion partial differential equation, which is the same as Spaninks's equation for a sphere,

$$\frac{\partial x}{\partial \Theta} = D_s\left[\frac{\partial^2 x}{\partial r^2} + \frac{2\partial x}{r\partial r}\right] \qquad (51.6)$$

Rasmuson and Neretnieks[7] state that the solution of Babcock et al.[9] is a "limiting solution for low values of D_A. For $D_A = 0$, the solution is identical to that given by Rosen."[5]

The initial and boundary conditions used by Rosen are stated in terms of an initially completely unsaturated bed and an inlet concentration step function.

$$X'(r,z,0) = 0 \text{ and } Y/Y_{in} = 0 \text{ for } \Theta = 0 \text{ and } = 1 \text{ for } \Theta > 0 \qquad (51.7)$$

Where X' is the concentration distribution of adsorbed material in the interior of the spheres. It is claimed that due to the linearity of the system the solution can be used for the case where the bed is initially at equilibrium, e.g., $X' = X_0$.

In the cases presented by Rasmuson and Neretnieks[7] and Babcock et al.,[9] the initial and boundary conditions are

$$Y(0,\Theta) = Y_0, \ Y(\infty,\Theta) = 0, \ Y(z,0) = 0 \text{ for } Z = 0 \qquad (51.8)$$

$$X'(r,z,0) = 0, X'(a,z,\Theta) = X_s(z,\Theta) \qquad (51.9)$$

$$\frac{\partial X}{\partial \Theta} = \frac{3k_Y(Y - mX_s)}{a} \qquad (51.10)$$

For extract concentration mass transfer in packed beds, Equation (51.5) can also be written as:

$$-V\varepsilon\left(\frac{\partial Y}{\partial z}\right) + D_A\varepsilon\left(\frac{\partial^2 Y}{\partial z^2}\right) + \left[\frac{s_p(1-\varepsilon)}{V_p}\right]k_Y(mX - Y)_{r=a} = \varepsilon\frac{dY}{d\Theta} \qquad (51.11)$$

For cases (as in food leaching) where Bi is large, it is usually preferable to replace $k_Y(mX - Y)_{r=a}$ by $-D_s(\partial X/\partial r)_{r=a}$.

When axial dispersion, D_A, is not negligibly small, the dimensionless PDE solutions for $S(t)$ should include the Peclet group, $Pe = 2aV/D_A$ [where V is the fluid velocity (m/s)] as a variable. Other types of contacting imperfections may also occur, and these may require the use of further dimensionless variables. When solving the auxiliary PDE, Equation (51.11) is converted, and thus its initial (IC) and boundary (BC) conditions, into dimensionless equations. Besides Pe (or $Pe' = Pe \{L/2a\}$), where L is the length of the packed column (m)), other groups may be important, such as $\alpha' = m\varepsilon/(1 - \varepsilon)$, $t = (\Theta - \Theta_f)/\Theta_f$, $T_f = D_s\Theta_f/a^2$ and Bi. Solutions of the PDE, IC, and BC provide values for $S(t) = (Y - mX_0)/(Y_{in} - mX_0)$ in terms of these groups.

The cause for lack of conformity appears to be neglect of axial dispersion in some DPDE formulations[5] and neglect of mass-transfer resistance ($1/k_Y$) in the liquid in other formulations. While, for most flows through fixed beds, the liquid-side, mass-transfer resistance in the solid (e.g., $Bi > 50$), it can be shown that even a relatively small liquid-side, mass-transfer resistance will tend to markedly limit the extract concentration at the leading edge of the entering extract. Because of this resistance, Y at the leading edge of the extract will not be in equilibrium with the solute concentration in the solid; i.e., will not equal mX_0, as assumed in many DPDE solutions for fixed beds. When $Y_{in} = 0$, Y at the leading edge of the liquid often lies between 0.2 to 0.35 mX_0 for the beds tested as shown by Flores.[10] Axial dispersion, by causing solute from the leading edge to diffuse back into the advancing extract, also tends to reduce the concentration at the leading edge. This seems to be confirmed by the somewhat higher values obtained in previously filled extraction beds when Y for the linear extraction region is extrapolated to $t = 0$.

Cowan et al.[11] preliminarily validated a computer code for design, performance prediction, optimization, and scaling-up for industrial adsorption equipment. They claim to have found good agreement among the results from an analytical solution for a packed bed adsorption column with experimental data for affinity adsorption,[12] as well as with those using the numerical methods code. They used a zero-axial-dispersion analytical solution but agree with the need to solve more-complex mass transfer rate models and to cover adsorption, washing, elution, and regeneration.

Arve and Liapis,[13] applying a PDE as Equation (51.11) when dealing with biospecific adsorption (affinity chromatography) in fixed beds, found that the breakthrough time of the adsorbate is significantly influenced by the rate of inter-action step between the adsorbate and the ligand. It is noteworthy that the concentrations used in these cases are very dilute, which explains in part the good results obtained.

The removal by suitable solvent of a desirable or undesirable solute or group of solutes associated with an insoluble solid matrix is classified as solid liquid extraction. Nowadays, it is used to produce many products: vegetable oil, sugar, decaffeinated coffee, protein concentrates, flavor extracts, hormones, antibiotics, etc. Sometimes the leaching process involves more than one solvent, as when obtaining carotenoid pigments. Recently, advances in extraction technology involve the use of supercritical CO_2 as solvent, e.g., in the production of hop extracts. Another example of such use is the removal of undesired caffeine from green coffee beans by means of supercritical CO_2.

Extraction behavior of vegetable oil from oilseed flakes by hexane or isopropanol cannot be modeled by DPDE standard solutions based on Fick's law, in accordance with Segado and Schwartzberg.[14] They claim this is due to structural nonuniformity.

Fick's law can be generalized by means of defining a matrix of Fick diffusivities, which can be applied to these mixtures. However, it normally holds only for dilute mixtures, and the elements of this matrix lose their simple meaning, as the non-diagonal elements can assume negative values. The limitations of Fick's law to describe molecular diffusion are well documented. It has been argued that the most convenient and general approach to model multicomponent mass transfer is the Maxwell–Stefan formulation. To describe intraparticle mass diffusion coupled to reaction, its equivalent, the dusty-gas model, should be used. Furthermore, the use of the Maxwell–Stefan approach generally leads to more complicated mathematical models; however, it should be applied to describe mass transfer in concentrated mixtures.

It is the aim of this chapter to discuss rigorous modeling and simulation for reaction, separation, and reactive-separation operations, using a heterogeneous description of a multiphase and multicomponent system. The conservation equations are written considering that heat and mass transport at the fluid phases interact through their common interface and are coupled, through fluid-solid interfaces, to reaction/adsorption and diffusion phenomena at the solid phase. In this way, models representing different processes have a common mathematical framework. This feature allows the performance of an *a priori* behavior analysis to introduce realistic simplifying assumptions and to select the appropriate numerical method for its solution.

51.2 MODELING OF REACTIVE AND SEPARATION PROCESSES

Models describing separation, reaction, and reactive-separation processes occurring in a heterogeneous food system (i.e., a multicomponent mixture contained in two-fluid phases interacting with a solid phase) could be described as members of the same class. The heterogeneous nature of the system requires from the mathematical model the description of heat and mass transport interactions between fluid phases occurring at the interparticle (reactor) level. This can be done via interphase transport, which couples them to intraparticle (reaction) diffusion, conduction and reaction/adsorption mechanisms.[15] This is schematically represented by Figure 51.1.

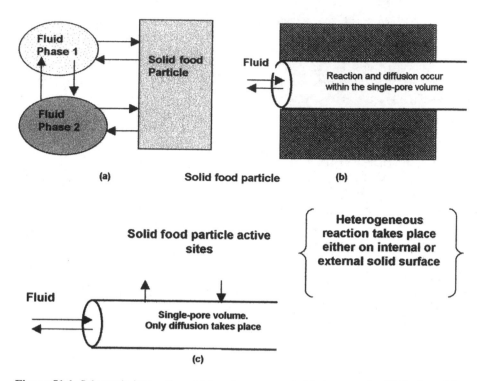

Figure 51.1 Schematic interactions taking place among fluid phases and solid food particle. (a) Both fluids may interact with the solid state, (b) the pseudo-homogeneous food particle, and (c) the pseudo-heterogeneous food particle.

When modeling such systems, two important aspects should be incorporated: the description of multicomponent diffusion mass transfer, in terms of the Maxwell–Stefan formulation[16–18] and the velocity of fluid fields by Darcy's law, which are not usually accounted for in most cases.

On the modeling issue, a remark by Aris[19] is pertinent at this point.

> A mathematical model is a representation, in mathematical terms, of certain aspects of a nonmathematical system. The arts and crafts of mathematical modeling are exhibited in the construction of models that not only are consistent in themselves and mirror the behavior of their prototype, but also serve some exterior purpose.

Therefore, although the mathematical structure of the class of systems considered in this work is the same, each model contains different ingredients, aiming to describe the particularities of the phenomena to be modeled.

Schwartzberg[20] stressed the importance of recognizing the occurrence of "solubilization" reactions thanks to which insoluble precursors yield new solutes; then, these created solutes transfer to the extract. Quality of the final product can be enhanced by controlling the extent and rate of solubilization stage. The product could be composed of several valuable solutes or, as in the production of gelatin from collagen, the former is the created solute itself. Sugar beets become too soft when processed at temperatures higher than 70°C, since protopectin, contained in the intercellular vegetable structure, reactively yields pectin, which in turn contaminates the extracted sugar. Heating promotes denaturation reactions of cell membranes and precipitate proteins when beet sugar is extracted. Solutes diffuse through denatured membranes faster; the precipitated proteins do not pass into the extract, so, by controlling this, the selectivity is controlled.

51.2.1 INTERPARTICLE FLUID PHASE CONSERVATION EQUATIONS FOR HEAT AND MASS

Two fluid phases interacting with a solid phase can be represented by a continuum model. It describes the conservation of heat, mass, and momentum coupled to reaction/separation, in a heterogeneous dynamic porous media. For the kth fluid-phase, the heat and mass transfer at its bulk, the general form is[18,21,22]

$$\frac{\partial(\rho_k \phi^k)}{\partial t} + \nabla \cdot (\rho_k \phi^k v_k) + \nabla \cdot \aleph_k = \psi_{jk} + \psi_{sk} \qquad (51.12)$$

where k stands for the number of fluid phases present. Equation (51.12) describes, for an arbitrary field quantity per unit volume of mixture in the kth phase, $\rho_k \phi^k$, the heat and mass accumulation, convection, nonconvective flux, $\aleph_k$, through external bounding surface σ, and fluid-solid, ψ_{sk}, as well as fluid-fluid, ψ_{jk}, interface transports, respectively.

The net rates of heat and mass at fluid-fluid and fluid-solid interfaces are given by

$$\psi_{jk} = n_i \cdot \{ \aleph_k + \rho_k \phi^k(v_k - v_i) - \aleph_j - \rho_j \phi^j(v_j - v_i) \} \qquad (51.13)$$

$$\psi_{sk} = n_s \cdot \{ \aleph_k + \rho_k \phi^k(v_k - v_s) - \aleph_s^i \} \qquad (51.14)$$

in which it is assumed that the velocities of the fluid phases move relative to each other, and to the solid phase.

Heat and mass transfer coupled to chemical reaction at the solid phase is

$$\frac{\partial(\rho_s \phi_s)}{\partial t} + \nabla \cdot \aleph_s = \Re(\phi_s) \qquad (51.15)$$

The second important suggestion when modeling is describing fluid velocity fields in fixed- or fluid-bed (porous medium) in terms of Darcy's law, or using the expressions suggested by Forchheimer, Brinkman, or Ergun.[21,23]

Flow stability and instability are of paramount importance when modeling industrial diffusion batteries where the hydrodynamic conditions of the extract percolating through the food solids can behave quite differently from what is expected. This has been studied in the previous decades. Based on the maximum concentration behavior of a tracer traveling at the mean velocity of a liquid flowing in a capillary tube, Taylor[24] developed a differential equation which governs longitudinal or axial dispersion:

$$D_A \frac{\partial^2 G'}{\partial z^2} = \frac{\partial G'}{\partial \Theta} \tag{51.16}$$

where G' is the maximum tracer concentration around which the spreading occurs symmetrically. This can be used as a stability criterion; e.g., the closer the appearance of the peak concentration to the mean residence time, the more stable the flow is.

The absence of viscous fingering and the occurrence of mixing caused only by dispersion processes implies flow stability.[25] Axial dispersion, flow instability, and ill distribution frequently occur when operating fixed bed extractors. Channelling is promoted by the presence of radial viscosity and temperature gradients, whereas displacement instability causes irregular nonuniform flows called fingering. Mass transfer from solid to fluid during extraction causes density and viscosity gradients in the extract. Studying the displacement of sugar liquors by water in packed columns, Hill[26] developed an equation for critical velocity, V_c, above which fingering occurs.

$$V_c = \frac{g[\rho_1 - \rho_2]}{k[\mu_1 - \mu_2]} \tag{51.17}$$

where g = acceleration due to gravity field
  k = the flow resistance of bed
  ρ_1, ρ_2 = the respective density of the fluid above and below the displacement interface
  μ_1, μ_2 = the corresponding viscosity

Displacement instability tends to be suppressed by cross-flow near the end of the fingers.[27] Such radial dispersion can convert fingering-induced concentration discontinuities into smooth axial dispersion, provided the radial nonuniformity is not too large.[28] Carbonell also suggests that one can detect flow ill distribution by measuring pulse responses at different radial positions in the bed.

Analysis of the up-flow extraction is important, since it provides severe instability, which serves as good reference situation. Up-flow is frequently used in leaching batteries. Analysis of extraction carried out in initially empty beds is also important, since this eliminates back-diffusion of solute from the liquid being displaced by the entering liquid. Solvent frequently advances through an empty bed in the first stage of belt extractors.[10]

When not the all of unstable displacement is converted into axial dispersion, it is possible that other kinds of treatment would be required, e.g., the parallel extractor model. Plotting velocity against the fraction of the flow area, in which the local velocity is less than the velocity value on the ordinate, two parameters can be obtained, which can be used to characterize displacement instability.[29]

A general approach is to use the momentum equation, in terms of Darcy's law for slow flows, as follows:

$$\nabla P = -\frac{\mu}{K}\mathbf{v} \qquad\qquad (51.18)$$

where μ = fluid viscosity
K = bed permeability
$\mathbf{v}$ = velocity vector

The use of Brinkman's extension for high Reynolds numbers gives

$$\nabla P = -\frac{\mu}{K}\mathbf{v} + \tilde{\mu}\nabla^2\mathbf{v} \qquad\qquad (51.19)$$

With the exception of reactive distillation, which is treated below given its particular continuous-discrete nature, one fluid phase systems are described herein, since most of the processes considered in this work belong to this case. Extensions to multiphase fluid systems are easily made.

51.2.2 SEPARATION AND REACTION IN FIXED-BED OPERATIONS

For a single fluid phase, conservation equations for heat and mass transfer in reaction or separation processes in heterogeneous dynamic fixed bed operations have a similar structure to Equations (51.12) through (51.14). Traditionally, a Fick and Fourier description of dispersion terms have been used, and the corresponding heat and mass conservation equations in dimensionless form are

$$\frac{1}{\Lambda}\frac{\partial \vartheta_n}{\partial t} + \frac{\partial \vartheta_n}{\partial z} - \frac{1}{P_{eA}}\left(\frac{\partial^2 \vartheta_n}{\partial z^2}\right) - \frac{1}{P_{eR}}\left(\frac{\partial^2 \vartheta_n}{\partial r^2} + \frac{1}{r}\frac{\partial \vartheta_n}{\partial r}\right) = \delta^i(\vartheta_{ns} - \vartheta_n) \qquad (51.20)$$

where the quantity $\vartheta_n(n = 1,\ldots, M + 1)$ represents a total of M mass or mole specie concentrations plus the temperature of the fluid phase; Λ is the ratio of a characteristic intraparticle diffusion time to fluid phase residence time; δ^i represents the dimensionless interfacial fluid-solid heat and mass transfer coefficients; and Pe_A and Pe_R are modified axial and radial Peclet dimensionless groups for heat and mass.

51.2.3 BOUNDARY CONDITIONS FOR THE INTERPARTICLE FLUID PHASE

Initial and boundary conditions used for a single fluid phase could be summarized as follows:

At $t = 0$, an initial steady-state condition exists for all mass species and temperature in the fluid phase within the bed.

$$\vartheta_n(0,r,z) = \vartheta_{n0}(r,z) \tag{51.21}$$

At bed entrance ($z = 0$) and exit ($z = 1$), the more general conditions given by Wehner and Wilhelm[30] could be used.

$$\text{at } z = 0 \qquad \vartheta_n\big|_{0-} - \frac{1}{Pe_A}\frac{\partial \vartheta_n}{\partial z}\bigg|_{0-} = \vartheta_n\big|_{0+} - \frac{1}{Pe_A}\frac{\partial \vartheta_n}{\partial z}\bigg|_{0+} \tag{51.22}$$

$$\text{at } z = 1 \qquad \vartheta_n\big|_{1-} - \frac{1}{Pe_A}\frac{\partial \vartheta_n}{\partial z}\bigg|_{1-} = \vartheta_n\big|_{1+} - \frac{1}{Pe_A}\frac{\partial \vartheta_n}{\partial z}\bigg|_{1+} \tag{51.23}$$

However, it has been shown[31] that, for mass transport, these conditions can be reduced to the Dankcwerts conditions.[32]

$$\text{at } z = 0 \qquad \vartheta_n\big|_0 = \vartheta_n\big|_{0+} - \frac{1}{Pe_A}\frac{\partial \vartheta_n}{\partial z}\bigg|_{0+} \tag{51.24}$$

$$\text{at } z = 1 \qquad \frac{\partial \vartheta_n}{\partial z}\bigg|_{1-} = 0 \tag{51.25}$$

Step changes or varying conditions at the feed as a function of time, as in the case of soybean oil leaching in multistage extractors, coffee extracted in counter-currently arranged batteries, and chromatographic separations, should also be taken into account.[33]

Among the relevant problems that must be faced when modeling leaching systems is dealing with changes in boundary conditions, which occur in most actual situations. Some workers have overcome these problems using superposition techniques.[10,34] The Duhamel's theorem helps to deal with computational difficulties caused by such BC changes, as well as to account for discrepancies due to flow instability and axial dispersion that occur during extraction. These superposition techniques, while very useful and powerful, are limited just to linear equations.

In unit operations and mass transfer traditional textbooks, despite the fact that all of them recognize the inapplicability of the equilibrium stage, it is common to assume that every stage of an extraction system is an ideal stage. That is, the solution concentration in the outflow of a stage is equal to that of the solution occluded in the insoluble solid matrix leaving the same stage.

In the radial direction, symmetry is normally considered at $r = 0$, whereas an impermeable wall for mass transport and adiabatic behavior, or heat transfer throughout the wall, are invoked at $r = 1$. These are described as[35] at $r = 0$

$$\frac{\partial \vartheta_n}{\partial r} = 0 \tag{51.26}$$

at $r = 1$

$$-\frac{\partial \vartheta_{N+1}}{\partial r} = Bi_w(\vartheta_{N+1} - \vartheta_b) \text{ for temperature} \tag{51.27}$$

$$\frac{\partial \vartheta_n}{\partial r} = 0 \text{ for mass, and adiabatic} \tag{51.28}$$

When axial dispersion is considered to be negligibly small, the following condition applies at bed inlet:

$$\text{at } z = 0 \quad \vartheta_n(t,0,r) = \vartheta_{n0}(t) \tag{51.29}$$

In many food and biomaterials bed processes, simplifying assumptions may apply. Some considerations allow that neglecting any of the dispersion terms (or other mechanisms in general) may be based on intuition (physical insight) about the particular phenomena under study. Mears[36] has summarized some criteria for these cases, based on the analysis of the relative magnitude of different transport and reaction terms.

Many solid-liquid contacting patterns are used. Mass transfer properties and economic limitations will dictate which kind of arrangement is best. Particle size, texture, and solid morphology, as well as solvent properties, play a key role in selecting equipment and in equipment design problems. Usually, the designing of new extraction plants is based on practical experience (the required equipment size is commonly overestimated), and the basis for selecting batch versus continuous systems is not self-evident. This is due to the lack of a clear and a well established engineering approach, which can be used in designing solid-liquid extraction processes.

For the processes considered in this work, some considerations are in order.

For solid-fluid extraction,[34] immobilized enzyme reactors[37] and chromatographic separations,[38,39] only mass balances are needed. But the appearance of moving fronts due to sharp changes in the boundary conditions at the feed, or the saturation of active sites in the stationary phase, requires that axial dispersion of mass be considered. However, radial dispersion may be comparatively less important.

The case for the dynamics of nonisothermal reactors with catalyst deactivation, the generation of activity profiles coupled to temperature dynamics, makes the inclusion of the axial dispersion terms essential.[40,41] Strong exothermic reactions with large radial temperature gradients around the hot-spot region make radial dispersion of heat also important. However, when describing nonisothermal, nonadiabatic reactors under steady-state conditions, radial dispersion may be the only term to be retained.[40]

The above leads us to consider four cases.

Neglecting radial dispersion gives a system of linear parabolic partial differential equations (PDE) of the type

$$\frac{\partial \vartheta_n}{\partial t} + \frac{\partial \vartheta_n}{\partial z} - \frac{1}{Pe_A}\left(\frac{\partial^2 \vartheta_n}{\partial z^2}\right) = \delta^i(\vartheta_{ns} - \vartheta_n) \tag{51.30}$$

for which initial and boundary conditions given by Equations (51.21 through 51.25) apply.

Neglecting axial dispersion leads to a system of linear parabolic PDE of the type

$$\frac{\partial \vartheta_n}{\partial t} + \frac{\partial \vartheta_n}{\partial z} - \frac{1}{Pe_R}\left(\frac{\partial^2 \vartheta_n}{\partial r^2} + \frac{1}{r}\frac{\partial \vartheta_n}{\partial r}\right) = \delta^i(\vartheta_{ns} - \vartheta_n) \tag{51.31}$$

where initial and boundary conditions given by Equations (51.21) and (51.26 through 51.29) apply.

Neglecting both axial and radial dispersions leads to a system of hyperbolic PDE of the type

$$\frac{\partial \vartheta_n}{\partial t} + \frac{\partial \vartheta_n}{\partial z} = \delta^i(\vartheta_{ns} - \vartheta_n) \tag{51.32}$$

where initial and boundary conditions given by Equations (51.21) and (51.29) apply.

Neglecting the accumulation terms of mass species or temperature, in either phase, comparison of the relative magnitudes of heat and mass accumulation terms in the coupled dynamics of fluid and solid-phases allows the application of the quasi-steady-state assumption.[42]

51.2.4 INTRAPARTICLE CONSERVATION EQUATIONS FOR HEAT AND MASS TRANSPORT

From the point of view of a pseudo-heterogeneous description, the conservation of adsorbed species on heterogeneous surfaces could be accounted for by linking them to those for the fluid phase within the porous particle, represented in terms of a single cylindrical pore, assuming that fluid and solid temperatures are equal. In this way, the conservation equations for the fluid and adsorbed species become, respectively,

$$\frac{\partial \vartheta_{ns}}{\partial t} + \nabla \mathbf{N}_{ns} = 0 \tag{51.33}$$

$$\frac{\partial \theta_{ns}}{\partial t} = \gamma_{ns} \nabla^2 \theta_{ns} + \Re_{ns}(\vartheta_{ns}, \theta_{ns}) \tag{51.34}$$

$$Le \frac{\partial \vartheta_{(N+1)s}}{\partial t} = \nabla \mathbf{q}_s + \sum_{j=1}^{R} \upsilon_j \beta_j \Re_{ns}(\vartheta_{ns}, \theta_{ns}) \tag{51.35}$$

Equation (51.33) gives the mass balance of species in the fluid within the porous particle. Conservation of adsorbed species is given by Equation (51.34) considering surface diffusion in terms of Fick's law. Alternatively, a description using Maxwell–Stefan formulation[43] could also be used. Finally, Equation (51.35) gives the heat balance for the solid particle considering heat conduction and heat generation associated with adsorption/reaction mechanisms.

The generation term $\Re_{ns}(\vartheta_{ns}, \theta_{ns})$ may take one of the following forms:

- Langmuir isotherms for multicomponent mixtures
- Langmuir–Hinshelwood kinetics
- Michaelis–Menten enzyme kinetics
- Lump schemes like those used in catalytic cracking reactions

For diffusion of binary or dilute mixtures or for that occurring in the Knudsen regime, Equation (51.33) may be expressed in terms of Fick's law.[44]

$$N_{ns} = -D_{neff} \nabla \vartheta_{ns} \tag{51.36}$$

For multicomponent mixtures, Maxwell–Stefan description should be preferred. Then, mass conservation equations become[16]

$$-C_T \nabla \vartheta_{ns} = \frac{N_{ns}}{D_{kn}} + \sum_{j \neq n}^{NC} \frac{\vartheta_{js} N_{ns} - \vartheta_{ns} N_{js}}{D_{nj}} + \vartheta_{ns} \left(\frac{1}{RT} + \frac{B_o C_T}{\mu D_{kn}} \right) \nabla p \tag{51.37}$$

where R = the general gas constant
T = the absolute temperature
C_T = heat capacity at T

The last term accounts for changes in total pressure due only to reaction.

In the case of the intraparticle heat conservation equation, this could be of the Fourier-type.[44]

$$q_s = -k_{eff} \nabla \vartheta_{(N+1)s} \tag{51.38}$$

The formalism of diffusion coupled to heterogeneous reaction, based on volume-averaged methods, has been discussed in detail by Withaker.[45]

51.2.5 BOUNDARY CONDITIONS FOR THE PSEUDO-HETEROGENEOUS SOLID PHASE

Similar to the interparticle fluid phase, at $t = 0$, initial conditions for all mass species and temperature within the solid phase are given by its initial steady-state.

$$\vartheta_{ns}(0,\varsigma,\xi) = \vartheta_{ns0}(\varsigma,\xi) \tag{51.39}$$

$$\theta_{ns}(0,\varsigma,\xi) = \theta_{ns0}(\varsigma,\xi) \tag{51.40}$$

At $\xi = 0$, i.e., the bottom of the cylindrical pore being an impermeable inner wall; therefore, boundary conditions for all mass species (neglecting the contribution of surface reaction) and temperature are given by

$$\left.\frac{\partial \vartheta_{ns}}{\partial \xi}\right| = 0 \tag{51.41}$$

At the fluid-solid interface, boundary conditions for all mass species and temperature specify the interfacial transport,

$$-\left.\frac{\partial \vartheta_{ns}}{\partial \xi}\right| = Bi_n(\vartheta_{ns} - \vartheta_n) \tag{51.42}$$

For all mass species and temperature, symmetry is invoked at radius $\varsigma = 0$,

$$\left.\frac{\partial \vartheta_{ns}}{\partial \varsigma}\right| = 0 \tag{51.43}$$

At the porous cylindrical wall, boundary conditions for all mass species and temperature describe the surface reaction coupled with radial diffusion and the corresponding heat generation.[46]

$$-\left.\frac{\partial \vartheta_{ns}}{\partial \varsigma}\right| = \Re_{ns}(\vartheta_{ns},\theta_{ns}) \tag{51.44}$$

It can be considered that reaction or adsorption could be expressed in a pseudo-homogeneous form, assuming they occur at fluid conditions. Therefore, heat and mass conservation equations at the solid phase interacting with the external fluid phase are

$$\frac{\partial \vartheta_{ns}}{\partial t} = \nabla \mathbf{N}_{ns} + \Re_n(\vartheta_{ns}) \tag{51.45}$$

$$\text{Le}\frac{\partial \vartheta_{(N+1)s}}{\partial t} = \nabla \mathbf{q}_s + \sum_{j=1}^{R} \upsilon_j \beta_j \mathfrak{R}_j(\vartheta_{ns}) \tag{51.46}$$

where mass diffusion and heat conduction take place and can be expressed as above. The generation terms in the RHS of Equations (51.45) and (51.46) account for (the linear or nonlinear) heat and mass effects due to adsorption and/or reaction processes. They are only a function of the fluid quantities within a food solid or catalytic pores. In this case, it can be assumed that radial variations of heat and mass inside the pore are less important than axial ones. Thus, the boundary conditions are reduced to Equations (51.41) and (51.42).

In many cases, isothermal particles can be considered for exothermic reactions occurring in porous catalysts. On the other hand, in all adsorption and immobilized enzyme processes, this assumption easily applies.

51.3 ON THE COMMON STRUCTURE OF GOVERNING EQUATIONS

In the approach presented in this chapter, the dynamics of fixed bed operations are given in terms of the governing equations, which describe interparticle heat and mass transport (a linear set of PDE). They are coupled to the respective equations representing intraparticle heat and mass transfer (a nonlinear set of PDE). The resulting class of models (normally hyperbolic or parabolic PDEs) describing different phenomena has a similar structure that could be used to analyze, *a priori*, its general features and select a successful numerical scheme for its solution. The presence of moving fronts in fluid and solid phases requires the use of an efficient numerical method, which should successfully deal with any degree of steepness presented by the mentioned moving front. This could be due, for instance, to catalyst deactivation, to combustion moving front, to saturation of active sites or to changes at feed conditions, etc. The magnitude of steepness of the front is determined by the relative contribution of convection and dispersion terms as well as by the (slower) response at the solid phase. A measure of this magnitude and the velocity at which it travels would suggest the scheme to be employed. Finlayson[47] describes in detail several numerical techniques (finite differences, finite element, collocation, Galerkin methods) to be used when convection terms dominate. Orthogonal collocation based methods have also been successfully used in the solution of this class of models.[48-50] In the case of hyperbolic PDEs, the natural choice for a numerical solution is the characteristics method,[51] which reduces the original PDE to a system of ordinary differential equations (ODE), which in turn could be solved by orthogonal collocation. The use of assumption 4 (p. 859), when possible, avoids dealing with stiff ODE systems.[42] In those cases, after making discrete the spatial variables, the model is reduced to a system of differential-algebraic equations (DAE), where a semi-implicit Runge–Kutta scheme, coupled with Broyden's method, works well.[40]

51.3.1 A NONEQUILIBRIUM MODEL FOR SIMULTANEOUS REACTION-SEPARATION PROCESSES

There are several food or biological materials separation processes that are good candidates for treatment with the approaches presented. Now is the turn of solid-liquid-vapor or gas systems. Consider the case of food dryers in which hot air passes throughout a bed of particulate solids to remove part of their moisture content thanks to the heat transferred from the air to them. Water and aromas are removed, leaving the solids suffering a series of complex physical and chemical changes, such as color, flavor, and textural reactions. Consider also the case of cooked or partially cooked food products where, for instance, the raw potato chips primarily receive heat from the hot vegetable oil, allowing the moisture to be removed in the form of steam bubbles. After part of the water has left, the oil partially penetrates the chips' structure but rapidly changes the external surface texture, making it impermeable to oil. It is this combination of heat and multicomponent mass transfer, producing complex reactions and phase changes characteristic of cooking and processing of food solids or liquids, that makes the use of this mathematical modeling suitable.

The above-mentioned processes can be represented, taking into account the arts and crafts of each case,[19] by a nonequilibrium model proper for a reactive distillation operation. A schematic illustration of this is provided in Figure 51.2. Vapor from

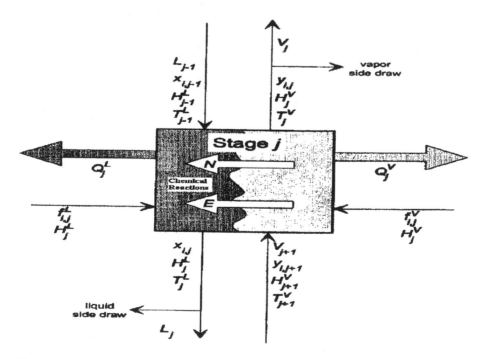

Figure 51.2 Schematic nonequilibrium stage. This stage represents a reactive tray in a trayed reactive column or a packing section in a reactive packed column.

the stage below is brought into contact with liquid from the stage above and allowed to exchange mass and energy across their common interface, represented in the diagram by the wavy line. At the same time, chemical reactions are occurring at the surface of the food solid or catalytic particle. Provision is made for vapor and liquid feed streams.

The nonequilibrium stage in Figure 51.2 may represent either a single tray or a section of solids in a packed column. The same equations are used to model both types of equipment. The only difference is that different expressions must be used for estimating the binary mass-transfer coefficients and interfacial areas. In each stage, the solids, food, or catalyst particles are suspended in a fluidized state. All the gas and liquid streams from adjacent stages are brought into contact counter-currently.

One or more relatively less volatile reactants are transferred from the gas mixture into the liquid phase. The most volatile products are stripped away by the gas stream from the liquid phases and are carried out at the top of the column.

In the equations modeling this nonequilibrium stage, the vapor and liquid phases flow rates leaving the jth stage are denoted by V_j and L_j, respectively. The mole fractions in these streams are y_{ij} and x_{ij}. The N_{ij} are the rates of mass transfer of species i on stage j.

The vapor and liquid phase temperatures are not assumed to be equal, and heat and mass transfer must be allowed to occur across the interface. E_j represents the rate of energy transfer across the phase boundary.

Equation (51.12) can be used to represent the conservation of mass and heat transfer occurring at stage j, provided its respective terms are substituted by the corresponding expressions in a discrete or staged form. This means that, if it is considered a steady-state operation, the convective term, $\nabla \cdot (\rho_k \phi^k \upsilon_k)$, can be replaced by the differences between the flow rates of species i in phase k entering and those leaving stage j [i.e., $(1 + S_j^V)V_j y_{ij} - V_{j+1} y_{ij} - f_{ij}^V$] for mass. For the energy case, the product differences of the total flow rates and its associated enthalpies are considered. The nonconvective flux, $\nabla \aleph_k$, through an external bounding surface, σ, is replaced by N_{ij}, where $N_{ij} = \int \aleph \, da$. With the above substitutions, Equation (51.12) can be used to represent the mass and energy balance around stage j.

51.3.2 MATERIAL BALANCE FOR COMPONENT i ON STAGE j

If a steady-state operation is assumed, the component material balance equations for the vapor phase may be written as follows:

$$M_{ij}^V = (1 + S_j^V)V_j y_{ij} - V_{j+1} y_{ij} - f_{ij}^V + N_{ij}^V = 0 \qquad i = 1,2,...,C \qquad (51.47)$$

where f_{ij}^V = the feed flow rate of component i to stage j in the vapor phase

M = the kg/h of component i accumulated on stage j

The component material balance for the liquid phases is

$$M_{ij}^L = (1 + S_j^L)L_j x_{ij} - L_{j-1} x_{ij} - f_{ij}^L + N_{ij}^L = 0 \quad i = 1,2,...C \quad (51.48)$$

where f_{ij}^L = the feed flow rate of component i to stage j in the liquid phase

L = kg/h of liquid phase

The last term in the left-hand side of Equations (51.47) and (51.48) represents the net loss or gain of component i on stage j due to interface transport.

51.3.3 ENERGY BALANCE FOR STAGE j

The energy balance for the vapor phase and for the liquid phase can be written as

$$E_j^V = (1 + S_j^V)H_j^V V_j - H_{j+1}V_{j+1} + q_j^V - H_j^{VF}\sum_{i=1}^{C} f_{ij}^V + E_j^V = 0 \quad (51.49)$$

$$E_j^L = (1 + S_j^L)H_j^L L_j - H_{j-1}^L L_{j-1} + q_j^L - H_j^{LF}\sum_{i=1}^{C} f_{ij}^L - E_j^L + E_j^S = 0 \quad (51.50)$$

where E = energy in kJ/h

H = enthalpy

It should be noted that the heat released by the chemical reaction is implicitly considered in the computation of the enthalpy for the liquid phase, taking as reference enthalpy the heat of formation of each component.

The rates of mass transfer from the gas bulk to the gas-liquid interface and from it to the bulk of the liquid can be written, respectively, as

$$R_{ij}^V = N_{ij}^V - \sum_{k=1}^{C-1} k_{ikj}^V a_j(y_{kj}^I - y_{kj}) - y_{ij}\sum_{k=1}^{C-1} N_{kj}^V = 0 \quad i = 1,2,...C \quad (51.51)$$

$$R_{ij}^L = N_{ij}^L - \sum_{k=1}^{C-1} k_{ikj}^L a_j(x_{kj}^I - x_{kj}) - x_{ij}\sum_{k=1}^{C-1} N_{kj}^L = 0 \quad i = 1,2,...C \quad (51.52)$$

The rates of mass transfer from the liquid bulk to the liquid-solid interface are

$$R_{ij}^S = r_{ij} - N_{ij}^S = r_{ij} - \sum_{k=1}^{C-1} k_{ikj}^S a_j^S(x_{kj} - x_{kj}^S) - x_{ij}\sum_{k=1}^{C} r_{kj} = 0 \quad i = 1,2,...C \quad (51.53)$$

The energy balance around the gas-liquid interface and at the liquid-solid interface is

$$E_j^L = E_j^V - E_j^L = h_j^V a_j (T_j^V - T_j^I) + \sum_{i=1}^{C} N_{ij}^V H_{ij}^V - h_j^L a_j (T_j^I - T_j^L) - \sum_{i=1}^{C} N_{ij}^L H_{ij}^L = 0$$

$$(51.54)$$

$$E_j^S = q_j^r - E_j^S = h_j^s a_j^s (T_j^L - T_j^s) - \sum_{i=1}^{C} r_{ij} H_{ij}^L = 0 \qquad (51.55)$$

The interface model is given by the phase equilibrium equations at gas-liquid interface.

$$Q_{ij}^I = K_{ij} x_{ij}^I - y_{ij}^I = 0 \qquad i = 1, 2, \ldots C \qquad (51.56)$$

When analyzing the set of Equations (51.47) through (51.56), it can be observed that the number of iteration variables and independent equations are $6C + 1$.

The governing equations of the model can be written in terms of equations for the computation of multicomponent mass-transfer coefficients, reaction rates, equilibrium ratios, enthalpies, and other physical properties.

Taylor and Krishna[18] provided a review for many different methods to calculate the multicomponent mass-transfer coefficients. For reactive distillation, it is recommended to calculate these coefficients through a Maxwell–Stefan equations solution for a mass-transfer film model. The binary volumetric mass transfer coefficients for the gas phase are usually calculated with the correlation proposed by Chan and Fair[52] for sieve tray columns. The binary liquid-solid mass transfer coefficients are calculated with the correlation of Nikov and Karamanev.[53]

The film model of simultaneous heat and mass transfer in a multicomponent system[18] is applied to calculate the finite flux heat transfer coefficient in the vapor phase, h^V, in which the zero flux heat transfer coefficients may be obtained from the Chilton–Colburn analogy. The heat transfer coefficients for the liquid phase can be obtained in the same way. The equilibrium ratios, K_{ij}, may be estimated using the modified UNIFAC method.

Even for the steady-state model presented here, it is difficult to obtain a solution with Newton methods. This is due to the high nonlinearity of the equations of nonequilibrium stage models.

The Newton–Homotopy continuation method is well known for being very robust numerical algorithms for solving difficult nonlinear problems and is preferred to solve sets of equations characteristic of a gas-liquid-solid separation.

When the residual terms are not zero, one must solve a set of differential-algebraic equations (DAE) which require, in general, much more effort than the set of nonlinear algebraic equations or single differential equations. A useful review of the methods for solving DAEs is presented by Brenan and Petzold.[54] As far as the authors are aware, there is only one paper in the literature in which a dynamic reactive distillation column considering nonequilibrium stage has been solved.[55] In this work, the method of lines is used for making discrete the spatial dimension of the partial differential equations of the model, and a set of DAEs is obtained. The ABACUSS software is used to integrate the differential-algebraic equations.

51.4 CONCLUSIONS

The rigorous modeling and simulation of separation, reaction, and reaction-separation processes have been discussed. The models developed consider the heterogeneous and multicomponent nature of food systems. By using Equation (51.12), it is noted that the reactive separation processes usually have a similar mathematical structure, even when the phenomena involved may be different. This common feature of different models allows making, when appropriate, simplifying assumptions, which leads to simpler models. Their appropriate numerical solution is then based on the analysis of the general features of its behavior. In the case of spatial continuum systems (packed beds), the conservation equations and the respective boundary conditions generate a set of partial differential equations that can be solved by several methods, notably orthogonal collocation (global or on finite elements). For hyperbolic systems, the characteristics method, along with orthogonal collocation, offers a good possibility. In the case of discrete or staged systems (reactive distillation columns), the conservation equations generate a set of nonlinear algebraic equations for the steady-state case, and a set of differential-algebraic equations for the dynamic case. In the former, due to no linearity of algebraic equations, a homotopy-continuation method is required. For the resulting differential-algebraic equations, the implicit one-step methods (e.g., Implicit Runge-Kutta), coupled with quasi-Newton (Broyden) methods, are strongly recommended.

REFERENCES

1. Spaninks, J. 1979. *Design Procedures for Solid-Liquid Extractors,* doctoral thesis, Agricultural University of the Netherlands, Wageningen.
2. Hougen, O. A. and W. R. Marshall. 1947. "Adsorption from a Fluid Stream Flowing Through a Stationary Granular Bed," *CEP,* 43: 197.
3. Bruniche-Olsen, H. 1962. *Solid Liquid Extraction,* Copenhagen: NYT Nordisk Forlag, Arnold Busck.
4. Thomas, H. C. 1951. "Solid Diffusion in Chromatography," *J. Chem. Phys.,* 19: 1213.
5. Rosen, J. B. 1952. "Kinetics of a Fixed Bed System for Diffusion into Spherical Particles," *J. Chem. Phys.,* 20: 387.
6. Rosen, J. B. 1954. "General Numerical Solution for Solid Diffusion in Fixed Beds," *Ind. Eng. Chem.,* 46: 1590.
7. Rasmuson, A. and I. Neretnieks. 1980. "Exact Solution of a Model for Diffusion in Particles and Longitudinal Dispersion in Packed Beds," *AIChE J.,* 26: 686.
8. Deisler, P. F. Jr. and R. H. Wilhelm. 1953. "Diffusion in Beds of Porous Solids; Measurements by Frequency Response Techniques," *Ind. Eng. Chem.* 45: 1919.
9. Babcock, R. E., D. W. Green, and R. H. Perry. 1966. "Longitudinal Dispersion Mechanisms in Packed Beds," *AIChE J.,* 12: 922.

10. Flores, S. I. 1985. *Analysis and Simulation of Countercurrent Crossflow Belt Extractors*, Ph.D. Dissertation, University of Massachusetts, Amherst.

11. Cowan, G. H., I. S. Gosling, J. F. Laws, and W. P. Sweetenham. 1986. "Physical and Mathematical Modelling to Aid Scale-Up of Liquid Chromatography," *Journal of Chromatography,* 56(37): 363.

12. Chase, H. A. 1984. "Prediction of the Performance of Preparative Affinity Chromatography," *Journal of Chromatography,* 297: 179.

13. Arve, B. H. and A. I. Liapis. 1988. "Biospecific Adsorption in Fixed and Periodic Countercurrent Beds," *Biotechnol and Bioengineering,* 32: 616–627.

14. Segado, R. R. and H.G. Schwartzberg. 1989. "Anomalous Behavior during Oil Extraction from Flaked Soybeans," Paper 43-9, presented at the Fifth International Congress on Engineering and Food, Cologne, West Germany, May 28–June 3.

15. Wilhelm, R. H. 1962. "Progress Towards the *a Priori* Design of Chemical Reactors," *Pure Appl. Chem.,* 5: 403.

16. Jackson, R. 1977. *Transport in Porous Catalysts*, Amsterdam: Elsevier.

17. Mason E. A. and A. P. Malinauskas. 1983. *Gas Transport in Porous Media: The Dusty-Gas Model*, Amsterdam: Elsevier.

18. Taylor, R. and R. Krishna. 1993. *Multicomponent Mass Transfer*, Wiley Interscience Series.

19. Aris R. 1999. *Mathematical Modeling—A Chemical Engineer's Perspective*, PSE vol. 1, London: Academic Press.

20. Schwartzberg, H. G. 1995. "Scale-Up of Solid-Liquid Extraction," Paper D.17 presented at Conference on Food Engineering, Nov. 2, Chicago.

21. Nield D. A. and A. Bejan. 1992. *Convection in Porous Media*. New York: Springer-Verlag.

22. Slattery, J. C. 1980. "Interfacial Transport Phenomena," *Chem. Eng. Commun.,* 4: 149.

23. Foumeny E. A and J. Ma. 1994. "Non-Darcian Non-Isothermal Compressible Flow and Heat Transfer in Cylindrical Packed Beds," *Chem. Eng. Technol.,* 17: 50.

24. Taylor, G. 1953. "Solid Diffusion in Chromatography," *J. Chem. Phys.,* 19: 1213.

25. Perrine, R. L. 1961. *Soc. Pet. Eng. J.,* March 9.

26. Hill, M. A. 1952. "Channeling in Packed Columns," *Chem. Eng. Sci.,* 1: 247.

27. Perkins, K., O. C. Johnston, and R. N. Hoffman. 1965. *Soc. Pet. Eng. J.,* Dec., 301.

28. Carbonell, R. G. 1980. "Flow Non-Uniformities in Packed Beds: Effect on Dispersion," *Chem. Eng. Sc.,* 35: 1347.

29. Schwartzberg, H. G. 1983. "Characterisation of Displacement Instability," personal communication.

30. Wehner and R. H. Wilhelm. 1956. "Boundary Conditions of Flow Reactors," *Chem. Eng. Sci.,* 6: 89–93.

31. Bischoff, K. B. 1961. "A Note on Boundary Conditions for Flow Reactors," *Chem. Eng. Sci.,* 16: 181–183.

32. Dankcwerts, P. V. 1953. "Continuous Flow Systems," *Chem. Engn. Sci.,* 2: 1–13.

33. Flores De Hoyos, S. I. 1986. "Problemas en el Diseño de Extractores Sólido-Líquido de Banda Transportadora," *Revista de la Academia Nacional de Ingenieria*, México, 5: 38.

34. Flores De Hoyos, S. I. and H. G. Schwartzberg, 1986; "Modelling of Countercurrent, Crossflow, Solid-Liquid Extractors and Experimental Verification," in *Food Engineering and Process Applications*, vol. 1, M. Le Maguer and P. Jelen, eds. London and New York: Elsevier Applied Science Publishers, 412–422.

35. Carberry, J. J. 1976. *Chemical and Catalytic Reaction Engineering*, McGraw-Hill, New York.

36. Mears, D. E. 1971. "Tests for Transport Limitations in Experimental Catalytic Reactors," *Ind. Eng. Chem. Process Develop.*, 10: 541.

37. Horstmann, B. J. and H. A. Chase. 1989. "Modelling the Affinity Adsorption of Immunoglobulin G to Protein A Immobilised to Agarose Matrices," *Chem. Eng. Res. Des*, 67: 243.

38. Luo, R. G. and J. T. Hsu. 1997. "Rate Parameters and Gradient Correlations for Gradient-Elution Chromatography," *AIChE J.*, 43: 464.

39. Joshi V. P., Karode S. K., Kulkarni M. G., and Mashekelkar. 1998. "Novel Separation Strategies Based on Molecularly Imprinted Adsorbents," *Chem. Eng. Sci.*, 53: 2271.

40. López-Isunza, F. 1992. "The role of Reversible Deactivation in Catalyst Activity in the Observed Multiple Steady States during Partial Oxidation Dynamics," *Chem. Eng. Sci.*, 47: 2817.

41. Puszynski, J. and V. Hlavacek. 1980. "Experimental Study of Traveling Waves in Nonadiabatic Fixed Bed Reactors for the Oxidation of Carbon Monoxide," *Chem. Eng. Sci.*, 35: 1769.

42. Ferguson, N. B and B. A. Finlayson. 1974. "Transient Modelling of a Catalytic Converter to Reduce Nitric Oxide in Automobile Exhaust," *AIChEJ.*, 20: 539.

43. Krishna, R. 1990. "Multicomponent Surface Diffusion of Adsorbed Species: a Description Based on the Generalised Maxwell-Stefan Equations," *Chem. Eng. Sci.*, 45: 1779.

44. Satterfield, C. N. 1970. *Mass Transfer in Heterogeneous Catalysis*, MIT Press.

45. Withaker, S. 1986. "Transport Processes with Heterogeneous Reactions," in *Concepts and Design Considerations of Chemical Reactors*, A. E. Cassano and S. Withaker, eds. New York: Gordon and Breach.

46. López-Isunza F. 1999. "Deactivation of FCC Catalysts—A Modelling Approach," Preprints, Symposium on The Advances in Fluid Catalytic Cracking, *Div. Pet. Chem.*, ACS, New Orleans, Aug. 22–26, 540.

47. Finlayson, B. A. 1991 *Numerical Methods for Problems with Moving Fronts*, New York: Hemisphere.

48. Yu, Q. and N.-H. L. Wang. 1989. "Computer Simulations of the Dynamics of Multicomponent Ion Exchange and Adsorption in Fixed Beds-Gradient-Directed Moving Finite Element Method," *Computers. Chem. Eng.*, 13: 915.

49. Michelsen, M. and J. Villadsen. 1978. *Solution to Differential Equation Models by Polynomial Approximation*, New Jersey: Prentice Hall.

50. Finlayson, B. A. 1980. *Nonlinear Analysis in Chemical Engineering*, New York: McGraw-Hill.

51. Aris, R. and N. R. Amundson. 1973. *Mathematical Methods in Chemical Engineering*, vol. 2, New Jersey: Prentice Hall.

52. Chan, H. and J. R. Fair. 1984 "Prediction of Point Efficiencies On Sieve Trays: I Binary Systems," *Ind. Eng. Chem. Process Des. Dev.*, 23: 187.

53. Nikov, I. and D. Karamanev. 1991. "Liquid-Solid Mass Transfer in Inverse Fluidized Bed," *AIChE Journal*, 37: 781.

54. Brenan, K. E., S.L. Campbell, and L. R. Petzold. 1989. *Numerical Solution of Initial-Value Problems in Differential-Algebraic Equations*, The Netherlands: North-Holland Elsevier.

55. Kreul, L. U., P. I. Barton and A. Gorak. 1998. "A Complex Rate-Based Model for the Simulation of Reactive Distillation Columns," paper presented at the *AIChE* Annual Meeting, Los Angeles CA.

52 Modeling Bean Heating during Batch Roasting of Coffee Beans

H. G. Schwartzberg

CONTENTS

Abstract

During batch roasting of coffee beans, parts of the batch are randomly exposed to a stream of hot gas and to metal surfaces heated by that gas. The heated beans then mix with and transfer heat to the rest of the batch, raising the average temperature of the entire charge. The exposure-heating-mixing cycle repeats many times per minute. When a selected measured bean temperature is reached, hot gas flow is turned off, and the beans are quickly cooled. Flavor-generating reactions take place as the beans roast. Reaction products provide the characteristic taste of roasted coffee. Extents of roasting reactions depend on both reaction rate and reaction duration; reaction rates depend on temperature and reactant concentrations. Some reactants used in reactions that occur at late stages of roasting are produced at earlier stages. Therefore, reproducible provision of desired product flavor requires control of bean temperature-time history during almost the entire roast. Most roasters do not yet use systems designed to achieve such control. Instead, they rely on monitoring bean temperature and provide a constant fuel flow rate or hot gas temperature and hot gas flow rate, or provide preselected step changes in these operating variables at a limited number of selected, measured bean temperatures. Methods for modeling bean heating and temperature vs. time histories during batch roasting of green coffee are presented and implemented by machine computation for a selected roaster. Good agreement is obtained between predicted and actual bean temperature vs. time behavior. Certain variable factors, such as average bean diameter, bean moisture content, gas flow rate, and bean load weight, markedly alter bean temperature vs. time history when normal control methods are used. Such changes can be suppressed by specifically controlling bean temperature vs. time history.

52.1 INTRODUCTION

Green coffee does not have an attractive taste or aroma. Desired coffee flavors, aromas, and color are produced by roasting.[1] These develop because of a multitude of complex thermally induced chemical reactions. At late stages of coffee roasting, strongly exothermic reactions occur and cause bean temperature to rise rapidly. These reactions generate CO_2. Part of the CO_2 escapes, and some is retained in cells in the beans, raising in-bean pressure, which reaches roughly 12 atm near the end of a roast. The roasting beans expand 50 to 120% in volume because of the internal pressure buildup. Some evidence tends to show that normal coffee flavor will not develop without internal CO_2 pressure buildup.

52.1.1 ROASTING REACTIONS

Roasting reactions determine how coffee flavor develops and the nature of that flavor. Some green coffee components—cellulose, lignin, caffeine, and inorganic compounds—do not react to a significant extent. Others—sucrose, reducing sugars, proteins, chlorogenic acid, trigonelline, and certain amino acids—react extensively and partly or almost wholly disappear.[2] These disappearances lead to wide ranges of reactions crucial to coffee flavor development.

More than 1000 aroma compounds have been found in roasted coffee. Large numbers of nonvolatile compounds are also produced. Since the number of product compounds is so great, many different reactions must be involved. Products from some reactions partly break down near the end of roasting or would break down if roasting continued. Other products are only produced to a significant extent if longer than normal roasting is used.

Some coffee components (e.g., trigonelline) break down over a wide range of temperatures and almost directly generate a substantial number of products.[2] Other components, such as sugars and amino acids, participate in networks of reactions, e.g., Maillard reactions,[2] involving series and parallel steps where reactions compete strongly with one another. Late-stage reactions in these networks can take place only if suitable intermediates are produced earlier. If different intermediates are produced or are produced in different proportions, different final products will be produced, and roasted coffee flavor will change.

Like all reactions, the extent of roasting reactions depends on both reaction rate and reaction duration. Reaction rates depend on temperature and reactant concentrations. Concentrations of some reactants used in late-stage reactions depend on early stage reactions. Therefore, reproducible provision of desired coffee flavor requires control of bean temperature-time history during almost the entire roast.

52.1.2 COFFEE ROASTING, RECENT HISTORY

Over the past 50 years, roaster manufacturers and users have developed ways to roast coffee more efficiently, rapidly, uniformly, safely, and controllably, and ways to greatly reduce roasting-caused air pollution.[3–5] Continuous roasters are now frequently used by large-volume roasters. Reflectance-color meters are now used almost universally to characterize roast completeness and character. Roasting techniques and pretreatments that upgrade the quality of robustas and other low-grade beans have been developed, and much information has been obtained about the chemistry of roasting reactions.

At the beginning of the twentieth century, heat was mainly transferred to beans during roasting by conduction,[3,4] with minor amounts being transferred by convection and radiation. As the century progressed, use of convective heat transfer progressively increased, and conduction's role in transferring heat to beans progressively declined.

Very rapid roasting is increasingly used to provide roasted beans that swell more and provide more soluble solids than slowly roasted beans.[3] Even when the end-of-roast temperatures and roast colors for both types of beans are the same, rapidly roasted, low-density coffees contain more chlorogenic acid, more free acid, more moisture, and greater residual amounts of CO_2.[3] When roasting speed changes, the chemical composition and bitterness of roasted coffee change,[6] as well as its acidity[7] and perceived cup strength.[8] Therefore, outcomes of coffee roasting reactions and roast coffee character depend on bean temperature history rather than on a single end-of-roast temperature or roast color.

52.1.3 MODERN ROASTERS

In modern roasters, hot gas passes through agitated beds or layers of coffee beans or through cascading or suspended streams of beans. Electrically heated gas is sometimes used by small custom roasters, but hot gas is usually produced by burning oil or fuel gas in air. Added air or recycled roaster gases are mixed with burned gas to provide gas temperatures low enough to not adversely affect the beans being roasted. Heat is mostly transferred to the roasting beans from the hot gas by convection. A smaller portion is transferred from the gas to roasting hardware and then from hardware surfaces to beans contacting or viewing those surfaces. A blower draws the gas through the roasting chamber. This keeps the chamber under a slightly negative gage pressure and prevents escape of roaster gases into the working environment. Gas leaving the roaster passes through a cyclone that removes the chaff released by beans as they roast. The gas then either (a) discharges directly into the atmosphere through a stack, (b) passes through an afterburner and then through the stack, or (c) is sent back to the roaster furnace.[3] In the latter case, part of the hot gas leaving the furnace is sent to the roaster and part is sent to the stack, either directly or via a catalytic afterburner.

Rotating-Drum Roasters

Horizontal, rotating drums are the most frequently used type of roasting chamber. Spiral flights in the drum lift the beans, which subsequently cascade downward as their dynamic angle of repose is exceeded. There are outer flights or sets of flights and oppositely pitched inner flights or sets of flights. The flights axially mix the beans. Thus, beans heated by hot entering gas mix with beans heated by cooler exiting gas. The drums rotate at a speed slightly lower than that at which centrifugal force would cause the beans to adhere to the drum wall. Some rotating-drum roasters have walls without perforations, whereas others have perforated walls.

Drum Walls without Perforations

Hot, e.g., 550°C, gas enters a large, central port at the rear of these roasters and flows axially down their length, then exits through a port at the front of the drum (see Figure 52.1). Beans falling from the flights drop into a tumbling pile of beans at the bottom of the drum where mixing occurs. The gas contacts (a) beans dropping from flights, (b) the outer surface of bean piles on flights and the drum bottom, and (c) exposed metal surfaces in the drum. Roasts are usually completed in 8 to 12.5 min; up to 20 min may be used in small-scale roasters.

   In some roasters of this type, the hot gas flows over the outer surface of the drum, heating it before entering the drum. Heat transfers to the beans by radiation from exposed metal surfaces in the drum, and by conduction at bean-covered metal surfaces. To prevent local scorching on beans because of excessive conduction heating, the inner and outer drum walls are separated by 6 mm of air.

Drums with Perforated Walls

Gas also enters the rear of some perforated drums, flows axially down the drum, and then flows radially outward through bean-covered and bean-free portions of the drum wall (Figure 52.2). Centrifugal force and suction used to draw gas through the drum cause beans to adhere to much of the drum wall. Beans drop from the wall near the top of the drum and fall on an axially oriented, upward-pointing, v-shaped sheet-metal spreader.

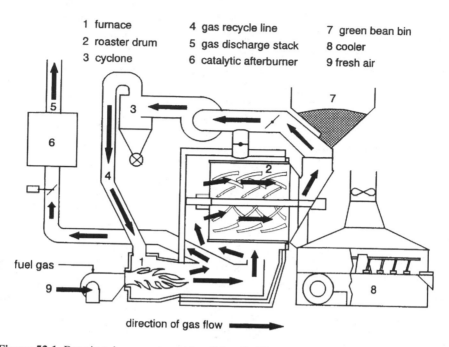

Figure 52.1 Rotating-drum roaster with solid wall. (Courtesy of Probat Werke GmbH.)

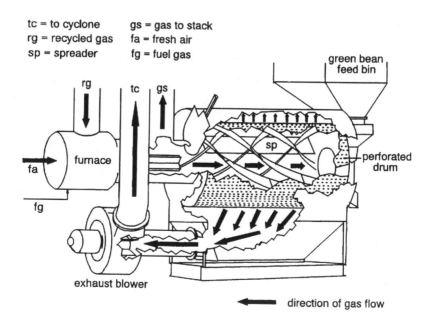

Figure 52.2 Rotating-drum roaster where gas flows outward radially through a perforated wall. (Courtesy of Jabez Burns, Inc.)

The spreader sends some beans backward, counter to the direction of drum rotation, and some forward, in the direction of drum rotation, causing mixing, which would not occur if the beans continued to adhere to the wall. Some forward-diverted beans strike, adhere to, and cover drum wall zones that would otherwise not be covered. Beans also fall into a tumbling pile of beans at the bottom of the drum, where added mixing occurs. Roasting times range between 8.5 and 15 min in these roasters.

In other perforated-wall drum roasters, hot gas enters through perforations along the top surface of the drum, passes downward through the drum and the tumbling pile of beans at the bottom of the drum, and flows out of perforations under the pile. External baffles ensure that hot gas enters only through uncovered portions of the drum wall and exits only through bean-covered portions of the wall. Roasting times range between 10 and 15 min.

Spouted-Bed Roasters

Fluidized-bed coffee roasters are described in patents and were once produced. They provide very large amounts of heat-transfer surface per unit weight of beans, but they are no longer produced and are very rarely used in industry. Coffee beans, because of their size and density, form poorly mixed (class D) fluidized beds. Asymmetric spouted bed roasters (Figure 52.3) provide good mixing, enough bean-air contact surface for very rapid roasting, and are used often in the coffee industry.

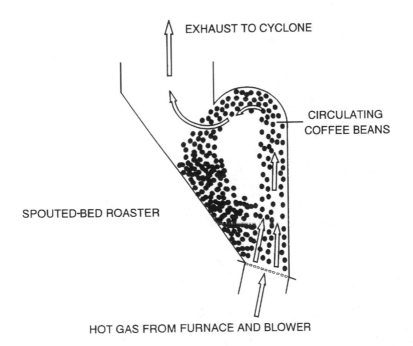

Figure 52.3 Spouted-bed roaster. (Courtesy of Neuhaus Neotec GmbH.)

Usually, 310 to 360°C inlet gas is provided. The roasting time is inversely proportional to the gas flow rate; 3 to 6 min roasting is normally used. Roasting can be completed in 1.5 min by using enough gas flow. Lower inlet gas temperatures of 232 to 276°C, and longer roasting times of 10 to 20 min, are used in some spouted-bed roasters.

Rotating-Bowl Roasters

In rotating-bowl roasters (Figure 52.4), beans are driven across the surface of a rotating heavy cast-iron bowl by centrifugal force and thrown upward at the rim of the bowl. The beans then strike a stationary circular cover, where flights direct the beans inward. Hot gas, drawn into the roaster through a large central circular duct in the top cover, is deflected by the beans moving across the bowl surface, spreads outward, and is drawn up into an annular duct located between the inlet duct and the rim of the cover. The beans move across the gas stream. Hot gas inlet temperatures between 480 and 550°C are used, and roasting times between 3 and 6 min are obtained.

In a modified form of this roaster, higher gas flow rates are used, and the entering gas passes upward through a concave perforated distributor plate covering the inner surface of the bowl. The beans slide outward across the distributor plate. The hot gas flows upward, first contacting the outwardly sliding beans and then contacting inward-moving beans. Roasting is usually completed in 3 min.

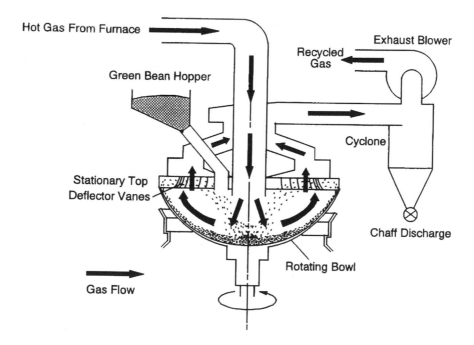

Figure 52.4 Rotating-bowl roaster. (Courtesy of Probat Werke GmbH.)

Scoop-Wheel Roasters

The body of this roaster is a trough with a hemicylindrical bottom and deep vertical side walls (Figure 52.5). Scoops mounted on a wheel rotate through the deep bed of beans in the trough. Hot gas flows tangentially downward into the trough through a long axially-oriented slit or perforated metal plate just below the junction of a side wall and the bottom of the trough. A blower draws the gas upward through the beans and a metal hood above the trough. The rotating scoops axially mix the bed of beans, strongly agitating the bed and throwing the beans upward into the head space. Inlet gas at 420°C is used. Again, though roasting times as short as 1.5 min can be obtained, 3 to 6 min is usually used.

Swirling-Bed Roasters

Swirling-bed roasters are vertical and circular in cross section, with walls that taper slightly outward in the upward direction (Figure 52.6). The walls are covered by an array of small, vertical, louver-like slits. Hot gas, drawn through the slits, spirals around the wall as it flows inward. The spiral gas flow causes a 30- to 50-mm thick layer of beans to spiral upward around the wall. Centrifugal force counteracts flow drag, causing the bean layer to adhere to the wall. After reaching the top of the wall, the beans drop onto a shallow upward-pointing cone at the bottom of the roaster. The beans then spiral outward across the cone and again spiral up the wall. The cycle takes roughly four seconds to complete. Increasing flow increases both centrifugal force and drag force proportionately. Therefore, very high hot gas flow rates can be used without

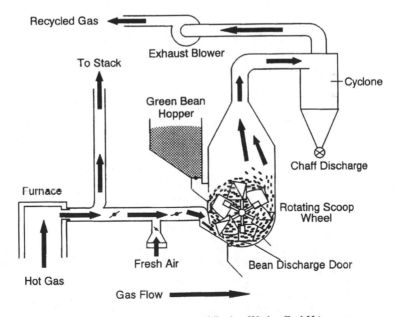

Figure 52.5 Scoop-wheel roaster. (Courtesy of Probat Werke GmbH.)

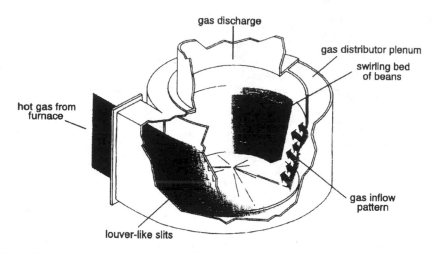

Figure 52.6 Swirling-bed roaster. (Courtesy of Jabez Burns, Inc.)

losing beans by entrainment. Inlet gas at 274 to 280°C is used. Roasting times are inversely proportional to the gas flow rate. Roasts can be completed in 1.5 min; 3-min roasting is usually used.

52.1.4 ROASTER CONTROL

Roaster operators measure and/or control the end-of-roast temperature, roasting time, roasted bean reflectance-color, overall weight loss, and dry matter loss. These control points or measurements serve to roughly indicate the broad overall extent of roasting reactions. However, coffee flavor does not simply depend on broad overall reaction extent. It depends on the relative extents of the many different reactions.

Fuel gas flow rates for small gas-fired roasters are usually manually adjusted at the factory or when the roaster is first installed. The gas is turned on manually or by means of a solenoid-actuated valve, mixes with excess air to ensure complete combustion, and is automatically ignited slightly before roasting starts. The hot gas produced then mixes with more air to reduce its temperature enough for roasting purposes, e.g., to 540°C. The proportion of burned gas and added air in the mix, and the consequent temperature of gas entering the roasting chamber is set by manual adjustment of dampers whose settings are rarely changed.

As bean temperature increases, the temperature difference between the beans and the roaster gas decreases. Since heat transfers to the beans at a rate proportional to this temperature difference, the rate of bean temperature rise progressively decreases. This situation continues until exothermic reactions near the end of roasting cause the bean temperature to rise more rapidly. When the beans reach the selected target end-of-roast temperature, the gas supply is turned off and cold water is immediately sprayed on the beans to evaporatively cool (quench) them enough to stop the roasting. The roasted beans are then transferred to a cooler where they are stirred while cool air, admitted through a perforated plate, is drawn through the beans to provide further cooling.

In some roasters, all sections of the fuel-gas burner remain on during the entire roast. Consequently, the gas inlet temperature for the roaster tends to remain constant. In spouted-bed, rotating-bowl, and swirling-bed roasters, software-based PID control maintains the inlet temperature constant at a selected set value.

In many rotating-drum roasters, the inlet roaster gas temperature drops twice in step-wise fashion during a roast. Either the first drop occurs when the bean temperature reaches a preset value and the second drop when a higher set temperature is reached, or the temperature drops occur at set elapsed times after roasting starts. In small roasters and some medium-size roasters, one section and later a second section of a three-section burner are turned off to reduce the inlet temperature. The second inlet temperature reduction helps decrease the temperature rise produced by heat-generating reactions late in the roast.

In many medium-size and most large roasters, roaster gas leaving the cyclone is sent back to the roaster furnace, where the recycled gas mixes with burned fuel gas and reduces its temperature. Part of the hot gas mixture leaving the furnace enters the roaster and part is sent to a gas-discharge system. The proportions of gas sent to these two destinations are controlled by damper settings. In some solid-wall rotating-drum roasters, step changes in damper settings are used to reduce gas flow through the roaster and cause a step drop in inlet gas temperature at two selected measured bean temperatures. Step reductions in inlet gas temperature or in gas flow rate prolong the end-of-roast period and increase the ease with which the end-of-roast can be controlled.

52.1.5 CONTROL ADJUSTMENTS

The prescribed end-of-roast temperature is adjusted upward if roasts consistently turn out be too light, i.e., have too large a reflectance color value, and downward if the opposite is the case. The inlet gas temperature is adjusted upward (by increasing the fuel gas delivery rate or reducing recycling of roaster gas) if roasting times are consistently too long and adjusted downward if roasting time is consistently too short. Total weight loss and dry matter loss are subsequently measured. These measurements are valuable cost accounting tools and also, somewhat belatedly, serve to alert roaster operators to abnormal roasting behavior. Occurrences of abnormally high or abnormally low total weight losses almost certainly indicate that green bean character has changed in some way or that something abnormal is happening during roasting.

Roast color measurements roughly indicate the extent of reactions that are the final step in a series of reactions. Thus, roast color provides an indicator of overall reaction extent. However, many different reaction paths lead to dark color production. Roast color does not tell us which paths are most favored. Therefore, roast color measurements, at least as currently implemented, do not tell us enough about flavor balance.

52.2 THEORY

Gas temperature, T_g, changes greatly in passing through a roaster. Therefore, changes in T_g in modeling roaster behavior will be discussed first. Usually, only a fraction of the beans in a roaster are exposed to hot gas flow at a given time, and the change in bean temperature, T_b, is small during the passage of a small amount of gas. If one can neglect gas-to-metal surface heat transfer,

$$-GC_g \frac{dT_g}{dZ} = h_e\left(\frac{dA_{gb}}{dZ}\right)[T_g - T_b]$$
(52.1)

where G = the gas mass-flow rate
C_g = its heat capacity
h_e = the effective gas-to-bean heat-transfer coefficient
Z = the length of the gas flow path across the beans
A_{gb} = the area across which gas-to-bean heat transfer occurs

There will be a distribution of T_b and $h_e A_{gb}$ values for the bean population exposed to the hot gas at any time. Therefore, T_b and $h_e A_{gb}$ represent average values for that population.

Integrating and rearranging, one obtains

$$(T_{go} - T_b) = (T_{gi} - T_b)\exp\left[-\frac{h_e A_{gb}}{GC_g}\right]$$
(52.2)

where T_{gi} and T_{go}, respectively, are the gas inlet and outlet temperature. Thus,

$$(T_{gi} - T_{go}) = (T_{gi} - T_b)\left(1 - \exp\left[-\frac{h_e A_{gb}}{GC_g}\right]\right)$$
(52.3)

Neglecting exothermic heat generation and evaporative cooling due to moisture evaporation during the roast, a heat balance yields

$$\frac{dT_b}{dt} = \frac{GC_g[T_{gi} - T_{go}]}{RC_b} = \frac{GC_g}{RC_b}(T_{gi} - T_b)\left(1 - \exp\left[-\frac{h_e A_{gb}}{GC_g}\right]\right)$$
(52.4)

where R = mass of beans in the roaster
C_b = the bean heat capacity
t = time

If T_{gi}, $(GC_g)/(RC_b)$, and $(h_e A_{gb})/(GC_g)$ remain constant, Equation (52.4), when integrated yields

$$\frac{GC_g t_r}{RC_b}\left(1 - \exp\left[-\frac{h_e A_{gb}}{GC_g}\right]\right) = \ln\left[\frac{T_{gi} - T_{bi}}{T_{gi} - T_{bf}}\right] \tag{52.5}$$

where T_{bi} and T_{bf} are, respectively, the initial and final bean temperature, and t_r is the roasting time. If one uses a modified C_b [i.e., 2.41 kJ/(kg·°C)] that roughly accounts for exothermic heat generation and evaporative cooling as well as sensible heat pickup by roasting beans, Equation (52.5) can be used to quickly estimate effects of changes in inlet gas temperature and bean load weight on roasting time in some roasters.

52.2.1 Heat Transfer from Hot Gas

C_b roughly doubles, and R experiences a 12 to 16% decrease during the course of roasting; evaporative cooling and exothermic heat generation occur over localized but different T_b ranges. Metal roaster parts transfer heat to or receive heat from roaster gas and/or beans and pick up an appreciable amount of thermal energy during the course of roasting. Therefore,

1. Equation (52.5), with t used in place of t_r, is not suitable for accurately predicting T_b vs. t behavior.
2. Equation (52.1) has to be modified for some types or roasters.

Although metal temperatures vary markedly from point to point in the roaster, to simplify modeling, we will assume that we can analyze heat transfer in terms of the mean metal temperature, T_m. For cases where heat transfer from gas to beans and to roaster metal occur along parallel paths, using an approach like that used to obtain Equations (52.2) and (52.3), we obtain

$$(T_{gi} - T_{go}) = \left[T_{gi} - \frac{T_b + FT_m}{1 + F}\right]\left(1 - \exp\left[-\frac{h_e A_{gb}(1 + F)}{GC_g}\right]\right) \tag{52.6}$$

where $F = (h_{gm}A_{gm})/(h_e A_{gb})$ and h_{gm} and A_{gm} are, respectively, the effective gas-to-metal heat-transfer coefficient and contact area. In rotating-bowl and scoop-wheel roasters, beans wholly or nearly wholly cover the metal surfaces, so $(h_{gm}A_{gm})$ and F are very small and can be neglected, and Equation (52.1) still applies. In spouted-bed and swirling-bed roasters, $(h_e A_{gb})$ is so large that F is negligibly small. Consequently, $\exp[-(h_e A_{gb})/GC_g] \approx 0$. Therefore, in those roasters, T_{go} and $T_m \approx T_b$.

Q_{gm}, the rate of heat transfer from gas to metal parts, is given by

$$Q_{gm} = \frac{F[h_e A_{gb}(T_b - T_m) + GC_g(T_{gi} - T_{go})]}{1 + F} \tag{52.7}$$

where $(T_{gi} - T_{go})$ is obtained from Equation (52.6). The rate of temperature rise of the beans is given by

$$\frac{dT_b}{dt} = \frac{GC_g[T_{gi} - T_{go}] - Q_{gm} + Q_{mb} + Rd(Q_r + \lambda dX/dt)}{R_d(1 + X)C_b} \tag{52.8}$$

where Q_{mb} = the rate of heat transfer from metal parts to beans

Q_r = the rate of heat production by roasting reactions

R_d = the dry weight of the beans

X = their dry-basis moisture content

dX/dt = the rate of moisture change per unit dry mass of beans (negative for evaporation)

λ = 2790 kJ/kg at typical bean moisture contents = the latent heat of vaporization and desorption of bean moisture

$Q_{mb} = h_{bm}A_{bm}(T_m - T_b)$, where h_{bm} and A_{bm}, are the effective coefficient and area for metal-to-bean heat transfer. Q_{mb} is positive when $T_m > T_b$ and negative when $T_b > T_m \cdot h_{bm}A_{bm} = h_rA_r + h_cA_c$, where h_r and A_r are the heat-transfer coefficient and area of metal surfaces transferring heat radiantly to beans, and h_c and A_c are the heat-transfer coefficient and area of metal surfaces transferring heat conductively and convectively to beans.

52.2.2 Moisture Loss

When X vs. T_b and t records for beans undergoing roasting are examined, one finds two broad $-dX/dt$ peaks: a large peak centered around $T_b \approx 160°C$, and a smaller one centered $T_b \approx 200°C$. The smaller peak occurs during the roasting exotherm and probably represents evaporative loss of water formed in roasting reactions (roughly 0.05 kg water/kg dry coffee). In modeling water evaporation, it was assumed that

1. Reaction-based water formation partly counterbalanced evaporative water loss, causing part of the $-dX/dt$ reduction at $T_b > 160°C$.
2. Evaporative loss of moisture originally contained in the beans continues at a high rate, even though on an overall basis $-dX/dt$ decreases above 160°C.

It is further assumed that cooling caused by evaporation of reaction-formed water can be accounted for in terms of the net heat of reaction. That is, the net heat of reaction equals the heat of reaction measured in a sealed vessel (where evaporation cannot occur) minus the evaporative cooling that occurs when water can escape from the beans.

Corrected X vs. t and T data obtained by subtracting estimated X increases caused by water formation by reaction were used to calculate corrected $-dX/dt$ vs. T_b data for evaporation of the bean's initial water content alone. Best modeling of the corrected data was obtained by assuming that

1. Moisture loss was diffusively regulated.

2. The diffusion coefficient's temperature dependence was governed by an Arrhenius-type equation.
3. The driving rate for mass-transfer was proportional to X.
4. The diffusion coefficient was also proportional to X.
5. Although d_p variation was not tested, $-dX/dt$ was inversely proportional to d_p^2, where d_p is the effective bean diameter.

The following equation (where T_b is in °C and d_p is in mm) agrees well with corrected $-dX/dt$ data if $d_p = 6$ mm:

$$-\frac{dX}{dt} = \frac{4.32 \times 10^9 \cdot X^2}{d_p^2} \exp\left[\frac{-9,889}{T_b + 273.2}\right] \tag{52.9}$$

52.2.3 Exothermic Roasting Reactions

Raemy[9] and Raemy and Lambelet,[10] using differential thermal analysis (DTA), found that, when evaporation of water from coffee beans is suppressed, 250 to 420 kJ of heat are generated between 150 and 230°C by exothermic roasting reactions. Allowing partial evaporation of water contained in the coffee diminishes the amount of energy released and the extent of diminution increases as the amount of evaporation increases. Roughly 0.05 kg water is produced and lost when exothermic reactions occur in roasters.[5] Correcting for evaporative cooling, less than 120 to 290 kJ/kg of heat is produced in roasters. Based on exotherm-induced T_b deviations in a published T_b vs. t curve,[1] estimated in-roaster exothermic heat production was only 107 kJ/kg for a 12-min roast, with a final $T_b = 240$°C.

In thermograms obtained at a 1°C/min heating rate, exothermic heating started at 150°C, peaked at 200°C, and ended at 240°C. The onset, peak and end of heat generation shift to lower temperatures when the heating rate is 0.5°C/min and the peak contains three poorly resolved sub-peaks. Separate heating of spent coffee grounds and coffee solubles indicates that the sub-peak occurring at the highest temperature is due to reactions involving the grounds and that the lower sub-peaks are due to reactions involving water-soluble coffee components.

Exothermic heat generation during roasting was modeled by assuming that

1. The rate of heat generation is proportional to the rate of reactions producing that heat.
2. The rates of these reactions are proportional to reactant concentration and to a coefficient governed by the Arrhenius equation.
3. Reactants are consumed in the reactions.
4. The concentrations of remaining reactants are proportional to $(H_{et} - H_e)/H_{et}$, where H_{et} is the total amount of reaction heat produced per kilogram of dry solids and H_e is the amount of heat produced thus far per kilogram of dry coffee.

Using these assumptions, the following equation was obtained:

$$Q_r = A \exp\left[-\frac{H_a}{R_g(T_b + 273.2)}\right]\left(\frac{H_{et} - H_e}{H_{et}}\right) \qquad (52.10)$$

where Q_r = the rate of exothermic heat production in (kJ/kg dry coffee)/s
$\quad H_a$ = the energy of activation
$\quad R_g$ = the perfect-gas-law constant
$\quad A$ = the Arrhenius equation prefactor times the amount of heat generated per unit amount of reaction, in (kJ/kg dry coffee)/s
$\quad T_b$ = in units of °C

In modeling calculations carried out later, $H_a/R_g = 5{,}500°$K, $A = 116{,}200$ (kJ/kg dry coffee)/s, and $H_{et} = 232$ kJ/kg.

52.2.4 Changes in Metal Temperature

The time rate of change of T_m is given by

$$\frac{dT_m}{dt} = \frac{Q_{gm} + h_{bm}A_{bm}(T_b - T_m) + Q_e}{M_m C_m} \qquad (52.11)$$

where Q_{gm} is given by Equation (52.7), M_m and C_m, respectively, are the metal mass and heat capacity, and Q_e is the net rate of heat transfer to roaster chamber metal from sources outside the roaster chamber. Q_e is negative for heat transferred to the environment or the insulated casing surrounding the roaster chamber and positive for heat transferred to chamber metal by hot gas. Q_e is important when hot gas heats a roaster drum's outer wall before entering the drum. Evaluation of Q_e is complicated in such cases because of the air gap between the drums inner and outer walls. Therefore, treatment of such cases will not be covered in this contribution. In most other cases, we can assume that Q_e is negligible.

52.2.5 Measured Bean Temperatures

Thermometers (usually thermocouples sheathed in stainless steel) inserted in dense, well mixed portions of bean loads are used to measure bean temperature. Coffee bean beds, even when well stirred, are poor conductors. Therefore, marked differences exist between T_b and measured bean temperatures, T_a.

$$\frac{dT_a}{dt} = \frac{h_t A_t}{M_t C_t}(T_b - T_a) = K_t(T_b - T_a) \qquad (52.12)$$

where h_t and A_t = the heat-transfer coefficient and contact area between the thermometer and bean bed

M_t and C_t = the mass and heat capacity of the thermometer

$$K_t = (h_t A_t)/(M_t C_t)$$

Accurate spot T_b vs. t values can be determined by off-line measurement, but such data are of little use for plant-performance monitoring and roaster control purposes. In swirling-bed roasters $T_b \approx T_{go}$. If gas bypassing does not occur, T_b should $\approx T_{go}$ for spouted-bed roasters.

52.3 SOLUTION OF MODELING EQUATIONS

Equations (52.6) to (52.11) were solved in step-wise fashion by machine computation using a program written in Quick Basic and one-second Δt steps. T_{gi} and initial T_a, T_b, and X were known. The initial T_m was estimated. Current T_b, T_m, X, and H_e were determined as described below. Equation (52.6) and estimated or known F and $(h_e A_{gb})/(G C_g)$ were used to obtain current $(T_{gi} - T_{go})$. Current $-dX/dt$ were obtained from Equation (52.9), current Q_{gm} from Equation (52.7), and current Q_r from Equation (52.10). Q_r, $(T_{gi} - T_{go})$, dX/dt, Q_{gm}, X, and $Q_{mb} = h_{bm} A_{bm} (T_m - T_b)$ were substituted in Equation (52.8) and used to obtain dT_b/dt. Equation (52.11) was used to determine dT_m/dt and Equation (52.12) to determine $dT_a/dt \cdot t_n = t_{n-1} + \Delta t$ and $(T_b)_n = (T_b)_{n-1} + (dT_b/dt) \cdot \Delta t$, where subscripts n and $n-1$ indicate the nth and $(n-1)$st time step, used to obtain t and the T_b used in the next times step. Similar equations were used to determine T_m, X, T_m, and T_a at the start of the next time step. H_e is the sum of $(Q_r)n \cdot \Delta t$ from $t = 0$ to the current t. Computation was stopped when the computed T_a reached the specified end-of-roast T_a. The main results obtained were predicted T_b and T_a vs. t records.

T_a values at ten-second intervals starting from $t = 0$, obtained from T_a vs. t records from a roasting plant, were listed as reference values in the computer program. Predicted T_a for the same t were compared with the plant T_a, and the nature and time trend of differences between these T_a values were noted. Estimated parameters, e.g., $h_e A_{gb}$, $h_{bm} A_{bm}$, and K_t, were revised until differences between experimental and plant data were minimized. Individual parameters most strongly affect different parts of predicted T_a vs. t curves. For example, the initial portion of the curve is sensitive to K_t variation. Therefore, minimizing differences between predicted and plant T_a vs. t data in a selected section of the T_a vs. t curve provides a fairly effective way of determining the value of the parameter affecting that section most strongly. Additional experimental or plant data, e.g., measured G, T_{go} vs. t data, permits added comparisons that increase the reliability of parameter determination.

52.3.1 C_b VARIATION

C_b is smaller for robustas than for arabicas. The limited C_b data available[10,11] show that C_b increases markedly as T_b and X increase. Therefore, C_b varies markedly during roasting. It is assumed that $C_b = (C_s + X \cdot C_w)/(1 + X)$, where C_s is the partial heat capacity of bean solids and C_w is the partial heat capacity of water in the beans. At moisture contents found in green coffee, $C_w = 5.0$ kJ/(kg·°C) and does not change much with temperature. Based on the available data, it appears that C_s in kJ/(kg·°C) = 1.099 + 0.0070 · T_b for a 50% arabica 50% robusta blend. In calculations used for this contribution, 1.099 + 0.0070 · T_b + 5.0 · X was used in place of $C_b(1 + X)$ in Equation (52.8).

52.3.2 G AND T_{gi} VARIATION

In some roasters, G or T_{gi} is automatically reduced when T_a reaches some preselected value, e.g., T_{gi} may be reduced from its initial value, 500°C, to 315°C when T_a reaches 145°C and then reduced again to 290°C when T_a reaches 170°C. G will also vary because of programmed changes in the extent of roaster gas recycling and slightly because of atmospheric pressure variation (±3.5% over the course of a year). T_{gi} may also vary because of variation in the heating value of the fuel used. Such variation has to be accounted for in solving Equations (52.6) through (52.11).

52.4 RESULTS

The modeling program was solved to provide predicted T_b and T_a vs. t data for a rotating-bowl roaster processing 400 kg of green beans per roast using a constant T_{gi} of 482°C. The end-of-roast T_a was 218°C and the roasting time was 332 s. The bowl weighed 1500 kg, and C_m = 0.418 kJ/(kg·°C). G was reduced in two steps— first, to 89% of its initial value when T_a reached 182°C, and then to 78% of its initial value when T_a reached 199°C. Based on T_{go} vs. t data, it appeared that initially $G \approx$ 43 kg/s. It was assumed that initially T_b = 18.3°C and X = 0.111 kg water/kg dry solids. As is usually the case, the roaster was preheated before beans were admitted. Therefore, it was assumed that the bowl temperature, T_m, initially = 121°C. Plant T_a vs. t data, show T_a = 253°C at the start of roasting. The bowl was completely covered by beans so $F = 0 · h_e A_{gb}$, $h_{bm} A_{bm} · K_t$, and G were adjusted until predicted T_a vs. t data closely agreed with plant operating data. Plant T_a vs. t and predicted T_a and T_b vs. t data are shown in Figure 52.7. T_a vs. t agreement is quite good over the entire t range. Except for a few points at the start of roasting, the maximum difference between the plant and predicted T_a is 2.0°C · T_a was markedly greater than T_b at the start of roasting and 21.4°C to 25.5°C lower than T_b for 150 s < t < 332 s.

The main purpose of modeling was to determine how much T_b vs. t behavior would be changed by expected variations in uncontrolled factors, e.g., atmospheric pressure or bean size, moisture content, initial temperature, and load weight. The predicted T_b vs. t changes were used to judge whether better control of bean-temperature vs. time behavior is needed. Figure 52.8 depicts effects of bean size (d_p) variation on T_b vs. t profiles. The surface area per unit volume of exposed beans is inversely proportional to $d_p · h_e \approx h/(1 + 0.3 · Bi)$ where h is the heat-transfer coefficient at the gas-bean surface. The Biot number, Bi, is used to correct the overall gas-bean heat-transfer coefficient for heat-transfer resistance in the bean. h is roughly proportional to $d_p^{-0.5}$. Therefore, $h_e · A_{gb}$ is roughly proportional to $d_p^{-1.5}$. Effects of d_p on T_a and T_b vs. t behavior will be smaller for roasters where

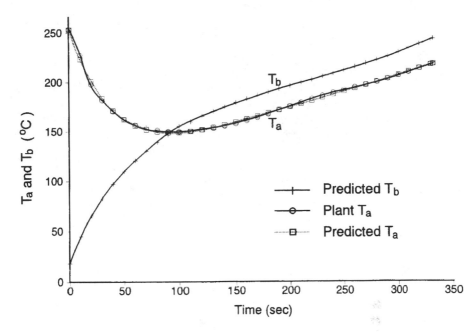

Figure 52.7 Experimental and predicted T_a and predicted T_b vs. time.

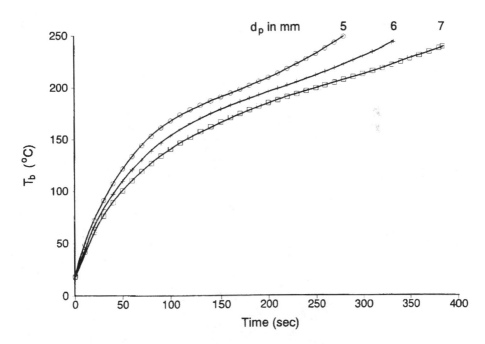

Figure 52.8 Effect of bean size (d_p) on predicted T_b vs. t behavior.

$(h_eA_{gb})/(GC_g)$ is markedly larger than in rotating-bowl roasters, i.e., in cases where $\exp[-(h_eA_{gb})/(GC_g)]$ approaches zero.

Figure 52.9 shows that initial moisture content variation greatly affects T_b vs. t behavior. Bean load weight (R) variation (caused by fully emptying feed bins at the end of a run) and G shifts caused by improper damper positioning also caused major changes in T_b vs. t behavior. Normal changes in atmospheric pressure caused only minor changes in T_b vs. t behavior, i.e., t_r varied by a few seconds. Changes in initial T_b affected T_b vs. t profiles moderately, but much less than changes in d_p, X, R, and G. A system[12] for controlling T_a vs. t behavior is now used in a number of roasting plants. The modeling was used to determine whether such control would improve the reproducibility of T_b vs. t behavior in rotating-bowl roasters. Reproducibility clearly would be improved. The modeling also permitted determination of how much control element variation was needed to eliminate T_b vs. t variation due to anticipated process upsets.

52.5 CONCLUSIONS

The model presented can successfully be used to describe bean temperature vs. time behavior for roasting of coffee in rotating-bowl roasters. It can also be used to determine how such behavior will be affected by variable factors encountered when roasting coffee. If correct h_eA_{gb}, G, R, M_mC_m, $h_{bm}A_{bm}$, and h_tA_t are used, the program used to implement the model can probably be used directly or with minor modifi-

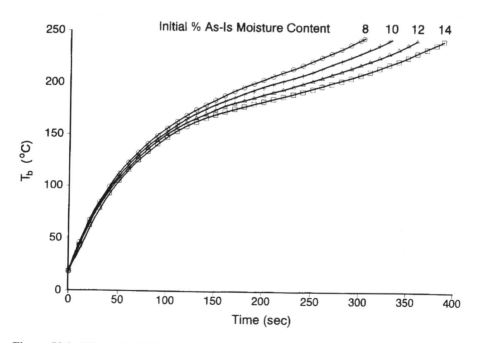

Figure 52.9 Effect of initial bean moisture content on predicted T_b vs. t behavior.

cation for other roasters where gas-to-metal heat-transfer is negligible or can be readily accounted for, e.g., spouted-bed roasters, scoop-wheel roasters, swirling-bed roasters, and drum roasters with both perforated walls and nonperforated walls where heat is not applied to the outside drum wall. Somewhat more complex modeling is required when heat is applied to the outside of a drum wall. Added factors, e.g., swelling of beans and loss in bean dry weight during roasting, should be and will be considered in more refined models and programs.

Ways to more accurately measure bean temperature should be developed. When $h_e A_{bg}/GC_g$ is larger than 6 and gas bypassing does not occur, T_b will be less than 0.5°C lower than T_{go}. Therefore, in such cases, T_{go}, rather than T_a as measured by normal means, will more accurately indicate how T_b is behaving, and T_{go}, not T_a, will be more suitable for controlling roaster operation.

REFERENCES

1. Viani R. 1985. "Coffee," in *Ullman's Encyclopedia of Industrial Chemistry*, 5th ed. vol. A7, W. Gerhatz, ed. Reinheim, Fed. Rep. of Germany: VBH, pp. 319–320, 325–327.

2. Clarke, R. J. and R. Macrae. 1985. *Coffee*. vol. 1, Chemistry. Barking, Essex, England: Elsevier Applied Science, pp. 127–132, 137–145, 223–265.

3. Rothfos, B. 1986. *Coffee Consumption*. Hamburg, Germany: Gordian-Max Rieck. GmbH, pp. 118–173.

4. Sivetz, M. and N. W. Desrosier. 1979. *Coffee Technology*. Westport, CT: AVI Publishing Co. pp. 226–264.

5. Clarke, R. J. and R. Macrae. 1987. *Coffee*. vol. 2, Technology. Barking, Essex, England: Elsevier Applied Science, pp. 72–97

6. Price R. E., R. F. Kussin, R. J. Fruhling, and M. S. Harris. 1991. U.S. Patent no. 4,988,590.

7. Mahlmann, J. P., M. S. Schecter, and L. Scher. 1985. U.S. Patent no. 4,501,761.

8. McAllister, R. V. and C. H. Spotholz. 1964. U.S. Patent no. 3,122,439.

9. Raemy, A. 1981. "Differential Thermal Analysis and Heat Flow Calorimetry of Coffee and Chicory Products," *Thermochimica Acta*, 43: 229–236.

10. Raemy, A. and F. Lambelet, 1982. *J. Food Technol.*, 17: 451–460.

11. Singh, P., R. Singh, S. Bhamidipati, S. Singh, and P. Barone. 1997. *J. Food. Proc. Eng.*, 20: 31–50.

12. Anon. 1995. Logofile. East Hanover, NJ: Praxis Werke Inc.

Part X

Food Biotechnology

53 New Frontiers for Food Processing: The Impact of Changes in Agrobiotechnology

R. Quintero-Ramírez

CONTENTS

Abstract

The commercialization of transgenic crops is a reality at the beginning of the new century. These new plant products have better traits and qualities, especially as related to human nutrition or for feeding purposes. In this chapter, the state of the art of agrobiotechnology and its future trends will be discussed and, based on these, some food processing changes will be presented and discussed, addressing not only the technical aspects but also the social concerns that transgenic crops have generated.

53.1 INTRODUCTION

The beginning of a new century is an exciting time, because it shows that the future that lies ahead is full of opportunities, challenges, and problems to be solved if we want to continue increasing our standards of living. One issue that has been discussed frequently regarding the next 20 years is food production. Some consider the present situation and its trends as a problem, and others view it as a wonderful opportunity. Let me present a brief analysis of the main concerns based on a recent study published by the World Bank.[1]

53.1.1 POPULATION

The world's population stands at 5.8 billion and is growing at about 1.5 percent per year. The wealthy industrial nations, including Japan and the nations of Europe and North America, have about 1.2 billion people. These nations are growing at a slow rate, roughly 0.1 percent a year. Population in the developing world is 4.6 billion and is expanding at 1.9 percent a year, a rate that has been decreasing somewhat in the past decade. The least developed nations, with a total population of 560 million, are growing at 2.8 percent a year. If they continue to grow at this rate, their population will double in 24 years. At present, about 87 million people are added to the world's population each year. Although fertility has been declining worldwide in recent decades, it is not known when it will decline to replacement levels. There is broad agreement among demographers that, if current trends are maintained, the world's population will reach about 8 billion by 2020, 10 billion by 2050, and possibly 12 to 14 billion before the end of the next century. Virtually all of the growth in coming decades will occur in the developing world.

53.1.2 NUTRITION

The wealthy nations have high levels of nutrition and little problem supplying all their citizens with adequate food. Indeed, well over one-third of world grain production is fed to livestock to enhance the supply of animal protein, which is consumed most heavily in the industrial world. In the developing world, matters are different. More than 1 billion people do not get enough to eat on a daily basis and live in what the World Bank terms *utter poverty;* about half a billion suffer from serious malnutrition. A minority of nations in the developing world are markedly improving their citizens' standard of living; in some 15 countries, 1.5 billion people have experienced rapidly rising incomes over the past 20 years. But in more than 100 countries, 1.6 billion people have experienced stagnant or falling incomes. To provide increased nutrition for a growing world population, it will be necessary to expand food production faster than the rate of population growth. Studies forecast a doubling in demand for food by 2025 to 2030. Dietary changes and the growth in nutritional intake that accompany increased affluence will contribute to making food demand larger than the projected increase in population.

53.1.3 AGRICULTURE

About 12% of the world's total land surface is used to grow crops, about 30% is forest or woodland, and 26% is pasture or meadow. The remainder, about one-third, is used for other human purposes or is unusable because of climate or topography. According to the Food and Agriculture Organization (FAO),[2] the increase in world population has exceeded the rate of increase in total cultivated land. In Figure 53.1, it is shown that, between 1961 and 1995, the amount of cultivated land has decreased from approximately 1.5 ha/person to 0.75 ha/person,[2] but, based on population projections, it will decrease to 0.15 by 2050. The rate of expansion of arable land is now below 0.2% a year and continues to fall. The bulk of the land best suited to rain-fed agriculture is already under cultivation, and the land that is being brought into cultivation generally has lower productivity. The FAO has projected that, over the next 20 years, arable land in the developing countries could be expanded by 12% at satisfactory economic and environmental costs, although such expansion would inflict major damage to the world's remaining biodiversity. The yields per hectare on this land would be less than on the land already in production. This expansion is to be compared with the 61% increase in food demand that is expected to occur in these countries during the same period, according to a scenario discussed by the FAO. The last major frontiers that can potentially be converted to arable land are the acid soil areas of the Brazilian cerrado, the llanos of Colombia and Venezuela, and the acid soil areas of central and southern Africa. Bringing these unexploited, potentially arable lands into agricultural production poses formidable but not insurmountable challenges.

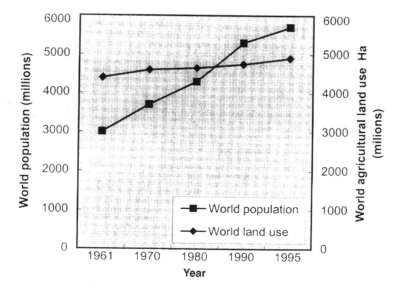

Figure 53.1 Increase in world population and the total global area used for agricultural purposes during the period 1961–1995.

Irrigation plays an important role in global food production. Of the currently exploited arable land, about 16% is irrigated, producing more than one-third of the world crop. Irrigated land is, on balance, over 2.5 times more productive than rain-fed land. The prospects for expanding irrigation, so critical to the intensification of agricultural productivity, are also troubling. The growth of irrigated land has been slowing since the 1970s, owing to problems such as siltation of reservoirs and environmental problems and related costs that arise from the construction of large dam systems.

There are additional factors that limit land for agriculture. Urbanization frequently involves the loss of prime agricultural land, because cities are usually founded near such land. Losses of prime land are often not counterbalanced by the opening of other lands to production, because the infrastructure generally required for market access is frequently lacking on those lands.

53.1.4 Food Demand

The task of meeting world food needs in 2010 by the use of existing technology may prove difficult, not only because of the historically unprecedented increments to world population that seem inevitable during the period, but also because problems of resource degradation and mismanagement are emerging.

The main challenge is to expand agricultural production at a rate exceeding population growth in the decades ahead. This goal must be accomplished in the face of a fixed or slowly growing base of arable land offering little expansion, and it must involve simultaneous replacement of destructive agricultural practices with more benign ones. One more problem is arising; there has been a reduction in yield increases. For example, in the area of cereals, the yield increase dropped from 7.2% in the period 1967–1983 to 1.5% in 1984–1994.[3] These results indicate that classical breeding is not paying off, and that other techniques should be used to fulfill the expected requirements.

The application of modern techniques based on biotechnology and genetic manipulation have been identified since the early 1980s as a key factor to achieve the needed improvements.[4] These techniques are a powerful new tool with which to supplement traditional approaches in agriculture.

53.2 AGROBIOTECHNOLOGY TODAY

Following the great success of the application of biotechnology to the health sector and to the pharmaceutical industry since the 1980s, it has been possible to genetically alter many species by the stable introduction of foreign DNA, creating so-called transgenic plants.[5] Well over 70 species have been transformed, and the number is growing continuously.[6]

Many of the first improvements achieved using transgenic plants have involved the transfer of input traits. These are traits that reduce the amount of agricultural inputs required to grow a crop. Input traits include such improvements as insect resistance, virus resistance, and herbicide tolerance. For example, the insertion of a gene that produces the crystalline toxin from *Bacillus thuringiensis* has led to the production of transgenic plants that are resistant to insects from the order *Lepidoptera*.[7] The transfer of coat protein genes from plant viruses has lead to the development of transgenic crops that are resistant to the virus from which the gene or genes were isolated. Various strategies have been developed that allow transgenic plants to tolerate applications of herbicides that allow for improved weed control. In addition to input traits, other strategies are now being employed that are directed at improving output traits. These include traits to enhanced product quality, shelf life, and ripening control.

The methods for genetic manipulation of the main crops are well established.[8] Of course, there is not one general and best technique to be applied to all plants and varieties, but in many cases high-efficiency transformation has been mediated by *Agrobacterium tumefasciens*.[9]

Today, we can classify the different types of transgenic crops into three classes: wide transfer, referring to the movement of genes from organisms of other kingdoms into plants; close transfer, referring to movements of genes between species of plants; and tweaking, which refers to the manipulation of levels or patterns of expression of genes already present in a plant's genome.

The potential of biotechnology for crop improvement will continue to expand over the coming decade. An explosion of information will inevitably result from the expanding efforts in the area of genomics and will provide germoplasm developers with a palette of traits never before seen in the history of plant improvement. These advances promise to allow for the production of high-quality, wholesome, and nutritious fruits and vegetables, which will bring value to growers, processors, and consumers.

The foreseeable impacts of agrobiotechnology are as follows:

- Extension to crop genetics of the level of intellectual property protection common to the chemical industry
- Much more extensive tailoring of crops to specific end-use requirements
- Consequent further segmentation of the markets for agricultural products
- Greater use of information technology in crop design, development, production, and marketing
- Significant reductions in the use of crop protection chemicals, with substitution of genetics as a means of pest and disease protection
- Increased use of agricultural systems for production of high-value compounds
- Widespread adoption of genetically modified crops around the world
- Continued reduction in the real prices of agricultural commodities

53.2.1 TRANSGENIC CROPS

In 1994, the first transgenic plant product sold in the U.S.A. was a tomato that contains an antisense polygalacturonase gene that slows fruit softening, thereby allowing the tomatoes to remain on the vine longer, resulting in an improved taste.[10]

This was a significant improvement over the current process of harvesting tomatoes at the green stage and gassing them with ethylene to develop red color, but which has no effect on taste.

Since that time, transgenic crops have been introduced into commercial agriculture; between 1996 and 1999, twelve countries (eight industrial and four developing) have contributed to more than a twenty-fold increase in the global area of transgenic crops (see Table 53.1.)[12] The U.S.A. continued to be the principal grower of transgenic crops in 1999, with a total of 28.7 million hectares, followed by Argentina and Canada, with 6.7 and 4.0 million hectares, respectively.

TABLE 53.1
Global Area of Transgenic Crops (1996–1999)

Year	Hectares (millions)	Acres (millions)
1996	1.7	4.3
1997	11.0	27.5
1998	27.8	69.5
1999	39.9	98.6

Of the four major transgenic crops grown in twelve countries in 1999, the two principal crops were soybeans and corn, with 54 and 28 percent respectively. The remaining 18% were equally shared by cotton and canola, as shown in Table 53.2.[12] Potato, squash and papaya occupied less than 1% of the global area of transgenic crops.

TABLE 53.2
Global Area of Transgenic Crops in 1998 and 1999

Crop	1998		1999		Increase
	Ha (millions)	%	Ha (millions)	%	
Soy	14.5	52	21.6	54	7.1
Corn	8.3	30	11.1	28	2.8
Cotton	2.5	9	3.7	9	1.2
Canola	2.4	9	3.4	9	1.0
Potato	<0.1	<1	<0.1	<1	<0.1
Squash	0.0	0	<0.1	<1	(–)
Papaya	0.0	0	<0.1	<1	(–)
Total	27.8	100	39.9	100	12.1

The distribution by trait was as follows: herbicide tolerant soybeans (54%), insect resistant maize (19%), herbicide tolerant canola (9%), combined insect resistant and herbicide tolerant maize (5%), herbicide tolerant cotton (4%), herbicide tolerant corn (4%), and the rest in transgenic cotton with different traits.

53.2.2 Public Response

One issue that many governments, multinational companies, and scientific research-ers have overlooked is how people will respond to the new transgenic foods. It was thought that the consumer was going to passively accept them as good, safe, and useful products. But now, all of them acknowledge that there are several concerns about genetically engineered products that were previously diminished; these sub-stantial concerns are not trivial; they do not have obviously self-evident answers, and they need careful and thoughtful consideration. The main concerns that have this far been identified are[11]

- Are genetically modified foods safe to eat?
- Is the regulatory process that each nation has in place sufficient to assure consumers of the safety of such foods?
- Will consumers have meaningful choices based on the kinds of informa-tion they want to have?
- Are these products safe for the environment? How are they going to affect biodiversity? How will they affect other plants, insects, and birds?
- How will the new technologies affect traditional agricultural practices, rural life, and, especially, organic farmers?
- Are we "playing God?" Do we collectively have the wisdom to use these technologies wisely?

If these questions are well answered, consumers will be satisfied with the progress of agrobiotechnology and will continue supporting its development.

Another public issue is related to intellectual property rights concerning trans-genic crops, its products, the access to genetic resources, how they will be commer-cialized, and who will benefit most from them.[13] In all these issues, there are several points of view, as well as conflicting economic and social interests. An illustrative case has been the nonacceptance of agrobiotechnology in Europe, which has given a bad image to large multinationals. In the U.S.A., so far, the negative response has been minimal.[14]

53.3 IMPACTS OF FOOD PROCESSING

In the previous section, it was demonstrated that new plant products are being developed; some have already reached the market. How will these new products impact food processing? The answer depends on the type of new trait or quality added. If the new product has a better nutritional value, then the composition of foods based on it will vary, thereby changing in some way its characteristics and properties for food processing. Another possibility is the appearance of new com-pounds with physicochemical properties that will change taste, aroma, consistency, etc., requiring a new approach not only in formulation but also in its processing.

A good example has been how, through the genetic manipulation of the ripening processes in fresh vegetables and fruits, harvest and post-harvest methods have changed.

Fruit ripening is a highly complex process, with marked variation in metabolism occurring between different fruits. Nevertheless, the process is characterized by a series of coordinated biochemical and physiological changes that lead to the development of a soft, edible fruit. Some of these changes include synthesis of secondary metabolites associated with flavor and aroma, synthesis of pigments, degradation of chlorophyll, alterations in organic acids and cell wall metabolism, and a softening of the fruit tissue (see Table 53.3).[15] These events occur in conjunction with a sharp autocatalytic increase in ethylene production and a peak of respiratory activity.

TABLE 53.3
Changes that May Occur during the Ripening of Fleshy Fruit

Color changes: synthesis of pigments

Degradation of chlorophyll

Softening: changes in composition of cell wall constituents

Production of flavor and aroma compounds

Changes in tissue permeability

Abscission

Changes in respiration rate

Changes in the rate of ethylene production

Changes in levels and composition of organic acids

Changes in gene expression

Synthesis of new proteins

Development of wax on skin

Based on the understanding of how ethylene is produced in plants[16] and/or by deleting genes of enzymes involved in degrading certain biopolymers, it has been possible to obtain fruits with extended shelf life, better taste, and fewer requirements for storage at cold temperature or controlled atmosphere for transportation.

Three examples of transgenic plants that produce new vegetable products with new or increased characteristics that will need changes in their processing will be described in detail.

53.3.1 INCREASE IN NUTRITION VALUE

Corn, which is a preferred animal feed because it is a low-cost energy source, is relatively poor in amino acid content, with lysine being the most limiting amino acid for the dietary requirements of many animals. Soybean meal, which is rich in lysine and other essential amino acids, is used to supplement corn-based animal feeds. In addition, about 200,000 tons of lysine are produced annually via fermentation, mostly for use as an animal feed additive. An increase in the lysine content of corn, soybeans, or other animal feed sources would reduce the need to supplement the seeds with crystalline lysine.

One approach to raising the lysine content of seeds is to increase its production by deregulating the biosynthetic pathway. Lysine, along with threonine, methionine, and isoleucine, are amino acids derived from asparte (Figure 53.2).[17] The first step in the pathway is the phosphorylation of aspartate by the enzyme aspartokinase (AK), and this enzyme has been found to be an important target for regulation, usually via end-product inhibition, in plants. The condensation of aspartyl β-semialdehyde with pyruvate catalyzed by dihydrodipicolinic acid synthase (DHDPS) is the first reaction committed to lysine biosynthesis. In plants, DHDPS is feedback inhibited by lysine and serves as the major regulator of the lysine branch of the pathway.

Recently, genetic engineered technology has been used to increase free lysine production in the leaves of plants. The lysine content in the seeds of canola and soybeans has been increased by circumventing the normal feedback regulation of two enzymes of the biosynthethic pathway, AK and DHDPS. Lysine-feedback insensitive bacterial DHDPS and AK enzymes encoded by the *Corynectaerium dapA* gene and a mutant *E. coli lysC* gene, respectively, were linked to a chloroplast transit peptide and expressed from a seed-specific promoter in transgenic canola and soybean seeds. Expression of *Corynebacterium* DHDPS resulted in a more than 100-fold increase in the accumulation of free lysine in the seed of canola; total seed lysine content approximately doubled. Expression of *Corynebacterium* DHDPS plus lysine-insensitive *E. coli* AK in soybean transformants similarly caused several hundred-fold increases in free lysine and increased total seed lysine content by as much as five-fold. Accumulation of α-amino adipic acid (AA) in canola and saccharopine in soybeans, which are intermediates in lysine catabolism, was also observed.[17]

53.3.2 NEW OIL COMPOSITION

Oil and fats, which, chemically, are glyceride esters of fatty acids (triacylglycerols), play an important role in human nutrition because of their high energy content. The

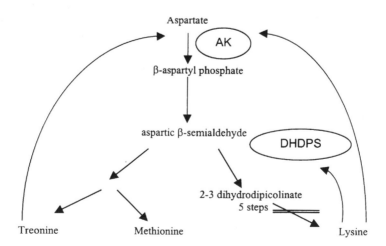

Figure 53.2 The aspartame family biosynthetic pathway.

chemical structures of the common fatty acids of plant storage oils are illustrated in Table 53.4.[18] Ninety percent of the vegetable oil produced is used for human consumption, predominantly in margarine, shortening, salad soils, and frying oils. The oil of the six major oil crops (soybeans, oil palm, rapeseed, sunflower, cotton seed, and groundnut), which account for 84% of worldwide vegetative oil production, consist mainly of palmitic, stearic, oleic, linoleic, and linolenic acid. The majority of the 210 known types of fatty acids synthesized by plants, however, are not available for economic use, because they occur in noncrop plants. Breeding efforts have traditionally focused on improving oils for these uses. Selected plants have considerably altered seed oil compositions, which indicate that they could tolerate a wide variation in fatty acid composition of storage lipids.

TABLE 53.4
Structures of Common Edible Oil Fatty Acids

Trivial name	Systematic name	Symbol
Lauric acid	Doecanoic acid	12:0
Myristic acid	Tetradecanoic acid	14:0
Palmitic acid	Hexadecanoic acid	16:0
Stearic acid	Octadecanoic acid	18:0
Oleic acid	cis-9-Octadecanoic acid	Δ-9,18:1
Linoleic acid	cis, cis, 9–12–Octadecadienoic acid	Δ-9,12–18:2
α-Linolenic acid	all-cis–9,12,15–Octadecatrienoic acid	Δ-9,12,15–18:3
γ-Linolenic acid	all-cis–6,9,12–Octadecatrienoic acid	Δ-6,9,12–18:3

Currently, only 10% of the vegetable oils produced are used in nonfood applications such as lubricants, hydraulic oil, biofuel or oleochemicals for coatings, plasticizers, soaps, and detergents. The economic requirements for industrial raw material are quite different from those for nutritional oils. The ideal oil for an oleochemical application would consist of a particular type of fatty acid that could be supplied constantly at a competitively low price, as compared to raw materials based on mineral oil products. Furthermore, such a fatty acid should have a reactive group in addition to the carboxyl function to provide an additional target for chemical modifications. It is expected that such fatty acids can be produced in plants. This challenge has been addressed using genetic engineering techniques.

By genetic manipulation, the vast capacity for the production of specific oils in agriculturally and economically attractive crop plants has been explored. For both food and nonfood uses, transgenic and traditional breeding approaches have been used to change the degree of desaturation and to reduce or increase the chain length of fatty acids. Tissue-specific promoter elements from genes encoding seed storage proteins can be used to direct gene expression to the desired storage tissue, thus avoiding possible deleterious effects that could result from expression of the transgene throughout the plant.[19]

The development of soybeans with increased concentrations of oleic acid also provides an instructive example of seed trait modification. Unsaturated fatty acids are healthier than saturated fatty acids, and the monounsaturated form, oleic acid (18:1), is also more stable in frying and cooking applications than are the polyunsaturated forms, linoleic (18:2) and linolenic (18:3). Chemical hydrogenation is currently used to improve oxidative stability by increasing the concentrations of 18:1 fatty acid, but hydrogenation also raises the concentration of trans-fatty acids, which have been linked to higher health risks. Biochemical methods were used to purify and sequence the relevant soluble proteins from soybeans, from which probes were created to isolate the genes that affect fatty acid composition.[20]

Several strategies have been tried in which genes are accessed and identified from a variety of plant and nonplant sources that have novel phenotypes.[21] For example, vernolic acid, an epoxide fatty acid, and ricinoleic acid, a hydroxy fatty acid, are oleic acid derivatives used as hardeners in paints and plastics (Figure 53.3).[20] Neither vernolic acid nor ricinoleic acids are made in soybean seeds. Complementary DNA libraries were made from Vernonia and castor bean seeds, plants known to produce the desired fatty acids. The genes that encoded the fatty acid-modifying enzymes were identified by DNA sequence homology to desaturase genes. After modification of the regulatory sequences, the genes were transferred to soybean, where the desired industrial oils were produced in seeds.

53.3.3 Provitamin A into Rice

Although half of the world's population eat rice daily and depend on it as their staple food, rice is a poor source of many essential micronutrients and vitamins. In Southeast Asia, 70% of children under the age of five suffer from vitamin A deficiency, leading to vision impairment and increased susceptibility to disease. UNICEF predicts that improved vitamin A nutrition could prevent 1–2 million deaths each year among children aged 1–4 years.

Mammals make vitamin A from β-carotene, which is one of the most abundant carotenoids found in plants. Carotenoids are yellow, orange, and red pigments that are essential components of the photosynthetic membranes of all plants. They serve as accessory light-harvesting pigments and as antioxidants that quench tissue-damaging free radicals such as single oxygen species. Rice in its milled form contains neither β-carotene nor any of its immediate precursors.

Carotenoids, along with a variety of other compounds including gibberellins, sterols, chlorophylls, and tocopherols, are derived from the general isoprenoid biosynthetic pathway (Figure 53.4).[22] Immature rice endosperm synthesizes the common precursor isopentenyl diphosphate (IPP) from which the carotenoid precursor geranyl geranyl diphosphate (GGPP) is obtained. To convert GGPP to β-carotene, the endosperm can be programmed to carry out the necessary additional enzymatic steps. Two molecules of GGPP must first be condensed to form phytoene, which is then desaturated to lycopene and finally cyclized to form β-carotene.

In the year 2000, one of the most exciting developments was the introduction of genes into rice that results in the production of the vitamin A precursor β-carotene

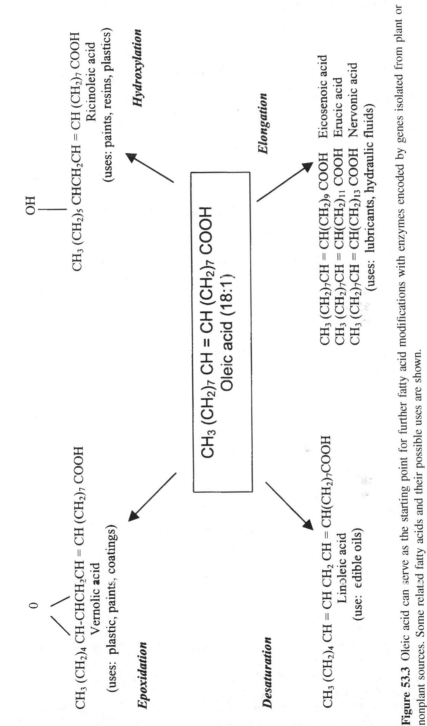

Figure 53.3 Oleic acid can serve as the starting point for further fatty acid modifications with enzymes encoded by genes isolated from plant or nonplant sources. Some related fatty acids and their possible uses are shown.

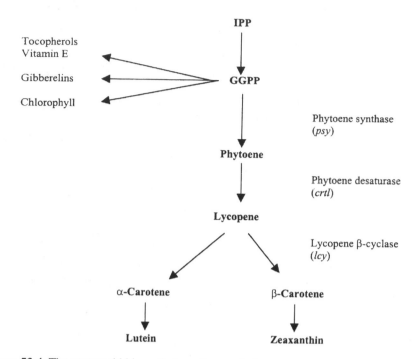

Figure 53.4 The carotenoid biosynthetic pathway of plants.

in the rice grain.[22] Traditional plant breeding has given us some plants that produce β-carotene in non-photosynthetic tissue, such as the roots of carrots, but, despite decades of searching, no rice mutants had been found that produce β-carotene in the grain, so conventional breeding was not an option.

To get the cells of the grain to produce β-carotene, genetic engineers introduced two of the required genes (encoding phytoene synthase and phytoene desaturase) on a construct that did not have a selectable marker. They achieved this by simultaneously introducing another construct, which carried the third gene of interest (lycopene β-cyclase) as well as a selectable antibiotic resistance gene. This cotransformation strategy should enable segregation of the antibiotic resistance gene away from the phytoene synthase and phytoene desaturase genes, thereby addressing one of the major concerns of opponents of genetically engineered crops. Such plants should still be able to produce β-carotene, because the authors have also shown that plants engineered with standard transformation procedures to express only the phytoene synthase and phytoene desaturase genes do not accumulate lycopene as predicted. Instead, these plants produce essentially the same end products (β-carotene, lutein, and zeaxanthin) as plants engineered to express all three carotenoid genes.

The grain of the transgenic rice has a light golden-yellow color and contains sufficient β-carotene to meet human vitamin A requirements from rice alone. This new rice offers an opportunity to complement vitamin A supplementation program, particularly in rural areas that are difficult to reach.

53.4 FUTURE DEVELOPMENTS

What will the new developments be in the next ten years? How will they affect food processing? One way to answer these questions is to analyze current research projects and to expect that most of them will be successful. If we base our estimation on past the performance of agrobiotechnology, the success rate will be high, and the time to reach commercialization will be faster than it has been in the past.

Most probably, the new transgenic products will come from a large variety of plants, and the new traits will be the result of having transferred more than one gene. From the scientific literature, the following selected examples illustrate potential new products:

- In tree crops, there is an effort to identify and modify the expression of endogenous genes/traits that enhance quality and/or nutritional value. Examples include the manipulation of sugar and ethylene metabolism in apples; the manipulation of oil/fat composition in walnuts; and, in almonds, there is a study on self-incompatibility.
- Through efforts in transformation, transgenic soybeans have been produced because, due to the high oil and protein content of the seed, soybeans have been modified with regards to seed amino acid profile and lipid composition.
- It has been possible to genetically engineer the constitutive accumulation of a resveratrol glucoside in transgenic alfalfa leaves and stems (this compound does not naturally accumulate in alfalfa). Resveratrol is proposed to have beneficial effects on human health (strong antioxidant, improvement of cardiovascular health, prevention of tumorigenesis in model systems), but there are few human dietary sources of resveratrol.
- Wheat is unique among cereals in that doughs prepared by mixing flour with water are extensible and strong enough to trap the gas bubbles made by leavening agents and, thus, to form light porous structures when processed into foods. This unique property of viscoelasticity is primarily due to the protein component of wheat seeds. Dough elasticity or strength is determined by the low- and high-molecular-weight glutenin proteins that form very large polymers, exceeding 10 million daltons in size, and linked by disulfide bonds. The amount and size of these polymers determine the suitability of flours for particular food end uses. For this purpose, transgenic wheat lines expressing the high-molecular-weight glutenin subunit 1Ax1 gene were obtained. The overexpressed protein 1Ax1 was shown to be incorporated into the insoluble gluten network, leading to increases in the amount of total insoluble gluten. A good correlation was found between the amount of insoluble gluten and dough strength.
- The genetic modification of plant seed storage proteins is one field of great interest and importance. The main groups of storage proteins present in cereals correspond to prolamins that account for about 40–50%, and glutelins for around 35–40% of total proteins from several seeds, like

wheat, barley, and rye, as shown in Table 53.5.[25] Apart from these two groups of storage proteins, cereals also contain other types such as albumins and globulins (glutelins), as indicated in the table. Prolamins are the major storage proteins in most cereal seeds except for oats and rice, where the main storage proteins have the solubility properties of globulins and glutelins, respectively, and prolamins are present at levels as low as 5–10% of total grain protein. One example is the increase of protein content in sweet potato using a synthetic storage protein gene.

TABLE 53.5
Comparison of Storage Protein Distribution from Some Cereals and Legumes

	Albumins	Globulins	Prolamin	Glutelins
Cereals				
Barley	3–4	10–20	35–45	35–45
Maize	2–10	10–20	50–55	20–45
Oats	5–10	50–60	10–15	5
Rice	2–5	2–8	1–5	85–90
Sorghum	Trace	Trace	60–70	30–40
Wheat	5–10	5–10	40–50	30–40
Pseudo-cereals				
Amaranth	50	16	2	30
Legumes				
Broad bean	20	60		15
Common bean	15	75		10
Pea	21	66		12
Peanut	15	70		10
Soybean	10	90		0

Another strong possibility is that new products will be obtained using transgenic plants like novel production systems or bioreactors. In this area, there are many interesting examples, among which are the following:

- Poly(hydroxyalkanoates) are natural polymers with thermoplastic properties. One polymer of this class with commercial applicability, poly(3-hydroxybutyrate-*co*-3-hydroxyvalerate) (PHBV), can be produced by bacterial fermentation, but the process is not economically competitive with polymer production from petrochemicals. *Arabidospis* and *Brassica* have been genetically modified to produce PHBV in leaves and seeds, respectively, by redirecting the metabolic flow of intermediates from fatty acid and amino acid biosynthesis.[23]
- Application of yeasts in traditional biotechnologies such as baking, brewing, distiller's fermentation, and wine making exposes them to numerous environmental stresses. Then, for yeast to perform efficiently in industrial

processes, it is necessary to make genetic manipulations that will allow them to respond to the different kind of stresses.[24]

- It has been demonstrated that plants can be utilized to produce high value proteins, such as hormones, antibodies ("plantibodies"), human serum albumin, human interferon and human and animal vaccine.[26]

Taking into account the public response toward transgenic agriculture in many countries, it is feasible that the field of nutraceuticals will grow stronger.[27] If one considers that, in the U.S.A., nutrition is the shoppers' second most important consideration when buying food (trailing taste, but ranked ahead of price and safety), then nutraceuticals will be favored, because they will be considered "natural" and healthier eating.[28]

53.4.1 AGROBIOTECHNOLOGY FOR DEVELOPING COUNTRIES

The application of advances in plant breeding, including tissue culture, marker-aided selection (which uses DNA technology to detect the transmission of a desired gene to a seedling arising from a cross), and genetic engineering, are going to be essential if farmers' yields and yield ceilings are to be raised, excessive pesticide use reduced, the nutrient value of basic foods increased, and farmers on less favored lands provided with varieties better able to tolerate drought, salinity, and lack of soil nutrients.[29]

For some time to come, tissue culture and marker-aided selection are likely to be the most productive uses of biotechnology for cereal breeding. However, progress is being made in the production of transgenic crops for developing countries. As in the industrialized countries, the focus has been largely on traits for disease and pest resistance, but genes that confer tolerance of high concentrations of aluminum have been added to rice and maize, and also two genes were added to rice that might help the plant tolerate prolonged submergence. There is also the possibility of increasing yield ceilings, through more efficient photosynthesis, for example, or by improved control of water loss from leaves through regulation of stomatal opening and closing.

In addition to generating new traits that enable the plant to grow better, which are useful to poor farmers, agrobiotechnology can also generate plants with improved nutritional features of benefit to poor consumers.[30]

Over the next decade, we are likely to see much greater progress in multiple gene introductions that focus on output traits or on difficult-to-achieve input characteristics, as indicated in Table 53.6.[31]

The potential benefits of plant biotechnology are considerable but are unlikely to be realized unless seeds are provided free or at nominal cost. This will require heavy public investment by national governments and donors, at times in collaboration with the private sector, both in the research and in the subsequent distribution of seed and technical advice. Breeding programs will also need to include crops such as cassava, upland rice, African maize, sorghum, and millet, which are food staples and provide employment for the 650 million rural poor who need greater stability and reliability of yield as much as increased yield.[31] In conclusion, food processing will certainly change as agrobiotechnology continues evolving.

TABLE 53.6
Biotechnology Research Useful in Developing Countries

Traits now in greenhouse or field test	Traits now in laboratory tests
Input traits	*Input traits*
Resistance to insects, nematodes, viruses, bacteria and fungi in crops such as rice, maize, potato, papaya and sweet potato	Drought and salinity tolerance in cereals
Delayed senescence, dwarfing, reduced shade avoidance and early flowering in rice	Seedling vigor in rice
	Enhanced phosphorus and nitrogen uptake in rice and maize
Tolerance to aluminum, submergence, chilling and freezing in cereals	Resistance to the parasitic week *Striga* in maize, rice and sorghum, to viruses in cassava and banana and to bacterial blight in cassava
Male sterility/restorer for hybrid seed production in rice, maize, oil-seed rape and wheat	Nematode resistance and resistance to the disease black sigatoka in banana
New plant types for weed control and for increased yield potential in rice	Rice with alternative C_4 photosynthetic pathway and the ability to carry out nitrogen fixation
Output traits	*Output traits*
Increased β-carotene in rice and oil-seed rape	Increased β-carotene, delayed post-harvest deterioration and reduced content of toxic cyanides in cassava
Lower phytates in maize and rice to increase bioavailable iron	Increased vitamin E in rice
Modified starch in rice, potato and maize and modified fatty-acid content in oil-seed rape	Apomixix (asexual seed production) in maize, rice, millet and cassava
	Delayed ripening in banana
Increased bioavailability protein, essential amino acids, seed weight and sugar content in maize	Use of genetically engineered plants such as potato and banana as vehicles for production and delivery or recombinant vaccines to humans
Lowered lignin content of forage crops	Improved amino-acids content of forage crops

REFERENCES

1. Kendall, H. W., R. Beachy, T. Eisner, F. Gould, R. Herdt, P. H. Raven, J. S. Schell, and M. S. Swaminathan. 1997. "Bioengineering of Crops: Report of the World Bank Panel on Transgenic Crops, Environmentally and Socially Sustainable Development Studies and Monographs," Series 23, The World Bank, Washington, D.C.
2. World Wide Web: www.faostat.html.

3. Izquierdo, J. 1999. "Rol Crítico de la Biotecnología en el Mejoramiento de Cutlvios Alimenticios: Perspectivas para el Próximo Milenio, Papel de REDBIO/FAO," presented at Workshop Balance de la Biotecnología Agrícola: a un Paso del Nuevo Milenio, National Council of Science and Technology, REDBIO/Venezuela, Caracas, Venezuela, October 19–20, 1999.

4. Quintero, R. 1996. "Tendencias Internacionales de la Aplicación de la Biotecnología en los Sectores Agroalimentario y Agroindustrial," in *Perspectivas de la Biotecnología en los Sectores Agrícola, Pecuario y Agroindustiral*, J. L. Solleiro, ed. CamBio-Tec/International Development Research Center/Centro para la Innovación Tecnológica, México, D.F., pp. 3–26.

5. Doyle, J. J. and G. J. Persley. 1996. "Enabling the Safe Use of Biotechnology: Principles and Practice," Environmentally Sustainable Development Studies and Monographs Series no. 10, The World Bank, Washington, D.C.

6. James, C. and A. F. Krattiger. 1996. "Global Review of the Field Testing and Commercialization of Transgenic Plants: 1986 to 1995," International Service for the Acquisition of Agri-biotech Applications, Briefs no. 1, Ithaca, New York.

7. Krattiger, A. F. 1997. "Insect Resistance in Crops: A Case Study of *Bacillus thuringiensis* (Bt) and its Transfer to Developing Countries," The International Service for the Acquisition of Agri-biotech Applications, Briefs no. 2, ISAAA, Ithaca, New York.

8. Vasil, I. K. 1996. "Milestones in Crop Biotechnology-Transgenic Cassava and *Agrobacterium*-Mediated Transformation of Maize," *Nature Biotechnology*, 14: 702–703.

9. Ishida, Y., H. Sato, S. Ohta, Y. Hiei, T. Komari, and T. Kumashiro. 1996. "High Efficiency Transformation of Maize (*Zea mays* L.) Mediated by *Agrobacterium tumefasciens*," *Nature Biotechnology* 14: 745–750.

10. Redenbaugh, K., L. Davis, J. Lindermann, and D. Emlay. 1995. "Commercialization of Biotechnology Products," presented at the North America Plant Protection Organization Meeting in Saskatoon, Sask., Canada, October, 1995.

11. Heylin, M. 1999. "Ag Biotech's Promise Clouded by Consumer Fear," *Chemical and Engineering News*, December: 73–82.

12. James, C. 1999. "Global Review of Commercialized Transgenic Crops: 1999," International Service for the Acquisition of Agri-biotech Applications, Briefs no. 12, Ithaca, New York.

13. Hawtin, G. and T. Reeves. 1997. "Intellectual Property Rights and Access to Genetic Resources in the Consultative Group on International Agricultural Research," presented at the Workshop on Intellectual Property Rights III-Global Genetics Resources: Access and Property Rights, Washington, D.C., June 4–7, 1997.

14. Gaskell, G., M. W. Bauer, J. Durant, and N. C. Allom. 1999. "Worlds apart? The Reception of Genetically Modified Foods in Europe and the US," *Science*, 285: 384–387.

15. Gómez-Lim, M. A. 1999. "Physiology and Molecular Biology of Fruit Ripening," in *Molecular Biotechnology for Plant Food Production*, O. Paredes-López, ed. Lancaster, PA: Technomic Publishing Co., Inc., pp. 303–342.

16. Ecker, J. R. 1995. "The Ethylene Signal Transduction Pathway in Plants," *Science*, 268: 667–675.

17. Falco, S. C., T. Guida, M. Locke, J. Mauvais, C. Sanders, R. T. Ward, and P. Webber. 1995. "Transgenic Canola and Soybean Seeds with Increased Lysine," *Biotechnology*, 13(6): 577–582.

18. Mackenzie, S. L. 1999. "Chemistry and Engineering of Edible Oils and Fats," in *Molecular Biotechnology for Plant Food Production*, O. Paredes-López, ed. Lancaster, PA: Technomic Publishing, Co., Inc., pp. 525–551.
19. Topfer, R., N. Martini, and J. Schell. 1995 "Modification of Plant Lipid Synthesis," *Science*, 268: 681–686.
20. Mazur, B. E., Krebbers and S. Tingey. 1999. "Gene Discovery and Product Development for Grain Quality Traits," *Science*, 285: 372–375.
21. Reddy, A. S. and T. L. Thomas. 1996. "Expression of a Cyanobacterial Δ^6-desaturase Gene Results in γ-linolenic Acid Production in Transgenic Plants," *Nature Biotechnology*, 14(5): 639–642.
22. Ye, X., S. Al-Babili, A. Kloti, J. Zhang, P. Lucca, P. Beyer, and I. Potrykus. 2000. "Engineering the Provitamina A (β-Carotene) Biosynthetic Pathway into (Carotenoid-Free) Rice Endosperm," *Science*, 287: 303–305.
23. Slater, S., T. A. Mitsky, K. L. Hoomiel, M. Hao, S. E. Reiser, N. B. Taylor, M. Tran, H. E. Valentin, D. J. Rodríguez, D. A. Tone, S. R. Padgette, G. Kishore, and K. J. Gruys. 1999. "Metabolic Engineering of *Arabidopsis* and *Brassica* for Poly (3-hydroxybutyrate-co-3-hydroxyvalerate) Copolymer Production," *Nature Biotechnology*, 17 (10): 1011–1016.
24. Attfield, P. V. 1997. "Stress Tolerance: the Key to Effective Strains of Industrial Baker's Yeast," *Nature Biotechnology*, 15(12): 1351–1357.
25. Segura, M. and R. Jiménez, 1999. "Genetic Modifications of Plant Seed Storage Proteins for Food Production," *in Molecular Biotechnology for Plant Food Production*, O. Paredes-López, ed. Lancaster, PA: Technomic Publishing Co., Inc., pp. 411–492.
26. Moffat, A. S. 1995. "Exploring Transgenic Plants As a New Vaccine Source," *Science*, 268: 658–660.
27. Craker, L. E., S. J. Herbert, and C. Barstaw. 1999. "Latest Trends in the US Herbal Market," presented at Regional Conference Prospects of Biotechnology and Biodiversity for the Caribbean Agro-industry, University of the West Indies, Organization of American States and National Council of Science and Technology, Kingston, Jamaica, November 29–30, 1999.
28. Inglett, G. E. 1999. "Nutraceuticals: the Key to Healthier Eating," *ChemTech*, October: 38–42
29. Serageldin, I. 1999. "Biotechnology and Food Security in the 21st Century," *Science*, 285: 387–389.
30. DellaPenna, D. 1999. "Nutritional Genomics: Manipulating Plant Micronutrients to Improve Human Health," *Science*, 285: 375–379.
31. Conway, G. and G. Toenniessen. 1999. "Feeding the World in the Twenty-First-Century," *Nature*, 402: C55–C58.

54 Novel and Potential Applications of α-Amylases in the Food Industry

G. Saab-Rincón

G. del Río

R. I. Santamaría

M. L. Díaz

X. Soberón

A. López-Munguía

CONTENTS

Abstract

This chapter presents a description of some properties of the α-amylases, as well as their current or potential application to production of tensoactive compounds, production of maltose or maltooligosaccharides, and in the corn tortilla industry.

54.1 INTRODUCTION

The worldwide energy supply and chemical industries are almost entirely based on fossil sources (coal, oil, and natural gas) that are used directly as energy sources or processed to more sophisticated products. They are also the base of primary and intermediate chemicals that are the building blocks of diversity of our chemical world.

However, since the 1970s, there has been a strong debate on the competition between renewable and nonrenewable sources, cellulose and starch being at the center of the discussion as the major products of carbon and solar energy storage through photosynthesis. Cellulose can be used directly as a source of energy, but, due to its various chemical and physical properties, starch is the world's most abundant commodity. This glucose polymer can be processed into products with a wide variety of applications; it is used as an energy carrier and in numerous industrial food or feed applications, including its transformation to glucose, which may be used as a feedstock for the fermentation industry. As a result of this versatility, millions of tons of starch are processed annually by corn wet millers worldwide and transformed into products such as glucose and fructose syrups, crystalline glucose, cyclodextrins, maltodextrins, dextrins, and modified starches. For instance, according to the Tropical Development Research Institute, in England, in 1997, 35 million tons of cornstarch, or about 6% of the world corn production, were extracted. Among other products, in the U.S.A. alone, 8 million tons of high fructose corn syrups (HFCS) and 3.5 million tons of glucose were produced. According to the Food and Agriculture Organization (FAO), the total world production of corn in 1998 was 604 million tons, with 41% being produced in the U.S.A.

Future expanded uses for starch may result from the emerging ability to modify the starch structure and to develop and improve transformation processes to obtain relatively inexpensive products. One of the main driving forces is the development of new enzymes and enzymes with new properties for biocatalytic processes. This is clearly shown by the fact that less than 20% of starch is used in its native form, while more than 60% is enzymatically transformed to HFCS, glucose syrups, and crystalline dextrose. Cyclodextrins, another enzymatic product obtained from starch, with a current production volume of several thousand tons per year, appeared on the market during last decade due to the development of an appropriate enzyme using genetic engineering to overproduce and improve it.[1] CGTases are unique transglucosidases that transform starch to a homologous series of cyclic nonreducing D-glucosyl polymers known as cyclodextrins, or Shardinger dextrins. The largest market of enzymes in the food industry is occupied by those involved in starch transformation, particularly α-amylase, glucoamylase, and glucose isomerase. Together, they accounted for global sales of $156 million in 1996.[2] As the demand for starch-based sweeteners grows, products such as high maltose syrups, extremely high maltose syrups, and high conversion syrups, as well as maltooligosaccharides, are gaining importance as industrial commodities for use as food ingredients. Among α-amylases, the thermostable liquefying enzyme from *B. licheniformis,* and the saccharifying fungal amylases, are the most commonly used. New amylases have been discovered, particularly in themophilic organisms,[3] but new applications and new properties are also emerging from advances in the structure-function knowledge of the already existing enzymes. Some of these advances and applications are discussed in this contribution.

54.2   STRUCTURE OF α-AMYLASES

As shown in Table 54.1, four well conserved regions are evident when comparing the amino acid sequence derived from almost 100 genes coding for α-amylases. These four regions are important for substrate binding and catalysis. These sequence signatures are also present in enzymes such as CGTase, neopullulanase, and isoamylase, the last two also known as debranching enzymes for their activity toward α(1–6) bonds in amylopectin. Three-dimensional structures have shown that most amylases are composed of three domains.

TABLE 54.1
Conserved Regions in α-Amylases and Related Enzymes

	Region I	Region II	Region III	Region IV
AMY *A. ory.*	[117]DVVANH	[202]GLRIDTVKH	[230]EVLD	[292]FVENHD
AMY *B. stearo.*	[101]DVVFDH	[230]GFRLDAVKH	[264]EYWS	[326]FVDNHD
AMY *B. amylo.*	[98]DVVLNH	[227]GFRLDAAKH	[261]EYWQ	[323]FVENHD
AMY *B. sub.*	[97]DAVINH	[172]GFRLDAAKH	[208]EILQ	[264]WVESHD
AMY Rat	[96]DAVINH	[190]GFRLDAAKH	[230]EVID	[292]FVDNHD
AMY Mouse, s.	[96]DAVINH	[193]GFRLDASKH	[233]EVID	[295]FVDNHD
AMY Human, p	[99]DAVINH	[196]GFRLDASKH	[236]EVID	[298]FVDNHD
AMY Barley	[101]DIVINH	[127]DGRLDWGPH	[218]EVWD	[299]FVDNHD
NPL *B. stearo.*	[242]DAVFNH	[324]GWRLDVANE	[357]EIWH	[419]LLGSHD
PUL *K. aero.*	[600]DVVYNH	[671]GFRFDLMGY	[704]EGWD	[827]YVSKHD
CGT *B. mace.*	[135]DFAPNH	[225]GIRFDAVKH	[258]EWFL	[324]FIDNHD
CGT *B. stearo.*	[131]DFAPNH	[221]GIRMDAVKH	[253]EWFL	[319]FIDNHD

Reprinted from Reference 4, courtesy of Marcel Dekker, Inc.

Domain A has a structural arrangement of $(\alpha/\beta)_8$ barrel. Domain B is inserted in domain A, between β-strand 3 and helix 3, forming an extended loop of 3-stranded antiparallel β-sheet (Taka amylase). Domain C is found in the C terminus, which folds into two, 4-stranded and one, 2-stranded antiparallel β-sheets (porcine pancreatic amylase) or 5-stranded antiparallel (barley α-amylase).[5] A model of the α-amylase from *B. stearothermophilus* based on the 3D structure of the enzyme from *B. licheniformis* is shown in Figure 54.1.

The active site of most α-amylases is located in a cleft at the C-terminal side of the β-strands of the $(\alpha/\beta)_8$ barrel of domain A. All the amino acid residues involved in the active site are found on the loops or β-strands. The two catalytic amino acid

Figure 54.1 Ribbon representation of α-amylase, showing Ala 289 in *B. Stearothermophilus.*
The three carboxylic acids are shown, as well as the proposed path for water.

residues in Taka amylase Glu 230 (acting as general acid) and Asp 297 (acting as
the nucleophilyc base) are highly conserved among amylases and are located in
conserved Regions III and IV respectively.

54.3 α-AMYLASE SPECIFICITY

α-amylases contain a substrate binding site formed by several subsites and strongly
related to its specificity. In Taka amylase from *A. niger,* seven subsites are present,
with bond cleavage occurring between subsites 4 and 5. A minimum of five (pancreatic α-amylase) and a maximum of ten (*B. amyloliquefaciens*) subsites have been
described.[5] Many additional amino acids involved in substrate binding have been
identified, particularly a highly conserved histidine His 210 (conserved Region II)
also involved in calcium binding, control of pH optimum, and activation by chloride
ion.

It is also widely accepted that the properties and mechanisms of action of α-
amylase are dependent on the source of the enzyme. As they are all endo-acting
enzymes hydrolyzing only α(1–4) linkages, they are classified as liquefying or
saccharifying according to their mechanism of starch degradation. A saccharifying
α-amylase produces an increase in reducing power about twice that of a liquefying

enzyme, while the latter rapidly reduces the viscosity of starch solutions. However, after prolonged digestion of starch, the action of α-amylases results in major end-products such as G6 (*B. amyloliquefaciens* and *B. subtilis*), G2, G3, G5 (*B. licheni-formis*), G2, G3, G4 (*Micrococcus halobius*); G2 and G1 are the major products of saccharifying enzymes.[6] We proposed that a more extensive degradation of starch of certain amylases (saccharifying) is the consequence of a hydrolytic activity complemented with transglycosydation, as exemplified in Figure 54.2 for the oli-gosaccharide G5.[12] Therefore, the end-product specificity is a complex function of subsite size and structural features affecting the reactivity water and the concurrence of alternative nucleophiles that may interact with the glucosyl covalent intermediate.

54.4 ALCOHOLYSIS: SYNTHESIS OF TENSOACTIVE COMPOUNDS FROM STARCH

The reaction catalyzed by α-amylases is formally a nucleophilic substitution at the saturated carbon of the anomeric center, which takes place with retention of the anomeric configuration. Many retaining glycosidases will transfer glycosyl residues to low-molecular-weight alcohols such as methanol, as well as to water.[7] This activity is never found in inverting glycosidases like β-amylase or amyloglucosidase. Fur-thermore, we recently demonstrated that liquefying α-amylases contain little or no activity toward methanol, while saccharifying enzymes such as α-amylases from *A. oryzae* or *A. niger* synthesize a series of methyl glycosides from starch, methyl maltoside being the most abundant product in α-amylase from *A. niger*.[8] Alcoholysis is therefore related to transglycosylation.

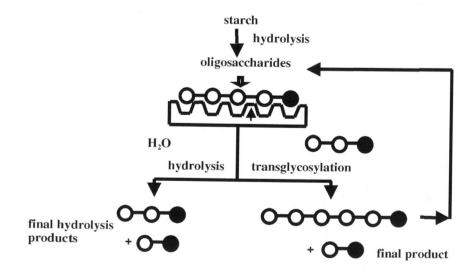

Figure 54.2 Example of the combination of transferase and hydrolase activities in the deg-radation of G5 (maltopentaose).

Alkyl polyglycosides are a new category of surfactants that meet the evolving requirements of the food industry. They belong to the nonionic surfactants group; among them, sucrose esters are now preferred in industry instead of polysorbates, due to the presence in the latter of the carcinogen 1–4 dioxane. In this context, alkyl-glucosides may also find interesting applications as biodegradable nontoxic tenso-active molecules.

A wide variety of retaining glycosidases has been reported for the synthesis of alkyl-glycosides. Best results been obtained with β-glucosidases,[9] β-galactosidases,[10] and xylanases[11] using cellobiose, lactose, and xylobiose as substrates, respectively. When high-molecular-weight alcohols are used as substrates, the resulting alkyl-glucoside contains a hydrophobic tail and a lypophilic sugar in the other extreme of the molecule, joined by glucoside type linkages. So far, no attempt has been made to produce alkyl-glucosides from starch and α-amylases, a simpler system than the ones already described. In Figure 54.3, the product profiles obtained when starch is used as substrate of *A. niger* α-amylase in the presence of one soluble (methanol) and two insoluble (butanol and hexanol) alcohols are shown. These results also include a reaction system containing water saturated with the alcohol (monophasic). High concentration of starch in the aqueous phase results in higher alcoholysis yields, probably as a result of reducing the available water for hydrolysis. In all cases, a series of alkyl-glycosides are produced as shown by the HPLC chromatograms, where it also may be observed that maltosyl-glycoside is the major end product (maltose is the main product of this α-amylase). Butanol has an inhibitory effect on the enzyme, but hexanol does not. As the products are extracted in the organic phase (which is also one of the substrates) the reaction may proceed to high conversions, provided a renewal of the water phase takes place to eliminate the hydrolysis by-products. In Table 54.2, the products are quantified and reported as the corresponding alkyl-glucoside, after treatment of all the alkyl glycosides with glucoamylase.

54.5 COMBINING TRANSFERASE AND HYDROLYTIC ACTIVITIES

In a recent publication, we proposed that transferase activity is likely to have evolved from an ancestral hydrolase.[12] Evidence of this hypothesis came from sequence analysis and by studying the effect of the combined action of transferase activity from CGTase from *B. macerans* and the liquefying α-amylase from *B. licheniformis*. The enzyme CGTase appears to catalyze at least three reactions: cyclization, coupling, and disproportionation. In Table 54.3, we further explore this idea, which may have important implications in the industrial production of glucose, maltose, or high conversion syrups. In this table, it is shown that a much more efficient liquefaction process was obtained when CGTase was added to the liquefying enzyme, producing 1.4 times more glucose, two times less maltotriose, and no maltotetraose. No cyclodextrins were formed. It is also interesting to observe that when the same combination was made with a saccharifying enzyme containing transglycosidase activity (*A.*

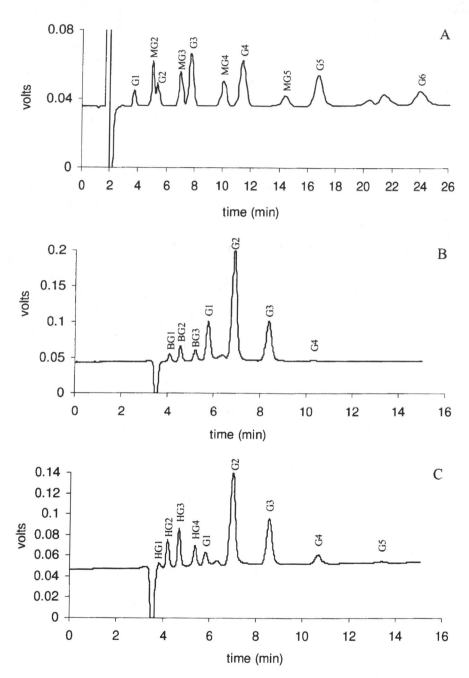

Figure 54.3 Alkyl glycosides produced by α–amylase from starch by alcoholysis, in the presence of methanol (A), butanol (B), and hexanol (C). G1, G2, G*n*... are glucose and the corresponding oligosaccharides, while HG*n*, BG*n*, and MG*n* are the corresponding alkyl glycosides, where *n* is the number of glucose molecules.

TABLE 54.2

Production of Alkyl-Glucosides from Starch by Alcoholysis with α-Amylase with Three Different Alcohols

n–alcohol	Alcohol/water ratio	Alky-glucoside (organic phase) mg/ml	Alkyl-glucoside (aqueous phase) mg/ml	Alkyl-glucoside (total) mg/ml
Methanol[†]	40% v/v			2.320
Butanol[‡]	monophasic[*]			0.663
	1:1	0.241	0.227	0.468
	2:1	0.079	0.094	0.253
	3:1	0.030	0.039	0.130
Hexanol[‡]	monophasic[*]			0.520
	1:2	0.808	0.126	0.530
	1:1	0.505	0.086	0.591
	2:1	0.229	0.041	0.499

[*]In the monophasic system, the water was saturated with the corresponding alcohol. In all cases, α-amylase from *A. niger* was used with 6%[†] w/v or 25%[‡] w/v starch; 60°C and pH4.

Alkyl-glycosides were indirectly quantified by addition of the glucoamylase to the alcoholysis reaction products (see Figure 54.3). All products were then hydrolyzed to glucose and the corresponding alkyl-glucoside; methyl-glucoside; butyl-glucoside; hexyl-glucoside.

niger), no significant differences were observed in the product profile. Due to the lower thermal stability of CGTase, the experiments shown in Table 54.3 were carried out at the optimal temperature of CGTase, so the application of these results still requires a higher thermal stability CGTase or the introduction of transglycosidase activity to the liquefying enzymes.

TABLE 54.3

Effect of Combining Liquefying (*B. licheniforms*) and Saccharifying (*A. niger*) α-Amylases with CGTase (*B. macerans*) in the Final Product Profile from Starch

	α-amylase (*B. licheniformis*)		CGTase + amylase (*B. licheniformis*)		α-amylase (*A. niger*)[*]		CGTase + amylase (*A. niger*)[*]	
	5 h	24 h	5 h	24 h	5 h	24 h	5 h	24 h
G1	26.5	36.7	36.8	50.7	25.8	38.1	26.2	40.4
G2	27.4	37.2	35.1	38.9	49.2	46.6	49.0	46.5
G3	23.1	19.9	14.3	9.4	25.0	15.4	24.8	13.2
G4	6.1	5.1	5.8	–	–	–	–	–
G5	17.0	–	8.0	–	–	–	–	–

[*]After two-hour treatment with the liquefying α-amylase from *B. licheniformis*.

54.6 IMPROVING THE TRANSFERASE PROPERTIES OF α-AMYLASE

As already stated, transferase activity is used in combination with hydrolysis for an efficient transformation of starch in saccharifying enzymes. However, the more robust, heat resistant α-amylase from bacteria (*B. licheniformis*, *B. stearothermophilus*, etc.) are liquefying, and, although new saccharifying amylases have been reported in these strains,[13] they also show optimal temperatures at 60°C or lower.

One structural explanation for the presence of transferase activity in certain amylases given by Holm et al.[14] is based on the longer loops in saccharifying enzymes like Taka amylase rather than those from *B. stearothermophilus* (which may block part of the subsite). Kuriki and coworkers[15] have suggested that a series of three residues near the active site of neopullulanase from *B. stearothermophilus* is responsible for controlling the entrance path of the catalytic water to the active site. One mutation (Y377F) increasing the hydrophobicity of this region also increased the transglycosidase/hydrolysis ratio with G3 as substrate from 0.75 in the wild type to 1.09 in the mutant enzyme.

Other changes in amino acid residues at this region, such as Met 375 and Ser 422, also affected this ratio. More recently, Cha et al.[13] reported that changes in Asn381 and Gly382 of a maltogenic α-amylase from *B. stearothermophilus* also increased transglycosylation. These amino acids are also located in this hydrophobic region. Recently, we were able to introduce transglycosylation activity in a liquefying α-amylase.[16] The strategy was based on the observation of another region of homology among amylases, found between the III and IVth conserved regions, as shown in Table 54.4.[16] It may be observed that, in this region, Tyr377 of neopullulanase or Phe284 of CGTases are substituted by small residues (Ala, Val, Ser) in hydrolytic (liquefying) amylases, as shown in the alignment.

Based on these observations, we tested two alternative residues, Tyr and Phe, to substitute Ala289 in the bacterial liquefying α-amylase from *B. stearothermophilus*. As shown in the TLC plates in Figure 54.4, the mutant Ala289Tyr substantially increased the transferase activity of the enzyme, as shown by the products of alcoholysis, absent in the wild-type enzyme from *B. stearothermophilus*. In the same figure, it is shown how introduction of transglucosidase activity in the enzyme changes the product profile when maltoheptaose (G7) is used as a substrate. In this case, higher-molecular-weight products are observed at the beginning of the reaction, and new end products appear such as glucose, G3, and G4. We have demonstrated that factors other than hydrophobicity, such as geometry and the electrostatic environment of the active site, may explain this change in activity and therefore in specificity.[16]

The stability of the enzyme is not affected by mutation, as the half-life at 80°C is reduced only from 1.7 to 1.6 h, while the pH profile, although not strongly modified, allows the assignment of three pKa values, most probably as a consequence of the electrostatic changes induced by the mutation.

These types of mutants with transglycosidase activity introduced in robust structures like α-amylases from *B. stearothermophilus* or *B. licheniformis*, combined

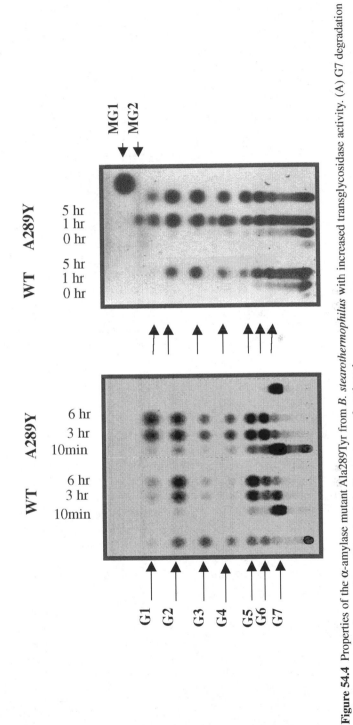

Figure 54.4 Properties of the α-amylase mutant Ala289Tyr from *B. stearothermophilus* with increased transglycosidase activity. (A) G7 degradation pattern and (B) reaction products from starch in the presence of methanol.

TABLE 54.4
A Fifth Conserved Region in α-Amylases Related to Specificity and Transferase Activity

AMY *A. oryzae*	[247]DGVLNYPIYYPLLNA
NPL *B. stearothermophilus*	[372] DAVMNYPFTDGVLRF
AMY *B. stearothermophilus*	[284]MSLFDAPLHNKFYTA
AMY *B. licheniformis*	[281]HSVFDVPLHYQFHAA
AMY *B. amyloliquefaciens*	[281]QSVFDVPLHFNLQAA
AMY *E. coli*	[285]TMLFDAPLQMKFHEA
CGT *B. macerans*	[279]MHLLDFAFAQEIREV
CGT *B. circulans*	[278]MSLLDFRFNSAVRNV
CGT *B. licheniformis*	[278]MSLLDFRFNSAVRNV
CGT *B. stearothermophilus*	[274]MSLLDFRFGQKLRQV

with already demonstrated stabilizing mutations,[17] may result in interesting industrial alternatives for the production of maltose or maltooligosaccharides in a one-step reaction from starch at high temperatures. Presently, maltose is produced by a combination of liquefying thermostable amylase, followed by a maltogenic saccharifying, but less stable, amylase.

54.7 USE OF AMYLASES IN THE CORN TORTILLA INDUSTRY

The *tortilla* industry represents one-fifth of the global food market in Mexico, with annual sales of $4 billion. Annual production is estimated at 11 million tons, 23% being produced from corn flour. The tortilla making process has been the subject of intensive research, and several reviews have been published regarding various aspects of this industry.[18] However, most studies dealing with additives are related to microbial spoilage, and less attention has been paid to the staling process. We have recently developed an α-amylase process to increase the shelf life of tortillas.[19] Extensively documented research has been carried out with the aim of extending the shelf life of bread. On standing, particularly at refrigerated temperatures, the swollen gelatinized starch granules become rigid by association of amylopectin, a phenomenon known as *retrogradation*. Therefore, a limited hydrolysis of amylopectin chains can inhibit and delay retrogradation with the concomitant extension of the shelf life. Although several chemicals have been studied to retard staling, the use of α-amylases has been widely accepted for some breads and has now reached commercial application in bakeries.[20] However, there are no reported applications concerning the use of α-amylases in the *tortilla* making process, in spite of huge losses due to tortilla staling.

Initial experiments with different fungal and bacterial enzymes added to corn dough before cooking have shown that concentration was a critical parameter. It was also evident that thermal deactivation during cooking was required to limit the reaction, avoiding damage to *tortilla* structure due to extensive hydrolysis. Therefore, thermostable enzymes were not useful. Successful treatments were easily obtained with activities of around 0.01 mU of α-amylases per gram of corn flour of thermolabile amylases during cooking. The enzymatic treatment decreased the rate of retrogradation by more than 50%. Clear evidence of the benefits of this process is given by the results of bending experiments: the rod diameter at which the treated tortilla fractures is smaller at any given time compared to a control. Treated tortillas and the corresponding controls were also analyzed in a texture analyzer. In all cases, the enzymatic treatment modified the rheological parameters, particularly reducing hardness and improving chewiness. Also, in the modified tortilla, the force required to penetrate (initial stress) was decreased. As no significant differences were found ($P < 0.01$) in the texture analysis results, it was concluded that tortillas treated with fungal α-amylases behave like fresh tortillas even after storage in conditions that favor staling. Significant changes in texture were, however, found at all levels for the control experiments in a sensory test, as 75 panelists were not able to differentiate (at a 5% level of significance) stored tortillas treated with α-amylases from fresh ones, with a clear rejection for the hardened samples. Industrial experiments were first conducted on one and two tons of flour experiments by MASECA S.A., and later scaled up, reaching commercial application in 1997. Future research in this area is related to the possibility of improving bacterial enzymes for this process by structural modifications to reduce thermal stability and to improve specificity.

54.8 ACKNOWLEDGEMENTS

Research grant IN212696 from DGAPA-UNAM.

REFERENCES

1. Schmid, G., 1989. "Cyclodextrin Glucosyl Transferases Production: Yield Enhancement by Overexpression of Cloned Genes," *Trends in Biotechnology,* 7: 244–248.
2. Crabb, W. D. and C. Mitchinson. 1997. "Enzymes Involved in the Processing of Starch to Sugars," *Trends in Biotechnology,* 15: 349–352
3. Brown, S. H., H. R. Constantino, and R. M. Kelly. 1990. "Characterization of Amylolytic Enzyme Activities Associated with the Hyperthermophilic Archaebacterium *Pyrococcus furiosus," Appl. Environ. Microbiol.,* 56: 1985–1991.
4. Imanaka, T. 1994. "Improvement of Useful Enzymes by Protein Engineering," in *Recombinant Microbes for Industrial and Agricultural Applications,* Y. Murooka and T. Imanaka, eds. N.York, NY: Marcel Dekker Inc., pp. 449–464.
5. Wong, D. W. S. 1995. *Food Enzymes.* New York, U.S.A.: Chapman & Hall.
6. Fogarty, W. M. 1983. "Microbial Amylases," in *Microbial Enzymes and Biotechnology,* W. M. Fogarty, ed. New York: Applied Sci. Pub., pp. 1–72.

7. Sinnott, M. L. 1990. "Catalytic Mechanisms of Enzymic Glycosyl Transfer," *Chem. Rev.*, 90: 1171–1202.
8. Santamaria, R. I., G. Del Rio, G. Saab, M. E. Rodríguez, X. Soberón, and A. López-Munguía. 1999. "Alcoholysis Reaction from Starch with α-Amylases," *FEBS Letters*. 452: 346–350.
9. Shinoyama, H., K. Takei, A. Ando, T. Fujii, M. Sasaki, Y. Doi, and T. Yasui. 1991. "Enzymatic Synthesis of Useful Alkyl β-Glucosides," *Agric. Biol. Chem.*, 55(6): 1676–1681.
10. Matsumura, S., H. Kubokawa, and S. Yoshikawa. 1991. "Enzymatic Synthesis of ω-Hydroxyalkyl and n-Alkyl β-D-Galactopyranosides by the Transglycosylation Reaction of β-Galactosidase," *Chemistry Letters*, 945–948.
11. Shinoyama, H., Y. Kamiyama, and T. Yasui. 1988. "Enzymatic Synthesis of Alkyl β-Xylosides from Xylobiose by Application of the Transxylosyl Reaction of *Aspergillus niger* Alkyl β-Xylosidase," *Agric. Biol. Chem.*, 51(9): 2197–2202.
12. Del Río, G., E. Morett, and X. Soberón. 1997. "Did Cyclodextrin Glycosyltransferases Evolve from α-Amylases?" *FEBS Letters*, 416: 221–224.
13. Cha, H. J., H. G. Yoon, Y. W. Kim, H. S. Lee, J. W. Kim, K. S. Kweon, B. H. Oh, and K. H. Park. 1998. "Molecular and Enzymatic Characterization of a Maltogenic Amylase that Hydrolyzes and Transglycosylates Acarbose," *Eur. J. Biochem.*, 253: 251–262.
14. Holm, L., A. K. Koivula, P. M. Lehtovaara, A. Hemminki, and J. K. C. Knowles. 1990. "Random Mutagenesis Used to Probe the Structure and Function of *B. stearothermophilus* α-Amylase," *Protein Engineering*, 3 (3): 181–191.
15. Kuriki, T., H. Kaneko, M. Yanase, H. Takata, J. Shimada, S. Handa, T. Takada, H. Umeyama, and S. Okada. 1996. "Controlling Substrate Preference and Transglycosylation Activity of Neopullulanase by Manipulating Steric Constraint and Hydrophobicity in Active Center," *J. Biol. Chem.*, 271(29): 17321–17329.
16. Saab-Rincón, G., G. Del Rio, R. I. Santamaría, A. López Munguía, and X. Soberón. 1999. "Introducing Transglycosylation Activity in a Liquefying α-Amylase," *FEBS Letters*, 453: 100–106.
17. Igarashi, K., Y. Hatada, K. Ikawa, H. Araki, T. Ozawa, T. Kobayashi. K. Ozaki, and S. Ito. 1998. "Improved Thermostability of a Bacillus α-Amylase by Deletion of an Arginine. Alycine Residue is Caused by Enhanced Calcium Binding," *Biochem. Biophys. Res. Comm.*, 248: 372–377.
18. Paredes-López O. and M. E. Saharopulos. 1983. "Maize: A review of Tortilla Production Technology," *Bakers Digest*, (13): 16–25.
19. Iturbe-Chiñas, F. A., R. M. Lucio, and A. López-Munguía. 1996. "Shelf Life of Tortilla Extended with Fungal Amylases," *Int. J. Food Sci. Technol.*, 31: 505–509.
20. Boyle, P. J. and R. E. Hebeda. 1990. "Antistaling Enzyme for Baked Goods," *Food Technol.*, (44): 129.

55 Modeling Conditions Affecting the Production of a Bacteriocin by *Enterococus faecium* UQ1 and the Kinetics of the Bacteriocin Antilisterial Activity

B. García-Almendárez
J. Ibarra-Silva
L. Mayorga-Martínez
J. Domínguez-Domínguez
C. Regalado

CONTENTS

Abstract

The antagonism of lactic acid bacteria (LAB) and/or their antimicrobial metabolites such as bacteriocins can be used in food biopreservation to control the growth of bacterial foodborne pathogens. Growth conditions have been associated with the production of bacteriocins, but the optimum must be determined experimentally for each bacteriocin, since the mechanism that activates their production is still unknown. A study of the antagonistic effect against related and pathogenic bacteria important to food safety was conducted on a bacteriocinogenic LAB isolated from kefir. It was identified as *Enterococcus faecium,* using biochemical tests and confirmed by ribotyping, and was labelled UQ1. The proteinaceous nature of the antimicrobial agent was confirmed by its sensitivity to proteolytic enzymes. Using the most sensitive strain, the activity was quantified by the agar diffusion method. Factorial designs and response surface methodology (RSM) were used to model the growth conditions leading to high bacteriocin activity. Using a 2^{6-2} fractional factorial screening design, six factors affecting bacteriocin production were evaluated. Significant factors were used for further optimization of bacteriocin production in a central composite design (2^{4-1}), using as a response variable a modified activity assay. The reduced second-order model obtained gave a good fit to experimental data. Conditions leading to maximum bacteriocin activity under static conditions were: yeast extract, 3g/l; beef extract, 6 g/l; peptone, 10 g/l; $MnSO_4$, 0.15 g/l; $MgSO_4$, 0.3 g/l; Tween 80, 2%; initial pH 6.25, temperature 25°C; 24 h; aerobiosis. Specific activity increased 17.5 times its value under initial conditions. After preliminary experiments, a 2^{5-1} factorial design was applied to optimize bacteriocin production using a shake culture, where a 30 times increase in specific activity was obtained after 5 h of incubation.

    L. monocytogenes viable counts in liquid medium decreased 6 log after 17 h using a bacteriocin concentration of 64 AU/ml. Using this activity the specific growth rate decreased to about a 1/3, while doubling time increased 2.7-fold.

55.1 INTRODUCTION

The antimicrobial effect of lactic acid bacteria (LAB) has been appreciated by mankind for centuries and has enabled the shelf life extension of many foods, mainly through fermentation processes. In recent years, the addition of chemical preservatives has been considered as a disadvantage by consumers looking for high quality, less processed, more natural foods, which in addition promote health. This search has resulted in the onset of a new generation of foods that are minimally processed and kept refrigerated.[1] Today, food preservation relies heavily on the use of multiple barriers known as "hurdle technology,"[2] to minimize the risk of microbial activity and therefore to ensure food safety.

LAB have a great potential to be exploited in biopreservation, because they are currently classified as GRAS by the Food and Drug Administration of the U.S.A. In addition, they can produce a variety of antimicrobial substances, including bacteriocins, and both microorganisms and their metabolites could be used as preservative hurdles. Bacteriocins usually have antimicrobial activity against taxonomically related species. In addition, they can inhibit bacterial foodborne pathogens such as *Clostridium botulinum, Bacillus spp., Listeria monocytogenes,* and *Staphylococcus aureus,* among others.

Bacteriocins are a heterogeneous group of antimicrobial compounds of proteinaceous nature, ribosomally produced by a wide group of bacterial species. They vary in their mode of action, activity spectrum, molecular weight, biochemical properties, and genetic determinants.[3] According to these criteria, they have been classified into four groups:

I. Lantibiotics, small peptides (<5 kDa) containing unusual aminoacids like lanthionin and b-methyl-lanthionin. They are subdivided into types A and B. Type A are elongated, cationic, pore-forming peptides such as nisin A[4] and subtilin, some of them consisting of two components.[5] Type B are enzyme inhibitors, immunologically active globular peptides;

II. Small, heat-stable nonlantibiotic peptides (<10 kDa). They are divided into four subgroups: IIA, peptides active against *Listeria,* with an N-terminal consensus sequence (YGNGVXC) such as bavaricin MN;[6] IIB, two peptide complexes, such as plantaricin S;[7] IIC, bacteriocins translated with secretion dependent leader peptides such as enterocin P;[8] and IID, bacteriocins that do not belong to the other subgroups: enterocin B.[9]

III. Heat labile proteins (>30 kDa), narrow antimicrobial spectra, such as Helveticin J.[10]

IV. Complex bacteriocins composed of an undefined mixture of proteins, lipids, and carbohydrates.

Bacteriocin production from enterococci is apparently a common feature for strains associated with food systems, especially those of *Enterococcus faecium.*[11] These bacteriocins have been shown active against Gram-positive and Gram-negative organisms, such as the foodborne pathogens *Staphylococcus* spp., *Clostridium* spp.,[12] and *Vibrio cholerae* 01,[13] respectively. *Ent. faecium* has also been used for fermentation of sugar cane bagasse to be used as a probiotic for animal feed.[14]

Many enterococci display antimicrobial activity against *Listeria monocytogenes,* a pathogen of great concern to public health.[15] This property is of big interest because of the involvement of this pathogen in many outbreaks related to soft cheeses and meat products, which have caused high mortality rates in recent years.[16] Therefore enterococci-producing bacteriocins and/or their bacteriocins have been successfully explored against *Listeria* spp. in various food products.[11,17] Enterocins activity against *L. monocytogenes* has been reported on cell free extracts,[15,18,19] but only a few enterocins have been purified and characterized.[9,20,21]

Enterococci are considered good indicators of faecal contamination in water, and epidemiological studies have shown a correlation between enterococci and disease risk.[22] However, Giraffa et al.[23] argued that there is a strong variability within the genus *Enterococcus* and important properties like antibiotic resistance and potential pathogenicity depend on the source of the isolate. Although enterococci and other lactic acid bacteria could occasionally be involved in clinical infections, many strains were concluded to be safe and useful microorganisms in food technology.

55.1.1 FACTORS AFFECTING BACTERIOCIN PRODUCTION

Commercial application of bacteriocins depends on the optimization of their production either by modification of genetic regulation or environmental growth parameters.[24] The cost of bacteriocin production can be reduced by determining the optimum parameters for its production. Few studies have been reported in which different media, the effect of media ingredients, or environmental factors for bacteriocin production are compared. Factors affecting the production of the lantibiotic nisin have been thoroughly studied;[25,26] apparently, carbon source regulation is a major control mechanism for its production.[25] However, De Vuyst[26] found that nisin biosynthesis had a strong dependence on a sulfur source that could be fulfilled by inorganic salts or sulfur aminoacids.

Reports about media ingredients affecting bacteriocins production are scarce. The effect of the ingredients of a complex medium (TGE) with supplementation or replacement of nutrients was evaluated on the production of pediocin AcH.[24] Glucose was the best carbon source; Tween 80 (0.2%) and Mn^{2+} (0.033 mM) produced optimal biomass (measured as OD) and pediocin AcH, while 0.02 mM Mg^{2+} had a stabilizing effect. Bacteriocin production using 16 strains of lactococci grown in M17 broth, BHI, milk, lactic broth, and a synthetic medium were compared by Geis et al.[27] where the highest titers were obtained using lactic broth. Pediocin production was increased by 27% when TGE broth was supplemented with yeast extract (1.5%), triptone (1.5%), pantotenic acid, and biotin.[28] They concluded that large biomass, when pH and growth temperature are optimal, is necessary for high levels of bacteriocin production, which is in agreement with other report.[29] However, plantaricin F specific activity was maximal when pH and temperature did not produce maximum cell yield.[30] A summary of growth factors having a significant influence on bacteriocin production is shown in Table 55.1.

Bacteriocin production is also dependent on the nature of the growth medium; higher yields have been obtained from a solid or semisolid than from a liquid medium.[31] Opposite results were found for plantaricin F, where 6.5 higher specific activity (AU/mg dry weight cells) was observed using liquid vs. agar-solidified MRS medium.[30]

Current understanding of bacteriocin regulation is inadequate for *a priori* selection or prediction of the optimum growth conditions for maximum production.[32] The traditional approach of varying one parameter at a time while keeping others constant

---

TABLE 55.1
Factors Associated with the Production of Bacteriocins

Factor	Level used	Reference
PH	5.2–7.9	35
	5–6	36
Temperature	30–45°C	37
	25–37°C	38
Yeast extract	0–2%	24
Oxygen	Aerobic, anaerobic	30
Media	TYT, MRS, M17, ELB	11
Tween 80	0.5–2%	29

TYT: triptone, yeast extract, Tween; MRS: Man-Rogosa-Sharpe; ELB: Eliker lactic broth.

---

is time consuming and may lead to incomplete understanding of the behavior of the system. The true optimum may not be detected, since interactions among the different factors may be missed.[33] A more efficient technique to establish significance of factorial interactions between variables to achieve the optimal conditions is response surface methodology (RSM).[34] The optimal conditions must be evaluated experimentally for each bacteriocin.[32]

The objective of this work was to evaluate the bacteriocinogenic potential of *E. faecium* UQ1 isolated from kefir and to assess the bacteriocin antagonism against foodborne pathogens and food spoilage microorganisms in liquid and solid media. We also deal with investigations using RSM to determine the most suitable growth conditions to maximize bacteriocin production in static and shake cultures.

55.2 MATERIALS AND METHODS

55.2.1 Microorganisms

E. faecium UQ1 was isolated from kefir after a screening made on 150 different foods such as meat, vegetables, fruits, and locally made fermented food products.[39] The producer strain was identified using the API 50 CHL system, and the results were analyzed through API LAB computer software. Identification was confirmed by sequence similarity of the 1500 nucleotide pairs of 16S rRNA (Midilab, DE, U.S.A.).

The strains used here (Table 55.2) were preserved at –70°C. Two previous cultures (1% v/v inoculum) were conducted at optimum growth conditions for each strain before the inoculum was ready for each experiment.

55.2.2 Media and Chemicals

The producing strain was grown on TGE broth, having the following composition:[40] peptone (Oxoid), 10 g/l; meat extract (Bioxon), 6 g/l; yeast extract (YE, Difco),

TABLE 55.2
Indicator Strains, Media and Incubation Temperatures Used

Strain	Medium	Temperature (°C)/time (h)
B. cereus ATCC 11778	STB	37/24
B. circulans NCIMB 11844	STB	37/24
B. subtilis NCIMB 08565	STB	37/24
E. coli EPEC (pozol)	STB	37/24
Lb. plantarum 5T	MRS	30/18
Lb. plantarum 6L	MRS	30/18
Lb. plantarum 14L	MRS	30/18
Lb. plantarum 9L	MRS	30/18
Lb. plantarum 18L	MRS	30/18
Lb. plantarum 20L	MRS	30/18
Lb. plantarum 34L	MRS	30/18
Lb. plantarum 35L	MRS	30/18
Lc. lactis NCDO 496	MRS	30/18
Le. mesenteroides NCDO 523	MRS	30/18
L. innocua	STB–0.6%YE	37/24
L. monocytogenes Scott A	STB–0.6%YE	37/24
M. luteus NCIB 8166	AM	30/48
P. acidilactici ATCC 8092	MRS	30/18
P. pentosaceus ATCC 8092	MRS	30/18
S. aureus ATCC 25923	STB	37/24

STB: Soya trypticasein broth; MRS: Man, Rogosa, Sharpe; YE: yeast extract; AM: assay medium.

4 g/l; glucose (Sigma), 10 g/l; $MnSO_4$ (Sigma), 0.05 g/l; $MgSO_4$ (Sigma), 0.10 g/l; Tween 80 (Sigma), 2% v/v. *Micrococcus luteus* was grown on assay medium having the following composition:[41] peptone, 10 g/l; meat extract, 3 g/l; yeast extract (YE), 1.5 g/l; bacteriological agar (Bioxon), 10 g/l; NaCl (Sigma), 3 g/l; demerara sugar (Dilis), 1 g/l.

The bacteriocin sensitivity was tested against the following enzymes: bacterial α-amylase (Merck), catalase solution from bovine liver (Boehringer), lipase type II from porcine pancreas, bacterial pronase E type XIV, trypsin from bovine pancreas, α-chymotrypsin type II from bovine pancreas, and proteinase K from *Tritirachium album* were purchased from Sigma.

Nisin was a gift from Applin & Barret, Ltd. (England). All other chemicals used were purchased from commercial suppliers and were of analytical grade or better.

Anaerobic conditions were created in an anaerobic jar added with a gas generating kit (Oxoid BR 38B, H_2 and 10% v/v CO_2), while, for microaerophilic conditions, a system envelope (campy pack BBL; 10% CO_2, 6% O_2, and H_2) was added to the jar.

55.2.3 Bacteriocin Activity Spectrum

The bacteriocin antagonic effect was conducted using the spot method.[42] Two milliliters of an overnight broth culture was spotted onto a soft TGE (0.8% agar), containing 150 U/ml of catalase, without glucose. After overnight incubation at 30°C, the plates were overlaid with 8 ml of TGE or STA (0.8% agar)–0.6% YE, and inoculated with 20 μl of the indicator strains to a level of 10^6–10^7 cfu/ml, respectively. The plates were incubated for 18–24 h at the optimum temperature of these strains (see Table 55.2). The presence of an inhibition zone around the spots was considered a positive antagonic effect.

55.2.4 Preparation of the Cell Free Extract (CFE)

Samples were taken at different time intervals from the fermentation vessel and were heated to 70°C for 30 min to avoid bacteriocin inactivation by proteases. Previous experiments showed that this heat treatment had no effect in the bacteriocin activity. Samples were adjusted to pH 2.0 using 1 M HCl, and later placed at 4°C for 2 h under mild stirring. Previous experiments conducted with cell cultures in the pH range 1–6, followed by cell removal, showed that bacteriocin activity was the highest in cultures adjusted to pH 2.0, according to the procedure depicted by Yang et al.[43] Cells were removed by centrifugation at 13000xg for 15 min at 4°C. The supernatant pH was adjusted to 6.5 with 1 M NaOH and the filter sterilized using a Millipore membrane (0.22 μm).

55.2.5 Bacteriocinogenic Activity

The activity of CFE was evaluated using the well diffusion assay.[44] Plates having a bottom layer of 10 ml AM (1.5% agar) were overlaid with 7 ml AM containing 0.8% agar and inoculated with 20 μl of *M. luteus* or *L. monocytogenes* (10^6–10^7 cfu/ml). A stainless steel cork borer was used to cut wells of 5 mm diameter. Twenty five μl of serial two-fold dilutions of CFE were added to the wells[45] and incubated at 30°C (*M. luteus*) or 37°C (*L. monocytogenes*). The antimicrobial activity was defined as the reciprocal of the highest dilution showing clear zones of inhibition, of at least 1 mm, and was expressed as arbitrary units per ml (AU/ml). Each assay was conducted in triplicate.

The bacteriocin sensitivity to different enzymes was determined by incubating 300 μl of CFE with 30 μl of the tested enzymes (10 mg/ml), kept at optimum pH (20 mM buffers) and temperature for 2 h.[46] The residual activity was evaluated as described above.

55.2.6 Growth Factors Affecting Bacteriocin Production

The experimental strategy sequence used to study the influence of six growth factors on increased bacteriocin production is shown in Figure 55.1.

Based on reports in the literature, a preliminary scrutiny was conducted using growth factors and levels that have been associated with the production of high amounts of bacteriocin activity. A fractional factorial design 2^{6-2} was selected to

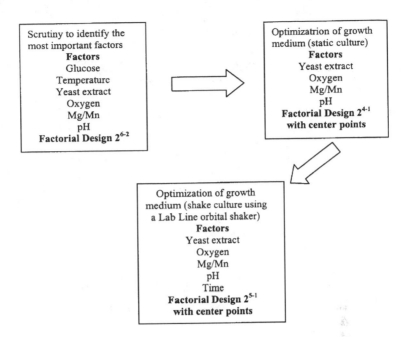

Figure 55.1 Sequence of the experimental strategy followed to optimize bacteriocin production starting from six growth conditions.

identify the significant factors. The studied factors and their corresponding low and high levels were pH (5.5–6.5); YE (3.0–6.0 g/l); Mg^{+2} (0.1–0.3 g/l); oxygen (anaerobic-aerobic); temperature (25–30°C) and glucose (0–5 g/l). The effect of Mg^{2+} and Mn^{2+} were evaluated as the ratio [Mg]/[Mn] having a constant value of 2.

Experiments were conducted in random order and in duplicate. The effect of the different factors on bacteriocin activity (response variable) was estimated using a second-order polynomial empirical model. This model allows a good approximation of the true behavior of a wide variety of multifactorial systems,[47] and can be expressed by

$$Y = a_o + a_1x_1 + a_2x_2 + a_3x_3 + a_4x_4 + a_{12}(x_1x_2) + a_{13}(x_1x_3) +$$
$$a_{14}(x_1x_4) + a_{23}(x_2x_3) + a_{24}(x_2x_4) + a_{34}(x_3x_4) + a_{11}(x_1)^2$$
$$+ a_{22}(x_2)^2 + a_{33}(x_3)^2 + a_{44}(x_4)^2$$

$$(55.1)$$

where
Y = estimated response
a_0 = regression coefficient at center point
a_1, a_2, a_3, a_4 = linear coefficients
x_1, x_2, x_3, x_4 = independent variables
a_{12}, a_{13}, a_{14} = second-order interaction coefficients
$a_{11}, a_{22}, a_{33}, a_{44}$ = quadratic coefficients

In this study, factors having significant effects on the response variable were selected from the screening factorial experiments. These factors were used in a fractional central composite design 2^{4-1} added with center points to increase the precision of the fitted model, in a further optimization of the bacteriocin activity. Equation (55.1) was applied to model the experimental results, and a reduced model was obtained by including only the most significant factors and their interactions, leading to a maximum predicted bacteriocin activity. Confirmatory tests were carried out to validate the model predictions for static conditions.

The same factors were also tested under shaking conditions, and, after preliminary experiments, significant factors affecting bacteriocin activity were detected. Using a new central composite design, 2^{5-1}, bacteriocin activity could be optimized under these conditions (Figure 55.1). Statistical analysis was carried out using Statgraphics v. 6.0 software. Contour and response surface plots were obtained using SAS software.[48]

55.3 RESULTS AND DISCUSSION

55.3.1 IDENTIFICATION OF THE ANTIMICROBIAL AGENT AS BACTERIOCIN

Spectrum of Activity

Activity spectrum of the bacteriocin produced by *E. faecium* UQ1 is shown in Table 55.3. *Listeria monocytogenes* Scott A and *Micrococcus luteus* NCIB8166 were among the most sensitive strains to this bacteriocin, along with other taxonomically related lactic acid bacteria. The sensitivity of *M. luteus* to this bacteriocin allowed activity quantitation in later experiments.

Sensitivity to Proteolytic and Other Enzymes

The proteinaceous nature of the bacteriocin was confirmed by its sensitivity to different proteolytic enzymes, where a total inactivation after treatment with pronase E was observed (Table 55.4). The activity losses observed after treatment with α-amylase and lipase suggest a complex composition of this bacteriocin. However, protease activity evaluated by the method of Kunitz,[49] using Hammarsten casein as substrate, was observed in the lipase used (as stated by the supplier), while a weak proteolytic activity was also found in the α-amylase sample. Therefore, traces of proteases were probably responsible for the observed effect.

55.3.2 GROWTH OF *E. faecium* UQ1 AND BACTERIOCIN PRODUCTION

The growth curve, bacteriocin production, biomass, and pH were evaluated using initial conditions in modified TGE medium[40] (Figure 55.2). Bacteriocin activity was detected at the late stationary phase, and achieved its highest, 320 AU/ml, after 18 h of incubation. Bacteriocin production appeared simultaneously with a sudden pH

TABLE 55.3
Activity Spectrum of the Antagonic Effect of the Bacteriocin Produced by *E. faecium UQ1*

Indicator microorganism	Sensitivity	Zone of Inhibition (mm)		
Pediococcus acidilactici ATCC 8092	+++	3.0	3.5	3.7
P. pentosaceus NCDO 990	+++	3.2	3.4	4.0
Lactococcus lactis NCDO 496	+	1.7	1.6	1.5
Leuconostoc mesenteroides NCDO 523	+	1.4	1.2	1.3
Lactobacillus plantarum 5T*	+	1.1	0.8	1.0
L. plantarum 6L*	+	1.0	0.9	1.0
L. plantarum 5L*	−			
L. plantarum 14L*	+	1.0	1.9	1.0
L. plantarum 9L*	+	1.0	0.9	0.9
L. plantarum 18L*	+	1.5	1.6	1.4
L. plantarum 20L*	−			
L. plantarum 34L*	−			
L. plantarum 35L*	−			
E. faecium UQ1	−			
Listeria monocytogenes Scott A	+++	3.6	4.4	3.1
Listeria innocua	+++	3.7	3.2	3.4
Escherichia coli (EPEC)	−			
Micrococcus luteus NCIB 8166	+++	3.3	3.5	3.7
Bacillus cereus ATCC 11778	−			
Bacillus subtilis NCIMB 08565	−			
Bacillus circulans NCIMB 11844	−			
Staphylococcus aureus ATCC 25923	−			

(+++) very sensitive; (+) sensitive; (−) not sensitive

*From culture collection of Facultad de Qúimica, UNAM, México.

decrease. The high bacteriocin activity remained for a short time, and a sharp decrease followed, probably due to the presence of extracellular proteases, as estimated by the method of Kunitz.[49]

55.3.3 OPTIMIZATION OF GROWTH MEDIUM FOR BACTERIOCIN PRODUCTION

Static Culture

The influence of growth conditions on bacteriocin production was carried out, initially studying six factors: pH (A), temperature (B), yeast extract (C), Mg (D), oxygen (E), and glucose (F), using two levels of each factor. This screening test was

TABLE 55.4
Sensitivity of the Bacteriocin Produced by *E. faecium* UQ1 to Different Enzymes

Enzyme	Sensitivity	Activity (AU/ml)	% Activity
Control		320	100
α-chymotrypsin	+	160	50
Trypsin	++	80	25
Pronase E	+++	0	0
Proteinase K	++	80	25
Lipase	+++	0	0
Catalase	–	320	100
α-amylase	+	160	50

conducted using a 2^{6-2} fractional factorial design, whose design matrix is shown in Table 55.5. Preliminary trials were conducted using selected experiments, sampling after 18, 20, and 24 h of fermentation time. In all cases, bacteriocin activity was highest after 24 h and was kept constant throughout the experiments. From analysis of variance, the statistically significant ($p < 0.05$) factors and interactions were pH, Mg, oxygen, and YE.

Temperature and glucose were not significant factors in the chosen levels, although the carbon source has been reported as a major control mechanism for nisin production.[25] Analyzing the effect of principal factors using the statgraphics

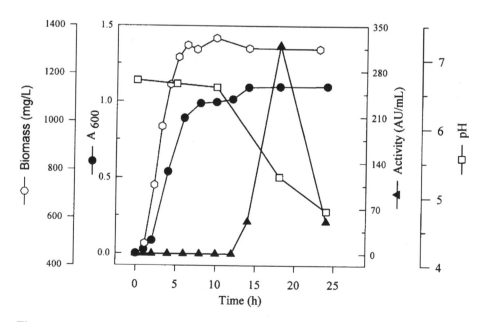

Figure 55.2 Initial growth conditions for *E. faecium* UQ1 in TGE medium at 30°C for 24 h.

TABLE 55.5
Experimental Design Showing Values of Factors and Response Variable (Bacteriocin Activity) in the Screening Fractional Factorial 2^{6-2}

Trial	pH	T	YE	Mg	Oxygen	Glucose	Activity
1	5.5	25	3	0.1	Anaerobic	0	80
2	6.5	25	3	0.1	Aerobic	0	160
3	5.5	30	3	0.1	Aerobic	5	0
4	6.5	30	3	0.1	Anaerobic	5	160
5	5.5	25	6	0.1	Aerobic	5	0
6	6.5	25	6	0.1	Anaerobic	5	80
7	5.5	30	6	0.1	Anaerobic	0	0
8	6.5	30	6	0.1	Aerobic	0	0
9	5.5	25	3	0.3	Anaerobic	5	640
10	6.5	25	3	0.3	Aerobic	5	320
11	5.5	30	3	0.3	Aerobic	0	0
12	6.5	30	3	0.3	Anaerobic	0	1280
13	5.5	25	6	0.3	Aerobic	0	0
14	6.5	25	6	0.3	Anaerobic	0	0
15	5.5	30	6	0.3	Anaerobic	5	80
16	6.5	30	6	0.3	Aerobic	5	80
17	5.5	25	3	0.1	Anaerobic	0	80
18	6.5	25	3	0.1	Aerobic	0	160
19	5.5	30	3	0.1	Aerobic	5	0
20	6.5	30	3	0.1	Anaerobic	5	160
21	5.5	25	6	0.1	Aerobic	5	0
22	6.5	25	6	0.1	Anaerobic	5	80
23	5.5	30	6	0.1	Anaerobic	0	0
24	6.5	30	6	0.1	Aerobic	0	0
25	5.5	25	3	0.3	Anaerobic	5	640
26	6.5	25	3	0.3	Aerobic	5	160
27	5.5	30	3	0.3	Aerobic	0	0
28	6.5	30	3	0.3	Anaerobic	0	1280
29	5.5	25	6	0.3	Aerobic	0	0
30	6.5	25	6	0.3	Anaerobic	0	0
31	5.5	30	6	0.3	Anaerobic	5	80
32	6.5	30	6	0.3	Aerobic	5	80

T: temperature; YE: yeast extract; MG: Magnesium [ratio MG/Mn (w/w) = 2].

software, the model suggested that pH and Mg contributed to higher response at their high levels, while the opposite was indicated for YE and oxygen. The normal probability and the residual plots suggested a completely random behavior (results not shown). A determination coefficient (R^2) of 0.87 indicated a good fit between the model and experimental data. Therefore, the optimization of the significant factors was further examined.

The most important factors influencing bacteriocin activity were arranged in a central composite design 2^{4-1}, having 16 trials plus 3 center points; the response variable was bacteriocin activity. Factors and levels used in this experimental design and the results obtained for the response variable are shown in Table 55.6. Final pH and biomass are also shown in this table. Fermentation time was kept constant at 24 h. Results showed little pH change, with a mean difference of acidification through lactic acid production, at the expense of a reduced growth. This can be observed by comparison of the low biomass values found (Table 55.6) with those shown in Figure 55.2.

TABLE 55.6

Fractional Central Composite Design 2^{4-1}, Showing Values of Factors and Response Variable (Bacteriocin Activity)

Trial	Initial pH	YE (g/l)	Mg (g/l)	Oxygen	Biomass (mg/l)	Activity (AU/ml)	Final pH
1	5.5	1	0.20	Anaerobic	302	160	6.04
2	7.0	1	0.20	Aerobic	289	160	6.97
3	5.5	5.0	0.20	Aerobic	385	80	5.47
4	7.0	5.0	0.20	Anaerobic	498	160	6.73
5	5.5	1	0.40	Aerobic	267	320	5.44
6	7.0	1	0.40	Anaerobic	295	80	6.96
7	5.5	5.0	0.40	Anaerobic	315	640	5.43
8	7.0	5.0	0.40	Aerobic	472	160	6.97
9	5.5	3.0	0.30	Microaerophilic	417	80	5.46
10	7.0	3.0	0.30	Microaerophilic	417	160	6.67
11	6.25	1.0	0.30	Microaerophilic	237	160	6.23
12	6.25	5.0	0.30	Microaerophilic	470	160	6.61
13	6.25	3.0	0.20	Microaerophilic	291	320	6.01
14	6.25	3.0	0.40	Microaerophilic	350	80	6.18
15	6.25	3.0	0.30	Anaerobic	341	160	6.24
16	6.25	3.0	0.30	Aerobic	350	1280	6.01
17	6.25	3.0	0.30	Microaerophilic	363	640	6.25
18	6.25	3.0	0.30	Microaerophilic	413	320	6.14
19	6.25	3.0	0.30	Microaerophilic	365	640	6.20

Analysis of variance showed that the quadratic effects of pH, Mg, and oxygen were the most significant ($p < 0.05$). However, the model gave a low fit to the experimental data. When the CFE was slightly concentrated by heating under vacuum (85°C for 30 min), the inhibition zone observed upon activity evaluation was larger than that of the unconcentrated sample. However, both samples reached the same critical dilution and thus had the same activity (AU/ml).

This observation was incorporated to develop a modified activity measurement, using the critical dilution method and the same sample volume (25 µl), but subtracting the final from the initial length of the inhibition zone (mm). This activity was defined as modified arbitrary units (MAU). There was a good correlation between these two methods of assessing activity ($R^2 = 0.87$), having the following regression equation:

$$\text{Activity (AU/mL)} = 227.4 \text{ Activity (MAU)} - 8.7 \tag{55.2}$$

When the response variable was expressed as MAU, analysis of variance showed that the factors affecting significantly bacteriocin production were pH, Mg, and oxygen, while YE and its interactions were not significant. Taking the statistically significant factors, a reduced model was obtained ($R^2 = 0.86$; $p < 0.05$). The fitted model gave good agreement with experimental data and is shown in Equation (55.3).

$$\text{Activity (MAU)} = 0.60 - 0.069 \, A + 0.0057 \, C + 0.0028 \, D + -0.44 \, A^2$$

$$-0.39 \, C^2 + 0.85 \, D^2 - 0.097 \, AC + 0.043 \, AD + 0.10 \, CD \tag{55.3}$$

where coded values for the factors are $A = (\text{pH} - 6.25)/0.75$; $C = (\text{Mg} - 0.3)/0.1$; Oxygen $= D$, discrete values were assigned ($-1 =$ anaerobic, $0 =$ microaerophilic, $1 =$ aerobic).

Conditions for optimal bacteriocin activity were obtained from contour plots (data not shown) and response surface analysis (Figure 55.3). Maximum activity could be achieved by using pH 6.2; $MgSO_4$, 0.32 g/l; aerobic; keeping constant: peptone, 10 g/l; YE, 3 g/l; meat extract, 6 g/l; and $MnSO_4$, 0.16 g/l, 25°C.

Confirmatory tests were carried out using these conditions, in triplicate, where activity, pH, and biomass were monitored as a function of time, and optimal activity was 5.67 ± 0.35 mm (1280 AU/ml) after 24 h of incubation (Figure 55.4). The specific bacteriocin activity at this optimum was 4275.2 AU/mg dry weight biomass. This value is 17.5 times larger than under initial conditions (245 AU/mg dry weight biomass; see Figure 55.2).

The relatively high pH and low biomass observed for optimized conditions in this study demonstrates that it is not necessary to have a low final pH or high biomass for increased bacteriocin production, which is in agreement with the results of Paynter et al.[30]

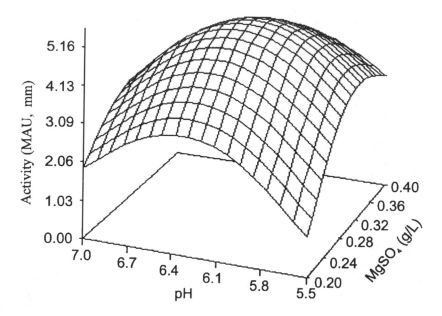

Figure 55.3 Response surface of the length of inhibition zone (MAU) calculated with Equation (55.3) as a function of pH and MgSO$_4$. Oxygen level was kept constant to 1 (aerobic conditions).

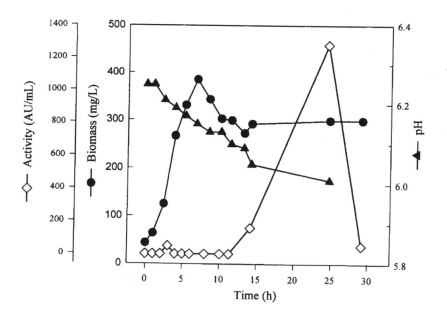

Figure 55.4 Growth of *Enterococcus faecium* UQ1 under optimized bacteriocin production conditions in a static culture at 25°C.

Shake Culture

A shake culture was used to evaluate the influence of higher oxygen tension under optimized parameters from the static culture as a function of time. Agitation speed was kept constant at 150 rpm, while temperature was 25°C. Preliminary experiments showed that bacteriocin production was strongly influenced by fermentation time, and the maximum activity under these conditions was detected after about 4 h. A fractional factorial design 2^{5-1}, added with four center points to increase precision of the model, was used to study the influence of four growth factors on bacteriocin production. These factors were found significant under static conditions, plus time. Time levels studied were 3, 4, and 5 h, while the other factors had the same levels of the 2^{4-1} design. The response variable was activity expressed as MAU. Analysis of variance showed a good fit of experimental data to the model, having an adjusted R^2 of 0.87. The residual and normal plots showed that experimental values distributed randomly with no unusual behavior. The lack of fit ($p = 0.88$) showed the adequacy of the model to the data. These data were adjusted to a first-order model and expressed as

$$\text{Activity (MAU)} = 1.27 + 0.37\ B - 0.54\ C + 0.70\ D + 0.47\ E - 0.27\ AB$$
$$+ 0.35\ AC + 0.37\ BD + 0.43\ BE - 0.43\ CD + 0.87\ DE \qquad (55.4)$$

where $B = (\text{YE} - 3)/2$; $E = (\text{time} - 4)/1$; other values are defined in Equation (55.3).

The bacteriocin activity predicted using Equation (55.4) increased as YE and fermentation time increased (Figure 55.5). The response can be expressed as AU/ml using Equation (55.2). The effect of pH was not significant, only its interactions,

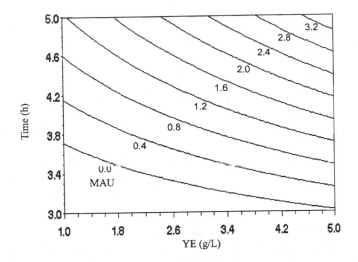

Figure 55.5 Contour plot of predicted bacteriocin production, according to Equation (55.4), as a function of pH and YE at 25°C. Factors fixed: pH, 6.25; Mg, 0.2 g/l. Values for MAU are shown inside.

and, for the confirmatory experiments (four replicates), it was kept constant at its medium level, while the low level for Mg and the highest YE and time levels were used. An example of such growth curves as a function of time is shown in Figure 55.6. The mean MAU was 5.16 ± 0.32 mm, while the predicted value using Equation (55.4) was 5.43 mm. According to a t-test, the MAU value is not different from the predicted maximum ($p < 0.05$). Further tests were carried out increasing YE and incubation time, as suggested by the rising ridge shown in Figure 55.5, but bacteriocin activities were always lower than the maximum previously found. Therefore, growth conditions found here were close to the optimum bacteriocin activity production: pH 6.25; YE 5 g/l; Mg^{2+} 0.2 g/l; Mn^{2+} 0.1; incubation time 5 h; peptone, 10g/l; meat extract, 6 g/l; 150 rpm, at 25°C.

The fact that bacteriocin activity obtained under static and shake cultures stopped at the same value (1280 AU/ml) indicates that a specific medium component necessary for synthesis or secretion of this bacteriocin is the limiting factor. Another possibility is a defense mechanism of the producer strain, which might be activated when a certain bacteriocin titer is achieved.[50] The specific activity found was 7333 AU/mg dry weight biomass, which is higher than initial and static conditions. Initially, the pH probably decreased because *E. faecium* UQ1 uses a carbon source from YE; when it was exhausted, alternative metabolic routes might have produced compounds that increased the pH.

55.3.4 KINETICS OF ANTILISTERIAL ACTIVITY

The optimized conditions from the shake culture were used to produce CFE that were challenged against growing cells of *Listeria monocytogenes* Scott A in STB-

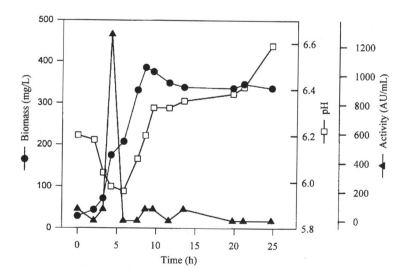

Figure 55.6 Growth of *E. faecium* UQ1 under optimized bacteriocin production conditions in a shake culture (150 rpm) at 25°C.

0.6% YE (Figure 55.7). After 17 h, there was a 6 log reduction using 64 AU/ml, while a 3 log reduction was observed using 19 AU/ml. However, using 19 AU/ml, the viable count after 23 h was similar to the control, indicating a bacteriostatic effect. On the other hand, using the high titer, a bactericidal effect was maintained throughout the experiment. The effect of this bacteriocin on the growth kinetics of *L. monocytogenes* is shown in Table 55.7. The bacteriocin effect on doubling time showed similar behavior. The use of higher concentrations might result in the aggregation of the peptide, owing to its hydrophobic nature, which may decrease the effectiveness of bactericidal effects.[37]

TABLE 55.7
***L. monocytogenes* Growth Kinetics at Different *E. faecium* Bacteriocin Titers, STB–0.6% YE at 37°C**

Bacteriocin activity (AU/ml)	*L. monocytogenes* Scott A	
	μ (h^{-1})	t_d (h)
0	1.07	0.64
19	0.58	1.19
64	0.39	1.75

μ = specific growth rate (h^{-1}); t_d = doubling time (h).

Therefore, the potential of this bacteriocin to control *L. monocytogenes* growth has been demonstrated; further experiments are needed in food systems to assess its use as an additional hurdle in food biopreservation applications.

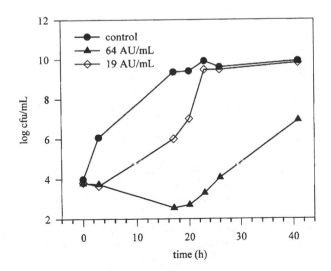

Figure 55.7 Effect of the bacteriocin from *E. faecium* UQ1 on *L. monocytogenes* Scott A growing cells. STB-0.6% YE at 37°C.

55.4 CONCLUSIONS

M. luteus and *L. monocytogenes* Scott A were found to be sensitive to the bacteriocin produced by *E. faecium* UQ1, while the activity was effective against *L. monocytogenes* on solid and liquid media. The proteinaceous nature of the antimicrobial compound was confirmed by partial or total inactivation by different proteolytic enzymes. Heat treatments at 70°C for 30 min did not cause any activity loss. Factorial designs and response surface methodology can be a valuable tool to determine significant factors affecting bacteriocin production and allow media optimization under static and shake culture conditions.

The models developed showed a good fit between experimental and predicted bacteriocin activities by varying four significant growth factors for static culture and five for shake culture. Specific bacteriocin activity was initially 245 AU/mg dry weight biomass; it increased 17.5 times under optimized static conditions and was 30 times higher under optimized shaking conditions. This type of modeling can be especially important to predict the effect of varying conditions for further scaling of *in vitro* and *in situ* bacteriocin production processes.

The kinetic parameters of *L. monocytogenes* growing cells were significantly affected by the bacteriocin produced by *Enterococcus faecium* UQ1 under optimized shaking conditions.

55.5 ACKNOWLEDGMENTS

This work was supported by grant no. 3145P-B9607, from CONACYT. Financial support through an MSc grant to J. I. S., and a BSc grant to L. M. M., from CONACYT, as well as through a visiting professorship to J. R. Whitaker is greatly appreciated.

REFERENCES

1. Stiles, M. E. 1996. "Biopreservation by Lactic Acid Bacteria," *Anton. Leeuw.,* 70: 331–345.
2. Leistner, L. and G. G. Gorris. 1995. "Food Preservation by Hurdle Technology," *Trends Food Sci. Tech.,* 6: 41–46.
3. Klaenhammer, T. R. 1993. "Genetics of Bacteriocins Produced by Lactic Acid Bacteria," *FEMS Microbiol. Rev.,* 12: 39–86.
4. Gross, E. and J. L. Morell. 1971. "The Structure of Nisin," *J. Am. Chem. Soc.,* 93: 4634–4635.
5. Navaratna, M. A. D. B., H. G. Sahl, and J. R. Tagg. 1998. "Two-Component Anti-*Staphylococcus aureus* Lantibiotic Activity Produced by *Staphylococcus aureus* C55," *Appl. Environ. Microbiol.,* 64: 4803–4808.
6. Kaiser, A. L. and T. J. Montville. 1996. "Purification of the Bacteriocin bavaricin MN and Characterization of its Mode of Action against *Listeria monocytogenes* Scott A Cells and Lipid Vesicles," *Appl. Environ. Microbiol.,* 62: 4529–4535.

7. Jiménez-Díaz, R., Barba-Ruiz, J. L., and D. P. Cathcart. 1995. "Purification and Partial Amino Acid Sequence of Plantaricin S, a Bacteriocin Produced by *Lactobacillus plantarum* LPCO10, the Activity of which Depends on the Complementary Action of Two Peptides," *Appl. Environ. Microbiol.*, 61: 4459–4463.

8. Cintas, L. M., L. S. Casaus, L. S. Håverstein, P. E. Hernández, and I. F. Nes. 1997. "Biochemical and Genetic Characterization of Enterocin P, a Novel Sec-Dependent Bacteriocin from *Enterococcus faecium* P13 with a Broad Antimicrobial Spectrum," *Appl. Environ. Microbiol.*, 63: 4321–4330.

9. Casaus, P., T. Nilsen, L. M. Cintas, I. F. Nes, P. E. Hernández, and H. Holo. 1997. "Enterocin B, a New Bacteriocin from *Enterococcus faecium* T136 which can Act Synergistically with Enterocin A," *Microbiol.*, 143: 2287–2294.

10. Joerger, M. C. and T. R. Klaenhammer. 1986. "Characterization and Purification of Helveticin J and Evidence from a Chromosomally Determined Bacteriocin Produced by *Lactobacillus helveticus* 481," *J. Bacteriol.*, 167: 439–446.

11. Parente, E. and C. Hill. 1992. "A Comparison of Factors Affecting the Production of Two Bacteriocins from Lactic Acid Bacteria," *J. Appl. Bacteriol.*, 73: 290–298.

12. Giraffa, G. 1995. "Enterococcal Bacteriocins: Their Potential Use as Anti-*Listeria* Factors in Dairy Technology," *Food Microbiol.*, 12: 291–299.

13. Simonetta, A. C., L. G. Moragues, and L. N. Frisón. 1997. "Antibacterial Activity of Enterococci Strains Against *Vibrio cholerae*," *Lett. Appl. Microbiol.*, 24: 139–143.

14. Iritani, S., M. Mitsuhashi, H. Chaen, and T. Miyake. 1997. "Biological Pure Strain of *Enterococcus faecium* FERM BP-4504," United States Patent 5,622,859.

15. Olasupo, N. A., U. Schillinger, C. M. A. P. Franz, and W. H. Holzapfel. 1994. "Bacteriocin Production by *Enterococcus faecium* NA01 from 'wara'—a Fermented Skimmed Cow Milk Product from West Africa," *Lett. Appl. Microbiol.*, 19: 438–441.

16. CDC. 1999. "Update: Multi-State Outbreak of Listeriosis," Centers for Disease Control and Prevention. Atlanta, GA. Office of communication. March 17, http://www.cdc.gov/od/oc/media/.

17. Mayorga-Martínez, L. 1999. *Estudio del Antagonismo y Potencial Bacteriocinogénico de una Cepa de Bacteria Láctica Aislada de Kefir.* BS Thesis. Faculty of Chemistry. UAQ. México.

18. Ennahar, D., D. Aoude-Werner, O. Assobhei, and C. Hasselmann. 1998. "Antilisterial Activity of Enterocin 81, a Bacteriocin Produced by *Enterococcus faecium* WHE 81 Isolated from Cheese," *J. Appl. Microbiol.*, 85: 521–526.

19. Lauková, A., S. Csikková, Z. Vasilková, P. Juris, and M. Marekova. 1998. "Occurrence of Bacteriocin Production Among Environmental Enterococci," *Lett. Appl. Microbiol.*, 27: 178–182.

20. Vlaemynck, G., L. Herman, and K. Coudijzer. 1994. "Isolation and Characterization of Two Bacteriocins Produced by *Enterococcus faecium* Strains Inhibitory to *Listeria monocytogenes*," *Int. J. Food Microbiol.*, 24: 211–225.

21. Hill, C. 1994. "Enterocin 1146, a Bacteriocin Produced by *Enterococcus faecium* DPC1146," in *Bacteriocins of Lactic Acid Bacteria*, L. De Vuyst and E. J. Vandamme, eds. London: Blackie Academic and Professional, pp. 515–528.

22. Wade, J. J. 1997. "*Enterococcus faecium* in Hospitals," *Eur. J. Clin. Microbiol. Infect. Dis.*, 16: 113–119.

23. Giraffa, G., D. Carminati, and E. Neviani. 1997. "Enterococci Isolated from Dairy Products: Review of Risks and Potential Technological Use," *J. Food Protect.*, 60: 732–738.

24. Biswas S. R., P. Ray, M. C. Johnson, and B. Ray. 1991. "Influence of Growth Conditions on the Production of a Bacteriocin, Pediocin AcH, by *Pediococcus acidilactici* H," *Appl. Environ. Microbiol.*, 57: 1265–1267.

25. De Vuyst, L. and E. J. Vandamme. 1992. "Influence of the Carbon Source on Nisin Production in *Lactococcus lactis* subsp. *lactis* Batch Fermentations," *J. Gen. Microbiol.*, 138: 571–578.

26. De Vuyst, L. 1995. "Nutritional Factors Affecting Nisin Production by *Lactococcus lactis* subsp. *lactis* NIZO 22186 in a Synthetic Medium," *J. Appl. Bacteriol.*, 78: 28–33.

27. Geis, A., J. Singh, and M. Teuber. 1983. "Potential of Lactic Streptococci to Produce Bacteriocin.," *Appl. Environ. Microbiol.*, 45: 205–11.

28. Yang, R. and B. Ray. 1994. "Factors Influencing Production of Bacteriocins by Lactic Acid Bacteria," *Food Microbiol.*, 11: 281–291.

29. Vignolo, G. M., M. N de Kairuz, A. Ruiz, and G. Oliver. 1995. "Influence of Growth Conditions on the Production of Lactocin 705, a Bacteriocin Produced by *Lactobacillus casei* CRL 705," *J. Appl. Bacteriol.*, 78: 5–10.

30. Paynter, M. J. B., K. A. Brown, and S. S. Hayasaka. 1997. "Factors Affecting the Production of an Antimicrobial Agent, Plantaricin F, by *Lactobacillus plantarum* BF001," *Lett. Appl. Microbiol.*, 24: 159–165.

31. Lyon, W. J. and Glatz, B. A. 1991. "Partial Purification and Characterization of a Bacteriocin Produced by *Propionibacterium thoenii*," *Appl. Environ. Microbiol.*, 57: 701–706.

32. Verna, C. M., G. Arendse, and J. W. Hastings. 1997. "Purification of Bacteriocins of Lactic Acid Bacteria: Problems and Pointers.," *Int. J. Food Microbiol.*, 34: 1–6.

33. Haaland, P. D. 1989. *Experimental design in Biotechnology*. New York: Marcel Dekker, pp. 9–12, 29–35.

34. Box, G. E. P., W. G. Hunter, and J. S. Hunter. 1978. *Statistics for Experimenters*. New York: John Wiley and Sons, pp. 510–537.

35. Lewus, C. B. and T. J. Montville. 1992. "Further Characterization of Bacteriocins Plantaricin BN, Bavaricin MN and Pediocin A," *Food Biotechnol.*, 6: 153–174.

36. Parente, E. and Ricciardi, A. 1994. "Influence of pH on the Production of Enterocin 1146 during Batch Fermentation," *Lett. Appl. Microbiol.,* 19: 12–15.

37. Lejeune, R., R. Callewaert, K. Crabbé, and L. De Vuyst. 1998. "Modelling the Growth and Bacteriocin Production by *Lactobacillus amylovorus* DCE 471 in Batch Cultivation," *J. Appl. Environ. Microbiol.*, 84: 159–168.

38. Batish, V. K., R. Lal, and S. Grover. 1990. "Studies on Environmental and Nutritional Factors on Production of Antifungal Substance by *Lactobacillus acidophillus* R.," *Food Microbiol.*, 7: 199–206.

39. Peña, L. R. 1996. *Actividad Bacteriocinogénica Contra Listeria monocytogenes Scott A de una Cepa de Pediococcus 6s, Aislada de Kefir.* MS thesis. Faculty of Chemistry. UAQ.

40. Elegado, F. B., W. J. Kim, and D. Y. Kwan. 1997. "Rapid Purification Partial Characterization, and Antimicrobial Spectrum of the Bacteriocin, Pediocin AcM, from *Pediococcus acidilactici* M," *Int. J. Food Microbiol.*, 37: 1–11.

41. Tramer, J. and G. G. Fowler. 1964. "Estimation of Nisin in Foods," *J. Sci. Food. Agric.*, 15: 522–528.

42. Lewus, C. B. and T. J. Montville. 1991. "Detection of Bacteriocins Produced by Lactic Acid Bacteria," *J. Microbiol. Meth.*, 13: 145–150.

43. Yang, R., M. C. Johnson, and B. Ray. 1992. "Novel Method to Extract Large Amounts of Bacteriocin from Lactic Acid Bacteria," *Appl. Environ. Microbiol.*, 58: 3355–3359.

44. Schillinger, U. and F. Lucke, F. 1989. "Antibacterial Activity of *Lactobacillus sake* Isolated from Meat," *Appl. Environ. Microbiol.*, 55: 1901–1906.

45. Mayr-Harting A., A. J. Hedges, and R. C. Berkeley. 1972. "Methods for Studying Bacteriocins," in *Methods in Microbiology,* vol. VII. J.R. Norris and D.W. Ribbons, eds. New York. Academic Press, pp. 316–422.

46. Coventry, M. J., J. B. Gordon, A. Wilcock, K. Harmark, B. E. Davidson, M. W. Hickey, A. J. Hillier, and J. Wan. 1997. "Detection of Bacteriocins of Lactic Acid Bacteria Isolated from Foods and Comparison with Pediocin and Nisin," *J. Appl. Microbiol.*, 83: 248–258.

47. Palasota, J. A and S. E. Deming. 1992. "Central Composite Experimental Designs," *J. Chem. Ed.,* 64: 560–563.

48. SAS. 1987. Statistical Analysis System Institute, Inc., North Carolina, U.S.A.

49. Colowick, S. P. and N. O. Kaplan. 1955. *Methods in Enzymology* Vol. 1. New York: Academic Press, pp. 33–34.

50. Callewaert, R. and L. De Vuyst. 1996. "Purification and Improved Production of Amylovorin L471, a Bacteriocin from *Lactobacillus amylovorus* DCE 471," *Mededelingen van de Faculteit Landbouwwetenschappen Universiteit Gent.*, 61/4a: 1565–1572.

Part XI

Environmental

56 Clean Technology in the Food Manufacturing Industry

P. J. Lillford
M. F. Edwards

CONTENTS

Abstract

The food industry appears to have minimal concerns with clean technology. After all, the raw materials are natural, the processing is mild, and the by-products are biodegradable and nontoxic. However, the delivery of beneficial and attractive products to consumers frequently requires separation, purification, and restructuring of agricultural materials, and each step yields unwanted by-products, which at present are disposed of as cheaply as possible. In future, the industry will need to consider all of its operations more critically. For example,

- Large-scale processing can produce by-products that threaten the local environment by overloading in the oxygen demand.
- Physical processes are a compromise between energy efficiency and unacceptable damage to product quality.
- Packaging is an integral part of the product because of the need for safety and quality maintenance during distribution. These materials are less

degradable than the foodstuff itself and represent an increasingly unacceptable source of environmental pollution.

- Examples of past and current successes will be given, along with the emerging challenges of redesign of processes and products.

56.1 INTRODUCTION

The food industry handles enormous volumes of material. For instance, in the UK, the annual expenditure by householders was £47.3 billion in 1994.[1] The industry's supply chain (see Figure 56.1) stretches from the production of raw materials through food processing in factories, to the distribution of products to the retail trade, and hence to the consumer. In terms of added value through the supply chain, agriculture contributes £9 billion, manufacture £14 billion, retailing and distribution £9.5 billion, and catering £10 billion.[2] Employment in the production sector is £0.5 million, with around 1 million in distribution and retail.

The complete supply chain of the food industry, from the production of raw materials, via food processing, to the consumption and disposal by the consumer, is complex (see Figure 56.1). Some of the environmental issues facing the industry can be identified in this diagram and are discussed below.

Raw materials for the food processing industry include fruits, cereals, vegetables, meats and poultry, fish, and food ingredients such as fats and oils, sugar, flavorings, thickeners, and emulsifiers. Production methods for these raw materials often involve the use of fertilizers and pesticides/fungicides, in addition to the consumption of plants, animal products, energy, water, etc. Obviously, there are several environmental issues in the production of the raw materials that should not be considered separately from the operations of the food processing industry.

One of the major materials supplied to the food processing industry is packaging, which is used to protect processed food from deterioration and/or contamination (primary packaging) and to provide physical protection through the distribution and retailing operations (secondary packaging). It is notable that the food industry, of all the manufacturing industries, makes the largest demand on packaging, whether it be paper/board (including laminates), plastics, glass, or metal. Indeed, the food industry is responsible for about two-thirds of the total industrial usage. Finding ways to reduce this packaging quantity and its subsequent waste is a demanding task. The issues surrounding the optimal strategies for the selection, recycling, re-use, and disposal of this packaging are extremely complex and cannot be resolved in isolation from other factors such as the product itself and its complete distribution chain.

The food processing industry is diverse, with a few major multinational companies at one extreme and a large number of small and medium-size enterprises at the other. This, along with the convenience of location close to sources of perishable raw materials, means that the factories in the industry are geographically scattered, with many small operations. In the limit, the "factory" becomes mobile, following the available raw material. Examples of this are factory ships and, to a degree, the combine harvester. This geographical spread, coupled with the biodegradability of waste foodstuffs, indicates a benign situation. However, the very large number of

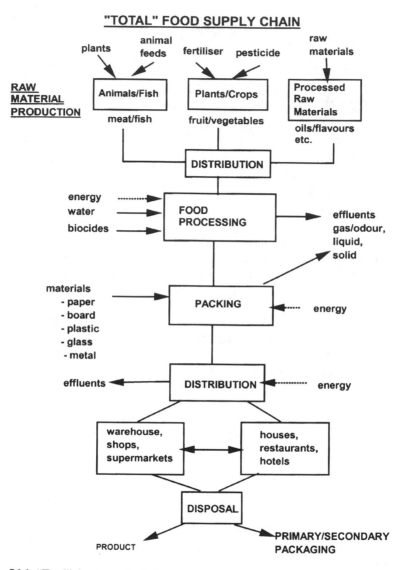

Figure 56.1 "Total" food supply chain.

factories means that the overall consumption of water and energy are significant, and it is not feasible to concentrate production into a few highly efficient processing units to gain the advantage of scale. Additionally, wastage from sectors of the food industry ranges from around 1.0% loss by weight for dairy and oil and fat processing, to 15% for certain confectionery operations, and in excess of 50% for some meat products.[3–4]

The food industry also includes many energy-intensive operations, such as sterilization/pasteurization and drying, and refrigeration of products during distribution, warehousing, and retailing. It is clear that there is a need for energy minimization as a key element in the industry's efforts to be environmentally friendly.

Finally, in the disposal from the consumer, some of the food is eaten and digested and goes to the sewage system. Some goes into garbage bins, along with much of the packaging material, for landfill or combustion.

It is clear from this brief introduction that the total supply chain of the food processing industry is complex, with several areas of environmental concern, e.g., fertilizers, pesticides, packaging, energy utilization, and water usage (Figure 56.1). As a result, it is not feasible to consider just the "food processing" operation in isolation from the full supply chain. For example, it is not reasonable to minimize the environmental impact in processing operations if this solution presents an even larger problem elsewhere in the chain. As another example, the energy requirements for processing and the amount of waste materials could be minimized by improvements to the raw materials during farming/agriculture.

Life cycle assessment (LCA) provides one way of taking an integrated approach to gaining environmental benefits across the total supply chain. The stages in LCA include drawing the system boundary, inventory (of raw materials, energy usage, effluents, etc.), assessment of environmental impact, followed by improvements to the system. However, the methodology for impact assessment is not fully agreed upon, and the database for the inventory is not complete, especially for many of the natural materials found in the food and agricultural industries. Nevertheless, LCA does provide a way to develop a framework for taking an integrated view of the environmental issues in food processing.

Referring to Figure 56.1, the LCA approach of defining the boundary, calculating the demands on raw materials and energy, evaluating the losses, etc., becomes apparent. However, since LCA is not yet a complete methodology, the following sections of this chapter focus only upon environmental case studies in food processing and the food supply chain. The role that clean technology can play is then highlighted and, finally, some of the research barriers that stand in the way are identified. However, any proposed solutions that use clean technology in the food industry to minimize environmental impact must ultimately be viewed in a full supply chain and LCA context.

56.2 CASE STUDIES

56.2.1 Case Study 1—White Fish Product

Fish Fillets

When we eat typical British fish and chips, we need to remember that the succulent white fillet of fish represents less than half the weight of the whole fish when it was swimming in the sea. The head, bones, skin, and digestive system (guts) are not normally eaten and must be disposed of in some way. Traditionally, large white fish such as cod or haddock would be gutted at sea. The guts would be thrown overboard, and thus not be a waste problem to the fisherman. The gutted fish could then be processed at sea—beheaded, filleted, and skinned—and the resultant waste also discarded overboard.

Nowadays, most fish are landed whole, and the processing steps (filleting, etc.) will be done on land. As a consequence, a significant amount of fish waste—skin and bones—must be disposed of in an acceptable way.

The yield of boneless fillet from the fish frame will depend on the size of the fish and the species. Typical filleting yields will be about 40% for cod, but as high as 60% for salmon because of the small size of both the head and digestive system of the salmon. Filleting can be carried out by hand or by machine. It is interesting to note that the yield from hand filleted fish is greater than that from machine filleting.

It is important that abundant fresh water be available during fish filleting and skinning steps so as to keep the fish flesh wet and clean. All of this water will end up as effluent from the factory and will be expensive to dispose of due to its high biological oxygen demand (BOD).

The most cost-effective way of disposing of fish waste is to convert it into fish meal, which can be used for animal feed or fertilizer. However, this waste is unstable and decomposes rapidly, not only rendering it unfit for use, but generating a highly unpleasant odor! The waste is of low value, which means that the fish meal factory must be close to the filleting factory to avoid excessive transport costs/times. Furthermore, fish meal production is not a very profitable process; therefore, the factories need to be large, taking in waste from more than one filleting factory, to be economically viable.

Thus, it can be seen that the production of 1 kg of fish fillet requires over 2 kg of starting fish, the disposal of more than 1 kg of nonfillet fish material, and the processing of at least 1 kg of high BOD wash liquor. The surimi process is even more demanding, using much higher levels of wash water.[5]

56.2.2 Case Study 2—Cheese Production[6]

Milk is a fluid consumed directly by almost all societies. In its liquid form, cow's milk is composed of 3.75% fat, 4.85% lactose, 3.30% protein, 0.72% minerals, and 87.5% water.

When consumed as a liquid, the solids are largely digested and the liquids disposed of by the individual. The effluent is not an issue for food manufacturers but only for the sewage industry. However, fermentation of milk by benign organisms has been practiced for at least 8,000 years, producing the interesting textures and flavors of cheeses and yogurts. Currently, about 11 million tons of cheese are produced annually and, for obvious reasons of economy of scale, process is centralized in increasingly large sites. The composition of cheeses varies but in every case is much higher in solids than the original milk (Table 56.1).

Irrespective of local tradition, the initial stages of the process produce a curd and liquid whey whose composition is 0.7% protein, 4.5% lactose, 0.6% salts, 0.6% acid, and 93.6% water. This represents an annual by-product of about 4.5 million tons of a liquid stream with very high BOD.

TABLE 56.1
Composition of Different Cheeses

	Protein %	Fat %
Cheddar	25.4	34.5
Edam	24.4	22.9
Parmesan	35.1	29.7

In the past, this was disposed of into local waterways or added directly to animal feed. Indiscriminate dumping has now been stopped in most countries, and a key factor in the economics of cheese manufacture is the disposal or resale costs of whey. However, this has become a success story, thanks largely to the introduction of ultrafiltration (UF) and reverse osmosis (RO).

Because whey is biologically unstable, its treatment has also been localized near or at cheese manufacturing plants. A combination of sterilization, RO evaporation, and drying produces a powder for use in animal feeds and calf milk food, and potable water. This accounts for almost 90% of the use of whey in the EU, and a cost analysis of a cheese plant producing 100,000 tons of whey per annum shows that the capital cost of RO equipment can be recovered within one year.[7]

There are even more advantages to be gained if hygiene is improved and processing is milder. The use of UF produces a protein concentrate, which, if not excessively heat treated, is of considerable functional value. Here, a compromise between energy efficiency and product value has to be reached. It is inefficient to transport and store whey in a liquid state, and the suspension is an almost perfect medium for microorganisms. As a result, whey is usually dried, by fluidized bed, roller, or spray drying. Engineers can calculate the most energy-efficient process, but these frequently require that the liquid whey reach temperatures of greater than 70°C. Unfortunately, this exceeds the thermal denaturation temperature of the constituent proteins. Protein denaturation is a highly cooperative, first-order process that is comparable with melting and, in concentrated solutions, is irreversible.[8] Thus, the most energy-efficient process is not optimal for the recovery of native protein. The whey proteins α-lactalbumin and β-lactoglobulin have similar solubility and heat setting properties as egg white (ovalbumin). Since much culinary practice in bakery, confectionery, etc. uses the properties of egg white, functional whey protein powders can command a comparable ingredient price of at least £2000 per ton. This whey product, therefore, is as valuable as the primary cheese.

56.2.3 Case Study 3—Instant Coffee[9]

For those who make coffee directly from ground beans, the volume of wet, spent grounds represents an inconvenience. However, a factory producing instant coffee is faced with a massive disposal problem. The grounds have considerable calorific value, but not at the solids contents recovered from the primary extraction. At the Nestlé factory in Staffordshire, instant coffee, equivalent to 5 billion cups per annum, is manufactured. Spent grounds are pressed, producing a dense cake and a viscous coffee oil. The cake provides a fuel source capable of providing up to 50% of the site's steam demand, but the pressed liquid is sprayed onto fields, since the BOD exceeds the capability of the municipal system.

Such procedures are scarcely environmentally friendly, and the liquid has considerable calorific value. Two-stage recovery of fines by centrifugation and coffee oil concentration produces further fuel, and saves 0.33 million pounds per annum. The next step will be to recycle potable water.

56.3 CLASSIFICATION OF CLEAN TECHNOLOGIES

Having outlined some particular case studies in the food industry, let us now take a generalized approach to clean technologies for the industry.

If one considers only the food processing part of the entire food supply chain, (Figure 56.1), we can generalize the unit operations of the industry to include conveying, washing, heating, cooling, extraction, emulsification, drying, etc. The equipment used in these processes must be cleaned (and often sterilized) and, in addition to the main products, the entire operation results in a further series of waste streams. These wastes are usually readily biodegradable by natural microorganisms and enzymes. However, there is a major role for clean technology to minimize the overall environmental impact by improving the raw materials, improving the unit operations, using reuse and recycling strategies and improving waste treatment.

The relevant clean technologies be classified into the following categories:

- *Reduction at source.* This covers improvement of raw materials. In the case of the food chain, these are agricultural products that have already been subject to breeding for increased yield. The next step is to design raw materials, particularly plants, that require less treatment. Short-stalk cereals, leafless peas, and ripening controlled tomatoes are examples that have already been introduced.

- *Recycling.* This includes procedures that permit recycling of energy, water, and potential waste streams back into earlier stages of the process. Formal techniques exist for energy minimization and water minimization.[10–11] However, attention must be paid to possible microbial contamination in any water minimization solutions involving water recycling/reuse.

- *Recovery.* This embraces technologies that allow materials previously regarded as waste to be recovered/treated, and thus appropriate for reuse or sale as a marketable product in its own right. The example of whey recovery was given above.

- *Conversion.* This refers to conversion of waste material streams into energy for improvement of the energy efficiency of the process or other processes on the site. For example, anaerobic digestion of some fat-laden aqueous streams can provide not only intensified treatment (taking up less space) and low levels of biomass production, but also generate methane, which can be burned to provide heat for use elsewhere in the processing operations. Coffee solids recycling is another example.

- *Treatment.* The subject covers the on-site treatment of waste streams to minimize environmental impact, e.g., odor removal from gaseous streams.
- *Disposal.* This covers the routes available off-site to dispose of waste streams via third parties; for example, fish meal processing.

The best response to increasing pressures from regulation and legislation is via an integrated environmental management system[12] with regular environmental audits providing a database from which improvements in performance can be measured. Most of the products from the food processing industry have a relatively low added value. Thus, the capital and running costs that the industry can tolerate for any new clean technology are severely constrained. To illustrate this pressure on costs, it is interesting to note that, in the UK in 1950, consumers spent approximately 25% of their income on food. By 1994, this figure had dropped steadily to about 11%[1] as food production became more efficient in response to continuous consumer pressure. Clean technology will be the next requirement.

56.4 RESEARCH CHALLENGES

As the above discussion illustrates, the food industry is not free from environmental problems. Thus, there is a constant search for improvements that can lead to fundamental changes in the industry's operations, leading to new processes and procedures. In parallel, incremental improvements to existing processes are also sought.

In this chapter, we highlighted some of the technical barriers to progress where research is required.

1. The development of a life cycle assessment methodology and database for the food processing supply chain. In particular, an agreed-upon approach to the assessment of environmental impact is needed. Only when such a procedure is available can a rational approach be taken to the evaluation of the significance of individual clean technologies.
2. Biotechnology and breeding techniques to modify animal and plant raw materials could facilitate waste and energy reductions in processing, as well as minimization of pesticides, etc., in agriculture.
3. The further development and application of energy and water minimization techniques in food processing.
4. Improved separation technology; for example, membrane processes for product recovery that have significantly lowered fouling and longer membrane life.
5. Techniques for improved waste treatment; for example, biological treatment with reduced sludge generation, super-critical oxidation, or improved odor removal from gaseous streams.
6. Improved techniques for cleaning in place of process equipment to reduce the amount of water, steam, surfactants, and biocides required.

56.5 ACKNOWLEDGEMENT

P. J. Lillford acknowledges with thanks the assistance of Dr. P. Harris in the assembly of the case studies.

REFERENCES

1. National Food Survey. 1994. MAFF.
2. Progress Through Partnership—Part 7. 1995. Food and Drink, HMSO.
3. Gorsuch, T. T., ed. 1986. *Food Processing Consultative Committee Report*, MAFF.
4. Niranjan, K. And N. C. Shilton. 1994. "Food Processing Wastes—Their Characteristics and an Assessment of Processing Options," *AIChE Symp. Ser.*, 90(300): 1–7.
5. Lanier and Lee, ed. 1992. *Surimi Technology*, Marcel Dekker.
6. Robinson, R. K. 1994. *Modern Dairy Technology*, vol. 1. Chapman & Hall.
7. Pepper, D. 1981. "Whey Concentration by Reverse Osmosis," *Dairy Industries Int.*, 46: 24, 26
8. Ruegg, M., U. Moor, and B. Blanc. 1975. "Hydration and Thermal Denaturation of Beta-Lactoglobulin-Calorimetric Study," *Biochimica et Biophysica Acta*, 400, 334–342.
9. Beck, M. 1996 *Feedback*, Food & Drinks Federation, 19.
10. Linhoff, B. 1993. "Pinch Analysis—A State of the Art Overview," *Trans. I. Chem. E.*, Part A, 71: 503–522.
11. Wang, Y. P. and R. Smith. 1994. "Waste Water Minimization," *Chem. Eng. Sci.*, 49: 981–1006.
12. Zaror, C. A. 1992. "Controlling the Environmental Impact of the Food Industry: An Integral Approach," *Food Control*, 3 (4): 190–199.

57 Pest Control in the Food Industry: Raw Materials Control

P. Valle-Vega

CONTENTS

Abstract

There are different risks associated with the wholesomeness of foods, in particular with the handling of raw materials. To maintain sanitary conditions at a food plant, a pest control program must be applied as a preventive measure; for that, it is necessary to identify the major pests that affect raw materials. In addition, different action levels must be at hand to stop a pest once it has been detected, while avoiding the use of chemicals, in particular where there is a food surface exposed or in a food area.

Pests must be prevented because of the shoddy image that they may impart to the packaging materials and because they are carriers of disease. In addition, the food itself must be free of rodent hair and insect fragments.

1-56676-963-9/02/$0.00+$1.50
© 2002 by CRC Press LLC

57.1 INTRODUCTION

To establish a pest control program, the pests' behavior must be known so as to exclude them from food processing areas. In food plants, rodents are one of the most troublesome pests, not only for their disgusting aspect but also because of the great damage they cause to the product, equipment, installations, and packaging materials. Different insects are considered as the main pests in neglected plants, although good manufacturing practices (GMPs) must be sufficient to halt their presence. And, more than occasionally, birds will nest and fly freely inside a plant, causing health risks resulting from *Salmonellae ssp.* contamination.[1,2]

There are some basic considerations to be aware of. Pests, like any living animal, require food, water, harborage, and a mate. Thus, a food plant may provide the ideal conditions for such needs.[3–7]

The damage produced by pest attacks can be

1. *Direct.* This is the most typical form of damage. It is usually measured as weight or volume loss, but this is not totally accurate, as there is an accumulation of insect bodies, exuviae, fecal matter, fragments, etc.
2. *Selective eating.* Some beetles, moth larvae, and mites show a feeding preference for the seed germ.
3. *Loss of quality.* This is the combination of all damage, the loss of nutrients, or if the food becomes unfit for consumption.
4. *Contamination.* The degree of contamination may be associated with cultural factors. In some parts of the world, pests are expected, dealt with, or the population has become accustomed to them. Sometimes, unethical considerations arise in food commerce, regardless of the risks associated with rat droppings, live or dead insects, larval and pupal exuviae, fecal matter, food fragments (frass), or unpleasant odors like urine. Another form of contamination is the spreading of pathogenic microorganisms, or promoting the growth of mold-produced mycotoxins. Insect contamination may also represent an increase in the temperature of stored grains as a result of their metabolism, creating hot spots and condensations, which in turn may help mold development. Silk webs or cocoons may clog ducts in food plants. Finally, through cross contamination, clean products may become infested or food damaged by pests.

57.2 GOOD MANUFACTURING PRACTICES (GMPs)

A reference in pest violations must be set in accordance to the health risk posed. They can be classified as: *critical* (representing a high risk to human health or product identity, such as product contamination or adulteration, or noncompliance with Federal or State Good Manufacturing Practices); *major* (representing a moderate risk to product contamination or adulteration); and *minor* (representing a low risk to product contamination or adulteration). The pest control operator (PCO) must observe criteria that are flexible, practical, and based on experience to determine what degree of risk is involved. Minor violations become critical if they are left unattended.[8,9] Table 57.1 presents a practical approach to pest control violations.

TABLE 57.1
Pest Control Violations

Critical	Major	Minor	Unclassified
Lack of pest control	Lack ILTs* or rodent stations	Incomplete pest registration	Improper station location
Active pests inside the plant	External pest activity	Dark areas Harborage areas	Lack of proper plant design
Bait usage in food areas	Lack of training	Obsolete equipment storage	No certified PCO
Detritus	Cracks and crevices	Hollow walls, false ceilings	Open plants
Chemicals stored close to a food area	Moisture	Improperly sealed junctions	Food residues
Dead pests	Lack of food protection (covers)	Lack of a 1 m perimeter	Pest access points
Pest attacks	Lack of good housekeeping	Pipe openings	
Nests inside the plant	Lack of bait rotation (exteriors)	Cleanliness	

*ILT (insect light trap or electrocution system).

The analysis of GMPs in different food suppliers, including raw and intermediate materials, may give an idea of the degree of pest control and the areas to be improved. Table 57.2 summarizes the behavior of 98 suppliers in different food areas, such as colorants, flavors, vinegar, sugar cane, slaughterhouses, antioxidants, dehydrated vegetables, frozen juices, tomato paste, canned food, aseptic processed concentrates, tomato powder, salt, lyophilized meat, rice, beans, flours, food enhancers, frozen egg yolk, dried milk powder, glucose syrup, and high-fructose corn syrup, among others.

To complete the analysis, it is necessary to know the kind of pest present in food plants. Table 57.3 presents their frequency, considering whether they are simultaneously present or alone.

57.3 COMMERCIAL PEST CONTROL OPERATORS

Another opportunity to improve food safety is in the PCOs' hands; they will be detecting at first glance potential human health risks. When a pest is detected, *corrective* action or chemical application must be discouraged; instead, *preventive*

TABLE 57.2
Pest Frequency (in %); the Amounts Given May not Add Up to 100 Because of Multiple Violations within a Single Plant or Repetitive Violation Among Pests

Insects		Rodents		Birds		Others pests and violations	
Type	%	Type	%	Type	%	Type	%
No ILTs	7	Alive	5	Feathers	1	Dogs and cats	3
Soiled ILTs	8	Dead	2	Inside birds	4	Weeds	11
ILT above food areas	3	Detritus	5	Externals birds	8	Gophers	1
Flies	3	No rodent stations	18	Nests	2	Molds	4
Ants	1	Stations damaged	3			Drainage damaged	18
Earwigs	2	No stations diagram	6			Open plant	25
Moths	2	Use of baits (interior)	32			Not audited	6
Death insects	9	No pest control	20			Structural damage	4
Crawling insects	8	No external control	3				
Webs	3						
Lack of fumigation	14						
Insecticide unknown	10						
Lack of insecticide storage area	1						

TABLE 57.3
Pest Frequency Detected by GMPs Audits

Pest kind or its combination	Frequency %
Insects	16.5
Rodents	16.5
Birds	3.5
Insects and rodents	29.4
Rodents and birds	2.4
Insects and birds	2.4
Insects, rodents and birds	17.6
No pest associated to GMPs audit	11.7

actions, such as GMPs, the use of external chemical barriers, integrated pest control systems, pheromones, food lures, cleanliness, and hermetic plants must be relied on more and more.[1,3,5,10]

In addition, they need to be trained in different areas such as food toxicology, pesticide recoveries (spills), pest identification, human physiology, plant sanitation, food law, first aids, local and federal regulations, pesticide chemistry, etc. With a strong background, the PCOs will help to control food pests.[11,12,13]

57.4 PESTICIDE USAGE

Different chemicals are employed to control pests (among them insecticides, rodenticides, and avicides), but their existence does not mean that they will be automatically applied in food contact areas.

According to the United Nations Environmental Program, the following chemicals are prohibited for application as pesticides: PCBs, Chlordane, Dioxin, Aldrin, Dieldrin, Furane, Myrex, Toxophene, DDT, Heptachlor, Endrin, and Hexachlorobenzene.[7,14-16] However, there is still some usage of banned pesticides, as their residues are detected through international commerce (Table 57.4).[20]

TABLE 57.4
**Detection of Banned Pesticides Present in
Vegetables Consumed in Mexico by a Private
Monitoring System**

Pesticide	Commodity	µg/Kg	Comments
Aldrin	Leek	0.13	Imported
Aldrin	Broccoli	0.14	Imported
Aldrin	Clove	0.50	National
Aldrin	Chili	<0.1	National
Pesticides	Mustard	Positive	Imported
Dieldrin	Clams	0.4	National
Aldrin	Mojarra *(Eugorres ssp.)*	0.4	National

There are some legal points to be expanded from the agricultural view to food processing plants; practically,

1. Different guidelines must be set for food plant fumigation.
2. Birds nesting inside food plants or silos must be considered as pests. Even though they may be protected by ecological considerations, the chance of salmonella cross contamination to foods is a real menace to public health.
3. Biomonitoring, pheromones, and insect growth regulators (IGR) need to be considered as a different category from pesticides.
4. Rodent control is barely covered in must of the cases. Therefore, a continuos educational program is needed to improve the level of the PCO's abilities.
5. Unusual pests must be removed from the facilities promptly, considering ecological aspects. Some unexpected pests are bats, amphibians, reptiles, foxes, etc.

Mexican food plants use different chemicals for pest control; some may be effective for short periods, and they are then replaced by a different category or the dosage is increased. Different chemicals employed for fumigation are methyl bromide, methyl dibromide (Brom-O-gas; MeBr; Profume), ethylene dibromide (DBE, EDB, Bromofume, Dowfume), dibromochloropropane (DBCP, Nemagon), dicloro-propane-dicloropropane (D-D, Telone), calcium, and aluminum or magnesium phos-phide (Delicia, Phostoxin).[7,8,17–22] However, the main point to emphasize is the GMPs as an alternative for pest control. Table 57.5 describes the common pesticides used in Mexican food plants.

TABLE 57.5
Some of the Most Commonly Applied Pesticides in Mexican Food Plants

Usage	Chemical	Comments/recommendations
Baits	Bromadiolone	Inside a rodent station
	Bradifacuom	Keep records
	Cumatetralyl	Not in food areas
	Colecalciferol	
Insect regulation growth (IGR)*	Triflumuron and benzophenyl-urea	Use mainly at the agricultural level
Pheromone lures*	Pheromones and food lures	Specific for insect species, such as red and confused beetle, cockroaches, flies, etc.
Herbicides	Glyphosate atrazine	Exteriors
Fumigants	Aluminum phosphide	To control beetles and moths
	Ethylene oxide	
	Etilenc bromidc	
Space spray food areas	Deltamethrin	Do not apply directly to food surfaces
	Cyfluthrin	Cover equipment
	Cypermethin	Keep GMPs
	Permetrhin	
	Deltamethrin w/piperonil butoxide	
Plant exteriors	Cholpirifos	Plant exteriors
	Cholpirifos and DDVP	Carefully applied and monitored
	Propoxur	
	Diazinon	
	Phoxim	

*Mexican law considers pheromones and lures as pesticides.

57.5 EXAMPLES OF PEST CONTROL

Some of the most common pest problems associated with raw materials and pro-cessed food are noted in Table 57.6, where the lack of GMPs or consumer abuse were in many cases the cause that allowed a pest to get into the food. In other cases, the lack of plant supervision was the origin for pest development. Biological diversity may produce unexpected pests, and some will be present as an incidental occurrence, such as earwigs in glass bottles. Other pests will be of no health risk but will impart

TABLE 57.6
Analysis of Pest Control Deviations in Raw and Finished Products

Product	Associated pest	Detection Point	Comments
Instant cereal	Beetles, insect borer (*Rhyzopertha ssp*)	Packaging materials perforated	Product is placed near grains and raw seeds at a public market
Rice	Insect borers (beetles)	From supplier warehouse	Use of photoxin as fumigant. Implement a 100% inspection
Beans	Rodents	Stored product was attacked under refrigeration	The refrigerated chambers did not have rodent protection. The product was destroyed
Corn flour corrugate	Book lice	The corrugate was kept under moisture, allowing the book lice to grow and reproduce	The corrugate supplier had to fumigate and remove all moisture from its warehouse; it had to eliminate molds
Glass bottles	Earwigs	Present inside the supplier plant: gardens, warehouse, railroad tracks, truck container	Lack of pest control, increased by the rainy season. Implementation of air blowing system for the bottles.
Flours (prepared)	Secondary beetles	Inside plastic bags, they were able to attack due to a bad seal.	Slow product rotation, stored close to feeds
Flour	Moths and rodents	Infested plant at the process area	Lack of GMPs, and plant safety. There were several access points
Bell peppers	Moths	Insects trapped inside the adhesive tape (on the corrugate)	Slow product rotation. Some eggs may survive,
Chili (*poblano*)	Moths	No visible external perforations; completely infested	The product was irradiated, but below the lethality dose against insects.
Toasted pasta, condimented	Insect borers	Consumer complaint, because some larvae were inside a finished product pouch	The food storage area was severely infested. The due date was past.
Intermediate materials warehouse	Birds	Open doors and damaged ceilings	Implement GMPs and maintenance
Finished products warehouse	Bats	Harborage spaces in ceilings and walls	Reinforcement of maintenance.
Condiments, catering presentation	Cockroaches	Consumer complained of cockroach presence inside the containers	After the facilities pest control audit, the problem was caused by consumer product abuse; the installations were close to a municipal market without any GMPs.
Mole paste	*Ephestia spp.* and *Tribolium spp* larvae	Larvae presence due to attack by moths and beetle larvae, the packing material is not insect resistant	External infestation at consumers house.

a negative effect on consumer product quality appreciation; e.g., book lice, which are able to reproduce and nurture from molds present in moistened corrugated paper.

57.6 INSECT CONTROL

Several factors may influence the presence of insects inside a food plant—mainly, poor housekeeping and ignorance of good product rotation (first in, first out, or FIFO). When the food storage is neglected, several pests may develop. Among the major insect pests are cockroaches, flies, moths, and beetles. They have different behaviors; some may be seen during daylight, others at night; some fly, others crawl. Some insects will be practically invisible to the human eye because of their small size, e.g., book lice.

Insects not only represent grain losses but also render food inedible due to extensive damage to the nutrients and sensory attributes. There are also health risks associated with flies and cockroaches, because they tend to wander around waste or fecal matter, posing a source for diseases (Table 57.7).

The housefly (*Musca domestica*), as well as other true flies that belong to the *Diptera* order, are of special interest, because they transmit bacteria (typhoid, dysentery, infantile paralysis, etc.), viruses, parasitic protozoa, and worms to plants and animals. Larvae accidentally swallowed in the food material sometimes survive in the human gut, causing intestinal myiasis, with symptoms of pain, nausea, vomiting etc. They cause food damage through eating, by fecal contamination and enzyme regurgitation.

To discuss insect control or other pests, the toxicity associated with insecticides has to be defined; Table 57.8 gives the different levels.

The Mexican Health Secretariat has assigned different colors to point out the toxicological level in the labels.

  I. Red, extremely toxic
 II. Blue, highly toxic
III. Yellow, moderately toxic
 IV. Green, slightly toxic

Since insects represent the most populous group of pests, it is difficult to classify them; based on their locomotive behavior, they may crawl or fly. If the way of feeding is considered, they can bite, chew, suck or bore. They also can have a diurnal or nocturnal activity.

The majority of insects are "cold blooded," which makes them susceptible to temperature changes. The ideal temperature is around 28°C and 55% relative humidity; the vast majority cannot grow at temperatures below 10°C.

In relation to the damage caused to stored grain, insects are considered as primary pests when they directly attack a grain or seed; in this category are some beetle larvae (rice borers) and moths (*Angousmois spp*). They spend their whole lives inside the grain and are unable to survive outside the infested kernels.

TABLE 57.7
Pathogens Associated with Cockroaches

Pathogen	Sickness/disease	Cockroach species
Bacteria		
Alcaligenes faecalis	Gastroenteritis, urinary tract infections, skin wounds	*Periplaneta americana* *Blatta orientalis*
Bacilus subtilis	Food poisoning, conjunctivitis	*Blaberus craniifer* *Blatta orientalis* *P. americana*
Bacillus cereus	Food poisoning	*Blaberus craniifer*
Clostridium perfringens	Gangrene (gaseous), food poisoning	*Blatta sp* *Periplaneta sp*
Campylobacter jejuni	Enteritis	*B. orientalis* *P. americana*
Clostridium novii	Gangrene (gaseous)	*B. orientalis*
Clostridium perfringens	Gangrene (gaseous), food poisoning	*B. orientalis*
Enterobacter aerogenes	Bacterimia	*B. germanica* *P. americana*
Escherichia coli	Diarrhea, wound infections	*B. orientalis* *B. germanica* *B. americana*
Klebsiella pneumoniae	Pneumonia, urinary tract infections	*Blatella sp* *Periplaneta sp*
Mycobacterium leprae	Leper	*B. germanica* *P. americana* *P. australasiae*
Proteous morganii	Wound infections	*P. americana*
Proteous rettgeri	Wound infections	*P. americana*
Proteous vulgaris	Gastroenteritis, wound infections	*P. americana*
Pseudomonas aeruginosa	Respiratory illness, gastroenteritis	*B. craniifer* *B. orientalis* *B. germanica* *P. americana*
Salmonella bredeny	Food poisoning, gastroenteritis	*P. americana*
Salmonella newport	Food poisoning	*P. americana*
Salmonella oranienburg	Food poisoning, gastroenteritis	*P. americana*
Salmonella panama	Food poisoning	*P. americana*
Salmonella paratyphi-B	Food poisoning, gastroenteritis	*P. americana*
Streptococcus pyogenes	Pneumonia	*B. orientalis*
Salmonella thyphi	Typhoid	*B. orientalis*
Salmonella typhimurium	Food poisoning, gastroenteritis	*B. germanica*
Serratia marcescens	Food poisoning	*B. orientalis* *B. germanica* *P. americana*

TABLE 57.7
Pathogens Associated with Cockroaches

Pathogen	Sickness/disease	Cockroach species
Shigella dysenteriae	Dysentery	*B. germanica*
Staphylococcus aureous	Wound infections, skin infections, internal organ infections	*B. craniifer* *B. orientalis* *B. germanica*
Streptococcus faecalis	Pneumonia	*B. orientalis* *P. americana*
Fungi		
Aspergillus niger	Pneumomycosis, otorhinomycosis	*P. americana*
Aspergillus fumigatus	Pneumomycosis, broncomycosis	*B. orientalis*
Helminths		
Ancylostoma duodenale	Hookworm	*P. americana*
Ascaris lumbricoides	Intestinal worms, ascarides	*P. americana*
Hymenolopsis sp	Tapeworm	*P. americana*
Trichuris trichuria	Trichinosis	*B. oriental* *B. germanica* *P. americana*
Molds		
Aspergillus niger	Otorhinomycosis	*B. orientalis*
Protozoa		
Entamoeba hystolytica	Amoeba	*B. orientalis* *B. germanica* *P. americana* *P. australasiae*
Virus		
Polyommyelitis	Polio	*B. germanica* *P. americana*

The primary insects (borers) lead the way for secondary pests that feed on the exposed flour or damaged grains; they may be present as larvae or in their adult stage. The presence of tertiary insects reflects extremely poor housekeeping practices, since they feed on mill wastes like flour or husks on the floor.

To control insects, there are different alternatives, varying from the use of chemicals to the insect electrocution trap (ILT). These provide excellent control of flying insects, but maintenance is a major concern, because they have to be cleaned, the dead insects have to be monitored and removed periodically, and they have to be hidden from the view of external insects; otherwise, they will work as attractants. The ILTs have evolved. Now they employ the same principle of UV light but, instead of electrocuting the insects, there is a glue board at the bottom of the equipment; the insect is guided to the UV light but trapped and left to die on a glue board instead of via electric shock. Thus, the spreading of insect fragments is avoided, as happens with ILTs.

TABLE 57.8
Pesticide Toxicity

Toxicity level	DL_{50}, rat, acute oral*	Acute mg/kg dermal exposure	CL_{50}-acute inhalation mg/kg, 1 h[†]	Example $(DL_{50}$, mg/kg) oral
I. Extreme	<5	<10	Up to 8.2	Aldicarb (0.93)
II. High	5–50	10–100	0.2–2	Azimfos (16)
III. Moderate	50–500	100–1000	2–20	Propoxur (98)
VI. Slight	>500	>100	>20	Permethrin (4000)
Practically nontoxic[‡]				Deltamethrin (9338)

*DL50 is the lethal dose to cause the death of 50% of the population.

[†]CL50 is the lethal concentration in air that causes a death rate of 50% for a given population. It is expressed as mg per liter (ppm). The exposure time is one hour.

[‡]This level is not considered as a class for Mexican regulation.

57.7 RODENTS

Rats and mice form part of the family *Muridae*; they are regular pests of food plants or stored food products. Rats and mice live in association with man in all parts of the world. They damage crops as well as stored products in warehouses and supermarkets. They need to regularly gnaw hard substances to control the size of their lower incisors, causing, on some occasions, extensive structural damage in pipes, wooden structures, electric ducts, etc.

Damage by rodents is primarily the eating of the produce, but contamination by feces, urine, fur, etc. is often the most worrisome. From a sanitary point of view, it implies extremely poor hygienic conditions; because of this, there are different legislations that control imports and exports. Rodent hair detection and tolerance are major issues in international trade as well as domestic evaluation of the food processing GMPs. Some products that are highly scrutinized so as to be accepted by another country (such as chili, ground spices, flours, etc.) are expected to be contaminated by rodent hairs; therefore, a limit is set to comply with food safety requirements.

Mus musculus (house mouse) is an extremely important urban pest, as any rodent can cause damage by gnawing. They have adapted so successfully to the human environment that some races can live at minus 20°C, inside freezers or refrigerators. The main damage is by direct contamination of foodstuffs with excreta, urine, and hair.

Rattus rattus (roof, ship, or black rat) is a major pest of several tropical tree crops (palms). It is also the main host of the tropical rat fly that is the vector of the plague organism.

Rattus norvergicus (sewer or common rat), totally omnivorous, is an extremely damaging animal to stored produce and includes almost anything edible in its diet.

57.7.1 CONTROL

1. *Chemical.* Natural or synthetic poisons (mainly anticoagulants and fumigants), sterilization chemicals.
2. *Biological.* Predators, bacteria.
3. *Physical.* Hermetic plant, electric shock system, sanitary rifle, mechanical traps, glue boards, ultrasound (not necessarily effective), intense light, explosions, high-volume and strident sounds.
4. *Cultural.* Environment control, good housekeeping, good storage practices and GMPs.

57.8 BIRDS

Sometimes, birds may become serious pests, especially when they carry disease, attack crops, damage buildings, and when they contaminate food.[23] Pigeons, starlings, and sparrows are the most common pest birds, affecting not only public buildings but food plants as well.

It is necessary to know their biology, habits, and behavior to understand why they are transformed from nice animals into pests inside a food plant.

Pigeons prefer smooth flat surfaces for feeding; they will feed from rooftops, because they like open feeding areas that permit speedy getaways. Their feeding range may be two or three kilometers. Common feeding sites include garbage areas from food plants. Other places may be of interest to pigeons, like parks, spilled grains, food loading docks, etc.

Roosting and resting sites include roofs, ledges, building corners, and different internal structures where they feel secure. Nests are usually located in open protected places; their nests are made of twigs, straw and other debris, feathers, and even droppings; with time, the nests will become hardened. The offspring mature rapidly, reaching full ground size in less than a month.

Pigeons in food warehouses, grain facilities, and processing plants can contaminate food. Accumulated pigeon droppings can lead to histoplasmosis and cryptococcis. Both are respiratory diseases, usually with no symptoms. Both diseases are spread by air when the droppings are disturbed, then workers inhale the disease organism.

Pigeons occasionally carry and transmit other diseases to people or livestock. Among the most important are pigeon ornithosis, encephalitis, Newcastle disease, and Salmonella food poisoning

Starlings are not native to America; they were brought from Europe and released initially in New York. They have spread to almost all of North America. Starlings nest in holes or cavities in trees or buildings. The young leave the nest when they are about three weeks old. They migrate in large flocks that sometimes exceed 1 million birds. They can shift their food from insects and fruits to grains, seeds, livestock rations, and food in garbage. They can fly up to 45 km daily to find food. They tend to roost in the evening in high places, which in turn may become risk sites because of the droppings.

Starlings have become responsible for public outbreaks of disease. The most serious is the fungal respiratory disease *histoplasmosis,* caused by a fungus growing in the soil beneath the roosts. The disease is spread when droppings are disturbed and the disease organism swirls into the air and is inhaled by workers. Starlings roost in and on food processing plants or storage areas and contaminate food.

Sparrows (*Passer domesticus*) were introduced to America around 1850, spreading efficiently throughout the continent. It is the most common bird around farm buildings. They can be pests in grain and bulk food storage. Contamination from their droppings causes much more damage than their actual feeding. Their droppings and feathers can make hazardous, unsanitary, and smelly waste inside and outside of buildings. They become a pest when one or more begin nesting inside a food plant, warehouse, mall, or atrium.

The sparrows tend to peck at rigid foam insulation and nest inside. They are also a fire hazard when nesting inside transformers, power stations, or traffic lights. Nests often plug rain gutters or jam power transformers. They feed preferentially on grain, fruits, seeds, and garbage.

Most sparrow pest problems are ectoparasites (mites and lice) and are rarely related to a public health concern. Probably the main public health issue is the potential food contamination in food plants, grain and seed storage, and outdoor restaurants.

57.8.1 CONTROL

There are some alternatives in bird control, but mainly they have to be restricted to GMPs because, in most places, birds are protected by law; therefore, chemicals are highly restricted. Birds tend to be seen as part of the ecology, but a pest is any living organism present in a place and time in which it causes or represents health risks, economic losses, or industrial problems. Some bird control alternatives are

1. Physical
   a. Hermetic plants
   b. Bird spikes
   c. Bird nets
   d. Bird barriers
   e. Stainless steel needle strips
   f. Electric shock systems
   g. Bird coils
   h. Nest removal
   i. Sanitary rifles
   j. Glue boards
   k. Ultrasound (not necessarily effective)
   l. Intense light
   m. Explosions
   n. High-volume and strident sounds
   o. Scarecrows (in different forms, e.g., eyes, balloons)
   p. Conical traps

2. Chemical
   a. Avicides
   b. Behavioral alteration chemicals
   c. Repellents
3. Biological
   a. Predators
4. Cultural
   a. Good housekeeping
   b. GMPs

57.8.2 SAFETY PRECAUTIONS

Safety precautions for removing accumulated bird droppings include

1. Wear a respirator (able to filter particles up to 0.3 microns).
2. Wear disposable protective gloves, hat, boots, coveralls, etc.
3. Wet down the droppings to keep spores and microorganisms from becoming airborne.
4. Put the wet droppings into sealed plastic garbage bags, and wet down the outside of the bag.
5. Carefully remove all protective wear; keep the respirator on until there are no signs of dust or airborne particles.
6. Dispose of trash bags by transporting whole and sealed to the local landfill; do not let them be taken by the local trash truck, where they can be broken.
7. Wash up or shower on the site; wash the respirator.

57.9 MEXICAN REGULATIONS IN RELATION TO PEST CONTROL

The Health Secretariat (Secretaria de Salud, SSA) is in charge of food plant surveillance as part of its social functions. When a GMP audit is performed, the following aspects are covered:

1. Lack of evidence for insect, rodent, or bird activity, as well as domestic animals.
2. Preventive equipment to control pests is in order and functioning (ILTs, meshes, baits, rodent traps, sewage, etc.).
3. There must be a written pest control plan and sanitation procedures. There must be evidence from the PCOs work.

57.10 OPPORTUNITIES IN THE FIELD FOOD PEST CONTROL

In conclusion, a different approach may provide a radical change in plant control, mainly oriented to cultural practices:

1. Get the CEO involved and compromise for continuous improvement through different quality philosophies.
2. Reach an ecological compromise; pests must be at a nondetectable level. The aim is not to extinguish any species, or unrelated or nontargeted species.
3. Change concepts: moisture presence or accumulation from washing floors does not imply a clean plant.
4. Implement integrated pest control management (IGRs, pheromones, bio-monitoring, biological controls, etc.).
5. The plant must be as hermetic as possible; avoid openings.
6. Keep up cleanliness and good housekeeping.
7. Implement chemical barriers.
8. Set an audit control plan.
9. Implement regulatory rules in relation to food plant pests.
10. Set up a continuous personnel improvement program and implement food plant auditors.
11. Use GMPs.
12. Keep records. Establish pest control records and procedures for fumigation and emergency action levels.
13. The PCO must be approved or certified by the government.
14. Assign personnel in charge and set time limits for follow-up.
15. Keep in mind that *preventive* action is far better than a corrective approach to pest control.

REFERENCES

1. American Institute of Baking (AIB). 1979. *Warehouse Sanitation Manual.* 1213 Bakers Way Manhattan, Kansas 66502.
2. ASTA. American Spice Trade Association. 1972. *The Paprika Manual.* 580 Sylvan Ave. Englewood, Cliffs, N.J. 07632. USA.
3. Imholte, T. J. 1984. *Engineering for Food Safety and Sanitation, Technical Institute of Food Safety.* Crystal, Minnesota, 55427. EUA.
4. Jamieson, M. and P. Jobber. 1987. *Manejo de los Alimentos.* México: Ed. Pax.
5. Lauhoff Grain Co. 1978. "A Guide to Good Manufacturing Practices for the Food Industry." Danville, Illinois, 61832. P.O. Box 5712, USA.
6. Schoenherr, W. H. and J. H. Rutledge. 1967. "Insects Pests of the Food Industry." Lauhoff Grain Co. Biological Control Department. PO Box 571, Danville, Ill. 61832. USA.
7. Valle-Vega, P. 1986. "Toxicología de Alimentos. Centro de Ecología Humana y de la Salud." Organización Panamericana de la Salud. Organización Mundial de la Salud. Metepec, Edo. de Mexico, Mexico.
8. Bayer, 1990. "ABC Productos Veterinarios," in *Manual Práctico del Hacendado,* 9th. ed., Mexico, D.F.
9. Brooks, J. E., G. K. Lavoie. 1990. "Rodent Control will Reduce Post-harvest Food Losses," *Agribusiness Worldwide,* November-December, 12 (7): 13.

10. Foulk, J. D. 1990. "Pest Occurrence and Prevention in the Foodstuffs Container Manufacturing Industry," *Dairy, Food and Environmental Sanitation.* 10(12): 725–730.

11. Bond, E. J. "Manual of Fumigation for insect Control," FAO paper 5. FAO, Rome, p. 432.

12. Centro de Ecología Humana y Salud. 1986. "Protocolo Estandarizado para Estudios de Campo Sobre Exposición a Plaguicidas," Centro Panamericano de Ecología y Salud; Organización Panamericana de la Salud; Organización Mundial de la Salud. Metepec, Edo. de México, Mexico.

13. Centro de Ecología Humana y Salud. 1986. "Plaguicidas. La Prevención de Riesgos en su Uso. Manual de Adiestramiento," Centro Panamericano de Ecología Humana y Salud; Organización Panamericana de la Salud; Organización Mundial de la Salud. Metepec Edo. de México, Mexico.

14. Centro de Ecología Humana y Salud. 1986. "Clasificación de Plaguicidas Conforme a su Peligrosidad, Recomendado por la Organización Mundial de la Salud," Centro Panamericano de Ecología Humana y Salud; Organización Panamericana de la Salud; Organización Mundial de la Salud. Metepec, Edo. de México, Mexico.

15. Diario Oficial de la Federación. 1988. "Catálogo de Plaguicidas," Tomo CDXIV, No. 10, México, D. F. *Commisión Nacional de Ecología. Subcomisión de Control de Agroquímicos.* SARH/SSA/SECOFI/SEDUE. Catálogo Oficial de Plaguicidas 1987. Diario Oficial, Primera Sección, Monday 14, march. Mexico, D.F.

16. EPA. Environmental Protection Agency, USA. 1979. "Suspended, Cancelled Pesticides. Office of Public Awareness (A-107)," Washington, D. C. 20460.

17. Kimball, A. C., J. Finkelman, A. Caracheo, and G. Molina. 1989. "Listado de Plaguicidas Restringidos y Prohibidos en Países de la Región de las Américas," Centro Panamericano de Ecología Humana y Salud; Organización Panamericana de la Salud; Organización Mundial de la Salud. Metepec, Edo. de México, Mexico.

18. Shell Agricultural Division. 1992. "Developing Safe Pesticides," *Agribusiness Worldwide* 14(6): 12.

19. Sagan, K. V. 1991. "Poison in your Backyard, The Pesticide Scandal," *Family Circle.* April 2nd: 59.

20. Alpuche, L. 1991. "Plaguicidas Organoclorados y Medio Ambiente. Ciencia y Desarrollo," *Conacyt* 16(96): 45.

21. González Avelar, R., J. J. Reyna García, M. Gil-Gutiérrez, and A. Ortiz Cornejo. 1990. *Manual de Fumigación*, Fascículo Primero. Centro Nacional de Investigación, Cerftificación y Capacitación. Uso de Bromuro de Metilo. Almacenes Nacionales de Depósito, S.A. ANSA. Dirección de Operación México, D.F.

22. Mueller, D. K. 1994. "Phosphine, Heat and Carbon Dioxide Deliver Death Punch to Insects," *Pest Control,* 62(3): 42.

23. Pinto, L. 2000. "Not Such Fine Feathered Friends," *Pest Control,* 68(3): 53.

Part XII

Space Missions

58 Storage Stability and Nutritive Value of Food for Long-Term Manned Space Missions

D. Zasypkin
T-C. Lee

CONTENTS

Abstract

Interplanetary manned space missions will require food with a shelf life of up to five years until a closed-loop bioregenerative life support system is developed. During this period, the nutritive value of such food components as proteins, lipids, and vitamins is reduced by 10–30% of the original value. Degradation of these components needs to be compensated for by increased consumption of some food items. The cost of supplying food is well above $11,000/kg. Therefore, accurate, quantitative assessment of nutritive degradation of the most critical food components is needed. Ample data exists on sensory acceptability of many food products in military rations during up to five years storage at ambient or elevated temperature. Numerous studies have been conducted on specific chemical reactions occurring in

1-56676-963-9/02/$0.00+$1.50
© 2002 by CRC Press LLC

important raw food materials or ingredients during long-term storage. Predictive models have been developed to evaluate shelf stability of some products. However, accurate, instrumental data on the nutritive value of the components in typical food products is fragmented for real-time ambient storage beyond two years. Therefore, physical, chemical, and nutritive changes in a variety of selected products should be traced during five years of ambient storage. The selected products are to be of high nutritive value, representative for their group of similar products and traditionally used in military and/or space menus. Multiple applications in menu items and lack of data on storage stability need to be considered in product selection. Minimal packaging could be explored along with most efficient packaging in vacuum-sealed meal-ready-to-eat (MRE) pouches developed by the U.S. Army Natick Laboratory. More specifically, physicochemical methods of analysis must be used to quantify the products of Maillard reaction and lipid oxidation, and to evaluate starch retrogradation and vitamin degradation. Protein bioavailability could be assessed using enzyme digestion methods.

58.1   INTRODUCTION

The first manned missions to Mars or other planets will most likely rely on ready-to-eat or easy-to-prepare food rather than on food processing on board. The current manned Mars mission scenario assumes that a cargo ship with food will first be sent into Mars orbit. Several months later, a crew is scheduled to arrive. The total time for the mission will approach three years, and the storage time for at least part of the food supply will exceed four years.[1] Alternative scenarios of food supply are under current consideration.[2-4] Some food can be prepared on board using primary ingredients, which are, as a rule, more stable than the final food products. In this case, the final products will be prepared fresh, and their acceptability is expected to be higher. The crew might enjoy cooking, which would also help to release psychological stresses. However, processing of ingredients into food is challenging under reduced gravity, and additional food processing equipment will be required. Even more complicated is a scenario assuming plant growing and food production "from scratch." At present, life support technologies for these alternative scenarios are not available or mature enough to provide reliable crew support. Therefore, a variety of food products with shelf stability from one to five years is needed. In addition, this variety should be large enough to provide a 28-day menu cycle as required by NASA specification.[2]

In 1987, the U.S. Army Natick Research Center completed a series of studies on long-term shelf stability of a group of products stored in a special MRE package for up to five years[5,6] (Table 58.1[6]). After five years of storage at 21°C, or three years at 30°C, or two years at 38°C, most of the products were found acceptable by a panel of 32 judges (Table 58.2). The number of products tested in these studies was 44, including 12 meat entrees, 3 varieties of beans, and 2 potato patties. The rest were mainly cakes, fruits, crackers, and condiments. This variety has to be significantly expanded to meet the 28-day menu cycle requirement. Most important, the conducted studies were limited to overall sensory evaluation (Table 58.2).[6] Very

TABLE 58.1
Items Studied for Sensory Acceptability during Five-Year Storage

Entrees	Pastries	Vegetables	Miscellaneous
Port sausage	Brownies	Beans	Cheese spread
Ham-chicken loaf	Cookies	Potato patty	Peanut butter
Beef patty	Pineapple cake		Jelly
Barbecued beef	Cherry cake	**Fruit**	Cocoa
Beef stew	Maple cake	Peaches	Coffee
Frankfurters	Fruit cake	Strawberries	Toffee
Turkey	Chocolate cake	Applesauce	Vanilla candy
Beef in gravy	Orange cake	Fruit mix	Catsup
Chicken			Crackers
Meatballs			Crackers and peanut butter
Ham slices			Crackers and cheese
Beef in sauce			Crackers and jelly

limited real-time data on chemical and physical changes or nutritional quality of food products is available about long-term storage (over two years). Studies conducted on model food systems or ingredients, and predictive models indicate that protein digestibility and bioavailability of essential amino acids decreases by 10–30% during three to five years storage at ambient temperature.

Inevitable nutritive degradation of food components during storage should be compensated for with specific additives and increased consumption of some items. Quantitative real-time analysis of chemical, physical, and nutritive changes is needed to supply sufficient food to the astronauts and reduce the cost of the missions. The estimated cost of sending 1 kg of food into a low Earth orbit is $11,000.[7] The cost will be significantly higher for the Mars or other planet missions. Therefore, it is critical to make accurate estimations of the needed amount of food to avoid either food shortage or oversupply. This data is essential to assure sufficient quantity, acceptability, nutritive value, and safety of food. Quantitative analysis of chemical, physical, and nutritional changes can also serve as a basis to improve the sensory acceptability and nutritional quality of the products and meet requirements for long-term storage stability. Microbiological and toxicological concerns should be addressed to assure the safety of food.

The number of products in a 28-day menu cycle is expected to exceed 100; it does not seem feasible to conduct the experiments on all specific products. Therefore, real-time quantitative tests of physical, chemical, and nutritive changes of some selected products representative of a group of products with similar composition, structure, and physical properties should be conducted. For example, one type of wheat bread may represent the group of bread-like bakery products in a viscoelastic state. Sensory evaluation will complement these studies. The following selection criteria can be applied: chemical composition, structure, and physical properties should be similar to the products in the group; highest nutritional value; priority of

TABLE 58.2
Average Hedonic Scores for Some Selected Products

Average score	After 60 months at 4°C	After 60 months at 21°C	After 36 months at 30°C	After 24 months at 38°C
7.4				
7.3	C			
7.2		C	C	C
7.1				
7.0				
6.9	PB			
6.8				PB
6.7		PB		
6.6				BN
6.5			BN	
6.4	CH, PP	BN		
6.3	BN			
6.2				
6.1				
6.0				
5.9	T			T
5.8		CH	T, CH	
5.7		T	BP	BP, CH
5.6	BP			
5.5		BP, PP		
5.4				
5.3			PP	
5.2				
5.1				
5.0				PP
4.9				

BP–Beef Patties; BN–Beans; C–Cookies; CH–Chicken; T–Turkey;
PP–Potato Patty, PB–Peanut Butter

traditional and palatable products; lack of data in the literature on nutritive value during more than two years of storage; presence, in significant amounts, of proteins and lipids, which are food components most susceptible to degradation; and potential application of the product in multiple menu items. In addition to the studied products, an extensive database can be created as to nutritive changes of most important food ingredients and palatability of the final products during storage of more than two years. This approach will allow for flexibility in product selection according to the specific mission requirements. Moreover, the obtained data is expected to affect storage and handling of food products in numerous applications on Earth, such as in distant outposts, military storage, and long-term expeditions.

58.2 SPECIFIC REACTIONS AFFECTING NUTRITIVE VALUE OF FOOD PRODUCTS AND CONSTITUENTS DURING STORAGE

Several reactions are responsible for most degradation of basic nutrients such as proteins, lipids, and carbohydrates. Besides major reactions, there are numerous reactions of minor components or the products of reactions involving the major components. Minor reactions in some cases may have a major effect on overall texture, flavor, acceptability, and safety.

In the case of proteins, nutritive deterioration is due to nonenzymatic Maillard browning that causes the loss of essential amino acids, primarily lysine; decreased digestibility due to protein cross-linking; color change; texture toughening; and off-flavor development.[8] The loss of protein nutritive value calculated from accelerated storage tests can be as high as 10% of the original value. This data needs to be verified in real-time experiments. Lysine loss can be up to 30%.[9] Physiological and safety effects of Maillard browning were studied by Lee et al.[10] and Kimiagar et al.[11] Experiments on feeding rats with browned egg albumin showed significant weight loss of the animals.[10]

Lipids are subject to hydrolysis (lipolysis), autoxidation, and enzymatic oxidation.[12] Lipolysis causes off-flavor development and changes in functional properties of fats and oils. Oxidation promotes development of rancid off-flavors, changes in both color and texture, and nutritional quality. Enzymatic oxidation leads to the same results as nonenzymatic oxidation, which is typical for nonheated products such as raw ingredients.

Starch carbohydrates contribute significantly to texture toughening of starch-containing food products.[13] The primary reason for this toughening is starch retrogradation that causes crystallization of amylose and amylopectin, accompanied by hydrogen bond formation. Water transfer between protein and carbohydrate-rich domains is another reason for quality deterioration related to carbohydrate content.[14] In this case, the phase loosing water is shifted toward the glassy state with a higher modulus of elasticity or firmness.

Below, we will consider the most typical studies conducted on long-term storage of food products and basic ingredients.

58.2.1 STORAGE OF FOOD PRODUCTS

Most of the data on the physical, chemical and nutritive changes in food products is limited to six-month storage. Other data are derived from accelerated storage tests conducted at elevated temperatures during several days or weeks. Prediction of the product quality was performed by extrapolation of the results, using evaluated temperature dependence of the rates of chemical reactions. An Arrhenius-type mathematical model was developed by Ross et al.,[5,6] to estimate the dependence of average consumer-acceptance scores on storage time and temperature, as well as the effect of temperature on shelf-life of MRE food products stored at 4, 21, 30, and 38°C for up to 60 months (Figure 58.1)[6] This approach, since it is predictive, may not be accurate for some products.

More specific studies have been done for some groups of products. Shelf life of canned pound cake was found to be less than 18 months at 38°C due to discoloration and development of off-flavors.[15] Breakfast cereals remain acceptable for two years at 15.6°C and 60% RH when stored in grease-proof bags.[16] Army-type biscuits have a 12-month shelf life at 38°C.[5] Spaghetti in fiberboard cartons has a 9-month shelf life at 32°C, showing rancidity, cracking, and mustiness after this time.[17] Berkowitz and Oleksyk[18] have developed an MRE pouch bread that is highly acceptable even after three years of storage at 27°C.

Most dried fruits and vegetables have a shelf life of one year at 38°C and one to two years at 21°C.[19] Darkening and loss of flavor are major types of deterioration. By using vacuum packaging in high-barrier material, the shelf life of freeze-dried food products can be extended from 30 days to 1 year.[20] Osmotically dehydrated blueberries, cherries, and cranberries mixed with sugar in a proportion of 3:1 or 4:1 had a predicted shelf life of 16 months at 25°C.[21]

Whole milk powder packaged in nitrogen with less than 1% oxygen and moisture under 2.5% has more than a two-year shelf-life at 38°C.[22] Rinschler[16] reported a

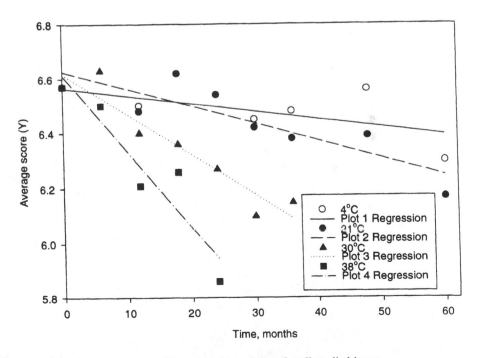

Figure 58.1 Average scores (Y) as functions of time for all studied items.

three-year shelf life at 15°C and 60% RH in grease-proof bags. A 15-month shelf life was found for dry coffee cream vacuum-packaged in aluminum foil laminated pouches.[23] Storage life of ice cream mixes is 12 months at 21°C. Margarine in cans has an estimated storage life of two years at 21°C.[17]

Canned green beans, yellow corn, peas, spinach, limas, and carrots had an 18-month shelf life, even at 38°C. Apricots in cans had less than a 12-month shelf life at 38°C.[24] Orange juice is acceptable up to 12 months at 21°C, while tomato juice is acceptable for 18 months at 32°C. Canned fruit juice will lose half of its vitamin C in 24 months at 27°C. Nitrogen-packaged dry meat was reported to have a shelf life over 12 months at 16°C.[25]

In summary, the studies conducted indicate that packaging under vacuum conditions with high barrier materials will significantly extend the shelf life of these products. Reformulation of the products, innovative preservation technologies, novel packaging materials, and antimicrobial additives could also markedly improve shelf-stability of these products. Most of the data on the shelf life of products is reported based on their sensory evaluation. In the following paragraphs, we review the results available for typical chemical reactions and bioavailability of long-stored food products.

The covalent cross-linking of proteins resulting from Maillard reaction in meat is expected to reduce the bioavailability of the proteins due to involvement of lysine residues and reduced digestibility by hydrolytic enzymes. Chicken meat was heated at 73°C for eight days in pure water and in the presence of 0.04 M lactose. The amount of tryptophan released by pronase from the browned chicken meat produced in the presence of lactose was approximately 12% less than that from the control chicken.[8]

Availability of lysine is an important index of protein quality. When the chicken meat was heated in the presence of 0.04 M lactose, the lysine loss increased to 13%.[8] When UHT milk was stored at 37°C for three years, the lysine loss reached 31%.[9]

Lipid oxidation is responsible for the short shelf life of many products; for example, instant potato powders, poultry, nuts, and dairy products.

58.2.2 STORAGE OF BASIC INGREDIENTS

More specific studies have been done on long term real-time storage of some industrially important raw materials and ingredients such as wheat grains and flour,[26–28] rice and soybeans,[29] sugar, dried milk or cream,[30] and others.

Zeleny[31] describes major changes occurring in wheat grains and flour. Wheat grains containing about 11% water showed a decrease in protein digestibility of 8% when stored in sealed jars at 25°C for two years. Under the same storage conditions, corn showed a decrease of 3.6% in protein digestibility. These changes, as well as changes in protein solubility, occur much more rapidly in the milled products of grain than in whole grain.

Whole bran wheat flour can be kept for a relatively short time in contact with air, because it readily becomes rancid, regardless of moisture content.[31] Detailed studies of lipid fractions during storage of cereals show that polar lipid fraction steadily increases during storage up to five years. The glycolipid peak ratios, particularly sitosteryl-beta-glucoside/sn-1,2-dilinolyl-1-3-galactosylglycerol, were shown to be practical indicators of not only the general lipid condition of wheat upon storage, but also in particular those lipids of interest in baking quality.[26]

Analysis of the data indicates that physical-chemical and nutritive changes were studied in detail for long-term storage (up to five years) of basic food ingredients. In the case of final food products, ample data exists on the changes during relatively short-term storage. In many studies, data were collected for up to several weeks at elevated temperatures. Sensory evaluation is available for a number of products for up to five years of storage. However, data is very limited as to quantitative real-time measurements of physical, chemical, and nutritive changes in food products.

Based on data availability, we consider the following studies need to be done:

1. Conduct physical and chemical tests of the products representative for groups of products of similar chemical composition, structure and physical properties during one to five years. Complement these studies with sensory evaluation. Texture changes, Maillard reaction, starch retrogradation, lipid oxidation, off-flavor development, microbial stability, and accumulation of toxic compounds should be the major issues of concern. HPLC, GC, DSC, texture analysis, and traditional chemical methods of food analysis can be used.
2. Quantitatively evaluate nutritional losses of the selected products and basic ingredients during real-time long-term storage.
3. Shelf stability of basic ingredients needs to be evaluated to explore the possibility of partial on-board food preparation.
4. Models predicting the shelf life of the products should be tested and corrected using real-time data on physical, chemical, and nutritive changes in the products.
5. Create a comprehensive database on nutritive changes and acceptability of most important food ingredients and final products during storage of more than two years. Recommendations and suggestions on product selection and menu development can then be derived for long-term manned missions.

58.3 FUTURE STUDIES

58.3.1 PRODUCT SELECTION

Since it may not be feasible to comprehensively test every product on the menu, we propose grouping the products and testing typical product(s) representing a group. For example, there are a number of bread-like products prepared mainly from wheat. Chemical composition, structure, and physical properties of the products within the group are similar. Therefore, it makes sense to test thoroughly one or several products of the entire group.

Products may be selected for the tests using the following criteria:

1. Chemical composition, structure, and physical properties similar to the products in the group
2. Highest nutritional value
3. Priority of traditional and palatable products
4. Lack of data in the literature on nutritive value during more than two years of storage
5. Presence of significant amounts of proteins and lipids most susceptible to degradation of food components
6. Multiple applications of the product in various menu items

Chemical composition of the products determines major chemical reactions occurring in the system. There are several reviews and experimental papers describing in detail major chemical reactions in systems containing proteins, sugars and lipids, separately or in combinations.[8, 2,13] The effect of physical state and structure of the products on their storage is less understood, although in a number of studies the critical role of these texture attributes has been demonstrated.[14,32] Based on this information, we will analyze the products according to their composition, i.e., protein, carbohydrate, and lipids content, and also combinations thereof. Types of typical components will be taken into account. For example, the type of protein can be meat protein, storage plant proteins, wheat protein, milk protein, etc. The type of carbohydrate can be neutral or charged, mono-, oligo-, or polysaccharide. Lipids can be classified as saturated, mono- and polyunsaturated, and phospholipids. The type of component will determine reactivity and nutritive value of the components during storage. Particular attention will be paid within a group of similar products to the products with the highest potential reactivity.

Structure can be classified as dense, porous, homogeneous, or multiphase and in relation to aggregation and physical state of the products. The aggregation state can be gas-like, liquid, or solid, while the physical state of the biopolymers can be glass, rubber, or fluid-like.

There is information available on sensory acceptability[6] or chemical and nutritional changes for some groups of products. As a rule, sensory information is available for some products stored up to five years, while chemical, physical, or nutritional data exists mainly for products stored up to six months.[21] Conclusions for the longer storage times were drawn from data extrapolation based on estimated temperature dependence of typical chemical reactions. This data can be used in product selection to avoid repetition and to make a target-oriented experimental design.

As an example, the following products or ingredients within the groups of the products can be used in long-term missions. The above-mentioned selection criteria are applied. The selected products can be vacuum-packed in MRE pouches and retorted for 1 h at 120°C for long-term storage studies.

Meat, poultry, and fish

- *Group:* beef sausages, beef patties, beef stew, frankfurters, turkey, chicken-a-la-king, ham slices, freeze-dried beef, dried and salted fish.
- *To be studied:* beef patties, beef jerky, cooked chicken, cooked whitefish.

Baked products

- *Group:* wheat bread, cakes, cookies, brownies, crackers, chips, pasta, flat bread, tortillas, noodles, breadsticks, waffles, breakfast cereals.
- *To be studied:* tortillas, sweet dough cake, cookies.

Dairy products

- *Group:* butter, cheese, freeze-dried ice cream, dry milk, concentrated milk with sugar.
- *To be studied:* milk fat.

Cereals and vegetables

- *Group:* beans, wheat flour, potato patties, potato granules, oat flakes, tofu, soy-based meat analogs, cheese analogs, soy margarine, soy milk powder, dried and freeze-dried vegetables, rice, corn, corn flakes, tortillas.
- *To be studied:* dried soybean curd.

Fruits

- *Group:* dried peaches, apples, prunes, apricots, cranberries, dried tomatoes, apple juice, orange juice, dry juice, fruit jelly, jams.
- *To be studied:* none.

Nuts

- *Group:* walnuts, peanuts, peanut butter, halvah.
- *To be studied:* walnut, peanut butter.

Sweets

- *Group:* chocolate, caramel, tomato candies, candies based on baked concentrated milk, syrup, honey, sherbet.
- *To be studied:* none.

Spices and condiments

- *Group:* coffee, tea, pepper, garlic, onion powder, dry parsley, starch, egg powder, mayonnaise, salad dressings.
- *To be studied:* none.

58.3.2 PACKAGING AND STORAGE

Several packaging methods can be applied. Some dried products in a glassy state will be sealed in plastic bags. Vacuum packaging is to be used for products prone to oxidation. Meal-ready-to-eat (MRE) pouches developed by the U.S. Army Natick Laboratory can be vacuum-sealed for this purpose. To reduce the quantity of packaging material, principles of minimal packaging should be applied.

The products might be stored at 25°C and 50% relative humidity in a controlled environmental chamber. Based on the literature, we assume that at lower temperatures shelf life will be longer. The temperature of 25°C is considered maximum for storage.

58.4 CONCLUSION

Analysis of the data available on degradation of food components during long-term storage (up to five years) indicates that quantitative real-time analysis of chemical, physical, and nutritive changes is needed to supply sufficient and palatable food to the astronauts, and reduce the cost of the missions. Microbiological and toxicological concerns should also be addressed to assure the safety of food.

Based on the literature data, we propose criteria for the selection of food products and ingredients to be tested for long-term missions. Led by these criteria, several groups of the products are suggested as an example for the prospective tests. Those most critical for storage physical and chemical reactions were discussed.

REFERENCES

1. NASA Special Publication 6107. 1998. *Human Exploration of Mars: The Reference Mission of the NASA Mars Exploration Study Team.* Johnson Space Center publication.
2. Evert, M. F., C. T. Bourland, K. D. Glaus-Late, and S. D. Hill. 1992. "Food Provisioning Considerations for Long Duration Space Missions and Planet Surface Habitation," SAE paper #961415.
3. Fu, B., P. E. Nelson, and C. Mitchell. 1995. "The Food Processing Subsystem for CELSS: A Conceptual Analysis," *Life Support Biosphere Sci.*, 2: 59–70.
4. Zasypkin, D. V. and T-C. Lee. 1999. "Food Processing on a Space Station: Feasibility and Opportunities," *Life Support & Biosphere Sci.*, 6: 39–52.
5. Ross, E. W., M. V. Klicka, J. Kalick, and M. E. Branagan. 1985. "Acceptance of a Military Ration After 24-Month Storage," *J. Food Sci.*, 50: 178–181, 208.
6. Ross, E. W., M. V. Klicka, J. Kalick, and M. E. Branagan. 1987. "A Time-Temperature Model for Sensory Acceptance of a Military Ration," *J. Food Sci.*, 52: 1712–1717.
7. Bozich, W. F. 1991. "Technology Requirements for Future Launch Vehicles: The Next 20 Years," presented at TMSA Space Requirements Conference, Los Angeles, CA, 1996.
8. Barnett, R. E. and H.-J. Kim. 1998. "Protein Instability," in *Food Storage Stability*, I. A. Taub & R. P. Singh, eds. New York: CRC Press, ch. 3, pp. 75–88.

9. Moller, A. B. 1981. "Chemical Changes in Ultra Heat Treated Milk During Storage," in *Progress in Food and Nutrition Science*, vol. 5, *Maillard Reactions in Food*, New York: Pergamon Press.

10. Lee, T-C., M. Kimiagar, S. J. Pintauro, and C. O. Chichester. 1981. "Physiological and Safety Aspects of Maillard Browning of Foods," *Prog. Fd. Nutr. Sci.*, 5: 243–256.

11. Kimiagar, M., Lee, T-C. and C.O. Chichester. 1980. "Long-term Feeding Effects of Browned Egg Albumin to Rats," *J. Agric. Food Chem.*, 28: 150–155.

12. Nawar, W. W. 1998. "Biochemical Processes: Lipid Instability," in *Food Storage Stability*, I. A. Taub & R. P. Singh, eds. New York: CRC Press, ch. 4, p. 89–104.

13. Gordon, J. and E. Davis. 1998. "Biochemical Processes: Carbohydrate Instability," in *Food Storage Stability*, I. A. Taub & R. P. Singh, eds. New York: CRC Press, ch. 5, pp. 105–124.

14. Chinachoti, P. 1998. "Water Migration and Food Storage Stability," in *Food Storage Stability*, I. A. Taub & R. P. Singh, eds. New York: CRC Press, ch. 9, pp. 245–268.

15. Mitchell, J. H., Jr. 1955. "Rate and Extent of Deterioration of Packaged Rations During Storage and Transportation," in *Establishing Optimum Conditions for Storage and Handling of Semiperishable Subsistence Items*, H. E. Goresline, N. J. Leinen, and E. M. Mrak, eds.

16. Rinschler, R. A. 1954. "Underground Storage Test," Phase I. Summary Report., Quartermaster Corps.

17. Anonymous. 1950. "Storage of Quartermaster Supplies," *Dept. of The Army Tech. Manual TM* 10-250. U.S. Gov't. Printing Office, Washington, DC.

18. Berkowitz, D. and L. Oleksyk. 1991. "Leavened Breads with Extended Shelf-Life," U.S. Patent no. 5,059,432.

19. King, J. 1948. "Scientific Problems in Feeding a Modern Army in the Field," *J. Soc. Chem. Ind.* 47: 739–743.

20. Duxbury, D. D. 1987. "Freeze-Dried Ingredients Use MAP to Extend Shelf Life 12-fold," *Food Proc.* 48: 28.

21. Yang, T. C. S. 1998. "Ambient Storage," in *Food Storage Stability*, I. A. Taub & R. P. Singh, eds. New York: CRC Press, ch.17, p. 435–458.

22. Coulter, J. S. 1947. "The Keeping Quality of Dry Whole Milk Spray-Dried in an Atmosphere of Inert Gas," *J. Dairy Sci.*, 30: 115.

23. Kemp, J. D., Ducker, A. J., Ballantyne, R. M., and G. G. Acheson. 1957. "A Study of the Flexible Packaging of Dry Cream and Potato Powders," *Def. Res. Med. Lab. Rpt.* 174-3. Toronto.

24. Brenner, S., Wodicka, V. O. and S. G. Dunlop. 1947. "Stability of Ascorbic Acid in Various Carriers," *Food Res.* 12: 253–269.

25. Sharp, J. G. and E. J. Rolfe. 1958. *Fundamental Aspects of the Dehydration of Foodstuff*, London: Society of Chemical Industry.

26. Warwick, M. J. and G. Shearer. 1980. "The Identification and Quantification of Some Non-Volatile Oxidation Products of Fatty Acids Developed During Prolonged Storage of Wheat Flour," *J. Sci. Food Agric.*, 31: 316–318.

27. Bothast, R. J., R. A. Anderson, K. Warner, and W. F. Kwolek. 1981. "Effects of Moisture and Temperature on Microbiological and Sensory Properties of Wheat Flour and Corn Meal During Storage," *Cereal Chem.*, 58: 309–311.

28. Galliard, T. 1986. "Hydrolytic and Oxidative Degradation of Lipids During Storage of Wholemeal Flour: Effects of Bran and Germ Components," *J. Cereal Sci.*, 4: 179–192.

29. Kermasha, S. and I. Alli. 1993. "Changes in Lipase and Lipoxygenase Activities and Fatty Acid Profile During the Storage of Unprocessed and Processed Full-Fat Soybeans," in *Developments in Food Science*. Amsterdam: Elsevier Scientific Publications, 33, pp. 1021–1035.
30. Hall, G. and H. Lingnert. 1986. "Analysis and Prediction of Lipid Oxidation in Foods," in *The Shelf Life of Foods and Beverages*, G. Charalambous, ed. Amsterdam: Elsevier Science Publishers B.V.
31. Zeleny, L. 1954. "Chemical, Physical, and Nutritive Changes During Storage," in *Storage of Cereal Grains and Their Products*, J.A. Anderson, and A.W. Alcock, eds., St. Paul MN: AACC, ch. 2, pp. 46–76.
32. Szczesniak, A. S. 1998. "Effect of Storage on Texture," in *Food Storage Stability*, I. A. Taub & R. P. Singh, eds. New York: CRC Press, ch. 8, p. 191–244.

59 Preservation Methods Utilized for Space Food

Y. Vodovotz
Ch. T. Bourland

CONTENTS

Abstract

Food for manned space flight has been provided by NASA–Johnson Space Center since 1962. The various mission scenarios and space craft designs have dictated the type of food preservation methodologies required to meet mission objectives. The preservation techniques used in space flight include freeze-dehydration, thermostabilization, irradiation, freezing, and moisture adjustment. Innovative packaging material and techniques have enhanced the shelf stability of the food items. Future space voyages may include extended duration exploration missions requiring new packaging materials and advanced preservation techniques to meet mission goals of up to five-year shelf-life foods.

59.1 INTRODUCTION

The National Aeronautics and Space Administration–Johnson Space Center (NASA–JSC) has been providing safe, nutritious, and palatable foods to crew members since the first extended manned mission to space in 1962. Each mission, from

Mercury to the International Space Station, had specific constraints that dictated the appropriate form for food preservation. The Mercury and Gemini missions confined crew members to a sitting position within the cabin and allowed for very limited food stowage volume. Additionally, the lack of adequate bathroom facilities promoted the development of low-fiber diets to reduce fecal output.

The Apollo era introduced special requirements due to the Extra Vehicular Activities (EVAs) on the moon's surface and a requirement by crew members for better quality and palatability of food. Skylab, the first space station, opened up a new era for the food system.

The availability of freezers, refrigerators, and a dining table with individual heated wells promoted an Earth-like dining experience. Additionally, the vast space (in comparison to previous missions) available on board the space station enabled a greater variety of food items.

The space shuttle, intended as a work vehicle for short-duration missions, does not include amenities such as a dining table or refrigerator/freezer for food consumption. Nonetheless, improvements in packaging and preservation techniques were and are a part of the shuttle program. Mir, the Russian space station, had limited capabilities for food preparation due to power limitations on board. Additionally, while astronauts were on board, half of the food on the Russian station was provided by the United States, while the other half remained Russian. Due to the incompatibility of the U.S. food with the Russian warmer, a suitcase heater was required, a situation that will continue on the International Space Station (ISS) during the assembly phase.

At the completion of the ISS, the habitation module (HAB) is planned to contain a full galley with refrigerators and freezers as well as a wardroom table. This hardware was included to simulate an Earth-like eating environment due to the longer mission duration expected on ISS.

Below, we describe the evolution of preservation techniques and packaging innovation used in the U.S. space program.

59.2 PRESERVATION TECHNIQUES

59.2.1 FREEZE DEHYDRATION

Freeze dehydration gained attention and use by the military for combat rations in the 1950s due to the increased shelf life of the products and their reduced weight.[1] With time, military and civilian use of freeze dehydrated products increased along with significant improvements from research stimulated by military and space applications. Freeze dried foods were first used by NASA on the Mercury 9 mission. However, only one-third of the items were eaten, due to problems with the food container and the water dispenser.[2]

Freeze drying allows for greater flavor retention and less vitamin loss of a product than any other drying method. The freeze drying process consists of processing the food to the ready to eat stage, freezing, placing in a freeze dryer vacuum chamber, and removing the moisture (final moisture content $\leq 3\%$) by sublimation. The dehydrated food is packaged immediately after removal from the freeze dryer to prevent reabsorption of moisture. Many foods for space are freeze dried in pre-designed forms (placing the processed food in a mold prior to freezing) to facilitate packaging. The final product is then in the shape of the mold. While all freeze-dried food for the space shuttle and the ISS is to be consumed after the addition of water, some of the early foods, such as ice cream and fruits, were made to eat directly. These were not popular with the astronauts.

Shrimp cocktail, the most popular food on the shuttle menu, is a freeze-dried product. The precooked freeze dried shrimp are mixed with a dried cocktail sauce and rehydrated together. Since the ISS will require the use of solar panels for electricity instead of fuel cells (which provide abundant water on liftoff), many of the advantages of freeze dried food will be negated.

59.2.2 Intermediate-Moisture Food

Reducing the moisture of food and the resultant lowering of water activity has been a common method for increasing the shelf life of food. The first intermediate moisture food was flown on Gemini. These bite-sized foods (such as bacon squares) were cooked, pressed into squares, then packaged (Figure 59.1). Intermediate mois-ture fruits (dried to 20–30% moisture) were introduced on Apollo 9 (1969) and are common items on shuttle flights. Dried beef (beef soaked in 3–5% brine then cooked and dried) was later added to the shuttle menu.

59.2.3 Freezing

Frozen foods were first available during the Skylab missions and presented a sig-nificant improvement to menu quality. Foods were cooked, placed in aluminum cans with a full-panel pullout lid, sealed under nitrogen atmosphere, and frozen. Prior to eating, the cans containing different foods (thermostabilized, dehydrated, frozen, etc.) were assembled in a food warmer/serving tray (Figure 59.2). Frozen foods were warmed from $-23.2 \pm 5.5°C$ to $65 \pm 3.3°C$ in a span of 2 h.[3] Frozen foods have not been used since Skylab but are planned for ISS. Foods will be cooked, placed into frozen food containers made of CPET (crystallized polyethylene terephthalate), sealed, and frozen. At ISS Assembly Complete (Habitation module becomes part of ISS), 50% of the food is planned to be frozen.

59.2.4 Thermostabilized Food

Initially, thermostabilized food was part of the contingency food supply on Apollo missions. This chocolate-flavored, nutrient-defined formula was packaged in flexible metal tubes with an attached pontube.[4] The tubes (Figure 59.3) were designed to provide nutrients to astronauts wearing a pressurized suit, such as required in emer-gency operations, and therefore were limited to the size of the helmet feedport (86 mm). On Apollo 8 (1968), thermostabilized meat products ("wet packs") were introduced into the menu when astronauts ate a wet pack of turkey and gravy as part of their Christmas meal.[5] These products were cooked then inserted into pouches

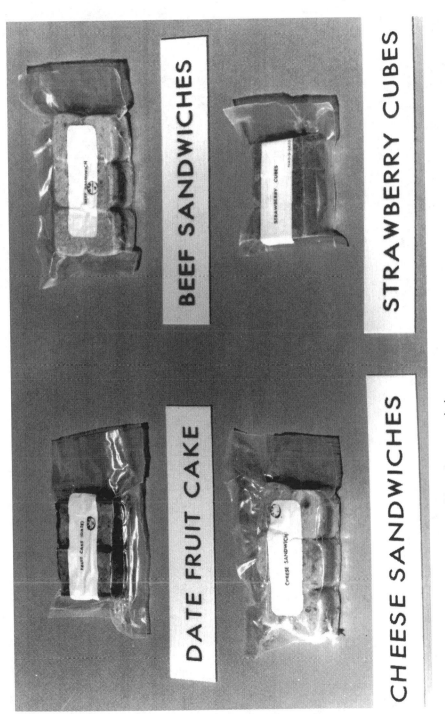

Figure 59.1 Examples of bite-size food used in early space missions.

Figure 59.2 Skylab warmer/serving tray with various food containers.

made from laminated polyester, aluminum foil, and polypropylene (Figure 59.4). Subsequently, pouches where induction sealed and retorted to commercial sterility. Sandwich spreads were initially packaged in aluminum tubes and retorted and later packaged in aluminum cans.

As the space program progressed, thermostabilized foods became an integral part of the menu. In Skylab, these foods were packaged in metal cans (as discussed in the packaging section), while the shuttle saw the reintroduction of the flexible metal pouches. Currently, eight different foods are thermostabilized in pouches in the shuttle program, and that number will increase significantly for ISS.

59.2.5 Food Irradiation

Irradiated foods significantly increase the shelf stability of products and are especially advantageous to NASA, since all missions have been without refrigerators or

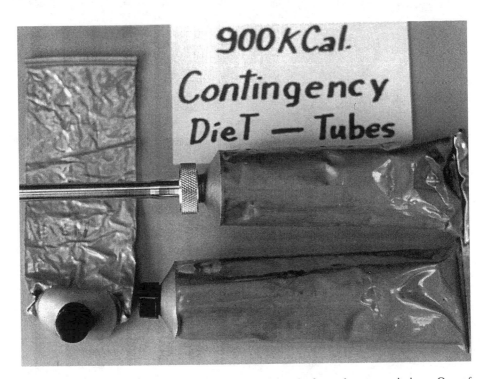

Figure 59.3 Tubes containing a contingency food supply for early space missions. One of the tubes has a pontube attached to it for feeding purposes.

freezers, with the exception of Skylab in 1972. Additionally, liquids can be controlled in irradiated products, thus reducing the difficulty of handling liquids in microgravity.

Irradiated food was first used in space when irradiation-sterilized ham was consumed on Apollo 17 in 1972. Later, irradiated beef steak, ham, turkey, and corned beef were included on the Apollo Soyuz Test Project (ASTP) in 1975, and irradiated flour (0.5 kG) was incorporated in bread for the Skylab missions.[6] Irradiated beef steak, corned beef, and smoked turkey were used on the early shuttle missions from 1981 to 1986. These irradiated products were provided by the U.S. Army Natick Research, Development and Engineering Center (RD&E Center). In 1988, NASA began the formal petition process for approval of sterilization doses (absorbed dose 44 kG) from the Food and Drug Administration (FDA). Prior to approval in 1995, the FDA issued a letter of no objection to NASA use of irradiated sterilized beef steaks on space missions.

The meats are irradiated by Food Technology Services, Inc., Mulberry, FL. Products are produced by Natick RD&E Center with production guides that outline formulas, inspection, and packaging procedures. The products are processed, i.e., cooked or grilled, and then heated for enzyme inactivation. Subsequently, products are packaged under vacuum (not less than 920 kPa) in foil pouches. Go/no-go dosimetry indicators are included with the pouches before they are placed into

Figure 59.4 Examples of thermostabilized food packaging used in space missions.

individual cardboard containers and shipped with dry ice to the irradiation facility. Upon receipt, the irradiation facility processes the packages at a temperature of between −30 and −40°C to an absorbed dose of 44 kG. After processing, the product is returned under ambient temperature to the Natick RD&E Center for standard microbial evaluations to determine commercial sterility. Those packages that pass testing are shipped to the Johnson Space Center, where the pouches are inspected for seal integrity and weighed. Pouches that do not meet the weight tolerance are used for training. The receiving inspection also includes a sensory evaluation. The remaining pouches are placed in bonded storage to await a flight assignment.

A total of 376 irradiated meats and 253 irradiated bakery products were sent on the first 24 shuttle missions (1981–1986). Irradiated meat has been on all shuttle missions since 1989 and was consumed by astronauts and cosmonauts on the Russian Mir space station (1995–1998).

The International Space Station (ISS) renewed interest in irradiated foods for space flight. More irradiated meat products are being developed by the U.S. Army Natick RD&E Center to supplement the available shelf stable food products. Irradiated BBQ beef brisket, pork sausage, and teriyaki beefsteak have been developed and are in use on shuttle and will be on the first food shipment to ISS. Products under development include beef enchiladas, beef sirloin tips with mushrooms, broiled lamb, teriyaki chicken, chicken dijon, and Thai chicken.

Astronaut candidates (Ascans) and astronauts use a nine-point Hedonic scale (1 being "dislike extremely" and 9 "like extremely") to evaluate new foods at the Space Food Systems Laboratory at JSC. Ratings for irradiated smoked turkey and beef steak are shown in Table 59.1 for 1995 through 1999. Smoked turkey was generally rated higher than beef steak; the Ascans' ratings are slightly lower than those of astronauts. In most cases, this is the first exposure to space food for the Ascans, whereas the astronauts may have been on several missions and are experienced space food consumers. The Mir cosmonauts rated both irradiated meats equal to or higher than the Ascans or crews.

TABLE 59.1
Astronaut Candidate (Ascan), Astronaut, and Cosmonaut Sensory Scores[*]
on Irradiated Smoked Turkey and Beef Steak

	Smoked Turkey		Beef Steak	
1995 Ascans	7.0 ± 1.4	$N = 21$	7.1 ± 1.8	$N = 22$
1998 Ascans	7.7 ± 0.8	$N = 10$	5.7 ± 2.0	$N = 37$
1999 Ascans	7.5 ± 1.1	$N = 22$	6.0 ± 1.6	$N = 22$
Ascans overall	7.2 ± 1.2	$N = 53$	6.2 ± 1.9	$N = 82$
Flight crews 1996-99	7.4 ± 1.4	$N = 46$	7.3 ± 1.6	$N = 45$
Overall	7.3 ± 1.3	$N = 99$	6.6 ± 1.9	$N = 126$
Mir cosmonauts	8.0 ± 1.0	$N = 15$	7.3 ± 1.8	$N = 16$

[*]1 = dislike extremely; 9 = like extremely. The crews represent all astronaut crews from 1996 to 1999 and include ISS crews. N is the number of crew members testing the product.

Irradiated meats have been a popular food for the astronauts and cosmonauts. Products like the beefsteak provide a whole-muscle item with a familiar texture that is difficult to duplicate with other preservation methods. Astronauts have never questioned the safety of the irradiation process. With the delay in getting freezers and refrigerators aboard the ISS, irradiation will be given further attention for processing a wider variety of foods. Fruits and vegetables may be irradiated instead of thermostabilized to improve texture qualities. Irradiation is under consideration for controlled atmosphere packaged fruits and vegetables that will be used on ISS when refrigerators are available. By irradiating the controlled atmosphere packaged products, potential pathogens can be reduced or eliminated.

59.2.6 PACKAGING

The first food consumed in space by a United States astronaut was applesauce packaged in an aluminum tube and eaten by John Glenn, over 30 years ago. The tubes were fitted with a pontube (Figure 59.3) for consumption directly into the mouth. The package design served the requirements well and did not allow any food to contaminate the cabin. Tubes continued to be used through the late Apollo missions in the early 1970s and are still being used by the Russian Space Agency.

However, tubed foods lacked odor, and the contents could not be seen. Additionally, the texture of the product was limited to the orifice of the tube and the tube-filling processing capabilities.

Cubed foods (Figure 59.1) were also part of the earlier space food systems. Cubes were formed from cereal-based products by subjecting them to high pressures. The bite-size cubes were covered with a gelatin coating to prevent crumbs from contaminating the cabin. The cubes were originally packaged in an aluminum foil and then later changed to a transparent plastic laminate. Cubed foods also suffered from some problems similar to the tubes. Although the starting ingredients were the same as familiar counterparts (e.g., sugar cookies pressed into sugar cubes), the engineered cubes did not have familiar texture and mouth feel. Even though popular in preflight taste tests, many cubes were returned uneaten.

The search for a transparent plastic material with high barrier properties began early in the development of space food packaging. A material developed in the late 1960s called SLP4 met these requirements. SLP4 was a laminate made of polyethylene, mylar, aclar, and polyethylene. The barrier properties, 2.5 cc oxygen and 0.0015 g water per 100 in² per 24 h, were more than adequate for space food and established a standard that proved to be difficult to meet for new materials. SLP4 did have a tendency to delaminate if not sealed properly and was a special-order film with a substantial setup charge.

Dehydrated foods were first consumed from a Zero G Feeder, a package that allowed insertion of water in one end and consumption through a large tube on the other end. Again, consumers were not able to smell what they were eating and the particle size was limited to the orifice of the mouthpiece.

Early in the Apollo program, the spoonbowl package (Figure 59.5) was developed as a solution to the problem of eating through pontubes and continued to be used through the first four shuttle missions. The spoonbowl was made with SLP4. Water was added through the one-way water port. After rehydration, the top of the package was cut open and the contents consumed with a spoon. Although this approached normal eating procedures, it required both hands: one to hold the package and one to hold the utensil. The consumer was committed to one package of food at a time, making eating more than one item at a time in microgravity very labor intensive.

A unique package developed for Apollo was the in-suit food bar made from compressed fruit leather and packaged with an edible starch film. It was used inside the astronaut's suit for consumption without the use of hands. The packaged bar fit into a sleeve that the astronaut could reach by mouth and bite off portions. The in-suit food bar was used on Apollo, Skylab, Apollo-Soyuz Test Project, and early Shuttle, but was later discontinued. The Apollo missions were also the first to use retort pouches ("wet packs"), as discussed earlier.

The Skylab program presented some new challenges in food packaging. All of the food was launched with the first mission, making it over two years old when the last crew consumed it. Skylab (as described above) was the first U.S. space mission to have freezers, refrigerators, and food warmers. Aluminum cans were chosen as the barrier package for the Skylab foods, even though round cans were

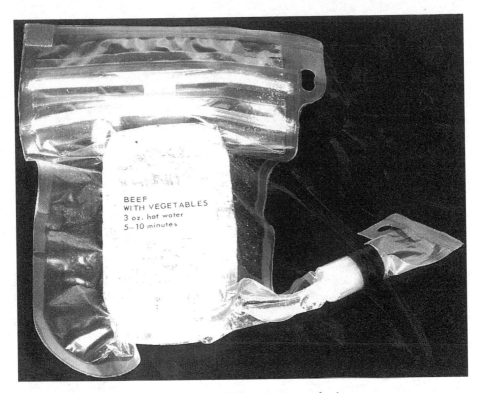

Figure 59.5 Spoonbowl package used for dehydrated space food.

not amenable to efficient storage. Three sizes of aluminum cans were used, with most of the entrees being packaged in 401 by 105 cans with full-panel pullout lids.

Frozen food that required heating, and some of the thermostabilized items, were packaged with a membrane under the lid to prevent spilling while heating and to facilitate opening in microgravity. Dehydrated foods were packaged in a thermoformed polyethylene inner package fitted to the can. The cans were subsequently sealed in a nitrogen atmosphere and placed in canisters designed to withstand the pressure changes between the ground and the spacecraft cabin pressure of 5 psia. The Skylab food maintained quality throughout the two-year period, despite cabin temperatures reaching 130°F early in the mission when a solar panel failed to deploy. The high temperature caused some browning, but the food was still edible, a testament to the high-barrier packaging.

With the planned short mission duration and lack of room and electrical power in the orbiter, the shuttle returned to a food system without freezers and refrigerators. The short-duration missions and 14.7-psia cabin pressure allowed some relaxation in the barrier properties and strength of the packaging material. The strict flammability and off-gassing requirements were still enforced.

A new rehydratable package was developed for the shuttle to permit consumption from a meal tray and to overcome the numerous package fabrication problems (over

30 steps) associated with the spoonbowl package. The new package base was injection molded from high-density polyethylene, and the lids were thermoformed from a saran-coated nylon polyextrusion laminate to adjust to the level of the package contents. A septum, which acts as a one-way valve, replaced the complex hand-assembled valve system used on previous packages.

As shuttle crew sizes increased and mission length was extended, food packaging began to receive attention again due to the accumulation of on-board trash. Waste is a major problem in the space craft environment due to minimal on-board storage space as it awaits return to ground. Trash from food containers contributed up to two-thirds of the trash volume. The rehydratable food and beverage packages were designed to nest to reduce volume, but they could not be compacted. The beverage package was replaced with a laminated foil package which can be compacted (top right in Figure 59.6), while the flexible rehydratable food package is constructed from a five-layer coextrusion of nylon/ethylene vinyl alcohol/nylon/tie layer of polyethylene/linear low-density polyethylene (bottom left in Figure 59.6).

Figure 59.6 is a display of all the types of packages used in the shuttle food system. Some selected MRE and commercial retort pouches are used. Weight and volume restrictions limit the number of retort pouches that can be used. Other types

Figure 59.6 Example of current space shuttle food on a meal tray. Velcro™ pads hold the food down in microgravity while magnets do the same for the eating utensils. Food packages from top left to bottom right: bite-size, cans, laminated foil beverage pouch, dehydratable, and thermostabilized.

of packaging include aluminum and bimetallic cans and single-service commercial condiment containers.

59.2.7 FUTURE TRENDS

The goal for ISS (assembly complete) is to have customized packages with one constant dimension that provide a restraint mechanism for storage, food preparation, and dining. A lip-lock concept will hold the food packages onto the dining tray (Figure 59.7). Most of the food will be packaged in single-service containers so as to provide real-time menu exchanges and to prevent the need to transfer food from one container to another in microgravity, which is usually a messy operation. The configuration of the packages has been determined through a series of tests and engineering evaluations that involved the interface with the storage facilities, oven, and serving trays.

Figure 59.7 Prototype of International Space Station assembly complete meal tray.

The material must

- protect the food in the intended storage environment
- be compatible with the food preparation equipment (microwave/convection oven)
- be easy to open and eat from
- interface with the restraint/storage system
- be particle free and easily compactible
- present no burn hazards
- comply with off-gassing and flammability specifications.

The current leading candidate material is CPET (crystallized polyethylene terephthalate).

Lunar base and Mars missions offer even greater challenges to food packaging. Shelf life of food must be extended to at least one year for a lunar base and up to five years for a Mars mission. Trash management will be a greater problem and, ideally, used package material will be easily converted to another use. The obvious solution to this application is the elusive edible package. There are some formulations for cabin components to be made from casein and starch and then used for food after their primary function is completed; however, their palatability is highly questionable.

On-base and in-flight packaging is also a distinct possibility if food is produced on a planetary base as part of an advanced life support system.[7] Excess food from the harvested crops may need to be packaged and preserved for storage and later use.

In many instances, the goals of space food packaging parallel those of the commercial packaging industry. NASA has utilized the innovations of the commercial packaging industry to get food into space, onto the moon and back, and in the near future to ISS. A concentrated joint effort will be required to process and package food for a journey to Mars.

REFERENCES

1. Hollender, H. A. 1969. "U.S. Army Food R & D Program," in *Aerospace Food Technology*. NASA SP-202, pp. 151–163.
2. Catterson, A. D., E. P. McCutcheon, H. A. Minners, and R. A. Pollard. 1963. "Aeromedical Observations," in *Mercury Project Summary Including Results of the Fourth Manned Orbital Flight May 15 and 16*, 1963, NASA-SP45, pp. 315.
3. Bourland, C. T., N. D. Heidelbaugh, C. S. Huber, P. R. Kiser, and D. B. Rowley. 1974. "Hazard Analysis of *Clostridium perfringens* in the Skylab Food System," *J. of Milk and Food Tech.*, 37(12): 624–628.
4. Smith, M. C., R. M. Rapp, C. S. Huber, P. C. Rambaut, and N. D. Heidelbaugh. 1974. *Apollo Experience Report—Food Systems*, NASA TN D-7720.
5. Smith, M. C., N. D. Heidelbaugh, P. C. Rambaut, R. M. Rapp, H. O. Wheeler, C. S. Huber, C. T. Bourland. 1975. "Apollo Food Technology," in *Biomedical Results of Apollo*, R. S. Johnston, L. F. Dietlein, and C. A. Berry, eds. Washington D.C.: National Aeronautics and Space Administration, pp. 437–468.

6. Hartung, T. E., L. B. Bullerman, R. G. Arnold, and N. D. Heidelbaugh. 1973. "Application of Low Dose Irradiation to a Fresh Bread System for Space Flight," *J. of Food Sci.*, 38: 129–132.

7. Vodovotz, Y., C. T. Bourland, and C. L. Rappole. 1997. "Advanced Life Support Food Development: a New Challenge," Presented at the 27th ICES, SAE paper number 972363.

60 NASA Food Technology Commercial Space Center

D. G. Olson

CONTENT

Abstract

The Food Technology Commercial Space Center (FTCSC) will develop foods and food processing technologies suitable for space travel and for commercial terrestrial markets. Commercial companies will be the primary developers of the products and technologies, and the marketers to the consumer market or food industry. The FTCSC will facilitate the interaction between NASA and food companies.

60.1 INTRODUCTION

NASA space program expansion to 90- to 120-day missions on the International Space Station and potential exploration flights to Mars increase the performance requirements of food for astronauts. For long-duration space exploration, food-crop production and food processing with minimal waste and efficient recycling of nutrients will be needed. Compared with production of small volumes of specialty foods for the space program, a more sustainable approach is for the food industry to develop foods and food-processing technologies that also have terrestrial commercial markets.

The NASA Food Technology Commercial Space Center would bring NASA and commercial food companies together to develop food products and food processing

1-56676-963-9/02/$0.00+$1.50
© 2002 by CRC Press LLC

technologies that can be used in space and are also commercially viable in terrestrial markets.

60.2 CENTER CONCEPT

The Food Technology Commercial Space Center (FTCSC) will develop foods and food processing technologies that are suitable for space travel and for commercial terrestrial markets. Commercial companies will be the primary developers of the products and technologies, and the marketers to the consumer market or food industry. The FTCSC will facilitate the interaction between NASA and food companies. This interaction will ensure that companies have a complete understanding of food specifications essential for space travel.

Companies, in their developmental activities, will need assistance from specialized facilities at NASA and Iowa State University. The FTCSC will facilitate the use of those facilities at Iowa State University. During the development of products and food processing technologies, scientific expertise will likely be needed. The FTCSC will help identify scientists who can provide their expertise as consultants.

When food products and food processing technologies or systems have been developed, the Center will assist companies in commercializing those products/technologies to the food industry and the terrestrial consumer market. A strong outreach program will be developed to inform the food industry and consuming public of the advances in food products and food processing developed in the FTCSC.

60.3 MISSION, OBJECTIVES, AND IMPLEMENTATION

60.3.1 Mission

The mission of the NASA FTCSC will be to lead a national effort in developing foods and food processing technologies that enhance space missions and advance commercial food products through cooperative efforts with NASA's scientists and technologists, commercial companies, and academic researchers.

60.3.2 Objectives

The following objectives will be used to develop projects in the center. All objectives will meet NASA requirements for nutrition and safety. The objectives will also take into consideration aspects such as convenience, eating pleasure, and space mission constraints of mass, power, volume, reliability, and crew time requirements.

Objective 1

Development of foods for 30- to 120-day space missions in support of the International Space Station (ISS) missions (minimum shelf life of one year).

Objective 2

Development of food and food processing technologies to support human exploration and development of space missions (up to five years). This would include stored foods systems for transit vehicles as well as food processing systems for site-grown crops.

Objective 3

Development of a terrestrial commercial production and marketing plan for products and processes developed for the space program.

Objective 4

Focused research in direct support of the development of food products, food production processes, waste processing, and product safety for space and terrestrial application.

60.3.3 IMPLEMENTATION

Development of Project Areas

The scoping committee has the responsibility to identify development projects, scope activities, and prioritize projects. The basis for identifying and prioritizing project areas will be from NASA space food needs, suggestions and recommendations from corporate partners and research scientists, and deliverables from previous projects. The FTCSC director will identify and solicit corporate or research scientists to work on the prioritized projects. The projects with the highest priorities and with scientists willing and able to perform the work will be implemented.

Project Development

The FTCSC director with the project scientists will develop for each project a statement of work, budget, time line, and deliverables. Sensitivities to intellectual property will be considered in the development of the statement of work. These statements of work will not necessarily be extensive but will have sufficient detail to measure progress and accomplishments. Approval of the projects will be by the FTCSC director with recommendations from the scoping committee.

Project Monitoring

FTCSC staff will monitor the progress of each project by monthly contact with the project leader, reviewing technical progress reports, schedule, and budgets. The monthly contacts may be by phone, e-mail, or a personal on-site visit. These monthly contacts will also be used to account for in-kind expenditures. A reporting form will be generated to provide details of the monthly contacts and will be placed in the project file. For projects that are lagging behind schedule for more than two months without adequate rationale, the project leader and the FTCSC director will determine appropriate actions or adjustments needed for the project.

Project Completion

A project will be considered completed when the objectives in the statement of work have been accomplished and the deliverables have been submitted to the FTCSC director, who will in turn submit them to the scoping committee. For projects that have products such as food or food related hardware as part of the deliverables, the products will be submitted to NASA JSC for evaluation.

Market Transition

Patents and licenses from FTCSC projects will be handled as detailed in Cooperative Agreement NCC9-100. Commercialization of intellectual property will be emphasized to the license holder. Technology development to bring the product to a commercialization stage may become a FTCSC project if recommended by the scoping committee. For new food products or food processing equipment developed in FTCSC projects, the FTCSC will facilitate, directly or through NASA field centers, marketing assistance to introduce products to the marketplace.

Education Outreach

To provide information to the public about the FTCSC and NASA food systems, several communication tools will be used. A web page will be developed and maintained containing information about the FTCSC and links to other NASA sites. Brochures and the annual report will be distributed internally and to the public. MSFC and other NASA outreach programs will be provided with information about the FTCSC outcomes. Booth display materials will be developed and shown at food related meetings and conferences when appropriate. Presentations to civic clubs, schools, consumer organizations, and other public groups will be given.

International

Due to the commitment of NASA to have international partners in the space program, the FTCSC will accept international corporate partners and corporate affiliates in the same manner as U.S. corporations. Government officials from other countries desiring to develop formal affiliation with the FTCSC will be referred to NASA for evaluation and possible implementation of that affiliation.

Part XIII

Education

61 Role of the Internet in Food Engineering Teaching

R.P. Singh

CONTENTS

Abstract

Instructional modules using Internet-assisted technologies may be used alone or as supplements to lecture materials to enhance learning. With careful preparation, including incorporation of features that have been used for decades in traditional teaching (e.g., Bloom's taxonomy), multimedia technologies may contribute significantly to teaching and learning. The use of text, self-quizzes, illustrations, animated figures, and spreadsheet calculations are some of the features that can enhance instructional units for delivery on the Internet.

61.1 INTRODUCTION

Teaching is undergoing a renaissance. During the last few years, there has been a dramatic increase in the availability of teaching aids to supplement the traditional podium lecture. While the twentieth century began with one-room schools equipped with a chalkboard, the classroom of the twenty-first century is a lecture hall with multimedia projectors and live Internet connections. The walls of this lecture hall extend well beyond the planet Earth, with live images and sound received from distant spacecraft. These extraordinary changes in the emerging tools now becoming available for teaching raise important questions. Are these technologies merely passing fads? Do these technologies improve teaching effectiveness? How can one best incorporate teaching aids as useful supplements to traditional lectures? There is considerable debate on the role of multimedia technologies in improving learning. Arguments supporting either side of the issue abound in the current literature. In this chapter, we will explore some selected aspects of using the Internet in teaching and issues related to its effectiveness in learning.

A variety of factors have caused an exponential rise in the use of the Internet in teaching. With increasing student enrollments, there is a growing need for resources and facilities to accommodate larger class sizes. For example, the University of California will be faced with 5,000 additional students each year for the next 10 years. This increase is due to what is being called the "Tidal Wave II"—the dramatically increasing number of high school students who will qualify for admission into the university. This type of extraordinary growth in student enrollments seriously challenges available resources. Similar increases in student enrollments are being experienced at universities around the world. Ironically, during the last several decades, funds for education have been shrinking, with capital investments for building infrastructure at woefully low levels. Many academic administrators therefore look at the potential for cost effectiveness of the Internet in meeting teaching needs.

Along with the increase in "traditional students," there is a growing number of life-long learners seeking higher education. As mid-career employees seek alternate job opportunities, new learning skills become necessary. It is not uncommon for today's students to pursue a part-time job to meet financial needs while enrolled in classes. For these nontraditional students, the availability of asynchronous learning offered via the Internet becomes highly appealing.

61.2 STUDENT ACCEPTANCE OF THE INTERNET

Today's freshman weaned on microwave ovens and television is computer-literate and has a high level of acceptance of technology. The introduction of computers at the kindergarten level during the past decade has made modern students conversant in computer use. For a student with a computer and network connection, the Internet offers access to unlimited sources of information. The global reach of the Internet has inspired many instructors around the world to provide course materials on the Web. With the emerging growth in e-business, many industries have started to provide educational materials relevant to their wares on the Internet. A quick search on the Internet reveals numerous examples of unusually rich course materials that are engaging and demanding.

Properly designed instructional units on the Internet incorporate a high level of interactivity not possible with other means such as books, radio, or television. The interactive lessons on the Internet make students active agents in the process of learning. An often-cited complaint that the Internet contains a lot of trivial information is in itself fortuitous. It forces a user to distinguish between trivial and substantial, thus making it a useful learning tool. Properly designed educational materials on the Web encourage the user to formulate thoughtful questions and seek new information. The use of e-mail, chat rooms, and net meetings provide an opportunity for students to work on team projects, unlimited by geographic location. The instructor's office hours may be extended to provide easy access through electronic conversation. Otherwise shy students, as well as students with physical disabilities, have increased opportunity to contact an instructor through an e-mail message. These features of the Internet provide an ease of communication not possible in the past. Students from different continents may be connected via the Internet to listen to an instructor and discuss questions in a live mode. This was recently demonstrated using the next generation of the Internet—Internet II—between two classrooms in Japan and the United States.[1] Similarly, remote hands-on experiments in food engineering have been conducted involving students who were continents apart.[2]

In exploring these opportunities offered by the Internet, it is easy to be swayed by the glitter. Therefore, we need to ask whether these technologies are truly effective in the student learning process. It is often noted that the medium offers a mere conveyance to the instructional materials, and that it alone cannot influence learning effectiveness. The widely tested and accepted learning theories of the past several decades must be kept in proper perspective as new media-assisted teaching materials are developed.

61.3 PEDAGOGICAL ISSUES AND THE INTERNET

The types of events necessary for learning comprise both internal and external processing by the learner. Internal processing involves temporary storage, encoding, and retrieval, whereas external events are about gaining attention, drill and practice, and feedback.[3] Bloom's cognitive taxonomy has been widely accepted since it was first presented in the mid-1960s.[4] According to Bloom's taxonomy, there are six learning objectives that increase in level of difficulty from basic (knowledge) to advanced (evaluation) (Table 61.1). The descriptors for each objective provide a

TABLE 61.1
Bloom's Cognitive Taxonomy

Objective	Descriptive	Common verbs
Knowledge	The learner must recall information	Define, list, state, identify
Comprehension	The learner understands what is being communicated by making use of the communication.	Describe, illustrate, substitute, give examples
Application	The learner uses abstractions (e.g., ideas) in particular and concrete situations.	Classify, compute, show, solve, predict
Analysis	The learner can break down a communication into its constituent elements or parts.	Associate, differentiate, discriminate, distinguish
Synthesis	The learner puts together elements or parts to form a whole.	Integrate, propose, theorize, compile
Evaluation	The learner makes judgments about the value of material or methods for a given purpose.	Contrast, critique, judge, evaluate, criticize, conclude

basis of how one may design an instructional module that challenges a student to advanced levels of learning. The types of questions a student is able to answer may determine the behavioral level achieved by a student at the completion of an instructional module. For example, questions to test knowledge seek answers to "Who?", "What?", "When?", and "Where?" whereas the objective pertaining to synthesis requires answers to questions such as "What would happen if?" and "How can we solve a given problem?"

For an instructor, the challenge in developing an instructional unit using the Web is how to incorporate the various elements of Bloom's taxonomy. The following is a description of a lecture unit from a course on heat transfer developed for partial use on the Web, which incorporates some of these concepts.

61.4 EXAMPLE OF A WEB-BASED INSTRUCTIONAL UNIT INCORPORATING BLOOM'S TAXONOMY

As shown in Figure 61.1, the lecture unit is presented in frames. This allows the user to view related items on the screen as desired. The right-hand frame contains

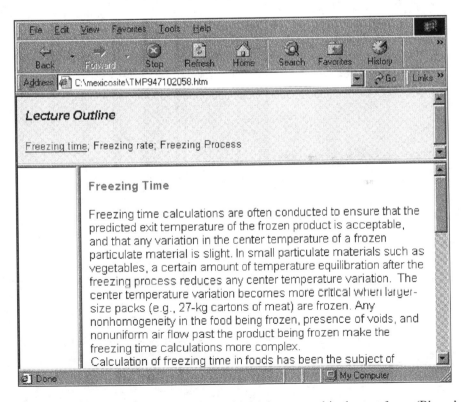

Figure 61.1 Text material of an instructional module presented in the text frame (Bloom's objective level: knowledge).

the textual content. This may include definitions, lists, facts, and other descriptions of a given concept. The text provides the knowledge base that a student is encouraged to learn and remember. After reading this material, a student should be able to recall the factual information. A self-quiz at the end of the section helps in this objective (Figure 61.2).

Certain terms in the textual content are highlighted and, with the use of links, clicking on such terms opens a new window that provides additional explanation, description, or examples. Similarly, illustrations may appear in the left frame either static or animated (Fig. 61.3). This step is aimed at increasing the comprehension of the text material so that the meaning of a term or concept may be grasped more easily.

The text for the lecture unit contains equations that are linked to short numerical problems to illustrate application of a given concept. The answers to those problems appear in the left-hand frame (Figure 61.4). Again, self-quizzes are included, which encourage the user to relate the learned material to other situations.

To encourage analysis of a concept, self-quizzes designed to analyze the problem require a student to select a correct diagram from several shown on the screen (Figure 61.5). Similar quizzes may be designed to promote discriminating between different descriptions, or determining the correct sequence of a process.

For the synthesis component, an interactive spreadsheet problem is provided in the left-hand frame (Figure 61.6). Analysis is encouraged by asking the user to

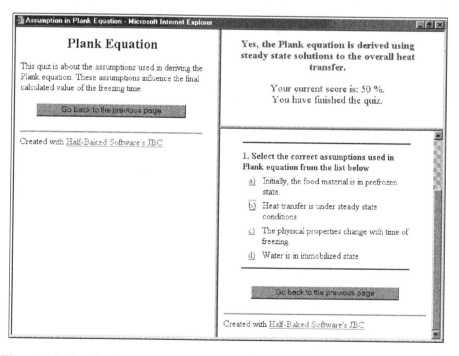

Figure 61.2 Use of self-quizzes to aid in learning the text material (Bloom's objective level: knowledge).

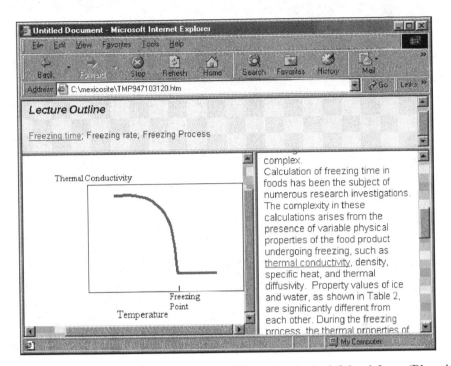

Figure 61.3 Use of highlighted text to show illustrations in the left-hand frame (Bloom's objective level: comprehension).

substitute different input values ("What if?") and observe the results. The results then may be used in creating a new design of equipment or a process.

The highest level of Bloom's taxonomy (evaluation) is included in the lesson plan as a question requiring a student to write a critique or an assessment that is then sent via e-mail to the instructor; for example, a report on current research on alternatives to ozone-depleting refrigerants from refrigeration systems. Samples of critiques may be included on the home page for students to obtain general directions.

In summary, it is possible to incorporate various aspects of Bloom's taxonomy into an instructional module destined for a Web page. Such attempts should improve the effectiveness of a given presentation. However, careful attention is required to encourage different levels of comprehension with liberal use of self-quizzes, illustrations (static and animated), descriptions, and glossary in the textual material. The use of frames provides an easy way for the user to navigate. Similarly, recent developments in integrated application programs, such as Microsoft 2000, offer opportunities to mix spreadsheets, text, and diagrams in an instructional unit.

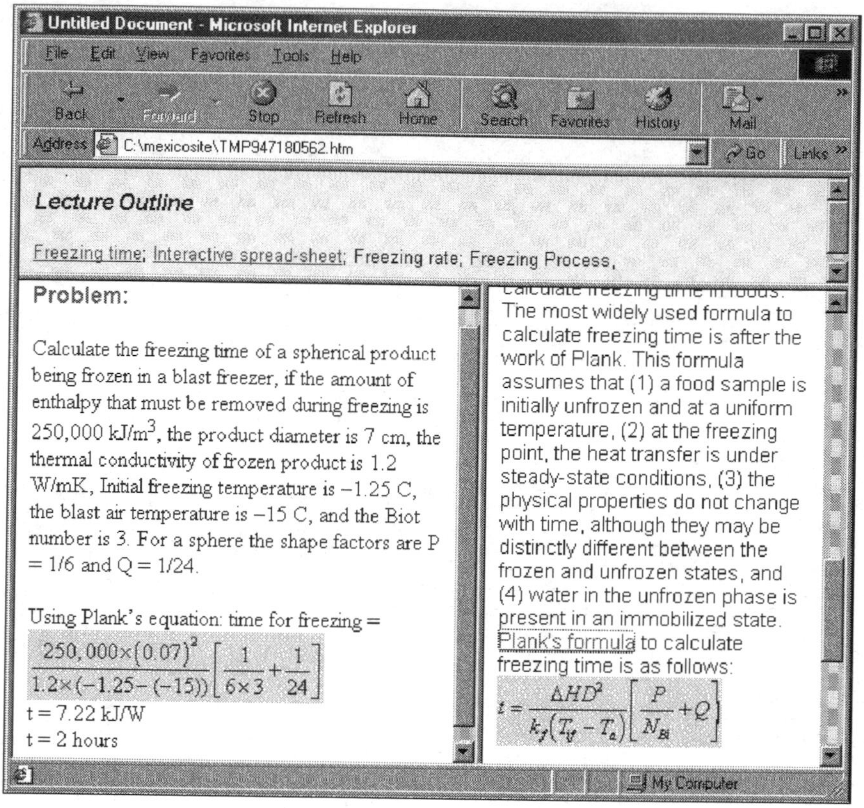

Figure 61.4 Linking equations with numerical examples (Bloom's objective level: application).

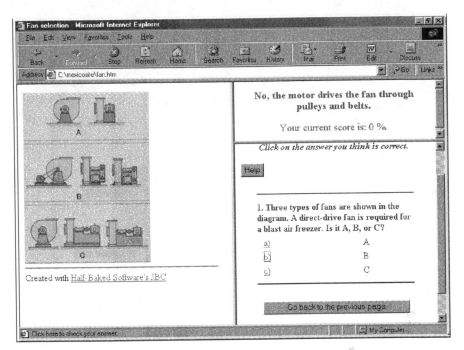

Figure 61.5 Use of self-quiz to distinguish among different designs (Bloom's objective level: analysis).

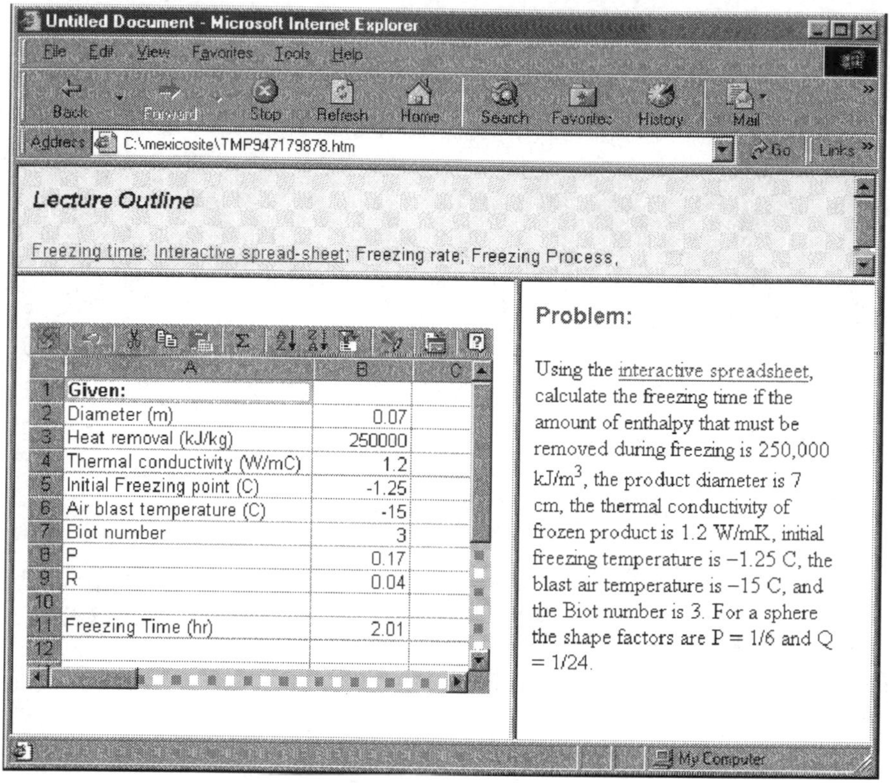

Figure 61.6 Use of interactive spreadsheet to seek answers to what-if questions (Bloom's objective level: synthesis).

REFERENCES

1. Guernsey, L. 1999. "Linking Classrooms, a World Apart, Via the Internet," *New York Times*, December 30, pp. D3.
2. Singh, R. P. and F. Courtois. 1999. "Conducting Laboratory Experiments Via the Internet," *Food Technology*, (53) 9: 54–59.
3. Gagne, R. M., L. J. Briggs, and W. W. Wager. 1992. *Principles of Instructional Design*, (4th ed.) Orlando, FL.: Harcourt Brace Jovanovich.
4. Bloom, B. S. 1968. *Taxonomy of Educational Objectives*. New York, N.Y.: David McKay, Publisher.

62 Teaching Food Engineering with the aid of Multimedia

M. López-Leiva

CONTENTS

Abstract

The explosive popularization of personal computers during the last decade, along with the appearance of the World Wide Web and of high-level programming techniques, has opened many new and exciting opportunities in the field of education.

In this chapter, a review of existing types of multimedia material used in education and of some of the authoring tools used in the production of this material is given, along with comments on their features and limitations.

A survey of higher education institutions working in food science and technology shows rather weak interest in this teaching technique. There is strong criticism against the high cost of development associated with such material. Through the study of some published work and our own experience, an attempt is made to explain this situation and suggest ways for improvement.

1-56676-963-9/02/$0.00+$1.50
© 2002 by CRC Press LLC

Finally, students in a master of science course present a multimedia program, developed at our department for teaching in membrane technology, with the results of its evaluation.

62.1 INTRODUCTION

Personal computers and the Internet have proved to be good tools for the transfer of knowledge, and we are currently witnessing increasing interest in producing and disseminating educational material with the aid of these instruments. At the same time, most of us are also experiencing tighter educational budgets (greater student/teacher ratios), which calls for new teaching techniques that can, at least in part, replace the need for face-to-face tutorials.

Being a relatively new way of teaching, we are still in the process of learning, among other things, the best pedagogical model to apply in each case or the best way of delivering the material.

The production of multimedia material for educational purposes requires expertise from at least three different areas: the subject itself (in our case food engineering), the pedagogical model (which is different from what teachers have been using to date), and expertise related to the multimedia tools themselves (programming, art design, sound design, etc.). This is an exciting and challenging area, with considerable potential for development. But, being a multidisciplinary task, the final product (courseware) will inherit characteristics from different sources, which will need to be compatible.

The production cost is almost zero, but the cost of development is considerable. This would not be an insurmountable problem if there were a sufficiently large market awaiting such products, but this is not the case. In addition, the very nature of teaching, where the teacher often puts a personal touch to the subject, means that there is little likelihood of a teacher using material developed by another teacher. Even books, in many cases, are trying to survive, and they are produced at a lower cost than multimedia products. The high development cost is almost certainly the main obstacle to more successful development of multimedia material for teaching, and the reason we have not seen a boom in this area.

Some of the features required in well designed educational software are

- Ease of use and simple function, so that a seamless interaction is produced between the student and the content of the program rather than with the program itself
- The possibility to choose the degree of difficulty to better accommodate the user's current level of knowledge on the subject
- A good degree of interactivity, allowing the student to influence events presented by the program (the ideal situation is for the program to mirror the student-teacher dialogue in conventional teaching)
- A "clear" program design (graphics design, sound, etc.) free from disturbing elements; it should be possible for the user to turn off nonessential elements (e.g., sound effects)

62.2 DELIVERY MULTIMEDIA

With the recent increase in the number of Internet connections, the concept of multimedia has often been equated with this technique. However, this is not the case; multimedia can exist in different forms, and delivery via the Internet is only one of them.

Multimedia material (courseware) for the purpose of education can be delivered in the following ways: Web-based material, CD-ROM, downloads from the Internet, by local network, and hybrid systems.

Web-based material has the advantage of being accessible worldwide (if a computer with an Internet connection is available), a low running cost, and easy actualization of the content, but it may run slowly.

Courseware delivered via CD-ROM always runs faster than its Web-based counterpart, is portable (can be accessed from any computer), is of better quality (e.g., pictures can be larger), but it is not easy to actualize its content, it is inexpensive to produce (U.S. $300 for a CD recorder; less than U.S. $1.00 per disk), and it is inexpensive to run.

Material delivered via a local network has the same advantages as those for a CD-ROM, with the additional advantage that new versions of the program will be accessible by the complete target group at the same time.

Hybrid delivery systems make use of the advantages of both Web-based material and CD-ROM. The student receives a CD-ROM containing the heaviest part of the course, while the rest, the "light" part, is delivered via the Internet. For example, in a course in periodontology, developed at Malmö University in cooperation with the European Dental Student Association, a CD-ROM with a complete set of cases (photographs) was delivered to students prior to the start of the course. When, during the course, a student asked a question via the Internet, the teacher's answer started the CD-ROM at the client site, where the necessary material could be obtained.[1]

The same applies for small programs that can be delivered, for example, on 1.44 MB disks as for CD-ROM (a 1.44 MB compressed file can correspond to a decompressed program of around 5 MB).

62.3 SOFTWARE FOR PROGRAMMING MULTIMEDIA

Several commercial programming tools are available for the production of multimedia programs. Some of them are listed in Table 62.1, classified according to the type of work for which they are suitable. In this list, the level of the tools decreases from top to bottom and from left to right. A higher level programming tool means that the tool is farther away from machine code (and consequently nearer the user) and is easier for the programmer to use, as many commonly used functions are predefined; but it also means that it offers a limited number of possibilities. Usually, more than one of these tools must be used in the production of multimedia courseware. High-level tools can be relatively easily used to produce a program with text, sound, hyperlinks, animations, etc. But a high degree of interactivity can be achieved only if a programming language is used.

TABLE 62.1
Examples of Software Used in the Development of Multimedia Programs for Teaching

Type of software	Examples
Reproduction programs	QuickTime, RealPlayer, RealAudio
Conferencing	CUSeeMe, NetMeeting, WebCT, Iparty
Aid to publishing a course on the Web	Luvit, Gentle
Text publisher	Acrobat, LaTex, Text processors
Authoring	PowerPoint, Authorware, Director, Toolbox
Graphics	Flash3, PhotoShop, Illustrator, Paint, FreeHand
Web authoring	FrontPage, HTML script
Help authoring	Help Workshop, raw code
Programming languages	Java, Visual Basic, Delphi, C++, Pascal

A multimedia program will usually contain one or more of the following components: text, figures, graphs, sound, tables, animations, simulations, calculations, training, assessment, glossary, links to other sources of information, etc.

We can distinguish between material produced for use in distance learning and material produced as an aid in conventional teaching. In the case of distance learning, the multimedia material can either be the only material available or an aid to the comprehension of accompanying written materials. In each of these cases, the design of the material must be specific to the task.

62.4 MULTIMEDIA COURSEWARE IN THE AREA OF FOOD SCIENCE AND TECHNOLOGY

Despite increasing interest in the application of multimedia to education, few multimedia courses or teaching aids in the area of food science and technology are described in the technical literature. This author conducted a small survey on the degree of usage of multimedia in the teaching of subjects related to food. Two questions were sent to 37 higher education institutions working in science and technology, asking if they used any computer-based teaching aids and, if so, what type of software they used. Twenty-six responses were received. Eleven stated that no such tools were in use at their institution, but many of those added that they would like to have access to this type of teaching material or to economical and easy means of publishing a course on the Internet.

From the responses received, it could be seen that the most popular tool was PowerPoint. This is easy to understand, since PowerPoint is simple to use and produces acceptable results with a minimum of effort. However, the images are too static to be considered a good tool for the production of educational material.

Publishing lectures and even complete courses on the Web has become popular for the delivery of distance learning courses. For example, the University of Guelph, Canada, has a complete program of distance learning food science courses.

Some colleagues stated that they used their own material, e.g., computer simulation and modeling of refrigeration and heat transfer in meat processing (University of Bristol, UK)[2] or simulations of food process applications (University of California–Davis, U.S.A.). One group has been experimenting with remote laboratory experiments (University of California–Davis, U.S.A., and ENSIA, France).[3]

Besides distance education courses, only two of the respondents mentioned complete courses: one in heat transfer (downloaded from the Internet; University of Alberta, Canada)[4] and one in nutrition (CD-ROM; University of Navarra, Spain).[5] Both were produced using Adobe Acrobat, which provides the possibility of hyperlinks and some degree of animation, but no interaction.

The best documented example of courseware, and the closest to food science found in the technical literature, is a course in weed biology developed at John Moores University in Liverpool.[6] This course was produced to replace part of the normal weed biology course in the BSc Applied Biology degree program. The courseware is stored on a CD-ROM and is composed of a total of 174 screens, 42 animations, 9 video clips, and 132 photographs.

The tool used in the development of this course was an authoring system developed at John Moores University known as SKYROS. In this system, a series of normalized screens are produced for interface with navigation controls built into the screens. It contains a help function, a notepad, and a search function for the user. This courseware was tested for the first time in 1994. Students' assessment indicated general satisfaction with this courseware, which can be summarized as follows:

- Students quickly and easily adjusted to the learning package, being able to concentrate on learning without needing to think about the technology (the package is considered to be "transparent").
- One main attraction was the flexibility of provision; they could study at their own pace.
- A degree of discipline is necessary on the part of the students, i.e., the necessity of organizing their time and using the required time for their own studies.
- Some structured lectures and tutorial support should be made available.

A critical analysis of this course is reported by David R. Clark.[7] His analysis is mainly centered on the problem of scaling up this educational pilot project to a real-life situation. The main conclusion is that the resources necessary for developing courseware like this are enormous, more than what could be considered reasonable. The author calculates that the development of this course cost as much as £200,000, and that this particular course replaces only 16 hours of normal lectures (of marginal value £500). This is a strong argument and, if proved true, it means that the use of multimedia in education is still far from an acceptable solution. Do we strive to lower the production costs, or do we instead find other niches for multimedia?

62.5 STUDY CASE: A COURSE ON MEMBRANE TECHNOLOGY TAUGHT WITH THE AID OF A MULTIMEDIA PROGRAM

For some years we have being offering a Master's program for foreign students. An introductory five-week course is given at the beginning of the program, including an introduction to membrane technology. In the 1997 version of the course, a written compendium on the subject, with an accompanying multimedia computer program, replaced the lecture time devoted to membrane technology.

The program was made available on the division's server. The students could access it at any time, and they were free to study the subject at their own pace. The only face-to-face contact with the lecturer during these five weeks was to answer general questions related to the computer program or the compendium; all other contact was via e-mail.

The 14 students were from 11 different countries (Ecuador 2, Russia 1, Thailand 1, Poland 1, China 3, Korea 1, Jordan 1, South Africa 1, Sweden 1, Latvia 1, and Turkey 1). They had varying skills and experience with computers, which was reflected in their degree of acceptance of this teaching technique.

Prior to the final examination the students were asked to evaluate the course by answering a written questionnaire.

Written final examinations were taken in each section of the course (Introduction to Food Engineering). One of these parts was the membrane technology section. No significant differences were seen in the results of this part, compared with the other traditionally taught parts. The average score was slightly lower (66%) than the average for the whole course (68%).

62.5.1 THE COMPUTER PROGRAM

The computer program was developed at Lund University within a project supported by the SOCRATES program of the European Community,[8] with Lund University as the general coordinator. The project was in progress during the years 1995–1997, with seven partner institutions: ENSIA, France, MAICH, Greece, Uppsala University and Lund University, Sweden, Technical University of Copenhagen, Denmark, Polytechnic University of Valencia, Spain, and University of Milan, Italy.

The courseware was constructed using three tools from Microsoft: a programming language (Visual Basic), Windows Help Compiler, and a bitmap-based drawing program (Paint).

The computer program was organized around the following sections (see Figure 62.1):

1. *Main Tutorial.* This part of the program was produced as a Windows Help file using simple help code with the aid of a word processor supporting Rich Text Format. The complete text of the program resides in this section. Graphics can be added directly to the text or referred to as links inside the Main Tutorial.

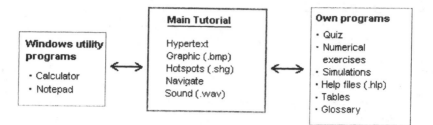

Figure 62.1 Structure of the program MembTech. Besides its own features, the Main Tutorial section can interact with externally produced computer programs; in this case, with Quiz, NumAssign, Simulations, Tables, Glossary, and help programs.

2. *MemQuiz.* A program with which the student can assess his/her comprehension of the subject.
3. *NumAssign.* A program designed to train students in solving numerical problems.
4. *Tables.* This series of tables can be consulted for the resolution of numerical problems.
5. *Glossary.* A list of terms used in membrane technology.
6. *Simulations.* Simulations of some key experiments (e.g., osmosis, ultrafiltration/diafiltration).

The five programs (Quiz, NumAssign, Tables, Glossary, and Simulations) were produced using Visual Basic and can be accessed directly from the Main Tutorial or from inside each other. Initially, sound effects were planned and incorporated into the program but were later removed.

Figure 62.2 is a reproduction of the screen viewed when the program NumAssign has been used by the student in the solution of a numerical assignment. Figure 62.3 presents the screen corresponding to the simulation of an ultrafiltration/diafiltration process.

A Windows Help file (.hlp) shows many of the characteristics common to many authoring tools:

- Production of hypertext that allows the user to jump to other parts of the text, or to produce "pop-up" texts to elucidate certain concepts or information.
- A search function that allows localization of pages or paragraphs where a certain word or concept is used or defined.
- To show graphics, both in bitmap and metafile format, for static images.
- To produce graphics in .shg format ("hypergraphs") for simple animated images containing so-called "hotspots."

The "visible" part of MemTech amounts to about 3.5 MB. Visual Basic requires a series of additional files (system files, an interpreter, dynamic link libraries, etc.), the size of which can be estimated to be an extra 3.5–5 MB.

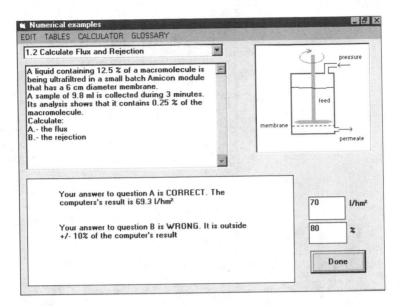

Figure 62.2 Computer screen of the program NumAssign. The computer has checked the answers given by the student.

It is difficult to determine the precise cost of producing this courseware, but it can be estimated to be of the order of 6 man-months (approximately 40,000 €)Since no professional programmer was engaged in the project, part of this time has been used learning to master the programming tool (Visual Basic) and the Help compiler.

In a later version of this program, the programming language Delphi was used instead of Visual Basic. Files produced using Delphi do not require any additional files; they are part of the executable file produced by the compiler. Only in special cases (e.g., when programs make use of a database) are some extra files required. Individual Delphi files are consequently larger than their Visual Basic counterparts, but Visual Basic will always require the use of an installation program (which will load any system and additional files needed to successfully run the program), no matter how small the program might be. Normally, Delphi does not need an installation program. Consequently, the executable files produced can be used in any compatible computer without the need for an installation procedure. In this case, it is also normal that the shipped files are smaller than the Visual Basic versions.

62.5.2 COURSE ASSESSMENT BY STUDENTS

Using a scale from 1 to 5 (1 = most negative, 5 = most positive), students were asked to rate the different parts of the program (Main Tutorial, Numerical Assignments, Quiz, Tables, Glossary, and Simulations) according to four criteria.

- It was easy to use.
- It covered the corresponding subject well.

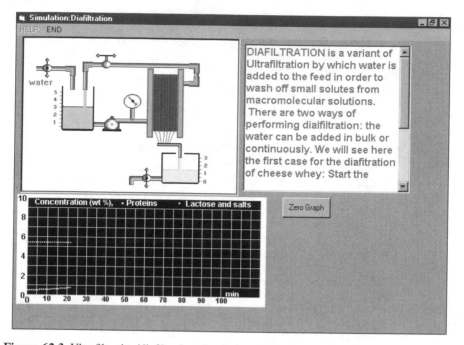

Figure 62.3 Ultrafiltration/diafiltration simulation. The process is running in batch, and the original volume of whey has been concentrated about 1.5 times. After a while, the process can be stopped, water can be added to the concentrated whey, and the process can be started again to complete the diafiltration cycle.

- It helped in the understanding of the subject.
- It had an adequate design.

The most positive responses were

- Easy to use: Main Tutorial = 4.7; Glossary = 4.5
- Covered the content: Glossary = 4.4, Quiz: 4.0
- Helped in the understanding of the subject: Quiz, Main Tutorial = 4.1
- Had an adequate design: Glossary = 4.4; Main Tutorial, Quiz = 4.3

The average of the marks assigned were as follows: Glossary: 4.4, Main Tutorial: 4.3, Quiz: 4.1, Simulations: 4.0, Tables: 3.5, and Numerical Exercises: 3.3. In relation to the section on Numerical Assignments, students specifically stated that this section should be more dynamic in the sense of offering the student help in learning how to solve particular types of numerical problems. Something similar could be implemented in the Quiz section where, if requested by the student, the program could give hints.

The students were also asked to report "bugs" and errors found in the program, as well as to make suggestions for its improvement. A total of 37 remarks/suggestions were received, distributed as follows:

- Bugs: 18 (referred to the same 5–6 bugs)
- Conceptual: 2
- Pedagogical: 5 (e.g., more help in the solution of problems, add a summary after each chapter, better way of navigating inside the Main Tutorial, etc.)
- Others: 12 (misspellings, different notations used for the same concept, grammatical errors, etc.)

Students were asked how much previous knowledge they had of membrane technology. For most, this was near zero. Only four of them declared having some degree of knowledge.

On average, students spent 13.8 h studying the notes (minimum 5; maximum 30) and stated that they worked with the computer program an average of 9.3 h (1–24). The obvious conclusion is that they preferred to read the notes rather than to use the software, but the fact that at least two of the students stated that they simply ignored the software (probably the same students who showed poor motivation/low skill in computers), makes the average number of hours spent on the software abnormally low.

62.6 CONCLUSION

The combination of a lower-level programming language (such as Visual Basic or Delphi) with the Windows help compiler has proved to be a good authoring tool. The programming itself is more complex than when using a higher level authoring package such as Director or Toolbox, but the results are much more rewarding, since high-quality interactivity can be added to the final product. Sound, in this level of multimedia teaching material, does not seem to be of much importance. The students do not miss it, and when it is present, it may become annoying, since this type of program will be open and run many times during a rather short period.

The development cost of multimedia educational programs is rather high. There is probably very little room for lowering these costs, but one may instead try to distribute the cost over several projects. This is possible if the multimedia software is produced using small, reusable modules that can be used in other courses with only minor changes. These modules could be similar to the Quiz, Numerical Assignments, or Glossary programs used in the program on membrane technology. If these small programs are written using Delphi, they will form independent units, with a size of 1–2 MB each, i.e., about 0.5 MB each when compressed. This is a reasonable size for downloading

REFERENCES

1. Aström, R., N. Matheos, and C. Bots. 1999. "A Virtual Classroom in Academic Learning," 21st. Annual EAIR Forum, Lund University, 22–25 August.
2. Anonymous. January, 1998. "BeefChil Computer Program," Newsletter of the Food Refrigeration and Process Engineering Research Center, University of Bristol, no. 19, January 1998.

3. Singh, R. P. and F. Courtois. 1999. "Conducting Laboratory Experiments Via the Internet," *Food Tech.*, 53(9): 54–59.
4. http://www.afns.ualberta.ca/foodeng/labs/acr-read.htm, University of Alberta, accessed on November, 1999.
5. Martínez, J. A. 1999. *Fundamentos Teórico-Prácticos de Nutrición y Dietética*, CD-ROM Interamericana-McGraw-Hill.
6. Lisewski, B. and C. Settle. 1999. http://www.cti.ac.uk, "Teaching with Multimedia: a Case Study in Weed Biology. Active Learning 3," accessed on November, 1999.
7. Clark, R. D. 1999. http://www.cti.ac.uk, "Interactive Media Programmes and the Problem of Scaling. Active Learning 4," accessed on November, 1999.
8. Network FIDEL: "Food Internet-based Distance Education Learning. SOCRATES programme," project Nr 35296-CP-2-96-1-SE-ODL-ODL.

63 A Low-Cost, Versatile Laboratory Experiment in Food Engineering Using the Internet

F. Courtois
R. P. Singh

CONTENTS

Abstract

A laboratory experiment for a class of 30 students was developed and made available over the Internet. Only free software and low-cost hardware were used for this purpose. Including the computer itself and a camera to continuously image the system, the overall price was less than $500. The laboratory experiment has been tested primarily for the teaching of the psychrometric properties of humid air.

63.1 INTRODUCTION

Teaching food engineering shouldn't be reduced to classroom sessions. Solving exercises with a pencil and a calculator cannot be called "practice in food engineering." Clearly, laboratory experiments must be included in a food-engineering syllabus. The resulting high cost per student of such experimental devices is usually a financial dilemma for universities. There are intermediate solutions between book exercises and process pilot experiments. Classically, one could use a simulator in place of a real experiment, which may be considered a virtual laboratory experiment. This is a financially interesting solution, compatible with large classrooms, although in many cases it may not convey a sense of reality. Experience shows that working with simulators may lead the students to a wrong perception of the reality behind the simulator.

Another solution is a real laboratory experiment shared on the Internet. The system is made available either directly in the laboratory or anywhere in the "Internet world." It performs real experiments in real time, without the need for the operator to be in the laboratory. The resulting cost per student is thus considerably reduced. This contribution describes such a system as developed and used at the University of California at Davis.

It must be emphasized that more and more web pages are now produced dynamically ("on the fly"), allowing the client to get close to real-time information. Thus, the Internet is compatible with real-time control of devices. Additionally, software for such development is becoming easier to use. It now takes just a few hours to program a dynamic page. Moreover, the decomposition of the programming into pages makes the sources intrinsically easy to read, understand, and maintain.

63.2 DESCRIPTION

The design of the experimental system was conducted according to the following constraints:

- *Low cost.* The complete system doesn't cost more than $500, with the use of standard components and low-cost or second-hand electronics.
- *Autonomy.* The system doesn't require human operation more than once a week.
- *Multiple locations.* The system is controllable from anywhere on the Internet, as well as from within the laboratory.
- *Multiple users.* The software allows many passive users simultaneously (only one active user at a time).
- *Versatility.* The device can be set up for different kinds of experiments.
- *Compatibility.* The only assumptions concern the internet connection. The basic hypothesis is that the bandwidth is at least 14,400 baud, and the browser is compatible with refresh tag and JPEG image format. No frames and no plug-ins are required.
- *Pleasurable.* The system/software is easy to use and makes the overall experience enjoyable for the students.

Figure 63.1 shows the schematics of the system and its instrumentation. In this configuration, the system can be called a dryer. A sphere sculpted in foam (Figure 63.2) is rewetted through use of a control valve and is dried with air heated by a set of two controlled 200-W bulbs and a fan. Four thermocouples give the temperatures of ambient air, hot air, sphere surface, and sphere center.

The system is composed of one DAQ (7 channels of 12-bit A/D, 8 channels of 8-bit D/A, and 1 digital output), 2 thermocouple signal amplifiers (2 K-type thermocouple channel amplification) obtained from BB-Electronics (P. O. Box 1040,

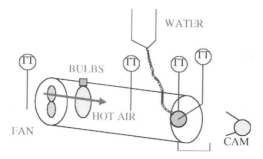

Figure 63.1 Overview of the system and its instrumentation (TT = temperature transmitter, QCAM = digital still camera).

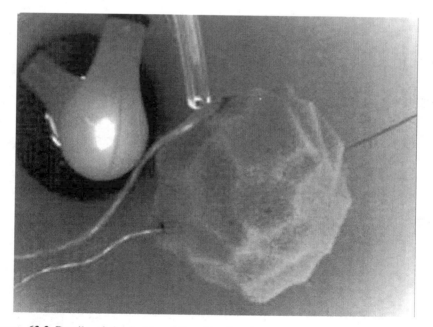

Figure 63.2 Details of the inside of the pipe showing two bulbs behind the sphere with thermocouples and plastic pipe (air filter has been removed).

Ottawa, IL 61350) and 2 dimmers obtained from Jameco (1355 Shoreway Road, Belmont, CA 94002) (see Figure 63.3 and 63.4).

The camera used for color imaging was the Color Quickcam II. This older, parallel-port version from CONNECTIX, is no longer distributed. Thus, it is available only as refurbished units from auction sites (at a very low price), with the advantage that it works on any workstation that has a parallel port.

The programs controlling the device were written entirely with free software (TCL/TK, VTCL, STAMP, CQCAM, CGI.TCL) on the LINUX free operating system and with the APACHE free web server. There are three modules as follows:

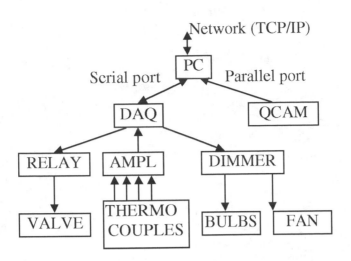

Figure 63.3 Overview of the hardware architecture (DAQ = data acquisition, QCAM = digital camera, AMPL = signal amplifier).

Figure 63.4 Details of the data acquisition system. The laptop is used as the main server.

1. *The laboratory experiment server:* communicates with the acquisition device and any network client (one could see it as an interface between software and hardware).
2. *The web server:* communicates as a network client with the above server and dynamically produces the web pages, displaying updated measures and images of the device.
3. *The client browser:* downloads and displays the dynamic pages and refreshes them every five seconds (see Figure 63.5).

These three entities can be in the same computer (when working in the laboratory itself) or in two (or three) separate computers for remote control (general case). The system is detailed by Singh and Courtois,[1] and Courtois et al.[2] and can be viewed at http://food.engr.ucdavis.edu.

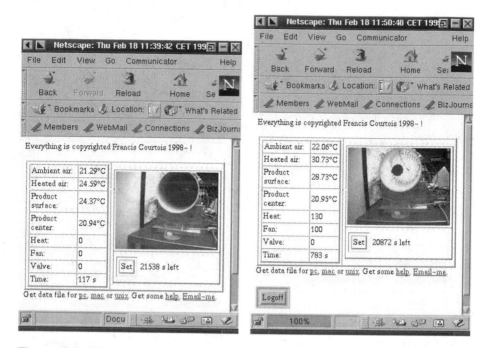

Figure 63.5 Web interface to the system with real-time data and image refresh. Users can change the settings for the heater, fan, and valve. Temperatures and image are updated every 5 s. Right-hand image shows the system with the heater on (50% full power) and the fan on (100% maximum rpm).

63.3 UTILIZATION

To teach psychrometrics to an undergraduate class at the University of California–Davis, the system was modified. The sphere was replaced by a cotton wick, which was submerged in water to measure the wet-bulb temperature of the hot air. A second cotton wick was used to measure the wet-bulb temperature of ambient air (Figure 63.6).

Students (30) were each given a login/password and had two weeks to perform their experiments. The system was configured to allow students only one hour of active connection (i.e., controlling the system), with an automatic disconnection after five minutes without any change in the actuators. After each experiment, students were asked to give their opinion (Figure 63.7). The system was reserved for them 24 hours a day, 7 days a week, for 2 weeks. They could connect to the server from their home or from any computer room.

Students were asked to connect to the system and collect real-time data of the dry- and wet-bulb temperatures of both ambient and heated air at different temperatures. After changing the heating power and retrieving the data, they could plot them (Figure 63.8) and locate them on a psychrometric chart.

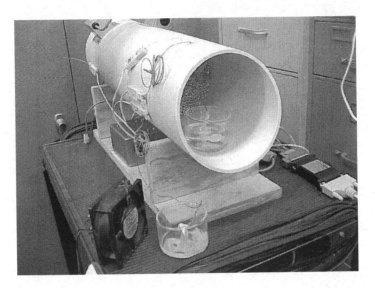

Figure 63.6 Experimental setup dedicated to teaching psychrometrics. Ambient and hot air, wet-bulb and dry bulb temperatures are measured.

From ThuLan N.
Likes best: I think it's great how we set up the apparatus and the fact that we are in charge of using the machine.
Likes least: I think the relay time is kind of distracting.
Has accomplished: Well, I learned how to use the apparatus.
Suggest changes: Actually, the procedures were very direct and easy to follow.
From Steve T.
Likes best: I liked the simplicity of the experiment. It does not take a computer-wiz to run it.
Likes least: I wanted to be a little more interactive in the experiment.
Has accomplished: I accomplished to run a food engineering experiment on the internet.
Suggest changes: Maybe give some definitions of the devices used in the experiment.

Figure 63.7 Some comments from the students (except from the feedback page).

It should be emphasized that each student generated an individual set of experimental data to analyze from a real experimental device.

63.4 CONCLUSION

The proposed system can cost as low as $500 and be built in a few days from standard parts, easy to find in local stores or on the Internet. The LINUX operating system combined with other free software is reliable, in this case operating 24 hours a day for months without a problem. The students at University of California–Davis who experienced the system for their classes gave a very positive feedback, showing a clear interest in this approach.

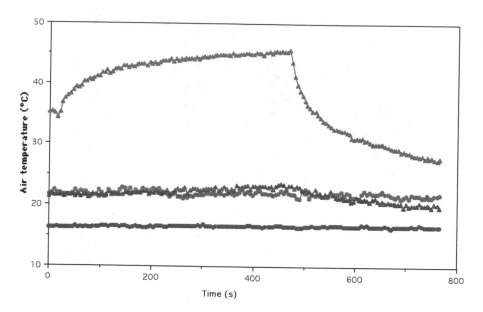

Figure 63.8 Plot of the four temperatures after a psychrometric experiment. Dry bulb temperatures are in light gray, wet-bulb temperatures in dark gray. Two different heater set points were used.

There are currently several other devices being developed at ENSIA and the University of California–Davis to broaden the range of unit operations that can be studied remotely on the Internet.

REFERENCES

1. Singh, R. and F. Courtois. 1999. "Conducting Laboratory Experiment via the Internet," *J. Food Tech.*, 53(9): 54–59.
2. Courtois, F., R. P. Singh, and N. Smith. 2000. "Laboratory Course in Food Engineering using Free Software and Low Cost Hardware," submitted to *Computers and Electronics in Agriculture*.

64 Toward a Virtual Cooperative Training Program in Food Engineering

E. Parada-Arias

G. F. Gutiérrez-López

C. Ordorica-Vargas

CONTENTS

Abstract

Modern sources of information and the development of communication technologies have opened a vast field of possibilities to training activities, including education in food engineering. Availability of syllabus and courses programs found in World Wide Web (WWW) sites, and the possibility of incorporating interactive multimedia and virtual reality capacities in teaching processes, are the basis for the design of long-distance, high-level education of strong social impact. Different approaches to the organization and use of these tools are already being applied in the educational systems of various countries. This chapter presents a number of syllabi for food engineering found on the WWW, as well as recent developments by groups working along the same approach, including those of an interactive and virtual reality nature in which the use of computing programs is stressed. A training program in food engineering is proposed based on a solid and novel application of computational tools, bringing together skills of groups in different countries that show capacities not only on the food engineering field, but also in the application of informational platforms and technologies to education.

64.1 FOOD ENGINEERING: A GENERAL SCOPE

The defining activity of engineering is design. Food engineering encompasses activities related to food processing and supply. This discipline requires a deep knowledge of mathematics, physics, transport phenomena, biological sciences, chemical reaction engineering, unit and combined operations, plant design and economics, and food chemistry.

Competitive food engineers demand a unique and continuous education, linking a strong background in the disciplines listed above with an understanding of the problems in the agricultural and biological industries and environmental areas.

A vast knowledge base of different fields, such as physical properties, microstructure, sensory properties, etc., most of a nonlinear nature, has a vital role on this subject and has opened many new fields of research, development, and, consequently, needs of training.[1]

Given the wide field of action of the food engineer and the rapidly changing methods, instruments, and processing techniques, on-line availability of multidisciplinary knowledge, ranging from processing technologies to education, on food engineering has become of key value within university-industry collaboration, including several forms and levels of formal education.

64.2 THE WWW: A NEW RESOURCE OF EDUCATIONAL AID

The widespread influence of on line information has entered into many fields of training in both university and industry. Approximate figures reveal that there has been a growth of web users, from 25 million in 1995 to a projected figure of 150 million in the year 2000.[2] Many of these users correspond to individuals seeking a different sort of academic information, which may range from an e-mail address to a complete syllabus of a graduate or undergraduate university course. In addition, a wide variety of electronic books and audiovisual tools are readily available from many libraries around the world. Although printed books will continue to have an important role in education, all sorts of electronic sources of information, including on- and off-line texts, are gaining attention within educational circles. For more information on electronic books go to http://www.softbook.com/, http://www.novomedia.com/ or to http://www.librius.com/

Web sites also give the opportunity of seeking for particular books and journals. The web side http://javeriana.educ.co/biblos/Elibros.html provides a good browser for the so-called new technologies. Internet II tools will not only speed up traditional on-line activities but also will allow selection of the required university sites.

The rise in the development of internet sites for academic purposes has been exponential over the past four years,[3] and these tools are becoming an excellent complement for traditional teaching and training. Computers will not replace lecturers but will change their role in front of a class (or a camera or a monitor).

New information technologies, including multimedia and telematics, along with the sophistication of microelectronic devices, allow old and difficult tasks to be carried out more efficiently.[4] Among these, the possibility of differentiation is one of the most important.[5] Individual students' needs, and the possibility of tracking personal progress to anticipate actions for improving learning processes, are only a few of the many advantages of application of new information technologies for computer aided training (CAT), including virtual training programs (VTPs). Elements such as collaborative environments of learning space like those developed by IBM (Macromedia) or by Lotus (Pathware) enable the integration of electronic resources in teaching aids. It is important to underline the potential use of the software: spaces for virtual learning (EVA, for the initials in Spanish). An expanded description of EVA is given in the section "Proposal of a Virtual Cooperative Training Program in Food Engineering," for the proposal given in this chapter as a platform for interactive activities.

Besides the above-mentioned items, internet tools make it possible to develop real-time interactive training schemes, with the aim of virtual cooperative training programs (VCTP), which will have the additional advantage of sharing and enriching teaching and learning. Within this frame, electronic libraries play a key role. A list of some international on-line libraries is given in Table 63.1.[6]

TABLE 64.1
Some Electronic Libraries Available on the WWW

Library	Web site
American Library Association	http://www.ala.org/
University of Virginia Library	http://www.lib.virginia/
The British Library Web Site	http://www.portico.bl.wk/
National Library of Canada	http://www.nic-bnc.ca/ehome.htm
Virtual Education Library (Australia)	http://www.csu.edu.au/education/library.htm
Latin-American Libraries	http://www.lanic.utexas.edu/la/region/
Spain National Library	http://www.bne.es/
Other Relevant Virtual Libraries	http://www.che.ofl.edu/www-CHE/topics/wwwvl.htm/

Other electronic tools, such as multimedia titles on CD-ROM, DVDs, videotapes, and supporting devices such as LCD projectors, digital cameras, scanners, and digitalization tools, among others, will also be of great help.[7]

Several education and research centers, including ours, have means of producing high-quality educational videos, CDs, and the like. The emphasis, however, has been on basic science and its diffusion, more than on technological applications. But given the availability of these electronic media, it should not be difficult to produce *ad hoc* materials.

In this chapter, a CAT unit on food engineering is proposed, which takes advantage of the already numerous sources of web information on the subject, stressing efforts made by individuals and institutions from the Ibero-American region.

64.3 INTERNET TOOLS ON FOOD ENGINEERING EDUCATION: BASIC AND ADVANCED RESOURCES, GEOGRAPHICAL DISTRIBUTION

A number of syllabi for students pursuing food engineering related degrees are available in the web. Levels covered are training, undergraduate, and postgraduate. Approximately 800 on-line syllabi on food engineering can be found (as of January 2000), and this number is expected to increase over the next several years.

Contents of courses on food engineering are also found; however, conventional searching engines like Yahoo and Altavista less frequently find these sites. A good example of a detailed food engineering course can be found at http://www.rci.rutgers.edu/%7Ebiorengg/whatfe.htm, which is a Rutgers University site.

The above-mentioned syllabi, courses, etc. have the advantage of being accessible to anyone having a computer linked to the web. Day and night consultation, and the possibility of interactivity by e-mail, are just some of these advantages. However, neither the permanence of sites nor the homogeneity of academic levels are guaranteed. Still, it is possible to identify a number of drawbacks for students seeking updated information. Some of these are

- Many sites are not updated for years.
- Some others are removed without any warning.
- Access to sites can be slow.
- Back-up notes/handouts are not frequently provided.
- Differences in content, scopes, and on-line tools may create confusion.
- So far, an in-depth approach to a systematic virtual training has not been found.

In addition to the long list of institutions found on the web offering subjects related to on-campus training on food engineering, it is possible to find a list of institutions offering food engineering-related syllabi. Of the 177 listed sites, 77 (43.5%) correspond to U.S. institutions; China, Canada, and the United Kingdom contribute with almost 8% each.[8] In total, 31 countries appear. A diagram showing comparative figures is presented in Figure 63.1.

A list of some important sites is shown in Table 63.2. Surprisingly, relatively few on-line training aids have been found, such as programs and problems on the subject. Singh and Courtois[3] offer a description of laboratory experiments via the internet. Users in different countries were given log-in password access to operate a wind tunnel with air heating elements and psychometric measuring devices. Psychometric properties of air at the inlet and exit of a heater can be monitored, and a full report downloaded for further analysis. Graphical and pictorial representations

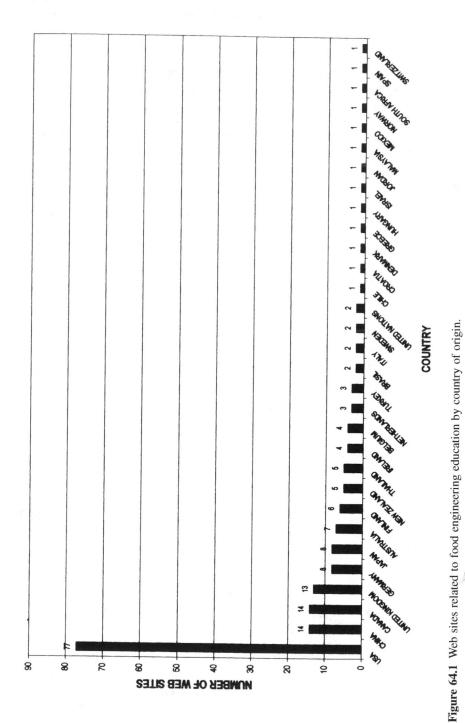

Figure 64.1 Web sites related to food engineering education by country of origin.

TABLE 64.2
Some Selected Web Sites Containing Food Engineering Syllabi

http://abe.www.ecn.purdue.edu/ABE/Undergrad/curr.fpe

http://www.chemeng.ntua.gr/courses/courintr.html

http://www.rci.rutgers.edu/~biorengg/whatfe.htm

http://www.agnr.und.edu/users/Bioreng/ugbre.htm

http://www.ume.maine.edu/~bre/undergra/under.htm

http://www.yatsushiro-nct.ac.jp/HOME/GAIYO/p2le.htm

http://www.psu.edu/registrar/progsumm/summaries/a_b_eb_s

http://www.twr.ac.za/subjectbiofo.htm

http://www.iyte.edu.tr/engweb/programs.htm

http://ageninfo.tamu.edu/programs/undergrad/food.html

http://aglab.ag.utk.edu/Undergrad/Facilities/Facilities.htm

http:/www.auth.gr/agro/eng/tt/

http://www.eng.ku.ac.th/~fe/feWelcom3.htm

http://www.eng.ohio-state.edu/student_info/ag_eng.html

http://perso.wanadoo.fr/patrick.perocheau

http://webapps.cc.umanitoba.ca/calendar/faculties/enginee...

http://www.ucd.ie/~food/degree.html

http://www.baeg.engr.uark.edu/curriculum/Food.htm

http://www.reading.ac.ur

http://www.ipn.mx

http://www.iztapalapa.uam.mx

http://www.fea.unicamp.br/

http://www.ing.pue.el

http://www.bsyse.usu.edu

http://custsv.univ-bpclermont.fr/enseignements/gb/agb2.html

http://www.bi.net.tr/~kunter/old_home

http://www.ics.uplb.edu.ph/catalog/ceat/ceat_bscel.html

http://www.metu.edu.tr/METU/Catalog/Depts/curriculum/fde_...

http://www.food.hacettepe.edu.tr/prog.htm

http://www.bae.ksu.edu/degrees/foodoption.html

http://www.engr.orst.edu/www/departments/bre/

http://www.toua-u.ac.jp/engineering/shokuhin/eguide.html

http://registrar.ucdavis.edu/UCDWebCatalog97_98/WebCatCrs...

http://www.ageng.ndsu.nodak.edu/undergrad.html

http://www.lib.rpi.edu/dept/core-eng/WWW/curriculums/chem...

http://www.up.ac.za/academic/bioagric/school-agric-rural/...

http://virtual.clemson.edu/groups/agbioeng/pages/students...

http://www.ncsu.edu/ncsu/provost/info/ugcatalog/programs/...

http://che.www.ecn.purdue.edu/ChE/Graduate/Courses/local/...

TABLE 64.2 (CONTINUED)
Some Selected Web Sites Containing Food Engineering Syllabi
http://ianrwww.unl.edu/ianr/bse/bsecours.htm
http://www.tu-dresden.de/mw/ilbv/english/estuplan.html
http://www.fsagx.ac.be/info_faculte/enseignement_an.html
http://www.ucc.ie
http://www.udlap.mx/ia_dept/
http://www.unam.mx
http://www.usach.cl/carreras/carrteca.htm#aline
http://www.upv.es/dtalim/

of the tunnel behavior are obtained within the set of results. Singh[9] also describes the use of internet tools to illustrate an on-line lecture entitled "Concentrating Liquid Foods by Evaporation." The home page is composed of different frames, and the content of the index include Introduction, Objectives, Process Description, Theoretical Considerations, Numerical Examples, Summary, Quiz, References, and Links to Other Resources. An example of a simple on-line food-engineering-related exercise is the pressure drop flow rate calculator by Paul Seamons, advertised in the web under the Spanish name *Programa de Reología* (Rheology Program). This simple program evaluates Reynolds number, Darcy friction factor, pressure drop per length, total pressure drop, and volumetric flow rate by the input these are obtained of the geometric characteristics of the system and the rheology constants of the problem fluid.

An outstanding resource found on the web related to food engineering is the collection of programs found in http://www.upv.es/dtalim/, under the Spanish title *Herramientas de Cálculo en Ingeniería de Alimentos* (Calculation Tools in Food Engineering). In this series, a thematic subject description of assorted software on food engineering is presented. The editors made an electronic recording of previously published material on the subject,[10–14] and produced a CD-ROM which has been already distributed throughout Ibero-America.[15]

This material has been produced bearing in mind its potential use for academic and industrial applications, and its success, measured in terms of usage, has met the editor's expectations.

The edition of this material has been coordinated by the Polytechnic University of Valencia (UPV-Spain) and the National Polytechnic Institute (IPN-Mexico). The comprehensive range of subjects included, as well as the architecture of the programs offered and supporting texts, may enable the edition of an important part of undergraduate or postgraduate food engineering syllabi. This collection is edited within the frame of the Ibero-American Program for Science and Technology for Development (acronym CYTED in Spanish), under Subprogram XI, Treatment and Preservation of Foods.

Another collection of data of potential use for interactive on-line training in food engineering may be provided by request from Dr. A. Sereno, of Catholic University of Porto, Portugal.[16] Data and information on prediction of physical properties of foods is available, including a proposal of a database on the subject. These resources are also part of the CYTED program.

A peculiar on-line training program on Introduction to Food Processing is found at http://www.open.uoguelph.ca/offerings, an introductory course on food processing and relationships between chemistry, microbiology, and engineering, prepared by the University of Guelph (Canada). The course is directed to government-employed food inspectors, quality assurance officers, food processing industry employees, food retailers, and public health officers, and it may not be taken for credit by students in specialized honors food science.

The previously mentioned, EVA learning space may constitute a solid software platform for interactivity. EVA is a multi-agent environment supporting Teaching/Learning processes, which contribute to[17–20]

- personal virtual assistants helping students to learn
- information, management, planning and organization of learning activities, including evaluation
- organization of working groups and group activities

Within EVA, students do not learn alone. EVA is a computer environment in which it is possible to obtain the necessary elements to learn. The learner navigates the space by routes (syllabi) suggested by the program and contacts with others to capture knowledge. EVA is a generator that guides, orients, and evaluates the *eva-nauta* (student) in the learning process, providing students with the physical, electronic, and human means required to learn. The constructors of EVA propose the development of information that allows the learner to

- individually process information and formulate needed studies
- obtain adequate orientation
- access efficient knowledge
- contact persons with the same interest to form working groups
- perform the required experimental work to learn effectively

EVA is structured into four essential elements, formed by four information sources, and the programs called Virtual Learning Spaces.

1. *knowledge*—all the necessary information to learn
2. *collaboration*—people who get together to learn
3. *consultation*—teachers or assessors who give the right direction for learning
4. *experimentation*—the work of students to obtain practical knowledge and abilities

EVA provides personalized learning material *polibooks* (Spanish: polilibros) to each student. EVA has access to a large collection of modules (chapters) written in several formats, e.g., Word, PowerPoint, etc. Knowing the current knowledge state and the desired final state, EVA arms the personalized polibooks. Then, EVA monitors the student's advancements through appropriate quizzes located at the end of each module. If the students provide their work schedule, EVA can program them with synchronous activities such as attending a teleconference on a fixed date on a proposed TV channel, watching an educational video on another channel, or joining a synchronous question-and-answer session.

Agglutination of the selected available material in food engineering found and to be generated within a CAT program can take advantage of the above-described EVA computer environment.

64.4 PROPOSAL FOR A VIRTUAL COOPERATIVE TRAINING PROGRAM IN FOOD ENGINEERING

It is important to point out that, even though there *is much* available on- and off-line electronic information on food engineering, materials such as tuition aid, methodologies, and technical information sources are spread out and disarticulated, and no proposal for agglutination has been made to date. This situation stresses the strong need for launching a plan toward a cooperative virtual training program on food engineering.

By means of the available information on the web and our own experience when dealing with many food engineering groups, mainly within the Ibero-American region, it has been considered that training in food engineering can be strongly supported by the use of electronic means and on-line resources.

General aspects to cover in training of this kind include

- the support and recognition of participating academic institutions
- faculty trained and convinced of the advantages and limitation of on-line resources
- available computing platforms for users, without administrative regulations
- previously agreed upon information platforms
- clear user rules (log-ins, passwords, controlled on line time, etc.)
- students willing to use and able to interact
- a solid and reliable software platform that allows the interaction of the different elements involved in the training

The core of the proposal is linked to the information on food engineering available at the site http://www.cyted.ipn.mx/, where this text, as well as links to web sites related to food engineering syllabi, e-mail addresses, and programs such as those previously described, are presented.

To help in the articulation of this cooperative virtual training program on food engineering, it is necessary to recall the elements proposed by Chang et al.,[21] who claimed that, to support educational programs and to manage their operation, it is necessary to count on important supports such as a virtual library, like those mentioned above; an intelligent learning system; and visualization and planning tools for the management of operations (EVA would be a good alternative for these last two points), and which will require a strong effort from the potential participant institutions.

Training might follow a direction given by the subject, as follows:

1. *Core of Training*
   Physical properties of foods
   Momentum transfer
   Heat and mass transfer
   Simultaneous heat and mass transfer
2. *Selective Subjects*
   Fermentations
   Use of low temperatures
   Product engineering
   Emerging technologies

Each of the subjects encompasses a collection of food engineering software constructed by Ibero-American authors who have always proposed to cooperate within a collaborative training program on the subject. Leaders of the groups are identified on the site, and e-mail addresses are also given for further contact.

It is proposed that long-distance training could be accomplished by interaction among groups belonging, in the first instance, to those mentioned in Table 63.3.

These groups can interact by means of software such as EVA, which may allow the device of a training scheme in which students would have access to specific subjects, consult expert faculty on subjects of interest, attend a lecture by means of teleconference, solve specific problems suggested by the tutors, use CYTED food engineering software as described above, or consult and enter virtual libraries on the field. Students could also use long distance laboratory facilities, like those reported by Singh.[3]

The above-depicted scope would allow having multidirectional interaction among students and tutors in a virtual environment, richer in many resources than the traditional bidirectional tutor-student relationship.

It is important to point out that the scope provided gives only a broad approach toward a virtual cooperative training program in food engineering, and the mentioned tools are only a starting point for developing a strong platform in which the activity of tutors and students can be carried out with a minimum of communication problems, and on which more and more specific pieces of electronic media can naturally be fitted. The revision presented does not attempt to be comprehensive but is illustrative of the many media available that can enrich the already discussed learning platforms.

Finally, it is noteworthy that the experiences of authors while dealing with multinational and multilateral program networks, R&D projects, and design and evaluation of curricula that are rich in interaction and (contrary to what was expected) relatively easy to handle, indicate that this program should be put into practice at once by undertaking specific tasks by specific groups, coordinated under a multina-

TABLE 64.3

Some Proposed Tutors for the Virtual Training Program and Their Affiliation

Country	Institution	Professor	E-Mail	Field
Argentina	CITA	Noemí Zaritzky	zaritzky@ing.unlp.edu.ar	Heat transfer
Argentina	PLAPIQUI	Jorge Lozano	allozano@criba.edu.ar	Modeling and simulation
Brazil	UNICAMP	Enrique Ortega	eortega@fea.unicamp.br	Education
Chile	USACH	Abel Guarda	aguarda@ussach.cl	Education
Chile	PUC	Jose Miguel Aguilera	jmaguile@ing.puc.cl	Microstructure
Mexico	IPN	Efrén Parada-Arias	leopardo@df1.telmex.net.mx	Education
Mexico	IPN	Gustavo Gutiérrez-López	gustavog@df1.telmex.net.mx	Dehydration, downstream processing
Mexico	IPN	Antonio Jiménez	aparici@redipn.ipn.mx	Physical properties
Mexico	IPN	Carlos Ordorica-Vargas	cvargas@redipn.ipn.mx	Modeling and simulation
Mexico	UNAM	Edmundo Brito	ebrito@servidor.unam.mx	Rheology
Mexico	UDLA-P	Jorge Welti-Chanes	jwelti@mail.udlap.mx	Hurdle technology
Portugal	UP	Alberto Sereno	sereno@up.te.pt	Physical properties of foods
Spain	UPV	Pedro Fito	pfito@tal.upv.es	Osmotic dehydration
Spain	UPV	Antonio Mulet	amulet@tal.upv.es	Modeling and simulation
Sweden	UL	Miguel López-Leiva	ltkk-mll@pop.lu.se	Education, membrane processes
UCD	USA	Paul Singh	rpsingh@ucdavis.edu	Education, physical properties
WSU	USA	Gustavo Barbosa-Cánovas	barbosa@mail.wsu.edu	Non thermal processing

tional framework scheme. In the short term, it is recommended that a strong platform of tutors and potential applications of this scheme be integrated into this effort.

The proposed program may be useful in several directions; the complete training can be directed toward obtaining credits in a formal university course coordinated by one of the leading centers and/or tutors within a software environment platform like EVA. Alternatively, students may choose among topics and tutors to improve their abilities in food engineering for an *ad hoc* purpose such as updating course work and diploma courses for lectures and students; e.g., training in industry, quality assurance, standardization, and inspections.

The configuration of a virtual training program in food engineering could be enhanced by the formation of international working networks in which several groups from different countries may participate. In the first instance, the different nodes of these networks would work under the international coordination of one of the groups, and the capabilities of each group would be directed toward the common goal of the program. Up to now, several institutions, not only from the Ibero-American region but also from other countries, have asked for participation and agree with the idea that has been put forward in this chapter. It is believed that the configuration of such networks will not only be of great importance toward the objective of training in food engineering but also a vehicle for improving capabilities in teaching, research and multilevel international cooperation.

REFERENCES

1. Welti-Chanes, J. and E. Parada-Arias. 1997. "Past, Present and Future of Food Engineering in México," COFE 97, New Frontiers in Food Engineering, *Proceedings of the 5th Conference of Food Engineering,* November 18-21, 1997, Barbosa-Cánovas, G., S. Lombardo, G. Nartsimhan, and M. S. Okos, eds. New York: *AIChE,* pp. 445–449.

2. IDC, Dataquest, 1996.

3. Singh, R. P. and F. Courtois. 1999. "Conducting Laboratory Experiments via the Internet," *Food Technol.,* 53(9): 54–59.

4. http://www.civila.com/universidades. Fernandez-Muñoz, R. 1998. "El Universo de las Nuevas Tecnologías: Información y Nuevas Tecnologías en la Enseñanza."

5. Lepeltak, J. and C. Verdidede. 1998. "Teaching in the Information Age: Problems and New Perspectives," in *Education for the Twenty-First Century,* Delors, J., ed. Issues and Prospects. UNESCO, France.

6. http:www.che.ufl.edu/WWW-CHE/topics/wwwvl.html. 2000. "Relevant Virtual Libraries."

7. http://lo.afs.udel.edu/Multimedia/engedu.htlm. Martin Lo, Y. 1998. "Using Multimedia in Food Engineering Education."

8. http://www.ucd.ie/~food/hotlink.html. 2000. "Agricultural & Food Engineering Around the World."

9. Singh, R. P. 1997. "Internet-Mediated Instruction in Food Engineering," in *Food Engineering Proceedings of the 5th Conference of Food Engineering, CoFE '97,* G. Barbosa-Cánovas, S. Lombardo, G. Narsimhan, and M. R. Okos, eds. New York: AIChE, New Frontiers, pp. 439–444.

10. Mulet, A., C. Ordorica-Vargas, and J. Bon, eds. 1995. *Herramientas de Cálculo en Ingeniería de Alimentos I,* Universidad Politécnica de Valencia, España.

11. Mulet, A., C. Ordorica-Vargas, J. Bon, and E. Ortega, eds. 1996. "Herramientas de Cálculo en Ingeniería de Alimentos II," *Anales del Primer Congreso Iberoamericano de Ingeniería de Alimentos,* vol. IV, Universidad Politécnica de Valencia, España.

12. Mulet, A., C. Ordorica-Vargas, and J. J. Benedito, eds. 1997. *Herramientas de Cálculo en Ingeniería de Alimentos III,* Instituto Politécnico Nacional, México, Universidad Politécnica de Valencia, España.

13. Mulet, A., C. Ordorica-Vargas, J. Bon, eds. 1998. *Herramientas de Cálculo en Ingeniería de Alimentos IV,* Universidad Politécnica de Valencia, España.

14. Fito, P., A. Mulet, C. Ordorica-Vargas, and J. Bon, eds. 1999. *Herramientas de Cálculo en Ingeniería de Alimentos V,* Universidad Politécnica de Valencia, España.

15. Fito, P., A. Mulet, C. Ordorica-Vargas, J. Bon, J. J. Benedito, E. Ortega, and M. A. Acedo, eds. 1999. *Herramientas de Cálculo en Ingeniería de Alimentos, CD ROM Versión 1.0,* Instituto Politécnico Nacional, México, Universidad Politécnica de Valencia, España.

16. Sereno, A. M. 1998. *PPFA-Computer Methods to Predict Physical Properties of Foods: an IberoAmerican Collaborative Project,* University of Porto, Porto, Portugal, sereno@fe.up.pt

17. Guzmán, A. A. and G. Nuñez-Esquer. 1998. "Virtual Learning Spaces in Distance Education: Tools for the EVA Project," *Expert Systems with Applications* 15: 205–210.

18. Núñez-Esquer, G., L. Sheremetov, J. A. Guzmán, and A. Albornoz. 1999. "The EVA Teleteaching Project—The Concept and the First Experience in the Development of

Virtual Learning Spaces," Personal Communication, Computer Science Research Center, Instituto Politécnico Nacional, México.

19. Sheremetov, L., G. Núñez-Esquer, and A. Guzmán. 1999. "Tecnologías de Inteligencia Artificial y de Agentes Computacionales en la Educación: El Proyecto EVA (Primera Parte)," *Revista Academia*, 23: 45–53.

20. Sheremetov, L., G. Núñez-Esquer, and A. Guzmán. 1999. "Tecnologías de Inteligencia Artificial y de Agentes Computacionales en la Educación: El Proyecto EVA (Segunda Parte)," *Rev. Academia*, 24: 58–63.

21. Chang, S. K., E. Hassanein, and C. Y. Hsieh. 1998. "A Multimedia Micro-University," *IEEE Multimedia 1070-986X/98,* 5(3): 60–68.

Index